AF616345

Applied Mathematical Sciences
Volume 84

Applied Mathematical Sciences

1. *John:* Partial Differential Equations, 4th ed.
2. *Sirovich:* Techniques of Asymptotic Analysis.
3. *Hale:* Theory of Functional Differential Equations, 2nd ed.
4. *Percus:* Combinatorial Methods.
5. *von Mises/Friedrichs:* Fluid Dynamics.
6. *Freiberger/Grenander:* A Short Course in Computational Probability and Statistics.
7. *Pipkin:* Lectures on Viscoelasticity Theory.
9. *Friedrichs:* Spectral Theory of Operators in Hilbert Space.
11. *Wolovich:* Linear Multivariable Systems.
12. *Berkovitz:* Optimal Control Theory.
13. *Bluman/Cole:* Similarity Methods for Differential Equations.
14. *Yoshizawa:* Stability Theory and the Existence of Periodic Solution and Almost Periodic Solutions.
15. *Braun:* Differential Equations and Their Applications, 3rd ed.
16. *Lefschetz:* Applications of Algebraic Topology.
17. *Collatz/Wetterling:* Optimization Problems.
18. *Grenander:* Pattern Synthesis: Lectures in Pattern Theory, Vol I.
20. *Driver:* Ordinary and Delay Differential Equations.
21. *Courant/Friedrichs:* Supersonic Flow and Shock Waves.
22. *Rouche/Habets/Laloy:* Stability Theory by Liapunov's Direct Method.
23. *Lamperti:* Stochastic Processes: A Survey of the Mathematical Theory.
24. *Grenander:* Pattern Analysis: Lectures in Pattern Theory, Vol. II.
25. *Davies:* Integral Transforms and Their Applications, 2nd ed.
26. *Kushner/Clark:* Stochastic Approximation Methods for Constrained and Unconstrained Systems
27. *de Boor:* A Practical Guide to Splines.
28. *Keilson:* Markov Chain Models—Rarity and Exponentiality.
29. *de Veubeke:* A Course in Elasticity.
30. *Sniatycki:* Geometric Quantization and Quantum Mechanics.
31. *Reid:* Sturmian Theory for Ordinary Differential Equations.
32. *Meis/Markowitz:* Numerical Solution of Partial Differential Equations.
33. *Grenander:* Regular Structures: Lectures in Pattern Theory, Vol. III.
34. *Kevorkian/Cole:* Perturbation methods in Applied Mathematics.
35. *Carr:* Applications of Centre Manifold Theory.
36. *Bengtsson/Ghil/Källén:* Dynamic Meteorology: Data Assimilation Methods.
37. *Saperstone:* Semidynamical Systems in Infinite Dimensional Spaces.
38. *Lichtenberg/Lieberman:* Regular and Stochastic Motion.
39. *Piccini/Stampacchia/Vidossich:* Ordinary Differential Equations in R^n.
40. *Naylor/Sell:* Linear Operator Theory in Engineering and Science.
41. *Sparrow:* The Lorenz Equations: Bifurcations, Chaos, and Strange Attractors.
42. *Guckenheimer/Holmes:* Nonlinear Oscillations, Dynamical Systems and Bifurcations of Vector Fields.
43. *Ockendon/Tayler:* Inviscid Fluid Flows.
44. *Pazy:* Semigroups of Linear Operators and Applications to Partial Differential Equations.
45. *Glashoff/Gustafson:* Linear Optimization and Approximation: An Introduction to the Theoretical Analysis and Numerical Treatment of Semi-Infinite Programs.
46. *Wilcox:* Scattering Theory for Diffraction Gratings.
47. *Hale et al.:* An Introduction to Infinite Dimensional Dynamical Systems—Geometric Theory.
48. *Murray:* Asymptotic Analysis.
49. *Ladyzhenskaya:* The Boundary-Value Problems of Mathematical Physics.
50. *Wilcox:* Sound Propagation in Stratified Fluids.
51. *Golubitsky/Schaeffer:* Bifurcation and Groups in Bifurcation Theory, Vol. I.
52. *Chipot:* Variational Inequalities and Flow in Porous Media.
53. *Majda:* Compressible Fluid Flow and Systems of Conservation Laws in Several Space Variables.
54. *Wasow:* Linear Turning Point Theory.

(continued following index)

Daniel D. Joseph

Fluid Dynamics of Viscoelastic Liquids

With 154 Figures in 206 Parts

Springer-Verlag
New York Berlin Heidelberg
London Paris Tokyo Hong Kong

Daniel D. Joseph
Department of Aerospace Engineering and Mechanics
University of Minnesota
Minneapolis, Minnesota 55455

AMS Subject Classifications: 35-02, 35Jxx, 35J60, 35Kxx, 35K55

Library of Congress Cataloging-in-Publication Data
Joseph, Daniel D.
Fluid dynamics of viscoelastic liquids / Daniel D. Joseph.
p. cm. — (Applied mathematical sciences ; v. 84)
Includes bibliographical references.
ISBN 0-387-97155-6
1. Viscous flow. 2. Liquids. 3. Differential equations, Partial.
I. Title. II. Series: Applied mathematical sciences (Springer
-Verlag New York Inc.) ; v. 84.
QA1.A647 vol. 84
[QA929]
510 s—dc20
[532'.0533] 89-21983

Printed on acid-free paper.

Text prepared by author in camera-ready form.
Printed and bound by R.R. Donnelley and Sons, Harrisonburg, Virginia.
Printed in the United States of America.

9 8 7 6 5 4 3 2 1

ISBN 0-387-97155-6 Springer-Verlag New York Berlin Heidelberg
ISBN 3-540-97155-6 Springer-Verlag Berlin Heidelberg New York

This volume is dedicated to Shifra Chana, Mike, Chuck and Samuel.

Preface

This book is about two special topics in rheological fluid mechanics: the elasticity of liquids and asymptotic theories of constitutive models. The major emphasis of the book is on the mathematical and physical consequences of the elasticity of liquids; seventeen of twenty chapters are devoted to this. Constitutive models which are instantaneously elastic can lead to some hyperbolicity in the dynamics of flow, waves of vorticity into rest (known as shear waves), to shock waves of vorticity or velocity, to steady flows of transonic type or to short wave instabilities which lead to ill-posed problems. Other kinds of models, with small Newtonian viscosities, give rise to perturbed instantaneous elasticity, associated with smoothing of discontinuities as in gas dynamics.

There is no doubt that liquids will respond like elastic solids to impulses which are very rapid compared to the time it takes for the molecular order associated with short range forces in the liquid, to relax. After this, all liquids look viscous with signals propagating by diffusion rather than by waves. For small molecules this time of relaxation is estimated as 10^{-13} to 10^{-10} seconds depending on the fluids. Waves associated with such liquids move with speeds of 10^5 cm/s, or even faster. For engineering applications the instantaneous elasticity of these fluids is of little interest; the practical dynamics is governed by diffusion, say, by the Navier-Stokes equations. On the other hand, there are other liquids which are known to have much longer times of relaxation. Polymers mixed in Newtonian solvents and polymer melts, like high viscosity silicone oils or molten plastics, are examples. The longest times of relaxation for these liquids are of practical interest; times we can read on our clock, of the order of milliseconds to minutes, or longer. The study of hyperbolic dynamics is complicated by the presence of many relaxation times. The limiting wave speed is determined by the fastest rather than by the slowest relaxation so that the instantaneous elastic response has already begun before the slow relaxation has begun. The fast relaxation of small molecules gives rise to an effective viscosity which smooths slow waves. If the total viscosity is much greater than the effective viscosity we may consider the theory of perturbed elasticity, with relatively small effective rigidities associated with the long lasting relaxations. The effective wave speeds are slow, ranging roughly from 1 to 1000 cm/sec.

It follows from what has been said that the models which are instantaneously elastic and give rise to hyperbolicity and change of type are precise only for times too short for applications. For the applications, the effective theory appears to work well but not all issues have been resolved. One question is what type of theory may be developed when the effective viscosity is not just a small part of the total. A second question is to what extent we may expect robust values of the effective quantities which are not dependent on flow conditions. It is probable that the successful resolution of these issues will depend more on experiments than on theory.

The contents of the seventeen chapters on the elasticity of liquids is taken from relatively recent papers not before collected into one volume.

The three chapters on asymptotic theory treat some well-known things in a new way. In 15 I review theories of fading memory and show how different theories will lead to different types of constitutive equations. Various types of perturbation theories are considered in detail in 16. In 17 I deal with second order theory emphasizing features which I consider fundamental like the balance of inertia and normal stress effects, the persistence of normal stress, the correlation between extensional viscosity and the intensity of secondary motion, the importance of nonelastic contributions to extensional and secondary motions and the general rheometrical problem of determining values of the quadratic constants.

I have tried to avoid repeating things which are well expressed in other books listed in the references. Since only special topics are treated, this book cannot be used as a general reference, but for many of the special topics treated it is effectively the only reference. Some complementary results for wave propagation in viscoelastic materials and many results about existence and uniqueness of solutions for one-dimensional models can be found in the book by Renardy, Hrusa, and Nohel [1987].

Finally I want to acknowledge the help I have received from Michael Renardy and Jean Claude Saut in joint works, separate works, and discussions, and Edward Fraenkel for his contributions to the solution of the problem of flow over a flat plate. Mark Ahrens, Kangping Chen, Howard Hu, Amitabh Narain, Luigi Preziosi, Oliver Riccius, and Claude Verdier helped me in different ways but especially doing the research reported in this book. Special thanks are due to Verdier and Hu for proofreading and to Hu for his help with the calculation in §5.8. Eric Scouten did the initial word processing of the manuscript and Lee Reynolds carried through revisions and formatted the text as it appears. My work has been supported for many years by the division of mathematics of the Army Research Office and division of fluid mechanics of the National Science Foundation.

Announcement of the grand opening of

RHEOLOGY DRUGSTORE

Our motto: "Fit The Data"
Proprietor: Daniel D. Joseph

"To make your experiment agree with your theory you should have the right fluids."

We carry many different fluids, corresponding to the thirty or forty models currently considered most realistic.

Standard brandname Fluids (well advertised):

Maxwell	Curtiss-Bird	Johnson-Segalman
Jeffreys	White-Metzner	Lodge's
BKZ	Phan Thien-Tanner	Green-Tobolsky
KBKZ	Newtonian	Oldroyd
Doi-Edwards	Reiner-Rivlin	Giesekus

Graded Fluids:

Single integral
Multiple order integral
1st, 2nd, 3rd order, etc.
Fluids of complexity 1, 2, 3, etc.

Composite Fluids:

With Springs and Dumbbells
With Beads and Chains
With Reptating Snakes

Retarded fluids with a strong backbone and fading memory

Mathematician's Delight:

Models with 1, 2, or 3 Fréchet derivatives
Less good fluids with only 1, 2, or 3 Gateaux derivatives

Less expensive fluids:

Liquid gold
Milky Way dust
Water with c=1 cm/sec

Table of Contents

CHAPTER 1

Models Like Maxwell's and Boltzmann's

1.1 The Maxwell element

We can build a very sophisticated nonlinear invariant theory from springs and dashpots. A spring and dashpot in series is called a Maxwell element (Figure 1.1). The spring constant is called G and the force σ_1 in the spring is $G\gamma_1$ where γ_1 is the displacement of the spring. The force σ_2 in the dashpot is $\eta(\partial\gamma/\partial t)_2$ where η is the viscosity and $(\partial\gamma/\partial t)_2$ is the velocity, the time rate of change of γ_2, and because they are in series, $\sigma_1=\sigma_2=\sigma$.

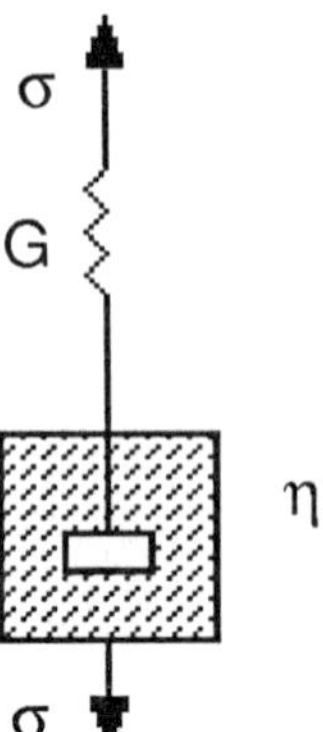

Figure 1.1. Maxwell element. The force σ_1 in the spring is the same as σ_2, the force in the dashpot.

We also have that the time rate of change of the total displacement is

$$\frac{\partial\gamma}{\partial t} = \left(\frac{\partial\gamma}{\partial t}\right)_1 + \left(\frac{\partial\gamma}{\partial t}\right)_2 = \frac{(\partial\sigma/\partial t)_1}{G} + \frac{\sigma_2}{\eta} = \frac{(\partial\sigma/\partial t)}{G} + \frac{\sigma}{\eta} . \tag{1}$$

Hence,

$$\lambda \frac{\partial\sigma}{\partial t} + \sigma = \eta \frac{\partial\gamma}{\partial t} \tag{2}$$

where $\lambda = \eta/G$ is a relaxation time. Another expression for σ is

$$\sigma = \frac{\eta}{\lambda} \int_{-\infty}^{t} \exp\left(\frac{-(t-\tau)}{\lambda}\right) \frac{\partial\gamma(\tau)}{\partial\tau} \, d\tau . \tag{3}$$

Obviously (2) is a differential equation model for the relation between force and deformation and (3) is an integral model showing that the present value of the force $\sigma(t)$ is determined by the history of γ.

1.2 Stress relaxation and instantaneous elasticity

Equation (2) is the constitutive equation for the Maxwell element. It will support a force with constant deformation $\gamma \neq 0$, $\partial\gamma/\partial t = 0$ and

$$\sigma = \sigma(0)\exp(-t/\lambda)\,. \tag{4}$$

Eventually σ relaxes to zero. We can regard $\sigma(x,t)$ as a stress and $\gamma(x,t)$ as a strain. We say that the Maxwell element has instantaneous elasticity. At $t=0$, we increase γ suddenly, producing a stress. Then we keep γ constant and the stress relaxes.

1.3 A one-dimensional model in the linearized case

Let

$$\gamma = \frac{\partial\xi(x,t)}{\partial x} \tag{5}$$

be the strain, where

$$\frac{\partial\xi}{\partial t} = u \tag{6}$$

is the velocity. Then

$$\frac{\partial\gamma}{\partial t} = \frac{\partial u(x,t)}{\partial x} \tag{7}$$

and

$$\lambda\frac{\partial\sigma}{\partial t} + \sigma = \eta\frac{\partial u}{\partial x}\,. \tag{8}$$

The equation of motion is

$$\rho\frac{\partial u}{\partial t} = \frac{\partial\sigma}{\partial x}\,. \tag{9}$$

We may combine (8) and (9)

$$\frac{\partial^2 u}{\partial t^2} + \frac{1}{\lambda}\frac{\partial u}{\partial t} = \frac{\eta}{\rho\lambda}\frac{\partial^2 u}{\partial x^2}\,. \tag{10}$$

Equation (10) is a telegraph equation. Recall that $G=\eta/\lambda$. The telegraph equation is a wave equation,

$$\frac{\partial^2 u}{\partial t^2} = c^2 \frac{\partial^2 u}{\partial x^2} \tag{11}$$

perturbed by

$$\frac{1}{\lambda}\frac{\partial u}{\partial t} \tag{12}$$

where

$$c = \sqrt{G/\rho} \tag{13}$$

is the wave speed. If $\partial u/\partial x$ is a longitudinal strain, c is the sound speed. If $\partial u/\partial x$ is a shear strain, c is the speed of a shear wave. For incompressible fluids, the sound speeds are infinite. In liquids, it is more interesting to consider shear waves.

Fluids with long times of relaxation are very nearly elastic; those with short times of relaxation are very nearly viscous.

Suppose $\lambda \to 0$, then

$$\sigma = \eta \frac{\partial u}{\partial x} \tag{14}$$

and

$$\rho \frac{\partial u}{\partial t} = \eta \frac{\partial^2 u}{\partial x^2}\,, \tag{15}$$

which are the equations of a Newtonian fluid. If, on the other hand, $\lambda \to \infty$, then

$$\frac{\partial \sigma}{\partial t} = G \frac{\partial u}{\partial x} \quad \text{or} \quad \sigma = G \frac{\partial \xi}{\partial x} \tag{16}$$

is elastic.

1.4 Hyperbolicity, characteristics

Equation (15) is parabolic and equation (11) is hyperbolic. Hyperbolic equations allow wave propagation of discontinuities without smoothing. Parabolic problems smooth discontinuities by diffusion.

The telegraph equation is hyperbolic. To see this, look at the system (8) and (9)

$$\lambda\sigma_t - \eta u_x = -\sigma ,$$
$$\rho u_t - \sigma_x = 0 \tag{17}$$

and ask if there are lines $\phi(x,t)=0$ across which the derivatives of u and σ are discontinuous with continuity for u and σ. We form equations for the jumps of the derivatives across these lines and get

$$\lambda[\![\sigma_t]\!] - \eta[\![u_x]\!] = -[\![\sigma]\!] ,$$
$$\rho[\![u_t]\!] - [\![\sigma_x]\!] = 0 . \tag{18}$$

According to our set-up, σ and u are continuous on $\phi=0$; hence,

$$[\![u]\!] = [\![\sigma]\!] = 0 \quad \text{on} \quad \phi(x,t) = 0 . \tag{19}$$

All the tangential derivatives of $[\![u]\!]=0$ and $[\![\sigma]\!]=0$ also vanish and the only derivatives of $[\![u]\!]=0$ and $[\![\sigma]\!]=0$ are normal to ϕ. Hence, for example,

$$[\![\sigma_t]\!] = [\![\sigma_\phi]\!]\phi_t .$$

Since $d\phi = \phi_t dt+\phi_x dx = 0$, we may reduce (18) to

$$\lambda[\![\sigma_\phi]\!]\frac{dx}{dt} + \eta[\![u_\phi]\!] = 0 ,$$
$$[\![\sigma_\phi]\!] + \rho[\![u_\phi]\!]\frac{dx}{dt} = 0 . \tag{20}$$

We can solve (20) for $[\![\sigma_\phi]\!]$ and $[\![u_\phi]\!]$ if

$$\det \begin{vmatrix} \lambda \frac{dx}{dt} & \eta \\ 1 & \rho \frac{dx}{dt} \end{vmatrix} = \lambda\rho \left(\frac{dx}{dt}\right)^2 - \eta = 0 . \tag{21}$$

There are two families of characteristics defined by (21)

$$x \pm ct = \text{const} \tag{22}$$

where c is the wave speed (13).

1.5 Linearized Maxwell models

We make a tensorial equation of (17) by declaring that σ is a symmetric tensor field

$$\sigma(\mathbf{x},t) = \sigma^T(\mathbf{x},t) , \tag{23}$$

$\mathbf{u}(\mathbf{x},t)$ is a solenoidal vector field satisfying

$$\text{div}\mathbf{u} = 0 \tag{24}$$

and $\partial u/\partial x$ now stands for

$$2\mathbf{D}[\mathbf{u}] \overset{\text{def}}{=} \mathbf{A}_1[\mathbf{u}]$$

where $\mathbf{D}[\mathbf{u}]$ stands for the symmetric part of $\mathbf{L}(\mathbf{x},t) \overset{\text{def}}{=} \nabla\mathbf{u}$ ($L_{ij}=\partial u_i/\partial x_j$). The constitutive equation of the rate type generalizing (17) is

$$\sigma = -p\mathbf{1} + \tau , \tag{25}$$

$$\lambda \frac{\partial\tau}{\partial t} + \tau = 2\eta\, \mathbf{D}[\mathbf{u}] \tag{26}$$

where τ is called the extra stress and p is the reaction pressure. The equation expressing the balance of mass is (24) and the balance of momentum is

$$\rho \frac{\partial \mathbf{u}}{\partial t} = -\nabla p + \text{div}\, \tau \tag{27}$$

where the reaction pressure p=p[**u**] may be determined as a functional of **u** by the solution of the four equations (27) and (24) for the three components of **u** and p[**u**]. The reaction pressure is not determined by a constitutive expression, it varies from problem to problem, as in an incompressible Newtonian fluid.

1.6 Nonlinear Maxwell models

By declaration, we shall call fluids which obey a constitutive equation of the rate type

$$\lambda \frac{D\boldsymbol{\tau}}{Dt} + \boldsymbol{\tau} = 2\eta \mathbf{D}[\mathbf{u}] \tag{28}$$

Maxwell models. These models are not unique; they differ in that various invariant derivatives $D\boldsymbol{\tau}/Dt$ can be defined. Up to now we did not consider invariance. In fact all the invariant nonlinear derivatives reduce to partial time derivatives when linearized at states of rest,

$$\frac{D\boldsymbol{\tau}}{Dt} \to \frac{\partial \boldsymbol{\tau}}{\partial t}\,. \tag{29}$$

1.7 Form invariance and frame indifference

Many people believe that because constitutive equations characterize the response of the material they ought to be independent of the observer; two observers on different planets, or on different turntables on the same planet should come up with, say, the same equation relating stress and deformation and their equation should not depend on the frame. There are actually two requirements stated here. The first is that equations should be form invariant under Euclidean transformations (30) representing the change of frame. The second is that the invariant form should be independent of frame.

Suppose that the frame of the star observer at $\mathbf{x}^*$ is in a rigid motion so that

$$\mathbf{x}^* = \mathbf{Q}(t)\mathbf{x} + \mathbf{b}(t) \tag{30}$$

where $\mathbf{b}(t)$ is a time dependent constant and $\mathbf{Q}(t)$ is an orthogonal tensor which rotates vectors*; $\mathbf{Q}^{-1}(t)=\mathbf{Q}^T(t)$, $\mathbf{Q}\mathbf{Q}^T=\mathbf{1}$ and $\mathbf{x}$ is the point you see in your frame.

Our * observer studies some material which we think satisfies (28), and we will think him clever if he finds that

$$\lambda \frac{D^*\boldsymbol{\tau}^*}{D^*t} + \boldsymbol{\tau}^* = 2\eta\mathbf{D}^*[\mathbf{u}^*] \tag{31}$$

where the * objects have the same form as in (28). For example,

$$2\mathbf{D}^* = \nabla^*\mathbf{u}^* + (\nabla^*\mathbf{u}^*)^T$$

$$\left(2D^*_{ij} = \frac{\partial u^*_i}{\partial x^*_j} + \frac{\partial u^*_j}{\partial x^*_i}\right). \tag{32}$$

Now we know that the rate of strain is an *indifferent* tensor: under a change of frame (30), it transforms like an ordinary tensor

$$\mathbf{D}^*[\mathbf{u}^*] = \mathbf{Q}(t)\mathbf{D}[\mathbf{u}]\mathbf{Q}^T(t)\,. \tag{33}$$

Actually,

$$\mathbf{L}^*(\mathbf{x}^*,t) = \mathbf{Q}(t)\mathbf{L}(\mathbf{x},t)\mathbf{Q}^T(t) + \Omega[\mathbf{u}]$$

where

$$\Omega = \frac{d\mathbf{Q}(t)}{dt}\mathbf{Q}^T(t) = -\,\Omega^T \tag{34}$$

is an antisymmetric tensor (because $\mathbf{Q}\mathbf{Q}^T=\mathbf{1}$), does not transform as an indifferent tensor; but $\mathbf{L}+\mathbf{L}^T$ does. We are going to *assume* that the extra stress is an indifferent tensor

---

* If $\mathbf{Q}(t)$ is a rotation around, say, x_3, then the components of $\mathbf{Q}(t)$ in coordinates (x_1,x_2,x_3) are given by

$$[\mathbf{Q}(t)] = \begin{vmatrix} \cos\theta & \sin\theta & 0 \\ -\sin\theta & \cos\theta & 0 \\ 0 & 0 & 1 \end{vmatrix}$$

where $\theta=\theta(t)$.

$$\boldsymbol{\tau}^*(\mathbf{x}^*,t) = \mathbf{Q}(t)\boldsymbol{\tau}(\mathbf{x},t)\mathbf{Q}^T(t)\,. \tag{35}$$

Hence, our constitutive equation will be form invariant if we can find some invariant derivatives satisfying

$$\frac{D^*\boldsymbol{\tau}^*}{Dt} = \mathbf{Q}(t)\,\frac{D\boldsymbol{\tau}}{Dt}\,\mathbf{Q}^T(t)\,. \tag{36}$$

If (35) and (36) hold, then

$$\begin{aligned} &\lambda\,\frac{D^*\boldsymbol{\tau}^*}{Dt} + \boldsymbol{\tau}^* - 2\eta\mathbf{D}^*[\mathbf{u}^*] \\ &= \mathbf{Q}(t)\left\{\lambda\frac{D\boldsymbol{\tau}}{Dt} + \boldsymbol{\tau} - 2\eta\mathbf{D}[\mathbf{u}]\right\}\mathbf{Q}^T(t) \\ &= 0 \end{aligned} \tag{37}$$

and all the observers are in agreement about the form of the constitutive equation. Indifferent constitutive equations are form invariant but they may depend on the frame (cf. Exercise 1.13).

The belief that constitutive equations should not depend on the frame is called the "principle of material frame indifference." Perhaps it is stated as a principle because many people believe in it strongly even though they can't prove it. It seems that the exact circumstances under which dynamics give rise to constitutive equations which are independent of frame is a largely unexplored topic at the foundation of continuum mechanics. For a fuller discussion of these issues, see Joseph and Preziosi, "Addendum to the paper 'Heat Waves.'"

1.8 Frame independent invariant derivatives

A properly invariant tensor rate can be obtained if we interpret the rate operator as the time derivative $\partial/\partial t$ with respect to a reference frame suitably fixed to the body*. Different choices of body-fixed frames in this interpretation yield different invariant tensor rates. We present two examples in Equations (40) and (46) below.

---

* Oldroyd, J.G. On the formulation of rheological equations of state. *Proc. R. Soc.*, London **A200**, 523-41. A complete exposition of Oldroyd's approach to constitutive modeling is developed by Lodge [1974].

1.9 Upper convected invariant derivatives

To facilitate the presentation of the body-fixed frames in these examples, we cover the body with a convected coordinate system ξ^i. (A convected coordinate system deforms with the body so that the coordinates (ξ^1,ξ^2,ξ^3) associated with a particular material point do not change with time.) Base vectors $\mathbf{g}_i$ at a material point are defined by

$$\mathbf{g}_i(\xi^1,\xi^2,\xi^3,t) = \frac{\partial}{\partial \xi^i}\mathbf{x}(\xi^1,\xi^2,\xi^3,t) . \tag{38}$$

where $\mathbf{x}(\xi^1,\xi^2,\xi^3,t)$ is the position vector at time t of the material point with convected coordinates (ξ^1,ξ^2,ξ^3). A 3-D tensor $\mathbf{M}$ may be expressed with respect to the basis $\mathbf{g}_i \otimes \mathbf{g}_j$

$$\mathbf{M}(\xi,t) = M^{ij}(\boldsymbol{\xi},t)\, \mathbf{g}_i(\xi,t) \otimes \mathbf{g}_j(\xi,t) \tag{39}$$

where $\boldsymbol{\xi}$ is shorthand for the convected label (ξ^1,ξ^2,ξ^3) of the material point.

The frame at the material point with basis $\{\mathbf{g}_1,\mathbf{g}_2,\mathbf{g}_3\}$ rotates and deforms with the neighborhood of the point. As our first choice of invariant tensor rate, we define the rate of tensor $\mathbf{M}$ to be the time derivative of $\mathbf{M}$ with respect to this body fixed frame; explicitly,

$$\frac{D}{Dt}\mathbf{M} \overset{\text{def}}{=} \left[\frac{\partial}{\partial t} M^{ij}(\boldsymbol{\xi},t)\right] \mathbf{g}_i(\boldsymbol{\xi},t) \otimes \mathbf{g}_j(\boldsymbol{\xi},t) . \tag{40}$$

The rate defined in Equation (40) differs from the material derivative $\frac{\partial \mathbf{M}}{\partial t}$ since it ignores the change of the base vectors $\mathbf{g}_i$ with time.

We now remove the dependence on convected coordinates from the rate definition, Equation (40). Taking the material derivative of Equation (39) gives

$$\frac{d\mathbf{M}}{dt} \overset{\text{def}}{=} \frac{\partial}{\partial t}\mathbf{M}(\boldsymbol{\xi}, t) = \left[\frac{\partial}{\partial t} M^{ij}(\boldsymbol{\xi},t)\right] \mathbf{g}_i(\boldsymbol{\xi},t) \otimes \mathbf{g}_j(\boldsymbol{\xi},t)$$

$$+ M^{ij}(\boldsymbol{\xi},t)\left[\frac{\partial}{\partial t}\mathbf{g}_i(\boldsymbol{\xi},t)\right] \otimes \mathbf{g}_j(\boldsymbol{\xi},t) + M^{ij}(\boldsymbol{\xi},t)\, \mathbf{g}_i(\boldsymbol{\xi}, t) \otimes \left[\frac{\partial}{\partial t}\mathbf{g}_j(\boldsymbol{\xi},t)\right] . \tag{41}$$

We note that

$$\frac{d\mathbf{g}_i}{dt} \overset{\text{def}}{=} \frac{\partial}{\partial t}\mathbf{g}_i(\xi,t) = \frac{\partial}{\partial t}\left[\frac{\partial}{\partial \xi^i}\mathbf{x}(\xi,t)\right] = \frac{\partial}{\partial \xi^i}\left[\frac{\partial}{\partial t}\mathbf{x}(\xi,t)\right]$$

$$= \frac{\partial}{\partial \xi^i}\mathbf{v}(\xi,t) = \frac{\partial}{\partial \mathbf{x}}\mathbf{v}(\mathbf{x},t)\,\frac{\partial}{\partial \xi^i}\mathbf{x}(\xi,t) = \mathbf{L}\mathbf{g}_i\,. \tag{42}$$

Use of this result and the definition (40), in (41) gives, after some arranging,

$$\frac{D}{Dt}\mathbf{M} = \frac{d\mathbf{M}}{dt} - M^{ij}\mathbf{L}\mathbf{g}_i \otimes \mathbf{g}_j - M^{ij}\mathbf{g}_i \otimes \mathbf{L}\mathbf{g}_j$$

$$= \frac{d\mathbf{M}}{dt} - \mathbf{L}(M^{ij}\mathbf{g}_i \otimes \mathbf{g}_j) - (M^{ij}\mathbf{g}_i \otimes \mathbf{g}_j)\mathbf{L}^T \tag{43}$$

$$= \frac{d\mathbf{M}}{dt} - \mathbf{L}\mathbf{M} - \mathbf{M}\mathbf{L}^T \overset{\text{def}}{=} \overset{\nabla}{\mathbf{M}}\,.$$

The definition of $\frac{D}{Dt}\mathbf{M}$ in Equation (40) is therefore seen to be the upper convected rate, here denoted by $\overset{\nabla}{\mathbf{M}}$.

1.10 Lower convected invariant derivatives

Other body-fixed frames besides the frame with base vectors $\mathbf{g}_j$ can be defined. Reciprocal base vectors $\mathbf{g}^i$ are defined by

$$\mathbf{g}^i \bullet \mathbf{g}_j = \delta^i_j\,, \tag{44}$$

and we may express the tensor $\mathbf{M}$ as

$$\mathbf{M}(\xi,t) = M_{ij}(\xi,t)\,\mathbf{g}^i(\xi,t) \otimes \mathbf{g}^j(\xi,t)\,. \tag{45}$$

As an alternative to Equation (40), we can define the rate of tensor $\mathbf{M}$ by

$$\frac{D}{Dt}\mathbf{M} \overset{\text{def}}{=} \left[\frac{\partial}{\partial t}M_{ij}(\xi,t)\right]\mathbf{g}^i(\xi,t) \otimes \mathbf{g}^j(\xi,t)\,. \tag{46}$$

This rate differs from the upper convected rate defined in Equation (40) since the vectors $\mathbf{g}^i$ deform in time differently than $\mathbf{g}_i$. In fact, the rate defined in Equation (46) is the lower convected rate $\overset{\Delta}{\mathbf{M}}$, given by

$$\overset{\Delta}{\mathbf{M}} = \frac{d\mathbf{M}}{dt} + \mathbf{L}^T\mathbf{M} + \mathbf{M}\mathbf{L} \,. \tag{47}$$

Useful in the transformation of equations (46) to (47) is the relation

$$\frac{d\mathbf{g}^i}{dt} \overset{\text{def}}{=} \frac{\partial}{\partial t}\mathbf{g}^i(\xi, t) = -\mathbf{L}^T\mathbf{g}^i \,, \tag{48}$$

which follows from Equation (42) and the material derivative of Equation (44).

1.11 Corotational invariant derivatives

The corotational rate $\overset{\circ}{\mathbf{M}}$, defined by

$$\overset{\circ}{\mathbf{M}} \overset{\text{def}}{=} \frac{1}{2}(\overset{\nabla}{\mathbf{M}} + \overset{\Delta}{\mathbf{M}}) = \frac{d\mathbf{M}}{dt} - \mathbf{W}\mathbf{M} + \mathbf{M}\mathbf{W} \,,$$

$$\mathbf{W} = \frac{1}{2}(\mathbf{L} - \mathbf{L}^T) \,, \tag{49}$$

corresponds to the time derivative of $\mathbf{M}$ with respect to a basis which shares the rotation of the neighborhood of the material point, but not its deformation.

1.12 Other invariant derivatives

Another invariant rate is

$$\overset{\square}{\mathbf{M}} \overset{\text{def}}{=} \frac{1}{2}(\overset{\Delta}{\mathbf{M}} - \overset{\nabla}{\mathbf{M}}) = \mathbf{D}\mathbf{M} + \mathbf{M}\mathbf{D} \,.$$

Hence, the four rates $\overset{\nabla}{\mathbf{M}}$, $\overset{\Delta}{\mathbf{M}}$, $\overset{\circ}{\mathbf{M}}$, and $\overset{\square}{\mathbf{M}}$ are included as special cases in the general rate

$$\frac{D}{Dt}\mathbf{M} \overset{\text{def}}{=} \frac{d\mathbf{M}}{dt} - \mathbf{W}\mathbf{M} + \mathbf{M}\mathbf{W} - a(\mathbf{D}\mathbf{M} + \mathbf{M}\mathbf{D}) \tag{50}$$

where a is constant. This rate satisfies invariance, expressed by (36), and preserves symmetry (in the sense that the rate of a symmetric tensor is necessarily symmetric) for all values of a. The choices a=1,–1, and 0 correspond to $\overset{\nabla}{\mathbf{M}}$, $\overset{\Delta}{\mathbf{M}}$, and $\overset{\circ}{\mathbf{M}}$, respectively.

Some properly invariant rates, such as

$$\frac{D}{Dt}\mathbf{M} \overset{\text{def}}{=} \left[\frac{\partial}{\partial t}M^{j}_{i}\right] \mathbf{g}^{i} \otimes \mathbf{g}_{j} = \frac{d\mathbf{M}}{dt} - \mathbf{LM} + \mathbf{ML}$$

and

$$\frac{D}{Dt}\mathbf{M} \overset{\text{def}}{=} \left[\frac{\partial}{\partial t}M^{i}_{j}\right] \mathbf{g}_{i} \otimes \mathbf{g}^{j} = \frac{d\mathbf{M}}{dt} + \mathbf{L}^{T}\mathbf{M} - \mathbf{M}\mathbf{L}^{T},$$

have the disadvantage that the rate of a symmetric tensor is not in general symmetric.

1.13 List of Maxwell models

$$\lambda\overset{\nabla}{\boldsymbol{\tau}} + \boldsymbol{\tau} = 2\eta\mathbf{D} \quad \text{(upper convected Maxwell model, UCM)} \quad (51)$$

$$\lambda\overset{\Delta}{\boldsymbol{\tau}} + \boldsymbol{\tau} = 2\eta\mathbf{D} \quad \text{(lower convected Maxwell model, LCM)} \quad (52)$$

$$\lambda\overset{\circ}{\boldsymbol{\tau}} + \boldsymbol{\tau} = 2\eta\mathbf{D} \quad \text{(corotational Maxwell Model, COM)} \quad (53)$$

$$\lambda\frac{D\boldsymbol{\tau}}{Dt} + \boldsymbol{\tau} = 2\eta\mathbf{D} \quad \text{(interpolated Maxwell model)} \quad (54)$$

where $\frac{D\boldsymbol{\tau}}{Dt}$ is given by (50).

1.14 Invariant derivatives of vectors

We may find time derivatives of vectors which are invariant under superposed rigid motions by Oldroyd's method. We may decompose the vector $\mathbf{a}$ in the basis $\mathbf{g}_i$, or its dual $\mathbf{g}^i$:

$$\mathbf{a} = a^{i}\mathbf{g}_{i} = a_{i}\mathbf{g}^{i}\,.$$

The invariant differentiation is again carried out in the body fixed frame. Then, using (42) and (48), we find that

$$\frac{d\mathbf{a}}{dt} = \overset{\nabla}{\mathbf{a}} + \mathbf{L}\mathbf{a} = \overset{\Delta}{\mathbf{a}} - \mathbf{L}^T\mathbf{a} \, .$$

1.15 Integral forms of Maxwell models

The upper convected Maxwell model is given by (50) with a=1. It may be written as an integral model

$$\boldsymbol{\tau} = \frac{\eta}{\lambda^2}\int_{-\infty}^{t} \exp[-(t-\tau)/\lambda][\mathbf{C}_t^{-1}(\tau) - \mathbf{1}]\, d\tau \qquad (55)$$

where $\mathbf{C}_t^{-1}(\tau)$ is a relative strain tensor, called the Finger tensor. It is the inverse of

$$\mathbf{C}_t(\tau) \overset{\text{def}}{=} \mathbf{F}_t^T(\tau)\mathbf{F}_t(\tau) = \mathbf{C}_t^T(\tau) \qquad (56)$$

which is another measure of strain, called the right relative Cauchy-Green tensor, defined in terms of the relative deformation gradient

$$\mathbf{F}_t(\tau) \overset{\text{def}}{=} \nabla\boldsymbol{\chi}_t(\mathbf{x},\tau) \, ,$$

$$\left(F_{ij} = \frac{\partial \chi_i}{\partial x_j}\right)$$

where

$$\boldsymbol{\chi} = \boldsymbol{\chi}_t(\mathbf{x},\tau) \qquad (57)$$

is a backward path line of a particle $\tau \leq t$ presently at

$$\mathbf{x} = \boldsymbol{\chi}_t(\mathbf{x},t) \, .$$

The lower convected Maxwell model is given by (50) with a=–1. It may be written as an integral model

$$\boldsymbol{\tau} = \frac{\eta}{\lambda^2} \int_{-\infty}^{t} \exp[-(t-\tau)/\lambda][\mathbf{1} - \mathbf{C}_t(\tau)]\, d\tau \,. \tag{58}$$

I do not know of integral expressions for any of the one parameter family of parameters model (50) except a=±1.

The eigenvalues of the tensor $\mathbf{C}_t(\tau)$ are real, because the tensor is symmetric; they are positive because $\mathbf{F}_t(\tau)$ and its transpose have the same eigenvalues. This is very important because it restricts the range of $\boldsymbol{\tau}$, which cannot then take all possible values in $\mathbb{R}^6$.

To prove (58), we shall show that the differential equation for the upper convected Maxwell model arises from differentiating (58). For this calculation, we need to calculate the partial t derivative of the Cauchy-Green tensor.

$$\mathbf{C}(\tau) = \mathbf{C}(t,\mathbf{x},\tau) \overset{\text{def}}{=} \mathbf{C}_t(\tau)\,, \quad \mathbf{C}_t(t) = \mathbf{1}.$$

We find that

$$\frac{\partial \mathbf{C}(\tau)}{\partial t} = -\left(\mathbf{u}(t)\frac{\partial}{\partial \mathbf{x}}\right)\mathbf{C}(\tau) - \mathbf{C}(\tau)\mathbf{L}(t) - \mathbf{L}^T(t)\mathbf{C}(\tau)\,, \tag{59}$$

where $\mathbf{u}\frac{\partial}{\partial \mathbf{x}} = (\mathbf{u}\bullet\nabla)$.

To prove (59), we note that if two points $(\mathbf{x}_1,t_1)$ and $(\mathbf{x}_2,t_2)$ are on the same path line and $\tau<t_1$ and $\tau<t_2$, then

$$\boldsymbol{\chi}(\mathbf{x}_2,t_2,\tau) - \boldsymbol{\chi}(\mathbf{x}_1,t_1,\tau) = 0\,. \tag{60}$$

Passing to the limit $(\mathbf{x}_1,t_1)\to(\mathbf{x}_2,t_2)$, we find

$$\frac{d\boldsymbol{\chi}_t(\mathbf{x},\tau)}{dt} = \frac{\partial\boldsymbol{\chi}_t(\mathbf{x},\tau)}{\partial t} + (\mathbf{u}(\mathbf{x},\, t)\bullet\nabla)\boldsymbol{\chi}_t(\mathbf{x},\tau) = 0\,. \tag{61}$$

The path line is invariant following the motion.

Now we compute

$$\frac{\partial \mathbf{C}(\tau)}{\partial t} = \frac{\partial}{\partial t}\left(\mathbf{F}_t^T(\tau)\mathbf{F}_t(\tau)\right) = \frac{\partial \mathbf{F}_t^T(\tau)}{\partial t}\mathbf{F}_t(\tau) + \mathbf{F}_t^T(\tau)\frac{\partial \mathbf{F}_t(\tau)}{\partial t}$$

$$\left(\frac{\partial C_{ij}}{\partial t} = \frac{\partial F_{ei}}{\partial t}F_{ej} + F_{ei}\frac{\partial F_{ej}}{\partial t}\right)$$

and, using (61)

$$\frac{\partial \mathbf{F}_t}{\partial t} = \frac{\partial}{\partial \mathbf{x}}\left(\frac{\partial \boldsymbol{\chi}_t(\mathbf{x},\tau)}{\partial t}\right) = -\frac{\partial}{\partial \mathbf{x}}\left[\left(\mathbf{u}(\mathbf{x},t)\frac{\partial}{\partial \mathbf{x}}\right)\boldsymbol{\chi}_t(\mathbf{x},\tau)\right].$$

$$\left(\frac{\partial F_{ei}}{\partial t} = -\frac{\partial}{\partial x_i}\left(u_m\frac{\partial \chi_e}{\partial x_m}\right) = -\frac{\partial}{\partial x_i}(u_m F_{em})\right)$$

Then, using components

$$\begin{aligned}\frac{\partial C_{ij}}{\partial t} &= -\left(\frac{\partial}{\partial x_i}u_m F_{km}\right)F_{kj} - F_{ki}\frac{\partial}{\partial x_j}(u_m F_{km}) \\ &= -u_m\left[\frac{\partial F_{km}}{\partial x_i}F_{kj} + F_{ki}\frac{\partial}{\partial x_j}F_{km}\right] \\ &\quad - L_{mi}F_{km}F_{kj} - F_{ki}L_{mj}F_{km}\,.\end{aligned} \tag{62}$$

Since

$$\frac{\partial F_{km}}{\partial x_i} = \frac{\partial F_{ki}}{\partial x_m},$$

we get

$$\frac{\partial F_{km}}{\partial x_i}F_{kj} + F_{ki}\frac{\partial F_{km}}{\partial x_j} = \frac{\partial (F_{ki}F_{kj})}{\partial x_m}, \tag{63}$$

and (62) and (63) prove (59).

Now we write (58) as

$$\boldsymbol{\tau} = \frac{\eta}{\lambda}\mathbf{1} - \hat{\boldsymbol{\tau}}, \tag{64}$$

$$\hat{\boldsymbol{\tau}} = \frac{\eta}{\lambda^2}\int_{-\infty}^{t}\exp[-(t-\tau)/\lambda]\mathbf{C}(\tau)d\tau\,. \tag{65}$$

Differentiating (64), we get

$$\frac{\partial \boldsymbol{\tau}}{\partial t} = -\frac{\partial \hat{\boldsymbol{\tau}}}{\partial t} .$$

Using (65), (59), and (64), we get

$$\frac{\partial \hat{\boldsymbol{\tau}}}{\partial t} = -\mathbf{u}\frac{\partial \hat{\boldsymbol{\tau}}}{\partial \mathbf{x}} - \hat{\boldsymbol{\tau}}\mathbf{L}(t) - \mathbf{L}^T(t)\hat{\boldsymbol{\tau}} - \frac{\hat{\boldsymbol{\tau}}}{\lambda} + \frac{\eta}{\lambda^2}\mathbf{1}$$

and

$$\frac{\partial \boldsymbol{\tau}}{\partial t} = -\mathbf{u}\frac{\partial \boldsymbol{\tau}}{\partial \mathbf{x}} - \boldsymbol{\tau}\mathbf{L}(t) - \mathbf{L}^T(t)\boldsymbol{\tau} + \frac{\eta}{\lambda}(\mathbf{L}(t) + \mathbf{L}^T(t)) - \frac{\boldsymbol{\tau}}{\lambda} ,$$

proving that the integral expression (58) is equivalent to the rate equation (52).

In the same way, using (58), we may show that the integral equation (55) is equivalent to the rate equation (51) for the upper convected Maxwell model.

1.16 Restrictions of the range of stresses of the upper and lower convected Maxwell models

The restrictions have their origin in the fact that $\mathbf{C}_t(\tau)$ and $\mathbf{C}_t^{-1}(\tau)$ are positive tensors; their eigenvalues are always positive. Since they are symmetric, they may be diagonalized at each and every point $(\mathbf{x},t)$ and all $\tau \le t$ and at each such representative point.

$$\boldsymbol{\xi}\cdot\mathbf{C}_t(\tau)\boldsymbol{\xi} = \xi_1^2\lambda_1 + \xi_2^2\lambda_2 + \xi_3^2\lambda_3 \ge 0 ,$$

$$\boldsymbol{\xi}\cdot\mathbf{C}_t^{-1}(\tau)\boldsymbol{\xi} = \xi_1^2\lambda_1^{-1} + \xi_2^2\lambda_2^{-1} + \xi_3^2\lambda_3^{-1} \ge 0 ,$$

for all vectors $\boldsymbol{\xi}$. Therefore,

$$\boldsymbol{\xi}\cdot\left(\boldsymbol{\tau} + \frac{\eta}{\lambda}\mathbf{1}\right)\boldsymbol{\xi} = \frac{\eta}{\lambda^2}\int_{-\infty}^{t} \exp[-(t-\tau)/\lambda]\boldsymbol{\xi}\cdot\mathbf{C}_t^{-1}(\tau)\boldsymbol{\xi}\, d\tau$$

is not negative and

$$\xi \cdot \left(\boldsymbol{\tau} - \frac{\eta}{\lambda} \mathbf{1} \right) \xi = -\frac{\eta}{\lambda^2} \int_{-\infty}^{t} \exp[-(t-\tau)/\lambda] \xi \cdot \mathbf{C}_t(\tau) \xi \, d\tau$$

is not positive;

$$\xi \cdot \left(\boldsymbol{\tau} + \frac{\eta}{\lambda} \mathbf{1} \right) \xi \geq 0 \quad \text{(UCM)} \tag{66}$$

$$\xi \cdot \left(\boldsymbol{\tau} - \frac{\eta}{\lambda} \mathbf{1} \right) \xi \leq 0 \quad \text{(LCM)} \tag{67}$$

for all $\xi \in \mathbb{R}^3$. Now choose ξ to lie on the principal diagonals of $\boldsymbol{\tau}$

$$\tau_{11} + \frac{\eta}{\lambda} \geq 0, \quad \tau_{22} + \frac{\eta}{\lambda} \geq 0, \quad \tau_{33} + \frac{\eta}{\lambda} \geq 0 \qquad \text{(UCM)}$$

$$\tau_{11} - \frac{\eta}{\lambda} \leq 0, \quad \tau_{22} - \frac{\eta}{\lambda} \leq 0, \quad \tau_{33} - \frac{\eta}{\lambda} \leq 0 \qquad \text{(LCM)}$$

We shall see that the restrictions (66) and (67) on the range of the stresses suffice to guarantee that the upper and lower convected Maxwell models never become catastrophically unstable to short waves; they are never Hadamard unstable. Similar results cannot be derived for the other Maxwell models; for example, the corotational model can become Hadamard unstable on certain solutions.

1.17 Quasilinear models like Maxwell's which differ in lower order terms

Maxwell models are quasilinear because the derivatives of $\boldsymbol{\tau}$ and $\mathbf{u}$ appear linearly and only linearly, a term like $\boldsymbol{\tau}\mathbf{L}$ is nonlinear but linear in derivatives of $\mathbf{u}$, $\mathbf{L}=\nabla\mathbf{u}$. Models of the form

$$\lambda \frac{D\boldsymbol{\tau}}{Dt} + \mathbf{f}(\boldsymbol{\tau}) = 2\eta \mathbf{D}[\mathbf{u}] \tag{68}$$

where $\mathbf{f}(\bullet)$ is a nonlinear function and D/Dt is the invariant derivative (50) for interpolated models. A model due to Giesekus [1982] is expressed in terms of the upper convected derivative

$$\lambda\overset{\nabla}{\boldsymbol{\tau}} + \boldsymbol{\tau} + c\boldsymbol{\tau}^2 = 2\eta\mathbf{D}[\mathbf{u}]\ , \tag{69}$$

where $c=\alpha\lambda/\eta$ and α is a mobility factor. The Giesekus model contains four parameters: a relaxation time λ, the solvent and polymer contributions to the zero-shear-rate viscosity, μ and η, and the dimensionless mobility factor α. In this case $\mu=0$.

Phan-Thien and Tanner [1977] proposed another model with a different nonlinear lower order term expressed as $Y(\mathrm{tr}\ \boldsymbol{\tau})\boldsymbol{\tau}$, where Y is a scalar function of the trace of the extra stress.

$$\lambda\frac{D\boldsymbol{\tau}}{Dt} + \boldsymbol{\tau}Y\ (\mathrm{tr}\ \boldsymbol{\tau}) = 2\eta\mathbf{D}[\mathbf{u}] \tag{70}$$

where D/Dt, given by (50), is derived from a molecular argument using a network which is allowed to be non-affine.

Various forms of the function $Y(\bullet)$ can be used for curve fitting to certain rheological data. Two such forms are

$$Y(\mathrm{tr}\ \boldsymbol{\tau}) = \exp\left\{\frac{\varepsilon\lambda}{\eta}\ \mathrm{tr}\ \boldsymbol{\tau}\right\}\ ,$$

$$Y(\mathrm{tr}\ \boldsymbol{\tau}) = 1 + c\ \mathrm{tr}\ \boldsymbol{\tau} \tag{71}$$

where ε and c are constants and $c=\varepsilon\lambda/\eta$ when the second form is a small argument approximation of the first.

1.18 Nonlinear models which are not quasilinear

The constitutive equations are expressed by rate equations, but in some place first derivatives of $\boldsymbol{\tau}$ and $\mathbf{u}$ appear nonlinearly. An example of this is the White-Metzner model in which the viscosity and the relaxation time depend on the second invariant $II = \frac{1}{2}\ \mathrm{tr}\ \mathbf{D}^2[\mathbf{u}]$ of the rate of strain.

$$\lambda(II)\frac{D\boldsymbol{\tau}}{Dt} + \boldsymbol{\tau} = 2\eta(II)\mathbf{D}[\mathbf{u}]\ . \tag{72}$$

Obviously $\lambda(\mathrm{II})\frac{\partial\tau}{\partial t}$ and $\eta(\mathrm{II})\mathbf{D}[\mathbf{u}]$ are not quasilinear. Many nonlinear models which are not quasilinear have been proposed.

1.19 Constitutive equations of Boltzmann's type

Ludwig Boltzmann [1874] introduced a linear integral representation of stress in viscoelastic solids. When applied to fluids in the one-dimensional case, Boltzmanns' equation becomes

$$\tau = \int_{-\infty}^{t} G(t-\tau') \frac{\partial u(x,\tau')}{\partial x} d\tau' . \tag{73}$$

Maxwell's equation is a special case of Boltzmann's with

$$G(t-\tau') = \frac{\eta}{\lambda} \exp[-(t-\tau')/\lambda] . \tag{74}$$

For steady flow $u(x,\tau')$ is independent of τ' and

$$\tau = \eta \frac{\partial u(x)}{\partial x} . \tag{75}$$

I call

$$\eta = \int_{0}^{\infty} G(s)\ ds$$

the elastic viscosity.

We call $G(s)$ the shear relaxation function and we assume that it is a positive, monotone-decreasing function which tends to zero for large values of s, like the exponential (74). We shall also assume that like the exponential, $G(s)$ is a convex function. These assumptions are the usual ones, they express what physical intuition demands of stress relaxation (§1.20) and they are necessary for the stability of the fluid when it is in a state of rest.

Equation (73) may be generalized in the same steps as Maxwell's model using the replacement

$$\frac{\partial G(t-\tau)}{\partial \tau} = \frac{\eta}{\lambda^2} \exp[-(t-\tau)/\lambda]$$

implied by (74). Then, for example, (55) may be written as

$$\boldsymbol{\tau} = \int_{-\infty}^{t} \frac{\partial G(t-\tau)}{\partial \tau} [\mathbf{C}_t^{-1}(\tau) - \mathbf{1}] \, d\tau$$

$$= - \int_{-\infty}^{t} G(t-\tau) \frac{\partial \mathbf{C}_t^{-1}(\tau)}{\partial \tau} \, d\tau \, . \tag{76}$$

BKZ models generalize this by allowing the relaxation function to depend on strain invariants. All these models reduce to

$$\boldsymbol{\tau}(\mathbf{x},t) = 2 \int_0^{\infty} G(t-\tau) \, \mathbf{D}[\mathbf{u}(\mathbf{x},\tau)] \, d\tau \tag{77}$$

when linearized on a state of rest. In this sense, G(s) is a model independent material function for incompressible fluids with instantaneous elasticity.

1.20 Wave speeds and stress relaxation for models of Boltzmann's type

We determine the rate at which τ changes with t for motions perturbing rest by differentiating (73):

$$\frac{\partial \tau}{\partial t} = G(0) \frac{\partial u(x,t)}{\partial x} + \int_{-\infty}^{t} G'(t-\tau') \frac{\partial u}{\partial x}(x,\tau') \, d\tau' \, . \tag{78}$$

This equation is the analogue of (17). The equation of motion is

$$\rho \frac{\partial u}{\partial t} = \frac{\partial \tau}{\partial x} \tag{79}$$

where τ is given by (73).

There is no pressure gradient in (78). We can think of a field of shear, there is one component of **u**, u(x) where x is perpendicular to u. To get the wave speed we use the method of simple jumps; u(x,t) and τ(x,t) are continuous but derivatives of u and τ jump across x = $x_0(t)$. Then

$$\left[\!\left[\frac{\partial\tau}{\partial t}\right]\!\right] = G(0)\left[\!\left[\frac{\partial u}{\partial x}\right]\!\right],$$

$$\rho\left[\!\left[\frac{\partial u}{\partial t}\right]\!\right] = \left[\!\left[\frac{\partial\tau}{\partial x}\right]\!\right].$$

The point here is that

$$\left[\!\left[\frac{\partial u(x,t)}{\partial x}\right]\!\right] = 0$$

at each $x \neq x_0(t)$ so that the jump of the last term of (78) vanishes. Following now the development after (18) we get

$$\left(\frac{dx_0}{dt}\right)^2 - c^2 = 0$$

where

$$c^2 = G(0)/\rho\,. \tag{80}$$

The instantaneous value G(0) of the relaxation function G(s) is the shear modulus or rigidity.

Later we shall learn that the multidimensional generalization of a shear wave is a wave of vorticity. Vorticity waves propagate at a speed c given by (80).

Suppose we now consider a sudden step in strain

$$\mathbf{C}_t^{-1}(\tau)-\mathbf{1} = -\mathbf{C}_t^{-1}(0)H(\tau) \tag{81}$$

where H(τ) is a step function

$$H(\tau) = \begin{pmatrix} 1\,, \ \tau > 0 \\ 0\,, \ \tau < 0 \end{pmatrix}.$$

To see how the stress relaxes after a step of the strain at t=0 we substitute (81) into (76) and integrate.

$$\begin{aligned} \boldsymbol{\tau} &= -\mathbf{C}_t^{-1}(0) \int_0^t \frac{\partial G(t-\tau)}{\partial \tau} H(\tau)\, d\tau \\ &= \mathbf{C}_t^{-1}(0)\, G(t)\,. \end{aligned}$$

The function G(t) describes the relaxation of the stress after the application of a constant strain, $\boldsymbol{\tau} \rightarrow 0$ as $t \rightarrow \infty$. The deformation (81) is actually impossible in a fluid, even in a viscoelastic fluid. The fluid will start to flow, the strain cannot be maintained.

We shall say the models which exhibit finite wave speeds and stress relaxation after a sudden strain possess instantaneous elasticity. It is enough to say that if the rigidity G(0), sometimes called the instantaneous shear modulus, is finite then the fluid possesses instantaneous elasticity. The idea of fading memory may be expressed already in what intuition would demand from stress relaxation. The effect of the initial t=0 change of strain in (81) should decay monotonically in time, with the most rapid decay at early times. This leads to viscosity, monotonicity and convexity of G(t).

Exercises

Exercise 1.1.
Point differences of two vectors are indifferent. (Indifferent tensors transform like tensors under a change of frame.)

Exercise 1.2.
The velocity **u** at a point $\mathbf{x}=\mathbf{x}(\xi,t)$, where ξ is any point of reference independent of t, is transformed to

$$\mathbf{u}^* = \mathbf{Q}(t)\mathbf{u} + \frac{d\mathbf{Q}}{dt}\mathbf{x} + \frac{d\mathbf{b}}{dt}\,. \tag{82}$$

Exercise 1.3.

Find a formula showing how the acceleration $\frac{d\mathbf{u}^*}{dt}$ transforms under a change of frame.

Exercise 1.4.

$$\mathbf{L}^* = \mathbf{Q}(t)\mathbf{L}\mathbf{Q}^T(t) + \Omega\,, \tag{83}$$

$$\mathbf{D}^* = \mathbf{Q}(t)\mathbf{D}\mathbf{Q}^T(t)\,,$$

$$\mathbf{W}^* = \mathbf{Q}(t)\mathbf{W}\mathbf{Q}^T(t) + \Omega\,, \tag{84}$$

$$\mathbf{W} = \text{skew symmetric part of } \mathbf{L}\,,$$

$$\mathbf{F}_t^*(\tau) = \mathbf{Q}(\tau)\mathbf{F}_t(\tau)\mathbf{Q}^T(t) \text{ is not indifferent}\,, \tag{85}$$

$$\mathbf{C}_t^*(\tau) = \mathbf{Q}(t)\mathbf{C}_t(\tau)\mathbf{Q}^T(t)\,.$$

Exercise 1.5.

The partial time derivative of a scalar $\phi^*=\phi(\mathbf{x},t)$ is not indifferent

$$\frac{\partial\phi^*}{\partial t^*} = \frac{\partial\phi}{\partial t} - \mathbf{Q}^T\left(\frac{d\mathbf{b}}{dt} + \frac{d\mathbf{Q}}{dt}\mathbf{x}\right)\nabla\phi\,. \tag{86}$$

Exercise 1.6.

The substantial time derivative of ϕ is indifferent (generalized Galilean invariance).

$$\frac{\partial\phi^*}{\partial t^*} + \mathbf{u}^*\frac{\partial\phi^*}{\partial\mathbf{x}^*} = \frac{\partial\phi}{\partial t} + \mathbf{u}\frac{\partial\phi}{\partial\mathbf{x}}\,. \tag{87}$$

Exercise 1.7.

Prove that the derivative

$$\overset{\nabla}{\mathbf{a}} = \frac{d\mathbf{a}}{dt} - \mathbf{L}\mathbf{a}$$

of an indifferent vector $\mathbf{a}$, $\mathbf{a}^*=\mathbf{Q}\mathbf{a}$, is indifferent.

Exercise 1.8.
Prove that the derivative

$$\overset{\nabla}{\mathbf{M}} = \frac{d\mathbf{M}}{dt} - \mathbf{L}\mathbf{M} - \mathbf{M}\mathbf{L}^T$$

of an indifferent tensor **M** is indifferent.

Exercise 1.9.
Show that

$$\frac{\partial \mathbf{C}_t^{-1}(\tau)}{\partial t} = -\mathbf{u}(t) \frac{\partial \mathbf{C}_t^{-1}(\tau)}{\partial \mathbf{x}} + \mathbf{L}(t)\mathbf{C}_t^{-1}(\tau) + \mathbf{C}_t^{-1}(\tau)\mathbf{L}^T(t) \,. \quad (88)$$

(Hint: $\dfrac{\partial \mathbf{C}_t^{-1}(\tau)}{\partial t} = -\mathbf{C}_t^{-1}(\tau) \dfrac{\partial \mathbf{C}_t(\tau)}{\partial t} \mathbf{C}_t^{-1}(\tau)$.)

Exercise 1.10.
Prove that the eigenvalues of $\mathbf{C}_t(\tau) = \mathbf{F}_t^T\mathbf{F}_t$ and $\mathbf{C}_t^{-1}$ are not negative.

Exercise 1.11.
(Upper convected Maxwell model with m times relaxation)

An upper convected Maxwell model with m times of relaxation may be defined by the equation

$$\boldsymbol{\tau}_n = \int_{-\infty}^{t} \frac{G_n}{\lambda_n} \left\{ \mathbf{C}_t^{-1}(\tau) - \mathbf{1} \right\} \exp\left[-(t-\tau)/\lambda_n\right] d\tau \,, \quad \boldsymbol{\tau} = \sum_{n=1}^{m} \boldsymbol{\tau}_n$$

Use (88) to reduce this expression for $\boldsymbol{\tau}$ to

$$\overset{\nabla}{\boldsymbol{\tau}} = 2\, G(0)\, \mathbf{D}[\mathbf{u}] + \sum_{n=1}^{m} \frac{\boldsymbol{\tau}_n}{\lambda_n} \quad (89)$$

where

$$G(0) = \sum_{n=1}^{m} G_n \ .$$

Equation (89) is a quasilinear equation whose principal part is identical to an UCM with a single time of relaxation and a rigidity G(0).

Exercise 1.12.

$\hat{O}$ is the origin of an inertial frame. O and O* are the origins of moving frames. **d** and **d*** are position vectors of $\hat{O}$ relative of O and O* and **x**, **x*** and $\hat{\mathbf{x}}$ are position vectors for a moving body in O, O* and $\hat{O}$, respectively.

We define a change of frame for **x*** and $\hat{\mathbf{x}}$, **x** and $\hat{\mathbf{x}}$ using orthogonal matrices **P*** and **P** as follows

$$\mathbf{x}^* - \mathbf{d}^* = \mathbf{P}^*\hat{\mathbf{x}} \ ,$$
$$\mathbf{x} - \mathbf{d} = \mathbf{P}\hat{\mathbf{x}} \ , \ (x_i = P_{ij}\hat{x}_j) \ . \tag{90}$$

Show that (30) holds with

$$\mathbf{Q} = \mathbf{P}^*\mathbf{P}^T \quad (Q_{ij} = P^*_{ie}P_{je}) \tag{91}$$

and

$$\mathbf{d}^* = \mathbf{Q}\mathbf{d} + \mathbf{b} \ . \tag{92}$$

Define an angular velocity tensor relative to the inertial frame

$$\hat{\Omega} = \dot{\mathbf{P}}\mathbf{P}^T \tag{93}$$

Show that this tensor is skew-symmetric

$$\hat{\Omega} = -\hat{\Omega}^T \tag{94}$$

and that it may be expressed in terms of the three components of an angular velocity vector $\boldsymbol{\omega}$

$$\hat{\Omega}_{ik} = -\varepsilon_{ijk}\omega_j$$

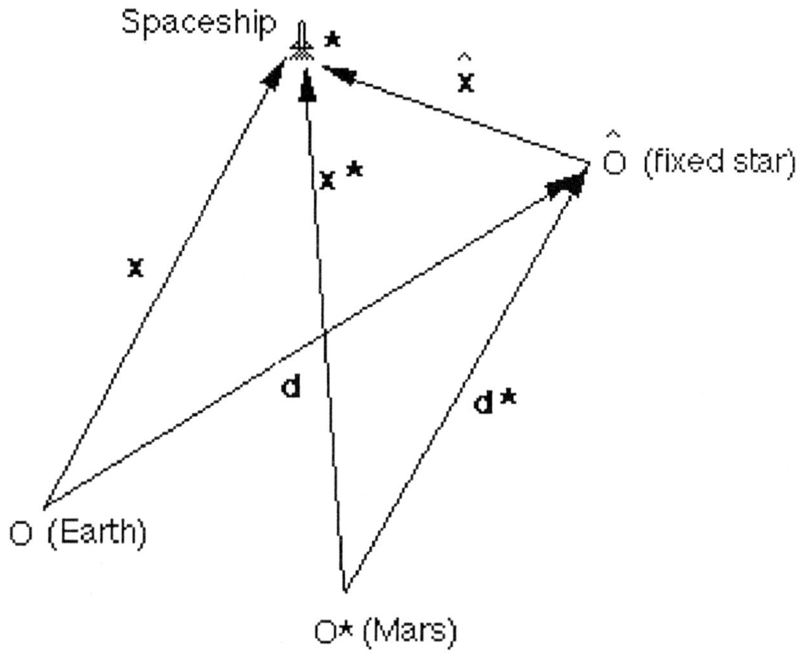

Figure 1.2 Vectors involved in the Euclidean transformation called change of frame. Let $\hat{e}_i$ be a fixed orthonormal basis in the inertial frame $\hat{O}$, $\mathbf{e}_i(t)$ are basis vectors fixed in O which rotate relative to $\hat{O}$, $\mathbf{e}_i^*(t)$ are similar. In the Figure $(x_i-d_i)\mathbf{e}_i=(x_i^*-d_i^*)\mathbf{e}_i^*=\hat{x}_i\hat{\mathbf{e}}_i$. Hence $(x_i-d_i)=P_{ij}\hat{x}_j$, $x_i^*-d_i^*=P_{ij}^*\hat{x}_j$, where $P_{ij}=\mathbf{e}_i\cdot\hat{\mathbf{e}}_j$, $P_{ij}^*=\hat{\mathbf{e}}_i\cdot\mathbf{e}_j^*$.

such that for any vector $\mathbf{v}$

$$\hat{\Omega}\mathbf{v} = -\boldsymbol{\omega}\wedge\mathbf{v}\ ,$$
$$\hat{\Omega}^2\mathbf{v} = \omega\wedge(\boldsymbol{\omega}\wedge\mathbf{v})\ . \qquad (95)$$

Show that $\hat{\Omega}$ transforms under (30) according to the rule

$$\hat{\Omega}^* = \mathbf{Q}\hat{\Omega}\mathbf{Q}^T + \dot{\mathbf{Q}}\mathbf{Q}^T\ . \qquad (96)$$

Exercise 1.13.
Let $\mathbf{e}_i(t)$ be an orthonormal cartesian basis in O, $\mathbf{e}_i^*(t)$ in O*. Further set

$$\begin{aligned} \mathbf{x} - \mathbf{d} &= \mathbf{y} = y_i\mathbf{e}_i(t) , \\ \mathbf{x}^* - \mathbf{d}^* &= \mathbf{y}^* = y_i^*\mathbf{e}_i^*(t) . \end{aligned} \tag{97}$$

$$\begin{aligned} \mathbf{v} &= \mathbf{e}_i v_i(\mathbf{x}, t) , \quad v_i = \dot{y}_i , \quad u_i(\mathbf{x}, t) = \dot{x}_i , \\ \mathbf{v}^* &= \mathbf{e}_i^* v_i^*(\mathbf{x}^*, t) , \; v_i^* = \dot{y}_i^* , \quad u_i^*(\mathbf{x}^*, t) = \dot{x}_i^* . \end{aligned} \tag{98}$$

Let

$$\begin{aligned} \mathbf{a} &\overset{\text{def}}{=} \ddot{\mathbf{y}} , \\ \mathbf{a}^* &= \ddot{\mathbf{y}}^* \end{aligned} \tag{99}$$

be the accelerations relative to a fixed star at $\hat{O}$ as seen by observers in the moving frames O and O* shown in Figure 1.2, respectively. Show that

$$\mathbf{a} = \mathbf{e}_i\left[\frac{\partial v_i}{\partial t} + \mathbf{u}\frac{\partial}{\partial \mathbf{x}} v_i\right] + 2\boldsymbol{\omega}\wedge\mathbf{v} + \boldsymbol{\omega}\wedge(\boldsymbol{\omega}\wedge\mathbf{y}) + \dot{\boldsymbol{\omega}}\wedge\mathbf{y} \tag{100}$$

and $\mathbf{a}^*$ is given by the same expression using $\mathbf{e}_i^*$, v_i^*, $\mathbf{u}^*$, $\mathbf{x}^*$, ω^*, $\mathbf{y}^*$. Show further that

$$\mathbf{a} = \mathbf{e}_i\left(\frac{\partial u_i}{\partial t} + \mathbf{u}\frac{\partial}{\partial \mathbf{x}}u_i\right) + 2\boldsymbol{\omega}\wedge\mathbf{u} + \boldsymbol{\omega}\wedge(\boldsymbol{\omega}\wedge\mathbf{u}) + \dot{\boldsymbol{\omega}}\wedge\mathbf{u} - \ddot{\mathbf{b}} , \tag{101}$$

and that

$$\mathbf{a}^* = \mathbf{Q}(t)\mathbf{a} \tag{102}$$

is indifferent, but frame dependent.

Suppose that the force $\mathbf{f}$ is indifferent. Conclude, the equations of motion

$$\rho^*\mathbf{a}^* - \mathbf{f}^* = \mathbf{Q}(\rho\mathbf{a} - \mathbf{f}) = 0 \tag{103}$$

are indifferent, but frame dependent.

Exercise 1.14.
Assume η in (21) is negative and $\lambda\to\infty$ and show that u satisfies Laplace's equation

$$\frac{\partial^2 u}{\partial t^2} + \frac{|\eta|}{\rho\lambda}\frac{\partial^2 u}{\partial x^2} = 0 .$$

Prove that there are no real characteristics.

Solutions for the exercises

Exercise 1.1.
Point differences of two vectors are indifferent

$$\mathbf{x}^*_\ell = \mathbf{Q}(t)\,\mathbf{x}_\ell + \mathbf{b} , \quad \ell = 1 \text{ and } 2$$

$$\mathbf{x}^*_1 - \mathbf{x}^*_2 = \mathbf{Q}(t)(\mathbf{x}_1 - \mathbf{x}_2) .$$

Exercise 1.2.

$$\mathbf{x}^*(\xi,t) = \mathbf{Q}(t)\mathbf{x}(\xi,t) + \mathbf{b} ,$$

$$\frac{d\mathbf{x}^*}{dt} = \mathbf{Q}(t)\frac{d\mathbf{x}}{dt} + \dot{\mathbf{Q}}\mathbf{x} + \dot{\mathbf{b}} ,$$

$$\frac{d\mathbf{x}^*}{dt} = \mathbf{u}^* , \quad \frac{d\mathbf{x}}{dt} = \mathbf{u} .$$

Exercise 1.3.

$$\mathbf{u}^*(\xi,t) = \mathbf{Q}(t)\mathbf{u}(\xi,t) + \dot{\mathbf{Q}}\mathbf{x} + \dot{\mathbf{b}} , \tag{104}$$

$$\frac{d\mathbf{u}^*}{dt} = \mathbf{Q}(t)\frac{d\mathbf{u}}{dt} + 2\dot{\mathbf{Q}}(t)\mathbf{u} + \ddot{\mathbf{Q}}\mathbf{x} + \ddot{\mathbf{b}} . \tag{105}$$

Exercise 1.4.
First, note that $\partial\mathbf{x}/\partial\mathbf{x}^* = \mathbf{Q}^T(t)$. Then from (104)

$$\frac{\partial \mathbf{u}^*}{\partial \mathbf{x}^*} = \mathbf{Q}(t)\,\frac{\partial \mathbf{u}}{\partial \mathbf{x}}\frac{\partial \mathbf{x}}{\partial \mathbf{x}^*} + \dot{\mathbf{Q}}\frac{\partial \mathbf{x}}{\partial \mathbf{x}^*} = \mathbf{QLQ}^T + \dot{\mathbf{Q}}\mathbf{Q}^T .$$

Hence,

$$\mathbf{L}^* = \mathbf{QLQ}^{\mathrm{T}} + \Omega\,, \quad \Omega = \dot{\mathbf{Q}}\mathbf{Q}^{\mathrm{T}} = -\,\mathbf{Q}\dot{\mathbf{Q}}^{\mathrm{T}} = -\Omega^{\mathrm{T}}, \tag{106}$$

$$\begin{aligned} 2\mathbf{D}^* &= \mathbf{L}^* + \mathbf{L}^{*\mathrm{T}} = \mathbf{Q}(\mathbf{L}+\mathbf{L}^{\mathrm{T}})\mathbf{Q}^{\mathrm{T}} + \Omega + \Omega^{\mathrm{T}} = 2\mathbf{QDQ}^{\mathrm{T}}\,, \\ 2\mathbf{W}^* &= \mathbf{L}^* - \mathbf{L}^{*\mathrm{T}} = \mathbf{QLQ}^{\mathrm{T}} - \mathbf{QL}^{\mathrm{T}}\mathbf{Q}^{\mathrm{T}} + \Omega - \Omega^{\mathrm{T}} \\ &= 2\mathbf{QWQ}^{\mathrm{T}} + 2\Omega\,, \end{aligned}$$

$$\boldsymbol{\chi}_t^*(\mathbf{x}^*,\tau) = \mathbf{Q}(\tau)\boldsymbol{\chi}_t(\mathbf{x},\tau) + \mathbf{b}(\tau) \quad \text{change of frame}\,,$$

$$\mathbf{F}_t^*(\tau) = \frac{\partial \chi_t^*(\mathbf{x}^*,t)}{\partial \mathbf{x}^*} = \mathbf{Q}(\tau)\frac{\partial \chi_t}{\partial \mathbf{x}}\frac{\partial \mathrm{x}}{\partial \mathbf{x}^*} = \mathbf{Q}(\tau)\mathbf{F}_t(\tau)\mathbf{Q}^{\mathrm{T}}(t)\,. \tag{107}$$

$$\begin{aligned} \mathbf{C}_t^* = \mathbf{F}_t^{*\mathrm{T}}\mathbf{F}_t^* &= \mathbf{Q}(t)\mathbf{F}_t^{\mathrm{T}}(\tau)\mathbf{Q}^{\mathrm{T}}(\tau)\,\mathbf{Q}(\tau)\,\mathbf{F}_t(\tau)\mathbf{Q}^{\mathrm{T}}(t) \\ &= \mathbf{Q}(t)\,\mathbf{F}_t^{\mathrm{T}}(\tau)\,\mathbf{F}_t(\tau)\mathbf{Q}^{\mathrm{T}}(t)\,. \end{aligned}$$

Exercise 1.5.

$$\dot{\phi} \overset{\text{def}}{=} \frac{\partial \phi}{\partial t} + (\mathbf{u}\bullet\nabla)\phi\,,$$

$$\phi_t \overset{\text{def}}{=} \frac{\partial \phi}{\partial t}\,.$$

Chain rule

$$\frac{\partial \phi(\mathbf{x},t)}{\partial t^*} = \frac{\partial t}{\partial t^*}\,\phi_t(\mathbf{x},t) + \frac{\partial \mathbf{x}}{\partial t^*}\frac{\partial \phi}{\partial \mathbf{x}} \quad (\mathbf{x}^* \text{ is fixed})\,,$$

$$\mathbf{x}^* = \mathbf{Q}(t)\mathbf{x} + \mathbf{b}(t)\,,$$

$$d\mathbf{x}^* = \mathbf{Q}(t)d\mathbf{x} + [\dot{\mathbf{Q}}\mathbf{x} + \dot{\mathbf{b}}]dt\,.$$

When $\mathbf{x}^*$ is fixed,

$$\mathbf{Q}(t)\frac{\partial \mathbf{x}}{\partial t} = -\,\dot{\mathbf{Q}}\mathbf{x} - \dot{\mathbf{b}}\,. \tag{108}$$

Hence,

$$\frac{\partial \phi^*(\mathbf{x}^*,t^*)}{\partial t^*} = \frac{\partial \phi}{\partial t} - [\mathbf{Q}^T(t)\dot{\mathbf{Q}}\mathbf{x} + \mathbf{Q}^T(t)\dot{\mathbf{b}}]\frac{\partial \phi}{\partial \mathbf{x}} . \tag{109}$$

Exercise 1.6.

$$\mathbf{u}^* \frac{\partial \phi^*}{\partial \mathbf{x}^*} = [\mathbf{Q}\mathbf{u} + \dot{\mathbf{Q}}\mathbf{x} + \dot{\mathbf{b}}]\frac{\partial \phi}{\partial \mathbf{x}}\frac{\partial \mathbf{x}}{\partial \mathbf{x}^*} = \mathbf{Q}^T(t)\,[\mathbf{Q}\mathbf{u} + \dot{\mathbf{Q}}\mathbf{x} + \dot{\mathbf{b}}]\frac{\partial \phi}{\partial \mathbf{x}}$$

$$= \mathbf{u}\frac{\partial \phi}{\partial \mathbf{x}} + \mathbf{Q}^T\left(\frac{d\mathbf{b}}{dt} + \frac{d\mathbf{Q}}{dt}\mathbf{x}\right)\nabla\phi . \tag{110}$$

Hence, using (109),

$$\frac{\partial \phi^*}{\partial t^*} + \mathbf{u}^*\frac{\partial \phi}{\partial \mathbf{x}^*} = \frac{\partial \phi}{\partial t} + \mathbf{u}\frac{\partial \phi}{\partial \mathbf{x}} \tag{111}$$

or

$$\frac{d^*\phi^*}{dt^*} = \frac{d\phi}{dt} .$$

Exercise 1.7.

$$\frac{d^*\mathbf{a}^*}{dt^*} = \frac{\partial \mathbf{Q}\mathbf{a}}{\partial t^*} + \mathbf{u}^*\frac{\partial \mathbf{Q}\mathbf{a}}{\partial \mathbf{x}^*} ,$$

$$\frac{\partial \mathbf{Q}\mathbf{a}}{\partial t^*} = \frac{\partial \mathbf{Q}\mathbf{a}}{\partial t} - \mathbf{Q}^T(t)\,(\dot{\mathbf{Q}}\mathbf{x} + \dot{\mathbf{b}})\frac{\partial \mathbf{Q}\mathbf{a}}{\partial \mathbf{x}} ,$$

$$\mathbf{u}^*\frac{\partial \mathbf{Q}\mathbf{a}}{\partial \mathbf{x}^*} = \mathbf{u}\frac{\partial \mathbf{Q}\mathbf{a}}{\partial \mathbf{x}} + (\mathbf{Q}^T\dot{\mathbf{Q}}\mathbf{x} + \mathbf{Q}^T\dot{\mathbf{b}})\frac{\partial \mathbf{Q}\mathbf{a}}{\partial \mathbf{x}} \tag{112}$$

$$\frac{d^*\mathbf{a}^*}{dt^*} = \frac{\partial \mathbf{Q}\mathbf{a}}{\partial t} + \mathbf{u}\frac{\partial \mathbf{Q}\mathbf{a}}{\partial \mathbf{x}}$$
$$= \dot{\mathbf{Q}}\mathbf{a} + \mathbf{Q}\frac{d\mathbf{a}}{dt} ,$$

$$\mathbf{L}^*\mathbf{a}^* = (\mathbf{Q}\mathbf{L}\mathbf{Q}^T + \Omega)\mathbf{Q}\mathbf{a} = \mathbf{Q}\mathbf{L}\mathbf{a} + \dot{\mathbf{Q}}\mathbf{Q}^T\mathbf{Q}\mathbf{a} = \mathbf{Q}\mathbf{L}\mathbf{a} + \dot{\mathbf{Q}}\mathbf{a} . \tag{113}$$

Hence,

$$\overset{\nabla}{\mathbf{a}}^* = \frac{d^*\mathbf{a}^*}{dt^*} - \mathbf{L}^*\mathrm{a}^* = \mathbf{Q}\left(\frac{d\mathbf{a}}{dt} - \mathbf{L}\,\mathbf{a}\right) = \mathbf{Q}\,\overset{\nabla}{\mathbf{a}}\,. \tag{114}$$

Exercise 1.8.

$$\frac{d^*\mathbf{M}^*}{dt^*} - \mathbf{L}^*\mathbf{M}^* - \mathbf{M}^*\mathbf{L}^{T*} = \frac{d^*}{dt^*}\mathbf{QMQ}^T$$
$$-(\mathbf{QLQ}^T + \Omega)\mathbf{QMQ}^T - \mathbf{QMQ}^T(\mathbf{QL}^T\mathbf{Q}^T + \Omega^T)$$

$$= \dot{\mathbf{Q}}\mathbf{MQ}^T + \mathbf{QM}\dot{\mathbf{Q}}^T + \mathbf{Q}\frac{d\mathbf{M}}{dt}\mathbf{Q}^T - \mathbf{QLMQ}^T$$
$$-\mathbf{QML}^T\mathbf{Q}^T - \Omega\mathbf{QMQ}^T - \mathbf{QMQ}^T\Omega^T$$

$$= \mathbf{Q}\left(\frac{d\mathbf{M}}{dt} - \mathbf{L}\mathbf{M} - \mathbf{M}\mathbf{L}^T\right)\mathbf{Q}^T. \tag{115}$$

Exercise 1.9.
Since $\mathbf{C}_t\mathbf{C}_t^{-1}=\mathbf{1}$,

$$\frac{\partial \mathbf{C}_t^{-1}}{\partial t} = -\,\mathbf{C}_t^{-1}\,\frac{\partial \mathbf{C}_t}{\partial t}\,\mathbf{C}_t^{-1}\,. \tag{116}$$

Now use (59).

Exercise 1.10.
The eigenvalues of $\mathbf{C}_t$ are $\lambda_1,\lambda_2,\lambda_3$ and are real because $\mathbf{C}_t=\mathbf{C}_t^T$.

The eigenvalues of $\mathbf{F}_t$ are $\alpha_1,\alpha_2,\alpha_3$ and $\mathbf{F}_t^T$ has the same eigenvalues. Hence,

$$\det \mathbf{C}_t = \lambda_1\,\lambda_2\,\lambda_3 = (\det \mathbf{F}_t)^2 = \alpha_1^2\alpha_2^2\alpha_3^2\,.$$

In fact, for incompressible material, $\det \mathbf{F}_t=1$. The other two invariants give

$$\lambda_1 + \lambda_2 + \lambda_3 = \alpha_1^2 + \alpha_2^2 + \alpha_3^2$$

and

$$\lambda_1\lambda_2 + \lambda_2\lambda_3 + \lambda_1\lambda_3 = \alpha_1^2\alpha_2^2 + \alpha_2^2\,\alpha_3^2 + \alpha_1^2\alpha_3^2\ .$$

Clearly, all these equations are solved by some permutation of scripts, which without loss of generality, are equivalent to

$$\lambda_1 = \alpha_1^2\ ,\quad \lambda_2 = \alpha_2^2\ ,\quad \lambda_3 = \alpha_3^2\ .$$

Exercise 1.11. (cf. Exercises 18.1 and 18.2)

Exercise 1.12.
Solve (90) for $\hat{\mathbf{x}}$.

$$\mathbf{P}^{*T}(\mathbf{x}^* - \mathbf{d}^*) = \mathbf{P}^T(\mathbf{x} - \mathbf{d}) \tag{117}$$

and premultiply this by P*. We get

$$\mathbf{x}^* = \mathbf{Q}\mathbf{x} - \mathbf{Q}\mathbf{d} + \mathbf{d}^*\ . \tag{118}$$

The tensor $\hat{\Omega} = -\hat{\Omega}^T$ is skew symmetric because the time derivative of $\mathbf{P}\mathbf{P}^T = \mathbf{1}$ vanishes. A skew symmetric tensor may always be associated with a vector as in (95). In fact, the components of this vector are given by

$$\omega_\ell = -\frac{1}{2} \in_{\ell k i} \Omega_{ik}\ . \tag{119}$$

To verify the transformation (96) we use (91) to find

$$\hat{\Omega}^* = \dot{\mathbf{P}}^*\mathbf{P}^{*T} = (\dot{\mathbf{Q}}\mathbf{P} + \mathbf{Q}\dot{\mathbf{P}})\mathbf{P}^T\mathbf{Q}^T = \dot{\mathbf{Q}}\mathbf{Q}^T + \mathbf{Q}\hat{\Omega}\mathbf{Q}^T\ . \tag{120}$$

Exercise 1.13.
To prove (100), use (99), noting that

$$\dot{\mathbf{e}}_i = \boldsymbol{\omega}\wedge\mathbf{e}_i\ ,\ \dot{\mathbf{e}}_i^* = \boldsymbol{\omega}^*\wedge\mathbf{e}_i^* \tag{121}$$

and

$$\dot{v}_i\,(\mathbf{x}, t) = \frac{\partial v_i}{\partial t} + \dot{\mathbf{x}}\frac{\partial}{\partial \mathbf{x}}\, v_i = \frac{\partial v_i}{\partial t} + \mathbf{u}\frac{\partial}{\partial \mathbf{x}}\, v_i\ . \tag{122}$$

To prove (101), note that $y_i = x_i - d_i$, $v_i = u_i - \dot{\mathbf{d}}_i$ and

$$\ddot{\mathbf{d}} = \mathbf{e}_i\ddot{d}_i + 2\boldsymbol{\omega}\wedge\mathbf{e}_i\dot{d}_i + \boldsymbol{\omega}\wedge(\boldsymbol{\omega}\wedge\mathbf{d}) + \dot{\boldsymbol{\omega}}\wedge\mathbf{d}\ . \tag{123}$$

To prove (102), we first use (95) to write

$$a_i^* = \frac{\partial v_i^*}{\partial t} + \mathbf{u}^* \frac{\partial}{\partial \mathbf{x}^*}\, v_i^* - 2\hat{\Omega}_{ij}^* v_j^* + \hat{\Omega}_{ij}^* \hat{\Omega}_{jk}^* y_k^* - \dot{\hat{\Omega}}_{ij}^* y_j^* \;. \tag{124}$$

Since

$$\mathbf{y}^* = \mathbf{x}^* - \mathbf{d}^* = \mathbf{Q}(\mathbf{x}-\mathbf{d}) = \mathbf{Q}\mathbf{y} \tag{125}$$

we have

$$\mathbf{v}^* = \mathbf{Q}\mathbf{v} + \dot{\mathbf{Q}}\mathbf{y} \tag{126}$$

and from (111)

$$\frac{\partial}{\partial t} + \mathbf{u}^* \frac{\partial}{\partial \mathbf{x}^*} = \frac{\partial}{\partial t} + \mathbf{u}\frac{\partial}{\partial \mathbf{x}} = \frac{d}{dt} \,.$$

Hence

$$\frac{dv_i^*}{dt} = \dot{v}_i^* = Q_{ij}\dot{v}_j + 2\dot{Q}_{ij} v_j + \ddot{Q}_{ij} y_j \,. \tag{127}$$

Applying (96), (125) and (126) now to the other terms in (124), we get

$$\hat{\Omega}_{ij}^* v_j^* = Q_{il}\hat{\Omega}_{lk} v_k + \dot{Q}_{il} v_l + Q_{il}\hat{\Omega}_{lk} Q_{mk}\dot{Q}_{mp} y_p + \dot{Q}_{ij} Q_{kj}\dot{Q}_{kl} y_l \,,$$

$$\hat{\Omega}_{ij}^* \hat{\Omega}_{jk}^* y_k^* = Q_{ij}\hat{\Omega}_{jk}\hat{\Omega}_{kl} y_l + \dot{Q}_{ij}\Omega_{jk} y_k + Q_{ij}\hat{\Omega}_{jk} Q_{lk}\dot{Q}_{lm} y_m + \dot{Q}_{ij} Q_{kj}\dot{Q}_{kl} v_l \,,$$

$$\dot{\hat{\Omega}}_{ij}^* y_j^* = Q_{ij}\dot{\hat{\Omega}}_{jk} y_k + \dot{Q}_{ij}\hat{\Omega}_{jk} y_k + Q_{ij}\hat{\Omega}_{jk}\dot{Q}_{lk} Q_{lm} y_m + \ddot{Q}_{ij} y_j + \dot{Q}_{ij}\dot{Q}_{kj} Q_{km} y_m \,.$$

After forming the sum in (124), we get

$$a_i^* = Q_{ij}\left[\frac{dv_j}{dt} - 2\,\hat{\Omega}_{jl} v_l + \hat{\Omega}_{jl}\hat{\Omega}_{lk} y_k - \dot{\Omega}_{jk} y_k\right] = Q_{ij} a_j \,,$$

proving (102). The expressions for **a** and **a*** depend on the frame, though they are indifferent.

CHAPTER 2

Models Like Jeffreys'

2.1 Voigt element

The Maxwell models were generated by a spring and a dashpot in series. The system may be displaced by increasing the displacement to a constant value. Initially all the displacement is taken up by the spring. The element then relaxes with friction proportional to velocity $\partial\gamma/\partial t$. Permanent deformation forces, as in fluids, are impossible. This system has instantaneous elasticity in that the initial resistance is due to the spring.

A Voigt or Kelvin model is a spring and dashpot in parallel, as in Figure 2.1. The force in the elastic element is $G\gamma$ and the force in the viscous element is $\eta_2 \frac{\partial\gamma}{\partial t}$; and when in parallel, they are additive.

$$\sigma = G\gamma + \eta_2 \frac{\partial\gamma}{\partial t} \; . \tag{1}$$

This element is instantaneously viscous because, by design, the deformation γ and the rate of deformation $\partial\gamma/\partial t$ of the spring and dashpot occur simultaneously. The dashpot must work in each and every deformation. If a constant force is applied, the deformation will be damped by viscosity and the system will come to equilibrium with $\sigma=G\gamma$, as in an elastic body. The Voigt model is "good" for viscoelastic solids and not for fluids.

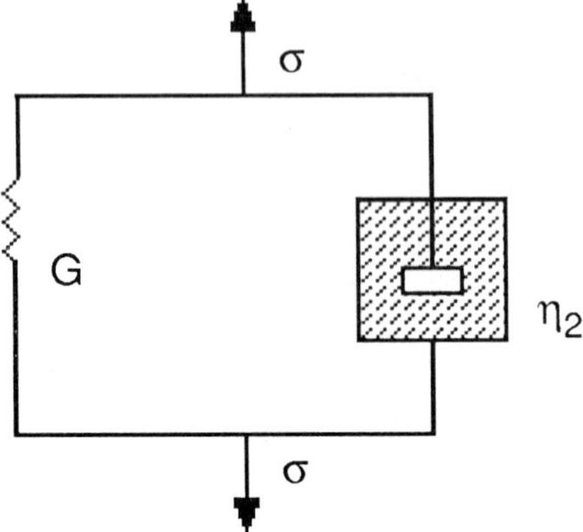

Figure 2.1. Voigt or Kelvin element. The viscosity of the dashpot is η_2.

2.2 Jeffreys element

The Jeffreys model [1929] is a dashpot and a Voigt model in series, as in Figure 2.2.

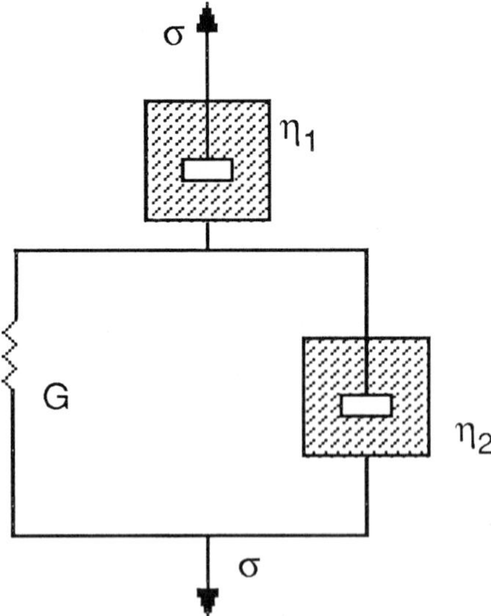

Figure 2.2. Jeffreys element.

The total displacement γ in a Jeffreys element is the sum of the strains in the dashpot η_1 and in the Voigt element. Hence,

$$\frac{\partial \gamma}{\partial t} = \frac{\partial \gamma_1}{\partial t} + \frac{\partial \gamma_2}{\partial t} . \tag{2}$$

The forces in both elements are the same;

$$\sigma = \eta_1 \frac{\partial \gamma_1}{\partial t} = G\gamma_2 + \eta_2 \frac{\partial \gamma_2}{\partial t} . \tag{3}$$

Equations (2) and $(3)_{1,2}$ are three equations for σ, γ, γ_1, and γ_2. After eliminating γ_1 and γ_2, we get

$$\frac{\eta_1+\eta_2}{G} \frac{\partial \sigma}{\partial t} + \sigma = \eta_1 \left(\frac{\partial \gamma}{\partial t} + \frac{\eta_2}{G} \frac{\partial^2 \gamma}{\partial t^2} \right) . \tag{4}$$

The Jeffreys model is viscous because purely elastic deformation of the element is impossible by design, as in the Voigt model. Viscosity is active in every deformation. The Jeffreys element cannot sustain a constant force in equilibrium; the force must relax. The Jeffreys model is good for fluids and not for solids.

We may define a relaxation time as

$$\lambda_1 = \frac{\eta_1+\eta_2}{G} \tag{5}$$

and a retardation time

$$\lambda_2 = \frac{\eta_2}{G} . \tag{6}$$

Then, after changing our interpretation of force and deformation to stress and strain, as in §1.3, we get

$$\lambda_1 \frac{\partial \sigma}{\partial t} + \sigma = \eta_1 \left[\frac{\partial u}{\partial x} + \lambda_2 \frac{\partial^2 u}{\partial x \partial t} \right] . \tag{7}$$

We recover the Maxwell model when $\lambda_2=0$ ($\eta_2=0$). On the other hand, Newtonian fluids are such that

$$\sigma = \eta_1 \frac{\partial u}{\partial x} ,$$

so that Newtonian fluids arise from Jeffreys' model when the relaxation and retardation times are equal;

$$\lambda_1 = \lambda_2 .$$

If the stress is put to zero, the strain $\partial u/\partial x$ will relax and its time of relaxation is the retardation time λ_2. The Jeffreys model allows for relaxation of stress and strain and, in general, they relax at different rates.

2.3 Tensorial generalization

The tensorial generalization (7) follows exactly along the lines laid down in §1.5. We have (23), (24), (25) of §1.5 and

$$\lambda_1 \frac{\partial \boldsymbol{\tau}}{\partial t} + \boldsymbol{\tau} = 2\eta_1 \left[\mathbf{D}[\mathbf{u}] + \lambda_2 \frac{\partial \mathbf{D}[\mathbf{u}]}{\partial t} \right] . \tag{8}$$

We may write (8) as

$$\boldsymbol{\tau} = \frac{2\eta_1\lambda_2}{\lambda_1} \mathbf{D}[\mathbf{u}(t)]$$

$$+ \frac{2\eta_1}{\lambda_1}\left(1 - \frac{\lambda_2}{\lambda_1}\right) \int_{-\infty}^{t} \exp[-(t-\tau)/\lambda_1]\, \mathbf{D}[\mathbf{u}(\tau)]\, d\tau . \tag{9}$$

For steady flow,

$$\boldsymbol{\tau} = 2\eta_1\, \mathbf{D}[\mathbf{u}] .$$

The static or zero shear viscosity is η_1. The viscosity η_1 is the value found in tables in books on fluid mechanics and in handbooks of material parameters. If the spring is very stiff, all the deformation will take place in the dashpot with the viscosity η_1.

It is convenient to define a Newtonian viscosity

$$\mu = \frac{\eta_1\lambda_2}{\lambda_1} = \frac{\eta_2\eta_1}{\eta_1+\eta_2} \tag{10}$$

and an elastic viscosity

$$\eta = \eta_1 \left(1 - \frac{\lambda_2}{\lambda_1}\right) = \frac{\eta_1^2}{\eta_1+\eta_2} . \tag{11}$$

Then, (9) may be written as

$$\boldsymbol{\tau} = \boldsymbol{\tau}_N + \boldsymbol{\tau}_E \tag{12}$$

where

$$\boldsymbol{\tau}_N = 2\mu \mathbf{D}[\mathbf{u}] \tag{13}$$

is the Newtonian contribution and

$$\boldsymbol{\tau}_E = \frac{2\eta}{\lambda_1} \int_{-\infty}^{t} \exp[-(t-\tau)/\lambda_1]\, \mathbf{D}[\mathbf{u}(\tau)]\, d\tau \tag{14}$$

is the viscoelastic contribution.

2.4 Integral form of Jeffreys model: Oldroyds models A and B [1950]

Now we are going to carry out a nonlinear upgrade of (14). The fruit of this upgrade are integral expressions of the upper convected type (Oldroyd B)

$$\boldsymbol{\tau}_E = \frac{\eta}{\lambda_1^2} \int_{-\infty}^{t} \exp[-(t-\tau)/\lambda_1][\mathbf{C}_t^{-1} - \mathbf{1}]\, d\tau \tag{15}$$

or the lower convected type (Oldroyd A)

$$\boldsymbol{\tau}_E = \frac{\eta}{\lambda_1^2} \int_{-\infty}^{t} \exp[-(t-\tau)/\lambda_1][\mathbf{1} - \mathbf{C}_t(\tau)]\, d\tau . \tag{16}$$

First, I wish to show you that (14) is uniquely the linearization of both (15) and (16) relative to the rest state $\chi_t(\mathbf{x},\tau)=\mathbf{x}$ for all τ's. To do this, we integrate (15) and (16) by parts

$$\boldsymbol{\tau}_E = -\frac{\eta}{\lambda_1}\int_{-\infty}^{t} \exp[-(t-\tau)/\lambda_1]\,\frac{\partial \mathbf{C}_t^{-1}(\tau)}{\partial \tau}\,d\tau \tag{17}$$

and

$$\boldsymbol{\tau}_E = \frac{\eta}{\lambda_1}\int_{-\infty}^{t} \exp[-(t-\tau)/\lambda_1]\,\frac{\partial \mathbf{C}_t(\tau)}{\partial \tau}\,d\tau\,. \tag{18}$$

Now

$$\frac{\partial \mathbf{C}_\tau}{\partial \tau} = \frac{\partial \mathbf{F}_t^T}{\partial \tau}\mathbf{F}_t + \mathbf{F}_t^T\frac{\partial \mathbf{F}_t}{\partial \tau} = \left[\frac{\partial \mathbf{u}(\boldsymbol{\chi},\tau)^T}{\partial \mathbf{x}}\right]\mathbf{F}_t + \mathbf{F}_t^T\left[\frac{\partial \mathbf{u}(\boldsymbol{\chi},\tau)}{\partial \mathbf{x}}\right] \tag{19}$$

where $\mathbf{u}(\chi,\tau)$ is the velocity of a particle at point $\chi=\chi_t(\mathbf{x},\tau)$ at time $\tau\le t$. In the state of rest, $\chi=\mathbf{x}$ for all $\tau\le t$. If ε is a small parameter forcing motion, then

$$\boldsymbol{\chi} = \mathbf{x} + O(\varepsilon)\,,$$

$$\mathbf{F}_t = \mathbf{1} + O(\varepsilon)\,, \tag{20}$$

$$\mathbf{u}(\boldsymbol{\chi},\tau) = \mathbf{u}(\mathbf{x},\tau) + O(\varepsilon)$$

where $\mathbf{u}(\mathbf{x},\tau)=\varepsilon\mathbf{v}(\mathbf{x},t)$ and $\mathbf{v}(\mathbf{x},t)$ is bounded from zero independent of ε.

Combining (19) and (20), we get

$$\frac{\partial \mathbf{C}_t}{\partial \tau} = \left[\frac{\partial \mathbf{u}(\boldsymbol{\chi},\tau)}{\partial \mathbf{x}}\right]^T + \frac{\partial \mathbf{u}(\boldsymbol{\chi},\tau)}{\partial \mathbf{x}} = 2\mathbf{D}[\mathbf{u}(\mathbf{x},\tau)]\,. \tag{21}$$

Similarly, using $\mathbf{C}_t\mathbf{C}_t^{-1}=\mathbf{1}$, we have

$$\frac{\partial \mathbf{C}_t^{-1}}{\partial \tau} = -2\mathbf{D}[\mathbf{u}(\mathbf{x},\tau)]\,. \tag{22}$$

This proves that (9) is the unique limit of different Oldroyd models with a single time of relaxation and a single time of retardation.

In general, we may write (12) with τ_N defined by (13) provided only that

$$\frac{D\boldsymbol{\tau}_E}{Dt} + \frac{\boldsymbol{\tau}_E}{\lambda_1} = \frac{2\eta_1}{\lambda_1}\mathbf{D}[\mathbf{u}] \tag{23}$$

where D/Dt is any invariant derivative, not just the upper and lower convected ones for which $\boldsymbol{\tau}_E$ has an integral form. In this case, (12), (13) and (23) imply that

$$\lambda_1 \frac{D\boldsymbol{\tau}}{Dt} + \boldsymbol{\tau} = 2\eta\left\{\mathbf{D}[\mathbf{u}] + \lambda_2 \frac{D}{Dt}\mathbf{D}[\mathbf{u}]\right\}. \tag{24}$$

Equation (9) is the rate form of (12) when $\boldsymbol{\tau}_N$ is given by (13) and $\boldsymbol{\tau}_E$ satisfies (23). More rate models may be written down which do not decompose along these lines. For example, different invariant derivatives could be used on the left and right of (24). The field of nonlinear constitutive equations is a nightmare of possibilities.

2.5 Lower order terms and higher order terms

We shall call (24) with D/Dt given by (1.50) a generalized Jeffreys model. Oldroyd's models A and B are special cases for a=±1. When $\lambda_2=0$, (24) reduces to an interpolated Maxwell model. Equation (24) represent six equations for the components of stress. These components appear linearly in (24) and the only derivatives of $\boldsymbol{\tau}$ are in the form

$$\frac{\partial\boldsymbol{\tau}}{\partial t} + (\mathbf{u}\bullet\nabla)\boldsymbol{\tau} = \frac{d\boldsymbol{\tau}}{dt}.$$

The operator

$$\lambda_1 \frac{D\boldsymbol{\tau}}{Dt} - 2\eta_1 \mathbf{D}[\mathbf{u}] \tag{25}$$

is linear in the first derivatives of $\boldsymbol{\tau}$ and linear in the first derivatives of $\mathbf{u}$. For example, using the upper convected derivative (1.43), (25) may be written as

$$\frac{\partial \boldsymbol{\tau}}{\partial t} + \mathbf{u}\boldsymbol{\cdot}\nabla\boldsymbol{\tau} - \mathbf{L}\boldsymbol{\tau} - \boldsymbol{\tau}\mathbf{L}^{\mathrm{T}} - \frac{\eta_1}{\lambda_1}(\mathbf{L}+\mathbf{L}^{\mathrm{T}}) = 0\,. \tag{26}$$

The operator in (26) is linear in the first derivatives of $\boldsymbol{\tau}$ and $\mathbf{u}$ but is nonlinear. This type of operator is called a quasilinear operator of first order. The operator

$$\frac{\mathrm{D}}{\mathrm{Dt}}\mathbf{D}[\mathbf{u}] \tag{27}$$

is quasilinear of second order, linear in second derivatives of $\mathbf{u}$. This higher order term is associated with the instantaneous viscous response of the Voigt model.

2.6 Generalizations of Jeffreys' model

Giesekus [1982] proposed a model like Jeffreys' in which the lower order term $\boldsymbol{\tau}$, linear in the stress in (23), is replaced by a nonlinear term, and the general invariant derivative by an upper convected derivative.

$$\boldsymbol{\tau}_{\mathrm{E}} + \lambda_1\overset{\nabla}{\boldsymbol{\tau}}_{\mathrm{E}} - \alpha\frac{\lambda_1}{\eta}\boldsymbol{\tau}_{\mathrm{E}}\boldsymbol{\tau}_{\mathrm{E}} = 2\eta\mathbf{D}[\mathbf{u}]\,. \tag{28}$$

The Giesekus model contains four parameters: a relaxation time λ_1, the solvent and polymer contributions to the zero-shear-rate viscosity, μ and η, and the dimensionless mobility factor α.

We may decompose the stress $\boldsymbol{\tau}=\boldsymbol{\tau}_{\mathrm{N}}+\boldsymbol{\tau}_{\mathrm{E}}$ as in (12) but with a more general interpretation of $\boldsymbol{\tau}_{\mathrm{N}}$ and $\boldsymbol{\tau}_{\mathrm{E}}$. Let $\boldsymbol{\tau}_{\mathrm{E}}$ stand for any expression giving rise to instantaneous elasticity; for example, we could use (15) or (16) or even more general integral models with instantaneous elasticity. The Newtonian part of the stress $\boldsymbol{\tau}_{\mathrm{N}}$ could be replaced by any invariant symmetric function depending on the present value of the velocity gradient of $\mathbf{L}(\mathbf{x},t)=\nabla\mathbf{u}(\mathbf{x},t)$ and not on its history.

$$\boldsymbol{\tau}_{\mathrm{N}} = \mathbf{f}(\mathbf{L}) = [\mathbf{f}(\mathbf{L})]^{\mathrm{T}}\,,\quad \mathbf{Q}\mathbf{f}(\mathbf{L})\mathbf{Q}^{\mathrm{T}} = \mathbf{f}(\mathbf{Q}\mathbf{L}\mathbf{Q}^{\mathrm{T}})\,.$$

The solution of this problem in the linear case is given by (13), **f** is a linear function of the symmetric part of **L**, independent of the skew symmetric part of **L**, independent of vorticity. (See Exercise 15.1.) The solution of the nonlinear problem is also that the invariant function cannot depend of the skew symmetric part of **L** (see, for example, Huilgol, 1975, p. 176). Then

$$\boldsymbol{\tau}_N = \mathbf{f}(\mathbf{D}[\mathbf{u}]) = \alpha(II,III)\mathbf{D}[\mathbf{u}] + \beta(II,III)\mathbf{D}^2[\mathbf{u}] \qquad (29)$$

where $\alpha(II,III)$ and $\beta(II,III)$ are essentially arbitrary functions, II and III are second and third invariants of $\mathbf{D}[\mathbf{u}]$ defined in §9 of Appendix A (I=0 where div $\mathbf{u}$=0) and the reduction to a quadratic polynomial follows from the Cayley-Hamilton theorem, see (§10, App. A). The expression (29) for $\boldsymbol{\tau}_N$ is known as the Reiner-Rivlin Fluid. Hence, we get a general representation of the form

$$\boldsymbol{\tau} = \alpha(II,III)\mathbf{D}[\mathbf{u}] + \beta(II,III)\mathbf{D}^2[\mathbf{u}] + \boldsymbol{\tau}_E \qquad (30)$$

where $\boldsymbol{\tau}_E$ is any expression, say an integral, which is instantaneously elastic. A general constitutive model is the direct sum of a Reiner-Rivlin fluid and an instantaneously elastic fluid.

Exercise 2.1

Write $\boldsymbol{\tau}=\boldsymbol{\tau}_N+\boldsymbol{\tau}_E$ according to the decomposition (12). Suppose $\boldsymbol{\tau}_E$ is given by (15). Then $\boldsymbol{\tau}$ is for Oldroyd B fluid. Show that $\boldsymbol{\tau}_E+\frac{\eta}{\lambda}\mathbf{1}$ is a positive definite tensor, has positive eigenvalues. The proof is as in §1.16. $\boldsymbol{\tau}_E$ is essential in the criteria which shall be developed for change of type. Strictly speaking, there will be no change of type when $\boldsymbol{\tau}_N \neq 0$, because of the smoothing action of viscosity. However, if the Newtonian viscosity μ is small compared to the elastic viscosity η, the smoothing can be expected to be small in the same way that a small viscosity smooths shocks in gas dynamics (see Exercise 13.1).

CHAPTER 3

Equations of Motion

In this chapter, we collect the equations expressing the balance of mass, energy and momentum, which together with the constitutive equation, determine the motion of fluids. I have tried to suppress details of derivations that I believe are adequately explained in many books on fluid dynamics.

3.1 Transport identities

The motion of a material volume V(t) is governed by

$$\frac{dJ}{dt} = J \operatorname{div} \mathbf{u}(\mathbf{x},t) \tag{1}$$

where J is the Jacobian of the transformation from $\xi(\mathbf{X},t)=\mathbf{x}$ to $\mathbf{X}$ (see Equation (C.1)). Equation (C.2)

$$\rho J = \rho_0 \tag{2}$$

expresses the notion that mass is conserved. Equations (1) and (2) imply that

$$\frac{d\rho}{dt} + \rho \operatorname{div} \mathbf{u} = 0\,. \tag{3}$$

Let V(t) be a material volume across which no mass passes and let $V(t_o)$ to locate this volume in a reference configuration with coordinate **X**. Let f(**x**,t) be a smooth field in V(t); then

$$\frac{d}{dt}\int_{V(t)} f\,dV = \int_{V(t_0)} \frac{d}{dt}(fJ)\,dV_0 = \int_{V(t)} \left(\frac{df}{dt} + f \operatorname{div} \mathbf{u}\right) dV \tag{4}$$

and since $(d\rho/dt) + \rho \operatorname{div} \mathbf{u} = 0$, we have

$$\frac{d}{dt}\int_{V(t)} \rho f\,dV = \int_{V(t)} \rho \frac{df}{dt}\,dV\,. \tag{5}$$

Now let f(**x**,t) be any field in V(t) which has a simple discontinuity across the surface Σ shown in Figure 3.1.

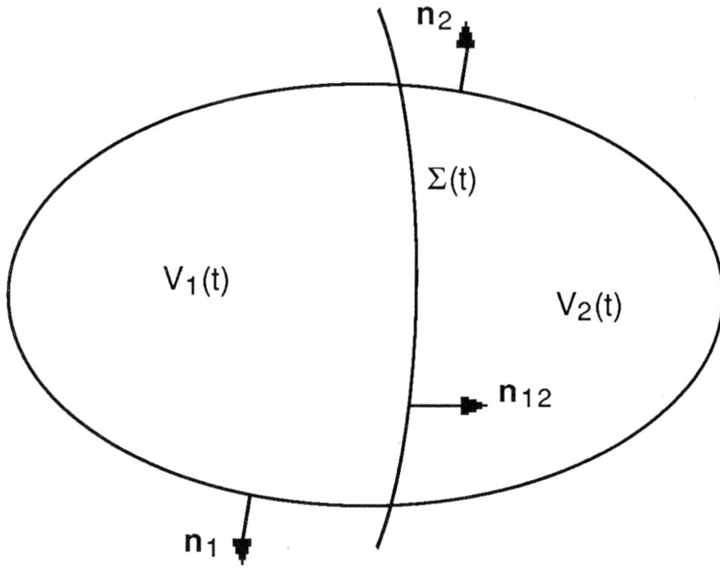

Figure 3.1 $V(t)=V_1(t)\cup V_2(t)$ is a material volume and $\Sigma(t)$ is a surface across which the otherwise smooth field f(**x**,t) has a simple discontinuity, with smooth derivatives on either side of Σ. $\mathbf{n}_1$ is the outward normal on ∂V_1 and $\mathbf{n}_2$ on ∂V_2. $\mathbf{n}_{12}$ is from V_1 to V_2 on Σ and $\mathbf{n}_{12}=\mathbf{n}_1$ there.

We consider limits in which V(t) tends to zero, while Σ is held fixed; that is, V is collapsed onto Σ. Define

$$[\![f]\!] = f_1(\mathbf{x},t) - f_2(\mathbf{x},t) \tag{6}$$

as the jump in f at the point $\mathbf{x} \in \Sigma$. There are two jump identities. The first one is

$$\lim_{V\to\Sigma} \int_V \mathbf{n} f \, dS = \int_\Sigma \mathbf{n}_{12} [\![f]\!] \, d\Sigma \tag{7}$$

where dS is an element on the surface ∂V. The proof of (7) is obvious from Figure 3.1. The second jump identity is

$$\lim_{V\to\Sigma} \frac{d}{dt} \int_{\partial V} f \, dV = \int_\Sigma [\![fv]\!] \, d\Sigma \tag{8}$$

where

$$v = \mathbf{n}_{12} \cdot (\mathbf{u}_\Sigma - \mathbf{u}) \,. \tag{9}$$

$\mathbf{u}(x,t)$ is the velocity at a point $\mathbf{x}$ on this or that side of Σ and $\mathbf{u}_\Sigma$ is the velocity of Σ at that point. In general, $\mathbf{u}_\Sigma \neq \mathbf{u}_\ell(\mathbf{x},t)$, $\mathbf{x} \in \Sigma$ where $\mathbf{u}_1$ is the particle velocity on the left side at Σ and $\mathbf{u}_2$ the particle velocity on the right. To prove (8), we note that

$$\frac{d}{dt} \int_{V_\ell} f(\mathbf{x},t) dV_\ell = \int_{V_\ell} \frac{\partial f}{\partial t} dV_\ell + \int_{\partial V_\ell} f \mathbf{n}_\ell \cdot \mathbf{v}_\ell \, dS_\ell \tag{10}$$

holds for $\ell=1$ and $\ell=2$, where $\mathbf{v}_\ell(\mathbf{x},t)$ is the velocity of points $\mathbf{x} \in \partial V_\ell$ and $\mathbf{v}_\ell(\mathbf{x},t) = \mathbf{u}(\mathbf{x},t)$ for $\mathbf{x} \in \partial V$ and $\mathbf{u}$ is the particle velocity and $\mathbf{v}_\ell(\mathbf{x},t) = \mathbf{u}_\Sigma(\mathbf{x},t)$ for $\mathbf{x} \in \Sigma$. Then

$$\int_{\partial V_1} f\mathbf{n}_1\text{•}\mathbf{v}_1\, dS_1 + \int_{\Sigma} f\mathbf{u}_1\text{•}\mathbf{n}_1\, dS_1 - \int_{\Sigma} f\mathbf{u}_1\text{•}\mathbf{n}_1\, dS_1$$

$$= \int_{\partial V_1} f\mathbf{u}\text{•}\mathbf{n}_1\, dS_1 + \int_{\Sigma} f(\mathbf{u}_\Sigma - \mathbf{u}_1)\text{•}\mathbf{n}_1\, d\Sigma \tag{11}$$

$$= \int_{V_1} \text{div}\,(f\mathbf{u})\, dV_1 + \int_{\Sigma} f(\mathbf{u}_\Sigma - \mathbf{u}_1)\text{•}\mathbf{n}_{12}\, d\Sigma\,.$$

We get the same equation (11) with subscript 2; then, we add the two equations (10), noting that $\mathbf{n}_1 = \mathbf{n}_{12} = -\mathbf{n}_{21}$ on Σ and get

$$\frac{d}{dt}\int_V f(\mathbf{x},t)\, dV = \int_V \left(\frac{\partial f}{\partial t} + \text{div}\, f\mathbf{u}\right) dV + \int_\Sigma [\![fv]\!]\, d\Sigma \tag{12}$$

leading to (9). We remind the reader that

$$\int_V (\text{•})\, dv = \int_{V_1} (\text{•})\, dv + \int_{V_2} (\text{•})\, dV.$$

Since

$$\frac{d}{dt}\int_V \rho\, dV = 0$$

for every material volume V, we find from (8) and (12) that

$$\rho_1 v_1 = \rho_2 v_2 \overset{\text{def}}{=} m\,. \tag{13}$$

Then, replacing f by ρf in (8), we get

$$\lim_{V\to\Sigma} \frac{d}{dt}\int_V \rho f\, dV = \int_\Sigma m[\![f]\!]\, d\Sigma \tag{14}$$

and

$$\frac{d}{dt}\int_V \rho f \, dV = \int_V \rho \frac{df}{dt} + \int_\Sigma m[\![f]\!] \, d\Sigma \,. \tag{15}$$

3.2 Balance of momentum

We balance the momentum of a material volume V(t):

$$\frac{d}{dt}\int_V \rho \mathbf{u} \, dV = \int_{\partial V} \mathbf{t} \, dS + \int_V \rho \mathbf{g} \, dV \tag{16}$$

change of momentum — tractions on the boundary — body forces

Now we replace the traction vector **t** (force/area) with a stress tensor **T** related to **t** by the Cauchy-Fourier formula

$$\mathbf{t} = \mathbf{T}^T \mathbf{n} \tag{17}$$

where **n** is the outward normal on ∂V. After applying (15) to (16), using (17), we get

$$\int_V \rho \frac{d\mathbf{u}}{dt} dV + \int_\Sigma m[\![\mathbf{u}]\!] \, d\Sigma = \int_{\partial V} \mathbf{T}^T \mathbf{n} \, dS + \int_V \rho \mathbf{g} \, dV \,. \tag{18}$$

It is easy to verify that

$$\int_{\partial V} \mathbf{T}^T \mathbf{n} \, dS = \int_{\partial V_1} \mathbf{T}^T \mathbf{n}_1 \, dS + \int_{\partial V_1} \mathbf{T}^T \mathbf{n}_2 \, dS$$

$$- \int_\Sigma [\![\mathbf{T}^T]\!] \mathbf{n}_{12} \, d\Sigma \,. \tag{19}$$

After using the divergence theorem, we get

$$\int_V \left\{\rho \frac{d\mathbf{u}}{dt} - \operatorname{div} \mathbf{T}^T - \rho \mathbf{g}\right\} dV$$

$$+ \int_\Sigma ([\![\mathbf{T}^T]\!]\mathbf{n}_{12} + m[\![\mathbf{u}]\!])\, d\Sigma = 0\,. \tag{20}$$

Standard arguments based on the fact (20) holds for any material volume and surface of discontinuity lead then to

$$\rho \frac{d\mathbf{u}}{dt} = \operatorname{div} \mathbf{T}^T + \rho \mathbf{g}\,, \quad \mathbf{x} \in V_1 \text{ or } V_2\,, \tag{21}$$

$$m[\![\mathbf{u}]\!] = -[\![\mathbf{T}^T]\!]\mathbf{n}_{12}\,, \quad \mathbf{x} \in \Sigma\,. \tag{22}$$

In the absence of couple stresses, $\mathbf{T}=\mathbf{T}^T$ is symmetric. The proof uses the balance of angular momentum of V.

3.3 Balance of energy

If we neglect radiative transfer and other body sources of energy, we have

$$\frac{d}{dt}\int_V \rho\left[e + \frac{|\mathbf{u}|^2}{2}\right] dV = -\int_{\partial V} \mathbf{q}\cdot\mathbf{n}\, dS + \int_{\partial V} \mathbf{u}\cdot\mathbf{t}\, dS + \int_V \rho\mathbf{u}\cdot\mathbf{g}\, dV$$

rate of change of the specific internal plus kinetic energy	heat in	power of tractions	power of the body force	(23)

Arguing, as in (19) and (20), we find, using (15), that

$$\int_V \left\{\rho \frac{d}{dt}\left(e + \frac{|\mathbf{u}|^2}{2}\right) + \operatorname{div}\mathbf{q} - \operatorname{div}(\mathbf{T}^T\mathbf{u}) - \rho\mathbf{u}\cdot\mathbf{g}\right\} dV$$

$$= \int_\Sigma \left\{-m\left[\!\!\left[e + \frac{|\mathbf{u}|^2}{2}\right]\!\!\right] + [\![\mathbf{q}\cdot\mathbf{n}_{12} - \mathbf{u}\cdot\mathbf{T}^T\mathbf{n}_{12}]\!]\right\} d\Sigma\,. \tag{24}$$

Standard arguments then lead us to

$$\rho \frac{d}{dt}\left(e + \frac{|\mathbf{u}|^2}{2}\right) + \operatorname{div} \mathbf{q} - \operatorname{div} (\mathbf{T}^T\mathbf{u}) - \rho\mathbf{u}\cdot\mathbf{g} = 0 \tag{25}$$

for $\mathbf{x} \in V_1$ or $\mathbf{x} \in V_2$, or using (21)

$$\rho \frac{de}{dt} + \operatorname{div} \mathbf{q} - \operatorname{tr} (\mathbf{L}\mathbf{T}^T) = 0 \tag{26}$$

and

$$-m\left[\!\!\left[e + \frac{|\mathbf{u}|^2}{2} \right]\!\!\right] + [\![\mathbf{q}]\!]\cdot\mathbf{n}_{12} - [\![\mathbf{u}\cdot\mathbf{T}^T\mathbf{n}_{12}]\!] = 0 \tag{27}$$

for $\mathbf{x} \in \Sigma$.

The quantity $\operatorname{tr} (\mathbf{L}\mathbf{T}^T)$ is the stress power, heating due to friction, and

$$\operatorname{tr} (\mathbf{L}\mathbf{T}^T) = \frac{1}{2} \operatorname{tr} (\mathbf{D}\mathbf{T}) \text{ when } \mathbf{T}=\mathbf{T}^T . \tag{28}$$

3.4 Boundary conditions

Consider the boundary between a solid and fluid or between immiscible liquids across which no mass flows. Then $m = 0$ and (22)

$$[\![\mathbf{T}^T]\!]\mathbf{n}_{12} = [\![\mathbf{t}]\!] = 0 \tag{29}$$

shows that the tractions are continuous. In fact, the usual condition for immiscible liquids is that the normal component of the jump $[\![\mathbf{t}]\!]$ is balanced by surface tension. To get this boundary condition, more assumptions are required; for example, we could say that there is a film between the two liquids. Using (29), (27) reduces to

$$[\![\mathbf{u}]\!]\cdot\mathbf{t} = [\![\mathbf{q}]\!]\cdot\mathbf{n}_{12} . \tag{30}$$

This says that the heat transported across a boundary to which the fluid adheres must vanish.

3.5 Incompressible fluids and the reaction pressure

By an incompressible material we mean that the volume is conserved.

$$J = \det \mathbf{F} = 1 \; . \tag{31}$$

For incompressible fluids,

$$\frac{d\rho}{dt} = 0 \quad \text{and} \quad \operatorname{div} \mathbf{u} = 0 \; . \tag{32}$$

The stress in an incompressible fluid is

$$\mathbf{T} = - p\mathbf{1} + \boldsymbol{\tau} \tag{33}$$

where p is the reaction pressure. We shall adopt the convention that $\boldsymbol{\tau}=\mathbf{0}$ in a fluid of rest so that

$$p = -\frac{1}{3} \operatorname{tr} \mathbf{T} \; . \tag{34}$$

The reaction pressure does no internal work. The pressure part of the stress power is

$$\operatorname{tr} [\mathbf{L} p \mathbf{1}] = p \operatorname{div} \mathbf{u} = 0 \; .$$

There is a universal relation between stress and deformation, a constitutive equation. There is no universal relation between the reaction pressure and the deformation. The reaction pressure is determined by the problem, and its relation to the motion varies from problem to problem.

3.6 Equations of motion of generalized Jeffreys models

We shall confine our attention to incompressible fluids. Then the equations of motion are

$$\lambda_1 \frac{D\boldsymbol{\tau}}{Dt} + \boldsymbol{\tau} - 2\eta\left(\mathbf{D}[\mathbf{u}] + \lambda_2 \frac{D}{Dt}\mathbf{D}[\mathbf{u}]\right) = 0 \; ,$$

$$\rho\left[\frac{\partial \mathbf{u}}{\partial t} + \mathbf{u}\bullet\nabla\mathbf{u}\right] - \operatorname{div} \boldsymbol{\tau} + \nabla p = 0 \; , \; \operatorname{div} \mathbf{u} = 0 \tag{35}$$

where, using a slightly different way of writing,

$$\frac{D\mathbf{M}}{Dt} = \frac{\partial \mathbf{M}}{\partial t} + (\mathbf{u}\cdot\nabla)\mathbf{M} + \frac{1-a}{2}(\mathbf{ML} + \mathbf{L}^T\mathbf{M}) - \frac{1+a}{2}(\mathbf{LM} + \mathbf{ML}^T) . \tag{36}$$

This is a system of ten equations for six components of $\boldsymbol{\tau}$, three for $\mathbf{u}$ and one for p. When $\lambda_1=\lambda_2=0$, or when $\lambda_1=\lambda_2$, we get a Newtonian fluid with $\tau=2\eta\mathbf{D}[\mathbf{u}]$, four equations in four unknowns.

In the next chapters, we are going to study the effects of instantaneous elasticity, $\lambda_2=0$, $\lambda_1=\lambda$, $\boldsymbol{\tau}=\boldsymbol{\tau}_E$. If then, we use an upper or lower convected Maxwell model, we may write $\boldsymbol{\tau}$ as the integral expression, as in (2.15) and (2.16). The stress may be eliminated from the momentum equation, say

$$\rho \frac{d\mathbf{u}}{dt} = -\nabla p + \frac{\eta}{\lambda^2} \int_{-\infty}^{t} \exp[-(t-\tau)/\lambda] \, \text{div}\mathbf{C}_t^{-1} \, d\tau \tag{37}$$

where $\chi_t(\mathbf{x},\tau)$ is related to $\mathbf{u}$ by

$$\frac{\partial \chi_t(x,\tau)}{\partial \tau} = \mathbf{u}(\chi_t(\mathbf{x},\tau),\tau) . \tag{38}$$

Together with div $\mathbf{u}(\mathbf{x},t)=0$, we have seven equations for three components of χ_t, three of $\mathbf{u}$ and p.

3.7 Quasilinear first order systems governing interpolated Maxwell models with different lower order terms

In the next two chapters, we shall study the effects of instantaneous elasticity for which the lower order terms are not important. We emphasize this feature by writing the constitutive equation for interpolated Maxwell models as

$$\lambda \frac{D\boldsymbol{\tau}}{Dt} - 2\eta\mathbf{D}[\mathbf{u}] = \text{lot} \tag{39}$$

where lot stands for lower order terms. For Maxwell models, these are $-\boldsymbol{\tau}$; for the Giesekus and Phan-Thien models, a more complicated expression for $\boldsymbol{\tau}$ appears on the right of (39).

The dynamical equations governing the motion of fluids like Maxwell's are quasilinear; they are nonlinear, but linear in derivatives. We can write the dynamical equations

$$\mathbf{G}(\mathbf{Q}_t + \mathbf{u}\cdot\nabla\mathbf{Q}) + \mathbf{H}\mathbf{Q}_x + \mathbf{J}\mathbf{Q}_y = \boldsymbol{\ell}[\mathbf{Q}] \tag{40}$$

where $\mathbf{G}$ is not invertible (e.g., there is no p_t in this system). $\mathbf{Q}$ is a system vector whose components are the velocity, stresses and pressure and $\mathbf{G},\mathbf{H},\mathbf{J}$, and $\boldsymbol{\ell}$ depend on $\mathbf{Q}$, but not on the derivatives of $\mathbf{Q}$. In two dimensions, we have velocity components $\mathbf{u}=(u,v)$ corresponding to x,y stress components

$$[\boldsymbol{\tau}] = \begin{vmatrix} \sigma & \tau \\ \tau & \gamma \end{vmatrix} \tag{41}$$

and

$$\mathbf{Q} = [\sigma,\gamma,\tau,u,v,p] \ . \tag{42}$$

There are six quasilinear equations for the six scalar fields $[\sigma,\gamma,\tau,u,v,p]$, linear in derivatives with lower order right-hand sides, ℓ_1, ℓ_2, and ℓ_3.

$$\begin{gathered}
\sigma_t + u\sigma_x + v\sigma_y + m_1u_x + m_2u_y + m_3v_x = \ell_1 \,, \\
\gamma_t + u\gamma_x + v\gamma_y + m_3u_y + m_2v_x + m_6v_y = \ell_2 \,, \\
\tau_t + u\tau_x + v\tau_y + m_7u_y + m_8v_x = \ell_3 \,, \\
\rho(u_t+uu_x+vu_y) - \sigma_x - \tau_y + p_x = 0 \,, \\
\rho(v_t+uv_x+vv_y) - \gamma_y - \tau_x + p_y = 0 \,, \\
u_x + v_y = 0
\end{gathered} \tag{43}$$

where

$$\begin{aligned}
m_1 &= -2a\sigma - 2G\,,\\
m_2 &= -(1+a)\tau\,,\\
m_3 &= (1-a)\tau\,,\\
m_6 &= -2a\gamma - 2G\,,\\
m_7 &= \frac{1}{2}(1-a)\,\sigma - \frac{1}{2}(1+a)\,\gamma - G\,,\\
m_8 &= -\frac{1}{2}(1+a)\sigma + \frac{1}{2}(1-a)\,\gamma - G\,.
\end{aligned} \tag{44}$$

Equations (43) are in the form (40) with

$$[\mathbf{G}] = \mathrm{diag}[1,1,1,\rho,\rho,0]\,, \tag{45}$$

$$[\mathbf{H}] = \begin{vmatrix} 0 & 0 & 0 & m_1 & m_3 & 0\\ 0 & 0 & 0 & 0 & m_2 & 0\\ 0 & 0 & 0 & 0 & m_8 & 0\\ -1 & 0 & 0 & 0 & 0 & 1\\ 0 & 0 & -1 & 0 & 0 & 0\\ 0 & 0 & 0 & 1 & 0 & 0 \end{vmatrix}\,, \tag{46}$$

$$[\mathbf{J}] = \begin{vmatrix} 0 & 0 & 0 & m_2 & 0 & 0\\ 0 & 0 & 0 & m_3 & m_6 & 0\\ 0 & 0 & 0 & m_7 & 0 & 0\\ 0 & 0 & -1 & 0 & 0 & 0\\ 0 & -1 & 0 & 0 & 0 & 1\\ 0 & 0 & 0 & 0 & 1 & 0 \end{vmatrix}\,. \tag{47}$$

3.8 Evolution of the vorticity

It is slightly tedious, but not at all difficult to show that the vorticity

$$\zeta = \mathrm{curl}\,\mathbf{u}$$

of the interpolated Maxwell models (1.50) satisfies an evolution equation of the form

$$\rho[\partial_{tt}\zeta_k + 2\mathbf{u}\cdot\nabla\partial_t\zeta_k + u_e u_j \zeta_{k,ej}] + \frac{1}{2}(a-1)\varepsilon_{kej}\,\tau_{jq}\,[\text{curl }\zeta]_{q,e}$$
$$-\frac{1}{2}(a+1)\tau_{mp}\zeta_{k,mp} - G\nabla^2\zeta_k = \ell_k \qquad (48)$$

where ℓ_k are the lower order terms. For the Maxwell models,

$$\ell_k = -\frac{1}{\lambda}\varepsilon_{kej}\,\tau_{jm,me}$$
$$+\rho\left[\left(\frac{\partial}{\partial t}+\mathbf{u}\cdot\nabla\right)(\zeta\cdot\nabla u_k) - \partial_t\mathbf{u}\cdot\nabla\zeta_k - u_e u_{j,e}\zeta_{k,j}\right]$$
$$+\frac{a+1}{2}\tau_{mp,m}\,\zeta_{k,p} + \frac{1-a}{2}\varepsilon_{kej}\tau_{jq,e}(\text{curl }\zeta)_q$$
$$+\varepsilon_{kej}\Big[\frac{a+1}{2}(\tau_{mp,e}u_{j,mp} + \tau_{mp,me}u_{j,p} - \tau_{jm,mp}u_{p,e}) \qquad (49)$$
$$+\frac{a-1}{2}(\tau_{mp,e}u_{p,jm} + \tau_{mp,me}u_{p,j} + \tau_{jp,me}u_{p,m}$$
$$+\tau_{jp,m}u_{p,me} + \tau_{jp,me}u_{m,p} + \tau_{jp,m}u_{m,pe})\Big]\,.$$

Moreover,

$$\varepsilon_{kej}\tau_{jm,me} = (\text{curl div }\tau)_k = \rho\left[\text{curl }\frac{d\mathbf{u}}{dt}\right]_k = \rho\left[\frac{d\zeta_k}{dt} + (\boldsymbol{\zeta}\cdot\nabla)u_k\right] \qquad (50)$$

Potential flow with $\boldsymbol{\zeta}=0$ can be a solution of (48) only if the last bracket in (49) vanishes when $\mathbf{u}=\nabla\phi$.

In general, the leading terms on the left of the equality in (48) are different for k=1,2,3. However, if a=1 (UCM), the coefficients of the leading terms in the equation (48) for ζ_1,ζ_2, and ζ_3 are the same. This means that criteria for change of type in the vorticity are the same for each component of the vorticity of an UCM, but a different criterion for change of type in the equations for ζ_1,ζ_2, and ζ_3 can be expected when $a\neq 1$.

In plane flow, there is one nonzero component of vorticity satisfying

$$\rho\frac{\partial^2\zeta}{\partial t^2} + 2\rho(\mathbf{u}\cdot\nabla)\frac{\partial\zeta}{\partial t} - A\frac{\partial^2\zeta}{\partial x^2} - 2B\frac{\partial^2\zeta}{\partial x\partial y} - C\frac{\partial^2\zeta}{\partial y^2} + \ell = 0\,. \qquad (51)$$

where ℓ is of lower order, and A,B,C are defined by

$$A = -\rho u^2 + G + \frac{1}{2}\sigma(1+a) - \frac{1}{2}\gamma(1-a) , \tag{52}$$

$$B = \tau - \rho uv , \tag{53}$$

$$C = -\rho v^2 + G - \frac{1}{2}\sigma(1-a) + \frac{1}{2}\gamma(a+1) , \tag{54}$$

$$G = \eta/\lambda . \tag{55}$$

Nonlinear models which are not quasilinear do not usually give rise to hyperbolic vorticity equations. In plane flow, and in axisymmetric flow, one may replace the vorticity equation with a fourth order quasilinear equation for the stream function. The fourth order equation may then be analyzed for type. Examples of this type of analysis will be given in §4.12 and §6.10. The analysis of the White-Metzner model given in §6.9 is very interesting because it shows that there is a big class of constitutive equations which are nonlinear and not quasilinear which are "more hyperbolic" than models like Maxwell's.

3.9 Vorticity equations in direct notation

We want to set up the vorticity equation in a direct notation. When expressed in this way, the vectors and tensors may be resolved in any coordinate system.

$$\rho[\partial_{tt}\boldsymbol{\zeta} + 2(\mathbf{u}\cdot\nabla)\partial_t\boldsymbol{\zeta} + \{\mathrm{tr}\,(\mathbf{u}\otimes\mathbf{u})(\nabla\otimes\nabla)\}\boldsymbol{\zeta}] - G\nabla^2\boldsymbol{\zeta}$$

$$+\frac{a-1}{2}\,\mathrm{curl}\,(\boldsymbol{\tau}\cdot\mathrm{curl}\,\boldsymbol{\zeta}) - \frac{a+1}{2}\,\mathrm{tr}\,\{\boldsymbol{\tau}(\nabla\otimes\nabla)\}\boldsymbol{\zeta} = \boldsymbol{\ell} . \tag{56}$$

To facilitate computations in any coordinate system we shall use a hybrid notation to express $\boldsymbol{\ell}$

$$\boldsymbol{\ell} = -\frac{1}{\lambda}\,\mathrm{curl}\,(\mathrm{div}\,\boldsymbol{\tau}) + \rho\Big[\Big(\frac{\partial}{\partial t} + \mathbf{u}\cdot\nabla\Big)$$

$$(\boldsymbol{\zeta}\cdot\nabla)\mathbf{u} - (\partial_t\mathbf{u}\cdot\nabla)\boldsymbol{\zeta} - (\{(\mathbf{u}\cdot\nabla)\mathbf{u}\}\nabla)\boldsymbol{\zeta}\Big]$$

$$+\frac{a+1}{2}(\mathrm{div}\,\boldsymbol{\tau}\cdot\nabla)\boldsymbol{\zeta} + \frac{1-a}{2}\,\mathrm{curl}\,\boldsymbol{\tau}\,\mathrm{curl}\,\boldsymbol{\zeta} \tag{57}$$

$$+\frac{a+1}{2}[\nabla\tau_{mp}\wedge\mathbf{u}_{,mp}+\nabla(\text{div}\,\boldsymbol{\tau})_p\wedge\mathbf{u}_{,p}+(\text{div}\,\boldsymbol{\tau})_{,p}\wedge\nabla\mathbf{u}_{,p}]$$
$$+\frac{a-1}{2}[\nabla(\text{div}\,\boldsymbol{\tau})_p\wedge\nabla u_p+\nabla\tau_{mp}\wedge\nabla u_{p,m}+\text{curl}\,(\boldsymbol{\Theta}+\boldsymbol{\Phi})]$$

where $(\text{curl}\,\boldsymbol{\tau})_{ij}=\varepsilon_{ikm}\partial_k\tau_{mj}$,

$$\boldsymbol{\Theta}=\frac{\partial\boldsymbol{\tau}}{\partial x_m}\frac{\partial\mathbf{u}}{\partial x_m}, \tag{58}$$

$$\boldsymbol{\Phi}=\frac{\partial\boldsymbol{\tau}}{\partial x_m}\nabla u_m\,. \tag{59}$$

Suppose we want to resolve these equations in a curvilinear coordinate system with coordinates α,β,γ. A term like $\nabla\tau_{mp}\wedge\nabla u_{p,m}$ generates certain terms proportional to, say

$$\frac{\partial\tau_{mp}}{\partial\alpha}\frac{\partial u_{p,m}}{\partial\beta}=\text{tr}\left[\frac{\partial\boldsymbol{\tau}}{\partial\alpha}\frac{\partial\nabla\mathbf{u}}{\partial\beta}\right] \tag{60}$$

which can be completely resolved in α,β,γ coordinates.

The term $\nabla\tau_{mp}\wedge\mathbf{u}_{,mp}$ generates terms of the form

$$\text{tr}\left[\frac{\partial\boldsymbol{\tau}}{\partial\alpha}\nabla\otimes\nabla u_\beta\right]. \tag{61}$$

From $\nabla(\text{div}\,\boldsymbol{\tau})_p\wedge\mathbf{u}_{,p}$, we get terms like

$$\frac{\partial\,\text{div}\,\boldsymbol{\tau}}{\partial\alpha}\bullet\nabla u_\beta \tag{62}$$

whereas $\nabla(\text{div}\,\boldsymbol{\tau})_p\wedge\nabla u_p$ leads to

$$\frac{\partial\,\text{div}\,\boldsymbol{\tau}}{\partial\alpha}\bullet\frac{\partial\mathbf{u}}{\partial\beta}. \tag{63}$$

The equations of motion for interpolated Maxwell models in polar spherical coordinates are given in §9.3 and in cylindrical coordinates in Exercise 10.1.

Exercises

Exercise 3.1.
(Joseph, Renardy and Saut [1985], §12; Joseph [1985], (2.9)) Suppose the extra stress $\boldsymbol{\tau}$ satisfies

$$\frac{\partial \boldsymbol{\tau}}{\partial t} + (\mathbf{u}\cdot\nabla)\boldsymbol{\tau} - \mathbf{ML} = \ell \tag{64}$$

where $\mathbf{M}$ is a fourth order tensor $(\mathbf{ML})=(\mathbf{ML})^T$ which may depend on $\mathbf{u}$ and $\boldsymbol{\tau}$, but not on the derivatives of $\mathbf{u}$ or $\boldsymbol{\tau}$, and ℓ is of lower order. Show that the equation governing the vorticity is

$$\begin{aligned}
&\rho[\partial_{tt}\zeta_k + 2\mathbf{u}\cdot\nabla\partial_t\zeta_k + u_e u_j \zeta_{k,ej}] - \varepsilon_{k\ell j}\, M_{jmpq} u_{q,pm\ell} \\
&= \rho[(\partial_t + \mathbf{u}\cdot\nabla)\varepsilon_{kei} u_{\ell,e} u_{i,\ell} - (\partial_t \mathbf{u}\cdot\nabla)\zeta_k - u_e u_{j,e}\zeta_{k,j}] \\
&\quad + \varepsilon_{kej}\, [M_{jmpq,e}\, u_{q,pm} + (M_{jmpq,m} u_{q,p})_{,e} \\
&\quad - u_{p,e}\, \tau_{jq,pq} - (u_{p,q}\tau_{jq,p})_{,e} + \ell_{jm,me}]\,.
\end{aligned} \tag{65}$$

where $(\bullet)_{,m} \overset{\text{def}}{=} \partial/\partial x_m(\bullet)$. Equation (65) reduces to a second order equation for the vorticity plus lower order terms whenever the last term on the left of (65) can be expressed in terms of second order derivatives of $\boldsymbol{\zeta}$. Potential flow is possible only when (65) is satisfied with $\boldsymbol{\zeta}=\mathbf{0}$, $\mathbf{u}=\nabla\phi$.

Exercise 3.2.
(Joseph, Renardy and Saut [1985], §12) Show that (64) may always be written

$$\frac{\partial \boldsymbol{\tau}}{\partial t} + \mathbf{u}\cdot\nabla\boldsymbol{\tau} - \mathbf{SD} - \mathbf{AW} = \ell \tag{66}$$

where $\mathbf{S}$ is a fourth order tensor symmetric with respect to $\mathbf{D}$, and $\mathbf{A}$ is a fourth order tensor skew symmetric with respect to $\mathbf{W} = \frac{1}{2}(\mathbf{L}-\mathbf{L}^T)$

$$\begin{aligned}
A_{ijk\ell} &= -A_{ij\ell k}\,, \\
S_{ijk\ell} &= S_{ij\ell k}\,.
\end{aligned} \tag{67}$$

Exercise 3.3.
Suppose that **S** is expressible as products of Kronecker's delta and any second order tensor **P**. Show that

$$\frac{d\boldsymbol{\tau}}{dt} = \mathbf{PD} + \mathbf{DP}^T + \mathbf{AW} + (\mathbf{AW})^T + \ell \ . \tag{68}$$

Show that the leading order terms in the vorticity equation associated with (68) are all linear in second derivatives of ζ.

Exercise 3.4.
Write the system of equations for models like Maxwell's in three dimensions as in (40) defining 10×10 matrices **G**,**H**,**J**, as in (45–47).

Exercise 3.5.
Determine the equations satisfied by characteristic surfaces $\phi(x,y,z,t)$ for the equations of Exercise 3.4.

Exercise 3.6.
Every equation of the form

$$A_{ij} \frac{\partial^2 \zeta_j}{\partial x_\ell \partial x_\ell} = \ell_i$$

where A_{ij} and ℓ_i may be written

$$\frac{\partial}{\partial x_\ell}\left(A_{ij} \frac{\partial \zeta_j}{\partial x_\ell} \right) = \ell_i \tag{69}$$

where ℓ_i is of lower order. Show that the vorticity equation for the upper-convected Maxwell model may be written as

$$\operatorname{div} \{(\boldsymbol{\tau} - \rho \mathbf{u} \otimes \mathbf{u}) \bullet \operatorname{grad} \boldsymbol{\zeta}\} = \ell \ . \tag{70}$$

Exercise 3.7.

Let $\hat{\boldsymbol{\tau}}$ be the normal to Σ in Σ and on $\partial\Sigma=\Sigma\cap\partial V$, see Figure 3.1. Suppose that Σ is an interface with surface tension which cannot resist bending or shearing. There is an additional force on $\partial\Sigma$ arising from the membrane stress

$$\int_{\partial\Sigma} \sigma\hat{\boldsymbol{\tau}}\, dS \tag{71}$$

on the right of (18). This force can be written (using the divergence theorem for surfaces) as

$$\int_{\Sigma} \nabla_{II}\sigma\, d\Sigma - \int_{\Sigma} (\nabla_{II}\bullet\mathbf{n}_{12})\, \sigma\mathbf{n}_{12}\, d\Sigma \tag{72}$$

where

$$\nabla_{II} = \nabla - \mathbf{n}_{12}\,(\mathbf{n}_{12}\bullet\nabla)$$

is a surface gradient and

$$\nabla_{II}\bullet\mathbf{n}_{12} = -2H = -\left(\frac{1}{R_1} + \frac{1}{R_2}\right)$$

where H is the mean curvature and R_1 and R_2 are principal radii of curvature. Following again the analysis just given we get the same momentum equation in the bulk fluid, but there are more terms on the interface

$$m[\![\mathbf{u}]\!] + [\![\mathbf{T}^T]\!]\,\mathbf{n}_{12} = \nabla_{II}\sigma + 2H\sigma\mathbf{n}_{12}\,. \tag{73}$$

The energy equation (23) must be modified by adding the rate of change of the stored energy of the surface

$$\frac{d}{dt}\int_{\Sigma} \sigma\, d\Sigma \tag{74}$$

on the left and the power of the surface tension traction

$$\int_{\partial\Sigma} \sigma \hat{\boldsymbol{\tau}} \cdot \mathbf{u}_\Sigma \, d\Sigma \tag{75}$$

where $\mathbf{u}_\Sigma$ is the velocity of points on the line $\partial\Sigma$, on the right.

The interface is known to satisfy*

$$\frac{d}{dt}\int_\Sigma \sigma \, d\Sigma - \int_{\partial\Sigma} \sigma \hat{\boldsymbol{\tau}} \cdot \mathbf{u}_\Sigma = \int_\Sigma \left[\frac{\partial\sigma}{\partial t} - \mathbf{u}\cdot\nabla_{II}\sigma - 2H\mathbf{u}\cdot\mathbf{n}_{12}\right]. \tag{76}$$

In this case (27) is replaced by

$$-m\left[\!\left[e + \frac{|\mathbf{u}|^2}{2} \right]\!\right] + [\![q]\!]\cdot\mathbf{n}_{12} - [\![\mathbf{u}\cdot\mathbf{T}^T]\!]\mathbf{n}_{12}$$

$$= \frac{\partial\sigma}{\partial t} - \mathbf{u}\cdot\nabla_{II}\sigma - 2H\sigma\mathbf{u}\cdot\mathbf{n}_{12}. \tag{77}$$

Equation (77) holds also at a phase change boundary which grows by mass accretion or depletion. In this case the surface velocity $\mathbf{u}_\Sigma \neq \mathbf{u}$ and $\mathbf{u}_\Sigma$ should replace the material velocity $\mathbf{u}$ on the right of (77).

Solutions for the exercises

Exercise 3.1.

$$\frac{\partial\boldsymbol{\tau}}{\partial t} + \mathbf{u}\cdot\nabla\boldsymbol{\tau} - \mathbf{ML} = \ell \ ,$$

or

* See, for example, Joseph, D.D. [1976] *Stability of Fluid Motions II*, Springer-Verlag: New York, pgs. 9 and 243)

$$\partial_t \tau_{ij} + u_m \tau_{ij,m} - M_{ijpq} u_{q,p} = \ell_{ij} ,$$
$$\partial_t \tau_{ij,j} + u_{m,j} \tau_{ij,m} + u_m \tau_{ij,mj} - M_{ijpq,j} u_{q,p} - M_{ijpq} u_{q,pj} = \ell_{ij,j} . \quad (78)$$

Momentum equation:

$$\rho(\partial_t u_i + u_m u_{i,m}) - \tau_{ij,j} + p_{,i} = 0 , \quad (79)_1$$
$$\rho(\partial_{tt} u_i + \partial_t u_m u_{i,m} + u_m \partial_t u_{i,m}) - \partial_t \tau_{ij,j} + \partial_t p_{,i} = 0 . \quad (79)_2$$

Eliminate $\partial_t \tau_{ij,j}$ in (78) and $(79)_2$ and take $\varepsilon_{kei}[\ldots]_{i,e}$ of equation $(79)_2$ to find that

$$\zeta_k \overset{\text{def}}{=} \varepsilon_{kei} u_{i,e} ,$$
$$\begin{aligned} \rho(\partial_{tt}\zeta_k + \varepsilon_{kei}\partial_t u_{m,e} u_{i,m} + \partial_t u_m \zeta_{k,m} + \varepsilon_{kei} u_{m,e} \partial_t u_{i,m} \\ + u_m \partial_t \zeta_{k,m}) + \text{``}\varepsilon_{kei}\partial_t p_{,ie}\text{''} + \varepsilon_{kei}[u_{m,je}\tau_{ij,m} \\ + u_{m,j}\tau_{ij,me} + u_{m,e}\tau_{ij,mj} + u_m \tau_{ij,mje}] - \varepsilon_{kei}[(M_{ijpq,j}\, u_{q,p}),_e \\ + M_{ijpq,e} u_{q,pj} + M_{ijpq} u_{q,pje}] = \varepsilon_{kei} \ell_{ij,je} \end{aligned} \quad (80)$$

where the term between quotation marks vanishes.

Now we take the curl of $(79)_1$ and find that

$$\varepsilon_{kei}\tau_{ij,je} = \rho(\partial_t \zeta_k + u_\ell \zeta_{k,\ell}) + \rho \varepsilon_{kei} u_{\ell,e} u_{i,\ell}$$
$$u_m \varepsilon_{kei} \tau_{ij,jem} = \rho u_m(\partial_t \zeta_{k,m}) + \rho u_m (u_\ell \zeta_{k,\ell})_{,m} +$$
$$\rho \varepsilon_{kei}(u_{\ell,e} u_{i,\ell})_{,m}\, u_m .$$

Therefore, (80) becomes:

$$\begin{aligned} \rho(\partial_{tt}\zeta_k + 2u_m \partial_t \zeta_{k,m} + u_m u_\ell \zeta_{k,\ell m}) + \rho[\varepsilon_{kei}\partial_t u_{m,e} u_{i,m} \\ + \partial_t u_m \zeta_{k,m} + \varepsilon_{kei} u_{m,e} \partial_t u_{i,m} + u_m u_{\ell,m} \zeta_{k,\ell} \\ + \varepsilon_{kei} u_m (u_{\ell,e} u_{i,\ell})_{,m}] - \varepsilon_{kei} M_{ijpq} u_{q,pje} \\ = \varepsilon_{kei}\ell_{ij,je} + \varepsilon_{kei}[(M_{ijpq,j} u_{q,p})_{,e} + M_{ijpq,e} u_{q,pj} \\ - (u_{m,j}\tau_{ij,m})_{,e} - u_{m,e}\tau_{ij,mj}] \end{aligned}$$

or

$$\begin{aligned} \rho(\partial_{tt}\zeta_k + 2u_m \partial_t \zeta_{k,m} + u_m u_\ell \zeta_{k,\ell m}) - \varepsilon_{kei} M_{ijpq} u_{q,pje} \\ = -\rho(\partial_t u_m \zeta_{k,m} + u_m u_{\ell,m} \zeta_{k,\ell}) + \varepsilon_{kei}[M_{ijpq,e} u_{q,pj} \\ + (M_{ijpq,j} u_{q,p})_{,e} - (u_{m,j}\tau_{ij,m})_{,e} - u_{m,e}\tau_{ij,mj} \\ + \ell_{ij,je}] - \rho\, \varepsilon_{kei}[\partial_t(u_{m,e} u_{i,m}) + u_m (u_{\ell,e} u_{i,\ell})_{,m}] \end{aligned}$$

and

$$\varepsilon_{kei}[\partial_t(u_{m,e}u_{i,m}) + u_m(u_{\ell,e}u_{i,\ell})_{,m}]$$
$$= [\partial_t + \mathbf{u}\cdot\nabla](\varepsilon_{kei}u_{\ell,e}u_{i,\ell}) .$$

Exercise 3.2.
Show that

$$\mathbf{ML} = \mathbf{SD} + \mathbf{AW}$$

where

$$\mathbf{D} = \frac{1}{2}(\mathbf{L} + \mathbf{L}^T) ,$$

$$\mathbf{W} = \frac{1}{2}(\mathbf{L} - \mathbf{L}^T) .$$

Actually, we have

$$S_{ijpq} = \frac{1}{2}(M_{ijpq} + M_{ijqp}) ,$$

$$A_{ijpq} = \frac{1}{2}(M_{ijpq} - M_{ijqp}) .$$

Thus,

$$S_{ijpq} = S_{ijqp} , \quad A_{ijpq} = -A_{ijqp} ,$$

and

$$M_{ijpq}L_{qp} = \frac{1}{2}(M_{ijpq} + M_{ijqp})\frac{1}{2}(L_{qp} + L_{pq})$$
$$+ \frac{1}{2}(M_{ijpq} - M_{ijqp})\frac{1}{2}(L_{qp} - L_{pq})$$
$$= S_{ijpq}D_{qp} + A_{ijpq}W_{qp} .$$

Exercise 3.3.

$$S_{ijk\ell} = \delta_{ij}P_{k\ell} + \delta_{ik}P_{j\ell} + \delta_{i\ell}P_{jk} + \delta_{jk}P_{i\ell} + \delta_{j\ell}P_{ik} + \delta_{k\ell}P_{ij} ,$$
$$S_{ijk\ell}D_{\ell k} = \delta_{ij}P_{k\ell}D_{\ell k} + P_{j\ell}D_{\ell i} + P_{jk}D_{ik}$$
$$+P_{i\ell}D_{\ell j} + P_{ik}D_{jk} + \text{``}P_{ij}D_{\ell\ell}\text{''}$$
$$= \delta_{ij}P_{k\ell}D_{\ell k} + 2(P_{j\ell}D_{\ell i}+P_{i\ell}D_{\ell j})$$
$$= \delta_{ij}p + 2(\mathbf{PD}+\mathbf{DP}^T)_{ij} .$$

where the term between quotation marks vanishes, and $p=P_{kl}D_{lk}$.

Let

$$2\mathbf{P} = \mathbf{P}' ,$$

then

$$\begin{aligned}\mathbf{ML} &= \mathbf{SD} + \mathbf{AW} \\ &= (\mathbf{P}'\mathbf{D} + \mathbf{DP}'^{T}) + \mathbf{AW} + p\mathbf{1}.\end{aligned}$$

Since $\boldsymbol{\tau}^T = \boldsymbol{\tau}$, we have $(\mathbf{ML})^T=(\mathbf{ML})$ and $(\mathbf{AW})^T=(\mathbf{AW})$. Therefore,

$$\mathbf{ML} = \mathbf{P}'\mathbf{D} + \mathbf{DP}'^{T} + (\mathbf{A}'\mathbf{W}) + (\mathbf{A}'\mathbf{W})^T + p\mathbf{1}$$

where $\mathbf{A}'=1/2\ \mathbf{A}$.

The leading terms in the vorticity equation

$$\rho[\partial_{tt}\zeta_k + 2\mathbf{u}\bullet\nabla\partial_t\zeta_k + u_e u_j \zeta_{k,ej}] - \varepsilon_{k\ell j} M_{jmpq} u_{q,pm\ell} \cdot$$

are third derivatives of $\mathbf{u}$. If the last term is expressible in terms of the second derivatives of ζ alone, we have it made.

We compute

$$\begin{aligned}\varepsilon_{klj} M_{jmpq} u_{q,pml} &= \varepsilon_{klj}\, P_{jq}(\text{curl }\zeta)_{q,l} + P_{mp}\zeta_{k,mp} \\ &\quad + \varepsilon_{klj}\, \varepsilon_{pqr}\, A_{jmpq}\zeta_{r,ml} + \text{lot} .\end{aligned}$$

This shows that all the third derivatives $\mathbf{u}$ are expressible as second derivatives of ζ.

Exercise 3.4.

Constitutive equation (interpolated Maxwell model):

$$\begin{aligned}\frac{\partial \boldsymbol{\tau}}{\partial t} + (\mathbf{u}\bullet\nabla)\boldsymbol{\tau} + \frac{1-a}{2}(\boldsymbol{\tau}\mathbf{L}+\mathbf{L}^T\boldsymbol{\tau}) - \frac{1+a}{2}(\mathbf{L}\boldsymbol{\tau}+\boldsymbol{\tau}\mathbf{L}^T)& \\ - 2G\mathbf{D}[\mathbf{u}] = \text{lot} .&\end{aligned} \tag{81}$$

Equations of motion:

$$\begin{aligned}\rho\left[\frac{\partial \mathbf{u}}{\partial t} + \mathbf{u}\bullet\nabla\mathbf{u}\right] - \text{div}\,\boldsymbol{\tau} + \nabla p = 0 ,& \\ \text{div}\mathbf{u} = 0 .&\end{aligned}$$

In the three-dimensional case,

$$\tau = \begin{vmatrix} \sigma & \tau & \alpha \\ \tau & \gamma & \theta \\ \alpha & \theta & \beta \end{vmatrix} , \quad \mathbf{u} = [u,v,w] .$$

Let $\mathbf{Q}=[\sigma,\gamma,\beta,\tau,\alpha,\theta,u,v,w,p]$. The dynamical equations are in the form

$$\mathbf{G}(\mathbf{Q}_t + \mathbf{u}\cdot\nabla\mathbf{Q}) + \mathbf{H}\mathbf{Q}_x + \mathbf{J}\mathbf{Q}_y + \mathbf{I}\mathbf{Q}_z = \ell[\mathbf{Q}] .$$

We may write (81) as

$$\frac{\partial\tau_{ij}}{\partial t} + (\mathbf{u}\cdot\nabla)\tau_{ij} + \frac{1-a}{2}\left(\tau_{ik}\frac{\partial u_k}{\partial x_j} + \frac{\partial u_k}{\partial x_i}\tau_{kj}\right)$$
$$-\frac{1+a}{2}\left(\frac{\partial u_i}{\partial x_k}\tau_{kj} + \tau_{ik}\frac{\partial u_j}{\partial x_k}\right) - G\left(\frac{\partial u_i}{\partial x_j} + \frac{\partial u_j}{\partial x_i}\right) = \text{lot} \qquad (82)$$

or

$$(\partial_t+\mathbf{u}\cdot\nabla)\sigma + (1-a)(\sigma u_x+\tau v_x+\alpha w_x)$$
$$-(1+a)(\sigma u_x+\tau u_y+\alpha u_z) - 2Gu_x = \ell_1 ,$$

$$(\partial_t+\mathbf{u}\cdot\nabla)\gamma + (1-a)(\tau u_y+\gamma v_y+\theta w_y)$$
$$-(1+a)(\tau v_x+\gamma v_y+\theta v_z) - 2Gv_y = \ell_2 ,$$

$$(\partial_t+\mathbf{u}\cdot\nabla)\beta + (1-a)(\alpha u_z+\theta v_z+\beta w_z)$$
$$-(1+a)(\alpha w_x+\theta w_y+\beta w_z) - 2Gw_z = \ell_3 ,$$

$$(\partial_t+\mathbf{u}\cdot\nabla)\tau + \frac{1-a}{2}(\sigma u_y+\tau v_y+\alpha w_y+\tau u_x+\gamma v_x+\theta w_x)$$
$$-\frac{1+a}{2}(\tau u_x+\gamma u_y+\theta u_z+\sigma v_x+\tau v_y+\alpha v_z) - G(u_y+v_x) = \ell_4 ,$$

$$(\partial t+\mathbf{u}\cdot\nabla)\alpha + \frac{1-a}{2}(\sigma u_z+\tau v_z+\alpha w_z+\alpha u_x+\theta v_x+\beta w_x)$$
$$-\frac{1+a}{2}(\alpha u_x+\theta u_y+\beta u_z+\sigma w_x+\tau w_y+\alpha w_z) - G(u_z+w_x) = \ell_5 ,$$

$$(\partial t+\mathbf{u}\cdot\nabla)\theta + \frac{1-a}{2}(\tau u_z+\gamma v_z+\theta w_z+\alpha u_y+\theta v_y+\beta w_y)$$
$$-\frac{1+a}{2}(\alpha v_x+\theta v_y+\beta v_z+\tau w_x+\gamma w_y+\theta w_z) - G(v_z+w_y) = \ell_6 ,$$

or organized into the following:

$$(\partial t+\mathbf{u}\cdot\nabla)\sigma + m_{11}u_x + m_{12}u_y + m_{13}u_z + m_{14}v_x + m_{15}w_x = \ell_1 ,$$

$$(\partial t+\mathbf{u}\cdot\nabla)\gamma + m_{21}u_y + m_{22}v_x + m_{23}v_y + m_{24}v_z + m_{25}w_y = \ell_2 ,$$

$$(\partial t+\mathbf{u}\cdot\nabla)\beta + m_{31}u_z + m_{32}v_z + m_{33}w_x + m_{34}w_y + m_{35}w_z = \ell_3 ,$$

$$(\partial t+\mathbf{u}\cdot\nabla)\tau + m_{41}u_x + m_{42}u_y$$
$$+ m_{43}u_z + m_{44}v_x + m_{45}v_y + m_{46}v_z + m_{47}w_x + m_{48}w_y = \ell_4 ,$$

$$(\partial t+\mathbf{u}\cdot\nabla)\alpha + m_{51}u_x + m_{52}u_y$$
$$+ m_{53}u_z + m_{54}v_x + m_{55}v_z + m_{56}w_x + m_{57}w_y + m_{58}w_z = \ell_5 ,$$

$$(\partial t+\mathbf{u}\cdot\nabla)\theta + m_{61}u_y + m_{62}u_z$$
$$+ m_{63}v_x + m_{64}v_y + m_{65}v_z + m_{66}w_x + m_{67}w_y + m_{68}w_z = \ell_6 ,$$

where

$$m_{11} = -2G - 2a\sigma , \quad m_{12} = -(1+a)\tau , \quad m_{13} = -(1+a)\alpha ,$$
$$m_{14} = (1-a)\tau , \quad m_{15} = (1-a)\alpha ,$$

$$m_{21} = (1-a)\tau , \quad m_{22} = -(1+a)\tau , \quad m_{23} = -2G - 2a\gamma ,$$
$$m_{24} = -(1+a)\theta , \quad m_{25} = (1-a)\theta ,$$

$$m_{31} = (1-a)\alpha , \quad m_{32} = (1-a)\theta , \quad m_{33} = -(1+a)\alpha ,$$
$$m_{34} = -(1+a)\theta , \quad m_{35} = -2G - 2a\beta ,$$

$$m_{41} = -a\tau , \quad m_{42} = -G + \frac{1-a}{2}\sigma - \frac{1+a}{2}\gamma , \quad m_{43} = -\frac{1+a}{2}\theta ,$$
$$m_{44} = \frac{1-a}{2}\gamma - \frac{1+a}{2}\sigma - G , \quad m_{45} = -a\tau , \quad m_{46} = -\frac{1-a}{2}\alpha ,$$
$$m_{47} = \frac{1-a}{2}\theta , \quad m_{48} = \frac{1-a}{2}\alpha ,$$

$$m_{51} = -a\alpha\,, \quad m_{52} = -\frac{1+a}{2}\theta\,, \quad m_{53} = -G + \frac{1-a}{2}\sigma - \frac{1+a}{2}\beta\,,$$
$$m_{54} = \frac{1-a}{2}\theta\,, \quad m_{55} = \frac{1-a}{2}\tau\,, \quad m_{56} = -G + \frac{1-a}{2}\beta - \frac{1+a}{2}\sigma\,,$$
$$m_{57} = -\frac{1+a}{2}\tau\,, \quad m_{58} = -a\alpha\,,$$

$$m_{61} = \frac{1-a}{2}\alpha\,, \quad m_{62} = \frac{1-a}{2}\tau\,, \quad m_{63} = -\frac{1+a}{2}\alpha\,, \quad m_{64} = -a\theta\,,$$
$$m_{65} = -G + \frac{1-a}{2}\gamma - \frac{1+a}{2}\beta\,, \quad m_{66} = -\frac{1+a}{2}\tau\,,$$
$$m_{67} = -G + \frac{1-a}{2}\beta - \frac{1+a}{2}\gamma\,,$$
$$m_{68} = -a\theta\,.$$

Therefore:

$$[\mathbf{G}] = \text{diag}\ [1,1,1,1,1,1,\rho,\rho,\rho,0]\,,$$

$$[\mathbf{H}] = \left|\begin{array}{cccccccccc} 0 & 0 & 0 & 0 & 0 & 0 & m_{11} & m_{14} & m_{15} & 0 \\ 0 & 0 & 0 & 0 & 0 & 0 & 0 & m_{22} & 0 & 0 \\ 0 & 0 & 0 & 0 & 0 & 0 & 0 & 0 & m_{33} & 0 \\ 0 & 0 & 0 & 0 & 0 & 0 & m_{41} & m_{44} & m_{47} & 0 \\ 0 & 0 & 0 & 0 & 0 & 0 & m_{51} & m_{54} & m_{56} & 0 \\ 0 & 0 & 0 & 0 & 0 & 0 & 0 & m_{63} & m_{66} & 0 \\ -1 & 0 & 0 & 0 & 0 & 0 & 0 & 0 & 0 & 1 \\ 0 & 0 & 0 & -1 & 0 & 0 & 0 & 0 & 0 & 0 \\ 0 & 0 & 0 & 0 & -1 & 0 & 0 & 0 & 0 & 0 \\ 0 & 0 & 0 & 0 & 0 & 0 & 1 & 0 & 0 & 0 \end{array}\right|\,,$$

$$[\mathbf{J}] = \left|\begin{array}{cccccccccc} 0 & 0 & 0 & 0 & 0 & 0 & m_{12} & 0 & 0 & 0 \\ 0 & 0 & 0 & 0 & 0 & 0 & m_{21} & m_{23} & m_{25} & 0 \\ 0 & 0 & 0 & 0 & 0 & 0 & 0 & 0 & m_{34} & 0 \\ 0 & 0 & 0 & 0 & 0 & 0 & m_{42} & m_{45} & m_{48} & 0 \\ 0 & 0 & 0 & 0 & 0 & 0 & m_{52} & 0 & m_{57} & 0 \\ 0 & 0 & 0 & 0 & 0 & 0 & m_{61} & m_{64} & m_{67} & 0 \\ 0 & 0 & 0 & -1 & 0 & 0 & 0 & 0 & 0 & 0 \\ 0 & -1 & 0 & 0 & 0 & 0 & 0 & 0 & 0 & 1 \\ 0 & 0 & 0 & 0 & 0 & -1 & 0 & 0 & 0 & 0 \\ 0 & 0 & 0 & 0 & 0 & 0 & 0 & 1 & 0 & 0 \end{array}\right|\,,$$

$$[\mathrm{I}] = \left|\begin{array}{cccccccccc} 0 & 0 & 0 & 0 & 0 & 0 & m_{13} & 0 & 0 & 0 \\ 0 & 0 & 0 & 0 & 0 & 0 & 0 & m_{24} & 0 & 0 \\ 0 & 0 & 0 & 0 & 0 & 0 & m_{31} & m_{32} & m_{35} & 0 \\ 0 & 0 & 0 & 0 & 0 & 0 & m_{43} & m_{46} & 0 & 0 \\ 0 & 0 & 0 & 0 & 0 & 0 & m_{53} & m_{55} & m_{58} & 0 \\ 0 & 0 & 0 & 0 & 0 & 0 & m_{62} & m_{65} & m_{68} & 0 \\ 0 & 0 & 0 & 0 & -1 & 0 & 0 & 0 & 0 & 0 \\ 0 & 0 & 0 & 0 & 0 & -1 & 0 & 0 & 0 & 0 \\ 0 & 0 & -1 & 0 & 0 & 0 & 0 & 0 & 0 & 1 \\ 0 & 0 & 0 & 0 & 0 & 0 & 0 & 0 & 1 & 0 \end{array}\right| ,$$

$$\mathbf{G}\left[\frac{\partial \mathbf{Q}}{\partial t} + \mathbf{u}\bullet\nabla\mathbf{Q}\right] + \mathbf{H}\mathbf{Q}_x + \mathbf{J}\mathbf{Q}_y + \mathbf{I}\mathbf{Q}_z = \ell[\mathbf{Q}] .$$

CHAPTER 4

Hadamard Instability and Frozen Coefficients

In this chapter, we are going to do a stability study of a special type in which we look for short wave instabilities with growth rates that increase without bound as the wave length tends to zero. Such instabilities are a catastrophe for numerical analysis; the finer the mesh, the worse the result.

4.1 Hadamard instability and Laplace's equation

A bad instability arises in the study of an initial value problem for Laplace's equation. This is Hadamard's model of an ill-posed initial value problem.

Consider the half space

$$D = \{x,t\ ;\ t>0,\ -\infty<x<\infty\}$$

appropriate for an initial value problem for Laplace's equation

$$\frac{\partial^2 u}{\partial t^2}+\frac{\partial^2 u}{\partial x^2}=0 \tag{1}$$

with initial values

$$u(0,x)=0\ ,$$

$$\frac{\partial u}{\partial t}(0,x)=f(x)\ . \tag{2}$$

We may choose an oscillating function

$$f(x)=\frac{1}{k^p}\sin kx\ ,\quad p>0\ , \tag{3}$$

which is bounded for all x and tends to zero for small wave lengths $2\pi/k\to 0$. The solution of this problem is

$$u(t,x)=\frac{1}{k^{1+p}}\sin kx\sinh kt\ . \tag{4}$$

Small data at t=0 leads to huge, unbounded oscillations for any small t>0 and the wave length tends to zero. This lack of continuous dependence of the solution on the data is called *Hadamard instability*. This type of instability gets worse as the wave length $\lambda=2\pi/k$ decreases, with ultimate catastrophe as $k\to\infty$.

Ill-posedness of the Cauchy problem can also be associated with the nonexistence of solutions of initial value (Cauchy) problem (1,2) for nonanalytic data. If u is a solution of (1) in t>0, it is analytic there and, by reflection, to all of $\mathbb{R}^2$. But then f(x) must be analytic. If f(x) is not analytic, no solution is possible, even in a small interval of t around zero. One can show that ill-posed problems generally cannot be solved outside the class of analytic initial data.

4.2 Backward heat equation

Another example giving rise to Hadamard instability is the backward heat equation

$$\frac{\partial \theta}{\partial t}=-\frac{\partial^2 \theta}{\partial x^2}\ . \tag{5}$$

We may regard θ as the temperature perturbing a uniform temperature in D. We may solve (5) with normal modes,

$$\theta = \theta_0 \exp [\sigma t + ikx] . \tag{6}$$

Hence,

$$\sigma = k^2 \tag{7}$$

grows without bound as $k^2 \to \infty$. This problem is the time reversal of the heat equation for which

$$\sigma = -k^2 \tag{8}$$

is always stable. Some implications of the forward and backward properties of the heat equation were beautifully described by Clark Maxwell [1891].

> If the body is originally heated in any arbitrary manner, Fourier shows how to express the original temperature as the sum of a series of harmonic distributions. When the body is left to itself the part depending on the higher harmonic rapidly dies away, so that after a certain time the distribution of heat continually approximates to that due to the fundamental harmonic, which therefore represents the law of cooling of a body after the process of diffusion of heat has gone on for a long time.
>
> Sir William Thompson has shown, in a paper published in the 'Cambridge and Dublin Mathematical Journal' in 1844, how to deduce, in certain cases, the thermal state of a body in past time from its observed condition at present.
>
> For this purpose, the present distribution of temperature must be expressed (as it always may be) as the sum of a series of harmonic distributions. Each of these harmonic distributions is such that the difference of the temperature of any point from the final temperature diminishes in a geometrical progression as the time increases in arithmetical progression, the ratio of the geometrical progression being the greater the higher the degree of the harmonic.
>
> If we now make t negative, and trace the history of the distribution of temperature up the stream of time, we shall find each harmonic increasing as we go backwards, and the higher harmonics increasing faster than the lower ones.
>
> If the present distribution of temperature is such that it may be expressed in a finite series of harmonics, the distribution of temperature at

> any previous time may be calculated; but if (as is generally the case) the series of harmonics is infinite, then the temperature can be calculated only when this series is convergent. For present and future time it is always convergent, but for past time it becomes ultimately divergent when the time is taken at a sufficiently remote epoch. The negative value of t for which the series becomes ultimately divergent, indicates a certain date in past time such that the present state of things cannot be deduced from any distribution of temperature occurring previously to that date, and becoming diffused by ordinary conduction. Some other event besides ordinary conduction must have occurred since that date in order to produce the present state of things.
>
> This is only one of the cases in which a consideration of the dissipation of energy leads to the determination of a superior limit to the antiquity of the observed order of things.

We can add stabilizing lower order terms to (5) without affecting the catastrophic instability at large k; for example, if

$$\frac{\partial \theta}{\partial t} = -b\frac{\partial^2 \theta}{\partial x^2} - a^2\theta \tag{9}$$

then

$$\sigma = bk^2 - a^2 . \tag{10}$$

If b is positive, the long waves are stabilized by the lower order terms, but the short waves are still Hadamard unstable. The lower order term is irrelevant for Hadamard instability.

It is instructive to imagine that b is a function of a system parameter, say U, and is such, that

$$b = B(U) - B(U_c) \tag{11}$$

is greater than zero when $U>U_c$ and is less than zero when $U<U_c$. As $U-U_c$ passes through zero, the short waves change from unboundedly damped to unboundedly amplified. This is a very singular type of stability behavior.

4.3 Frozen coefficients

The method of frozen coefficients gives deep results for nonlinear problems cheaply. Suppose that

$$\frac{\partial\theta}{\partial t} = g(\theta)\frac{\partial^2\theta}{\partial x^2} + f(\theta) \tag{12}$$

where $f(\theta)$ and $g(\theta)$ are nonlinear functions of θ and

$$\theta(0,t) = 0\ , \quad \theta(1,t) = 1\ . \tag{13}$$

We may choose $g(\bullet)$ and $f(\bullet)$ so that $\theta_0(x)$ satisfies (12) and (13). For example, if $(g,f)= -(b,a^2)$ are constants, then

$$\theta_0(x) = \frac{\sin\frac{a}{\sqrt{b}}\,x}{\sin\frac{a}{\sqrt{b}}} \tag{14}$$

To study the stability of $\theta_0(x)$, we linearize for small ϕ where ϕ satisfies

$$\frac{\partial\phi}{\partial t} = g(\theta_0)\frac{\partial^2\phi}{\partial x^2} + \{g'(\theta_0)\theta_0'' + f'(\theta_0)\}\phi\ . \tag{15}$$

In general, the solution of (14), subject to (15), cannot be carried in terms of normal modes because (14) has coefficients which vary with x. However, if we look at very short waves, so short that $k\to\infty$, the coefficients of (14) will not vary appreciably over one wave length $2\pi/k$. So locally, in the neighborhood of some point $x=x_0$, (14) has constant coefficients and we can look at normal modes

$$\phi = e^{\sigma t}e^{ik(x-x_0)}\ . \tag{16}$$

Of course, we cannot directly satisfy the boundary conditions (13) with (16), but we are local, infinitely far from boundaries. So the variable coefficients and the boundaries are both lost on short waves.

Combining now (15) and (16), we find a dispersion relation

$$\sigma = -g(\theta_0)k^2 + g'(\theta_0)\theta_0'' + f'(\theta_0)\ . \tag{17}$$

Clearly,

$$\lim_{k^2 \to \infty} \frac{\sigma}{k^2} = -g(\theta_0) , \tag{18}$$

stable if $g(\theta_0)>0$ and Hadamard unstable if $g(\theta_0)<0$. Again, we lose the lower order terms.

Now we have found a very deep result: if the steady solution $\theta_0(x)$ enters into a forbidden region for which $g(\theta_0)<0$, we will get a catastrophic instability which will frustrate any attempt to calculate $\theta_0(x)$ numerically.

4.4 Hadamard instability of interpolated Maxwell models

For the moment, it is useful to think of how we might carry out an analysis of stability. First, we suppose that $\mathbf{Q}_0$ is a solution of (3.40). It could even be an unsteady solution. Then we write

$$\mathbf{Q} = \mathbf{Q}_0 + \mathbf{q} \tag{19}$$

and suppose that $\mathbf{q}$ is small. The linearized equations are

$$\mathbf{G}(\mathbf{q}_t + \mathbf{u}_0 \bullet \nabla \mathbf{q}) + \mathbf{H}_0 \mathbf{q}_x + \mathbf{J}_0 \mathbf{q}_y = \ell_0 \mathbf{q} \tag{20}$$

where $\mathbf{H}_0$ and $\mathbf{J}_0$ are $\mathbf{H}$ and $\mathbf{J}$, given by (3.46) and (3.47) when $\mathbf{Q}=\mathbf{Q}_0$. Some lower order terms which appeared in the linearization have been moved to the right side of (20). We could imagine trying to solve (20) plus boundary conditions as an initial value problem for the stability of $\mathbf{Q}_0$. If $\mathbf{Q}_0$ is steady, we could write $\mathbf{q}=e^{\sigma t}\hat{\mathbf{q}}(\mathbf{x})$ and determine stability from eigenvalues σ. To get the eigenvalues, we would have to solve a complicated set of partial differential equations over the whole field of flow, satisfying boundary conditions.

4.5 Frozen coefficients on short waves

The coefficients of (20) are constant on a small neighborhood of each point $\mathbf{x}$, and we may try for a solution in terms of normal modes

$$\mathbf{q} = \mathbf{a} \exp \{-i\omega(t-t_0) + i\alpha(x-x_0) + i\beta(y-y_0)\} \tag{21}$$

where $\mathbf{a}(\mathbf{x}_0)$ is an amplitude, ω a frequency and α and β are wave numbers. Of course, we cannot hope to satisfy boundary conditions with a solution of this form. It is a strange form for the solution because it applies at each and every point $\mathbf{x}_0$, so we may find stability at some points and instability at others.

To set some notations, we define a wave vector.

$$\mathbf{k} \overset{\text{def}}{=} \mathbf{e}_x\alpha + \mathbf{e}_y\beta \ , \tag{22}$$

$$|\mathbf{k}| = \sqrt{\alpha^2 + \beta^2} \overset{\text{def}}{=} k \ . \tag{23}$$

Since $\exp(-i\omega t) = \exp(\omega_i t) \exp(-i\omega_r t)$ where $\omega=\omega_r+i\omega_i$, we may define a growth rate $\sigma=\omega_i$.

4.6 Instability to short waves

After putting normal modes (21) into (20), we get

$$[(-\omega+\mathbf{u}_0\cdot\mathbf{k})\mathbf{G} + \alpha\mathbf{H}_0 + \beta\mathbf{J}_0]\mathbf{a} = \frac{1}{i}\,\ell_0\mathbf{a} \ . \tag{24}$$

We divide (25) by $|\mathbf{k}|$ and let $|\mathbf{k}|\to\infty$. Since ℓ_0 is independent of $\mathbf{k}$,

$$\frac{\ell_0\mathbf{a}}{|\mathbf{k}|} \to \mathbf{0} \tag{25}$$

as $|\mathbf{k}|\to\infty$. So when $\mathbf{k}$ is very large, we may solve (24) if the determinant of the coefficients of $\mathbf{a}$ vanishes

$$\Delta \overset{\text{def}}{=} \det\,[(-\omega + \mathbf{u}_0\cdot\mathbf{k})\mathbf{G} + \alpha\mathbf{H}_0 + \beta\mathbf{J}_0] = 0 \tag{26}$$

where

$$[(-\omega+\mathbf{u}_0\cdot\mathbf{k})\mathbf{G} + \alpha\mathbf{H}_0 + \beta\mathbf{J}_0] =$$

$$\begin{vmatrix} v_0 & 0 & 0 & \alpha m_1^0+\beta m_2^0 & \alpha m_3^0 & 0 \\ 0 & v_0 & 0 & \beta m_3^0 & \alpha m_2^0+\beta m_6^0 & 0 \\ 0 & 0 & v_0 & \beta m_7^0 & \alpha m_8^0 & 0 \\ -\alpha & 0 & -\beta & \rho v_0 & 0 & \alpha \\ 0 & -\beta & -\alpha & 0 & \rho v_0 & \beta \\ 0 & 0 & 0 & \alpha & \beta & 0 \end{vmatrix} \tag{27}$$

and

$$v_0 \overset{\text{def}}{=} -\omega + \mathbf{u}_0\cdot\mathbf{k} \ . \tag{28}$$

After expanding, we find that

$$\Delta = k^2 v_0^2 \left\{ 2\alpha\beta\tau_0 + \left[G + \frac{a}{2}(\gamma_0+\sigma_0) \right] k^2 + \frac{1}{2}(\alpha^2-\beta^2)(\sigma_0-\gamma_0) - v_0^2\rho \right\} = 0 \ . \tag{29}$$

The eigenvalue v_0 is found from putting the bracket in (29) to zero. Now we define

$$c^2 = v_0^2 / |\mathbf{k}|^2$$

and

$$\xi_1 = \alpha/|\mathbf{k}|,$$
$$\xi_2 = \beta/|\mathbf{k}| \ .$$

Then

$$\rho c^2 = f_0 \overset{\text{def}}{=} 2\xi_1\xi_2\tau_0 + G + \frac{a}{2}(\gamma_0+\sigma_0) + \frac{1}{2}(\xi_1^2 - \xi_2^2)(\sigma_0-\gamma_0) \ .$$

Since $\xi_1^2 + \xi_2^2 = 1$, another way to write this is

$$f_0 = 2\xi_1\xi_2\tau_0 + \frac{1}{2}\sigma_0(a+1)\xi_1^2 + \frac{1}{2}\sigma_0(a-1)\,\xi_2^2$$

$$+ G(\xi_1^2 + \xi_2^2) + \frac{1}{2}\gamma_0(a-1)\xi_1^2 + \frac{1}{2}\gamma_0(a+1)\xi_2^2 = \boldsymbol{\xi}\cdot\mathbf{b}\boldsymbol{\xi} \qquad (30)$$

where $\mathbf{b}$ is a 2×2 symmetric matrix with components

$$b_{11} = \frac{1}{2}(\sigma_0 - \gamma_0) + \frac{a}{2}(\sigma_0 + \gamma_0) + G\,,$$

$$b_{12} = b_{21} = \tau_0\,,$$

$$b_{22} = \frac{1}{2}(\gamma_0 - \sigma_0) + \frac{a}{2}(\sigma_0 + \gamma_0) + G\,. \qquad (31)$$

Since $\mathbf{b}$ is symmetric, it has real eigenvalues.

$$\lambda_1 = \frac{1}{2}(b_{11}+b_{22}) + \frac{1}{2}\{(b_{11}-b_{22})^2 + 4b_{12}^2\}^{1/2}\,,$$

$$\lambda_2 = \frac{1}{2}(b_{11}+b_{22}) - \frac{1}{2}\{(b_{11}-b_{22})^2 + 4b_{12}^2\}^{1/2}\,. \qquad (32)$$

We may diagonalize $\mathbf{b}$; then

$$\rho c^2 = f_0 = \lambda_1\,\xi_1^2 + \lambda_2\xi_2^2\,, \quad \xi_1^2 + \xi_2^2 = 1\,. \qquad (33)$$

The condition that c is real-valued means that λ_1 and λ_2 must be positive,

$$\lambda_1\lambda_2 = \det\mathbf{b} = b_{11}b_{22} - b_{12}^2 > 0\,.$$

Equation (33) defines an ellipse whose semi-major and semi-minor axes define the maximum and minimum wave speeds

$$\rho c_1^2 = \lambda_1\,, \quad \rho c_2^2 = \lambda_2\,. \qquad (34)$$

The formulas may be simplified slightly by choosing a coordinate system in which $2\tau_0 = b_{12} = 0$.

We may interpret $c=\sqrt{f_0/\rho}$ as the speed of a wave provided that $f_0>0$. In fact, $\omega/|\mathbf{k}|$ is a wave speed and c is the speed relative to the particle speed $\mathbf{u}_0$ at $\mathbf{x}_0$ in the direction $\mathbf{n} = \mathbf{k}/|\mathbf{k}|$.

If $f_0<0$, then c^2 is negative and

$$\frac{v_0}{|\mathbf{k}|} = \pm\, i \sqrt{|f_0|/\rho}$$

which together with (28) implies that

$$-\frac{\omega_i}{|\mathbf{k}|} = -\frac{\sigma}{|\mathbf{k}|} = \pm\sqrt{|f_0|/\rho}\ .$$

and $\omega_r=\mathbf{u}_0\bullet\mathbf{k}$. This shows that there is Hadamard instability at all places at which $f_0<0$. Another way to say this is that the flow becomes unstable at points at which the wave speed c=0 and the flow is not Hadamard unstable where c^2 is real and positive. If the solution enters into the region $f_0<0$ of forbidden stress, a very ugly instability will occur.

The upper convected Maxwell model cannot be Hadamard unstable. When a=1, (30) may be written as

$$f_0 = \boldsymbol{\xi}\bullet(\boldsymbol{\tau}_0+G\mathbf{1})\boldsymbol{\xi} = \sigma_0\xi_1^2 + 2\tau_0\, \xi_1\xi_2 + \gamma_0\xi_2^2 + G\ . \tag{35}$$

According to (1.66), f_0 given by (35) is always positive. Similarly, for the lower convected Maxwell model (a=–1), we find that

$$f_0 = -\xi\bullet(\boldsymbol{\tau}_0-G\mathbf{1})\xi \tag{36}$$

which, according to (1.67), is always positive.

The other Maxwell models, with $a\neq\pm1$ can and do have $f_0<0$ on the range of stresses allowed by the constitutive equation. Models like Phan-Thien's, Tanner's and Giesekus', which differ from Maxwell's in the lower order terms, can and do develop Hadamard instabilities. It is an open question as to whether the change in lower order terms, from linear to nonlinear stresses, can induce Hadamard instability in generalized models with upper or lower convected stable principal parts.

Exactly the same sort of analysis can be carried out in the three-dimensional case and was first carried out by Rutkevich [1970] for

Maxwell models with a=[1,–1,0]. The forms (35) and (36) for f_0 hold also in the three-dimensional case when $\boldsymbol{\tau}_0$ has six rather than three components and a=±1. Rutkevich did not notice the conditions (1.66) and (1.67) which protect the UCM and LCM from Hadamard instability. The type of analysis introduced by Rutkevich may be applied to the interpolated models. We introduce "normal modes"

$$[\mathbf{u},\boldsymbol{\tau},p] = [\mathbf{U},\mathbf{T},P] \exp\{i[\mathbf{k}\cdot(\mathbf{x}-\mathbf{x}_0) - \omega(t-t_0)]\}$$

where the ten amplitudes **U**,**T**,P depend on (x_0,t_0) through the basic flow. In the limit $k\rightarrow\infty$, we can solve ten amplitude equations

$$cT_{ij} + \frac{1}{2}\,\tau^0_{i\lambda}\,[(1-a)U_\lambda\xi_j - (1+a)U_j\xi_\lambda]$$

$$+ \frac{1}{2}\,[U_i\xi_\lambda(1+a) - U_\lambda\xi_i(1-a)]\,\tau^0_{\lambda j} - G(\xi_jU_i + \xi_iU_j) = 0\ ,$$

$$\rho cU_i + P\xi_i - T_{ij}\xi_j = 0\ ,$$
$$U_i\xi_i = 0\ ,$$

if the determinant

$$\Delta = \left[-\rho c^2 + G - \frac{1}{2}\tau^0_{22}(1-a) + \frac{1}{2}\tau^0_{11}(1+a)\right]$$
$$\times\left[-\rho c^2 + G + \frac{1}{2}\tau^0_{11}(1+a) - \frac{1}{2}\tau^0_{33}(1-a)\right] \qquad (37)$$

vanishes. In the derivation of (37), special coordinates are selected in which $\xi_1=1$, $\xi_2=\xi_3=0$, $\tau^0_{23}=0$ and $\tau^0_{22}>\tau^0_{33}$. Equation (37) shows that there are two waves in the three-dimensional case.

4.7 Hadamard instability of the White-Metzner model

We will start this analysis by writing the equations of motion for the White-Metzner model, the continuity equation, the momentum equation and the stress equation

$$\boldsymbol{\tau} + \lambda(\mathbf{II_D})\left(\frac{\partial \boldsymbol{\tau}}{\partial t} + (\mathbf{u}\cdot\nabla)\boldsymbol{\tau} - \nabla\mathbf{u}\boldsymbol{\tau} - \boldsymbol{\tau}\nabla\mathbf{u}^T\right) = 2\eta(\mathbf{II_D})\,\mathbf{D}[\mathbf{u}]\,.$$

One can see that the nonlinearity comes from the fact that λ and η depend on the second invariant of the rate of strain tensor $\mathbf{D}$, where $\mathbf{II_D}$ is defined by

$$\mathbf{II_D} = \frac{1}{2}\,\mathrm{tr}(\mathbf{D}^2) = \frac{1}{2}\,\mathbf{D}{:}\mathbf{D}\,, \quad \mathbf{D}{:}\mathbf{D} = D_{i\lambda}D_{\lambda i}\,.$$

The following analysis follows the paper by Verdier and Joseph [1989b].

To determine whether loss of evolution occurs, we perturb a basic solution designated with capital letters (rather than a subscript 0), with a perturbation, designated with primes; that is,

$$(\mathbf{u},\boldsymbol{\tau},p) = (\mathbf{U},\mathbf{T},P) + (\mathbf{u}',\boldsymbol{\tau}',p')\,.$$

We will now drop the primes on the perturbed quantities. Under linearization, we have

$$\mathbf{II_{D+d}} = \mathbf{II_D} + \mathbf{D}{:}\mathbf{d}$$

where $\mathbf{D} \overset{\text{def}}{=} \mathbf{D}[\mathbf{U}]$ and $\mathbf{d} \overset{\text{def}}{=} \mathbf{D}[\mathbf{u}]$.

We now freeze the coefficients

$$\mathrm{div}\,\mathbf{u} = 0\,,$$

$$\rho\left(\frac{\partial \mathbf{u}}{\partial t} + \mathbf{U}\cdot\nabla\mathbf{u}\right) + \nabla p - \mathrm{div}\,\boldsymbol{\tau} = \mathbf{lot}\,,$$

$$\begin{aligned}\lambda(\mathbf{II_D})\mathbf{L}_1 + (\mathbf{D}{:}\mathbf{d})\lambda'(\mathbf{II_D})\mathbf{L}_2 - 2\eta(\mathbf{II_D})\mathbf{d}\\ - 2(\mathbf{D}{:}\mathbf{d})\,\eta'(\mathbf{II_D})\mathbf{D} = \mathbf{lot}\,.\end{aligned}$$

where **lot** stands for lower order terms, λ' and η' are the derivatives of λ and η with respect to the second invariant $\mathbf{II_D}$, and $\mathbf{L}_1$ and $\mathbf{L}_2$ are tensors defined by

$$\begin{cases} \mathbf{L}_1 = \dfrac{\partial \boldsymbol{\tau}}{\partial t} + \mathbf{U} \cdot \boldsymbol{\tau} - \nabla \mathbf{u} \cdot \mathbf{T} - \mathbf{T} \cdot \nabla \mathbf{u}^T , \\ \mathbf{L}_2 = \dfrac{\partial \mathbf{T}}{\partial t} + \mathbf{U} \cdot \mathbf{T} - \nabla \mathbf{U} \cdot \mathbf{T} - \mathbf{T} \cdot \nabla \mathbf{U}^T . \end{cases}$$

Following Rutkevich, we now look for perturbations of the form:

$$(\mathbf{u}, \boldsymbol{\tau}, p) = (\omega_i, \sigma_{ij}, q) \exp i(\mathbf{k} \cdot \mathbf{x} - \omega t)$$

We define

$$\mathbf{n} = \frac{\mathbf{k}}{|\mathbf{k}|}$$

and introduce the velocity of propagation of wave fronts for short waves c:

$$c = \frac{\omega}{|\mathbf{k}|} - \mathbf{U} \cdot \mathbf{n} .$$

We next let $k = |\mathbf{k}| \to \infty$ since we are looking at short waves. This leads to the following system for the amplitudes:

$$\omega_i n_i = 0 ,$$

$$-\rho c \omega_i = -q n_i + \sigma_{ij} n_j ,$$

$$\lambda(c\sigma_{ij} + T_{\ell j}\omega_i n_\ell + T_{i\ell}\omega_j n_\ell) + \eta(\omega_i n_j + \omega_j n_i)$$
$$+ D_{k\ell}\,(\omega_k n_\ell + \omega_\ell n_k) V_{ij} = 0$$

where we have defined the tensor **V** by

$$\mathbf{V} = \mu' \lambda \, \mathbf{D} + \frac{\lambda'}{2\lambda} \mathbf{T}$$

where

$$\mu = \frac{\eta(\mathbf{II_D})}{\lambda(\mathbf{II_D})} = \mu(\mathbf{II_D})$$

and

$$\mu' = \mu'(\mathbf{II_D}) , \quad \lambda' = \lambda'(\mathbf{II_D}) .$$

This gives a system of 10 equations and 10 unknowns. For a non-trivial solution, the determinant has to vanish.

We can select a special system of coordinates such that

$$n_1 = 1, \quad n_2 = 0, \quad n_3 = 0\,, \quad \text{and} \quad T_{23} = 0\,.$$

The determinant Δ of the 10×10 system can then be factored. We find that

$$\Delta = -(\lambda c)^4(\rho\lambda c^2 - A)\,(\rho\lambda c^2 - A - 2D_{12}V_{12} - 2D_{13}V_{13})$$

where A is defined by

$$A = \lambda T_{11} + \eta\,.$$

There are therefore two waves with velocities c_+ and c_- given by

$$c_+^2 = \frac{\lambda T_{11}+\eta}{\rho\lambda}\,,$$

$$c_-^2 = \frac{\lambda T_{11}+\eta+2D_{12}V_{12}+2D_{13}V_{13}}{\rho\lambda}\,.$$

The condition for well-posedness is that the two squares of the propagation speeds have to be positive

$$\lambda T_{11}+\eta \geq 0\,,$$

$$\lambda T_{11}+\eta+2D_{12}V_{12}+2D_{13}V_{13} \geq 0\,.$$

In the 2D case we find that there is only one root:

$$c^2 = \frac{\lambda T_{11}+\eta+2D_{12}V_{12}}{\rho\lambda} \geq 0\,.$$

The first condition is the same as the one obtained by Rutkevich for an upper convected Maxwell model, whereas the second one is an additional condition which is needed for well-posedness. It is not possible to conclude anything on the sign of $D_{12}V_{12} + D_{13}V_{13}$ since it depends on the way that λ and η vary as functions of the second invariant $\mathbf{II_D}$. We can guarantee the two conditions by requiring that

$$\text{Min}\,(\lambda T_{11}+\eta, \lambda T_{11}+\eta+2D_{12}V_{12}+2D_{13}V_{13}) \geq 0\,.$$

It is of interest to carry out this kind of analysis for basic flows like shear flow or elongational flow.

- For shear flow the condition becomes:

$$\eta(1+2\lambda^2\kappa^2) + \frac{\kappa^2}{2}\eta' \geq 0 .$$

  This condition for well-posedness is satisfied always when the fluid is shear-thickening $\eta'>0$ or not too large. Obviously there are shear-thinning fluids for which shear flow is ill-posed.

- For elongational flow we find that the condition is satisfied when

$$\frac{1}{1-2\lambda s} \geq 0 \quad \text{or} \quad s \leq \frac{1}{2\lambda} .$$

  This condition is the same one as for elongational flow of an upper convected Maxwell model. There is ill-posedness whenever the inequality is not satisfied.

4.8 Catastrophic short wave instability and the loss of well-posedness

Suppose that $f_0<0$, then

$$\sigma = \pm |\mathbf{k}|\sqrt{|f_0|\rho} . \tag{38}$$

This is a strange formula. When $f_0=0$, $\sigma=0$; but when $f_0<0$ and small, $\sigma\to\infty$ with $|\mathbf{k}|$. We can get stability to short waves at some points and instability at others, depending on the values of γ_0 and σ_0 through frozen coefficients. Many well-known problems of mechanics and physics can undergo catastrophic instabilities to short waves. Some examples are the Kelvin-Helmholtz instability, the Rayleigh-Taylor instability, the Taylor instability of a flat interface for oil displacement in a porous medium, magnetohydrodynamics with the Hall effect, two-fluid models of bicomponent media when viscosity is neglected and the flow of some models of viscoelastic liquids with instantaneous elasticity (see Joseph and Saut, [1986]).

G. Birkhoff [1954] has called attention to the fact that the mathematical definition of ill-posed initial value problems requires the prior specification of a set of functions. He argues convincingly that the class of functions possessing Fourier transforms is appropriate for this discussion, because in this class, it is easy to make a direct connection between Hadamard instabilities and ill-posed initial value problems. In this class, he defines regular and irregular growth rates, "$\sigma(k)$ is regular if and only if it is bounded as $|\mathbf{k}|=k\rightarrow\infty$. Otherwise it is irregular." Problems with irregular eigenvalues are ill-posed as initial value problems in the class of functions with Fourier transforms. There is no uniform continuity with respect to initial values; the short wave components of the disturbance will amplify with asymptotically unbounded growth rates. This kind of instability would not lead to bifurcation as in the case of regular eigenvalues. When the model giving rise to irregular eigenvalues is physically reasonable, the short wave instabilities are associated with fibril structures leading to fingering rather than bifurcation.

It is possible to have short wave instabilities with $k\rightarrow\infty$ without ill-posedness. This is the case when the growth rates are bounded with a maximum growth rate for the shortest waves: $k\rightarrow\infty$. An example of a regular short wave instability of a shear flow is superposed liquids with different viscosities when the surface tension is neglected (Hooper and Boyd, [1983]).

Another connection between ill-posedness and irregular growth rates can be made in the class of functions having Fourier transforms. This connection states that problems with irregular growth rates have no solutions for any class of initial data more general than analytic. For example, the Cauchy problem on $\mathbb{R}^2$, which follows from freezing the coefficients, has no solution with initial data in the C^m class. This result is well known for the Laplace equation treated in the celebrated example by Hadamard. The solution of Laplace's equation is analytic, say in the half (x,y) plane with $x>0$, and it can be extended to $x<0$ by reflections. Hence, the initial data on $y=0$ must also be analytic. This "nonexistence" result is valid generally. A sketch of the proof is given below. Suppose that the coefficients in (20) are constant (as will, in fact, be implied by frozen coefficients) and that $\mathbf{q}$ lies in a Fourier transform class:

$$\mathbf{p}(\alpha,\beta,t) = \frac{1}{2\pi} \int_{-\infty}^{\infty} \int_{-\infty}^{\infty} e^{-i(\alpha x+\beta y)}\, \mathbf{q}(x,y,t)\; dx\, dy \tag{39}$$

is the transform of $\mathbf{q}(x,y,t)$ and $\mathbf{p}(\alpha,\beta,0)$ is the transform of the Cauchy initial data $\mathbf{q}(x,y,0)$ for $\mathbf{q}(x,y,t)$ and

$$\mathbf{q}(x,y,t) = \frac{1}{2\pi} \int_{-\infty}^{\infty} \int_{-\infty}^{\infty} e^{i(\alpha x+\beta y)}\, \mathbf{p}(\alpha,\beta,t)\; d\alpha\, d\beta\,. \tag{40}$$

The transform $\mathbf{p}(\alpha,\beta,t)$ satisfies the following ordinary differential equation:

$$\mathbf{G}\left[\frac{\partial \mathbf{p}}{\partial t} + i(\mathbf{u}_0\bullet\mathbf{k})\mathbf{p}\right] + i(\alpha\mathbf{H}_0+\beta\mathbf{J}_0)\mathbf{p} = \mathfrak{L}_0\mathbf{p} \tag{41}$$

with a prescribed $\mathbf{p}=\mathbf{p}(\alpha,\beta,0)$ at $t=0$. The solution of this initial value problem is given by a vector valued function in $\mathbb{R}^6$

$$\mathbf{p}(\alpha,\beta,t) = \mathbf{p}(\alpha,\beta,0)\,\exp(-i\omega t)$$

where $\omega=\omega_r(\alpha,\beta)+i\omega_i(\alpha,\beta)$ is an eigenvalue of (26) and $\sigma(\alpha,\beta) \overset{\text{def}}{=} \omega_i(\alpha,\beta)$ is the largest growth rate among the six eigenvalues, giving instability when positive. It follows now that

$$\mathbf{q}(x,y,t) = \frac{1}{2\pi} \int_{0}^{\infty} \int_{0}^{\infty} e^{i(\alpha x+\beta y-\omega t)}\, \mathbf{p}(\alpha,\beta,0)\; d\alpha\, d\beta\,. \tag{42}$$

The crucial factor under the integral is

$$e^{\sigma(\alpha,\beta,t)}\,\mathbf{p}(\alpha,\beta,0)\,. \tag{43}$$

If the growth rate $\sigma(\alpha,\beta)$ is irregular, then

$$\lim_{\sqrt{\alpha^2+\beta^2}\to\infty} \sigma(\alpha,\beta) \to \infty ,$$

and the integral defining $\mathbf{q}(x,y,t)$ can exist only if $\mathbf{p}(\alpha,\beta,0)$ decays exponentially as $k\to\infty$ with an exponent dominating $\sigma(\alpha,\beta,t)$.

Our investigation of the existence of solutions with irregular eigenvalues has now been reduced to the study of the asymptotic properties of the transform of the initial data $\mathbf{q}(x,y,0)$. Suppose the second x derivative of $\mathbf{q}(x,y,0)$ is discontinuous at $x=x^*$, that $\mathbf{q}(x,y,0)$ tends to zero at large x, and y is as required for functions in the Fourier transform class. Then, after integrating by parts, we find

$$2\pi\mathbf{p}(\alpha,\beta,0) = -\int_{-\infty}^{\infty} dy\, e^{-i\beta y} \left\{ \int_{-\infty}^{\infty} \frac{1}{i\alpha^3}\, e^{-i\alpha x}\, \frac{\partial^3\mathbf{q}(x,y,0)}{\partial x^3}\, dx \right\}$$

$$+ \frac{1}{i\alpha^3} \int_{-\infty}^{\infty} e^{-i\beta y}\, e^{-i\alpha x^*} \left[\!\left[\frac{\partial^2\mathbf{q}}{\partial x^2} \right]\!\right] (y)\, dy \tag{44}$$

where

$$\left[\!\left[\frac{\partial^2\mathbf{q}(x^*,y)}{\partial x^2} \right]\!\right] = \left[\!\left[\frac{\partial^2\mathbf{q}}{\partial x^2} \right]\!\right] (y) \tag{45}$$

is the jump in the second derivative of $\mathbf{q}(x,y)$ at $x=x^*$. It follows that, in general, $\mathbf{p}(\alpha,\beta,0)$ decays like $1/\alpha^3$ for large α when (45) holds. In this case (43) is unbounded as $\alpha\to\infty$ whenever $\sigma(\alpha,\beta)>0$ is irregular.

The proof that (43) is unbounded when $\mathbf{q}(x,y,0)$ is $C^\infty(x,y)$ is given by Joseph and Saut [1990]. Ill-posedness means nonexistence for all but analytic data.

4.9 Some further comments about frozen coefficients

Analysis of short waves on frozen coefficients has the following useful properties.

1. It leads us to linear equations. Richtmeyer and Morton [1967] note that "Indeed, it is in checking the 'local' stability of linearized equations obtained from truly nonlinear equations that the constant coefficient theory is mainly of use."

2. The short waves allow one to freeze the coefficients; the coefficients do not vary on a sufficiently small region, so the linearized problem has constant coefficients.

3. Since derivatives of the quasilinear system on short waves become unboundedly large, the lower order terms are increasingly unimportant and they may be neglected. This leads to a homogeneous system, to a set of linear homogeneous equations which can be solved only if a determinant of constants vanishes. What could be easier?

4. The reduced homogeneous system is now in $\mathbb{R}^2$ and boundary conditions can be neglected. The short wave instabilities start as a local phenomenon.

The foregoing analysis of short waves on frozen coefficients was formal and not rigorous. The connection to ill-posedness can be framed as in §4.8, but more needs to be done. Kreiss [1962] gives an example of an equation where the problem with variable coefficients is properly posed, yet all the corresponding constant coefficient problems are improperly posed; on the other hand, he gives an example in which the constant coefficient problems are all properly posed, yet the variable coefficient problem is not. So Kreiss' examples show that, in general, local stability is neither necessary nor sufficient for the overall stability of the variable coefficient problem. However, as Strang [1966] points out, if a quasilinear system of first order equations is properly posed, then all frozen coefficient problems are also properly posed. So first order systems like (20) must be properly posed locally, in the sense of frozen coefficients.

4.10 Regularization of ill-posed problems

The catastrophic loss of stability for short waves cannot be an acceptable mathematical description of any physical process. Some regularizing feature of physics, negligible in smoother motions, is brought into action by the rapid variations associated with exponentially growing short waves. A small viscosity or surface tension can be important in controlling short waves.

There is a temptation to discard constitutive equations which can lead to ill-posed initial value problems. However, constitutive equations which always have well-posed problems may be bad and those which may be ill-posed can be good. It is necessary to study how correct models which allow solutions that enter into domains of ill-posedness can be regularized. This type of study is a realization of the "axiom" of physics, which is "there is always a cut-off", or of medicine, "the bleeding always stops".

The addition of Newtonian viscosity to the constitutive model for fluids with instantaneous elasticity is the natural way to regularize ill-posed problems. Many models, like Oldroyd B, already have a Newtonian contribution, expressed by a retardation time. Physically, we expect a Newtonian viscosity to arise from the decay of rapidly decaying modes associated with small molecules. All this puts forward the rheometrical problem of measurements of an "effective" Newtonian viscosity. This new problem needs a solution.

4.11 Hadamard instability of phase change models based on reclining S shaped curves and the regularization of this instability by viscosity

Stress induced phase changes are very important in rheological fluid mechanics, but they are not well understood. One kind of phase change is called complexation. This is believed to be a stress induced networking of polymers.

Mathematical models of phase changes are sometimes based on material behavior based on reclining S shaped curves (Figure 4.1). A classical example of this behavior is the phase diagram for the Van der Waals gas. A theory for phase changes in solids based on non-monotone reclining S shaped stress-strain curves was introduced by Ericksen [1975]. Slemrod [1983] has applied these ideas to various problems involving fluids.

The one-dimensional isothermal motion of a compressible elastic fluid or solid can be described in Lagrangian coordinates by the coupled system

$$\frac{\partial u}{\partial t} = \frac{\partial \rho(\upsilon)}{\partial x},$$

$$\frac{\partial \upsilon}{\partial t} = \frac{\partial u}{\partial x} \tag{46}$$

where $\rho(\upsilon)$ lies on a reclining S shaped curve shown in Figure 4.1.

The curve $\rho(\upsilon)$ is a constitutive assumption giving, say, the stress as a function of strain rate. Equations (46) are a first order quasilinear system which is equivalent to a single second order equation.

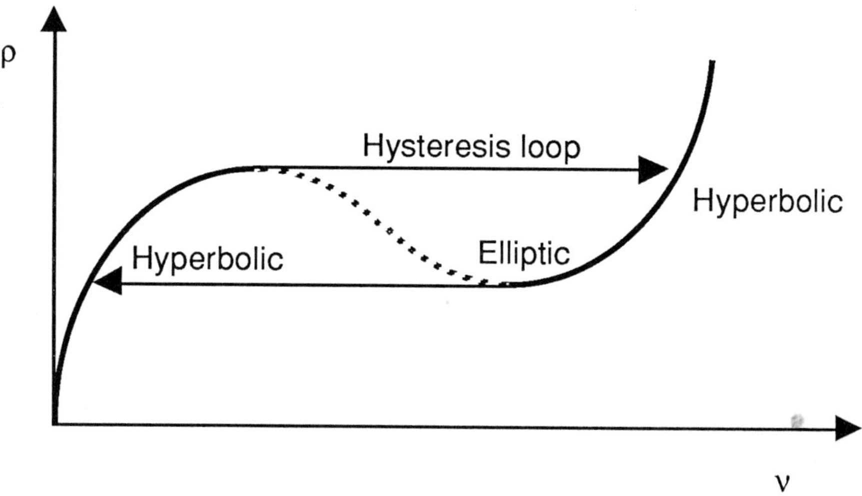

Figure 4.1 The system (46) is hyperbolic when $\rho' > 0$. The elliptic branch is unstable in the sense of Hadamard. This region is called "spinoidal".

$$\frac{\partial^2 \upsilon}{\partial t^2} = \frac{\partial^2 \rho(\upsilon)}{\partial x^2} \tag{47}$$

It is clear from (47) that wherever $\rho'(\upsilon)$ is negative (47) is like Laplace's equation and is Hadamard unstable.

Let us do something a little artificial to how Hadamard's instability arises on frozen coefficients of a stability problem for (47). We could solve

$$\frac{\partial^2 \rho(\upsilon)}{\partial x^2} = 0$$

for $\upsilon(x)$ such that $\upsilon(0)=0$ and $\upsilon(1)=1$.

Call this solution υ_0 and let ω be a small perturbation of υ_0. For ω we get

$$\frac{\partial^2 \omega}{\partial t^2} = 2\rho''(\upsilon_0)\frac{\partial \upsilon_0}{\partial x}\frac{\partial \omega}{\partial x} + \rho'(\upsilon_0)\frac{\partial^2 \omega}{\partial x^2} + \left[\rho'''(\upsilon_0)\left(\frac{\partial \upsilon_0}{\partial x}\right)^2 + \rho''(\upsilon_0)\frac{\partial^2 \upsilon_0}{\partial x^2}\right]\omega \tag{48}$$

where

$$\omega = 0 \text{ at } x = 0,1. \tag{49}$$

In general (48) has variable coefficients and doesn't admit stability studies using normal modes, but if the wave length of a disturbance is short enough, υ_0 will be nearly constant over the whole length of the wave. This leads us to freeze the coefficients in (48), treating υ_0 as a constant in a small neighborhood of each and every point.

In each such neighborhood, we write

$$\omega = \text{const } e^{\sigma t + ikx} \tag{50}$$

and we get

$$\sigma^2 = -\rho'(\upsilon_0)k^2 + 0(|k|) .$$

Of course, we cannot satisfy the boundary condition (49) with (50), the analysis of stability here is purely local and it may hold only if $k \to \infty$, for short waves. Hence

$$\frac{\sigma^2}{k^2} = -\rho'(\upsilon_0)\,. \tag{51}$$

asymptotically and we have Hadamard instability wherever υ_0 is such that $\rho'(\upsilon_0)<0$.

In another version of this problem we could try to regularize with a Newtonian viscosity μ. Then we would consider the system

$$\frac{\partial u}{\partial t} = \frac{\partial \rho(\upsilon)}{\partial x} + \mu\,\frac{\partial^2 u}{\partial x^2}\,,$$
$$\frac{\partial \upsilon}{\partial t} = \frac{\partial u}{\partial x} \tag{52}$$

which can be reformulated into one third order equation.

$$\frac{\partial^2 \upsilon}{\partial t^2} = \frac{\partial^2 \rho(\upsilon)}{\partial x^2} + \mu\,\frac{\partial^3 \upsilon}{\partial x^2 \partial t}\,. \tag{53}$$

Suppose $\upsilon_0(x-ct)$ is a travelling wave solution of (52). To study the stability of this travelling wave, we write

$$\upsilon = \upsilon_0 + \omega$$

where ω satisfies

$$\frac{\partial^2 \omega}{\partial t^2} = 2\rho''(\upsilon_0)\,\upsilon_0'\,\frac{\partial \omega}{\partial x} + \rho'''(\upsilon_0)\upsilon_0'^2\omega$$
$$+\,\rho'(\upsilon_0)\,\frac{\partial^2 \omega}{\partial x^2} + \rho''(\upsilon_0)\omega\,\upsilon_0'' + \mu\frac{\partial^3 \omega}{\partial x^2 \partial t}$$

We freeze the coefficients and use normal modes

$$\omega = e^{\sigma t + ikx}.$$

This leads to

$$\sigma^2 = 2ik\upsilon_0'\rho''(\upsilon_0) + \rho'''(\upsilon_0)\,\upsilon_0'^2 + \rho''(\upsilon_0)\,\upsilon_0''$$
$$-\,\rho'(\upsilon_0)\,k^2 - \mu\sigma k^2\,.$$

Divide this by k^2, letting $k \to \infty$ to find

$$\frac{\sigma}{k} = -\frac{1}{2}\mu k \pm \frac{1}{2}\sqrt{\mu^2 k^2 - 4\rho'}\,.$$

Clearly we have stability when $\rho'>0$. If $\rho'=-|\rho'|<0$, then the most dangerous mode is given by

$$\frac{\sigma}{k}=\frac{\mu k}{2}\left\{-1+\sqrt{1+4\,|\rho'|/\mu^2k^2}\right\}\,. \tag{54}$$

For large μk this is given by

$$\sigma \sim |\rho'|/\mu\,.$$

This shows that the travelling wave is unstable but not Hadamard unstable. Hadamard instability "occurs" when

$$\mu^2k^2 \ll |\rho'|\,,\quad \sigma \sim k|\rho'|\,,$$

$$\text{and } k\to\infty\,.$$

So in this case we get a formal regularization of the equations, but the Hadamard instability persists in a practical sense if μ is small enough.

Of course, there is some continuity between ill-posed problems and regularized ill-posed problems as the regularizing parameter tends to zero.

Returning now to Figure 4.1 and (47) we note that the solid lines, where $\rho'(\upsilon)>0$, lead to a hyperbolic equation and the dashed line leads to an elliptic equation. The elliptic portion is rejected because it will exhibit Hadamard instabilities; and actual solutions are required to operate only on the hyperbolic parts of the curve. This leads to spatially segregated solutions, separated by lines of discontinuity, each part operating on a different hyperbolic branch of the curve. There is hysteresis and abrupt transitions in the response of such models. These features are all present in the study of Hunter and Slemrod [1983], which attempts to explain some observations of Tordella [1969] of a type of melt fracture called ripple. This phenomenon shows hysteresis loops, double-valued shear rates at certain stresses and spatially segregated flow regimes.

4.12 Hadamard instability for some non-Newtonian fluids based on analysis of the fourth order equation for the stream function

Regirer and Rutkevich [1968] have considered fluids of the Reiner-Rivlin type which exhibit change of type. Their constitutive law is

$$\mathbf{T} = -p\mathbf{1} + \eta f\,(\mathrm{II})\mathbf{D}\ ,$$

where $\mathbf{D} = \frac{1}{2}\,(\nabla\mathbf{u}+(\nabla\mathbf{u})^{T})$, $\mathrm{II} = \mathrm{tr}\,\mathbf{D}^2$. Written in terms of a stream function

$$\mathbf{u} = (u,v) = (\psi_y, -\psi_x)\ ,$$

the equation governing steady two-dimensional flows is as follows,

$$L\psi \overset{\text{def}}{=} a_1\left[\frac{\partial^4\psi}{\partial x^4} + \frac{\partial^4\psi}{\partial y^4}\right] + 2a_2\,\frac{\partial^4\psi}{\partial x^2\partial y^2} + 4a_3\left(\frac{\partial^4\psi}{\partial x\partial y^3} - \frac{\partial^4\psi}{\partial x^3\partial y}\right) = H(\psi) \tag{55}$$

where $H(\psi)$ is a nonlinear third order operator and the coefficients a_j are nonlinear functions of the second derivatives of ψ. The characteristic curves $y(x)$ are solutions of

$$a_1 y_x^4 + 4a_3 y_x^3 + 2a_2 y_x^2 - 4a_3 y_x + a_1 = 0\ . \tag{56}$$

All plane flows can be formulated in terms of a quasilinear fourth order problem for the stream function even when the generating first order problem is nonlinear and not quasilinear (cf. §6.10). Quartics like (56) are critically important for the analysis of plane flows. In the Regirer-Rutkevich problem, there are three cases:

(i) $f + 2\mathrm{II}f' > 0$ (no real roots, elliptic),

(ii) $f + 2\mathrm{II}f' = 0$ (parabolic),

(iii) $f + 2\mathrm{II}f' < 0$ (four real roots, hyperbolic).

We can evalute the shear stress component τ of $\mathbf{T}$ when II is given as a function of κ, the shear rate and

$$\tau = 2\eta f(\kappa)\kappa\,. \tag{57}$$

The hyperbolic regions are those where the shear stress decreases as a function of shear rate, and the elliptic regions are those where it increases. Many constitutive equations are like this one; for example, the corrotational Maxwell model has a maximum shear stress at a finite rate of shear.

The unsteady problem corresponding to (55) is

$$\rho\frac{\partial \Delta\psi}{\partial t} = L(\psi)\,. \tag{58}$$

When the right of (58) is elliptic, the steady flow is elliptic, the problem is parabolic and well-posed. So an increasing shear stress does not lead to instability. However, when the shear stress starts to decrease, $f+2\mathrm{II}f\,'<0$, the unsteady equation is like a backward heat equation and Hadamard unstable stress occurs. This situation is analogous to the instability associated with reclining S-shaped curves.

Another interesting problem of the type just discussed is the stability analysis of Couette flow by Akbay, Becker, Krozer, and Sponagel [1980a, b]. In order to obtain a manageable equation, they introduce the "short memory approximation." This means that, in the memory integrals occurring in the equation for the disturbances, only terms of first order in the relaxation time of the fluid are kept. Proceeding in the manner described in this text following (16.16), they find

$$\left(\frac{\partial}{\partial t} + \rho\gamma\kappa\frac{\partial}{\partial x}\right)\Delta\Psi = L\Psi \tag{59}$$

where κ is the rate of shear of the Couette flow

$$\begin{aligned} a_1 &= \tau'\,, \\ 2a_2 &= -2\tau' + \frac{4\tau}{\kappa}\,, \\ 4a_3 &= N_1{}' - \frac{N_1}{\kappa} \end{aligned} \tag{60}$$

and $\tau(\kappa)$, $N_1(\kappa)$ are the shear stress and first normal stress functions of the rate of shear.

The problem is posed on the strip

$$-\infty<x<\infty\ ,\quad 0<y<h\ ,$$

$$\psi=\frac{\partial\psi}{\partial y}=0 \quad \text{on } y=0,h\ .$$

The paper by Ahrens, *et al.* [1984] reports a study of the stability of viscometric flow using the type of short memory introduced by Akbay, *et al.* [1980a, b]. The instability found by Akbay, *et al.* can be identified as a loss of evolution leading to the catastropic short wave instability of Hadamard type whenever

$$\frac{[(N_1(\kappa)/\kappa)']^2\kappa^3}{\tau(\kappa)\tau'(\kappa)}>16\ . \tag{61}$$

The problem and its solution are discussed in detail in Chapter 16.

Exercises

Let $N=N(\psi,\psi_x,\psi_y)$ be an analytic function of ψ and the first derivatives of ψ. Use the method of frozen coefficients to prove the following results:

Exercise 4.1.
The initial value problem for

$$\frac{\partial^2\psi}{\partial t^2}=\nabla^2\psi+N$$

is Hadamard stable.

Exercise 4.2.
The initial value problem for

$$\frac{\partial^2\psi}{\partial t^2}=\frac{\partial^2\psi}{\partial x^2}-\frac{\partial^2\psi}{\partial y^2}+N$$

is Hadamard unstable in a certain sector of the x,y plane. Identify the sector.

Exercise 4.3.

$$\frac{\partial^2\psi}{\partial t^2} = A\frac{\partial^2\psi}{\partial x^2} + B\frac{\partial^2\psi}{\partial y^2} + N$$

Suppose A,B are analytic functions of ψ and the first derivatives of ψ and that the steady equation is hyperbolic where $AB<0$ and is elliptic where $AB>0$. (You may prove rather than assume this by using the theory of classification of the type of a 2nd order equation set down in §6.1). Show that the unsteady problem is hyperbolic, hence, well-posed, if and only if $AB>0$ and $A>0$. The unsteady problem is elliptic and Hadamard unstable when the steady equation is hyperbolic and $A<0$.

Exercise 4.4.
Consider the first order quasilinear system

$$\frac{\partial u}{\partial t} = \frac{\partial\sigma(v)}{\partial x},$$

$$\frac{\partial v}{\partial t} = \frac{\partial u}{\partial x}$$

where $\sigma(\bullet)$ is the reclining S curve shown in Figure 4.1.

Show that this system is Hadamard unstable where $\sigma'(v)<0$.

Exercise 4.5.
Write (30)

$$f_0 = b_{11}\xi_1^2 + 2b_{12}\xi_1\xi_2 + b_{22}\xi_2^2$$

where

$$b_{11} = G + \frac{1}{2}(\sigma_0-\gamma_0) + \frac{a}{2}(\sigma_0+\gamma_0),$$
$$b_{22} = G - \frac{1}{2}(\sigma_0-\gamma_0) + \frac{a}{2}(\sigma_0+\gamma_0),$$
$$b_{12} = \tau_0 .$$

Show that there are always values $-\infty<\xi_1<\infty$, $-\infty<\xi_2<\infty$ such that

$$f_0 < 0 \text{ if } b_{12}^2 - b_{11}b_{22} > 0 .$$

Moreover,

$$f_0 > 0 \text{ if } b_{12}^2 - b_{11}b_{22} > 0 \text{ and } b_{11} > 0 .$$

Solutions for the exercises

Exercise 4.1.

$$\frac{\partial^2\Psi}{\partial t^2} = \nabla^2\Psi + N$$

Let ψ^0 be the basic solution, ψ be the perturbation,

$$\Psi = \psi^o + \psi .$$

Solving ψ with normal modes,

$$\psi = \psi_0 \exp\left[\sigma t + i\mathbf{k}\bullet\mathbf{x}\right] .$$

Therefore,

$$\sigma^2 = -|\mathbf{k}|^2 + 0(|\mathbf{k}|) .$$

As $|\mathbf{k}|\to\infty$, $\sigma\to\pm i|\mathbf{k}|$. Therefore, the problem is Hadamard stable.

Exercise 4.2.

$$\frac{\partial^2\Psi}{\partial t^2} = \frac{\partial^2\Psi}{\partial x^2} - \frac{\partial^2\Psi}{\partial y^2} + N .$$

Introduce normal modes

$$\psi = \psi_0 \exp\left[\sigma t + i(\alpha x + \beta y)\right] .$$

Therefore,

$$\sigma^2 = -(\alpha^2 - \beta^2) + \text{lot} = \beta^2\left(1 - \left(\frac{\alpha}{\beta}\right)^2\right) + \text{lot}$$

As $\alpha,\beta \to \infty$,

$$\sigma \to \pm|\beta| \sqrt{1-(\alpha/\beta)^2} .$$

If $|\alpha| \geq |\beta|$: the problem is Hadamard stable ,
$|\alpha| < |\beta|$: the problem is Hadamard unstable .

Exercise 4.3.

$$\frac{\partial^2 \Psi}{\partial t^2} = A \frac{\partial^2 \Psi}{\partial x^2} + B \frac{\partial^2 \Psi}{\partial y^2} + N .$$

When AB<0: the steady equation is hyperbolic;
When AB>0: the steady equation is elliptic.

Suppose Ψ^0 is a solution of the problem, we could write

$$\Psi = \Psi^0 + \psi$$

and try a solution of normal modes

$$\psi = \psi_0 \exp\left[\sigma t + i\alpha x + i\beta y\right] .$$

Then,

$$\sigma^2 = -(A_0\alpha^2 + B_0\beta^2) + \text{lot} ,$$
$$A_0 = A(\Psi^0,\ldots) , \quad B_0 = B(\Psi^0,\ldots) .$$

If $A_0>0$ and $B_0>0$, the unsteady problem is Hadamard stable, hence, well-posed.

If the unsteady problem is Hadamard stable, we must have $\sigma^2<0$, as $\alpha,\beta \to \infty$, thus, $A_0\alpha^2 + B_0\beta^2 > 0$, or, $A_0>0$, $A_0B_0>0$.

Therefore, when the steady equation is elliptic, the unsteady problem is hyperbolic (well-posed) if and only if $A_0>0$, $A_0B_0>0$.

If $A_0<0$, $B_0>0$ ($A_0B_0<0$, the steady equation is hyperbolic), then

$$\sigma^2 = (-A_0)\left(\alpha^2 - \frac{B_0}{(-A_0)}\beta^2\right) + \text{lot} .$$

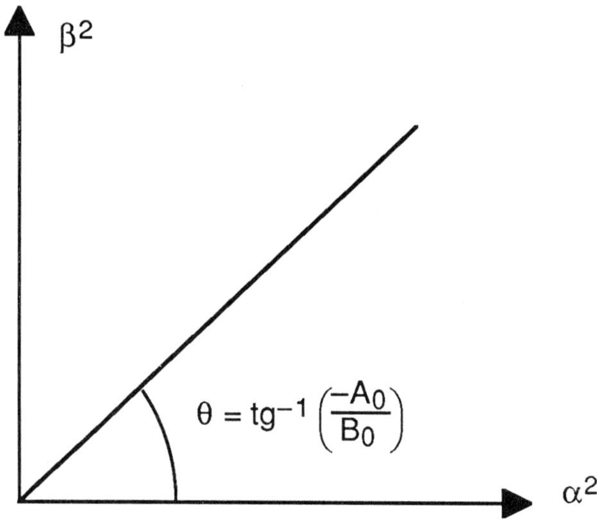

Figure 4.2. In the region $\alpha^2 > \frac{B_o}{(-A_o)}\beta^2$, the unsteady problem is Hadamard unstable (elliptic).

Exercise 4.4.

$$\frac{\partial u}{\partial t} = \frac{\partial \sigma(v)}{\partial x} ,$$

$$\frac{\partial v}{\partial t} = \frac{\partial u}{\partial x} .$$

Since

$$\frac{\partial u}{\partial t} = \frac{\partial \sigma}{\partial x} = \frac{\partial \sigma}{\partial v} \frac{\partial v}{\partial x} ,$$

we have

$$\frac{\partial^2 v}{\partial t^2} = \frac{\partial}{\partial x}\left(\frac{\partial \sigma}{\partial v} \frac{\partial v}{\partial x}\right) = \frac{\partial \sigma}{\partial v} \frac{\partial^2 v}{\partial x^2} + \text{lot} .$$

For solutions in normal modes

$$v = v_0 \exp\left[\omega t + i\alpha(x - x_0)\right]$$

and

$$\omega^2 = \left(\frac{d\sigma}{dv}\right)_0 (-\alpha^2) + \text{lot} .$$

Therefore, when $\sigma'(v)<0$, this system is Hadamard unstable.

Exercise 4.5.

$$f_0 = b_{11}\xi_1^2 + 2b_{12}\xi_1\xi_2 + b_{22}\xi_2^2 = (\xi_1,\xi_2)\begin{vmatrix} b_{11} & b_{12} \\ b_{12} & b_{22} \end{vmatrix}\begin{vmatrix} \xi_1 \\ \xi_2 \end{vmatrix} ,$$

$$-\infty<\xi_1<\infty , \quad -\infty<\xi_2<\infty .$$

Therefore, for positive definite f_0, we must have

$$b_{11} > 0 , \quad b_{11}b_{22} - b_{12}^2 > 0 .$$

If $b_{11}b_{22} - b_{12}^2 < 0$, there are always values of ξ_1,ξ_2 such that $f_0<0$. If $b_{11}<0$, then let

$$\xi_2 = 0, \quad f_0<0 .$$

If $b_{11}>0$, let

$$\xi'_1 = \xi_1 + \frac{b_{12}}{b_{11}}\xi_2 , \quad \xi'_2 = \xi_2 .$$

$$f_0 = b_{11}\xi'^2_1 + (b_{22} - \frac{b_{12}^2}{b_{11}})\xi'^2_2 , \quad \text{let } \xi'_1 = 0 \text{ then } f_0 < 0 .$$

CHAPTER 5

Characteristics and Classification of Type

In §1.4, we studied a first order system equivalent to a telegraph equation and determined lines $\phi(x,t)=0$ across which the system variables were continuous, with simple discontinuities of derivatives. In multidimensional systems with three or more independent variables, the characteristic lines are replaced with characteristic surfaces.

5.1 Characteristic surfaces

The method of discontinuous derivatives can be used to determine the characteristic surfaces for multidimensional flows of models, like Maxwell's (3.40). In the two-dimensional case, we look for characteristic surfaces $\phi(x,y,t)=0$ across which the system vector $\mathbf{Q}$ of the quasilinear system (3.40) is continuous with bounded jumps of the first derivatives. We first note that the continuity of $\mathbf{Q}$ implies $[\![\ell(\mathbf{Q})]\!]=0$, so that lower order terms vanish. All the tangential derivatives of $[\![\mathbf{Q}]\!]$ vanish, the only nonzero derivative is $[\![\mathbf{Q}_\phi]\!]$ normal to $\phi=0$. Hence, if discontinuities of the proposed type exist, they satisfy

$$\left\{\left(\frac{\partial\phi}{\partial t}+\mathbf{u}\cdot\nabla\phi\right)\mathbf{G}+\frac{\partial\phi}{\partial x}\mathbf{H}+\frac{\partial\phi}{\partial y}\mathbf{J}\right\}[\![\mathbf{Q}_\phi]\!]=0\,. \tag{1}$$

These are six equations in six unknowns which can be satisfied if and only if the determinant of the coefficients vanishes. We already computed the determinant in our analysis of Hadamard instability and the result is given by (4.29). After making the obvious replacements, we find a solution for our problem if and only if

$$\Delta = |\nabla\phi|^2 \left(\frac{\partial\phi}{\partial t} + \mathbf{u}\cdot\nabla\phi\right)^2 \left\{ 2\frac{\partial\phi}{\partial x}\frac{\partial\phi}{\partial y}\tau + \left[G + \frac{a}{2}(\gamma+\sigma)\right] |\nabla\phi|^2 \right.$$

$$\left. + \frac{1}{2}\left[\left(\frac{\partial\phi}{\partial x}\right)^2 - \left(\frac{\partial\phi}{\partial y}\right)^2\right](\sigma-\gamma) - \rho\left(\frac{\partial\phi}{\partial t} + \mathbf{u}\cdot\nabla\phi\right)^2\right\} = 0\,. \qquad (2)$$

The criterion (2) for characteristic surfaces is a general one, and it applies to the nonlinear problem without approximation.

5.2 Analysis of the roots

There are six "roots" for equation (2), appearing in three quadratic pairs:

$$|\nabla\phi|^2 = 0\,, \qquad (3)$$

$$\left(\frac{\partial\phi}{\partial t} + \mathbf{u}\cdot\nabla\phi\right)^2 = 0\,, \qquad (4)$$

$$2\frac{\partial\phi}{\partial x}\frac{\partial\phi}{\partial y}\tau + \left[G + \frac{a}{2}(\gamma+\sigma)\right] |\nabla\phi|^2$$

$$+ \frac{1}{2}\left[\left(\frac{\partial\phi}{\partial x}\right)^2 - \left(\frac{\partial\phi}{\partial y}\right)^2\right](\sigma-\gamma) - \rho\left(\frac{\partial\phi}{\partial t} + \mathbf{u}\cdot\nabla\phi\right)^2 = 0\,. \qquad (5)$$

There are no real roots of (3). This situation is characteristic of elliptic problems (Appendix D). Equation (4) says that ϕ is conserved following the motion, $d\phi/dt=0$. This means that particle trajectories are characteristic; in fact, they are doubly characteristic because $(d\phi/dt)^2=0$. Equation (5) may have real roots or complex roots, depending on the values of the unknown dependent variables

u,v,γ,σ,τ. It is possible, and frequently happens, that there are real roots in some parts of the flow and complex roots in other parts.

Our quasilinear system is not hyperbolic and is not elliptic. It is said to have roots of composite type. It is important to sort out the roots and to associate physical quantities with roots of different type.

5.3 One-dimensional unsteady flow

It is of interest to consider one-dimensional unsteady flows, like shear flows, which depend only on one-space variables, say x, so that u,σ,τ,γ depend on x and t. Then we seek characteristic surfaces $\phi(x,t)=0$. On $\phi=0$, we have

$$d\phi = \frac{\partial \phi}{\partial x} dx + \frac{\partial \phi}{\partial t} dt = 0 , \tag{6}$$

and (2) reduces to

$$\Delta = \left(\frac{\partial \phi}{\partial x}\right)^6 \left(u - \frac{dx}{dt}\right)^2 \left\{ G + \frac{a}{2}(\gamma+\sigma) + \frac{1}{2}(\sigma-\gamma) - \rho\left(u - \frac{dx}{dt}\right)^2 \right\} = 0 . \tag{7}$$

We get real characteristic surfaces, from the bracket, when

$$G + \frac{a}{2}(\gamma+\sigma) + \frac{1}{2}(\sigma-\gamma) > 0 . \tag{8}$$

In the linearized case treated in §1.4, we put u, γ, and σ to zero and find $\rho(dx/dt)^2=G$, corresponding to the straight characteristic lines (1.22).

5.4 Two-dimensional steady flows

We suppose that the six-field variables are independent of time t. Then the characteristic surfaces are of the form $\phi(x,y)=0$,

$$d\phi=\frac{\partial \phi}{\partial x} dx + \frac{\partial \phi}{\partial y} dy = 0 , \tag{9}$$

on such a surface and

$$\Delta = \left(\frac{\partial\phi}{\partial y}\right)^6\left[\left(\frac{dy}{dx}\right)^2+1\right]\left[-u\frac{dy}{dx}+v\right]^2\left\{-2\frac{dy}{dx}\tau\right.$$
$$+\left[G+\frac{a}{2}(\gamma+\sigma)\right]\left[1+\left(\frac{dy}{dx}\right)^2\right]+\frac{1}{2}\left[\left(\frac{dy}{dx}\right)^2-1\right](\sigma-\gamma) \tag{10}$$
$$\left.-\rho\left[v-u\,\frac{dy}{dx}\right]^2\right\}=0\,.$$

In steady two-dimensional flow, the streamlines are doubly characteristic.

Of course, the factor $(dy/dx)^2=-1$ does not give rise to real characteristics. These roots can be identified with a stream function ψ. $u=\partial\psi/\partial y$, $v=-\partial\psi/\partial x$. The vorticity is given by

$$\zeta = \frac{\partial v}{\partial x}-\frac{\partial u}{\partial y} = -\frac{\partial}{\partial x}\left(\frac{\partial\psi}{\partial x}\right)-\frac{\partial}{\partial y}\left(\frac{\partial\psi}{\partial y}\right),$$
$$0 = \frac{\partial u}{\partial x}+\frac{\partial v}{\partial y} = \frac{\partial}{\partial x}\left(\frac{\partial\psi}{\partial y}\right)-\frac{\partial}{\partial y}\left(\frac{\partial\psi}{\partial x}\right). \tag{11}$$

Let us see if jumps in the second derivatives of ψ are possible when the vorticity is continuous $[\![\zeta]\!]=0$. Using (11), we find

$$[\![\psi_x]\!]_x + [\![\psi_y]\!]_y = 0\,,$$

$$[\![\psi_y]\!]_x - [\![\psi_x]\!]_y = 0\,.$$

Then,

$$[\![\psi_x]\!]_\phi\phi_x + [\![\psi_y]\!]_\phi\phi_y = 0\,,$$

$$[\![\psi_y]\!]_\phi\phi_x - [\![\psi_x]\!]_\phi\phi_y = 0\,. \tag{12}$$

To solve for the jumps in (12), we must have

$$\left(\frac{\partial\phi}{\partial x}\right)^2+\left(\frac{\partial\phi}{\partial y}\right)^2=0$$

or

$$\left(\frac{dy}{dx}\right)^2=-1\,. \tag{13}$$

So the jumps in the derivatives of ψ are not possible. In fact,

$$\nabla^2\psi = -\zeta \tag{14}$$

shows that ψ satisfies one elliptic equation.

To understand the pair of interesting roots in the third factor of Δ, we next study Equation (3.48) governing the evolution of the vorticity ζ.

5.5 Characteristic surfaces for the vorticity

Let us suppose that the second derivatives of ζ, satisfying (3.51), are discontinuous, with continuity of the first and lower derivatives. First derivatives of ζ involve third derivatives of u and v, which are assumed to be continuous across $\phi(x,y,t)=0$. We also assume that τ, γ, and σ are continuous. After forming the jumps, we get

$$\rho\left[\!\left[\frac{\partial^2\zeta}{\partial t^2}\right]\!\right] + 2\rho\left[\!\left[(\mathbf{u}\cdot\nabla)\frac{\partial\zeta}{\partial t}\right]\!\right] - A\left[\!\left[\frac{\partial^2\zeta}{\partial x^2}\right]\!\right] - 2B\left[\!\left[\frac{\partial^2\zeta}{\partial x\partial y}\right]\!\right] - C\left[\!\left[\frac{\partial^2\zeta}{\partial y^2}\right]\!\right] = 0\,. \tag{15}$$

Since the tangential derivatives of first derivatives of ζ are continuous across $\phi=0$ only second derivatives with respect to ϕ can appear in (15); for example,

$$\left[\!\left[\frac{\partial^2\zeta}{\partial t^2}\right]\!\right] = \left[\!\left[\frac{\partial^2\zeta}{\partial t\partial\phi}\right]\!\right]\phi_t = [\![\zeta_{\phi\phi}]\!]\phi_t^2\,. \tag{16}$$

The equation governing the characteristic surfaces for $\phi(x,y,t)=0$ of the vorticity is

$$\rho\phi_t^2 + 2\rho u\phi_x\phi_t + 2\rho v\phi_y\phi_t - A\phi_x^2 - 2B\phi_x\phi_y - C\phi_y^2 = 0\,. \tag{17}$$

After taking account of the definition of A, B, and C, we reduce (17) to

$$\rho(\phi_t+\mathbf{u}\cdot\nabla\phi)^2 = 2\phi_x\phi_y\,\tau + \left\{ G + \frac{a}{2}(\gamma+\sigma) \right\} |\nabla\phi|^2 + \frac{1}{2}[\phi_x^2 - \phi_y^2](\sigma-\gamma) . \qquad (18)$$

Comparing (18) with (2), we see that the roots of (2), which depend on the values of the solution, are associated with the vorticity.

5.6 Wave speeds and hyperbolicity

On a characteristic surface

$$\phi_t = -\phi_x \frac{dx}{dt} - \phi_y \frac{dy}{dt} . \qquad (19)$$

After introducing the wave velocity

$$\hat{\mathbf{c}} \overset{\text{def}}{=} \mathbf{e}_x \frac{dx}{dt} + \mathbf{e}_y \frac{dy}{dt} , \qquad (20)$$

we may rewrite (18) using

$$\phi_t + \mathbf{u} \cdot \nabla\phi = (\mathbf{u} - \hat{\mathbf{c}}) \cdot \nabla\phi . \qquad (21)$$

Equations (18) and (21) form a quadratic equation for $\hat{\mathbf{c}}$, given $\mathbf{u},\phi_x,\phi_y,\sigma,\gamma,\tau$. Typically, there are two real-valued $\hat{\mathbf{c}}$'s or two complex conjugate roots.

The vorticity equation is hyperbolic if the two roots are real and elliptic if the two roots are complex. The vorticity equation is hyperbolic if there are real wave speeds and elliptic if the wave speeds are complex.

Equation (18) is homogeneous and of degree two in ϕ. Therefore, without losing generality, we may normalize ϕ so that

$$|\nabla\phi|^2 = 1 . \qquad (22)$$

We may then write (18) and (21) as

$$\rho\{(\mathbf{u}-\hat{\mathbf{c}}) \cdot \nabla\phi\}^2 = f(u,v,\sigma,\gamma,\tau,\phi_x,\phi_y) \qquad (23)$$

where f_o given by (4.30) is immediate.

$$f_o = f(u_o, v_o, \sigma_0, \tau_0, \xi_1, \xi_2) \; . \tag{24}$$

The vorticity is, therefore, Hadamard unstable if f<0; in this case, the vorticity equation is elliptic.

- The quasilinear system (3.40) is Hadamard unstable if f<0 for any choice of ϕ_x, ϕ_y ($f_o<0$ for any choice of ξ_1, ξ_2).

- The quasilinear system (3.40) is ill-posed if it is Hadamard unstable.

- The quasilinear system is Hadamard unstable wherever the vorticity is elliptic.

- The initial value problem for the quasilinear system (3.40) is well-posed for all solutions such that the vorticity is hyperbolic.

5.7 How to convert a nonlinear system into a quasilinear one

It is always possible to define systems of partial differential equations of any order as a system of first order equations in more unknowns. Without losing generality we restrict attention to systems of first order partial differential equations which may be quasilinear or nonlinear. For example

$$\frac{\partial u}{\partial t} + u^2 \frac{\partial u}{\partial x} = f(u)$$

is quasilinear, linear in $\frac{\partial u}{\partial t}$ and $\frac{\partial u}{\partial x}$ while

$$\frac{\partial u}{\partial t} + u\left(\frac{\partial u}{\partial x}\right)^2 - f(u) = 0 \tag{25}$$

is nonlinear. We may write (25) as

$$F \stackrel{\text{def}}{=} p + uq^2 - f(u) = 0\,, \quad p = \frac{\partial u}{\partial t}\,, \quad q = \frac{\partial u}{\partial x}\,.$$

We can reduce a system of N first order nonlinear partial differential equations in γ independent variables to a system of $(\gamma+1)N$ quasilinear equations. A general system of N first order PDE's in two independent variables can be expressed as

$$F_i(x,y,u_1,\ldots,u_n,p_1,\ldots,p_n,q_1,\ldots,q_n) = 0 \quad i = 1, \ldots, n \tag{26}$$

where

$$p_i = \frac{\partial u_i}{\partial x}\,, \quad q_i = \frac{\partial u_i}{\partial y}$$

are introduced as additional unknowns. We have 3N variables and 3N equations but one of the equations is nonlinear rather than quasilinear. The system can be reduced to a quasilinear one by differentiation but the reduction is not unique. One symmetric reduction is: $F_i=0$ is an identity in x and y jointly, hence

$$\frac{dF_i}{dy} = \frac{\partial F_i}{\partial y} + \frac{\partial F_i}{\partial u_j} q_j + \frac{\partial F_i}{\partial q_j}\frac{\partial q_j}{\partial y} + \frac{\partial F_i}{\partial p_j}\frac{\partial p_j}{\partial y} = 0\,,$$

$$\frac{dF_i}{dx} = \frac{\partial F_i}{\partial x} + \frac{\partial F_i}{\partial u_j} p_j + \frac{\partial F_i}{\partial q_j}\frac{\partial q_j}{\partial x} + \frac{\partial F_i}{\partial p_j}\frac{\partial p_j}{\partial x} = 0\,.$$

Equation (26) implies that

$$\frac{\partial F_i}{\partial p_j} p_j + \frac{\partial F_i}{\partial q_j} q_j = \frac{\partial F_i}{\partial p_j}\frac{\partial u_j}{\partial x} + \frac{\partial F_i}{\partial q_j}\frac{\partial u_j}{\partial y}\,. \tag{27$_1$}$$

We put this system into a symmetric form by writing $\partial p_j/\partial y = \partial q_j/\partial x$. Then we put the principal part on the right and the lower order terms on the left

$$-\left(\frac{\partial F_i}{\partial y} + \frac{\partial F_i}{\partial u_j} q_j\right) = \frac{\partial F_i}{\partial p_j}\frac{\partial q_j}{\partial x} + \frac{\partial F_i}{\partial q_j}\frac{\partial q_j}{\partial y}\,, \tag{27$_2$}$$

$$-\left(\frac{\partial F_i}{\partial x} + \frac{\partial F_i}{\partial u_j} p_j\right) = \frac{\partial F_i}{\partial p_j}\frac{\partial p_j}{\partial x} + \frac{\partial F_i}{\partial q_j}\frac{\partial p_j}{\partial y}\,. \tag{27$_3$}$$

Equations $(27)_1$, $(27)_2$, and $(27)_3$ are 3N equations for the 3N unknowns.

The principal parts of each of the equations $(27)_1$, $(27)_2$, and $(27)_3$ are identical. Each one determines the same characteristic directions. We have

$$\frac{\partial F_i}{\partial p_j}\frac{\partial a_j}{\partial x}+\frac{\partial F_i}{\partial q_j}\frac{\partial a_j}{\partial y}=\text{lot}\,.$$

Hence the characteristics $\lambda = dy/dx$ are determined from

$$\det\left[\lambda\frac{\partial F_i}{\partial p_j}-\frac{\partial F_i}{\partial q_j}\right]=0\,. \tag{28}$$

(The value of λ for (25) is $\lambda=2uq$.) Equation (28) has N roots. The N nonlinear first order PDE's give rise to N characteristic roots for the quasilinear system arising from differentiating the nonlinear system once with respect to each independent variable.

The White-Metzner model discussed in §6.9 gives rise to a nonlinear system which is not quasilinear.

If we generate (33) by the method of simple jumps we can state that real characteristic directions are the loci for discontinuities in the derivatives of p_i and q_i. This means that the second derivatives of u_i suffer jumps in the nonlinear case and first derivatives jump in the quasilinear case. The first derivatives are smooth when second derivatives jump so that we get one more derivative of smoothness in the nonlinear case.

It appears that a more far reaching conclusion following along lines of the last paragraph may be true. Compare quasilinear and nonlinear first order systems which allow blow up in finite time. The solutions are smooth before the blow up time. To find blow up we look for intersecting characteristics as in Chapter 20. First derivatives blow up in quasilinear systems, second derivatives in nonlinear systems. This conjecture is true for some special one-dimensional models of flow of a viscoelastic fluid which have been studied by M. Slemrod [1985] and Renardy, Hrusa, and Nohel [1987].

5.8 Weak compressibility

Some effects of compressibility have been considered by Phelan, Malone and Winter [1989] and Edwards and Beris [1990]. They allow the viscoelastic fluid to be compressible with an analytic density-pressure relation $\rho=f(p)$ at constant temperature. They then approximate this relation with a one-term Taylor expansion

$$\rho = \rho_0 \, [1 + (p-p_0)\varepsilon] \tag{29}$$

where ε, the isothermal compressibility, is small when the liquid is nearly incompressible. In this case the isothermal sound speed is huge

$$\left.\frac{\partial p}{\partial \rho}\right|_T = \frac{1}{\varepsilon\rho_0} \overset{\text{def}}{=} c_s \, . \tag{30}$$

The continuity equation $\dot{\rho} + \rho \, \mathrm{div}\,\mathbf{u} = 0$ reduces to $\varepsilon\rho_0\dot{p} + \rho_0[1+(p-p_0)\varepsilon]\mathrm{div}\mathbf{u} = 0$, which may be approximated by

$$\frac{dp}{dt} + \frac{1}{\varepsilon}\,\mathrm{div}\mathbf{u} = 0 \, . \tag{31}$$

We may therefore regard the case of slight compressibility as generated by a small pertubation of $\mathrm{div}\mathbf{u}=0$ which cannot be neglected in regions where $\dot{p}$ is as large as $O(\varepsilon^{-1})$. In the analysis of type, such rapid variations are all-important.

We now record how the basic first order quasilinear system is modified by the introduction of (29). We first remark that in the decomposition of the stress into a pressure and an extra stress $\boldsymbol{\tau}$

$$\mathbf{T} = -p(\rho)\mathbf{1} + \boldsymbol{\tau}$$

the p is the thermodynamic pressure satisfying (29) and (30) and $\boldsymbol{\tau}$ is given by the same interpolated Maxwell model used in the incompressible case. Of course, it is not true that $\boldsymbol{\tau}$ is traceless so that when there is motion, the mean normal stress $\frac{1}{3}\,\mathrm{tr}\mathbf{T}$ does not equal $-p$. With this understanding, the governing quasilinear system is again like (3.43), with ρ replaced by ρ_0, the third equations $(3.43)_3$ replaced by

$$\tau_t + u\tau_x + v\tau_y + m_9u_x + m_7u_y + m_8v_x + m_9v_y = \ell_3$$

where $m_9 = -a\tau$ and m_7, m_8 are as in (3.44), and the fifth equation $(3.43)_5$ is replaced by

$$\varepsilon(p_t + up_x + vp_y) + u_x + u_y = 0\ .$$

The new quasilinear system matrix form is again (3.40) with $[\mathbf{G}]=\text{diag}[1,1,1,\rho_0,\rho_0,\varepsilon]$,

$$[\mathbf{H}] = \begin{vmatrix} 0 & 0 & 0 & m_1 & m_3 & 0 \\ 0 & 0 & 0 & 0 & m_2 & 0 \\ 0 & 0 & 0 & m_9 & m_8 & 0 \\ -1 & 0 & 0 & 0 & 0 & 1 \\ 0 & 0 & -1 & 0 & 0 & 0 \\ 0 & 0 & 0 & 1 & 0 & 0 \end{vmatrix}\ ,$$

and

$$[\mathbf{J}] = \begin{vmatrix} 0 & 0 & 0 & m_2 & 0 & 0 \\ 0 & 0 & 0 & m_3 & m_6 & 0 \\ 0 & 0 & 0 & m_7 & m_9 & 0 \\ 0 & 0 & -1 & 0 & 0 & 0 \\ 0 & -1 & 0 & 0 & 0 & 1 \\ 0 & 0 & 0 & 0 & 1 & 0 \end{vmatrix}\ .$$

We note that in the case of weak compressibility, **G** is invertible. Characteristic surfaces are again generated from (5.1) by putting the determinant $\Delta=0$ where

$$\Delta = \det\left\{ (\phi_t+\mathbf{u}\cdot\nabla\phi)\mathbf{G}+\phi_x\mathbf{H}+\phi_y\mathbf{J} \right\} = \frac{1}{2}(\phi_t+\mathbf{u}\cdot\nabla\phi)^2\left\{ A- \right.$$

$$\left. B(\phi_t+\mathbf{u}\cdot\nabla\phi)^2 + C(\phi_t+\mathbf{u}\cdot\nabla\phi)^4 \right\} = 0$$

where

$$A = A_0 + \varepsilon\left\{ [4G^2+2aG(\sigma+\gamma) + 2a(a-1)(\sigma\gamma-\tau^2)]\,(\phi_x^2+\phi_y^2)^2 \right.$$

$$+ 2G(\sigma-\gamma)\left(\phi_x^4-\phi_y^4\right) + 2a(a+1)\left(\sigma\phi_x^2 + \gamma\phi_y^2 + 2\tau\phi_x\phi_y\right)^2$$

$$+ 4aG\left(\phi_x^2+\phi_y^2\right)\left(\sigma\phi_x^2 + \gamma\phi_y^2 + 2\tau\phi_x\phi_y\right)$$

$$+ 8G\tau\,\phi_x\phi_y\left(\phi_x^2+\phi_y^2\right)\Big\}$$

$$\stackrel{\text{def}}{=} A_0 + \varepsilon A_1$$

$$B = B_0 + \varepsilon\rho_0\Big[6G\left(\phi_x^2+\phi_y^2\right) + (\sigma-\gamma)\left(\phi_x^2-\phi_y^2\right) + a\left(\tau\phi_y^2+\gamma\phi_x^2\right)$$

$$+ 5a\left(\sigma\phi_x^2+\gamma\phi_y^2\right) + 4(1+a)\tau\phi_x\phi_y\Big]$$

$$\stackrel{\text{def}}{=} B_0 + \varepsilon B_1$$

$$C = 2\varepsilon\rho_0^2$$

and

$$A_0 = \left(\phi_x^2+\phi_y^2\right)\Big[2G\left(\phi_x^2+\phi_y^2\right) + (\sigma-\gamma)\left(\phi_x^2+\phi_y^2\right)$$

$$+ a(\sigma+\gamma)\left(\phi_x^2+\phi_y^2\right) + 4\tau\phi_x\phi_y\Big]$$

$$B_0 = 2\rho_0\left(\phi_x^2+\phi_y^2\right)$$

are the values of A and B in the incompressible case in which $\varepsilon=0$. There are six roots of the equation $\Delta=0$. Two of them are particle trajectories. The other four are determined by

$$A - B\,(\phi_t + \mathbf{u}\cdot\nabla\phi)^2 + C(\phi_t + \mathbf{u}\cdot\nabla\phi)^4 = 0\,, \tag{32}$$

hence

$$(\phi_t + \mathbf{u}\cdot\nabla\phi)^2 = \frac{B\pm\sqrt{B^2-4AC}}{2C}$$

Since $\varepsilon \ll 1$, we have

$$\sqrt{B^2-4AC} = \sqrt{(B_0+\varepsilon B_1)^2 - 4(A_0+\varepsilon A_1)(\varepsilon 2\rho_0^2)}$$

$$= (B_0+\varepsilon B_1)\sqrt{1-\frac{4(A_0+\varepsilon A_1)(\varepsilon 2\rho_0^2)}{(B_0+\varepsilon B_1)^2}}$$

$$\cong (B_0+\varepsilon B_1)\left(1-\frac{2(A_0+\varepsilon A_1)(\varepsilon 2\rho_0^2)}{(B_0+\varepsilon B_1)^2}\right)$$

$$\cong B_0+\varepsilon B_1 - \varepsilon\frac{4\rho_0^2}{B_0}\left[A_0 + \varepsilon A_1 - \varepsilon\frac{A_0}{B_0}B_1\right],$$

$$\frac{B+\sqrt{B^2-4AC}}{2C} \cong \frac{2B_0 + 2\varepsilon B_1 - \varepsilon\frac{4\rho_0^2}{B_0}A_0}{2\varepsilon 2\rho_0^2} = \frac{B_0}{2\varepsilon\rho_0^2} + \frac{B_1 - 2\rho_0^2\frac{A_0}{B_0}}{2\rho_0^2}$$

and

$$\frac{B-\sqrt{B^2-4AC}}{2C} \cong \frac{\varepsilon\frac{4\rho_0^2}{B_0}\left(A_0+\varepsilon A_1-\varepsilon\frac{A_0}{B_0}B_1\right)}{2\varepsilon 2\rho_0^2} = \frac{A_0}{B_0} + \varepsilon\frac{1}{B_0}\left(A_1 - \frac{A_0}{B_0}B_1\right).$$

It follows now that to leading order, the quartic (32) has four roots

$$\pm\sqrt{\frac{B_0}{2\varepsilon\rho_0^2}}, \pm\sqrt{\frac{A_0}{B_0}}$$

the last pair of roots are just those associated with (5), for the incompressible case and if $A_0<0$, then the quasilinear system is ill-posed. Recall that if $a= \pm 1$, then $A_0>0$ and the incompressible problem is well-posed.

The first pair of roots are new and real because

$$\frac{B_0}{2\varepsilon\rho_0^2} = \frac{1}{\varepsilon\rho_0}\left(\phi_x^2+\phi_y^2\right) = c_s^2\left(\phi_x^2+\phi_y^2\right) \tag{33}$$

where c_s is the isothermal sound speed. These roots disappear entirely in the incompressible case $c_s \to \infty$. One effect of weak compressibility is to replace "infinitely" fast waves with finitely fast waves which would not be expected to strongly influence the slower waves which are associated with viscoelasticity.

Another way to classify the type of a general quasilinear system like (3.40) in x,y,t is to study stability to short waves as we did in §4.4. This study leads us to (4.24) which we may write as

$$(c\mathbf{G} + \alpha\mathbf{H} + \beta\mathbf{J})\,\mathbf{a} = \mathbf{0}$$

where α and β are wave numbers,

$$c = -\omega + \mathbf{u} \cdot \mathbf{k}$$

is an eigenvalue and $\mathbf{G}$, $\mathbf{H}$, $\mathbf{J}$ are system matrices, say for the weakly compressible system just studied. The correct condition for the solvability is that the determinant Δ of the coefficients of $\mathbf{a}$ must vanish for all α, β. This system will be hyperbolic if and only if the six roots $c(\alpha, \beta)$ of $\Delta(c, \alpha, \beta)=0$ are real for all values of α and β.

Exercises

Exercise 5.1.
Suppose that the velocity field is given and $\mathbf{L}$ is continuous and you want to solve for the stresses (3.38). Show that the analysis of simple jumps in the extra stress yields

$$\left(\frac{\partial\phi}{\partial t} + \mathbf{u}\cdot\nabla\phi\right)[\![\boldsymbol{\tau}_\phi]\!] = 0\,.$$

In two dimensions, the constitutive equation considered separately has the real characteristics which are all particle trajectories.

Exercise 5.2.
The equations governing motions of Jeffreys' model (2.23) with a general invariant derivative may be written as

$$\frac{D\boldsymbol{\tau}_E}{Dt} = -\frac{\boldsymbol{\tau}_E}{\lambda_1} + 2\frac{\eta_1}{\lambda_1}\mathbf{D}[\mathbf{u}] ,$$
$$\mu\nabla^2\mathbf{u} - \nabla p + \text{div}\,\boldsymbol{\tau}_E = \rho\frac{d\mathbf{u}}{dt} , \tag{34}$$
$$\text{div}\,\mathbf{u} = 0 .$$

We want to write out a quasilinear first order system for these equations but second derivatives of $\mathbf{u}$ are on the left of (34). It follows then that the first derivative of $\mathbf{u}$, the components of $\mathbf{L}$ are of lower order. The variables of this quasilinear system are the 6 components of $\boldsymbol{\tau}_E$, one pressure p and the 9 components of $\mathbf{L}=\nabla\mathbf{u}$, 16 in all. For these variables, we have 6 stress equations

$$\frac{\partial\boldsymbol{\tau}_E}{\partial t} + (\mathbf{u}\cdot\nabla)\boldsymbol{\tau}_E = \mathbf{lot} ,$$

3 momentum equations

$$\mu\,\text{div}\,\mathbf{L} - \nabla p + \text{div}\,\boldsymbol{\tau}_E = \text{lot} ,$$

one continuity equation

$$\text{tr}\,\mathbf{L} = 0 ,$$

and 9 compatibility equations

$$\frac{\partial L_{ij}}{\partial x_\ell} = \frac{\partial L_{i\ell}}{\partial x_j}$$

of which 6 are independent.

In two dimensions, we have

$$\left|\begin{aligned} &\frac{\partial\tau_E}{\partial t} + (\mathbf{u}\cdot\nabla)\tau_E = \boldsymbol{\lambda}_1 , \\ &\frac{\partial\sigma_E}{\partial t} + (\mathbf{u}\cdot\nabla)\sigma_E = \boldsymbol{\lambda}_2 , \\ &\frac{\partial\gamma_E}{\partial t} + (\mathbf{u}\cdot\nabla)\gamma_E = \boldsymbol{\lambda}_3 , \end{aligned}\right. \tag{35}$$

$$\left|\begin{array}{l} \mu\left[\frac{\partial L_{11}}{\partial x}+\frac{\partial L_{12}}{\partial y}\right]-\frac{\partial p}{\partial x}+\frac{\partial \sigma_E}{\partial x}+\frac{\partial \tau_E}{\partial y}=\lambda_4 , \\ \\ \mu\left[\frac{\partial L_{21}}{\partial x}+\frac{\partial L_{22}}{\partial y}\right]-\frac{\partial p}{\partial y}+\frac{\partial \tau_E}{\partial x}+\frac{\partial \gamma_E}{\partial y}=\lambda_5 , \end{array}\right. \tag{36}$$

$$\left|\begin{array}{l} \frac{\partial L_{11}}{\partial y}-\frac{\partial L_{12}}{\partial x}=0 , \\ \\ \frac{\partial L_{21}}{\partial y}-\frac{\partial L_{22}}{\partial x}=0 , \end{array}\right.$$

$$L_{11}+L_{22}=0 .$$

There are 8 equations for $(\tau_E,\sigma_E,\gamma_E,p,L_{11},L_{12},L_{21},L_{22})$. After eliminating L_{22}, we get the three equations (35) and

$$\mu\left[\frac{\partial L_{11}}{\partial x}+\frac{\partial L_{12}}{\partial y}\right]-\frac{\partial p}{\partial x}+\frac{\partial \sigma_E}{\partial x}+\frac{\partial \tau_E}{\partial y}=\lambda_4 , \tag{37$_1$}$$

$$\mu\left[\frac{\partial L_{21}}{\partial x}-\frac{\partial L_{11}}{\partial y}\right]-\frac{\partial p}{\partial y}+\frac{\partial \tau_E}{\partial x}+\frac{\partial \gamma_E}{\partial y}=\lambda_5 , \tag{37$_2$}$$

$$\frac{\partial L_{11}}{\partial y}-\frac{\partial L_{12}}{\partial x}=0 , \tag{37$_3$}$$

$$\frac{\partial L_{21}}{\partial y}+\frac{\partial L_{11}}{\partial x}=0 . \tag{37$_4$}$$

Define $\mathbf{Q} = [\tau_E,\sigma_E,\gamma_E,L_{11},L_{12},L_{21},p]^T$ and write the system (35) and (37) as a matrix equation

$$\mathbf{G}[\mathbf{Q}_t+\mathbf{u}\cdot\nabla\mathbf{Q}]+\mathbf{H}\mathbf{Q}_x+\mathbf{J}\mathbf{Q}_y=\boldsymbol{\lambda} \tag{38}$$

where

$$\mathbf{G} = \text{diag}\ [1,1,1,0,0,0,0]$$

Find $\mathbf{H}$ and $\mathbf{J}$. Use the method of simple discontinuities to show that characteristic surfaces $\phi(x,y,t)=0$ must satisfy

$$(\phi_t+\mathbf{u}\cdot\nabla\phi)^3[\phi_x^2+\phi_y^2]^2=0 .$$

The only real characteristic surfaces are particle trajectories and these have a multiplicity of three. How does this compare with the characteristic surfaces of Maxwell models?

Exercise 5.3.
Suppose that at some point the stress τ is referred to as principal axes in which the shear stress τ vanishes. Show that (23) may be written as

$$\{\rho(u-\hat{c})^2 - b_{11}\}\phi_x^2 + \{\rho(v-\hat{c})^2 - b_{22}\}\phi_y^2$$

$$+ 2\rho(u-\hat{c})(v-\hat{c})\phi_x\phi_y = 0 .$$

where b_{11} and b_{22} are defined by (4.31). This may be solved for real ϕ_x and ϕ_y if and only if

$$\frac{\rho(v-\hat{c})^2}{b_{22}} + \frac{\rho(u-\hat{c})^2}{b_{11}} \geq 1 .$$

Exercise 5.4.
Hulsen, [1986]

Prove that the stream function, the velocity and the shear stress along streamlines for the perturbed motion of a Maxwellian fluid in the 2-D case satisfies

$$L\Delta\Psi = \text{lot} , \tag{39}$$

$$L\Delta u = \text{lot} , \tag{40}$$

$$L\Delta v = \text{lot} , \tag{41}$$

$$L\Delta\tau_{12} = \text{lot} \tag{42}$$

where the shear stress along streamlines, τ_{12}, is defined by

$$\left(U^2 + V^2\right) \tau_{12} = (U^2 - V^2)\tau + UV(\gamma - \sigma)$$

and the operator L is defined by

$$L = B_{xx}\frac{\partial^2}{\partial x^2} + 2B_{xy}\frac{\partial^2}{\partial x \partial y} + B_{yy}\frac{\partial^2}{\partial y^2} = \mathbf{B}:\nabla\nabla$$

and

$$\mathbf{B} = \lambda\mathbf{T} + \eta\mathbf{1} - \lambda\rho\mathbf{v}\cdot\mathbf{v} = \mathbf{A} - \lambda\rho\mathbf{v}\cdot\mathbf{v} .$$

Use (42) to conclude that $L\tau_{12}$ is elliptic and τ_{12} is not a variable which changes type.

Solutions for the exercises

Exercise 5.1.
Choose the interpolated Maxwell models for $\boldsymbol{\tau}$

$$\frac{\partial \boldsymbol{\tau}}{\partial \tau} + (\mathbf{u}\cdot\nabla)\boldsymbol{\tau} + \frac{1-a}{2}(\boldsymbol{\tau}\mathbf{L} + \mathbf{L}^T\boldsymbol{\tau})$$
$$-\frac{1+a}{2}(\mathbf{L}\boldsymbol{\tau} + \boldsymbol{\tau}\mathbf{L}^T) - 2G\mathbf{D}[\mathbf{u}] = -\frac{1}{\lambda}\boldsymbol{\tau}\,. \tag{43}$$

The velocity $\mathbf{u}$ is supposed to be known. Assume the stress $\boldsymbol{\tau}$ is continuous but the derivatives of $\boldsymbol{\tau}$ have a jump across characteristic surface $\phi(x,y,t)=0$. Then

$$\left[\!\left[\frac{\partial \tau}{\partial t}\right]\!\right] + [\![\mathbf{u}\cdot\nabla\boldsymbol{\tau}]\!]=0\,,\ [\![\boldsymbol{\tau}]\!]=0\,,\ [\![\mathbf{D}[\mathbf{u}]]\!]=0\,,\ [\![\mathbf{L}]\!]=0\,. \tag{44}$$

Hence,

$$[\![\boldsymbol{\tau}_\phi]\!]\frac{\partial \phi}{\partial t} + \mathbf{u}\cdot\nabla\phi[\![\boldsymbol{\tau}_\phi]\!] = 0 \text{ or } \left(\frac{\partial \phi}{\partial t} + \mathbf{u}\cdot\nabla\phi\right)[\![\boldsymbol{\tau}_\phi]\!] = 0\,. \tag{45}$$

In the two-dimensional case, $\left(\frac{\partial \phi}{\partial t} + \mathbf{u}\cdot\nabla\,\phi\right)[\![\boldsymbol{\tau}_\phi]\!] = 0$ can be written as

$$\begin{vmatrix} \frac{\partial \phi}{\partial t}+\mathbf{u}\cdot\nabla\phi & 0 & 0 \\ 0 & \frac{\partial \phi}{\partial t}+\mathbf{u}\cdot\nabla\phi & 0 \\ 0 & 0 & \frac{\partial \phi}{\partial t}+\mathbf{u}\cdot\nabla\phi \end{vmatrix} \begin{vmatrix} [\![\sigma_\phi]\!] \\ [\![\gamma_\phi]\!] \\ [\![\tau_\phi]\!] \end{vmatrix} = 0\,. \tag{46}$$

The characteristics are determined by

$$\det \begin{vmatrix} \frac{\partial\phi}{\partial t}+\mathbf{u}\cdot\nabla\phi & 0 & 0 \\ 0 & \frac{\partial\phi}{\partial t}+\mathbf{u}\cdot\nabla\phi & 0 \\ 0 & 0 & \frac{\partial\phi}{\partial t}+\mathbf{u}\cdot\nabla_{\phi} \end{vmatrix} = 0\ .$$

Therefore,

$$\left(\frac{\partial\phi}{\partial t}+\mathbf{u}\cdot\nabla\phi\right)^3 = 0 \quad \text{or} \quad \left(\frac{d\phi}{dt}\right)^3 = 0\ .$$

Three real characteristics are particle trajectories.

Exercise 5.2.
The coefficients in (38) are

$$\mathbf{H} = \begin{vmatrix} 0 & 0 & 0 & 0 & 0 & 0 & 0 \\ 0 & 0 & 0 & 0 & 0 & 0 & 0 \\ 0 & 0 & 0 & 0 & 0 & 0 & 0 \\ 0 & 1 & 0 & \mu & 0 & 0 & -1 \\ 1 & 0 & 0 & 0 & 0 & \mu & 0 \\ 0 & 0 & 0 & 0 & -1 & 0 & 0 \\ 0 & 0 & 0 & 1 & 0 & 0 & 0 \end{vmatrix}\ , \qquad (47)$$

$$\mathbf{J} = \begin{vmatrix} 0 & 0 & 0 & 0 & 0 & 0 & 0 \\ 0 & 0 & 0 & 0 & 0 & 0 & 0 \\ 0 & 0 & 0 & 0 & 0 & 0 & 0 \\ 1 & 0 & 0 & 0 & \mu & 0 & 0 \\ 0 & 0 & 1 & -\mu & 0 & 0 & -1 \\ 0 & 0 & 0 & 1 & 0 & 0 & 0 \\ 0 & 0 & 0 & 0 & 0 & 1 & 0 \end{vmatrix}\ . \qquad (48)$$

For simple jumps across the characteristic surfaces $\phi(x,y,t)=0$,

$$\left[\mathbf{G}\left(\frac{\partial\phi}{\partial t}+\mathbf{u}\bullet\nabla\phi\right)+\mathbf{H}\,\frac{\partial\phi}{\partial x}+\mathbf{J}\,\frac{\partial\phi}{\partial y}\right][\![\mathbf{Q}_\phi]\!]=0\,. \tag{49}$$

Hence,

$$\det\left[\mathbf{G}\left(\frac{\partial\phi}{\partial t}+\mathbf{u}\bullet\nabla\phi\right)+\frac{\partial\phi}{\partial x}\mathbf{H}+\frac{\partial\phi}{\partial y}\mathbf{J}\right]=0\,, \tag{50}$$

$$\det\begin{vmatrix} \nu_0 & 0 & 0 & 0 & 0 & 0 & 0 \\ 0 & \nu_0 & 0 & 0 & 0 & 0 & 0 \\ 0 & 0 & \nu_0 & 0 & 0 & 0 & 0 \\ \frac{\partial\phi}{\partial y} & \frac{\partial\phi}{\partial x} & 0 & \mu\frac{\partial\phi}{\partial x} & \mu\frac{\partial\phi}{\partial y} & 0 & -\frac{\partial\phi}{\partial x} \\ \frac{\partial\phi}{\partial x} & 0 & \frac{\partial\phi}{\partial y} & -\mu\frac{\partial\phi}{\partial y} & 0 & \mu\frac{\partial\phi}{\partial x} & -\frac{\partial\phi}{\partial y} \\ 0 & 0 & 0 & \frac{\partial\phi}{dy} & -\frac{\partial\phi}{\partial x} & 0 & 0 \\ 0 & 0 & 0 & \frac{\partial\phi}{\partial x} & 0 & \frac{\partial\phi}{\partial y} & 0 \end{vmatrix}=0\,,$$

where $\nu_o=\frac{\partial\phi}{\partial t}+\mathbf{u}\bullet\nabla\phi$

This reduces to

$$\nu_0^3\left[\left(\frac{\partial\phi}{\partial x}\right)^2\mu\left(\frac{\partial\phi}{\partial x}\right)^2-\frac{\partial\phi}{\partial y}\left(-\mu\frac{\partial\phi}{\partial y}\left(\frac{\partial\phi}{\partial x}\right)^2-\mu\left(\frac{\partial\phi}{\partial y}\right)^3-\mu\frac{\partial\phi}{\partial y}\left(\frac{\partial\phi}{\partial x}\right)^2\right)\right]=0 \tag{51}$$

or

$$\nu_0^3\,\mu\left[\left(\frac{\partial\phi}{\partial x}\right)^2+\left(\frac{\partial\phi}{\partial y}\right)^2\right]^2=0\,. \tag{52}$$

Therefore, the characteristic surfaces satisfy

$$(\phi_t+\mathbf{u}\bullet\nabla\phi)^3(\phi_x^2+\phi_y^2)^2=0\,. \tag{53}$$

The particle paths are triply characteristic for Jeffreys' model, or when the velocity is given, and only doubly characteristic for Maxwell's model.

Exercise 5.3.

We start with the perturbed equations of motion, where the terms in capital letters correspond to the basic flow, $\mathbf{T}$ is the stress in the basic flow, $\mathbf{t}$ is the perturbed stress .

$$\rho\mathbf{V}\cdot\text{grad }\mathbf{v} + \text{grad p} - \text{div }\mathbf{t} = -\rho\mathbf{v}\cdot\text{grad }\mathbf{V}\,,$$
$$\text{div }\mathbf{v} = 0\,,$$
$$\lambda\mathbf{V}\cdot\text{grad }\mathbf{t} - \boldsymbol{\lambda}\mathbf{A} - \mathbf{A}\boldsymbol{\lambda}^{T} = -\mathbf{t} - \lambda\mathbf{v}\cdot\text{grad }\mathbf{T} + \lambda(\mathbf{Lt}+\mathbf{tL}^{T})$$

where $\mathbf{A}$, $\mathbf{L}$, and $\boldsymbol{\lambda}$ are defined by

$$\mathbf{A} = \lambda\mathbf{T} + \eta\mathbf{1}\,,$$
$$\mathbf{L} = \text{grad }\mathbf{V}\,,$$
$$\boldsymbol{\lambda} = \text{grad }\mathbf{v}\,.$$

Then, in index form, we have

$$\rho V_j \frac{\partial v_i}{\partial x_j} + \frac{\partial p}{\partial x_i} - \frac{\partial \tau_{ij}}{\partial x_j} = \text{lot}_i{}^{(1)}\,,$$
$$\frac{\partial v_i}{\partial x_i} = 0\,,$$
$$\lambda V_k \frac{\partial \tau_{ij}}{\partial x_k} - \frac{\partial v_i}{\partial x_k} A_{kj} - A_{ik}\frac{\partial v_j}{\partial x_k} = \text{lot}_{ij}{}^{(2)}$$

where

$$\text{lot}_i{}^{(1)} = -\rho v_k \frac{\partial V_i}{\partial x_k}\,,$$
$$\text{lot}_{ij}{}^{(2)} = -\tau_{ij} - \lambda v_k \frac{\partial T_{ij}}{\partial x_k} + \lambda\left(\frac{\partial V_i}{\partial x_k}\tau_{kj} + \frac{\partial V_j}{\partial x_k}\tau_{ik}\right).$$

Following now the procedure used in Chapter 3, we find the vorticity equation

$$(A_{jk}-\lambda\rho V_j V_k)\frac{\partial^2 \zeta_b}{\partial x_j \partial x_k} =$$
$$-\varepsilon_{bmi}\left(\lambda V_k \frac{\partial^2 \text{lot}_i{}^{(1)}}{\partial x_k \partial x_m} - \frac{\partial^2 \text{lot}_{ij}{}^{(2)}}{\partial x_k \partial x_m}\right) + \text{lot}_b{}^{(3)} = \text{lot}_b{}^{(4)}$$

where

$$\mathrm{lot}_b^{(3)} = \varepsilon_{bmi}\Bigg[\frac{\partial}{\partial x_m}\left(\lambda\frac{\partial V_k}{\partial x_j}\frac{\partial \tau_{ij}}{\partial x_k} - \frac{\partial v_i}{\partial x_k}\frac{\partial A_{kj}}{\partial x_j} - \frac{\partial A_{ik}}{\partial x_j}\frac{\partial v_i}{\partial x_k}\right)$$
$$+\lambda\left(\frac{\partial V_k}{\partial x_m}\frac{\partial^2 \tau_{ij}}{\partial x_j \partial x_k} - \frac{\partial^2 v_i}{\partial x_j \partial x_k}\frac{\partial A_{kj}}{\partial x_m}\right)$$
$$+\lambda\rho\left(V_k\frac{\partial}{\partial x_k}\left(\frac{\partial V_j}{\partial x_m}\frac{\partial v_i}{\partial x_j}\right) + V_j\frac{\partial V_j}{\partial x_k}\frac{\partial^2 v_i}{\partial x_j \partial x_m}\right)\Bigg].$$

We are looking at the 2-D case so there is only one component for the vorticity ζ

$$(\mathbf{B}{:}\nabla\nabla)\zeta = \left(B_{xx}\frac{\partial^2}{\partial^2 x} + 2B_{xy}\frac{\partial^2}{\partial x \partial y} + B_{yy}\frac{\partial^2}{\partial^2 y}\right)\zeta$$
$$= B_{jk}\frac{\partial^2 \zeta}{\partial x_j \partial x_k} = L\zeta = \mathrm{lot}_3^{\,(4)}.$$

Then we introduce the stream function Ψ

$$u = \frac{\partial \Psi}{\partial y}, \quad v = -\frac{\partial \Psi}{\partial x}.$$

Therefore

$$\Delta\Psi = -\zeta.$$

It follows that Ψ satisfies

$$L\Delta\Psi = -\mathrm{lot}_3^{\,(4)}. \tag{54}$$

Deriving the next equations for u and v is straightforward. Using $\frac{\partial u}{\partial x} + \frac{\partial v}{\partial y} = 0$ and (54), we get

$$L\Delta u = -\frac{\partial \mathrm{lot}_3^{\,(4)}}{\partial y}, \tag{55}$$

$$L\Delta v = \frac{\partial \mathrm{lot}_3^{\,(4)}}{\partial x}. \tag{56}$$

In what follows we will use the following notations for the stresses

$$\mathbf{t} = \begin{pmatrix} \sigma & \tau \\ \tau & \gamma \end{pmatrix}.$$

We now differentiate the 3 stress equations with respect to x and y to get

$$\lambda U \frac{\partial^2\sigma}{\partial x^2} + \lambda V \frac{\partial^2\sigma}{\partial x\partial y} - 2A_{xx}\frac{\partial^2 u}{\partial x^2} + 2A_{xy}\frac{\partial^2 v}{\partial y^2} = \frac{\partial \mathrm{lot}_{11}{}^{(2)}}{\partial x} + \mathrm{lot}_1{}^{(5)} = \mathrm{lot}_1{}^{(6)} ,$$

$$\lambda U \frac{\partial^2\sigma}{\partial x\partial y} + \lambda V \frac{\partial^2\sigma}{\partial y^2} + 2A_{xx}\frac{\partial^2 v}{\partial y^2} - 2A_{xy}\frac{\partial^2 u}{\partial y^2} = \frac{\partial \mathrm{lot}_{11}{}^{(2)}}{\partial y} + \mathrm{lot}_2{}^{(5)} = \mathrm{lot}_2{}^{(6)} ,$$

$$\lambda U \frac{\partial^2\gamma}{\partial x^2} + \lambda V \frac{\partial^2\gamma}{\partial y\partial x} - 2A_{xy}\frac{\partial^2 v}{\partial x^2} + 2A_{yy}\frac{\partial^2 u}{\partial x^2} = \frac{\partial \mathrm{lot}_{22}{}^{(2)}}{\partial x} + \mathrm{lot}_3{}^{(5)} = \mathrm{lot}_3{}^{(6)} ,$$

$$\lambda U \frac{\partial^2\gamma}{\partial x\partial y} + \lambda V \frac{\partial^2\gamma}{\partial y^2} + 2A_{xy}\frac{\partial^2 u}{\partial x^2} - 2A_{yy}\frac{\partial^2 v}{\partial y^2} = \frac{\partial \mathrm{lot}_{22}{}^{(2)}}{\partial y} + \mathrm{lot}_4{}^{(5)} = \mathrm{lot}_4{}^{(6)} ,$$

$$\lambda U \frac{\partial^2\tau}{\partial x^2} + \lambda V \frac{\partial^2\tau}{\partial x\partial y} - A_{xx}\frac{\partial^2 v}{\partial x^2} + A_{yy}\frac{\partial^2 v}{\partial y^2} = \frac{\partial \mathrm{lot}_{12}{}^{(2)}}{\partial x} + \mathrm{lot}_5{}^{(5)} = \mathrm{lot}_5{}^{(6)} ,$$

$$\lambda U \frac{\partial^2\tau}{\partial x\partial y} + \lambda V \frac{\partial^2\tau}{\partial y^2} + A_{xx}\frac{\partial^2 u}{\partial x^2} - A_{yy}\frac{\partial^2 u}{\partial y^2} = \frac{\partial \mathrm{lot}_{12}{}^{(2)}}{\partial y} + \mathrm{lot}_6{}^{(5)} = \mathrm{lot}_6{}^{(6)}$$

where $\mathrm{lot}_i{}^{(5)}$ are given by

$$\mathrm{lot}_1{}^{(5)} = -\lambda \frac{\partial U}{\partial x}\frac{\partial\sigma}{\partial x} - \lambda \frac{\partial V}{\partial x}\frac{\partial\sigma}{\partial y} - 2\frac{\partial A_{xx}}{\partial x}\frac{\partial u}{\partial x} - 2\frac{\partial A_{xy}}{\partial x}\frac{\partial u}{\partial y} ,$$

$$\mathrm{lot}_2{}^{(5)} = -\lambda \frac{\partial U}{\partial y}\frac{\partial\sigma}{\partial x} - \lambda \frac{\partial V}{\partial y}\frac{\partial\sigma}{\partial y} - 2\frac{\partial A_{xx}}{\partial y}\frac{\partial u}{\partial x} - 2\frac{\partial A_{xy}}{\partial y}\frac{\partial u}{\partial y} ,$$

$$\text{lot}_3{}^{(5)} = -\lambda \frac{\partial U}{\partial x}\frac{\partial \gamma}{\partial x} - \lambda \frac{\partial V}{\partial x}\frac{\partial \gamma}{\partial y} - 2\frac{\partial A_{xy}}{\partial x}\frac{\partial v}{\partial x} - 2\frac{\partial A_{yy}}{\partial x}\frac{\partial v}{\partial y},$$

$$\text{lot}_4{}^{(5)} = -\lambda \frac{\partial U}{\partial y}\frac{\partial \gamma}{\partial x} - \lambda \frac{\partial V}{\partial y}\frac{\partial \gamma}{\partial y} - 2\frac{\partial A_{xy}}{\partial y}\frac{\partial v}{\partial x} - 2\frac{\partial A_{yy}}{\partial y}\frac{\partial v}{\partial y},$$

$$\text{lot}_5{}^{(5)} = -\lambda \frac{\partial U}{\partial x}\frac{\partial \tau}{\partial x} - \lambda \frac{\partial V}{\partial x}\frac{\partial \tau}{\partial y} - \frac{\partial A_{xx}}{\partial x}\frac{\partial v}{\partial x} - \frac{\partial A_{yy}}{\partial x}\frac{\partial u}{\partial y},$$

$$\text{lot}_6{}^{(5)} = -\lambda \frac{\partial U}{\partial y}\frac{\partial \tau}{\partial x} - \lambda \frac{\partial V}{\partial x}\frac{\partial \tau}{\partial y} - \frac{\partial A_{xx}}{\partial y}\frac{\partial v}{\partial x} - \frac{\partial A_{yy}}{\partial x}\frac{\partial u}{\partial y}.$$

Now we combine the last six equations and find that

$$\lambda U\left(-\frac{\partial^2\sigma}{\partial x^2} + \frac{\partial^2\gamma}{\partial x^2} - 2\frac{\partial^2\tau}{\partial x\partial y}\right) + \lambda V\left(-\frac{\partial^2\sigma}{\partial x\partial y} + \frac{\partial^2\gamma}{\partial x\partial y} - 2\frac{\partial^2\tau}{\partial y^2}\right)$$
$$+ 2A_{yy}\Delta u - 2A_{xy}\Delta v = \text{lot}_1{}^{(7)}, \qquad (57)_1$$

$$\lambda U\left(\frac{\partial^2\sigma}{\partial x\partial y} - \frac{\partial^2\gamma}{\partial x\partial y} - 2\frac{\partial^2\tau}{\partial x^2}\right) + \lambda V\left(\frac{\partial^2\sigma}{\partial y^2} - \frac{\partial^2\gamma}{\partial y^2} - 2\frac{\partial^2\tau}{\partial x\partial y}\right)$$
$$- 2A_{xy}\Delta u + 2A_{xx}\Delta v = \text{lot}_2{}^{(7)} \qquad (57)_2$$

where

$$\text{lot}_1{}^{(7)} = -\,\text{lot}_1{}^{(6)} + \text{lot}_3{}^{(6)} - 2\,\text{lot}_6{}^{(6)},$$

$$\text{lot}_2{}^{(7)} = \text{lot}_2{}^{(6)} - \text{lot}_4{}^{(6)} - 2\,\text{lot}_5{}^{(6)}$$

which shows how the equations are combined.

Taking the divergence and the curl of the momentum equation, we find that

$$\Delta p = \frac{\partial^2\sigma}{\partial x^2} + 2\frac{\partial^2\tau}{\partial x\partial y} + \frac{\partial^2\gamma}{\partial y^2} + \text{lot}_1{}^{(8)}, \qquad (58)_1$$

$$-\rho U\Delta v + \rho V\Delta u + \frac{\partial^2(\gamma-\sigma)}{\partial x\partial y} + \frac{\partial^2\tau}{\partial x^2} - \frac{\partial^2\tau}{\partial y^2} = \text{lot}_2{}^{(8)} \qquad (58)_2$$

where

$$\mathrm{lot}_1{}^{(8)} = \frac{\partial \mathrm{lot}_1{}^{(1)}}{\partial x} + \frac{\partial \mathrm{lot}_2{}^{(1)}}{\partial y} - \rho \frac{\partial U}{\partial x}\frac{\partial u}{\partial x} - \rho \frac{\partial V}{\partial x}\frac{\partial u}{\partial y}$$
$$- \rho \frac{\partial U}{\partial y}\frac{\partial v}{\partial x} - \rho \frac{\partial V}{\partial y}\frac{\partial v}{\partial y},$$

$$\mathrm{lot}_2{}^{(8)} = \frac{\partial \mathrm{lot}_1{}^{(1)}}{\partial y} - \frac{\partial \mathrm{lot}_2{}^{(1)}}{\partial x} - \rho \frac{\partial U}{\partial y}\frac{\partial u}{\partial x} - \rho \frac{\partial V}{\partial y}\frac{\partial u}{\partial y}$$
$$+ \rho \frac{\partial U}{\partial x}\frac{\partial v}{\partial x} + \rho \frac{\partial V}{\partial x}\frac{\partial v}{\partial y}.$$

Combining (57) and (58) we find that

$$\lambda U\Delta(\gamma - p) - \lambda V\Delta\tau + (2A_{yy} - \lambda\rho V^2)\Delta u$$
$$+ (-2A_{xy} + \lambda\rho UV)\Delta v = \mathrm{lot}_1{}^{(9)}, \qquad (59)_1$$

$$-\lambda U\Delta\tau + \lambda V\Delta(\sigma - p) + (-2A_{xy} + \lambda\rho UV)\Delta u$$
$$+ (2A_{xx} - \lambda\rho U^2)\Delta v = \mathrm{lot}_2{}^{(9)} \qquad (59)_2$$

where

$$\mathrm{lot}_1{}^{(9)} = -\mathrm{lot}_1{}^{(6)} + \mathrm{lot}_3{}^{(6)} - 2\,\mathrm{lot}_6{}^{(6)} - \lambda U\,\mathrm{lot}_1{}^{(8)} - \lambda V\,\mathrm{lot}_2{}^{(8)},$$

$$\mathrm{lot}_2{}^{(9)} = \mathrm{lot}_2{}^{(6)} - \mathrm{lot}_4{}^{(6)} - 2\,\mathrm{lot}_5{}^{(6)} + \lambda U\,\mathrm{lot}_2{}^{(8)} - \lambda V\,\mathrm{lot}_1{}^{(8)}.$$

We combine the last two equations (V times the first – U times the second) and keep only the leading order terms. This gives

$$\Delta[\lambda(U^2 - V^2)\tau + \lambda UV(\gamma - \sigma) + (2A_{yy} - \lambda\rho V^2)Vu$$
$$+ (-2A_{xy} + \lambda\rho UV)Vv - (-2A_{xy} + \lambda\rho UV)Uu$$
$$- (2A_{xx} - \lambda\rho U^2)Uv] = \mathrm{lot}^{(10)}$$

where

$$\mathrm{lot}^{(10)} = V\,\mathrm{lot}_1{}^{(9)} - U\,\mathrm{lot}_2{}^{(9)} + \lambda\tau\Delta(U^2 - V^2) + \lambda(\gamma - \sigma)\Delta(UV)$$
$$+ u\Delta[V(2A_{yy} - \lambda\rho V^2)] + v\Delta[V(-2A_{xy} + \lambda\rho UV)]$$
$$- u\Delta[U(-2A_{xy} + \lambda\rho UV)] - v\Delta[U(2A_{xx} - \lambda\rho U^2)].$$

Finally we multiply by L and use the definition of τ_{12}

$$\|V^2\|\,\tau_{12} = (U^2 - V^2)\tau + UV(\gamma - \sigma),$$

where $\|V^2\| = U^2 + V^2$, and find that

$$L\Delta(\lambda\|V^2\|\tau_{12}) = \text{lot}^{(11)}$$

where

$$\begin{aligned}\text{lot}^{(11)} = \text{lot}^{(10)} &- uL\Delta\,[V(2A_{yy}-\lambda\rho V^2) - U(-2A_{xy} + \lambda\rho UV)] \\ &- vL\Delta\,[V(-2A_{xy} + \lambda\rho UV) - U(2A_{xx} - \lambda\rho U^2)] \\ &- \frac{\partial \text{lot}_3{}^{(4)}}{\partial y}\,[V(2A_{yy} - \lambda\rho V^2) - U(-2A_{xy} + \lambda\rho UV)] \\ &+ \frac{\partial \text{lot}_3{}^{(4)}}{\partial x}\,[V(-2A_{xy} + \lambda\rho UV) - U(2A_{xx} - \lambda\rho U^2)]\,.\end{aligned}$$

Finally we let

$$\text{lot} = \frac{\text{lot}^{(11)}}{\lambda\|V^2\|}$$

so that

$$L\Delta\tau_{12} = \text{lot}\,.$$

To within lot we may commute L with Δ so that

$$\Delta L\tau_{12} = \text{lot}$$

and$L\tau_{12}$ is elliptic.

CHAPTER 6

Hyperbolicity and Change of Type in Steady Flow

Now we put the time derivatives to zero. This means that we have left behind ill-posed problems and short wave instabilities. The problem now is to find the regions of steady flow in which the vorticity equation gives rise to real characteristic directions. In general, the analysis of the characteristics of (5.10) when the time derivatives vanish will lead to conditions for the emergence of transcritical flow, like transonic flow in aerodynamics, elliptic in some regions of flow and hyperbolic in others.

6.1 Two kinds of change of type

The characteristics for the vorticity in plane steady flow are determined by (5.18) with ϕ_t=0. Using (5.9), this criterion may be reduced by putting the last factor in (5.10) to zero. An expression for the slope of the characteristics which follows from this criterion may be written as

$$\frac{dy}{dx} = \frac{B}{A} \pm \sqrt{\frac{B^2 - AC}{A^2}} \tag{1}$$

where A,B,C are defined under (3.52-54). Clearly, there are real characteristics wherever the discriminant

$$\begin{aligned} B^2 - AC = -G^2 + \rho\left[G + \frac{a}{2}(\sigma+\gamma)\right](u^2+v^2) + \frac{1}{2}\rho(\gamma-\sigma)(u^2-v^2) \\ + \tau^2 + \frac{1}{4}(\sigma-\gamma)^2 - \frac{a^2}{4}(\sigma+\gamma)^2 - Ga(\sigma+\gamma) - 2\rho\tau vu > 0\,, \end{aligned} \tag{2}$$

and the vorticity equation is elliptic wherever $B^2-AC<0$.

It is important to draw attention to the fact that the criterion (2) for change of type depends on the velocity and the stress, while the criterion $f<0$ for Hadamard instability depends only on the stress. There are two kinds of change of type, one which leads to a mixed vorticity field in steady flow and another which leads to instability and ill-posedness of the initial value problem. The first kind of "transonic" change of type does not necessarily lead to instability, and it may give a correct description of flows.

The two kinds of change of type are related mathematically but not physically. The mathematical relationship comes from considering the criterion (2) for an inertialess fluid, $\rho=0$. In this case,

$$\begin{aligned} B^2 - AC &= b_{12}^2 - b_{11}b_{22}\,, \\ b_{11} &= \frac{1}{2}(\sigma-\gamma) + G + \frac{a}{2}(\gamma+\sigma)\,, \\ b_{22} &= \frac{1}{2}(\gamma-\sigma) + G + \frac{a}{2}(\gamma+\sigma)\,, \\ b_{12} &= \tau\,. \end{aligned} \tag{3}$$

The results given in exercise 4.5 relate the sign of f_0 to $b_{12}^2-b_{11}b_{22}$. We get Hadamard instability $f_0<0$ if $b_{12}^2-b_{11}b_{12}>0$. The classification of the unsteady vorticity equation is related to the sign of f_0 through f of (5.23), and it is elliptic if $b_{12}^2-b_{11}b_{22}>0$. We get Hadamard instability when the unsteady vorticity equation is elliptic and the steady vorticity equation with $\rho=0$ is hyperbolic. The "stable" kind of change of type is often associated with high velocities and $\rho \neq 0$.

Summarizing, we have proved that *if the vorticity equation for steady flow is hyperbolic when* $\rho=0$, *then the unsteady vorticity equation is elliptic and the quasilinear system is ill-posed. Conversely, if the vorticity of an inertialess steady flow is elliptic and* $b_{11}>0$, *the quasilinear system is well-posed.*

A large number of examples of problems which change type using different constitutive models and different flows will be discussed in the sequel. Some models are always evolutionary (well-posed) and do not change type in unsteady flow. The vorticity equation for steady flow of such models can and does change type. Other models can become ill-posed and undergo bad instability to short waves. Some flows of all these models, like simple shear and Poiseuille flow, are always well-posed while other flows, like plane extension or sink flow, can become ill-posed. Sink flows of upper convected and lower convected Maxwell models change type in steady flow but cannot be ill-posed. On the other hand, sink flows of corotational Maxwell models change type in steady flow and are also ill-posed. Nearly every possibility is realized for some flow of some model.

6.2 Linearized problems and change of type

Up to now our study of steady problems has been exact and fully nonlinear. We may advance our understanding of the problems of vorticity of changing type by considering simpler problems which arise under linearization.

We first identify a class of motions, say motions in the plane or axisymmetric motions. We find some special exact solution of all the equations which fall in the given class. Usually this special solution is featureless, like uniform flow, flow into a sink, flow in a channel, flow between cylinders, extensional flow and so on. The featureless solution does not exhibit the unusual features of change of type. We then perturb all the equations around the special one with perturbations in the given class, and we linearize. This leads us to linearized problems with variable coefficients, depending alone on the special solution and not on the perturbation. We analyze the linear perturbed problem for hyperbolicity and change of type.

Something more than linearization is involved in the study of linearized equations for hyperbolicity and change of type. If we use a subscript zero for unperturbed quantities and a prime for perturbed ones, we may note that new lower order terms will appear as a result of linearization. Under linearization for steady flow, the first term of the vorticity equation (3.51) becomes

$$A\frac{\partial^2\zeta}{\partial x^2} = A_0\frac{\partial^2\zeta'}{\partial x^2} + A'\frac{\partial^2\zeta_0}{\partial x^2}\,.$$

The second term on the right is of lower order. Hence, our linearized vorticity equation is of the form

$$A_0\frac{\partial^2\zeta'}{\partial x^2} + 2B_0\frac{\partial^2\zeta'}{\partial x\partial y} + C_0\frac{\partial^2\zeta'}{\partial y^2} = \text{lot}\,. \tag{4}$$

The equations for the characteristics are now given by (1) with A,B,C replaced by A_0,B_0,C_0, the same functions evaluated on the unperturbed flow

$$\frac{dy}{dx} = \frac{B_0}{A_0} \pm \sqrt{\frac{B_0^2 - A_0C_0}{A_0^2}}\,. \tag{5}$$

The procedures of linearization are such that the characteristic directions are the same for each and every linear problem, perturbing the basic flow. The characteristic surfaces for linearized problems are a gift, since nothing beyond the special solution is needed to compute them. We can know which are the elliptic regions and which are the hyperbolic regions of flow cheaply; but, to find other properties of the perturbed flow, say isovorticity and stream lines, we are obliged to solve linear PDE's.

6.3 Perturbation of uniform flow, the viscoelastic Mach number

Let

$$\mathbf{U}_0 = \mathbf{e}_x U_0$$

for a constant U_0. The stresses on this flow vanish and

$$\begin{aligned} A_0 &= -\rho U_0^2 + G , \\ B_0 &= 0 , \\ C_0 &= G . \end{aligned} \tag{6}$$

Hence,

$$B_0^2 - A_0 C_0 = [\rho U_0^2 - G]G \tag{7}$$

is positive if

$$\frac{U_0^2}{c^2} > 1$$

where

$$c = \sqrt{G/\rho}$$

is the speed of vorticity waves into rest. Now we define a "Mach" number

$$M_o = U_o/c . \tag{8}$$

The vorticity is hyperbolic everywhere if $M_0>1$ and is elliptic if $M_0<1$. The characteristics for all small steady motions perturbing uniform motion are straight lines

$$x = \pm \frac{1}{\sqrt{M_0^2-1}} \; y + \text{const} . \tag{9}$$

6.4 Perturbation of simple shear flow

(a) Maxwell models

The simple shear flow (C.21) of a Maxwell model with

$$[u,v,w] = [\kappa y,0,0]^* \tag{10}$$

---

* Here and henceforth we will designate the shear rate by κ; that is, $\dot{\gamma} \overset{\text{def}}{=} \kappa$.

satisfies (3.43) with $(\ell_1,\ell_2,\ell_3) = -\frac{1}{\lambda}(\sigma,\gamma,\tau)$ if the stresses take on constant values

$$\tau = \frac{\eta\kappa}{1+\kappa^2\lambda^2(1-a^2)} \overset{\text{def}}{=} \frac{\eta\kappa}{D}, \tag{11}$$

$$\sigma = \lambda\kappa(a+1)\tau , \tag{12}$$

$$\gamma = \lambda\kappa(a-1)\tau . \tag{13}$$

where κ is the shear rate where $-1 \le a \le 1$ (see Exercise 6.1). We find (omitting the subscript zero) that

$$\begin{aligned} A &= \frac{\eta}{\lambda} - \rho\kappa^2 y^2 + \frac{1}{2}\sigma(1+a) - \frac{1}{2}\gamma(1-a) , \\ B &= \tau , \\ C &= \frac{\eta}{\lambda} - \frac{1}{2}\sigma(1-a) + \frac{1}{2}\gamma(1+a) . \end{aligned} \tag{14}$$

Hence,

$$\frac{dy}{dx} = \frac{B}{A} \pm \frac{1}{A}\sqrt{B^2-AC} = \frac{\omega \pm \sqrt{DM^2-1-\omega^2}}{-DM^2+1+2\omega^2}$$

where

$$M^2 = \frac{\kappa^2y^2}{c^2}, \quad \omega = \lambda\kappa .$$

After simplifying,

$$\frac{dy}{dx} = \frac{1}{(\omega \pm \sqrt{M^2D-1-\omega^2})} . \tag{15}$$

The steady vorticity equation is hyperbolic in a strip exterior to the origin defined by

$$M^2 = \frac{\kappa^2y^2}{c^2} > \frac{1+\omega^2}{D} = \frac{1+\lambda^2\kappa^2}{1+\lambda^2\omega^2(1-a^2)} . \tag{16}$$

We have just shown that the vorticity of steady shear flows of generalized Maxwell models will always change type where $y^2 = (1+\omega^2)c^2/\kappa^2 D$. These flows are never Hadamard unstable because $c^2 \to \infty$ as $\rho \to 0$ so that the inequality (16) cannot be satisfied when $\rho = 0$. Moreover, since $A>0$ for this flow, the steady vorticity equation with $\rho=0$ is always elliptic so that the unsteady vorticity is hyperbolic.

(b) A Giesekus model $(a=1,\ \ell = -\dfrac{1}{\lambda}(\tau + (\alpha/G)\tau^2))$

The system (3.43) has the form

$$\sigma_t + u\sigma_x + v\sigma_y - 2\sigma u_x - 2\tau u_y - 2Gu_x = -\frac{\alpha}{\eta}(\sigma^2+\tau^2) - \frac{\sigma}{\lambda},$$

$$\tau_t + u\tau_x + v\tau_y - \gamma u_y - \sigma v_x - G(u_y + v_x) = -\frac{\alpha}{\eta}(\sigma\tau+\gamma\tau) - \frac{\tau}{\lambda},$$

$$\gamma_t + u\gamma_x + v\gamma_y - 2\tau v_x - 2\gamma v_y - 2Gv_y = -\frac{\alpha}{\eta}(\tau^2+\gamma^2) - \frac{\gamma}{\lambda},$$

$$\rho(u_t+uu_x+vu_y) + p_x - \sigma_x - \tau_y = 0, \tag{17}$$

$$\rho(v_t+uv_x+vv_y) + p_y - \tau_x - \gamma_y = 0,$$

$$u_x + v_y = 0.$$

In simple shear flow, we find $u=\kappa y$, $v=0$, $\tau=\eta\kappa/(1+\kappa^2\lambda^2)$, $\sigma=\lambda\kappa\tau$, $\gamma=-\sigma$ when $\alpha=1$.

The steady vorticity equation for the linearized flow is hyperbolic in a strip outside the origin defined by

$$\rho\kappa^2 y^2 > G.$$

This inequality cannot be satisfied for $\rho=0$ and the steady equation for vorticity is always elliptic in inertialess flows. Moreover, $A = G+\sigma = G+\lambda\eta\kappa^2/(1+\lambda^2\kappa^2) > 0$, and the linear system perturbing shear flow is well-posed, not Hadamard unstable.

6.5 Poiseuille flow of a Maxwell model

In this example, $p_x = -K$, $K>0$ is a prescribed constant and

$$\begin{aligned} \tau_y &= -K , \\ \tau(1+a)u_y &= \sigma/\lambda , \\ \tau(1-a)u_y &= -\gamma/\lambda , \\ \left[\frac{1}{2}(\sigma-\gamma) - \frac{1}{2}a(\sigma+\gamma) - \frac{\eta}{\lambda}\right]u_y &= -\frac{\tau}{\lambda} . \end{aligned} \tag{18}$$

Putting $u_y = \kappa = \kappa(y)$, we find that

$$\begin{aligned} \tau &= \frac{\eta\kappa}{\lambda^2(1-a^2)\kappa^2+1} , \\ \sigma &= \frac{\lambda\eta\kappa^2(1+a)}{\lambda^2(1-a^2)\kappa^2+1} , \\ \gamma &= \frac{\lambda\eta\kappa^2\,(a-1)}{\lambda^2(1-a^2)\kappa^2+1} . \end{aligned} \tag{19}$$

Flows perturbing Poiseuille flow of a Maxwell fluid will become unstable to short waves and lose evolutionarity wherever the steady vorticity equation with $\rho=0$ is hyperbolic. This condition may be expressed as

$$B^2-AC = -\frac{\eta^2}{\lambda^2} + \tau^2 + \frac{1}{4}(\sigma-\gamma)^2 - \frac{a^2}{4}(\sigma+\gamma)^2 - \frac{\eta}{\lambda}a(\sigma+\gamma) > 0 . \tag{20}$$

In the present case, this condition reads

$$\tau^2[1+\lambda^2\kappa^2(1-a^4)] - 2a^2\eta\kappa\tau - \frac{\eta^2}{\lambda^2} > 0 , \tag{21}$$

where τ is given by (19). After a short computation, the left-hand side of (21) is evaluated as a positive coefficient times $-\lambda^2\kappa^2-1$, hence the inequality (21) cannot be satisfied for any $a\in(-1,1)$.

The flows perturbing Poiseuille flow of a Maxwell model are always evolutionary; it is only necessary to verify that

$$A = \left[\frac{\eta}{\lambda}\right]\left[\frac{1+2\lambda^2\kappa^2}{1+\lambda^2(1-a^2)\kappa^2}\right] > 0\,.$$

6.6 Extensional flow

(a) Maxwell models

We find that

$$[u,v,\tau,\sigma,\gamma] = \left[sx,-sy,0,\frac{2\eta s}{p},-\frac{2\eta s}{q}\right]$$

where $p=1-2a\lambda s$, $q=1+2a\lambda s$. We take $s\geq 0$ and small enough so that p and q are bounded from above, $s^2 < 1/4\ a^2\lambda^2$.

The steady vorticity of motions perturbing extensional flow is hyperbolic when

$$\frac{s^2x^2}{c^2}\left[1+\frac{2\lambda s(2a^2\lambda s-1)}{pq}\right]+\frac{s^2y^2}{c^2}\left[1+\frac{2\lambda s(1+2a^2\lambda s)}{pq}\right]$$

$$> \left[1+\frac{2\lambda s(1+2a^2\lambda s)}{pq}\right]\left[1+\frac{2\lambda s(2a^2\lambda s-1)}{pq}\right]. \tag{22}$$

If the second factor on the right is positive, the region outside an ellipse is hyperbolic. We put $\rho=0$ for inertialess flow. Then we get hyperbolicity for the vorticity with $\rho=0$ when

$$0 > \left[1+\frac{2\lambda s(1+2a^2\lambda s)}{pq}\right]\left[1+\frac{2\lambda s(2a^2\lambda s-1)}{pq}\right]. \tag{23}$$

Therefore we lose evolution when

$$pq + 2\lambda s(2a^2\lambda s - 1) < 0\,,$$

that is, when $s>1/2\lambda$. This value of s is in the allowed range $s^2 < 1/4a^2\lambda^2$ provided $a\in(-1,1)$. The linear systems governing flows which perturb extensional flow are unstable to short waves whenever $s>1/2\lambda$. Steady flows with inertia change type in the manner specified by (2) and are evolutionary when $s<1/2\lambda$. (It is easily verified that A with $\rho=0$ is equal to $G+2\eta s/pq+4a^2\eta\lambda s^2/pq > 0$.)

(b) A model of Phan-Thien and Tanner (for a=0)
The analysis given under (a) above applies here also because tr $\boldsymbol{\tau}=0$ for extensional flow. This model is ill-posed when $s>1/2\lambda$.

(c) A model of Giesekus

The system (17) is satisfied by the following extensional flow: $u=sx$, $v=-sy$, $\tau=0$ where σ and γ are given by

$$\frac{\alpha}{\eta}\sigma^2 + \sigma\left(\frac{1}{\lambda} - 2\,s\right) - 2sG = 0\ ,$$

$$\frac{\alpha}{\eta}\gamma^2 + \gamma\left(\frac{1}{\lambda} + 2\,s\right) + 2sG = 0.$$

It follows that the stresses in extensional flow are given by

$$\sigma_{\pm} = \frac{\eta}{\alpha}\left[\frac{2s\lambda-1}{\lambda} \pm \left[\left(\frac{1-2s\lambda}{\lambda}\right)^2 + \frac{8s\alpha}{\lambda}\right]^{1/2}\right],$$

$$\gamma_{\pm} = \frac{\eta}{\alpha}\left[-\frac{2s\lambda+1}{\lambda} \pm \left[\left(\frac{1+2s\lambda}{\lambda}\right)^2 - \frac{8s\alpha}{\lambda}\right]^{1/2}\right]. \tag{24}$$

Since $0\leq\alpha\leq 1$, $[(1+2s\lambda)/\lambda]^2 - 8s\alpha/\lambda$ is positive and the stresses double valued.

Let us note now, following Giesekus [1982], that the configuration tensor $\mathbf{C}=\mathbf{1}+\boldsymbol{\tau}/G$ is positive definite. Then $1+\sigma/G>0$ and $1+\gamma/G>0$. These inequalities cannot be satisfied for the negative roots of (24) when $s<1/2\lambda$. To show this we set

$$\alpha\left[1+\frac{\sigma_-}{G}\right] = \alpha+2s\lambda-1-[(1-2s\lambda)^2 + 8s\alpha\lambda]^{1/2} \overset{\text{def}}{=} f(\alpha)\ .$$

One has $f(0)=2(2s\lambda-1)<0$ and $f(1)=-1$. On the other hand

$$f'(\alpha) = 1 - \frac{4s\lambda}{[(1-2s\lambda)^2 + 8s\alpha\lambda]^{1/2}}$$

and

$$f'(0) = \frac{1-6s\lambda}{1-2s\lambda}, \quad f'(1) = \frac{1-2s\lambda}{1+2s\lambda} > 0\,.$$

Since $f'(\alpha)=0$ has one and only one zero, we see that $f(\alpha)<0$ for $\alpha\in[0,1]$.

It is easy to verify that $\alpha[1+\sigma_+/G]>0$ for $\alpha\in[0,1]$ and σ_+ is admissible under the restriction on the eigenvalues of the configuration tensor.

Similarly the reader can easily verify that $\alpha[1+\gamma_-/G]<0$ for $\alpha\in[0,1]$, and that $\alpha[1+\gamma_+/G]>0$ for $\alpha\in[0,1]$ provided that $s<1/4\gamma$. It follows that under the assumed restrictions $\sigma=\sigma_+$ and $\gamma=\gamma_+$ for extensional flow.

The steady vorticity of motions perturbing extensional flow is hyperbolic when

$$\rho s^2x^2(\gamma_+ + G) + \rho s^2y^2\,(\sigma_+ + G) - (\sigma_+ + G)(\gamma_+ + G) > 0. \qquad (25)$$

Since $G+\gamma_+$, $G+\sigma_+$ are positive, (25) describes the exterior of an ellipse.

When $\rho=0$, (25) reduces to

$$(\sigma_+ + G)(\gamma_+ + G) < 0,$$

which is impossible for $\alpha\in[0,1]$ and $0\le s<1/4\lambda$: the linear system perturbing extensional flow is always evolutionary in the Giesekus model whenever the configuration tensor is positive. It is perhaps necessary to note that unlike the upper and lower convected Maxwell models, the Giesekus model does not restrict the range of stresses to an evolutionary domain; an extra condition, perhaps inconvenient for numerical analysis has to be imposed.

6.7 Rigid rotation

We linearize around $\Omega r\mathbf{e}_\theta$ where Ω is an angular velocity and r is the distance from the axis of rotation. Let us consider the problem of finding the characteristic surfaces in the three dimensions using the linearized theory. We write the velocity as

$$\Omega r\mathbf{e}_\theta + \mathbf{u}$$

and the vorticity as

$$2\Omega\mathbf{e}_z + \boldsymbol{\zeta}$$

The stress associated with rigid motion vanishes, only the perturbed stresses are not zero. The linearized vorticity equation for the component $\zeta=\zeta_\theta$ of $\boldsymbol{\zeta}$ is given by (11.28):

$$c^2\left[\frac{\partial^2\zeta}{\partial r^2}+\frac{\partial^2\zeta}{\partial z^2}\right]+\left[\frac{c^2}{r^2}-\Omega^2\right]\frac{\partial^2\zeta}{\partial\theta^2}=\text{lot} \tag{26}$$

where $c=G/\rho$ is the wave speed. Using the method of simple jumps across characteristic surfaces $\phi(r,\theta,z)=0$, we find that

$$\phi_r^2+\phi_z^2+\left(\frac{1}{r^2}-\frac{\Omega^2}{c^2}\right)\phi_\theta^2=0\,. \tag{27}$$

A solution of this is

$$\phi = a\theta + bz + g(r)$$

where g satisfies

$$g' = \pm\sqrt{a^2\left(\frac{\Omega^2}{c^2}-\frac{1}{r^2}\right)-b^2}$$

and is real if and only

$$a^2\left(\frac{\Omega^2}{c^2}-\frac{1}{r^2}\right)-b^2>0\,.$$

It follows that when

$$r > r_c = \frac{1}{\sqrt{\frac{\Omega^2}{c^2} - \frac{b^2}{a^2}}} \tag{28}$$

we have real characteristic surfaces in the form of spirals shown in Figures 11.17 and 11.18.

6.8 Nonlinear ordinary differential equations along characteristics

Let us write the quasilinear system (3.40) in the plane, for steady flow of interpolated Maxwell models as

$$\hat{\mathbf{H}}\mathbf{q_x} + \hat{\mathbf{J}}\mathbf{q_y} = \boldsymbol{\ell} \tag{29}$$

where

$$\mathbf{q} = \begin{pmatrix} u \\ v \\ \sigma \\ \gamma \\ \tau \\ p \end{pmatrix}, \quad \boldsymbol{\ell} = -\frac{1}{\lambda}\begin{pmatrix} \sigma \\ \tau \\ \gamma \\ 0 \\ 0 \\ 0 \end{pmatrix} \tag{30}$$

and

$$\hat{\mathbf{H}} = \begin{pmatrix} -2(a\sigma+G) & \tau(1-a) & u & 0 & 0 & 0 \\ 0 & (1-a)\frac{1}{2}\gamma-(1+a)\,\frac{1}{2}\sigma-G & 0 & 0 & u & 0 \\ 0 & -(1+a)\tau & 0 & u & 0 & 0 \\ \rho u & 0 & -1 & 0 & 0 & 1 \\ 0 & \rho u & 0 & 0 & -1 & 0 \\ 1 & 0 & 0 & 0 & 0 & 0 \end{pmatrix},$$

$$\hat{\mathbf{J}} = \begin{pmatrix} -\tau(1+a) & 0 & v & 0 & 0 & 0 \\ \frac{1}{2}\sigma(1-a)-\frac{1}{2}\gamma(1+a)-G & 0 & 0 & 0 & v & 0 \\ \tau(1-a) & -2(G+a\gamma) & 0 & v & 0 & 0 \\ \rho v & 0 & 0 & 0 & -1 & 0 \\ 0 & \rho v & 0 & -1 & 0 & 1 \\ 0 & 1 & 0 & 0 & 0 & 0 \end{pmatrix}.$$

The characteristic equation for (29) is

$$\Delta = \det\,[-\alpha\hat{\mathbf{H}}+\hat{\mathbf{J}}] = 0 \tag{31}$$

where α=dy/dx is the slope of the characteristic lines y=f(x), Δ is given explicitly by (5.10) and

$$-\alpha\hat{\mathbf{H}}+\hat{\mathbf{J}} =$$

$$\begin{pmatrix} 2\alpha(a\sigma+G)-\tau(1+a) & -\tau\alpha(1-a) & -\alpha u+v & 0 & 0 & 0 \\ \frac{\sigma}{2}(1-a)-\frac{1}{2}\gamma(1+a)-G & -\alpha\left[\frac{1}{2}\gamma(1-a)-\frac{1}{2}\sigma(a+1)-G\right] & 0 & 0 & -\alpha u+v & 0 \\ \tau(1-a) & \alpha\tau(1+a)-2(G+a\gamma) & 0 & -\alpha u+v & 0 & 0 \\ \rho(-\alpha u+v) & 0 & \alpha & 0 & -1 & -\alpha \\ 0 & \rho(-\alpha u+v) & 0 & -1 & \alpha & 1 \\ -\alpha & 1 & 0 & 0 & 0 & 0 \end{pmatrix} \tag{32}$$

There are two real characteristics associated with streamlines and two real characteristics for the vorticity given by (1) when $B^2-AC > 0$.

The left eigenvector $\mathbf{Y}_\alpha$ belonging to one of the four roots α is given by

$$\mathbf{Y}_\alpha[-\alpha\hat{\mathbf{H}} + \hat{\mathbf{J}}] = 0\,. \tag{33}$$

Of course, $\mathbf{Y}_\alpha$ is unique up to an arbitrary mulitplicative constant. We may use this eigenvector to generate nonlinear ODE's along characteristics. We have

$$\mathbf{Y}_\alpha(\hat{\mathbf{H}}\mathbf{q}_x + \hat{\mathbf{J}}\mathbf{q}_y) = \mathbf{Y}_\alpha\boldsymbol{\lambda}\,. \tag{34}$$

Now after writing $\mathbf{m}=\mathbf{Y}_\alpha\hat{\mathbf{H}}$, using (33), we get

$$\mathbf{m}(\mathbf{q}_x+\alpha\mathbf{q}_y) = \mathbf{Y}_\alpha\boldsymbol{\lambda} . \tag{35}$$

The characteristic belonging to α is $y=f(x)$. Hence (35) may be written as ODE on $y=f(x)$,

$$\mathbf{m}(x)\frac{d\mathbf{q}(x, f(x))}{dx} = \mathbf{Y}_\alpha\boldsymbol{\lambda}(x) . \tag{36}$$

Now we seek to implement (36) when α is one of the two roots (1) for the vorticity. Let $\mathbf{Y}_\alpha = (y_1,y_2,y_3,y_4,y_5,y_6)$. Then

$$\begin{aligned}
y_1 &= \frac{-\alpha}{-\alpha u+v} ,\\
y_2 &= \frac{1-\alpha^2}{-\alpha u+v} ,\\
y_3 &= -y_1 ,\\
y_4 &= 1 ,\\
y_5 &= \alpha ,\\
y_6 &= \frac{\alpha a(\gamma-\sigma)}{-\alpha u+v}
\end{aligned} \tag{37}$$

where α is either of the two roots (6.1). Now we compute

$$\mathbf{m} = \mathbf{Y}_\alpha\hat{\mathbf{H}} = (m_1,m_2,m_3,m_4,m_5,m_6)$$

where

$$\begin{aligned}
m_1 &= \frac{1}{-\alpha u+v}\{\rho u(-\alpha u+v) + 2\alpha G+\alpha a(\sigma+\gamma)\} ,\\
m_2 &= \frac{1}{-\alpha u+v}\left\{\rho v(-\alpha u+v)-2\left(a\frac{\gamma+\sigma}{2}+G\right)\right\} ,\\
m_3 &= -\frac{v}{-\alpha u+v} ,\\
m_4 &= \frac{\alpha u}{-\alpha u+v} ,\\
m_5 &= \frac{u-\alpha v}{-\alpha u+v} ,\\
m_6 &= 1.
\end{aligned} \tag{38}$$

Using (30), (37), and (38) we may write (36) in explicit form

$$
\begin{aligned}
&[\rho u\,(-\alpha u+v) + 2\alpha G+\alpha a\,(\sigma+\gamma)]\,\frac{du(x,f(x))}{dx} \\
&+ [\rho v(-\alpha u+v) - 2G - a\,(\sigma+\gamma)]\,\frac{dv(x,f(x))}{dx} \\
&- v\,\frac{d\sigma(x,f(x))}{dx} + \alpha u\,\frac{d\gamma(x,f(x))}{dx} \\
&+ (u-\alpha v)\,\frac{d\tau(x,f(x))}{dx} + (v-\alpha u)\,\frac{dp(x,f(x))}{dx} \\
&= \alpha\frac{\sigma}{\lambda} - (1-\alpha^2)\,\frac{\tau}{\lambda} - \alpha\frac{\gamma}{\lambda}
\end{aligned}
\tag{39}
$$

where α is either one of the two roots of (1).

We turn next to the computation of ODE's along streamlines. The streamlines $v-\alpha u$ are double, semi-simple roots of the characteristic equation. We put $\alpha=v/u$ in (32) and compute

$$
\mathbf{Y}_\alpha = (y_1, y_2, y_3, 0, 0, y_6) \tag{40}
$$

where y_1 and y_2 arbitrary

$$
\begin{aligned}
y_3 &= \frac{A_1y_1+A_2y_2}{2\alpha(a\gamma+G) - \tau(1-a) - \alpha^2\tau(1+a)}\,, \\
A_1 &= 2\alpha(a\sigma+G) - \tau(1+a) - \alpha^2\tau(1-a)\,, \\
A_2 &= (\alpha^2-1)\left(a\frac{\gamma+\sigma}{2} + G\right) + (\alpha^2+1)\frac{\sigma-\gamma}{2}
\end{aligned}
$$

and

$$
y_6 = \alpha\tau(1-a)y_1 + \alpha\left[\frac{\gamma-\sigma}{2} - a\,\frac{\gamma+\sigma}{2} - G\right]y_2 - [\alpha\tau(1+a) - 2(G+a\gamma)]y_3\,.
$$

We get two linearly independent vectors by alternatively setting $y_1=0$ and then $y_2=0$, while setting the other (y_2 in the first case, y_1 in the second) equal to some convenient value; for example, the denominator of y_3. After simplifying y_6, we find that

$\mathbf{Y}_{\alpha 1}$:

$$\begin{aligned} y_1 &= \tau(1-a) + \alpha^2\tau(1+a) - 2\alpha(a\gamma+G) , \\ y_2 &= 0 , \\ y_3 &= -2\alpha(a\sigma+G) + \tau(1+a) + \alpha^2\tau(1-a) , \\ y_4 &= 0 , \\ y_5 &= 0, \\ y_6 &= -4a\alpha\tau^2 + 2(1+a)\tau(a\gamma+G) + 2\alpha^2(1+a)\tau(a\sigma+G) \\ &\quad - 4\alpha(a\gamma+G)(a\sigma+G) \end{aligned} \tag{41}$$

and

$\mathbf{Y}_{\alpha 2}$:

$$\begin{aligned} y_1 &= 0 , \\ y_2 &= \tau(1-a) + \alpha^2\tau(1+a) - 2\alpha(a\gamma+G) , \\ y_3 &= (1-\alpha^2)(a\frac{\gamma+\sigma}{2} + G) - (\alpha^2+1)\frac{\sigma-\gamma}{2} , \\ y_4 &= 0 , \\ y_5 &= 0 , \\ y_6 &= \left(\frac{\sigma-\gamma}{2} - a\frac{\sigma+\gamma}{2} - G\right)(\alpha\tau(1+a) - 2(a\gamma+G)) \\ &\quad + \alpha\tau(1-a)\left(\frac{\gamma-\sigma}{2} - a\frac{\sigma+\gamma}{2} - G\right) \end{aligned} \tag{42}$$

where $\alpha = \frac{v}{u}$ in both (41) and (42). Of course, any linear combination of $\mathbf{Y}_{\alpha 1}$ and $\mathbf{Y}_{\alpha 2}$ are also eigenvectors.

The corresponding $\mathbf{m}=\mathbf{Y}_\alpha\hat{\mathbf{H}}$ are given by

$\mathbf{m}_{\alpha 2}$:

$$\begin{gathered} y_6 , \\ (\sigma-\gamma)(a\tau+\alpha(G+a\gamma)) + (a(\gamma+\sigma)+2G)(\alpha(G+a\gamma)-\tau) , \\ 0 , \\ uy_3 , \\ uy_2 , \\ 0 , \end{gathered} \tag{43}$$

with y_2, y_3, and y_6 from (42) and

$\mathbf{m}_{\alpha 1}$:

$$\begin{gathered} 2a\tau(a(\sigma+\gamma)+\gamma-\sigma+2G-2\alpha\tau) , \\ 2a\tau(\alpha a(\sigma+\gamma) + \alpha(\sigma-\gamma+2G)-2\tau) , \\ uy_1 , \\ uy_3 , \\ 0 , \\ 0 \end{gathered}$$

with y_1 and y_3 from (41).

The canonical form (36) gives rise to two ODE's, one for each eigenvector.

$$m_1 \frac{du}{dx}(x,f(x)) + m_2 \frac{dv}{dx}(x,f(x)) + m_3 \frac{d\sigma}{dx}(x,f(x)) + m_4 \frac{d\gamma}{dx}(x,f(x)) + m_5 \frac{d\tau}{dx}(x,f(x)) + m_6 \frac{dp}{dx}(x,f(x)) - \frac{y_1\sigma}{\lambda} - \frac{y_2\tau}{\lambda} - \frac{y_3\gamma}{\lambda} = 0. \tag{44}$$

Equations (39) and (44) represent a total of four equations for the six unknowns: $u,v,\sigma,\gamma,\tau,\rho$. Two more equations from the elliptic part of the governing system seem to be required for a complete description of the problem. The ODE (39) and (44) may be useful for numerical computations of the solution. The numerical algorithms required are not yet known.

6.9 Analysis of type in a nonlinear system which is not quasilinear: the White-Metzner model

First we write the system of six nonlinear equations

$$\begin{aligned}
\sigma + \lambda_{\mathbf{II_D}}(u\sigma_x + v\sigma_y - 2\sigma u_x - 2\tau u_y) - 2\eta_{\mathbf{II_D}} u_x = 0\,,\\
\tau + \lambda_{\mathbf{II_D}}(u\tau_x + v\tau_y - \gamma u_y - \sigma v_x) - \eta_{\mathbf{II_D}}(u_y + v_x) = 0\,,\\
\gamma + \lambda_{\mathbf{II_D}}(u\gamma_x + v\gamma_y - 2\gamma v_y - 2\tau v_x) - 2\eta_{\mathbf{II_D}} v_y = 0\,,\\
\rho(uu_x + vu_y) + p_x - \sigma_x - \tau_y = 0\,,\\
\rho(uv_x + vv_y) + p_y - \tau_x - \gamma_y = 0\,,\\
u_x + v_y = 0
\end{aligned} \tag{45}$$

where we wrote $\lambda_{\mathbf{II_D}}$ and $\eta_{\mathbf{II_D}}$ to keep in mind that λ and η are not constants.

Here we have

$$\mathbf{II_D} = \frac{1}{2}\mathbf{D}{:}\mathbf{D} = \frac{1}{2}u_x^2 + \frac{1}{2}v_y^2 + \frac{1}{4}(u_x + v_y)^2\,.$$

We introduce

$$p_i = \frac{\partial u_i}{\partial x} ,$$
$$q_i = \frac{\partial u_i}{\partial y} \tag{46}$$

where $\mathbf{u} = (u,v,p,\sigma,\tau,\gamma)$. Of course, we have the compatibility relation

$$\frac{\partial p_i}{\partial y} = \frac{\partial q_i}{\partial x} . \tag{47}$$

So we can rewrite the system as

$$\begin{gathered}
u_4 + \lambda_{\mathbf{II_D}}(u_1p_4+u_2q_4-2u_4p_1-2u_5q_1) - 2\eta_{\mathbf{II_D}}p_1 = 0 , \\
u_5 + \lambda_{\mathbf{II_D}}(u_1p_5+u_2q_5-u_6q_1-u_4p_2) - \eta_{\mathbf{II_D}}(q_1+p_2) = 0 , \\
u_6 + \lambda_{\mathbf{II_D}}(u_1p_6+u_2q_6-2u_6q_2-2u_5p_2) - 2\eta_{\mathbf{II_D}}q_2 = 0 , \\
\rho(u_1p_1+u_2q_1) + p_3 - p_4 - q_5 = 0 , \\
\rho\ (u_1p_2+u_2q_2) + q_3 - p_5 - q_6 = 0 , \\
p_1 + q_2 \ = 0 .
\end{gathered}$$

We differentiate the 6 equations with respect to x to get, after using (47),

$$\mathbf{A}\frac{\partial \mathbf{p}}{\partial x} + \mathbf{B}\frac{\partial \mathbf{p}}{\partial y} = \mathbf{lot}$$

where the components of $\mathbf{p}$ are defined by (46),

$$\mathbf{A} = \begin{pmatrix}
A_1 & A_2 & u & 0 & 0 & 0 \\
A_3 & A_4 & 0 & 0 & u & 0 \\
A_5 & A_6 & 0 & u & 0 & 0 \\
\rho u & 0 & -1 & 0 & 0 & 0 \\
0 & \rho u & 0 & 0 & -1 & 0 \\
1 & 0 & 0 & 0 & 0 & 0
\end{pmatrix}$$

and

$$\mathbf{B} = \begin{pmatrix} B_1 & B_2 & v & 0 & 0 & 0 \\ B_3 & B_4 & 0 & 0 & v & 0 \\ B_5 & B_6 & 0 & v & 0 & 0 \\ \rho v & 0 & -1 & 0 & 0 & 0 \\ 0 & \rho v & 0 & 0 & -1 & 0 \\ 1 & 0 & 0 & 0 & 0 & 0 \end{pmatrix}.$$

The coefficients A_i's and B_i's are defined as follows:

$$\begin{aligned} A_1 &= -2\frac{D_{11}V_{11}}{\lambda} - 2\sigma - 2G, \\ A_2 &= -2\frac{D_{12}V_{11}}{\lambda}, \\ A_3 &= -2\frac{D_{11}V_{12}}{\lambda}, \\ A_4 &= -2\frac{D_{12}V_{12}}{\lambda} - \sigma - G, \\ A_5 &= -2\frac{D_{11}V_{22}}{\lambda}, \\ A_6 &= -2\frac{D_{12}V_{22}}{\lambda} - 2\tau, \end{aligned} \tag{48}$$

$$\begin{aligned} B_1 &= -2\frac{D_{12}V_{11}}{\lambda} - 2\tau, \\ B_2 &= -2\frac{D_{22}V_{11}}{\lambda}, \\ B_3 &= -2\frac{D_{12}V_{12}}{\lambda} - \gamma - G, \\ B_4 &= -2\frac{D_{22}V_{12}}{\lambda}, \\ B_5 &= -2\frac{D_{12}V_{22}}{\lambda}, \\ B_6 &= -2\frac{D_{22}V_{22}}{\lambda} - 2\gamma - 2G. \end{aligned} \tag{49}$$

where the components V_{ij} of the tensor $\mathbf{V}$ are defined by

$$\mathbf{V} = \frac{\lambda'}{2\lambda}\boldsymbol{\tau} + G'\lambda\mathbf{D}\,. \tag{50}$$

The characteristic curves are determined by

$$\det(\alpha\mathbf{A} - \mathbf{B}) = 0$$

where $\alpha = dy/dx$ whenever α is real. After calculating the determinant we find that

$$\begin{aligned}&(\alpha u - v)^2\lambda^2\big((-A_1\alpha^2 + B_1\alpha - A_2\alpha^3 + B_2\alpha^2 + A_3\alpha(1-\alpha^2)\\ &+ B_3(\alpha^2-1)) + A_4\alpha^2(1-\alpha^2) + B_4\alpha(\alpha^2-1) + A_5\alpha^2 - B_5\alpha\\ &\quad + A_6\alpha^3 - B_6\alpha^2 - \rho\lambda(1+\alpha^2)(\alpha u - v)^2\big) = 0\,.\end{aligned} \tag{51}$$

Then we use (48) and (49). We find, after some calculations that (51) becomes

$$\begin{aligned}&(\alpha u - v)^2\lambda^2\,\Big\{(1+\alpha^2)((\sigma+G-\rho u^2)\alpha^2 - 2(\tau-\rho uv)\alpha + \gamma + G - \rho v^2)\\ &\frac{2}{\lambda}(D_{12}\alpha^2+(D_{11}-D_{22})\alpha-D_{12})(V_{12}\alpha^2+(V_{11}-V_{22})\alpha-V_{12})\Big\}\\ &= 0\,.\end{aligned} \tag{52}$$

First we note that the streamlines $\frac{dy}{dx} = \frac{v}{u}$ are twice solutions of (52) so there are at least two real roots, but they are double. The system is not strictly hyperbolic in any case.

It is also of interest to note the symmetric role of the tensors **D** and **V** in the second part of this formula.

The formula (52) shows that we add an extra part $Q(\alpha)$, to the usual polynomial $P(\alpha)$, where

$$P(\alpha) = (1+\alpha^2)((\sigma+G-\rho u^2)\alpha_2 - 2(\tau-\rho uv)\alpha + \gamma + G - \rho v^2)$$

and

$$\begin{aligned}Q(\alpha) = \frac{2}{\lambda}&(D_{12}\alpha^2 + (D_{11}-D_{22})\,\alpha - D_{12})\\ &(V_{12}\alpha^2 + (V_{11}-V_{22})\alpha - V_{12})\,.\end{aligned}$$

The polynomial $P(\alpha)$ is indeed the one that corresponds to the upper convected Maxwell model.

In addition to that, the extra part $Q(\alpha)$ only has real roots since the discriminants of each quadratic from $Q(\alpha)$ are positive, i.e.

$$\Delta_1 = (D_{11}-D_{22})^2 + 4D_{12}^2 \geq 0 \ ,$$

$$\Delta_2 = (V_{11}-V_{22})^2 + 4V_{12}^2 \geq 0 \ .$$

There are two limiting cases that can be considered:

- λ and η do not depend very strongly on $\mathbf{II_D}$.

In this case the coefficients V_{ij} are small and we can neglect $Q(\alpha)$ part; the roots are the roots of $P(\alpha)$

$$\alpha = \pm i \ ,$$

$$\alpha = \frac{\tau-\rho uv \pm \sqrt{(\tau-\rho uv)^2-(\sigma+G-\rho u^2)(\gamma+G-\rho v^2)}}{\sigma+G-\rho u^2} \ .$$

We have two imaginary roots, and the two others depend on the sign of $\Delta=(\tau-\rho uv)^2-(\sigma+G-\rho u^2)(\gamma+G-\rho v^2)$. We may have a change of type.

- λ and η depend very strongly on $\mathbf{II_D}$.

Assuming we can neglect $P(\alpha)$, then the roots are easy to find

$$\alpha = \frac{D_{22}-D_{11} \pm \sqrt{(D_{11}-D_{22})^2+4D_{12}^2}}{2D_{12}}$$

$$\alpha = \frac{V_{22}-V_{11} \pm \sqrt{(V_{11}-V_{22})^2+4V_{12}^2}}{2V_{12}}$$

These roots are always *real* and it does not even matter how λ and η depend on the second invariant $\mathbf{II_D}$.

It is hard to say how the roots change by studying the whole polynomial. Since the streamlines are always doubly characteristic, we have to study the roots of a polynomial of order 4. There are some standard techniques to determine the roots of a quartic but all of them are very lengthy. The only thing that needs to be said is that each fourth order polynomial can be factored into a product of two quadratics. Therefore there are either

- Four real roots
- Two real roots + a pair of complex conjugates
- Two pairs of complex conjugates

To these we append the usual two real roots corresponding to the streamlines.

We note that all the roots may be obtained in analytic form by using symbol manipulators on the computer. The expressions for the general case are very long and are not given here. Also we can determine the roots for small λ' and η' by perturbation.

6.10 Analysis of type for the fourth order stream function equation for flow of the White-Metzner model

The method of deriving the vorticity equation is well known. For the White-Metzner model, we find, for the components of the vorticity $\boldsymbol{\zeta} = \zeta_b \mathbf{e}_b$ that

$$\rho\left(\frac{\partial^2\zeta_b}{\partial t^2} + 2(\mathbf{u}\bullet\nabla)\frac{\partial\zeta_b}{\partial t} + (\mathbf{u}\bullet\nabla)^2\zeta_b\right) - (\boldsymbol{\tau}:\nabla\nabla)\zeta_b - G(\mathbf{II_D})\nabla^2\zeta_b$$
$$- \varepsilon_{bmi}V_{ij}D_{kl}\left(\frac{\partial^3 u_k}{\partial x_l\partial x_j\partial x_m} + \frac{\partial^3 u_l}{\partial x_k\partial x_j\partial x_m}\right) = \text{lot}\,.$$

where V_{ij} are the components of the tensor $\mathbf{V}$ defined by (50). Note one more time the importance of the tensor $\mathbf{V}$. The last terms in the previous equation are "spoilers" because they cannot be expressed in terms of the vorticity components.

In the 2D case we introduce a stream function ψ defined by

$$u = \frac{\partial\psi}{\partial y}, \quad v = -\frac{\partial\psi}{\partial x}$$

Inserting the stream function , we determine a fourth order equation for ψ. We find that

$$\left(\sigma+G-\rho u^2+2\frac{D_{12}V_{12}}{\lambda}\right)\frac{\partial^4\psi}{\partial x^4}$$
$$+2\left(\tau-\rho uv+\frac{(D_{22}-D_{11})V_{12}}{\lambda}+\frac{D_{12}(V_{22}-V_{11})}{\lambda}\right)\frac{\partial^4\psi}{\partial x^3\partial y}$$
$$+\left(\sigma+G-\rho u^2+\gamma+G-\rho v^2+2\frac{(D_{11}-D_{22})(V_{11}-V_{22})}{\lambda}-4\frac{D_{12}V_{12}}{\lambda}\right)\frac{\partial^4\psi}{\partial x^2\partial y^2}$$
$$+2\left(\tau-\rho uv+\frac{(D_{11}-D_{22})V_{12}}{\lambda}+\frac{D_{12}(V_{11}-V_{22})}{\lambda}\right)\frac{\partial^4\psi}{\partial x\partial y^3}$$
$$+\left(\gamma+G-\rho v^2+2\frac{D_{12}V_{12}}{\lambda}\right)\frac{\partial^4\psi}{\partial y^4}=\text{lot}$$

where lot represents lower order terms including derivatives up to the third order of ψ.

We can get the quartic equation obtained in §6.9 for the characteristic curves directly from the fourth order equation* . We will now introduce different notations for the coefficients

$$A=\sigma+G-\rho u^2\ ,$$

$$B=\tau-\rho uv\ ,$$

$$C=\gamma+G-\rho v^2\ ,$$

$$D=\frac{D_{12}(V_{11}-V_{22})}{\lambda}\ ,$$

$$E=\frac{D_{12}V_{12}}{\lambda}\ ,$$

---

* If a function ψ satisfies the equation

$$a_1\frac{\partial^4\psi}{\partial x^4}+a_2\frac{\partial^4\psi}{\partial x^3\partial y}+a_3\frac{\partial^4\psi}{\partial x^2\partial y^2}+a_4\frac{\partial^4\psi}{\partial x\partial y^3}+a_5\frac{\partial^4\psi}{\partial y^4}=\text{lot}$$

then the characteristic curves are given by

$$a_1\left(\frac{dy}{dx}\right)^4-a_2\left(\frac{dy}{dx}\right)^3+a_3\left(\frac{dy}{dx}\right)^2-a_4\frac{dy}{dx}+a_5=0\ .$$

$$F = \frac{(D_{11}-D_{22})V_{12}}{\lambda} ,$$

$$G = \frac{(D_{11}-D_{22})(V_{11}-V_{22})}{\lambda} .$$

Note also that DF=EG. This allows us to write

$$(A+2E)\frac{\partial^4\psi}{\partial x^4} + 2(B-D-F)\frac{\partial^4\psi}{\partial x^3\partial y} + (A+C+2G-4E)\frac{\partial^4\psi}{\partial x^2\partial y^2}$$

$$2(B+D+F)\frac{\partial^4\psi}{\partial x\partial y^3} + (C+2E)\frac{\partial^4\psi}{\partial y^4} = \text{lot} .$$

To see the structure of this operator, we shall rewrite it in terms which give rise to the polynomials $P(\alpha)$ and $Q(\alpha)$. Thus

$$\left[\left(A\frac{\partial^2}{\partial x^2} + 2B\frac{\partial^2}{\partial x\partial y} + C\frac{\partial^2}{\partial y^2}\right)\left(\frac{\partial^2}{\partial x^2} + \frac{\partial^2}{\partial y^2}\right)\right.$$

$$\left. + 2E\left(\frac{\partial^2}{\partial x^2} - \frac{F}{E}\frac{\partial^2}{\partial x\partial y} - \frac{\partial^2}{\partial y^2}\right)\left(\frac{\partial^2}{\partial x^2} - \frac{D}{E}\frac{\partial^2}{\partial x\partial y} - \frac{\partial^2}{\partial y^2}\right)\right]\psi = \text{lot}$$

We now let

$$L = A\frac{\partial^2}{\partial x^2} + 2B\frac{\partial^2}{\partial x\partial y} + C\frac{\partial^2}{\partial y^2} ,$$

$$\Delta = \frac{\partial^2}{\partial x^2} + \frac{\partial^2}{\partial y^2} \quad \text{(Laplacian Operator)} ,$$

$$L_1 = \frac{\partial^2}{\partial x^2} - \frac{F}{E}\frac{\partial^2}{\partial x\partial y} - \frac{\partial^2}{\partial y^2} ,$$

$$L_2 = \frac{\partial^2}{\partial x^2} - \frac{D}{E}\frac{\partial^2}{\partial x\partial y} - \frac{\partial^2}{\partial y^2} ,$$

so that we get

$$(L\Delta + 2EL_1L_2)\psi = \text{lot} .$$

The first part of this operator $L\Delta$ is of composite type and depends on the sign of B^2-AC, and the second operator L_1L_2 is the product of two hyperbolic operators, therefore it is purely hyperbolic. This corresponds to what we found in the study of the polynomial.

This shows how strong the effect of hyperbolicity is in the flow of viscoelastic fluids. Quasilinear models like Maxwell's are already endowed with a fairly rich hyperbolic structure. The White-Metzner model could be considered to be one of the easiest nonlinear generalizations of Maxwell models, representative of an even larger class of nonlinear models. These nonlinear models can be seen to be even "more hyperbolic" than models like Maxwell's.

Calderer, Cook, and Schleiniger [1989] have analyzed the Bird-Deaguiar model of polymer melts for a change of type. They considered the case of instantaneous elasticity, in which the solvent viscosity is set to zero, and worked in the range of parameters in which the model may be valid. The system for this model is quasilinear. The time independent, two dimensional system does change type. For certain ranges of parameters all the eigenvalues are real, an interesting result showing that fully hyperbolic systems are possible for certain viscoelastic models.

6.11 Numerical simulations and analysis of type

The type of a system of equations is actually a statement about the algorithms which can be used to construct solutions. You should not use the same algorithms, say, for elliptic and hyperbolic problems. We have already seen in Chapter 4 that elliptic equations are ill-posed as initial value problems. If you try to compute you will come up against Hadamard instability leading to a failure of numerical simulation; mesh refinement worsens the instability because it leads to shorter waves. On the other hand, elliptic problems are well-posed as boundary value problems and are relatively easy to simulate numerically. It is equally true that boundary value problems for the wave equation

$$\phi_{tt} = \phi_{xx}$$

with periodic conditions in x with period $2\pi/\alpha$ and, say, prescribed boundary data $\phi(x,0)=g_1(x)$, $\phi(x,1)=g_2(x)$ is ill-posed in another sense, loss of uniqueness. There are in fact an infinite number of solutions because $\phi=\sin n\pi t\, e^{i\alpha x}$ is a solution of the homogeneous problem with $g_1=g_2=0$ for each and every $\alpha=n\pi$. This problem can be solved uniquely on a rectangle with $\phi(0,t)=\phi(L,t)=0$ provided that there are not homogeneous solutions, $\pi n \neq \frac{m\pi}{L}$, that is, L is not a rational fraction. Even when this is the case we have a problem of small divisors.

Serious attention to numerical simulation of flows of polymer solutions and liquid polymers for processing applications began in the 1970s (Crochet, Davies and Walters, [1984]). Extrusion, film blowing, coating, fiber spinning, and injection molding are among the processing flows needing simulation. It had been thought that extensions of Newtonian algorithms would be sufficient for solving viscoelastic flow problems, but the field has been found much more difficult than expected. It was found that even when inertia was neglected, the simulation would fail when an elasticity parameter, say the Weissenberg number $W = \lambda U/L$ representing the ratio of material to processing times was greater than a number of order one whose particular value depended on the details of the simulation. The problem did not appear to depend on the choice of numerical method; finite elements, differences and spectral methods were all plagued by the same disease.

Over the course of years it became clear that there were several reasons for the failure of numerical simulations and all of them involved appeared to be associated with high stress, particularly with apparent corner singularities. Every two years for about the last dozen there is a meeting to discuss the high Weissenberg number (or high Deborah number problem.) In the Lake Morrey, Vermont meeting in the summer of 1983, I suggested that the problem was associated with hyperbolicity and change of type of the governing equations. I also suggested that type dependent upwinding schemes like those used in transonic flow might be useful. At the time, it was thought that limit points might be responsible for the failure of simulations. It appears now that limit points are common in the numerical approximation but usually disappear after more careful analysis.

A finite difference numerical simulation of the flow of an upper convected Maxwell fluid through a planar 4:1 contraction using a type dependent upwinding scheme for the vorticity was recently reported by Song and Yoo [1987]. A similar method, with the addition of an artificial viscosity and using a Giesekus model, was employed by Choi, Song, and Yoo [1987]. The introduction of these type dependent methods gives improved numerical results at higher values of Deborah number but they do not eliminate the problem.

The problem of numerical simulations is evidently associated with hyperbolicity, not so much for "transonic" steady problems as I suggested originally but with short wave length instabilities of the type associated with ill-posed problems. The first important results on ill-posed problems for Maxwell models can be found in a series of papers in the 1970s by Rutkevich; Joseph, Renardy and Saut showed that vorticity was the key variable and demonstrated that type changes could be expected in many, maybe nearly all flows. The paper by Dupret, Marchal, and Crochet [1985] is important in that it indicated that discretization errors could introduce short wave instabilties even for models, like the upper convected Maxwell model, which are always well-posed for smooth solutions. Discretization for such problems allows one to step into forbidden and otherwise inaccesible regions of the added stress. A recent work by Marchal and Crochet [1987] seems to have partially solved the problem of "false" discretization induced instability by upwinding on streamlines. They introduced technical improvements and the addition of an artificial diffusivity which goes to zero with mesh refinement. This code has been applied by Crochet and Delvaux [1990] to the problems of supercritical flow in a channel bounded by wavy walls (Chap. 12), by Delvaux and Crochet [1989] to the problem of delayed die swell, and to flow around a cylinder (Chap. 7) in the two papers last mentioned.

Michael Renardy [1988a, b] and M.A. Hulsen [1986] have used the analysis of characteristics to obtain useful results for numerical analysis of steady flow in an open strip, say a channel in the subcritical case. Steady flows of viscoelastic fluids cannot be uniquely determined by imposing boundary conditions for the velocities as in the Newtonian case. The reason for this is that the fluids have memory and therefore the flow inside the domain is affected by what happened before the fluid entered the domain. This leads to the need for extra boundary conditions at an inflow boundary. The nature of these inflow boundary conditions depends on the choice of a constitutive equation. In steady flow the data at an open boundary is carried into the computational domain along streamlines. We have seen that the streamlines are characteristic and their multiplicity is always greater than one. This suggests that the number of extra conditions, beyond the prescription of the value of the velocity components at an inflow boundary, is equal to the multiplicity M of the streamlines. For models like Maxwell's this multiplicity M given by (5.4) as M=2 for plane flow, as M=3 by (see Exercise 10.2) for axisymmetric flow and as M=4 by (11.4) for three-dimensional flow.

Renardy [1988a] proved that the multiplicity number M is also the number of extra conditions that must be added to the prescribed values of velocity (u,v in 2-D; u,v,w in 3-D) at the inflow boundary to problems perturbing uniform flow in strips. This suprised rheologists who believed that it was possible to prescribe the extra stress at an inflow boundary. Many flows are in open domains, like channels pipes. The domain is usually truncated for numerical studies. The prescription of all of the components of the extra stress overdetermines the problem for fluids like Maxwell's and errors must be expected from this procedure unless the artificial boundary is chosen far enough out. Renardy showed that prescription of the two normal components of the stress would be sufficient for solutions in the plane, but a more complicated prescription was required for the three-dimensional strip. Hulsen [1986] derived boundary conditions for a linearized upper convected Maxwell model.

> It appears that the shear stress tangential to the streamlines behaves elliptic and that it is not necessary to specify all the stresses at the inflow boundary...Moreover it is not sufficient to specify only a level for the pressure, a level for pressure gradients in streamline direction must be specified too.

For Jeffreys' model (5.53) show that M=3. We leave it to the reader to show that in three dimensions M=6. The rule of thumb and the results of Renardy [1988b] and Hulsen [1986] show that for models like Jeffreys' one can prescribe all components of the extra stress as missing conditions at the inflow boundary.

The state of the art in numerical simulations of complex flows of viscoelastic fluids is an active state of development (see Proceeding of the 5th Workshop on Numerical Methods in Non-Newtonian

Fluid Mechanics, *J. Non-Newtonian Fluid Mech.* **29**, 1988). Yet another idea has been put forward by Phelan, Malone, and Winter [1989]. They develop a purely hyperbolic model for unsteady, two-dimensional flow of the Giesekus model (1.69) for the extra stress and a pressure-density relation in the continuity equation to obtain a dynamic equation for the pressure. The small degree of compressibility allowed by the model appears to be negligible for practical purposes but it changes the unsteady governing equations from composite type to purely hyperbolic. This allows one to use hyperbolic codes which may be better suited to problems at high Weissenberg number where the elasticity of the fluid and hyperbolicity in the equations are important. In fact the advantages of fully hyperbolic code in the singularly perturbed problem for weak compressibility discussed in §5.8 may be more apparent than real. In certain cases discussed at the end of §6.10 we can get purely hyperbolic systems even when the fluid is strictly incompressible.

There is no well developed rigorous theory of existence and uniqueness for flows of viscoelastic fluids. The theory is only recently active (Renardy, [1989a]) with results for special models in special subcritical flows and for small perturbations of uniform supercritical flow of fluids of Maxwell's type (Renardy, [1989b]). There is no theory whatsoever for flows which undergo a change of type.

There are two papers which establish the existence of solutions of the initial-boundary value problem for different models, with different conditions on the data. Guillopé and Saut [1988], [1990], and Renardy [1988b] treat models of Jeffreys' type, and Renardy [1988a] treats hypoelastic models which include all the interpolated Maxwell models. Both papers are for the case when the motion of the boundary is entirely tangential.

Exercises

Exercise 6.1.

Show that simple shear flows of interpolated Maxwell models have singular stresses if $a^2>1$, are shear thinning when $a^2<1$, have a constant shear viscosity and strictly quadratic dependence on the shear rate κ when $a^2=1$.

Exercise 6.2.
Calculate the characteristics which cover the hyperbolic regions of the vorticity in steady flow for $a=\pm1$ and $a=0$. (You must integrate (15); numerical integration is acceptable.)

Exercise 6.3.
Carry out the calculation, giving the results of the analysis of perturbations of simple shear of a Giesekus model.

Exercise 6.4.
Derive equation (22).

Exercise 6.5.
The general solution of the wave equation

$$\phi_{xx} = c^2\phi_{yy}$$

can be expressed as

$$\phi(x,y) = \phi_1(r) + \phi_2(s)$$

where $\phi_1(\bullet)$ and $\phi_2(\bullet)$ are arbitrary differentiable functions and

$$r = y-cx\ , \quad s = y+cx$$

are characteristic coordinates. If we introduce r and s as coordinates, then the wave equation may be written in canonical form

$$4c^2\phi_{rs} = 0\ .$$

consider (3.51) for steady flow and show that when $B^2-AC>0$, (3.51) may be written as

$$4\frac{B^2-AC}{A}r_y s_y \frac{\partial^2\xi(r,s)}{\partial r\partial s} = \hat{\ell}(r,s) \tag{53}$$

where

$$\xi(r,s) = \zeta(x,y)\ ,$$
$$\hat{\ell}(r,s) = \ell(x,y)$$

and (r,s) are characteristic coordinates satisfying

$$Ar_x^2 + 2Br_xr_y + Cr_y^2 = 0 \ ,$$

$$As_x^2 + 2Bs_xs_y + Cs_y^2 = 0 \ . \tag{54}$$

Solutions for the exercises

Exercise 6.1.

For simple shear flow we have

$$[u,v,w] = [\kappa y,0,0] \ ,$$

$$\mathbf{L} = \begin{pmatrix} 0 & \kappa & 0 \\ 0 & 0 & 0 \\ 0 & 0 & 0 \end{pmatrix} .$$

This must satisfy the constitutive equation

$$\frac{\partial \tau}{\partial t} + (\mathbf{u}\bullet\nabla)\boldsymbol{\tau} + \frac{1-a}{2}(\boldsymbol{\tau}\mathbf{L}+\mathbf{L}^T\boldsymbol{\tau}) - \frac{1+a}{2}(\mathbf{L}\boldsymbol{\tau}+\boldsymbol{\tau}\mathbf{L}^T) - 2G\mathbf{D}[\mathbf{u}] + \frac{\boldsymbol{\tau}}{\lambda} = 0 \ ,$$

that is,

$$\frac{1-a}{2}\begin{pmatrix} 0 & \kappa\sigma & 0 \\ \kappa\sigma & 2\kappa\tau & 0 \\ 0 & 0 & 0 \end{pmatrix} - \frac{1+a}{2}\begin{pmatrix} 2\kappa\tau & \kappa\gamma & 0 \\ \kappa\gamma & 0 & 0 \\ 0 & 0 & 0 \end{pmatrix} - G\begin{pmatrix} 0 & \kappa & 0 \\ \kappa & 0 & 0 \\ 0 & 0 & 0 \end{pmatrix} + \frac{1}{\lambda}\begin{pmatrix} \sigma & \tau & 0 \\ \tau & \gamma & 0 \\ 0 & 0 & 0 \end{pmatrix}$$
$$= 0 \ .$$

This leads to the following expression for the components of the extra stress:

$$-\frac{1+a}{2}2\kappa\tau+\frac{\sigma}{\lambda}=0\,,$$

$$\frac{1-a}{2}\kappa\sigma-\frac{1+a}{2}\kappa\gamma-G\kappa+\frac{\tau}{\lambda}=0\,,$$

$$\frac{1-a}{2}2\kappa\tau+\frac{1}{\lambda}\gamma=0\,.$$

Therefore

$$\tau=\frac{\lambda G\kappa}{1+(1-a^2)\kappa^2\lambda^2}=\frac{\eta\kappa}{1+(1-a^2)\kappa^2\lambda^2}\,,$$
$$\sigma=(1+a)\lambda\kappa\tau\,,$$
$$\gamma=(a-1)\lambda\kappa\tau\,.$$

If $a^2>1$, then $1+(1-a^2)\kappa^2\lambda^2 = 1-(a^2-1)\kappa^2\lambda^2$ may be zero for sufficiently large κ, therefore these models have singular stresses if $a^2>1$.

If $a^2<1$, as κ increases, τ decreases. This is a shear thinning effect.

If $a^2=1$ then $\tau=\eta\kappa$ has a constant shear viscosity η and normal stresses

$$\sigma=2\lambda\eta\kappa^2\,,$$
$$\gamma=0.$$

Exercise 6.2.

The formula for the characteristic equation is

$$\frac{dy}{dx}=\frac{1}{\omega\pm\sqrt{M^2D-1-\omega^2}}$$

where

$$\omega=\lambda\kappa\,,\quad D=1+\kappa^2\lambda^2(1-a^2)\,,$$

$$M^2=\frac{\kappa^2y^2}{c^2}\,,\quad c^2=\frac{G}{\rho}\,,$$

$$\frac{dx}{dy} = \omega \pm \sqrt{\frac{\kappa^2}{c^2} D y^2 - (1+\omega^2)} \ ,$$

$$x = c_1 + \omega y \pm \frac{\kappa}{c}\sqrt{D} \int \sqrt{y^2 - \frac{1+\omega^2}{\kappa^2 D} c^2} \quad dy$$

$$= c_1 + \omega y \pm \frac{\kappa}{c}\sqrt{D}\left[\frac{y}{2}\sqrt{y^2 - \frac{1+\omega^2}{\kappa^2 D} c^2}\right.$$

$$\left. - \frac{1+\omega^2}{2\kappa^2 D} c^2 \ln\left(y + \sqrt{y^2 - \frac{1+\omega^2}{\kappa^2 D} c^2}\right)\right] .$$

Let

$$d^2 = \frac{1+\omega^2}{\kappa^2 D} c^2 \ ,$$

then the characteristics are given by

$$x = \omega y \pm \frac{\kappa}{c}\sqrt{D}\frac{1}{2}\left[y\sqrt{y^2 - d^2} - d^2 \ln\left(y + \sqrt{y^2 - d^2}\right)\right] + c_1$$

where c_1 is an arbitrary constant.

Exercise 6.3.
A Giesekus Model:

$$\frac{\partial \boldsymbol{\tau}}{\partial t} + (\mathbf{u}\cdot\nabla)\boldsymbol{\tau} - (\mathbf{L}\boldsymbol{\tau} + \boldsymbol{\tau}\mathbf{L}^T) - 2G\mathbf{D}[\mathbf{u}] = -\frac{1}{\lambda}\left(\boldsymbol{\tau} + \frac{\alpha}{G}\boldsymbol{\tau}^2\right)$$

For simple shear flow,

$$(u,v) = (\kappa y, 0) \ ,$$

$$\mathbf{L} = \begin{pmatrix} 0 & \kappa \\ 0 & 0 \end{pmatrix}, \quad \boldsymbol{\tau} = \begin{pmatrix} \sigma & \tau \\ \tau & \gamma \end{pmatrix} .$$

Hence,

$$-\begin{pmatrix} 2\kappa\tau & \kappa\gamma \\ \kappa\gamma & 0 \end{pmatrix} - G\begin{pmatrix} 0 & \kappa \\ \kappa & 0 \end{pmatrix} = -\frac{1}{\lambda}\begin{pmatrix} \sigma & \tau \\ \tau & \gamma \end{pmatrix} - \frac{\alpha}{\eta}\begin{pmatrix} \sigma^2+\tau^2 & \tau(\sigma+\gamma) \\ \tau(\sigma+\gamma) & \tau^2+\gamma^2 \end{pmatrix}$$

or

$$2\kappa\lambda\tau = \sigma + \frac{\alpha}{G}(\sigma^2+\tau^2) ,$$

$$\kappa\lambda\gamma + \lambda G\kappa = \tau + \frac{\alpha}{G}\tau(\sigma+\gamma) ,$$

$$0 = \gamma + \frac{\alpha}{G}(\tau^2+\gamma^2) .$$

If $\alpha=1$, then $\tau=\dfrac{\eta\kappa}{1+\kappa^2\lambda^2}$, $\sigma=\lambda\kappa\tau$, $\gamma=-\lambda\kappa\tau$ is a solution of the above equation.

$$A = -\rho u^2 + G + \frac{1}{2}\sigma(1+a) - \frac{1}{2}\gamma(1-a) = -\rho\kappa^2 y^2 + G + \sigma ,$$

$$B = \tau - \rho uv = \tau ,$$

$$C = -\rho v^2 + G - \frac{1}{2}\sigma(1-a) + \frac{1}{2}\gamma(1+a) = G + \gamma ,$$

$$\begin{aligned}
B^2-AC &= \tau^2 - (-\rho\kappa^2y^2+G+\sigma)(G+\gamma) \\
&= \tau^2 + \rho\kappa^2y^2\left(G - \frac{\lambda\kappa^2\eta}{1+\kappa^2\lambda^2}\right) - (G^2 - \lambda^2\kappa^2\tau^2) \\
&= (1+\lambda^2\kappa^2)\frac{\eta^2\kappa^2}{(1+\kappa^2\lambda^2)^2} - \rho\kappa^2y^2\frac{\lambda\eta\kappa^2}{1+\kappa^2\lambda^2} + G(\rho\kappa^2y^2-G) \\
&= \frac{\eta^2\kappa^2}{(1+\kappa^2\lambda^2)}(1-\rho\kappa^2y^2\frac{1}{G}) + G(\rho k^2y^2-G) \\
&= (\rho\kappa^2y^2-G)\left(G - \frac{\frac{1}{G}\eta^2\kappa^2}{1+\kappa^2\lambda^2}\right) = (\rho\kappa^2y-G)\frac{G}{1+\lambda^2\kappa^2} .
\end{aligned}$$

$B^2-AC > 0$ requires that

$$\rho\kappa^2y^2 > G .$$

The steady vorticity equation for the linearized flow is hyperbolic in the strip

$$|y| > \frac{c}{\kappa} .$$

For inertialess flow $\rho = 0$

$$b_{11} = G + \sigma ,$$
$$b_{12} = \tau ,$$
$$b_{22} = G + \gamma ,$$

$$b_{12}^2 - b_{11}b_{22} = \tau^2 - (G^2 - \lambda^2\kappa^2\tau^2)$$
$$= \frac{\eta^2\kappa^2}{1+\kappa^2\lambda^2} - G^2 = \frac{\lambda^2\eta^2\kappa^2 - (1+\kappa^2\lambda^2)\eta^2}{(1+\kappa^2\lambda^2)\lambda^2}$$
$$= -\frac{\eta^2}{(1+\kappa^2\lambda^2)\lambda^2} < 0 .$$

and

$$b_{11} = G+\sigma = \frac{\eta}{\lambda} + \frac{\lambda\eta\kappa^2}{1+\kappa^2\lambda^2} > 0 .$$

Therefore, the vorticity of the inertialess steady flow is elliptic and $b_{11}>0$ and the quasilinear system is well-posed.

Exercise 6.5.
Change variables to r and s, using (54). This leads to

$$2[Ar_xs_x + B(s_xr_y + r_xs_y) + Cr_ys_y]\, \xi_{rs} + \hat{\ell} = 0 . \tag{55}$$

On r = const $y_x = -r_x/r_y$. On s = const $y_x = -s_x/s_y$. Hence for both r and s we have

$$Ay_x^2 - 2By_x + C = 0 .$$

There are two roots, y_x^+ for r=const, y_x^- for s=const and

$$y_x^+ y_x^- = C/A\ ,$$

$$y_x^+ + y_x^- = 2B/A\ . \tag{56}$$

Since $r_x = -r_y y_x^+$, $s_x = -s_y y_x^-$, we may write (55) as

$$2r_y s_y[A y_x^+ y_x^- + B(-y_x^- - y_x^+) + C]\ \xi_{rs} + \hat{\ell} = 0\ .$$

After simplifying, using (56), we prove (53).

CHAPTER 7

Supercritical Flow Past Bodies

We are now going to study the problem of uniform flow past a body; this is the viscoelastic analogue of the exterior problem of aerodynamics. Two typical problems are uniform flow past a body or steady motion of a body in a fluid at rest. These two problems are equivalent because of Galilean invariance.

7.1 Linearization around uniform flow

A stationary body is held fixed in an infinite sea of fluid which moves with a constant velocity

$$\mathbf{u} = \mathbf{e}_x U$$

at points far away from the body. The stress vanishes on uniform motion, more generally in rigid body motions (see (19) of Appendix C)

$$\mathbf{C}_t(\tau) = \mathbf{F}_t^T(\tau)\mathbf{F}_t(\tau) = \mathbf{Q}^T(\tau)\mathbf{Q}(\tau) = \mathbf{1}$$

so that the strain for rigid body motions is identical to that for bodies at rest. Since the extra stress vanishes in a body at rest, it must also vanish for a body in rigid motion. The stress in an isotropic viscoelastic liquid in motion, which perturbs rigid motion is

$$\tau_{\ell m} = \int_{-\infty}^{t} G(t-\tau) A_{\ell m}[\mathbf{u}(\boldsymbol{\chi},\tau)]\, d\tau \tag{1}$$

where

$$A_{\ell m}[\mathbf{u}(\boldsymbol{\chi},\tau)] \stackrel{\text{def}}{=} \frac{\partial u_\ell(\boldsymbol{\chi},\tau)}{\partial \chi_m} + \frac{\partial u_m(\boldsymbol{\chi},\tau)}{\partial \chi_\ell}$$

and $\boldsymbol{\chi}(\mathbf{x},\tau) = \mathbf{Q}(\tau)\mathbf{x}+\mathbf{C}(\tau)$ is the pathline of a particle presently at $\mathbf{x}$ following the rigid motion. We might use (1) to describe the flow of a body rotating about a fixed stationary point.

It is useful in the analysis of hyperbolicity to express (1) as a first-order quasilinear rate equation with a principal part like Maxwell models. To this end, we differentiate (1) with respect to t, holding $\mathbf{x}$ fixed

$$\frac{\partial \tau_{\ell m}}{\partial t} = G(0) A_{\ell m}[\mathbf{u}(\mathbf{x},t)] + \int_{-\infty}^{t} G'(t-\tau)\left[\frac{\partial u_\ell}{\partial \chi_m} + \frac{\partial u_m}{\partial \chi_\ell}\right] d\tau$$

$$+ \int_{-\infty}^{t} G(t-\tau)\frac{\partial}{\partial t}\left[\frac{\partial u_\ell}{\partial \chi_m} + \frac{\partial u_m}{\partial \chi_\ell}\right] d\tau\,. \tag{2}$$

This model is instantaneously elastic if the rigidity G(0) is finite. If $G(s) = \frac{\eta}{\lambda} e^{-s/\lambda}$, $s = t-\tau$, then the second term on the right is $-\tau/\lambda$ and the model reduces to a Maxwell model, as we shall show in the course of the reduction of the last term in (2).

$$\frac{\partial}{\partial t}\left(\frac{\partial u_\ell(\boldsymbol{\chi},\tau)}{\partial \chi_m}\right) = \frac{\partial \boldsymbol{\chi}}{\partial t}\frac{\partial}{\partial \boldsymbol{\chi}}\left(\frac{\partial u_\ell}{\partial \chi_m}\right) = -(\mathbf{U}\cdot\nabla)\frac{\partial u_\ell}{\partial \chi_m} \tag{3}$$

where we have $d\boldsymbol{\chi}/dt=0$ (see (1.61)) and $\boldsymbol{\chi}$ is the pathline for rigid motion with velocity

$$\mathbf{U} = \dot{\mathbf{Q}}(t)\mathbf{x} + \mathbf{U}_0\,, \quad \mathbf{U}_0 = \dot{\mathbf{C}}\,. \tag{4}$$

It follows now from (3) and (1) that the last term of (2) is

$$-(\mathbf{U}\cdot\nabla)\tau_{\ell m} \tag{5}$$

and (2) may be written as

$$\frac{\partial \tau_{\ell m}}{\partial t} + (\mathbf{U}\cdot\nabla)\tau_{\ell m} = G(0)A_{\ell m}[\mathbf{u}(\mathbf{x},t)] + \int_{-\infty}^{t} G'(t-\tau)\left(\frac{\partial u_\ell}{\partial \chi_m} + \frac{\partial u_m}{\partial \chi_\ell}\right) d\tau\,. \tag{6}$$

This reduces to Maxwell's model when G is an exponential.

7.2 Waves of vorticity

We append

$$\operatorname{div} \mathbf{u} = 0 \tag{7}$$

and

$$\rho\left\{\frac{\partial \mathbf{u}}{\partial t} + (\mathbf{U}\cdot\nabla)\mathbf{u}\right\} = -\nabla p + \operatorname{div} \boldsymbol{\tau} \tag{8}$$

to (6). In the analysis of characteristic surfaces for this system, we note first that the last term of (6), the integral, is of lower order, equal to $-\boldsymbol{\tau}/\lambda$ if G(s) is an exponential. Simple jumps of derivative under the integral vanish almost everywhere according to the argument given in (§1.20). Roughly speaking, we can say that the integral is of zero order, differentiation with respect to $\mathbf{x}$, followed by integration on τ.

Having identified lower order terms, we can proceed with analysis of characteristic surfaces. It is easier to proceed directly to the vorticity equation. Let **IN** stand for the integral on the right of (6). Then, defining

$$\frac{d}{dt} = \frac{\partial}{\partial t} + (\mathbf{U}\cdot\nabla)\,, \tag{9}$$

we have

$$\text{curl div}\left(\frac{d\boldsymbol{\tau}}{dt}\right) = \frac{d}{dt}\,\text{curl div}\,\boldsymbol{\tau} + \mathbf{S}\,, \tag{10}$$

$$\frac{d}{dt}\,\text{curl div}\,\boldsymbol{\tau} - G\nabla^2\zeta = \text{curl div}\,\mathbf{IN} - \mathbf{S} \stackrel{\text{def}}{=} \mathbf{lot}_1\,, \tag{11}$$

$$\text{curl div}\,\boldsymbol{\tau} - \rho\frac{d\zeta}{dt} = \mathbf{lot}_2 \tag{12}$$

where the ith component of $\mathbf{S}$ is $\varepsilon_{im\ell}(\dot{Q}_{pj}\partial_m\partial_p\tau_{\ell j}+\dot{Q}_{pm}\partial_p\partial_j\tau_{\ell j})$ and of $\mathbf{lot}_2$ is $\rho\varepsilon_{ijk}\,\dot{Q}_{\ell j}\partial_\ell u_k$. Hence,

$$\rho\,\frac{d^2\zeta}{dt^2} - G(0)\nabla^2\zeta = \mathbf{lot}_3\,. \tag{13}$$

From equation (13), we may deduce that *the speed of propagation of vorticity waves relative to a particle in rigid motion is given by*

$$c = \sqrt{G(0)/\rho}\,. \tag{14}$$

The important result just stated holds for a very general class of fluids in motions which perturb rigid ones. Of course, the same result holds for Maxwell models with $G(0) = \eta/\lambda$.

7.3 "Mach" cones

It is perhaps not unreasonable to think that far from the body we have only a small perturbation of uniform motion with constant velocity $U\mathbf{e}_x$ in the direction x. For this flow, $\dot{\mathbf{Q}}=0$ and the pathlines are given by

$$\boldsymbol{\chi} = \left|\begin{array}{l} x-U(t-\tau) \\ y \\ z \end{array}\right. \quad . \tag{15}$$

There is now no difference between derivatives with respect to **x** or $\boldsymbol{\chi}$, $d\boldsymbol{\chi}$=dx. Equation (6) reduces to

$$\frac{\partial \boldsymbol{\tau}}{\partial t} + U \frac{\partial \boldsymbol{\tau}}{\partial x} = G(0)\ \mathbf{A}[\mathbf{u}(\mathbf{x},t)] + \int_{-\infty}^{t} G'(t-\tau)\mathbf{A}[\mathbf{u}(\boldsymbol{\chi},\tau)]\ d\tau\ . \quad (16)$$

The vorticity equation (13) becomes

$$\rho\left(\frac{\partial}{\partial t} + U \frac{\partial}{\partial x}\right)^2 \boldsymbol{\zeta} - G(0)\nabla^2\boldsymbol{\zeta}$$

$$= \text{curl div} \int_{-\infty}^{t} G'(t-\tau)\mathbf{A}[\mathbf{u}(\boldsymbol{\chi},\tau)]\ d\tau = \int_{-\infty}^{t} G'(t-\tau)\nabla^2\boldsymbol{\zeta}(\boldsymbol{\chi},\tau)\ d\tau\ . \quad (17)$$

For steady flow, these equations reduce to

$$U \frac{\partial \boldsymbol{\tau}}{\partial x} = G(0)\mathbf{A}[\mathbf{u}(\mathbf{x})] + \int_{-\infty}^{t} G'(t-\tau)\mathbf{A}[\mathbf{u}(\boldsymbol{\chi})]\ d\tau \quad (18)$$

and

$$\rho U^2 \frac{\partial^2 \boldsymbol{\zeta}}{\partial x^2} - G(0)\nabla^2\boldsymbol{\zeta} = \int_{-\infty}^{t} G'(t-\tau)\nabla^2\boldsymbol{\zeta}(\boldsymbol{\chi})\ d\tau \quad (19)$$

and (1) reduces

$$\boldsymbol{\tau} = \int_{-\infty}^{t} G(t-\tau)\mathbf{A}[\mathbf{u}(\boldsymbol{\chi})]\ d\tau\ . \quad (20)$$

The integrals do not collapse because $\boldsymbol{\chi}$ given by (15) depends on τ.

We can write (19) in a form familiar to students of gas dynamics,

$$(M^2-1)\frac{\partial^2 \boldsymbol{\zeta}}{\partial x^2} - \left(\frac{\partial^2}{\partial y^2} + \frac{\partial^2}{\partial z^2}\right)\boldsymbol{\zeta} = \mathbf{lot}\ . \quad (21)$$

The vorticity satisfies a hyperbolic equation when

$$M = \frac{U}{c} > 1\ , \quad c^2 = G(0)/\rho\ , \quad (22)$$

elliptic when M<1. Steady flows which perturb uniform flow do not have transcritical regions of vorticity. Either the whole field is elliptic (M<1) or hyperbolic (M>1).

The characteristic surfaces $\phi(x,y,z)=0$ for the wave equation on the left of (21) when M>1 are the family of cones

$$\frac{1}{M^2-1}(x-x_0)^2-(y-y_0)^2-(z-z_0)^2=0\,. \tag{23}$$

The apex of the cone is arbitrary, and the axis of the cone is parallel to the flow direction (the x-axis) with the cone semi-angle θ given

$$\tan\theta=(M^2-1)^{-1/2}\,. \tag{24}$$

We can borrow the name "Mach cone" from gas dynamics to describe (23), and it is strictly accurate in the same way to regard θ as the vertex angle of this cone.

7.4 Change of variables

I once asked Julian Cole how he arrived at the change of variables which allowed him to integrate the initial value problem for Burger's equation. He thought for a while and then replied, "Harry Bateman was my teacher, and I learned from him that no variable was good enough."

It does not seem natural to have the time variable τ in the equations for steady flow, so we shall change variables in the integrals

$$\chi=x-U(t-\tau)\,,\quad d\chi=Ud\tau\,. \tag{25}$$

We then have

$$\boldsymbol{\tau}=\frac{1}{U}\int_{-\infty}^{x}G\left[\frac{x-\chi}{U}\right]A[\mathbf{u}(\chi)]\,d\chi\,, \tag{26}$$

$$A_{ij}=\frac{\partial u_i(\chi)}{\partial\chi_j}+\frac{\partial u_j(\chi)}{\partial\chi_i}\,,$$

$$\rho U^2 \frac{\partial \zeta}{\partial x} = \int_{-\infty}^{x} G\left[\frac{x-\chi}{U}\right] \left\{\frac{\partial^2}{\partial \chi^2} + \nabla_2^2\right\} \zeta(\chi)\, d\chi\,, \tag{27}$$

$$\nabla_2^2 = \frac{\partial^2}{\partial y^2} + \frac{\partial^2}{\partial z^2}\,, \tag{28}$$

$$U \frac{\partial \boldsymbol{\tau}}{\partial x} = G(0)\, \mathbf{A}[\mathbf{u}(\mathbf{x})] + \frac{1}{U} \int_{-\infty}^{x} G'\left(\frac{x-\chi}{U}\right) \mathbf{A}[\mathbf{u}(\chi)]\, d\chi\,, \tag{29}$$

$$\rho U^2 \frac{\partial^2 \boldsymbol{\zeta}}{\partial x^2} - G(0)\nabla^2 \boldsymbol{\zeta} = \frac{1}{U} \int_{-\infty}^{x} G'\left(\frac{x-\chi}{U}\right)\left[\frac{\partial^2}{\partial x^2} + \nabla_2^2\right] \boldsymbol{\zeta}(\boldsymbol{\chi})\, d\chi\,. \tag{30}$$

7.5 Maxwell models

For Maxwell models $G' = -\lambda^{-1}G$ and

$$G\left[\frac{x-\chi}{U}\right] = \frac{\eta}{\lambda} \exp\left[\frac{-(x-\chi)}{\lambda U}\right]. \tag{31}$$

Using (26), (29) and (31) we get

$$U \frac{\partial \boldsymbol{\tau}}{\partial x} + \frac{\boldsymbol{\tau}}{\lambda} = \frac{\eta}{\lambda} \mathbf{A}[\mathbf{u}(\mathbf{x})]\,. \tag{32}$$

Using (27), (30) and (31) we get

$$\rho U^2 \frac{\partial^2 \boldsymbol{\zeta}}{\partial x^2} - \frac{\eta}{\lambda} \nabla^2 \zeta + \frac{\rho U}{\lambda} \frac{\partial \boldsymbol{\zeta}}{\partial x} = 0\,. \tag{33}$$

7.6 Dimensionless parameters

Let d be the scale of length, U the velocity scale, η/λ the scale of stress. Then (32) may be written in dimensionless variables as

$$\frac{\partial \boldsymbol{\tau}}{\partial x} + \frac{\boldsymbol{\tau}}{W} = \mathbf{A}[\mathbf{u}(\mathbf{x})] \tag{34}$$

and (33) becomes

$$\frac{\partial^2 \zeta}{\partial x^2}\left(1 - \frac{1}{M^2}\right) + \frac{1}{W}\frac{\partial \zeta}{\partial x} = \frac{1}{M^2}\nabla_2^2 \zeta \tag{35}$$

where

$$M = U/c \tag{36}$$

and

$$W = U\lambda/d \tag{37}$$

is the Weissenberg number.

The dynamics of linearized Maxwell models is determined by two dimensionless parameters expressed as M and W in (36) and (37). Actually there are four dimensionless parameters of which only two are independent

$$M, W, R, E$$

where

$$R = Ud\rho/\eta \tag{38}$$

is the Reynolds number and

$$E = \eta\lambda/\rho d^2 \tag{39}$$

is the elasticity number. The relations among these numbers are

$$RW = M^2, \quad W = RE, \quad R = M/\sqrt{E}, \quad W = M\sqrt{E}. \tag{40}$$

The choice of dimensionless parameters is somewhat arbitrary. However we may be assured that M is one important parameters in studies of change of type. When $M \ll 1$ it is usual to see (R,W) as the canonical basis. In this case (35) is a parabolic equation and it may be written as

$$R\frac{\partial \zeta}{\partial x} = (1-RW)\frac{\partial^2 \zeta}{\partial x^2} + \nabla_2^2 \zeta\ . \tag{41}$$

Certainly R is a fine parameter when $RW \ll 1$ because of all the intuition we have developed in studies of fluid mechanics of Newtonian fluids. The parameter $W=U\lambda/d$ is a measure of elasticity; we think of a fluid with large λ as elastic because it relaxes slowly. We may regard W as the ratio of times

$$W = \lambda(d/U) = \frac{\text{relaxation time}}{\text{process time}}\ . \tag{42}$$

In general, viscoelastic fluids do not have one time of relaxation, small molecules and polymer chains relax at different rates and the choice of time scale is more delicate. In Chapter 18 we shall treat the kind of physics in which an internal clock defined by one of many relaxation times is useful.

For studies of hyperbolicity and change of type it may be most useful to use (M,E) as the canonical basis. We have (34) in the form

$$\frac{\partial \boldsymbol{\tau}}{\partial x} + \frac{\boldsymbol{\tau}}{M\sqrt{E}} = \mathbf{A}[\mathbf{u}(\mathbf{x})] \tag{43}$$

and (35) in the form

$$\frac{\partial^2 \zeta}{\partial x^2} + \frac{M}{(M^2-1)\sqrt{E}}\frac{\partial \zeta}{\partial x} = \frac{1}{M^2-1}\nabla_2^2 \zeta\ . \tag{44}$$

For strongly supercritical flow, $M \gg 1$, (44) reduces to

$$\frac{\partial^2 \zeta}{\partial x^2} + \frac{1}{M\sqrt{E}}\frac{\partial \zeta}{\partial x} = \frac{1}{M^2}\nabla_2^2 \zeta\ . \tag{45}$$

when E is fixed and $M\rightarrow\infty$, the decay in the stress (43) and telegraph equation (45) tend to zero and $\frac{\partial \boldsymbol{\tau}}{\partial x} \sim \mathbf{A}[\mathbf{u}(\mathbf{x})]$ is a completely elastic response. Moreover, (45) makes sense now only if the variation of vorticity transverse to the flow direction are large. Then

$$\frac{\partial^2 \zeta}{\partial x^2} \sim \frac{1}{M^2} \nabla_2^2 \zeta \,. \tag{46}$$

The elasticity number $E=W^2/M^2=W/R$ is independent of the flow velocity and it has different interpretations. The most common interpretation

$$E = \frac{W}{R} = \frac{\text{elastic forces}}{\text{viscous forces}} \tag{47}$$

is perhaps useful for strongly subcritical $M \ll 1$ flows. For studies of change of type it is better to write

$$E = \frac{(\upsilon^2/d^2)}{(\upsilon/\lambda)} = \frac{(\upsilon/d)^2}{c^2}$$

where $\upsilon = \eta/\rho$ is the kinematic viscosity and

$$\sqrt{E} = \frac{\upsilon/d}{c} = \frac{\text{diffusion velocity}}{\text{wave velocity}} \,. \tag{48}$$

Another way to express E is

$$E = \frac{\upsilon}{\lambda} \frac{\lambda^2}{d^2} = \frac{c^2}{(d/\lambda)^2} \,.$$

If c is fixed, E is larger when λ is larger. For fixed λ, E is larger if c is. The fluids which have slow wave speeds for fixed υ are more elastic.

In steady flow it is perhaps better to speak of length scales, rather than velocities. The scale length d can be interesting. We write (48) as

$$\sqrt{E} = \frac{\lambda(\upsilon/d)}{\lambda c} = \frac{\text{length influenced by diffusion over a time } \lambda}{\text{length travelled by a wave in time } \lambda} \,.$$

This interpretation of the elasticity parameter may first appear to be strange, but it appears to explain well otherwise mysterious results which will appear in various examples in the sequel. In the problems which change type regions in which the vorticity is elliptic when $M>1$ is fixed grow with E. In the smaller hyperbolic regions the dynamics of the vorticity become more and more elastic. At the same M>1 we see elliptic regions decrease in size with increasing E and at the same time elastic behavior in hyperbolic regions is highly damped.

Astarita and Marrucci [1974] argue that the elasticity parameter E arises when inertia and pressure forces in a boundary layer are balanced by normal stresses. They write the boundary layer equation of motion

$$\rho\left[u\frac{\partial u}{\partial x}+v\frac{\partial u}{\partial y}-U\frac{dU}{dx}\right]=\frac{\partial \tau}{\partial y}-\frac{\partial}{\partial x}(\sigma-\gamma)\ .$$

They look at an "elastic" boundary layer for large W and argue that the first term on the right $\partial\tau/\partial y$ is negligible. The left side of the equation is estimated by $\rho U^2/x$ where x is the distance along the layer and the right side by $\lambda\eta U^2/x\delta^2$ where δ is the boundary layer thickness, a diffusion distance. Equating the left and right they get

$$\rho \sim \frac{\lambda\eta}{\delta^2}\ .$$

This estimate says that E is approximately one when the diffusion distance is used. Normal stresses can be expected to be largest in regions near the body.

D. F. James and O. Gupta [1971] have shown (see Figure 7.5) that for sufficiently large velocities, the drag also becomes independent of the velocity U, it becomes independent of Reynolds number. The critical condition does not seem to have been investigated. They found that the drag data in the regions where drag was independent of Reynolds number did not correlate with E so perfectly as the heat transfer data. They say that "A thorough review of the heat transfer experiments reveals a high degree of agitation in the preparation of the polymer solutions; ... and probably resulted in a negligible high weight tail in the molecular weight distribution for each grade." The polymer degradation of course makes it difficult to get a good correspondence between the critical values measured in the heat transfer experiments and the wave speeds measured on fresh solutions. We may have here a stiuation in which more than one relaxation time is required.

7.7 Flow around stationary bodies

There are many interesting and unexplained phenomena which occur in steady flow of viscoelastic fluids over bodies. Maybe some of these are associated with a change of type. In this chapter, we will try to interpret some of the phenomena using linearized theory. This certainly limits the type of results which can be obtained. The advantage and disadvantage of the linearized theory is that the material description is relatively simple, because the range of responses for which linearization is valid is so small.

In the linearized theory, the flow around a body is entirely subcritical if $U<c$ and entirely supercritical if $U>c$; transcritical flows with $M>1$ in some regions and $M<1$ in others does not occur. The linearization relative to $\mathbf{e}_x U$ must fail near the body where the velocity vanishes. It follows that a more exact statement of the flow around stationary bodies, which does not use linearized theories, would probably give a different result; even if the flow away from the body is supercritical, the flow near the body may always be subcritical, and the underlying problem is really "transcritical".

Perhaps the flow of an elastic fluid over a body is something like the ones we understand in gas dynamics, but the vorticity rather than the velocity is important. One possibility is the flow shown in Figure 7.1.

Vorticity is created at the surface of the body due to the no-slip condition, but this vorticity cannot propagate upstream into the region of uniform flow where the velocity U is larger than the speed c of vorticity waves into rest. In a linear theory, the first changes in the vorticity would occur across the leading characteristic surface which, in the axisymmetric case, forms a cone like the Mach cone of gas dynamics. The undisturbed region in front of the cone is like the region of silence in gas dynamics, the vorticity is zero there and it is an irrotational (rather than silent) region of flow. The linearization of the equations around uniform flow is not valid near the surface of the body on which the velocity must vanish. We would expect that a more exact treatment of the problem, linear or nonlinear, which takes into account both the no-slip condition at the body and the uniform flow at infinity, would give rise to transcritical flow with a subcritical region near the body as shown in Figure 7.1. I further expect that this region would grow, possibly to a limiting size, as the elasticity number increases at a fixed M and, at the same time, the vorticity in the hyperbolic cone behind the body would not decay rapidly, responding elastically at high E.

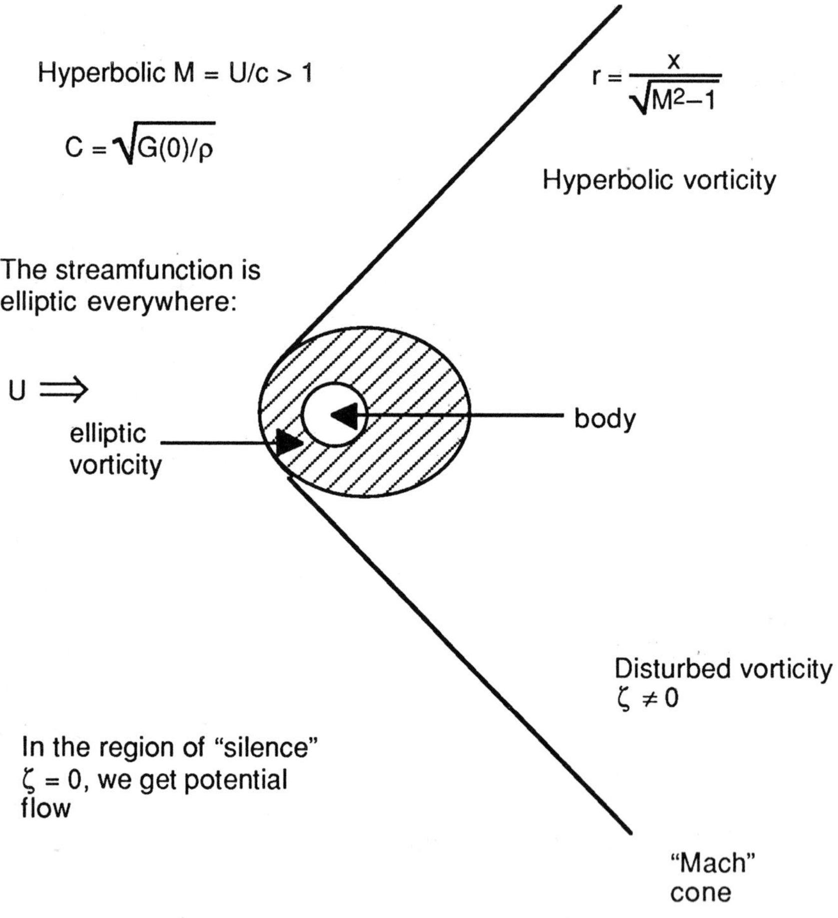

Figure 7.1. Possible distribution of the vorticity in the uniform flow over an axisymmetric body in the supercritical case.

7.8 Critical phenomena in heat and mass transfer from cylindrical waves in cross-flow

D.F. James [1967], in his thesis, and James and Acosta [1970] have reported a critical phenomenon in the laminar flow of dilute aqueous polyox solutions across small wires. There appears to be a critical speed $U=U_c$ for this problem; for $U<U_c$, the Nusselt number increases with U, as in a Newtonian fluid; for $U>U_c$, the Nusselt number is independent of U. The Nusselt number is

$$N_u = hd/k \tag{49}$$

where d is the cylinder diameter, k the thermal conductivity of the liquid and

$$h = Q/(T_0 - T_\infty)$$

is the average heat transfer coefficient, T_0 is an average temperature of the cylinder, T_∞ is the average temperature far from the cylinder and Q is the experimentally measured average heat flux. The drag on these cylinders also appears to become independent of U when $U>U_c$. Some of the original data taken by James [1967] is displayed in Figures 7.2–4. The transition at $U=U_c$ is not sharp, but a U_c may perhaps be identified and it appears to be independent of the diameter d of the cylinder. James and Acosta [1970] presented data giving the Nusselt vs. Reynolds number for three grades of dilute solutions of aqueous Polyox (WSR-301, 201 and coagulant); for each fluid, data is given for three different wires, d = 10^{-2}, 2×10^{-2}, 6×10^{-2} inches, and for each wire, dilute solutions of at least six different concentrations were tested. We replotted all their data in a Nusselt number vs. velocity plane and these replots are shown in Figures 7.2–4. The following observations are relevant here:

1. The critical condition is more nearly a speed than a Reynolds number.

2. The critical speed decreases with concentration for each and every wire and for all three polymers tested.

3. In the region $U>U_c$, where the Nusselt number is independent of U, the Nusselt number seems to correlate with the elasticity

number. The values of E in the experiments vary between 10^{-4} and 1. There appears to be an asymptote for the Nusselt number, slightly greater than one, for large E>1/10.

James and Acosta argued that the local accelerations of the fluid are everywhere small and, on this basis, they put forward the idea that, in this regime, the dilute solutions could be modeled by a Maxwell model with a single time of relaxation. They next wrote (U,d,η,ρ,λ) as the list of dynamical variables, noting that when $U>U_c$, the average Nusselt number should be a function of the remaining four variables. The only dimensionless group that can be formed from these four variables is the elasticity number $E=\eta\lambda/\rho d^2$, and they wrote

$$N_u = f(1/E)$$

for some $f(\bullet)$ when $U>U_c$. This argument could be incorrect in a case where more than one relaxation time is important, as may be the case in drag experiments of James and Gupta [1971].

A very similar set of data, for mass transfer rather than heat transfer, was presented by Ambari, Deslouis, and Tribollet [1984]. Their results are summarized in Figure 7.6. The current I in the Figure is a measure of the mass transfer obtained by an electrochemical technique. They found that

1. There is a large decrease in the mass transfer from the wire which takes place in the neighborhood of some critical value of the velocity. The critical velocity is not sharply defined but is a decreasing function of concentration.

2. As soon as deviation from the Newtonian response, indicated by the straight line in Figure 7.6, is observed, fluctuations of the current I take place and their rms value increases up to the maximum of the average current and then decreases.

Ambari, *et al.* did not run their experiments for different diameter wires, so we do not know if U_c is independent of d. However, the critical values are roughly the same as those observed by James and Acosta [1970] for heat transfer in aqueous coagulant polyox solutions of similar concentration. They say that

According to the data, the critical value of the velocity, which marks the onset of non-Newtonian behaviors is $U_C \cong 0.8$ cm/sec for the smaller concentrations 50 ppm and 100 ppm.

Inspection of Figure 7.6 reveals that the maximum I is found at about U=2.5 cm/sec for 50 ppm and 1.5 cm/sec for 100 ppm.

Figure 7.2. Heat transfer from heated wires in cross-flow of WSR-301 in water for different concentrations: (a) D=0.001 in, (b) D=0.002 in, (c) D=0.006 in. The critical U is a decreasing function of concentration.

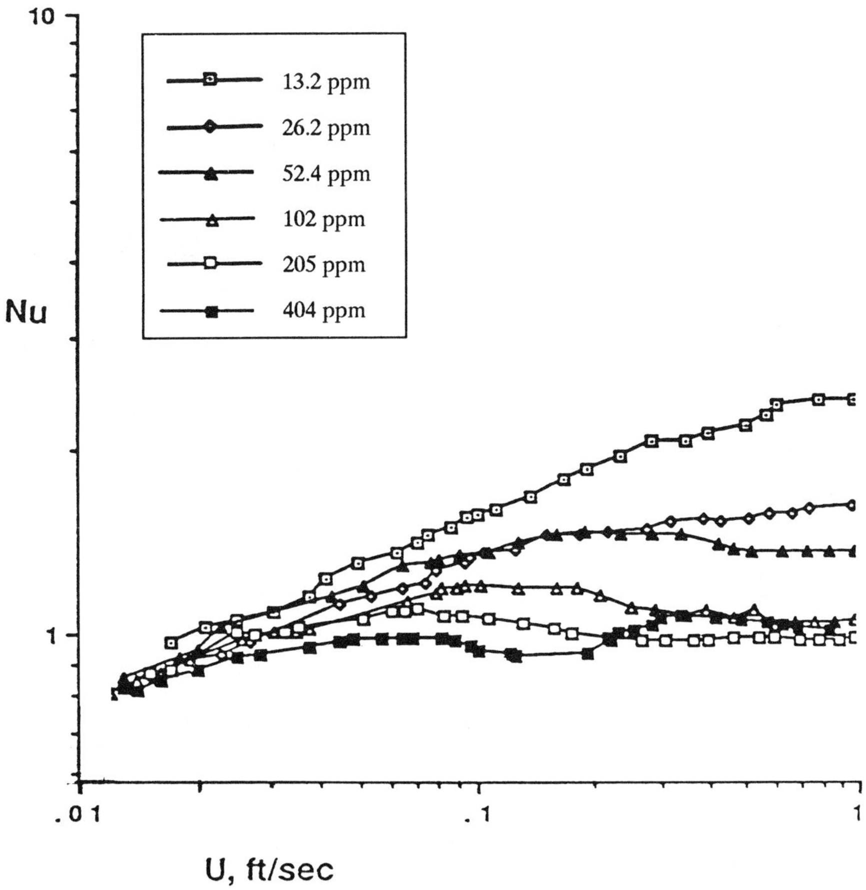

Figure 7.2(a). DIA=0.001 in / Polyox WSR-301.

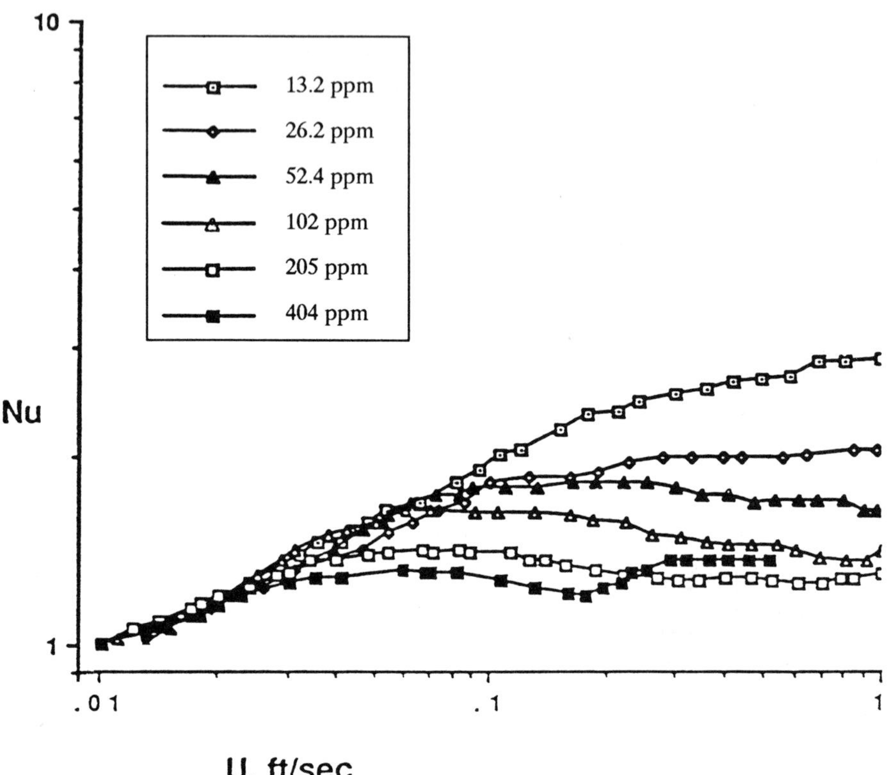

Figure 7.2(b). DIA=0.002 in / Polyox WSR-301.

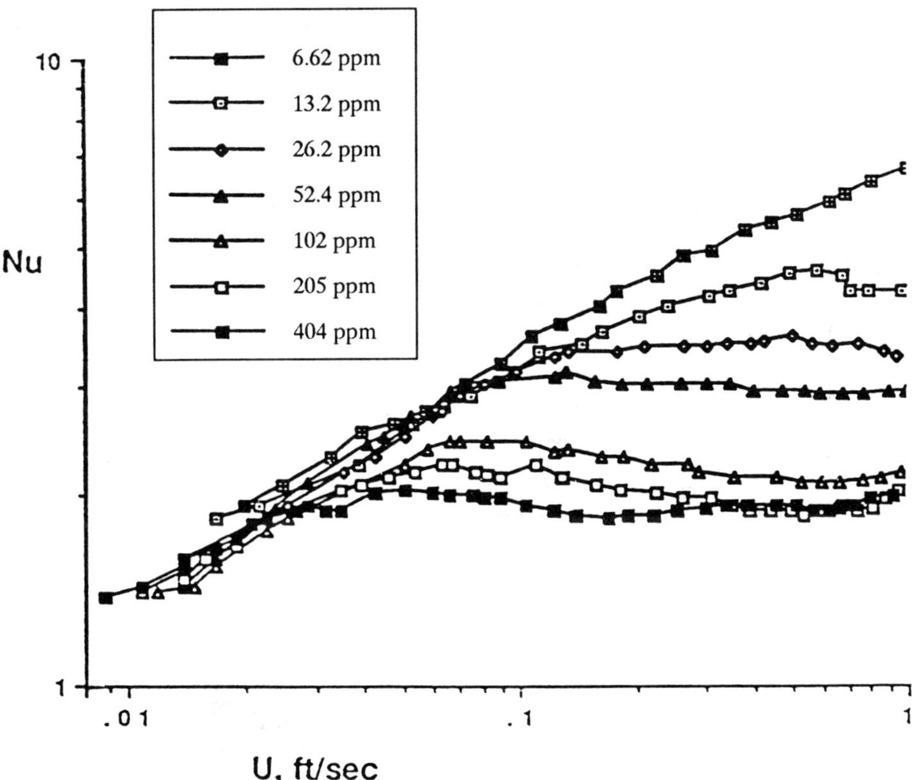

Figure 7.2(c). DIA=0.006 in / Polyox WSR-301.

Figure 7.3. Heat transfer from heated wires in cross-flow of WSR-205 in water for different concentrations: (a) D=0.001 in, (b) D=0.002 in, (c) D=0.006 in. The critical U is a decreasing function of concentration.

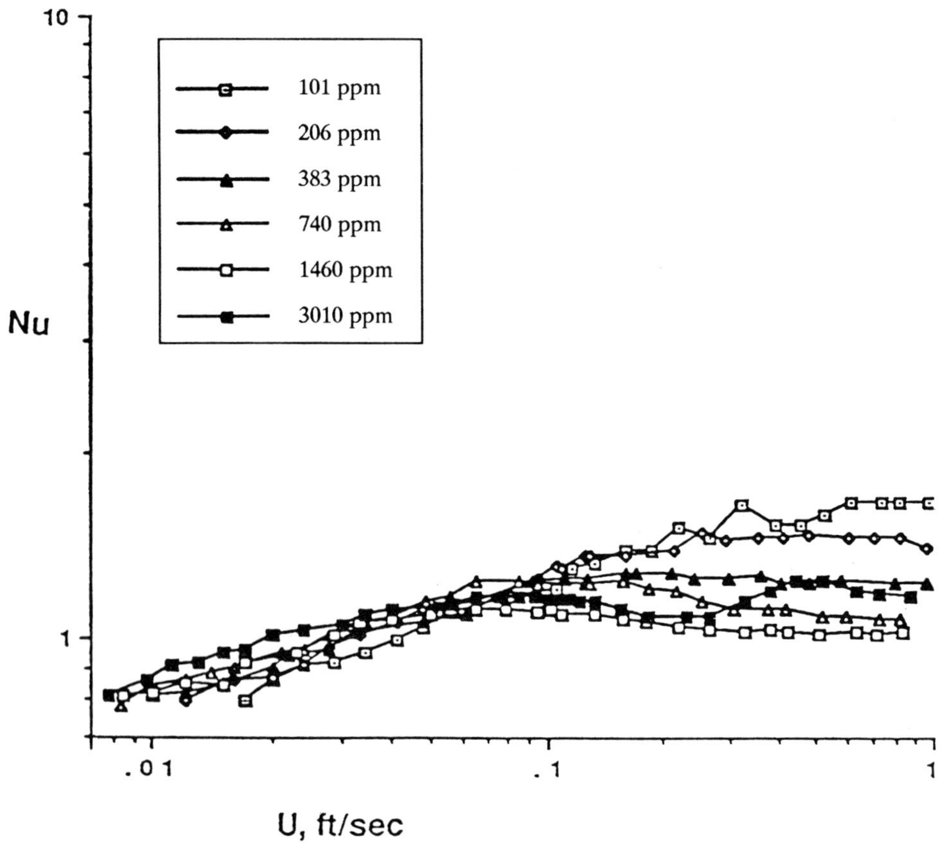

Figure 7.3(a). DIA=0.001 in / Polyox WSR-205.

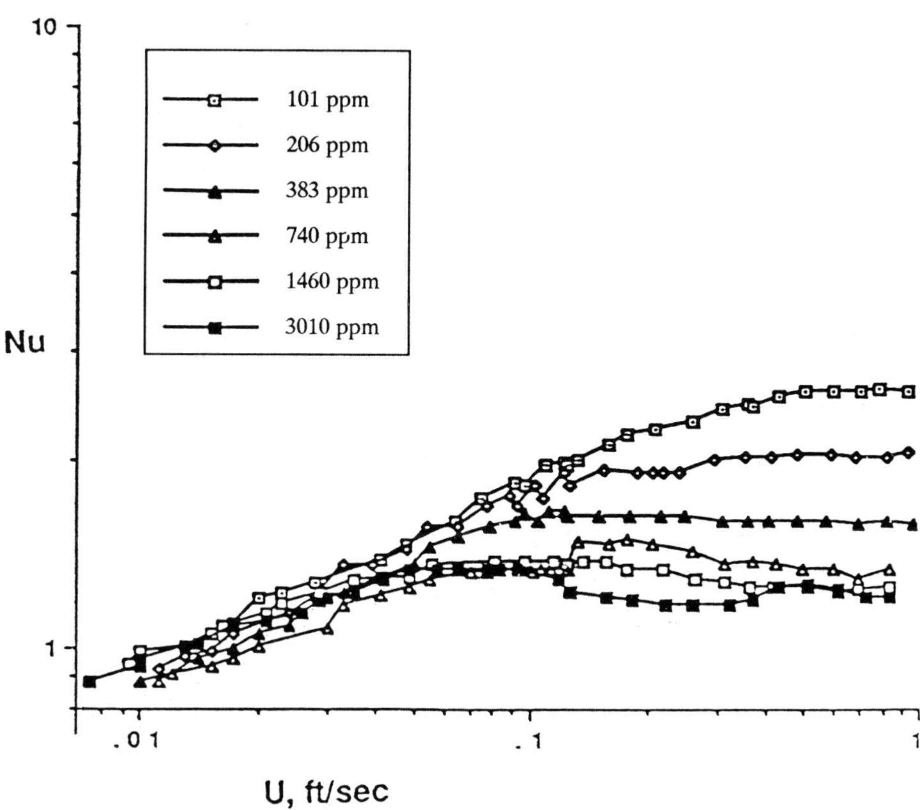

Figure 7.3(b). DIA=0.002 in / Polyox WSR-205.

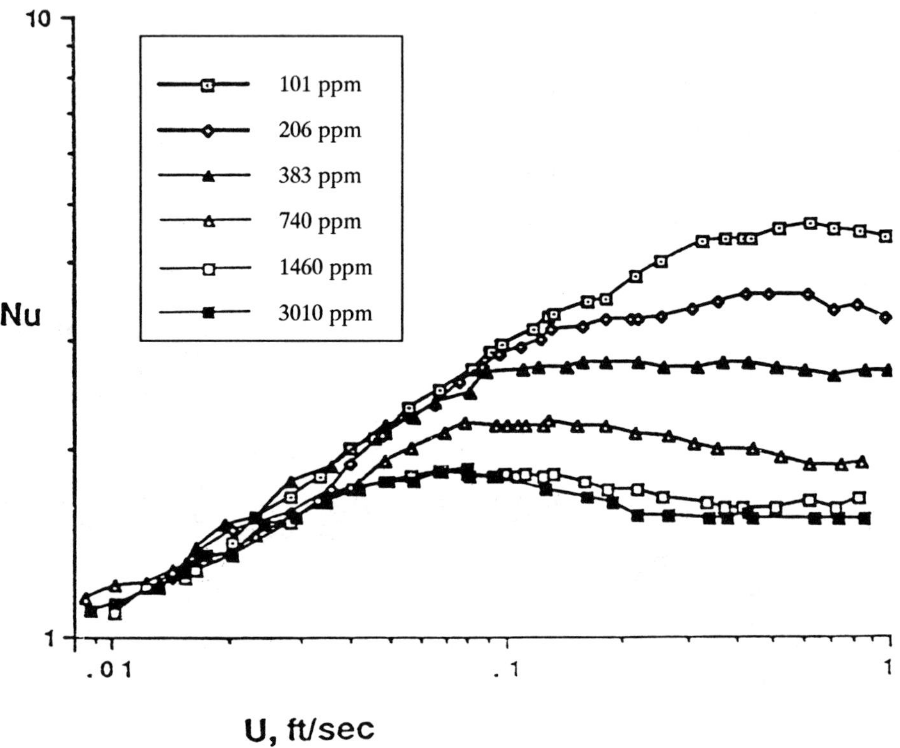

Figure 7.3(c). DIA=0.006 in / Polyox WSR-205.

Figure 7.4. Heat transfer from heated wires in cross-flow of Polyox Coagulant for different concentrations: (a) D=0.001 in, (b) D=0.002 in, (c) D=0.006 in. The critical U is a decreasing function of concentration.

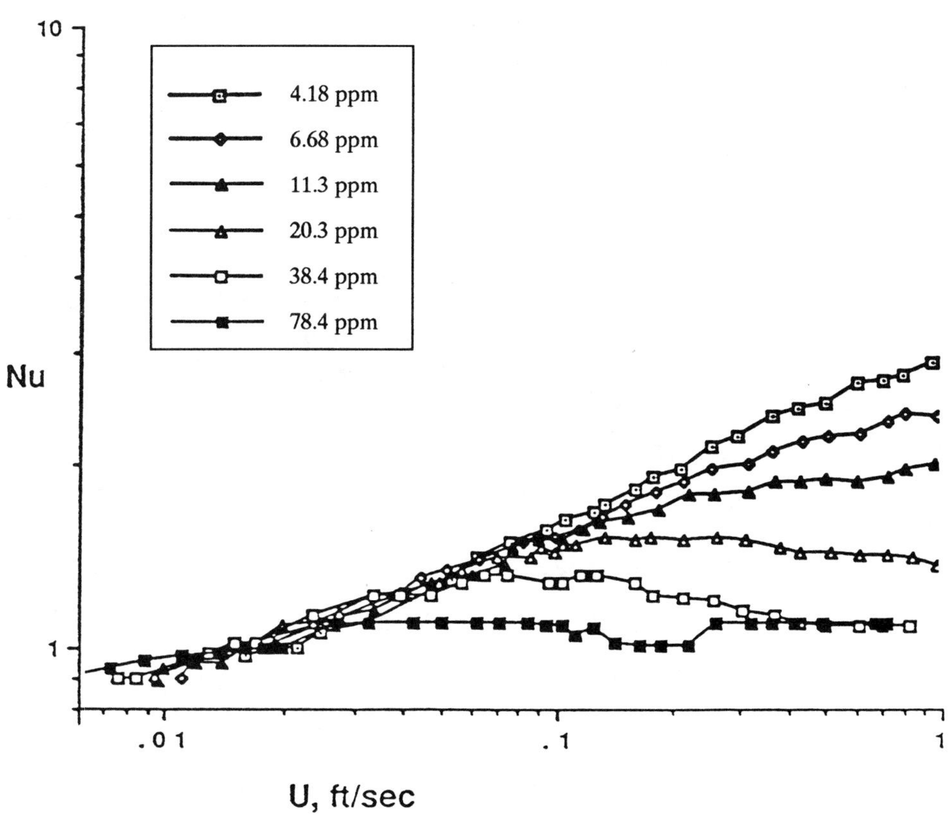

Figure 7.4(a). DIA=0.001 in / Polyox Coagulant.

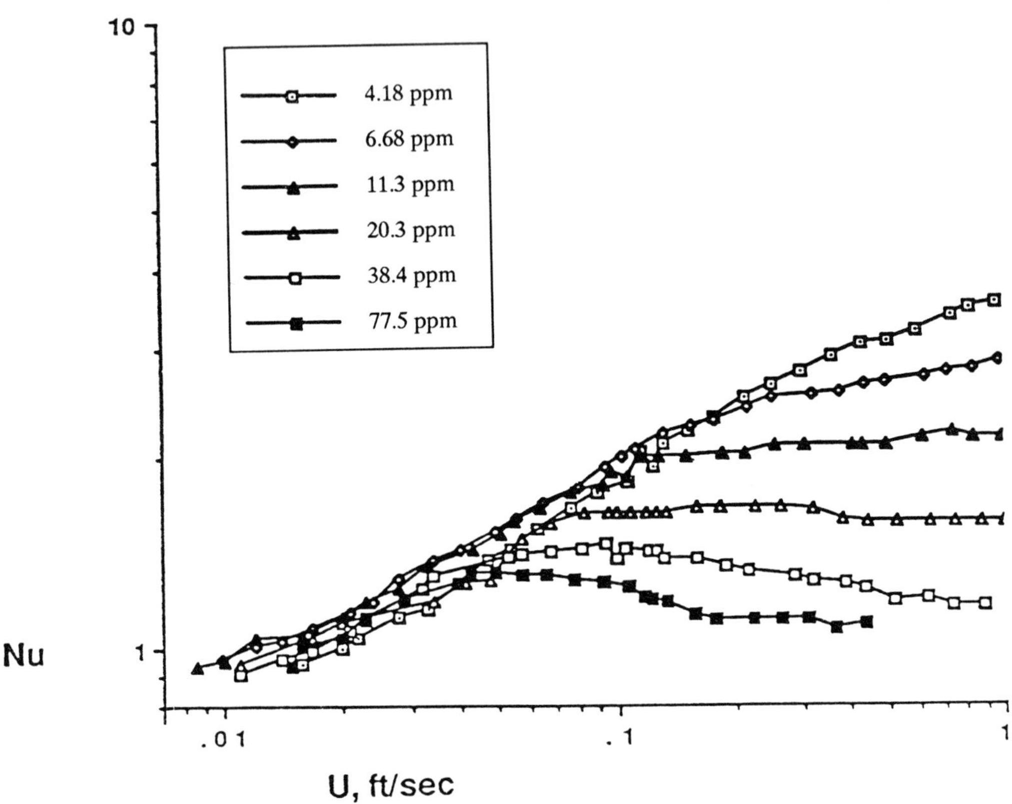

Figure 7.4(b). DIA=0.002 in / Polyox Coagulant.

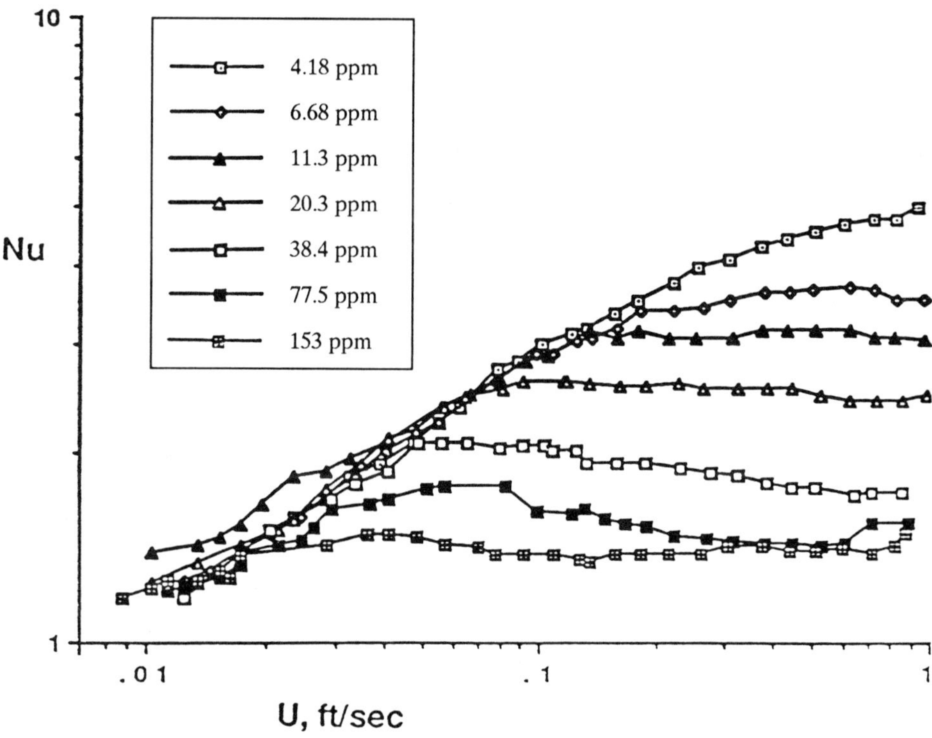

Figure 7.4(c). DIA=0.006 in / Polyox Coagulant.

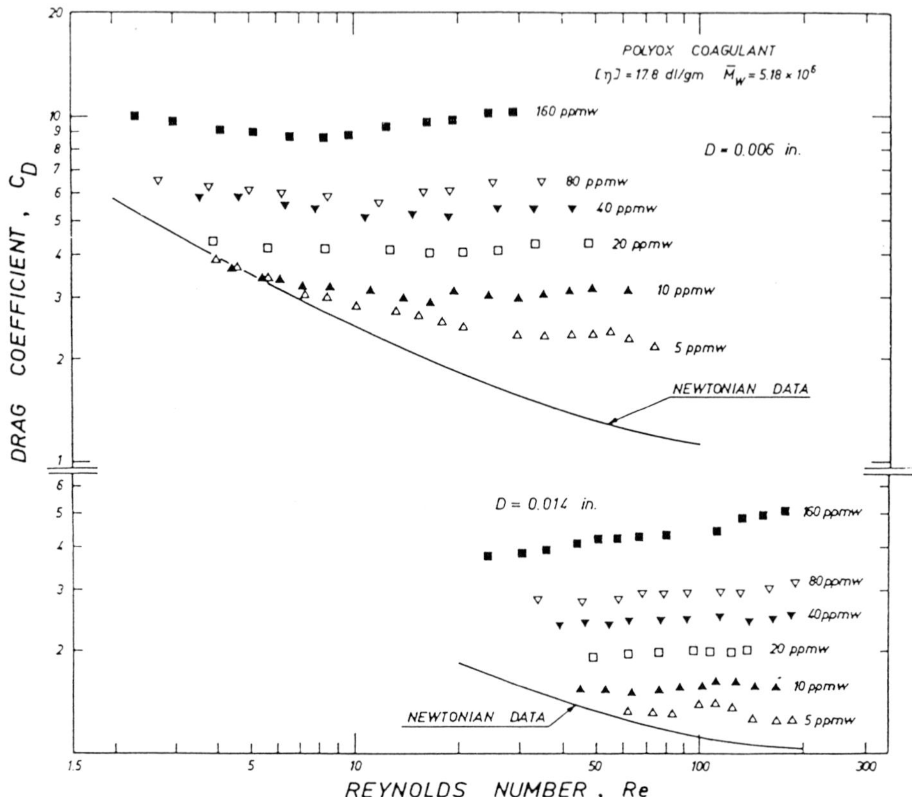

Figure 7.5. Drag measurements of circular cylinders in solutions of polyox coagulant (after James and Gupta, [1971]).

Figure 7.6. Comparison of heat transfer data (a) of James and Acosta [1970] with mass transfer data (b) of Ambari, Deslouis, and Tribollet [1980] for Polyox Coagulant solutions in water over 50 micron wires.

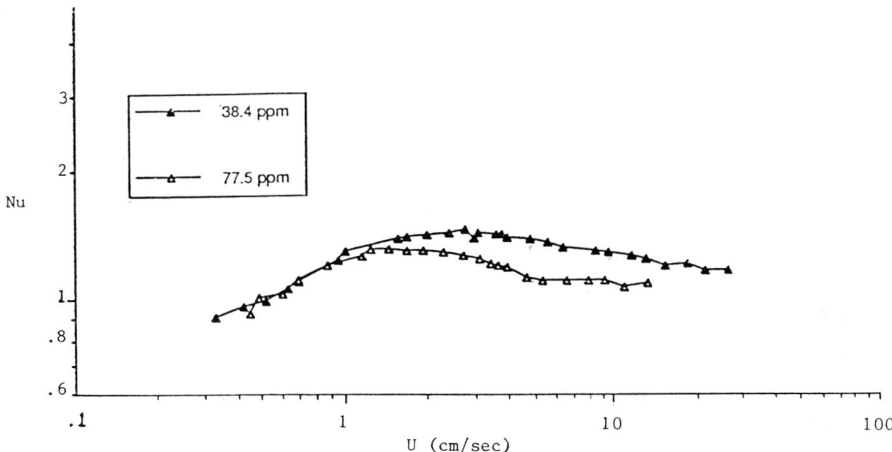

Figure 7.6(a). Heat transfer data.

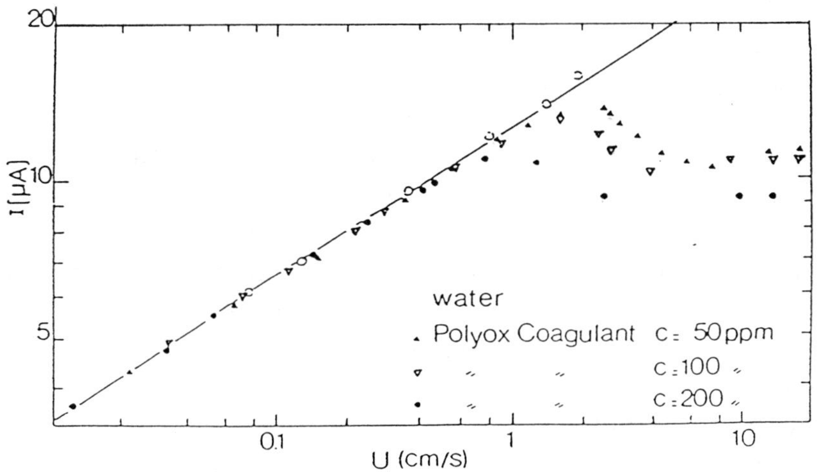

Fig 7.6(b). Mass transfer data.

7.9 The work of Ultman and Denn

Ultman and Denn [1970] tried to explain the critical phenomena reported by James as a change of type, with $U_c=c$, the speed of shear waves. This idea was motivated by the work of Coleman, Gurtin and Herrera [1965] on the propagation of acceleration waves. Ultman and Denn linearized the equations around uniform flow. They wrote down (32) and

$$\rho U \frac{\partial \mathbf{u}}{\partial x} + \nabla p = \operatorname{div} \boldsymbol{\tau} . \tag{50}$$

Eliminating div $\boldsymbol{\tau}$, they find that

$$\left[\frac{\lambda U}{\eta}\frac{\partial}{\partial x} + \frac{1}{\eta}\right]\nabla p + \frac{\rho U}{\eta}\frac{\partial \mathbf{u}}{\partial x} = \left(1 - \frac{U^2}{c^2}\right)\frac{\partial^2 \mathbf{u}}{\partial x^2} + \frac{\partial^2 \mathbf{u}}{\partial y^2} \tag{51}$$

where $c = \sqrt{\eta/\lambda\rho}$.

$$\mathbf{v} = \mathbf{u} + \mathbf{e}_x U$$

vanishes on the body. They note that div $\mathbf{u} = 0$ implies that

$$\frac{\lambda U}{\eta}\frac{\partial p}{\partial x} + \frac{1}{\eta}p$$

is harmonic and write down the equation for the x component of (51) noting that

> If the free stream velocity exceeds the wave velocity, the "supercritical" case, equation (51) becomes hyperbolic and admits discontinuities in the solution. This latter case is formally analogous to the development of standing shock waves in gas dynamics. It is proposed that this is the point at which the heat transfer and drag coefficient occurs.

The theory of Ultman and Denn is based on a discussion of velocity. Actually, the velocity is not governed by a hyperbolic equation, and, the velocity field does not change rapidly in regions where the vorticity changes rapidly. The theory requires that flow on the forward side of the cylinder be uniform up to some point of distance λU in front of stagnation.

The evidence for the interpretation presented by Ultman and Denn is in three parts.

1. The use of the Maxwell models to obtain Equation (51).

2. They use the molecular theories of Rouse and Bueche to get estimates for η and λ. From a theory they attribute to Bueche, they get

$$\lambda_B = 12(\eta-\eta_w)M/\pi^2\phi R_g T\,, \tag{52}$$

η_w is viscosity of water in poise,
M,ϕ is the solute molecular weight and concentration in parts per million,
R_g is the gas constant ,
T is absolute temperature .

They get

$$\eta-\eta_w = 10^{-6}\eta_w\phi(10^3+12\phi) \tag{53}$$

from an empirical fit from the experimental data on viscosity and concentration presented by James. They compare the prediction for the critical Reynolds number

$$R = cd\,\rho/\eta(= 1/\sqrt{E})$$

at the wave velocity $c = \sqrt{\eta/\rho\lambda}$, using $\lambda=\lambda_B$ from (52) and η from (53), and they get reasonable agreement with the experiments. They then note that the best fit can be obtained using

$$\lambda \sim 0.7\lambda_B \tag{54}$$

which lies between theoretical values of Rouse and Bueche. When (52) and (53) are combined we get

$$\lambda_B = \frac{12M\eta_w(10^3+12\phi)}{10^6\pi^2 R_g T}\ .$$

This shows that λ_B is finite when $\phi=0$; it is the time of relaxation for a single polymer in a sea of solvent and is not relevant to the dynamics of dilute solutions.

3. They give a heuristic argument to establish that there is a basis in the energy equation for the independence of the Nusselt number on velocity. The argument is that most of the heat transported is in the neighborhood of the stagnation point. Their result depends on assuming that uniform flow is achieved in a distance

$$x = U\lambda \tag{55}$$

in front of the stagnation point. A similar argument with less details was presented by Metzner and Astarita [1967].

The argument about the energy is as follows. First, the equations are made dimensionless with d, U, T_0 where T_0 is the temperature of the liquid on the wire. The dimensionless velocity is $\mathbf{v}$ and the temperature is θ. Then the energy equation

$$\mathbf{v}\bullet\nabla\theta = RP\nabla^2\theta \tag{56}$$

where $R=Ud/\nu$ and $P=c_p\mu/k$ is the Prandtl number,

$$\begin{aligned} &\mathbf{v} = 0\,, \quad \theta = 1 \text{ on the wire}\,, \\ &\mathbf{v} = \mathbf{e}_x\,, \quad \theta = 0 \text{ at infinity}\,. \end{aligned} \tag{57}$$

The Nusselt number given by $Nu=dh/k$, where $h=Q/T_0$ is the heat transfer coefficient and Q is the average heat flux; hence, in terms of the dimensionless variable,

$$Nu = -\frac{1}{S}\int_S \mathbf{n}\bullet\nabla\theta \; ds \tag{58}$$

where **n** is the outward normal. In the stagnation point approximation, one solves a one-dimensional problem in which x points against the stream, outward at the stagnation point and

$$v(x)\frac{d\theta}{dx} = RP\frac{d^2\theta}{dx^2}\,, \tag{59}$$

$$\begin{aligned} v = 0\,,\quad \theta = 1 \text{ at } x = 0\,,\\ v = -1\,,\quad \theta = 0 \text{ at } x = \infty \end{aligned} \tag{60}$$

and

$$\begin{aligned} Nu &= -\frac{d\theta}{dx}(0)\\ &= \left\{\int_0^{\infty} \exp\left[RP\int_0^{\zeta} v(\xi)\,d\xi\right]d\zeta\right\}^{-1}. \end{aligned} \tag{61}$$

Ultman and Denn assumed that

$$v \sim -x/W = -x/M\sqrt{E}\,. \tag{62}$$

Suppose we satisfy (60) by saying that (62) holds when $x<W = M\sqrt{E}$ and $v=-1$ when $x>W$. Then we evaluate (61) as

$$Nu = \left(\frac{\pi E}{2P}\right)^{1/2} \operatorname{erf}\left[M\frac{P}{2}\right] + \frac{\sqrt{E}}{PM}\exp\left[-PM^2\right]. \tag{63}$$

We get the result given by Ultman and Denn

$$Nu = \left(\frac{\pi E}{2P}\right)^{1/2} \tag{64}$$

for large values of M. The result (64) is completely independent of U. For large values of M>1, we get only a weak dependence of Nu on U in agreement with experiments.

7.10 The experiments of Koniuta, Adler, and Piau [1980]

The velocity field of an aqueous WSR-301 polyox solution across a cylinder of diameter 0.0206 cm using laser-doppler anemometry. The main result was that for all but the smallest speeds there was a stagnant or nearly stagnant region whose size increases with speed. The results of Koniuta., *et al.* are expressed in terms of a Reynolds number; this is equivalent to a speed U_∞ since the diameter d and fluid viscosity are given. A photograph showing the stagnation region is shown in Figure 7.7. Figure 7.8 shows how measured values of velocity vary with distance of the center line in front of the cylinder. Their measurements indicate the existence of a critical velocity, around 3.6 cm/sec, after which the stagnation region begins to develop.

Their results are completely consistent with the idea that the development of the stagnation is a critical event associated with a change of type. The formation for the stand-off distance in the left hand corner of Figure 7.8 begins at a value of velocity close to the value 3.65 cm/sec which we will suggest in the paragraph to follow is an estimate of the wave speed for the 1000 ppm polyox solution. The formation of a nearly stagnant region around the sphere is enough to explain that the heat transfer does not depend on velocity, because nearly all the heat is being transported away from the cylinder by conduction. Thus the results of Koniuta, *et al.*, James and Acosta, Ambari et al all fit together with the hypothesis that the critical events are associated with a change of type.

Now I shall explain the procedure used in computing the data exhibited in the caption of Figure 7.8. Figure 3 in the paper Koniuta, *et al.* corresponds to our 7.8 and they give six Reynolds numbers for flow of aqueous 1000 ppm Polyox WSR-301 across the 0.0206 cm wire. In order to convert this data to a velocity U, we need a kinematic viscosity ν. This is found from plotting data in the thesis of James [1967]. Unfortunately, the highest concentration of WSR-301 was 40 ppm. To interpolate we also used a viscosity for aqueous polyox WSR-301 of 2000 ppm which is listed in the tables of Joseph, Riccius, and Arney [1986]. We plotted this data and used a quadratic interpolation to obtain a kinematic viscosity of 0.0675 cm^2/s.

Figure 7.7 (after Koniuta, Adler, and Piau, 1980). Photograph showing the disturbance to the flow field by a 0.206 mm cylinder in a 500 wppm solution of PEO WSR-301. The fluid motion here is from left to right. R is equal to 17 ($U_\infty \approx 17$ cm/sec). The circular cylinder is not viewed exactly end-on which explains the slight asymmetry of the flow.

To obtain a Mach number M, we interpolated the wavespeed data from Joseph, Riccius, and Arney [1986] and obtained c=3.65 cm/s. Then, we computed the relaxation time λ and elasticity number using the following formulas

$$\lambda = \frac{\nu}{c^2}, \quad E = \frac{\nu^2}{c^2 d^2} .$$

Finally, we computed a dimensionless displacement number using the formula

$$\frac{x}{d} = \frac{\lambda U}{d} .$$

The values of the displacement number is indicated by the vertical bars in the left hand corner of Figure 7.8. This number evidently does not give the stand off distance exactly, but the velocities are relatively small for $x < \lambda U$, as was assumed by Metzner and Astarita [1967] and Ultman and Denn [1970] in their analysis of the heat transport problem.

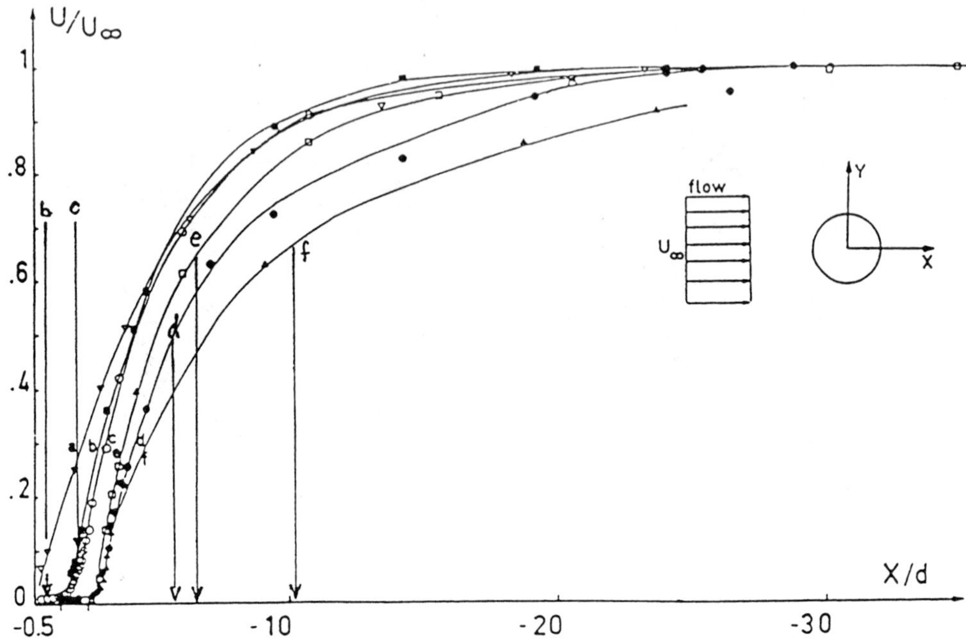

Figure 7.8 (after Koniuta, Adler, and Piau, 1980). Upstream velocity as a function of the distance from the cylinder center. Data are for a 1000 wppm solution. The vertical bars give $\lambda U/d$. The different curves correspond to the different Reynolds numbers: a(0.4), b(1.1), c(2.5), d(7.0), e(7.9), and f(12.7). The free stream velocity (U_∞), the Mach number (M), and the dimensionless displacement ($\lambda U/d$) corresponding to these Reynolds numbers are given in Table 7.1.

	Re	U_∞ (cm/s)	M	$\lambda U/d$
a	0.4	1.3	0.36	0.32
b	1.1	3.6	0.99	0.89
c	2.5	8.2	2.2	2.0
d	7.0	24	6.3	5.6
e	7.9	26	7.1	6.4
f	12.7	41.6	11.4	10.2

Table 7.1. The data in this table was computed for 1000 wppm PEO WSR-301 by the method described in the text. The wave speed is c=3.65 cm/s, the elasticity number is E=0.806, and the relaxation time is λ=0.0051.

7.11 Further remarks about critical phenomena and change of type

Ambari, *et al.* [1984] have suggested that the explanation of the results described in §7.7 can be ascribed to the fast stretching of the polymer coils which occurs when the elongational gradient in the upstream vicinity of the wire, equals the reciprocal of the polymer relaxation time. This is called a "coil-stretch" transition. The explanation of Ultman and Denn [1970] involving wave speeds and supercritical flow was also examined by Ambari *et al.* and they put forward analogies between the two approaches. The oscillations observed by Ambari, *et al.* [1984] in the neighborhood of criticality were explained as an instability associated with sharp gradients of viscosity but perhaps may even be more readily understood as an elastic response associated with change of type.

There are a number of features which support an explanation based on hyperbolicity and change of type of the critical phenomena observed in the experiments on heat and mass transfer from small cylinders in the cross flow of dilute aqueous polyox solutions.

1. There is a critical speed U_c, more or less. The transport of heat or mass with $U>U_c$ is independent of U, more or less.

2. The critical speed U_c seems to be independent of d, more or less. Hence, the critical speed is a material parameter independent of the flow and the wire.

3. The speed of transition in the case considered by Ultman and Denn, can be estimated from Figure 7.2 as about 2.9 cm/sec. This speed is consistent with the estimate of Ultman and Denn [1970] of the wave speed from a molecular theory. However, the molecular theories are perhaps irrelevant for the case of extreme dilution because they give a finite time of relaxation for zero concentration.

4. The maximum value of the transport curve in Figure 7.4 for the 50 ppm Polyox Coagulant occurs roughly at the same value, between 2 and 3 cm/sec.

5. Joseph, Riccius and Arney [1986] measured the wave speed of a 50 ppm WSR-301 solution directly, using a wave speed meter, and found a value c=2.48 cm/sec. We are still uncertain about the values of c in cases of extreme dilution (≤2000 ppm).

6. The data of James and Acosta [1970] and Ambari, *et al.* [1984] show that the wave speed decreases with concentration in the limit of extreme dilution. The viscosity η of of very dilute solutions is approximately the viscosity η of the solvent and it is not sensitive to the change of concentration under conditions of extreme dilution (see (53)). On the other hand, the relaxation time is a strong increasing function of concentration under the same conditions (see (52)). This means that under the conditions in the experiments of James and Acosta [1970] and Ambari *et al.* [1984] the quantity

$$\sqrt{\frac{\eta}{\rho\lambda}} = c$$

is a decreasing function of the concentration ϕ for small ϕ. This behavior is what one observes for the critical speed U_c=c.

7. The computation of Ultman and Denn [1970] giving a U independent expression for the Nusselt number is a suggestive one. It would be of interest to see the results of a more rigorous analysis.

8. The experiments of Koniuta, Adler, and Piau [1980] using laser-doppler anemometry to find the velocity field in the flow of a dilute solution around a circular cylinder are consistent with the explanation of the critical phenomenon as a change of type. The stagnation region around the cylinder, which would lead to the independence of heat transfer (N_u) on the velocity U_∞ appears just when the viscoelastic Mach number exceeds some number in the neighborhood of one.

9. The experiments of James and Gupta [1970] show that the drag also becomes independent of velocity, but critical values of U are not given. The drag data does not correlate as well as the heat transfer data with the elasticity number. This means other factors, not taken into account by Maxwell models are involved.

There are many issues raised by the study of the flow around bodies which need clarification. One question is, "At what degree of polymer concentration is a dilute solution well approximated by constitutive equations with instantaneous elasticity?" This question could be framed as a contest between Maxwell and Jeffrey models. The data of James and Acosta [1970] suggest some sort of effective Newtonian viscosity should be used in Polyox solutions of a few parts per million. The dilute solutions in the drag reduction range; say, 20 ppm already appear to be quite elastic.

More experiments, using other polymers and numerical simulations of the nonlinear problem of flow around a cylinder, should be done.

7.12 Numerical computation of the flow of an upper convected Maxwell model past a cylinder

The material in this chapter has stimulated very interesting calculations which aim for understanding of the nonlinear effects which arise in flow over wires. The results of these calculations will be summarized in this section.

Crochet and Delvaux [1990] report the results of a numerical simulation of flow past a cylinder using an upper convected Maxwell model. They replaced the infinite domain outside the cylinder of radius a with a finite domain of breadth 50a, length 250a, 200a downstream, 50a upstream. A uniform velocity is prescribed over the boundary of the rectangular domain. Their calculation exhibits all the features conjectured at the end of §7.7 to be true of the nonlinear supercritical problem. In addition to these features which were already anticipated in §7.7, they showed that vorticity could become hyperbolic in certain regions even when the Mach number is slightly less than one.

This is seen clearly in Figure 7.9. At M=0.95, a hyperbolic region has developed between the cylinder and the wall, whereas when M=1.5 and 1.73 the cylinder is surrounded by an elliptic region as in Figure 7.1. The existence of a vorticity shock asymptotically tangent to the line $y=x(M^2-1)^{-1/2}$ is shown in Figure 7.10. This Figure exhibits features anticipated in Figure 7.1. One of the most striking features is shown in Figure 7.11 where it is shown how the vorticity and forward velocity u vary with y on the line x=0 through the forward stagnation point on the cylinder. The major difference between the solution at M=0.95 on the one hand and M=1.5 and 1.73 on the other is that there is an essentially discontinuous slope at the position of the vorticity shock when M=1.5 and M=1.73, but not when M=0.95. The problem has also been studied for the same values of M at a higher value of E=W/R now four times as high. Figure 7.12 shows that the extent of the downstream elliptic region is now definitely larger. Again Figure 7.13 demonstrates an abrupt change of vorticity.

Delvaux and Crochet [1990] have also treated the heat transfer problem numerically. Their results are summarized in Figures 7.14 and 7.15 giving the Nusselt number as a function of Reynolds number for different Prandtl numbers and E=0 and 2 and the drag coefficient as a function of Reynolds number for three different values of E, respectively. These graphs can be compared with results of experiments by James and Acosta [1967] and James and Gupta [1971]. The Prandtl number for water is about 7. The elasticity number in the experiments varies from 10^{-4} to about 5. In the experiments the drag coefficient increases as the wire diameter is decreased at a fixed ϕ. The Nusselt number increases as the wire diameter is increased at a fixed ϕ. This agrees with the computation which shows that C_D increases with E (Figure 7.15) and Nu decreases with E (Figure 7.14). If we assume that $E=\nu\lambda/d^2$ is an increasing function of concentration ϕ at a fixed wire diameter d, then the experiments and computation agree with respect to ordering with ϕ.

The numerical simulation shows clearly that M=1 is the center of the neighborhood where the transition from Newtonian to viscoelastic response occurs in the flow of dilute solutions across waves.

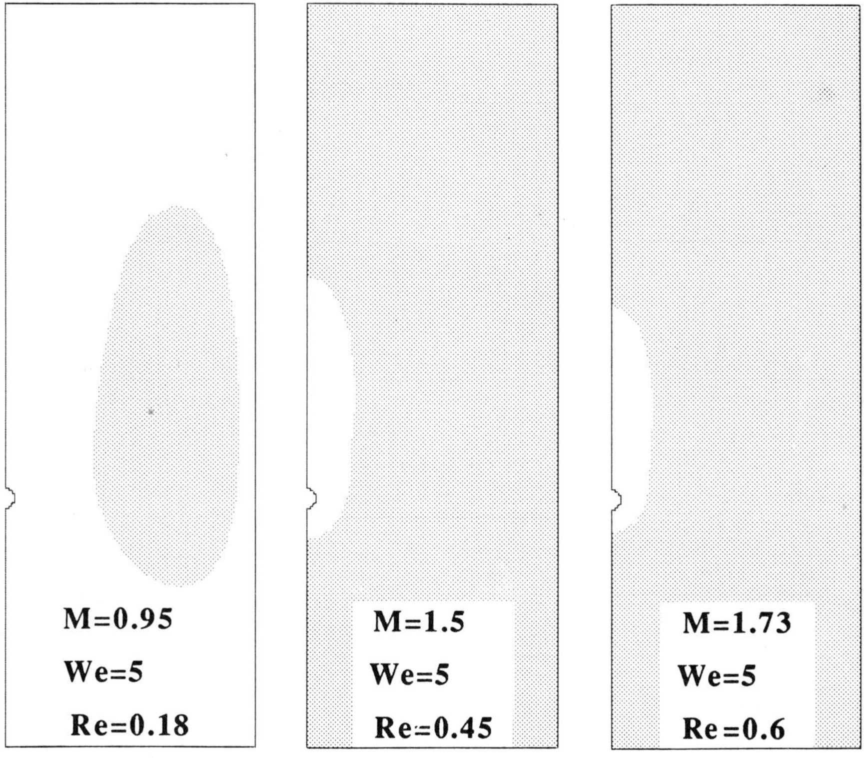

Figure 7.9. Hyperbolic (shaded) and elliptic (white) regions of vorticity in the flow of an upper-convected Maxwell fluid past a circular cylinder (a) (M, W, R) = (0.95, 5, 0.18), (b) (1.5, 5, 0.45), (c) (1.73, 5, 0.6).

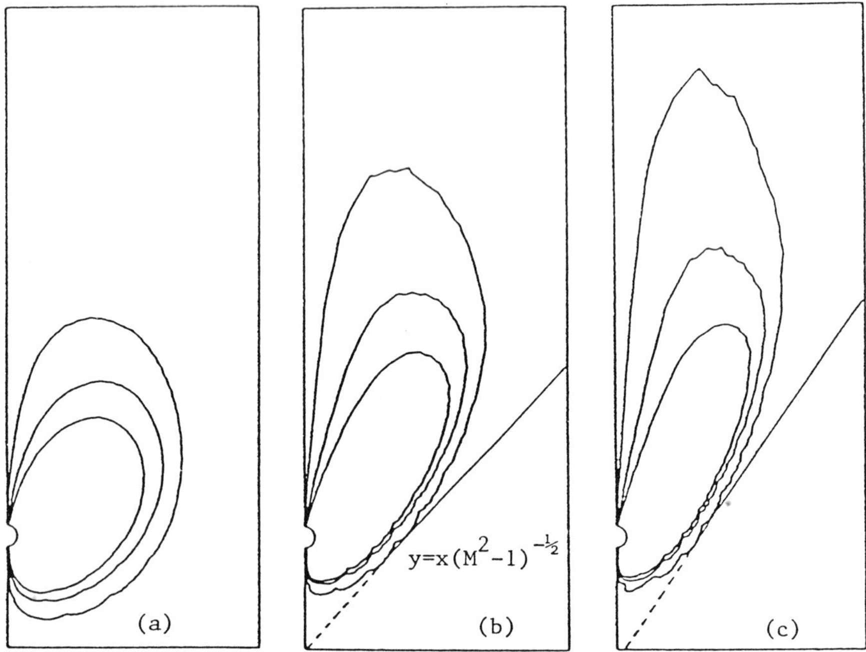

Figure 7.10. Vorticity contours for viscoelastic flow. (a) (M, W, R)=(0.95, 5, 0.18), (b) (1.5, 5, 0.45), (c) (1.73, 5, 0.6).

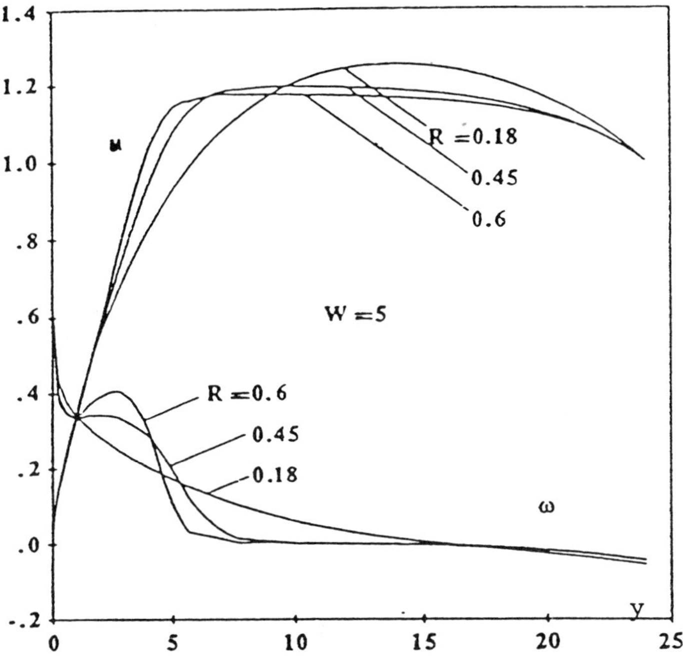

Figure 7.11. Axial velocity u/U and vorticity profile $\zeta d/2U$ along the axis x=0 for viscoelastic flow, W=5.

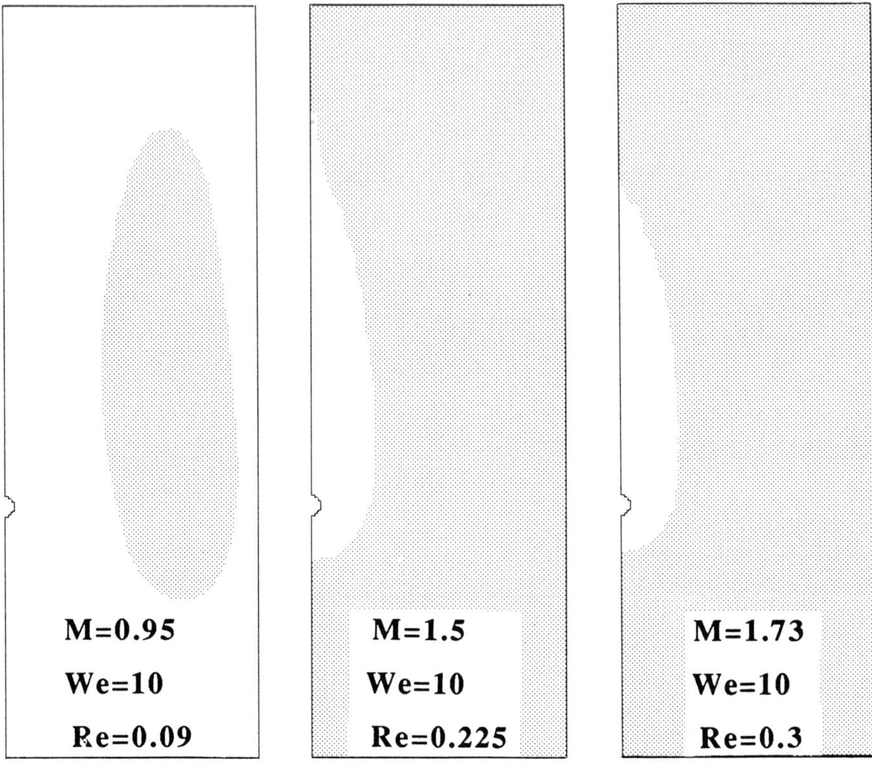

Figure 7.12. Region of hyperbolic vorticity indicated by the shaded area. (a) (M, W, R)=(0.95, 10, 0.09), (b) (1.5, 10, 0.225), (c) (1.73, 10, 0.3). The extent of the hyperbolic region increases with E=W/R, now four times larger than in Figure 7.9.

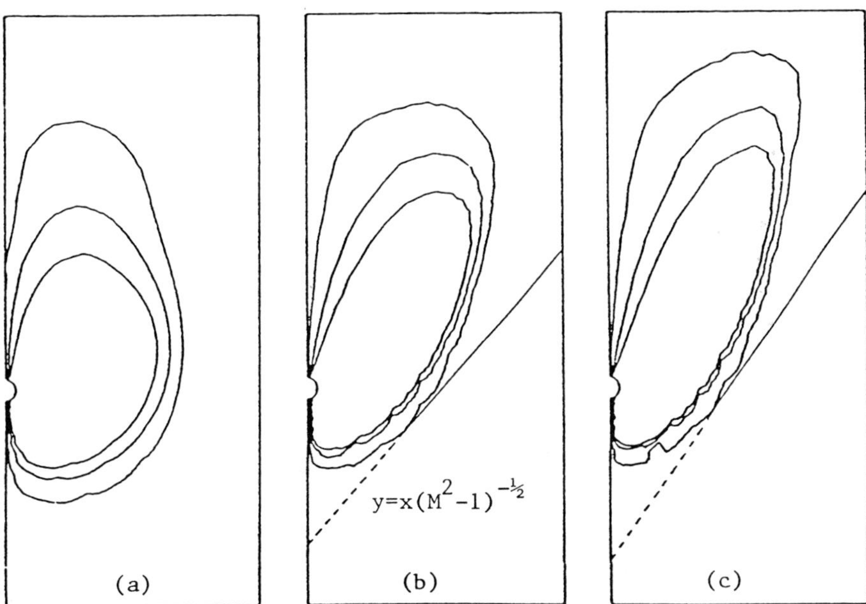

Figure 7.13. Vorticity contours for viscoelastic flow, W=10. (The values are the same as in Figure 7.12) (a) M=0.95, (b) M=1.5, (c) M=1.73.

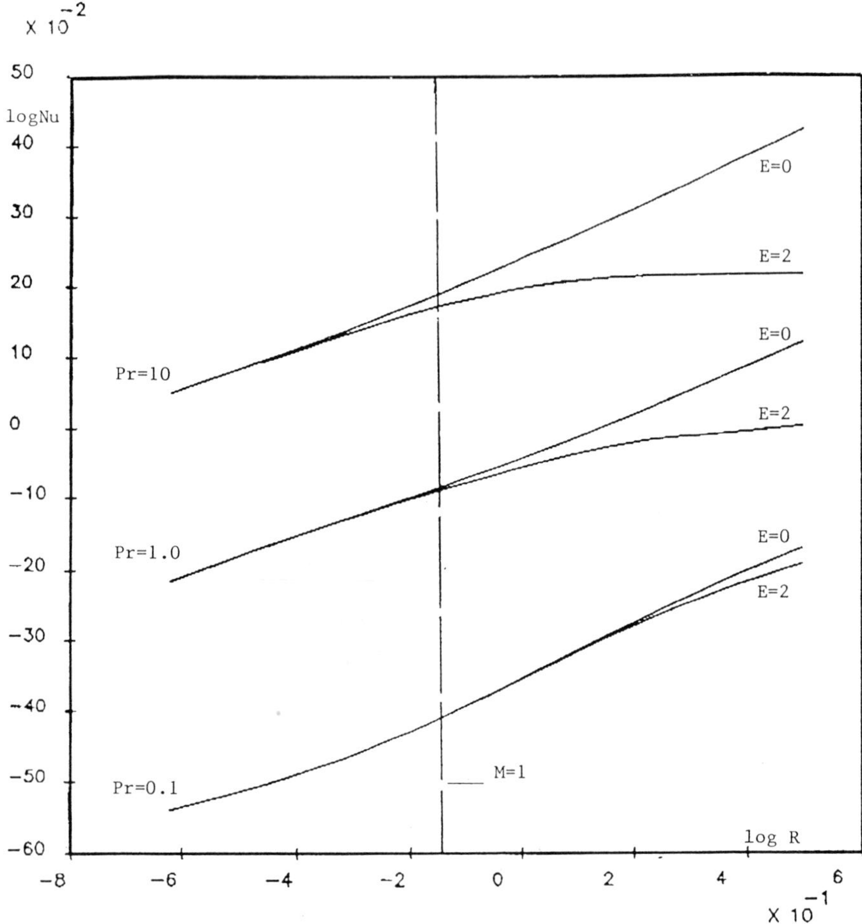

Figure 7.14 (after Delvaux and Crochet, 1990). Nusselt number for heat transfer from a hot cylinder across the flow of an UCM by numerical simulation ($R=Ud/\nu$, $E=\nu\lambda/d^2$)

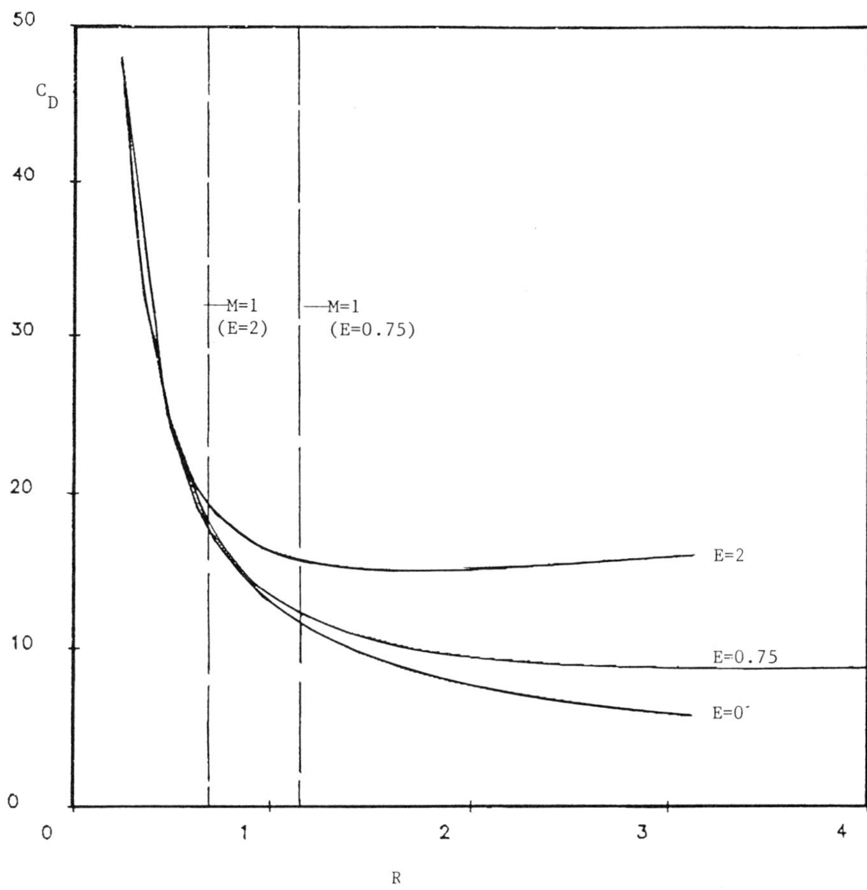

Figure 7.15. Drag coefficient for flow of an UCM over a circular cylinder by numerical simulation ($R=Ud/\nu$, $E=\nu\lambda/d^2$)

7.13 Drag reduction

The more dilute polyox solutions used in the experiments on flow over wires are well-known as drag reducers. We are obliged to ask if the hyperbolicity of the response which appears in these fluids when M>1 is also important for drag reduction. My ideas about this are only vaguely formed. We know that drag reduction is important only for turbulent flows; in fact the drag of viscoelastic and Newtonian liquids of the same viscosity in laminar flow is greater in the viscoelastic liquids (see Figures 7.5, 7.15 and 8.9). It appears that the reduction in drag is associated with turbulent polymer interactions in the buffer layer leading to the suppression of small eddies. Frequency spectra of wall shear stress fluctuations and of velocity fluctuations indicate relative attenuation of the high frequency components. This would suggest that elasticity rather than viscosity might play a major role in drag reduction, with the material wave speed providing a natural cut-off criterion for velocities which fluctuate at high frequencies. In fact the fluctuation velocities which are observed in turbulent flow of extremely dilute aqueous drag reducing solutions are of the right order for this, a few centimeters per second (see, for example, the Master's thesis of K. Harder [1989] done in the laboratory of W. Tiederman).

CHAPTER 8

Mach Wedges and Upstream Influence in the Partly Hyperbolic Flow Over a Flat Plate

The theory of slender bodies in aerodynamics is a perturbation of uniform flow which is perturbed less by slender bodies than by fat ones. The aerodynamic theory works well, but only because the fluid is required to slip along the slender body. Therefore a result so good as in slender body aerodynamic theory cannot be expected from the linearized Navier-Stokes theory or from the linearized Maxwell models derived in chapter 7. In the Navier-Stokes theory, linearization around uniform flow leads to Oseen's approximation. We can say that we are making an Oseen approximation to the flow of a viscoelastic fluid around a flat plate.

It is always very important to see what linearized theory can reveal even when the form of the nonlinear theory is known, as in Navier-Stokes theory. In the present case, there is a jungle of nonlinear equations, all subject to controversy. The material description of linearized models possessing instantaneous elasticity is not so problematic; for example, the expression (7.1) is fully general as a small perturbation of rigid motion and (7.33) is a form which applies uniformly to the linearization of any Maxwell model.

The problem of supercritical flow of a viscoelastic fluid over a flat plate was first studied by Joseph [1985]. In that study, the vorticity field was given by an integral expression when the distribution of the vorticity along the plate surface is prescribed. The stream function was represented in a double integral form using a Green function which vanishes on the plate surface. It was proposed that the correct vorticity distribution along the plate surface could be obtained by satisfying the no slip condition for velocity on the plate. This proposal was successfully implemented by H. Hu and reported here (§8.6). He used a less singular Green function which vanishes at $y=0$, not only on the plate $x>0$, but for $x<0$.

The problem of flow of a viscoelastic fluid over a flat plate was solved by L.E. Fraenkel [1988] using the Fourier transform method of Wiener and Hopf. He followed procedures used by Lewis and Carrier [1949] to solve Oseen's problem for the flow of a Newtonian fluid over a flat plate. Fraenkel's problem is more difficult. He had to do complex integrations with many branch cuts. His detailed derivations will not be given here. His solution has many interesting properties never before seen in the rheological literature or even in other branches of fluid mechanics.

There is a "Mach" wedge of vorticity attached to the leading edge of the plate (see Figure 8.1) with irrotational flow in front of the wedge and rotational flow behind. We could think of corresponding cases in gas dynamics in which the region in front of the wedge cannot be disturbed because the sound is convected by the flow faster than it can propagate. The region of "silence" in the viscoelastic case is the region of zero vorticity, but the velocity is not uniform there; it is determined as a potential flow varying from point to point which attains uniform flow only far upstream. The flow behind the wedge is rotational, with nonzero vorticity. Therefore the Mach wedge forms shock lines for the vorticity. The velocity and stresses are continuous across the shock. The first derivative of the velocity component normal to the shock is continuous across the shock. The two stresses normal to the shock line are continuous and have continuous first derivatives across the shock. The first normal derivative of the velocity component tangential to the shock, and the first normal derivative of the shear stress parallel to the shock, jump across the shock. Perhaps the most interesting property of the solution is that only vorticity vanishes before the shock; the irrotational parts of the velocity and the associated stresses are not zero there, so that even though the normal derivatives of the component of velocity tangent to the shock and the associated shear stress jump across the shock, the vorticity is the key hyperbolic variable. The peculiar mix of elliptic and hyperbolic fields is characteristic of problems of composite type.

In fully hyperbolic problems, and even in parabolic ones, the upstream influence is negligible, even completely suppressed. In problems of composite type, like the flat plate problem studied here, this mode of thought is incorrect; there is a huge upstream influence because the harmonic part of the velocity is elliptic. Students of the fluid mechanics of viscoelastic fluids should always take care in drawing analogies taken from Newtonian fluid mechanics.

8.1 Green function solution

The plate is on the half line y=0, x≥0 and is moving in negative x direction with velocity U. The velocity components corresponding to (x, y) are (u, v) and there is only one component of vorticity

$$\zeta = \frac{\partial v}{\partial x} - \frac{\partial u}{\partial y}\,. \tag{1}$$

The velocity components

$$(u, v) = \left(\frac{\partial \psi}{\partial y}, -\frac{\partial \psi}{\partial x}\right) \tag{2}$$

may be obtained from a stream function ψ.

Following the approach in Joseph [1985], the vorticity of the perturbed flow for the Maxwell model satisfies the equation

$$(M^2 - 1)\frac{\partial^2 \zeta}{\partial x^2} - \frac{\partial^2 \zeta}{\partial y^2} + \frac{\rho U}{\eta}\frac{\partial \zeta}{\partial x} = 0 \tag{3}$$

where $M = \frac{U}{c}$ is the Mach number, $c = \sqrt{\frac{\eta}{\lambda\rho}}$ is the wave velocity for propagation of vorticity waves in a Maxwell fluid, η is the viscosity, λ is the relaxation time and ρ is the density.

The stream function and vorticity are related by a second order equation

$$\frac{\partial^2\psi}{\partial x^2} + \frac{\partial^2\psi}{\partial y^2} = -\zeta \,. \tag{4}$$

We may write these equations in dimensionless form. We scale the length with ℓ, velocity with U, vorticity with $\frac{U}{\ell}$ and stream function with $U\ell$, and use the same symbols for the dimensionless variables. In the dimensionless form (3) becomes

$$\frac{\partial^2\zeta}{\partial x^2} + \frac{M^2}{\beta^2 W}\frac{\partial\zeta}{\partial x} - \frac{1}{\beta^2}\frac{\partial^2\zeta}{\partial y^2} = 0 \tag{5}$$

where $\beta^2 = M^2 - 1$ and $W = \frac{U\lambda}{\ell}$ is a Weissenberg number based on ℓ. The equation (4) is unchanged.

The vorticity ζ and stream function ψ must vanish far away from the plate. The no slip condition on the plate surface for velocity requires

$$\frac{\partial\psi}{\partial x} = 0 \qquad \text{on } y=0,\ x\geq 0, \tag{6}$$

and

$$\frac{\partial\psi}{\partial y} = -1 \qquad \text{on } y=0,\ x\geq 0. \tag{7}$$

If $\zeta(x,y)$ is prescribed on the plate surface

$$\begin{aligned} \zeta(x,\,0) &= g(x) \quad \text{for } x>0 \,, \\ \zeta(x,\,0) &= 0 \quad \text{for } x<0 \,. \end{aligned} \tag{8}$$

then the solution may be written as (see Carslaw and Jaeger [1963], p.202-203)

$$\zeta(x,y) = g(x-\beta y)\,\exp\left(-\frac{M^2}{2\beta W}\,y\right) +$$

$$\frac{M^2}{2\beta W}\,y\int_{\beta y}^{x} g(x-\tau)\;\exp\!\left(-\frac{M^2}{2\beta^2 W}\,\tau\right)\frac{I_1\!\left(\dfrac{M^2}{2\beta^2 W}\sqrt{\tau^2-\beta^2y^2}\right)}{\sqrt{\tau^2-\beta^2y^2}}\,d\tau \qquad (9)$$

where I_1 is the modified Bessel function of the first kind. On the bottom of the plate we require that

$$\zeta(x,-y) = -\zeta(x,y)\,. \qquad (10)$$

This solution reveals that there is a "Mach" wedge $x=\beta y$, outside this wedge the vorticity is zero (irrotational flow), and inside the wedge the vorticity is nonzero (rotational flow). This solution, with a constant $g=c$, was given by B. Caswell in a study of the effects of a leading edge singularity in the flow over a flat plate. A solution like (9) can be obtained for Boltzmann's model of a linear viscoelastic fluid using a Duhamel integral given by Narain and Joseph [1983a].

Equation (4) for the stream function can now be solved with a Green function

$$\psi(x,y) = \int\!\!\int_R G(x, x_0|\, y, y_0)\,\zeta(x_0, y_0)\;dx_0\,dy_0 \qquad (11)$$

where R is the complement of the upper half plane to the line $y=0$. The Green function for this domain, vanishing on the line $y=0$, is given on p.1180 of Morse and Feshbach [1953]

$$G(x, x_0|\, y, y_0) = \frac{1}{4\pi}\,\ell n\left[\frac{(x-x_0)^2 + (y+y_0)^2}{(x-x_0)^2 + (y-y_0)^2}\right].$$

The condition (7) takes the form

$$-1 = \frac{1}{\pi}\int_R \frac{y_0}{(x-x_0)^2+y_0^2}\,\zeta(x_0,y_0)dx_0dy_0 \quad \text{for } x\geq 0\,. \qquad (12)$$

Substituting vorticity (9) into (12) we get an equation for g(x). We have to choose the correct form of the prescribed plate vorticity g(x) to satisfiy (12).

Given g(x) solving (12), we can obtain the vorticity $\zeta(x,y)$ from (9), the stream function $\psi(x,y)$ from (11) and the velocities from (2). The stresses are determined by the following rate equations

$$\boldsymbol{\tau} = \begin{pmatrix} \sigma & \tau \\ \tau & \gamma \end{pmatrix}, \tag{13}$$

$$\frac{\partial \sigma}{\partial x} + \frac{\sigma}{W} = \frac{2}{M^2}\frac{\partial u}{\partial x}, \tag{14}$$

$$\frac{\partial \tau}{\partial x} + \frac{\tau}{W} = \frac{1}{M^2}\left(\frac{\partial u}{\partial y} + \frac{\partial v}{\partial x}\right) \tag{15}$$

or, in integral form

$$\sigma(x, y) = \frac{2}{M^2}\left[u(x, y) - \int_0^\infty u(x-\tau, y)\exp\left(-\frac{\tau}{W}\right)d\tau\right], \tag{16}$$

and

$$\tau(x,y) = \frac{2}{M^2}\left[v(x,y) - \int_0^\infty v(x-\tau, y)\exp\left(-\frac{\tau}{W}\right)d\tau\right] - \frac{1}{M^2}\int_0^x \zeta(x-\tau, y)\exp\left(-\frac{\tau}{W}\right)d\tau. \tag{17}$$

The relation $\gamma=-\sigma$ follows from the rate equation satisfied by γ and $v_y=-u_x$.

8.2 Fraenkel's solution

Fraenkel [1988] approached the supercritical flat plate problem from a different point of view. He notes that Lewis and Carrier [1949] solved the Oseen problem for the flow of Newtonian fluid past a flat plate by means of the Fourier transform and the Wiener-Hopf technique; later it was found that a formulation in terms of parabolic coordinates yields an elementary derivation (Kaplun [1954], Van Dyke [1975]) and an infinite number of different solutions (Olmstead [1975]). For a general viscoelastic fluid, satisfying (7.27) in the plane with dimension coordinates x, y, the presence of the convolution

$$K*\zeta = \int_{-\infty}^{X} K(X-\xi)\zeta(\xi,\mathcal{Y})\,d\xi = \int_{0}^{\infty} K(s)\zeta(X-s,Y)\,ds \qquad (18)$$

where

$$K(X) = \frac{1}{\rho U^2} G\left(\frac{X}{U}\right),$$

$$K(0) = \frac{1}{M^2} \qquad (19)$$

seems to dictate the use of Fourier transforms and the structure of the boundary value problem,

$$\nabla^2\Psi = -\zeta \qquad \text{upstream and downstream of the shock}, \quad (20)_1$$

$$\frac{\partial\zeta}{\partial X} = \nabla^2(K*\zeta) \qquad \text{downstream}, \quad (20)_2$$

$$\zeta = 0 \qquad \text{upstream}, \quad (20)_3$$

$$\left(\frac{\partial\Psi}{\partial Y}, -\frac{\partial\Psi}{\partial X}\right) = (-U,0) \qquad \text{on the plate}, \quad (20)_4$$

$$\left(\frac{\partial\Psi}{\partial Y}, -\frac{\partial\Psi}{\partial X}\right) \to (0,0) \qquad \text{at infinity upstream} \quad (20)_5$$

requires the use of the Wiener-Hopf technique. Parabolic coordinates cannot help in the viscoelastic problem and, surprisingly, uniqueness of solutions is recovered (Fraenkel [1988]).

Equation (20.2) implies that

$$\left\{(M^2-1)\frac{\partial^2}{\partial X^2}-\frac{\partial^2}{\partial Y^2}\right\}(K*\zeta)-M^2\frac{\partial}{\partial X}(K'*\zeta)=0 \qquad (20)_6$$

which differs from a hyperbolic differential equation only in the occurrence of K′ in the last term.

The solution of (20) naturally decomposes into a rotational part, designated with a superscript R, and an irrotational part associated with the harmonic part of the stream function, designated with the superscript H.

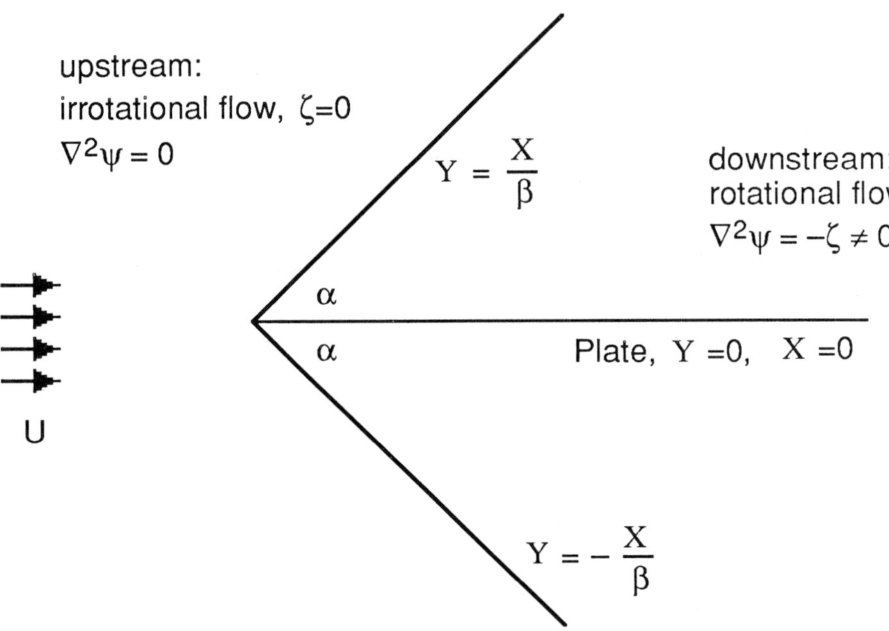

Figure 8.1. Supercritical linearized flow over a flat plate; $M=U/c>1$, $c=\sqrt{G(0)/\rho}$. The Mach wedge for the vorticity is a shock line across which the vorticity increases discontinuously from zero to infinity, the wedge angle is α and $\tan\alpha=1/\beta$, $\beta=\sqrt{M^2-1}$.

We write the decompositions

$$\Psi = \Psi^H + \Psi^R\,, \tag{21}$$

$$\nabla^2\Psi^H = 0 \qquad \text{downstream and upstream}\,, \tag{22}$$

$$\Psi^R \equiv 0 \qquad \text{upstream}\,, \tag{23}$$

$$\left(\Psi^H_Y, -\Psi^H_X\right) \to (0,0) \qquad \text{at infinity upstream}\,, \tag{24}$$

$$\nabla^2\Psi^R = -\zeta \qquad \text{downstream}\,, \tag{25}$$

$$\left(\Psi^H_Y + \Psi^R_Y, -\Psi^H_X - \Psi^R_X\right) = (-U,0) \qquad \text{on the plate}\,. \tag{26}$$

The first step of the Wiener-Hopf technique is to take Fourier transforms of the governing problem, say (20); $(20)_1$ and $(20)_2$. Thus

$$\left[\frac{\partial^2}{\partial Y^2} - p^2\right]\tilde{\Psi} = -\tilde{\zeta} \quad \text{for} \quad \operatorname{Im} p = 0\,, \quad p \neq 0\,, \qquad (27)_1$$

$$\left[\frac{\partial^2}{\partial Y^2} + \frac{ip}{\tilde{K}(p)} - p^2\right]\tilde{\zeta} = 0 \quad \text{for} \quad \operatorname{Im} p \geq 0\,, \quad p \neq 0 \qquad (27)_2$$

where $\tilde{\zeta} = \int_{-\infty}^{\infty} e^{ibx}\zeta(x,y)dx$ and $(K*\zeta)(x,y)$ transforms to

$\tilde{K}(p)\tilde{\zeta}(p,Y)$ and $\tilde{K}(0) = \int_0^{\infty} K(x)\,dx > 0$, $\tilde{K}(p) = \dfrac{i}{M^2 p} + 0(p^{-1-\omega})$, $\omega > 0$.

The interested reader will find a readable and thorough exposition of all the fine details of the solution in the various papers of Fraenkel.

After deriving a solution to the transformed problem, Fraenkel gets the solution of (18) by inverting the transforms. Fraenkel has derived an explicit asymptotic solution which he gets by expanding the inversion formulae for large values of p, and solutions far from the shock are obtained by expanding for small values of p. We are not going to present Fraenkel's results for general kernels. It could be important that his solution is so generally defined, since various kinds of solutions have been and can be investigated. It would be interesting to investigate this problem using two modes of relaxations, simulating conditions that lead to an effective viscosity.

8.3 Fraenkel's solution for Maxwell models

All of the interpolated Maxwell models collapse to (7.33) when $G(s) = \frac{\eta}{\lambda} e^{-s/\lambda}$. We introduce the dimensionless coordinate (x,y) defined by $\ell = NU\lambda$, $N = \beta^2/M^2$, $\beta^2 = M^2 - 1 > 0$ and $W = 1/N$. Velocities are made dimensionless with U and stresses are measured in units of ρU^2. The telegraph equation (3) for ζ may be written using dimensionless coordinates as

$$\frac{\partial^2 \zeta}{\partial x^2} - \frac{1}{\beta^2}\frac{\partial \zeta}{\partial y^2} + \frac{\partial \zeta}{\partial x} = 0 . \tag{28}$$

Equation (28) replaces (22) and the decompositions (21) through (26) are still valid. The equations governing the Fourier transforms are given with

$$\tilde{K}(p) = \int_0^\infty e^{ipx} K(x)\, dx \tag{29$_1$}$$

where

$$K(x) = \frac{1}{M^2} \exp\left(-\frac{\beta^2}{M^2} x\right) . \tag{29$_2$}$$

The extra stress with component σ, $-\sigma$, τ, now in dimensionless coordinates introduced above is related to the velocity by (14) and (15) with $W=1/N$. Only two stresses are needed, the normal stress σ and the shear stress τ. The vorticity is related to these two stresses by

$$\frac{\partial \zeta}{\partial x} = -\frac{\partial^2 \sigma}{\partial x \partial y} - \left(\frac{\partial^2}{\partial x^2} - \frac{\partial^2}{\partial y^2}\right)\tau . \tag{30}$$

It will be convenient to refer the velocity $\mathbf{u}$ and the stress $\boldsymbol{\tau}$ to coordinates tangent to and normal to the shock surface.

Let u', v' be the velocity components along x' and y' and σ', $\gamma' = -\sigma'$ and τ' are the corresponding normal and shear stresses in a coordinate system with $\mathbf{e}_x \bullet \mathbf{e}_x' = \cos\alpha$, $\mathbf{e}_x \bullet \mathbf{e}_y' = \cos(\alpha + \pi/2)$, where α is the semi-vertex angle of the Mach wedge. Then

$$\begin{aligned} u' &= u\cos\alpha + v\sin\alpha \ , \\ v' &= -u\sin\alpha + v\cos\alpha \ , \end{aligned} \tag{31}$$

$$\begin{aligned} \sigma' &= \sigma(\cos^2\alpha - \sin^2\alpha) + 2\tau\cos\alpha\sin\alpha \ , \\ \tau' &= -2\sigma\sin\alpha\cos\alpha + \tau(\cos^2\alpha - \sin^2\alpha) \ . \end{aligned} \tag{32}$$

The stresses may also be decomposed into harmonic and rotational parts. The rotational parts are identically zero before the shock and satisfy

$$\frac{\partial\sigma^R}{\partial x} + N\sigma^R = \frac{2}{M^2}\frac{\partial u^R}{\partial x} \ , \tag{33}$$

$$\frac{\partial\tau^R}{\partial x} + N\tau^R = \frac{2}{M^2}\frac{\partial v^R}{\partial x} - \frac{\zeta}{M^2} \ . \tag{34}$$

In deriving (34), we used the equation

$$\zeta = \frac{\partial v^R}{\partial x} - \frac{\partial u^R}{\partial y} \ . \tag{35}$$

The harmonic stresses σ^H and τ^H satisfy

$$\frac{\partial\sigma^H}{\partial x} + N\sigma^H = \frac{2}{M^2}\frac{\partial u^H}{\partial x} \ , \tag{36}$$

$$\frac{\partial\tau^H}{\partial x} + N\tau^H = \frac{2}{M^2}\frac{\partial v^H}{\partial x} \tag{37}$$

and are non-zero except possibly at discrete points.

8.4 Asymptotic expressions for the velocity and stress near the shock

8.4.1 The vorticity near the shock

First we recall that $\zeta(x,y)$ is zero in the region upstream from the shock. The vorticity near the shock on the downstream side, $\xi \stackrel{\text{def}}{=} x-\beta y>0$, $y>0$, is given by

$$\zeta(x,y) = \frac{C_1}{\xi^\gamma} \exp\left(\frac{-\beta y}{2}\right)\left[1 + C_2 \xi \log \frac{1}{\xi} + 0(\xi)\right] \tag{38}$$

where

$$\gamma = \frac{1}{2} + \frac{1}{\pi} \arctan \beta \,, \gamma \in \left(\frac{1}{2}\,, 1\right). \tag{39}$$

and C_1 and C_2 are defined in Table 8.1. The vorticity ζ given in (38) is infinite at $x=\beta y$, but integrable. It decays exponentially away from the plate. The functional dependence of the constants in Table 8.1 as they depend on M is given in Table 8.2.

8.4.2 Continuity of velocity across the shock

The velocity is continuous across the shock, with integrable singular derivatives at the shock. We could regard u and v as determined by ζ through the equations

$$\frac{\partial u}{\partial x} + \frac{\partial v}{\partial y} = 0 \,,$$

$$\frac{\partial v}{\partial x} - \frac{\partial u}{\partial y} = \zeta \,. \tag{40}$$

The harmonic parts of the velocity are smooth, actually analytic, across the shock and elsewhere. The continuity of the velocity is achieved because the rotational velocities vanish at the downstream, as well as the upstream, side, even though the first derivatives are singular on the downstream side.

$$\mu = \frac{1}{2} - \frac{\alpha}{\pi} \in (0,1/2)\,, \quad \alpha<\pi/2$$

$$E(0) = \exp\left\{\frac{1}{\pi}\int_1^\infty \left[\tan^{-1}\beta - \tan^{-1}\left(\beta\sqrt{1-\frac{1}{\rho}}\right)\right]\frac{d\rho}{\rho}\right\},$$

$x!\,,\quad$ x factorial ,

$$\mathcal{F}(x) = \frac{d}{dx}\log(x!)\,,\quad \text{digamma function},$$

$$E_1 = \frac{\beta}{2\pi M^2}\,,$$

$$E_2 = \frac{1}{\pi}\int_1^\infty \left[\tan^{-1}\beta - \tan^{-1}\left(\beta\sqrt{\frac{\rho-1}{\rho}}\right) - \frac{\pi E_1}{\rho}\right] d\rho$$

$$= E_1\left(1 + \log 4 - \frac{2\tan^{-1}\beta}{\beta}\right),$$

$$C_1 = \frac{(M^2-1)}{\mu!\; M^2\, E(0)}\,,$$

$$C_2 = \frac{E_1}{\mu+1}\,,$$

$$C_3 = C_2\mathcal{F}(\mu+1) + \frac{1}{\mu+1}\left[1 - \mu - \frac{\beta^2}{M^2} + E_2\right],$$

$$C_3' = C_3 - \frac{1}{2\mu+2}\,,$$

$$C_4 = \frac{1}{8\mu+8}\,.$$

Table 8.1 Constants of the asymptotic solution.

M	μ	$\mu!$	$G[\mu+1]$	E(0)	E_1	E_2
1.00	0.50000	0.88623	0.70316	1.00000	0.00000	0.00000
1.40	0.25325	0.90574	0.57436	1.12486	0.07956	0.06396
1.80	0.18749	0.92159	0.53710	1.12985	0.07352	0.07899
2.00	0.16667	0.92772	0.52502	1.12691	0.06892	0.08112
2.40	0.13680	0.93752	0.50743	1.11830	0.06028	0.08080
2.80	0.11625	0.94500	0.49515	1.10910	0.05309	0.07775
3.00	0.10817	0.94811	0.49028	1.10472	0.05002	0.07582
∞	0.00000	1.00000	0.42278	1.00000	0.00000	0.00000

M	C_1	C_2	C_3	C_3'	C_4	C_5
1.00	0.00000	0.00000	0.33333	0.00000	0.08333	0.50000
1.40	0.48075	0.06348	0.29253	-0.10643	0.09974	1.99959
1.80	0.66397	0.06191	0.20179	-0.21927	0.10526	1.76272
2.00	0.71739	0.05907	0.17197	-0.25660	0.10714	1.50000
2.40	0.78822	0.05303	0.13036	-0.30947	0.10996	0.77278
2.80	0.83241	0.04756	0.10333	-0.34460	0.11198	-0.17510
3.00	0.84866	0.04514	0.09320	-0.35799	0.11280	-0.72222
∞	1.00000	0.00000	0.00000	-0.50000	0.12500	-∞

Table 8.2. (Ahrens, 1989) Values of various constants as functions of M.

8.4.3 The rotational part of the velocity near the shock

The rotational part Ψ^R of the stream function $\Psi=\Psi^R+\Psi^H$ is also zero upstream and satisfies the equation $\nabla^2\Psi^R=-\zeta$ downstream. Near the shock, $\xi\to 0$.

$$\Psi^R(x,y) = \frac{C_1}{\beta} e^{-\beta y/2} \left\{ \frac{\xi^{\mu+1}}{\mu+1} + C_2 \frac{\xi^{\mu+2}}{\mu+2} \left(\log\frac{1}{\xi} + \frac{1}{\mu+2} \right) + (C_3'+C_4\beta y)\frac{\xi^{\mu+2}}{\mu+2} + 0\left(\xi^{\mu+3}\left[\log\frac{1}{\xi}\right]^2 \right) \right\} . \qquad (41)$$

where μ, C_1, C_2, C_3, C_3', and C_4 are defined in Table 8.1. The rotational part of the y component of velocity is

$$v^R(x,y) = -\frac{C_1}{\beta}\xi^\mu e^{-\beta y/2}\left\{1+C_2\xi\log\frac{1}{\xi}\right.$$
$$\left.+(C_3'+C_4\beta y)\xi+0\left(\left[\xi\log\frac{1}{\xi}\right]^2\right)\right\}. \tag{42}$$

The rotational part of the x component of velocity is

$$u^R(x,y) = -C_1\xi^\mu e^{-\beta y/2}\left\{1+C_2\xi\log\frac{1}{\xi}\right.$$
$$\left.+(C_3+C_4\beta y)\xi+0\left(\left[\xi\log\frac{1}{\xi}\right]^2\right)\right\} \tag{43}$$

The two rotational components of velocity are continuous, zero before the shock and at the shock, but the x and y derivatives of u^R and v^R are both singular at the shock.

Now we form the equations expressing the components of velocity u'^R and v'^R parallel and normal to the shock. Using (31), with $\beta=\cos\alpha/\sin\alpha$, we find that u'^R has a singular first derivative across the shock but the ξ derivative of

$$v'^R = C_1\xi^\mu e^{-\beta y/2}\sin\alpha\left\{[C_3-C_3']\xi+0\left(\left[\xi\log\frac{1}{\xi}\right]^2\right)\right\} \tag{44}$$

vanishes at $\xi=0$. The normal component v'^R of the velocity normal to the shock is continuous with a continuous first derivative across the shock. The component u'^R of the velocity parallel to the shock

$$u'^R = -\frac{C_1}{\cos\alpha}\xi^\mu e^{-\beta y/2}\left\{1+C_2\xi\log\frac{1}{\xi}+C_4\beta y\xi\right.$$
$$\left.+C_3\xi-\frac{\sin^2\alpha}{2(\mu+1)}\xi+0\left(\left[\xi\log\frac{1}{\xi}\right]^2\right)\right\} \tag{45}$$

has a singular derivative with respect to ξ at $\xi=0$.

8.4.4 The rotational part of the stresses near the shock

We can obtain σ^R and τ^R from (33) and (34), using the formulas (38) for ζ, (43) for u^R, and (42) for v^R.

$$\sigma^R - \frac{2}{M^2}u^R = -\frac{2N}{M^2}e^{-Nx}\int^x e^{Nx'}u^R(x',y)\,dx'$$

$$= \frac{2NC_1e^{\frac{-\beta y}{2}}}{M^2(\mu+1)}\xi^{\mu+1} + o(\xi^{\mu+1}), \tag{46}$$

$$\tau^R - \frac{2}{M^2}v^R = \frac{e^{-Nx}}{M^2}\int^x e^{Nx'}\zeta(x',y)\,dx'$$

$$-\frac{2N}{M^2}e^{-Nx}\int^x e^{Nx'}v^R(x',y)\,dx'$$

$$= \frac{C_1e^{\frac{-\beta y}{2}}}{M^2}\left\{\frac{1}{1-\gamma}\xi^{1-\gamma} + \frac{2N}{\beta(\mu+1)}\xi^{\mu+1}\right\} + o(\xi^{1-\gamma}) \tag{47}$$

It is easy to establish that

- σ^R and τ^R are bounded and continuous at the shock.
- The derivative of τ^R and σ^R are singular but integrable immediately after the shock, at $\xi=0^+$. (48)
- $\sigma^R \to 2u^R/M^2$ after the shock, as $\xi \to 0^+$, through positive values.

$$\tau^R = \frac{C_1}{M^2\beta}e^{-\beta y/2}\left\{(\beta^2-1)\left[\xi^{\mu}+C_2\xi^{\mu+1}\log\frac{1}{\xi}\right.\right.$$

$$\left.\left.+(C_3+byC_4)\xi^{\mu+1}\right] + C_5\frac{\xi^{\mu+1}}{\mu+1} + 0\left(\xi^{\mu+2}\left[\log\frac{1}{\xi}\right]^2\right)\right\} \tag{49}$$

where

$$C_5 = \frac{[M^4 - 2(M^2-1)(M^2-2)]}{2M^2} \ . \tag{50}$$

Now, using (49), (46), and (43), we may evaluate (32), σ' and τ', the normal stress and the shear stress at the shock. We find that σ' and τ' are continuous across the shock, $\partial\tau'/\partial\xi\to\infty$ at the shock, but, to leading order

$$\begin{aligned}\sigma'^R &= \frac{2u^R}{M^2}(\cos^2\alpha - \sin^2\alpha) + 2\tau^R \sin\alpha\cos\alpha \\ &= \sin^2\alpha\left\{\frac{2u^R}{M}(\beta^2-1) + 2\beta\tau^R\right\} \\ &= \sin^2\alpha\,\frac{2C_1}{M^2}\left\{\sin\alpha\cos\alpha\,\frac{\xi^{\mu+1}}{\mu+1} + 0\left(\xi^2\left[\log\frac{1}{\xi}\right]^2\right)\right\}.\end{aligned} \tag{51}$$

The first derivative of σ'^R across the shock is zero, and the first derivative of $\sigma'=\sigma'^R+\sigma'^H$ is continuous across the shock.

8.4.5 The harmonic part of the velocity near the shock

The total velocity $\mathbf{u}=\mathbf{u}^R+\mathbf{u}^H$ is composed of a rotational part $\mathbf{u}^R$, given near the shock by (42) and (43), and a harmonic part $\mathbf{u}^H$ which arises from a stream function $\Psi^H(x,y)$ satisfying Laplace's equation. Of course, the harmonic fields cannot be discontinuous; they are necessarily analytic and can be described by analytic functions of a complex variable $z=x+iy$. There is a function $F(z)$, of z alone such that

$$1 + u^H(x,y) - iv^H(x,y) = F(z) \ . \tag{52}$$

The data which drives this flow is the velocity -1 of the plate. Equation (26) shows that the harmonic and rotational fields are coupled at the plate, $x\geq 0$

$$\begin{aligned}1 + u^H(x,0) + u^R(x,0) &= 0 \ , \\ v^H(x,0) + v^R(x,0) &= 0 \ .\end{aligned} \tag{53}$$

If $u^R(x,0)$ and $v^R(x,0)$ were known at $x\geq 0$, then the harmonic problem could be formulated as a classical potential flow problem. Fraenkel [1988] notes that

$$u^R(x,0) - iv^R(x,0) \tag{54}$$

is a linear combination of terms

$$x^\rho\,(1 - i \tan \rho\pi)\,, \tag{55}$$

and

$$x^\rho \left\{\log\frac{1}{x}(1-i\tan\rho\pi) + i\pi\,(\sec\rho\pi)^2\right\} \tag{56}$$

for various constants $\rho>0$ such that ρ is not equal to half an odd integer. These are limiting values as $\arg z\to 0^+$ of

$$f_\rho(z) \overset{\text{def}}{=} \frac{(ze^{-i\pi})^\rho}{\cos\rho\pi}\,, \tag{57}$$

$$g_\rho(z) = \frac{(ze^{-i\pi})^\rho}{\cos\rho\pi}\left\{\log\frac{1}{ze^{-i\pi}} - \pi\tan\rho\,\pi\right\} \tag{58}$$

where $0<\arg z<2\pi$. The desired approximation to $1+u^H-iv^H$ is the linear combination of functions f_ρ and g_ρ implied by (53)

$$\begin{aligned} 1 + u^H(x,y) - iv^H(x,y) &= C_1 f_\mu(z) + C_1C_2 g_{\mu+1}(z) \\ &+ C_1C_3 f_{\mu+1}(z) + 0\left(z^{\mu+2}\left(\log\frac{1}{z}\right)^2\right), \end{aligned} \tag{59}$$

$$\begin{aligned} 1 + u^H(x,y) &= \frac{C_1}{\sqrt{N}}\Big\{r^\mu\cos\mu\,(\theta-\pi) \\ &+ C_2 r^{\mu+1}[-\cos\,(\mu+1)(\theta-\pi)]\left(\log\frac{1}{r} - \frac{\pi}{\beta}\right) \\ &- (\theta-\pi)\sin\,[(\mu+1)(\theta-\pi)] - C_3 r^{\mu+1}\cos\,[(\mu+1)(\theta-\pi)]\Big\}\,, \end{aligned} \tag{60}$$

$$\begin{aligned} v^H(x,y) &= -\frac{C_1}{\beta}\Big\{r^\mu[b\sin\mu\theta - \cos\mu\theta] \\ &+ C_2 r^{\mu+1}[\beta\sin(\mu+1)\theta - \cos\,(\mu+1)\theta]\left(\log\frac{1}{r} - \frac{\pi}{\beta}\right) \\ &- (\theta-\pi)\,[\beta\cos\,(\mu+1)\theta + \sin\,(\mu+1)\theta]\Big\}\,, \end{aligned} \tag{61}$$

$$y + \Psi^H(x,y) = \frac{C_1}{\beta}\left\{\frac{r^{\mu+1}}{\mu+1}[\beta \sin(\mu+1)\theta - \cos(\mu+1)\theta]\right.$$
$$+ C_2\frac{r^{\mu+2}}{\mu+2}\left([\beta \sin(\mu+2)\theta - \cos(\mu+2)\theta]\left(\log\frac{1}{r} + \frac{1}{\mu+2} - \frac{\pi}{\beta}\right)\right.$$
$$\left. - (\theta-\pi)[\beta \cos(\mu+2)\theta + \sin(\mu+2)\theta\right) \tag{62}$$
$$\left. + C_3\frac{r^{\mu+2}}{\mu+2}[\beta \sin(\mu+2)\theta - \cos(\mu+2)\theta]\right\}.$$

The harmonic part of the shear stress is governed by

$$\frac{\partial \tau^H}{\partial x} + N\tau^H = \frac{2}{M^2}\frac{\partial v^H}{\partial x}. \tag{63}$$

This can be written as

$$\frac{\partial}{\partial x}\left[\tau^H - \frac{2\beta^2}{M^4}\Psi^H - \frac{2}{M^2}v^H\right] + \frac{2N^2}{M^2}\Psi^H$$
$$-\frac{2N^3}{M^2}\int_{-\infty}^{x} e^{-N(x-\chi)}\,\Psi^H(\chi,y)\,d\chi = 0. \tag{64}$$

This does not lead to a local approximation since it depends globally on Ψ^H. However, $\Psi^H(\chi,0)=0$, $\tau^H(x,y)$ can be determined locally at the wall and on the zero streamline upstream

$$\tau^H(x,0) - \frac{2\beta^2}{M^4}\Psi^H(x,0) - \frac{2}{M^2}v^H(x,0) = 0. \tag{65}$$

After some manipulation we get $\tau(x,0)$, $x>0$, the total shear stress $\tau=\tau^R+\tau^H$ at the wall,

$$\tau(x,0^+) = \frac{C_1}{\beta}\left\{x^{\mu}+C_2x^{\mu+1}\log\frac{1}{x} + C'_3\,x^{\mu+1} + 0\left(x^{\mu+2}\left[\log\frac{1}{x}\right]^2\right)\right\}. \tag{66}$$

8.4.6 Comparison of the wall shear stress near the leading edge for the flow of a Newtonian and viscoelastic fluid with the same viscosity

The dimensional shear stress $\bar{\tau}^N$ of a Newtonian fluid flowing over a flat plate in the linearized (Oseen) approximation is given by (see, for example, Fraenkel [1988])

$$\bar{\tau}^N = \mu \frac{\partial \bar{u}}{\partial Y} = \frac{\mu U}{\sqrt{\pi}} \sqrt{\frac{U}{\nu X}}, \tag{67}$$

When written in dimensionless variables, $\tau^N = \bar{\tau}^N / \rho U^2$ and $x = X/U\lambda N$, we get

$$\tau^N = \frac{1}{\sqrt{\pi}} \frac{1}{\beta} x^{-1/2}. \tag{68}$$

The Newtonian stress is singular like $x^{-1/2}$ at the leading edge of the plate, then it decreases. The viscoelastic wall shear stress (66) starts at zero, increases, and then decreases. Eventually the shear stress will decay like a Newtonian fluid with the same viscosity. Figure 8.9 shows that the force of a Maxwell model on a flat plate is greater in a viscoelastic fluid than a Newtonian one of the same viscosity.

8.5 The far field

Fraenkel found that at large distances from the leading edge of the plate, the flow is scarcely distinguishable from that of a Newtonian fluid with viscosity $\eta = \int_0^\infty G(s)\, ds$. Near the plate, there is the familiar boundary layer of parabolic shape; elsewhere in the far field, the dominant feature is the irrotational velocity due to the displacement effect of the retarded fluid in the boundary layer.

The region near the plate may be described as $x \to \infty$ with Y bounded where

$$\frac{y^2}{4Lx} = \hat{Y}^2 , \quad L = \frac{1}{M^2 N} , \qquad \text{x and y are dimensionless ,} \qquad (69)$$

$$\frac{Y^2}{4U\lambda X} = \hat{Y}^2 , \qquad \text{X and Y are dimensional .} \qquad (70)$$

1. Near the plate, there is a boundary layer in which the dominant velocity is given by

$$1 + u^R = \operatorname{erf} \hat{Y}^2 \qquad (71)$$

This is the linearized version of the velocity found by Blasius for the boundary layer on a flat plate. The thickness of the layer grows like $x^{1/2}$, in view of the definition of $\hat{Y}$. Moreover

$$v \sim v^R = -\left[\frac{L}{\pi x}\right]^{1/2} e^{-\hat{Y}^2} . \qquad (72)$$

2. The rotational velocity is exponentially small outside the boundary layer ($x \to \infty$, $\hat{Y} \to \infty$) and inside the Mach wedge.

3. The boundary layer on the plate may be described mathematically as the region $x>0$ and y is either bounded or tends to infinity only logarithmically. The dominant disturbance velocity outside the boundary layer is given asymptotically, as $r \to \infty$, as

$$u^H(x,y) = -\left[\frac{L}{\pi r}\right]^{1/2} \sin \theta/2 , \qquad (73)$$

$$v^H(x,y) = -\left[\frac{L}{\pi r}\right]^{1/2} \cos \theta/2 . \qquad (74)$$

where $(x,y)=r(\cos \theta, \sin \theta)$

This is the disturbance velocity for irrotational flow, with velocity (1,0) at infinity, past the parabola

$$y^2 = \frac{4L}{\pi}\left[x + \frac{L}{\pi}\right] \tag{75}$$

which represents (for $x \to \infty$) the displacement effect of the retarded fluid in the boundary layer. Note that $1+u^H+u^R$ and v^H+v^R are exactly zero on the plate.

8.6 Numerical solution of H. Hu

The Fourier transform solution is very useful for asymptotic results when x is either large or small. Hu used a numerical method to compute the Green function solution given in §8.1. After substituting (9) into (12) and using the polar coordinates $(x_o, y_o) = r(\cos\phi, \sin\phi)$, one finds that

$$-\pi = \int_0^{\phi_0}\int_0^{\infty} \frac{r^2\sin\phi}{r^2-2xr\cos\phi+x^2}\exp(-k\beta r\sin\phi)\, g[r(\cos\phi-\beta\sin\phi)]\, dr d\phi$$

$$+ k\beta \int_0^{\phi_0}\int_0^{\infty} \frac{r^3\sin^2\phi}{r^2-2xr\cos\phi+x^2} \int_{\beta r\sin\phi}^{r\cos\phi} g(r\cos\phi-\tau)e^{-k\tau} \tag{76}$$

$$\frac{I_1\left(k\sqrt{\tau^2-\beta^2r^2\sin^2\phi}\right)}{\sqrt{\tau^2-\beta^2r^2\sin^2\phi}}\, d\tau dr d\phi,$$

where $k = \dfrac{M^2}{2\beta^2 W}$ and $\phi_0 = \arctan\left(\dfrac{1}{\beta}\right)$ is the angle of the Mach wedge.

To solve the integral equation (76) for g(x), x>0, the domain $\Omega=[0,\infty]$ is divided into n small elements $\Omega=[0, x_1]\,\mathbf{U}\{\mathbf{U}_{i=1}^{n-1}[x_i, x_{i+1}]\}$

$U[x_n, \infty)$. g(x) is assumed to be piece-wise linear inside each element; that is,

$$g(x) = \sum_{i=1}^{n} g_i N_i(x) \qquad \text{for } i = 1,2,\ldots,n \tag{77}$$

where $g_i = g(x_i)$, and the shape function $N_i(x)$ is defined as

$$N_i(x) = \begin{cases} \dfrac{x-x_{i-1}}{x_i-x_{i-1}}, & x_{i-1} \le x \le x_i, \\[2ex] \dfrac{x_{i+1}-x}{x_{i+1}-x_i}, & x_i \le x \le x_{i+1}, \\[2ex] 0, \text{ elsewhere}, & \text{for } i=2, 3, \ldots, n-1, \end{cases} \tag{78}$$

while

$$N_1(x) = \begin{cases} \left(\dfrac{x}{x_1}\right)^{-\gamma}, & 0 < x \le x_1, \\[2ex] \dfrac{x_2-x}{x_2-x_1}, & x_1 \le x \le x_2, \\[2ex] 0, \text{ elsewhere}, \end{cases} \tag{79}$$

and

$$N_n(x) = \begin{cases} \dfrac{x-x_{n-1}}{x_n-x_{n-1}}, & x_{n-1} \le x \le x_n, \\[2ex] \left(\dfrac{x}{x_n}\right)^{-0.5}, & x \ge x_n, \\[2ex] 0, \text{ elsewhere}, \end{cases} \tag{80}$$

where $\gamma = \dfrac{1}{2} + \dfrac{1}{\pi}\arctan(\beta)$. In (79) and (80) we used Fraenkel's asymptotic results for small x and large x.

Then substituting (77) into (76), and integrating the equation on the domain Ω using the same $N_i(x)$ as weighting functions, one finds a system of linear equations for the g_i $(i=1,2,\ldots,n)$. Each coefficient in the system is expressed by multiple integrals. The integrations over domain Ω with weighting function $N_i(x)$ can be obtained analytically. The other integrations are handled numerically. For each weighting function $N_i(x)$ the innermost integration over τ in (76) is built up into a two dimensional array for different values of ϕ in $[0, \phi_o]$ and r in $[0, r_c]$, where r_c is a cut-off value of r. The value of this innermost integral is calculated later by quadratic interpolation from the data stored in the array. In building up the array the subroutine CADRE (**C**autious **Ad**aptive **R**omberg **E**xtrapolation) by D.D. Boor [1971] is used. The numerical integrations over ϕ and r are handled with an Adaptive Gaussian Method (AGM) by I. Robinson [1976]. The largest relative error criteria (for the integration over ϕ) used in the program is 10^{-4}.

8.7 Graphs of vorticity, velocity and stress

In the numerical computation, $M^2=2$ and $W=1$ (λU is taken as length scale ℓ), thus $\beta=1$, $k=1$ and the Mach wedge angle is $\phi_o = \frac{\pi}{4}$.

The vorticity along the plate surface g(x) is presented in Figure 8.2 where numerical results using both n=21 and n=51 are plotted. Fraenkel's asymptotic results for small and large x are also presented for comparison. The agreement is very good; g(x) tends to $-\infty$ as $x\to 0$ as $x^{-\gamma}=x^{-0.75}$, and is therefore integrable. When x is large, g(x) decays like $x^{-0.5}$.

Using this g(x) we computed the vorticity field from (9). Figure 8.3 gives a three dimensional plot of the vorticity field. As we approach the plate from upstream, the vorticity jumps from zero to $-\infty$ at the shock x=y, then drops very quickly like $(x-y)^{-0.75}$ when x>y. Far away from the shock the vorticity decays quite slowly, like $x^{-0.5}$.

A similar numerical technique was used for all integrals, the stream function (11) and its derivatives were integrated numerically. Figure 8.4 shows the horizontal component of velocity u(x,y) for different values of y. Figure 8.4(a) shows that there is a huge upstream influence. At $x = -100$, the value of this component of velocity is about −0.04, or 4% of the maximum velocity. Figure 8.4(b) shows details close to the shock, the magnitude of the velocity u first increases, then decreases smoothly in front of the shock, and increases suddenly after the shock. Away from the shock, far downstream, this velocity approaches −1, the velocity of the plate, very slowly. The first derivative of the velocity $\frac{\partial u}{\partial x}$ is singular at the shock.

Figure 8.5 presents the vertical component of velocity v(x,y) for different values of y. The velocity is zero at the plate surface y=0. This velocity has a single peak at the shock. The peak widens for large y. The height of the peak first increases, then decreases as y increases. The first derivative of the velocity $\frac{\partial v}{\partial x}$ is also singular at the shock.

Thus we see that the velocities are continuous across the shock. The first derivative of the velocity at the shock is singular. If we decompose the velocity into two components, tangential and normal to the shock, (u', v'), then

$$u' = u \cos\phi_o + v \sin\phi_o \,,$$
$$v' = -u \sin\phi_o + v \cos\phi_o \,.$$

For $\phi_o = \frac{\pi}{4}$, $u' = \frac{\sqrt{2}}{2}(u + v)$ and $v' = \frac{\sqrt{2}}{2}(-u + v)$. The first derivative of the velocity component tangential to the shock u' is singular at the shock, just like u and v. But for v', the velocity component normal to the shock, as shown in Figure 8.6, has a smooth slope at the shock, actually the peaks for u and v just cancel.

The stresses are shown in Figure 8.7 and Figure 8.8. Both the overall view and closer view near the shock are presented. There is a large stress overshoot of both σ and τ. The stresses are continuous across the shock. The normal stress has a peak with a singular first derivative at the shock. The shear stress is smooth with a continuous first derivative across the shock. Since the shock line is 45° from the horizontal, the stress normal to the shock line is τ and the stress tangential to the shock line is $-\sigma$. It follows then that the stress normal to the shock line is continuous and has continuous first derivative across the shock, and the shear stress on the shock

Figure 8.9 shows how the shear stress on the plate for a viscoelastic fluid compares with that for a Newtonian fluid with the same viscosity. The Newtonian stress is singular like $x^{-1/2}$ at the leading edge of the plate, then it decreases. The viscoelastic wall shear stress starts at zero, increases, and then decreases. Eventually the shear stress will decay as in a Newtonian fluid with the same viscosity. This Figure shows that the force on a flat plate is greater in the Maxwell model than in a Newtonian fluid of the same viscosity.

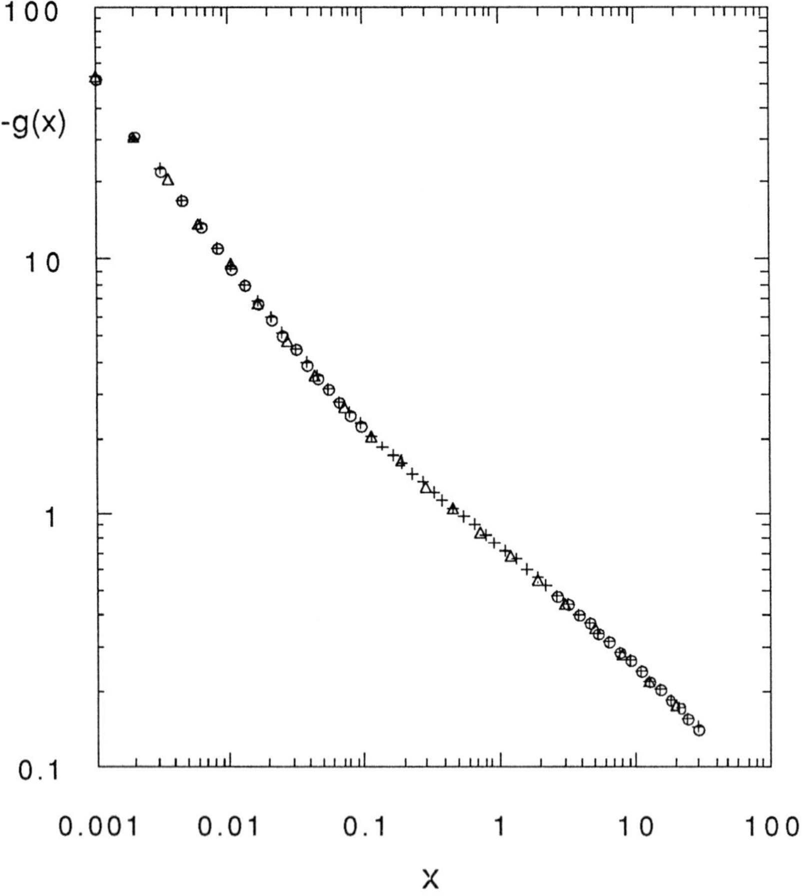

Figure 8.2. Vorticity along the plate surface g(x). Δ: numerical results with n=21; +: numerical results with n=51; o: Fraenkel's asymptotic results for small and large x. $g(x)\sim x^{-\gamma}=x^{-0.75}$ for small x and $g(x)\sim x^{-0.5}$ for large x.

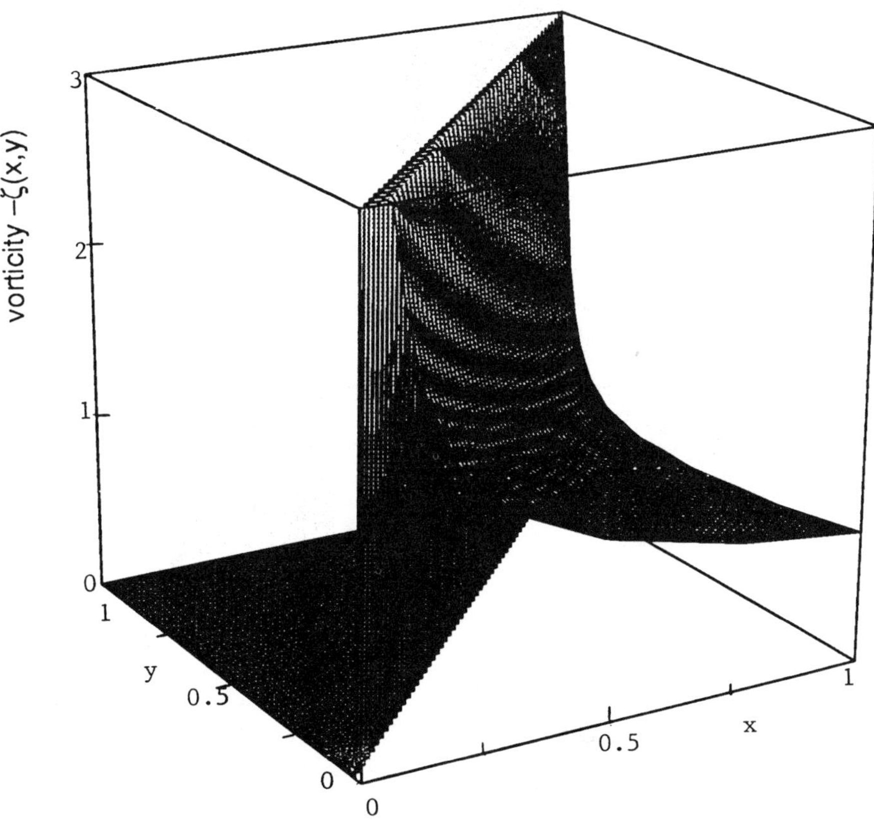

Figure 8.3. Three dimensional plot of the vorticity field $\zeta(x,y)$. The vorticity jumps from zero to $-\infty$ at the edge of the shock x=y, then rises quickly like $(x-y)^{-0.75}$ when x>y. Far away from the shock the vorticity decays slowly like $x^{-0.5}$.

Figure 8.4. The horizontal component of velocity (u(x,y) for different values of y. (a) Overall view of u(x,y). (b) Close view near the shock. In (a) we see a huge upstream influence in the form of a slow decay of velocity. The magnitude of the velocity decreases slowly and smoothly before the vorticity shock, after the shock there is a sudden rise in the vorticity. The first derivative of the velocity $\frac{\partial u}{\partial x}$ is singular at the shock.

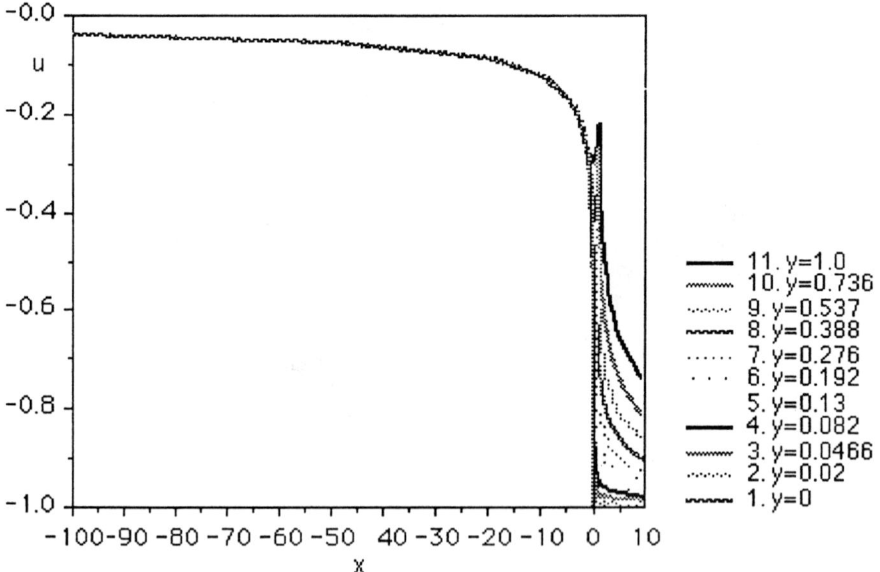

8.4 (a)

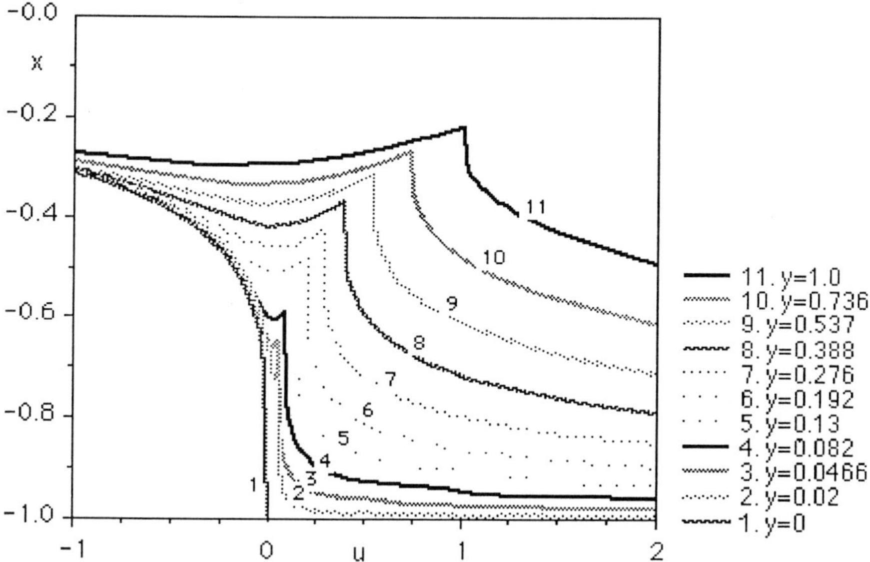
-0.0
x
-0.2
-0.4
-0.6
-0.8
-1.0
-1
0
u
1
2
11
10
9
8
7
6
5
4
3
2
1
11. y=1.0
10. y=0.736
9. y=0.537
8. y=0.388
7. y=0.276
6. y=0.192
5. y=0.13
4. y=0.082
3. y=0.0466
2. y=0.02
1. y=0

8.4 (b)

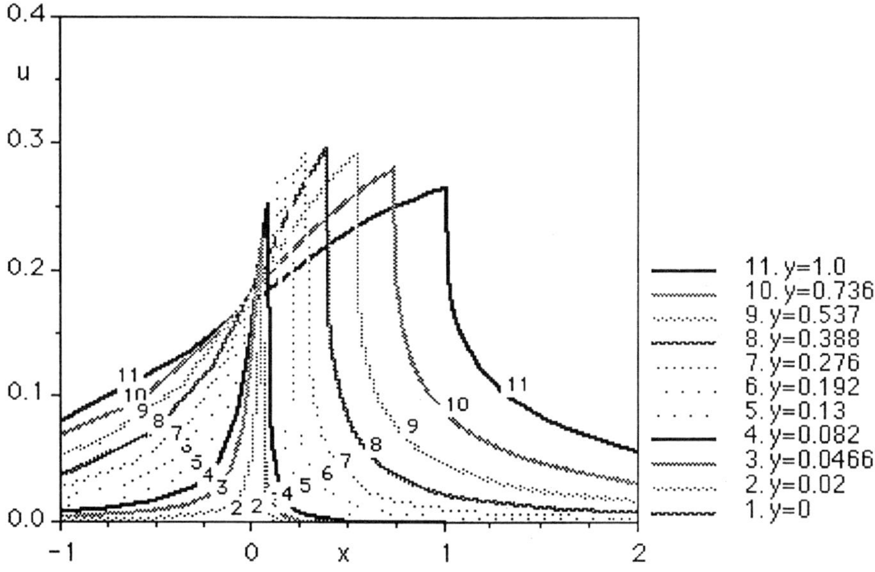

Figure 8.5. The vertical component of velocity v(x,y) for different values of y. The velocity is zero at the plate surface y=0 with a single peak at vorticity shock. The peak widens for large y. The height of the peak first increases, then decreases as y increases. The first derivative of the velocity $\frac{\partial v}{\partial x}$ is also singular at the shock.

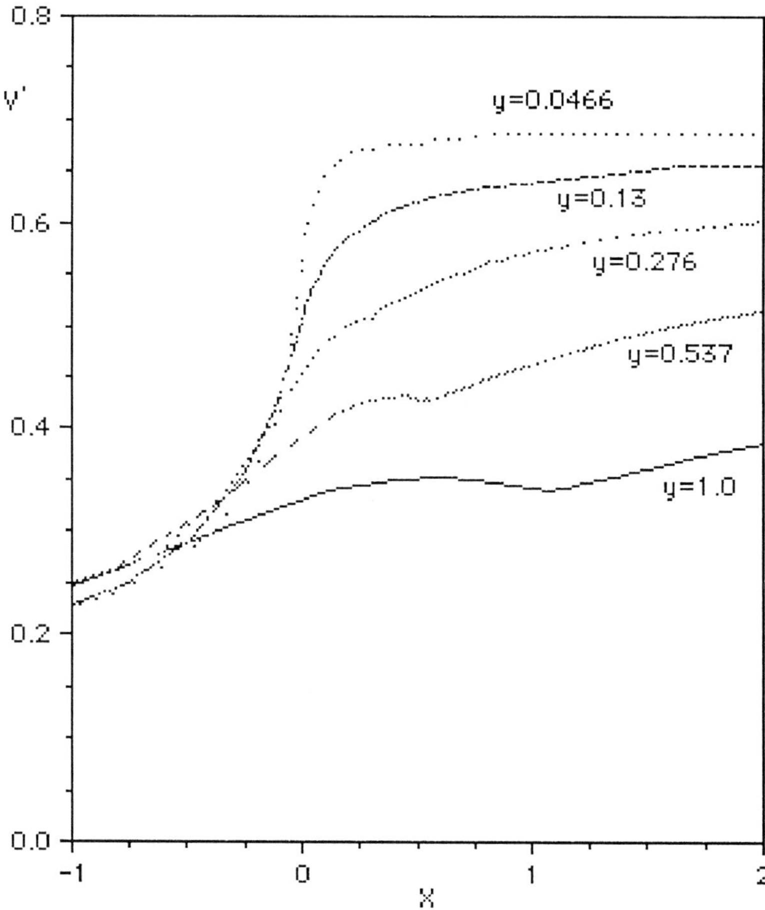

Figure 8.6. The component of velocity normal to the shock $v'(x,y)$ for different values of y. This velocity is smooth across the shock. The first derivative of the velocity $\frac{\partial v'}{\partial x}$ is continuous across the shock.

Figure 8.7. The normal stress $\sigma(x,y)$ for different values of y. (a) Overall view of $\sigma(x,y)$. (b) Close view near the shock.

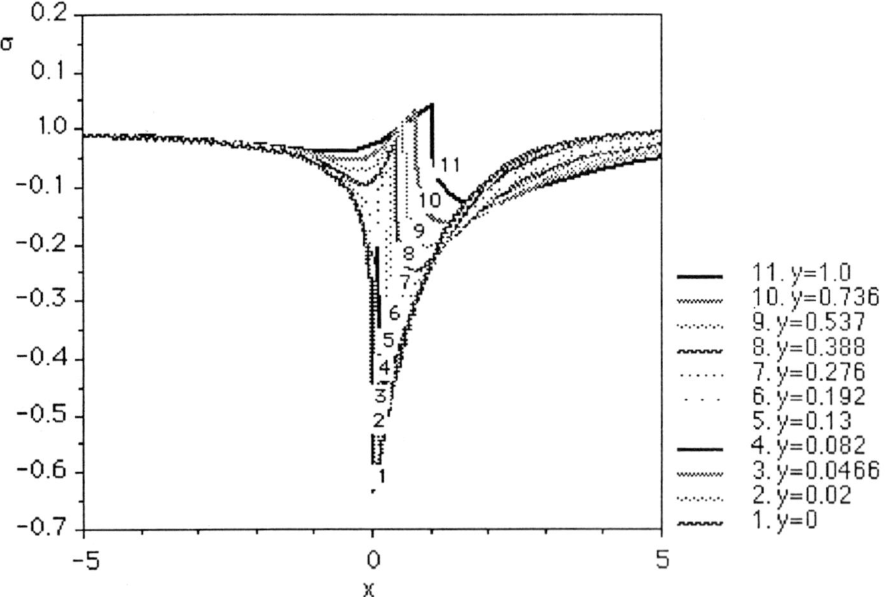

8.7 (a)

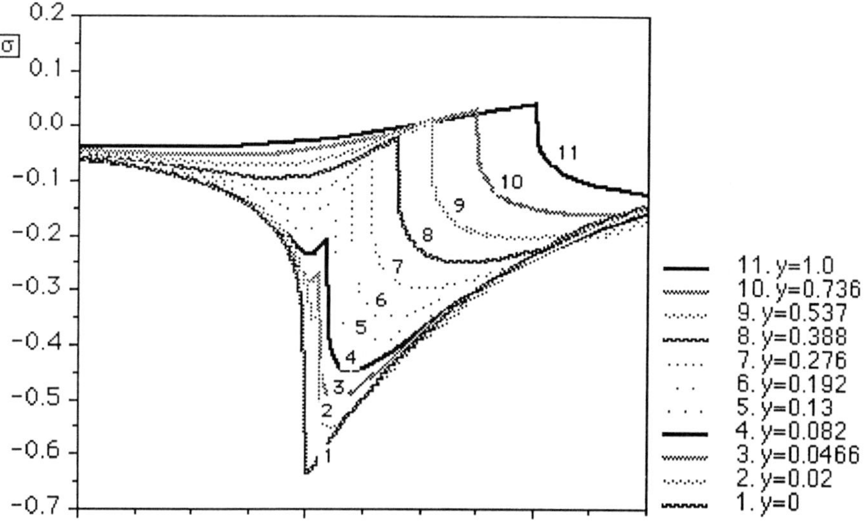
σ
0.2
0.1
0.0
-0.1
-0.2
-0.3
-0.4
-0.5
-0.6
-0.7
11. y=1.0
10. y=0.736
9. y=0.537
8. y=0.388
7. y=0.276
6. y=0.192
5. y=0.13
4. y=0.082
3. y=0.0466
2. y=0.02
1. y=0

8.7 (b)

Figure 8.8. The shear stress $\tau(x,y)$ for different values of y. (a) Overall view of $\tau(x,y)$. (b) Close view near the shock.

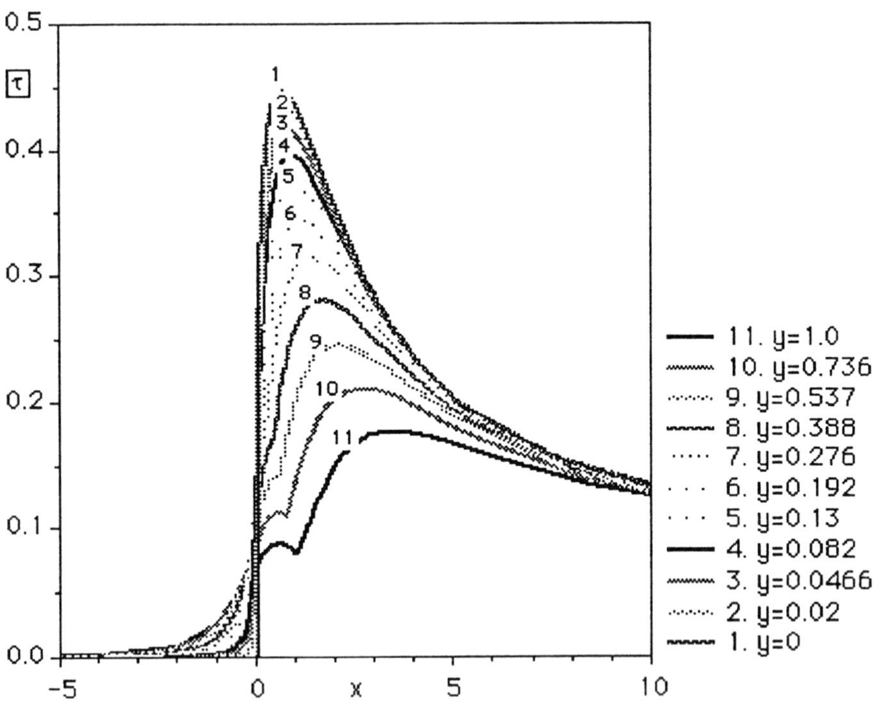

8.8 (a)

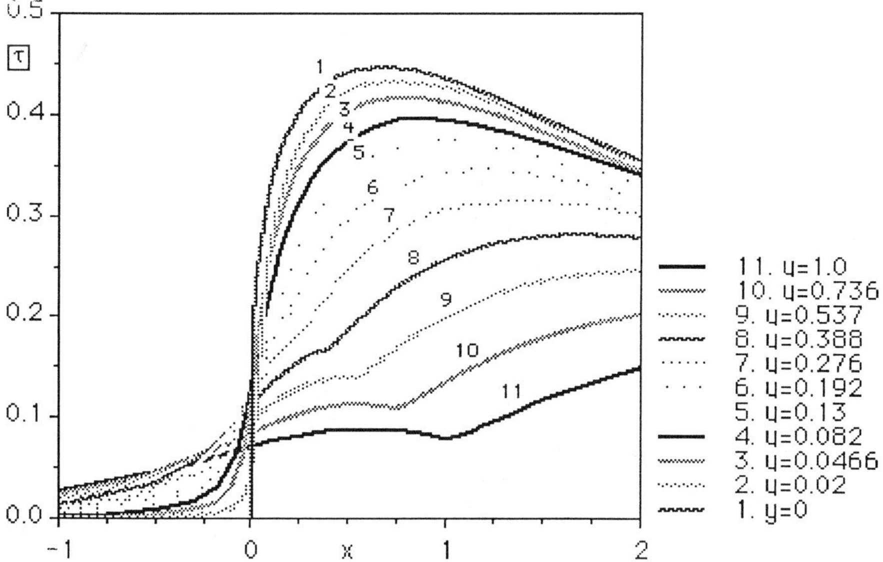
0.5
τ
0.4
0.3
0.2
0.1
0.0
-1
0
x
1
2
1
2
3
4
5
6
7
8
9
10
11
11. y=1.0
10. y=0.736
9. y=0.537
8. y=0.388
7. y=0.276
6. y=0.192
5. y=0.13
4. y=0.082
3. y=0.0466
2. y=0.02
1. y=0

8.8 (b)

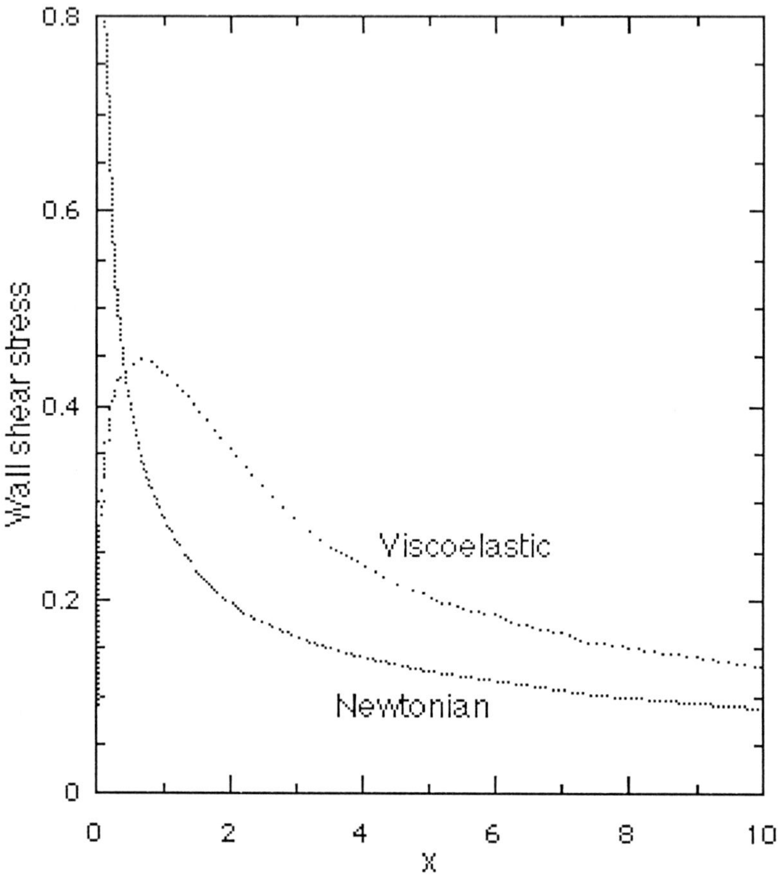

Figure 8.9. Comparison of the wall shear stress for a Maxwell model and a Newtonian fluid with the same viscosity. The force on a flat plate is greater in the Maxwell model.

8.8 Experiments of Hermes and Fredrickson

The experiments reported in the Ph.D. thesis of Hermes [1965] and in the paper of Hermes and Fredrickson [1967] were inspired by classical studies of the flow of a Newtonian fluid past a flat plate which is used to study boundary layers. Their stated goal was to characterize the flow patterns and measure the drag force for flow of viscoelastic fluid past a flat plate. They say that the experiments were undertaken to elucidate the qualitative effects of fluid elasticity and not intended to provide highly accurate data for uniform flow past a plate. A jet, falling freely through 3 cm from a $1\frac{1}{2}$ inch diameter pipe onto a 6 cm long flat plate with long sides along the stream, was used to simulate the boundary layer flow. This leads to an accelerated flow, rather than to uniform velocity, and the effects of finite jet boundaries also complicate the interpretation of their results.

Their working fluid was a 1.3% aqueous solution of grade 7H carboxymethylcellulose (CMC) powder made by the Hercules Powder Company. The wave speed was not measured by Hermes and Fredrickson; relaxation times are not given, so that estimates of the wave speed for their fluid cannot be given. They measure viscosity, apparently carefully and by two methods, and obtain a value of $\eta_0=17.2$ poise. This value disagrees with the manufacturer's data, which lists the viscosity of a 1% solution as between 16 to 30 poise. A value in excess of 100 poise would be expected of 1.3% solution. The effect of the small amount of alcohol and other additives used in their solutions is difficult to assess.

Riccius, Joseph and Arney [1987] measured the wave speed and zero shear rate viscosity of an aqueous 1.3% CMC solution (MN=160,000) manufactured by Louisiana Chemical Polymers in Baton Rouge, LA. They found that

$$c = 48 \text{ cm/s}, \quad \eta_0 = 108 \text{ poise}, \quad \lambda = \frac{\eta_0}{\rho c^2} = 0.047 \text{ sec}. \qquad (81)$$

We are going to pretend that these are the material parameters for the CMC solution used in the experiments. Therefore, our comparison may be closer to fantasy than reality.

Other warnings must be placed before readers. Velocities were measured by photographing tracer particles of hydrogen bubbles with a camera whose maximum framing speed was 90 frames/sec. This framing "... speed was not sufficiently fast to give an accurate determination of velocity in regions where it changes rapidly." In our Figure 8.10, we have reproduced a schematic diagram of the experiments discussed by Hermes and Fredrickson [1967]. The values shown in the Figure are the terminal y position of the tracer particles so that the curves may be said to give the u velocity of the fluid on different streamlines, to the extent that streaklines for hydrogen bubbles follow streamlines and give the fluid velocity. The sketch indicates that on all streamlines the velocity is U at the leading edge. This would be the case if there were no upstream influence, as in the Newtonian case. If there was no upstream influence, all the streamlines for $x<0$ would be straight and parallel. This is inconsistent with the data of Hermes [1965] on which their Figure 6 is based. For example, in Table A.7 of the thesis of Hermes, we find that $y=0.07$ cm at $x=0$ and $y=0.04$ cm for $x=-1.85, -0.90$, and -0.17 cm. The presence of upstream influence is consistent with the solutions of this chapter and Hermes' data in raw form, but not with their Figure 6.

Now we are going to assume that their aqueous, 1.3% CMC solution has a wave velocity and static viscosity close to the values given by (81). They report flow speeds of U = 104 cm/sec. Hence our assumption (81) gives rise to

$$M = \frac{U}{c} = 2.2 , \quad N = \frac{\beta^2}{M^2} = 0.77 . \tag{82}$$

The flow in the experiments was supercritical. The material length

$$U\lambda = 4.7 \text{ cm} . \tag{83}$$

is consistent with the observations in so much as the abrupt decreases in velocity shown in Figure 8.10 occurred in an interval of 3 or 4 cm from the leading edge. The graphs in §8.6 are based on Fraenkel's scale length

$$X = Lx , \quad L = U\lambda N = 3.63 \text{ cm} . \tag{84}$$

All of the events reported in Figure 8.10 are for $y \leq 0.80/3.63$ in the Fraenkel dimensionless coordinates used in §8.6.

Unfortunately the Hermes experiment could not resolve the existence of a shock of vorticity because the camera speed was too slow to determine "... velocity in regions where it changes rapidly." However, the observed behavior of the velocity is consistent with our theory in the following features:

1. The observed flow is supercritical and thus falls in the domain of the solution.

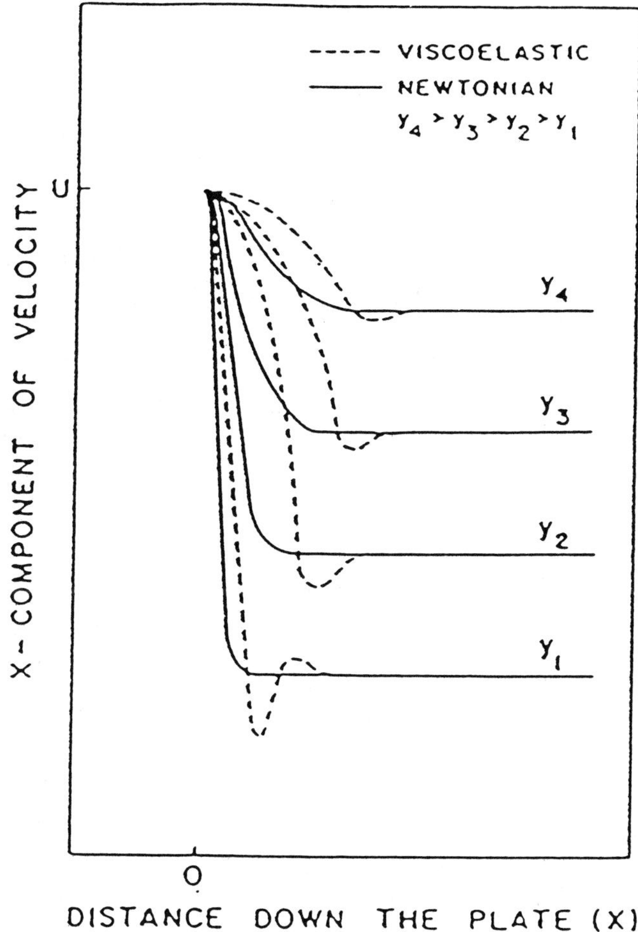

Figure 8.10. (Figure 6 of Hermes and Fredrickson, [1967]). Schematic diagram of observed kinematic behavior for viscoelastic fluids. The values of y_j, $j=1,2,3,4$ correspond to the downstream value of y on a streamline. Let $y_j=y_{jd}$ be the downstream y value of a certain streamline and y_{ju} be the upstream value on the same streamline. Then according to data given in Figure 5 of Hermes and Fredrickson $(y_{1u},y_1)=(0.04,0.11)$, $(y_{2u},y_2)=(0.10,0.26)$, $(y_{3u},y_3)= (0.25,0.60)$, $(y_{4u},y_4)=(0.45,0.80)$ where the y's are in centimeters.

2. Hermes and Fredrickson concluded that "elasticity is manifested in an oscillation in the velocity along a streamline." The theory gives rise to a shallow minimum before the shock

and the observed abrupt drop after the shock. The velocity is decidedly less monotonic for the Maxwell fluid that it is for a Newtonian one.

3. There is evidence in the raw data of Hermes [1965] for strong upstream influence.

4. Where the velocities of the two fluids differ most, it is the viscoelastic one that has the greater velocity.

5. The x intervals above the plate in which the viscoelastic fluid is significantly faster have lengths that increase (more or less) linearly with y in the experiments as they do in theory; these lengths are of the same order in the experiments and in the theory.

6. The observation that the drag force of a viscoelastic fluid on a flat plate is greater than in a Newtonian fluid of the same viscosity.

There surely are discrepancies between the theory and experiments, but they are probably impossible to evaluate because of various uncertainties. Likewise, the points of consistency between theory and experiment could vanish for more carefully constructed and documented experiments. The effects of nonlinearity and especially of a Newtonian viscosity, as in Jeffreys' model, ought to be considered.

CHAPTER 9

Hyperbolicity and Change of Type in Sink Flow

This chapter is based on the [1985] paper of Yoo, Ahrens, and Joseph. We consider the problem of steady fast flow of interpolated Maxwell models into a hole, and show that the flow field is partitioned into elliptic (subcritical) and hyperbolic (supercritical) regions. We analyze the characteristics and show that the vorticity changes type as in the experiments of Metzner, Uebler, and Fong [1969].

9.1 Introduction

In §15.3 we show that models which are much more general than Maxwell's have vorticity changing type, and that the vorticity of all motions of all models with instantaneous elasticity change type in small motions perturbing rigid motion. But it appears that not all models that change type are such that it is precisely the vorticity that changes type. We do not know if a given fluid fits some one constitutive model in all of its possible motions. Probably it never does. But if we are to progress we must certainly try to verify that some real fluids do exhibit the more striking properties implied by the model. A change of type to hyperbolic vorticity is one such striking property.

We are going to give a theory that leads to vorticity of changing type in the perturbation of sink flow. We apply this theory to the problem of flow into a hole, which could conceivably be regarded as perturbing sink flow in some sense. Many experiments on flow into a hole have been reported, but those by Metzner, Uebler, and Fong [1969] are notable because they exhibit a vorticity of changing type.

The plan of this chapter is as follows. In §9.2 we specify a constitutive model. We consider the interpolated Maxwell models. These models contain the upper and lower convected and the corotational Maxwell models, and they exhibit vorticity of changing type. The equations are analyzed for type in §9.3 and the characteristics are identified. The analysis of the quasilinear problem in §9.3 is without approximation. In §9.4 we linearize the quasilinear problem and show how the characteristics for any axisymmetric motion perturbing a basic one may be computed. In §9.5 we give the formulae for the characteristics of the axisymmetric motion that perturbs sink flow. In §9.6 we stretch our imaginations and imagine that flow into a hole is a small perturbation of sink flow, and we apply the results of such imaginative thinking to explain the curious experimental results of Metzner *et al.* [1969].

9.2 Interpolated Maxwell models with instantaneous elasticity

The interpolated Maxwell models are given by

$$\lambda \frac{\mathrm{D}\boldsymbol{\tau}}{\mathrm{Dt}} + \boldsymbol{\tau} = 2\eta \mathbf{D}\ , \tag{1}$$

where $\mathbf{D}$ is the symmetric part of the velocity gradient, $\boldsymbol{\tau}$ is determinate stress, λ is a relaxation time and D/Dt is an invariant derivative given by (2), can be said to have instantaneous elasticity. The fluids (1) have instantaneous elasticity when $\lambda \neq 0$, and the elastic response is more persistent when λ is large. Viscous fluids (Newtonian) have no elasticity, $\lambda=0$. The invariant derivative is given by

$$\frac{D\boldsymbol{\tau}}{Dt} = \frac{\partial\boldsymbol{\tau}}{\partial t} + (\mathbf{u}\cdot\nabla)\boldsymbol{\tau} + \boldsymbol{\tau}\mathbf{W} - \mathbf{W}\boldsymbol{\tau} - a(\mathbf{D}\boldsymbol{\tau}+\boldsymbol{\tau}\mathbf{D}) , \tag{2}$$

where $\mathbf{W} = \frac{1}{2}(\nabla\mathbf{u}-\nabla\mathbf{u}^T)$ and $-1 \le a \le 1$. All these derivatives are frame-indifferent. The Jaumann derivative has a=0. An upper convected Maxwell model arises from (2) when a=+1, while a=–1 corresponds to the lower convected Maxwell model and a=0 to the corotational model. These models may be evaluated in viscometric flows. Comparison of this evaluation with observations suggests that, of the three distinguished values a=–1,0,1, a=1 is best. More generally it is believed that the good a's are positive and slightly less than one (Astarita and Marrucci, [1974]).

We are interested in the quasilinear dynamical system associated with (1), (2) and

$$\operatorname{div}\mathbf{u} = 0 ,$$

$$\rho\left[\frac{\partial\mathbf{u}}{\partial t} + \mathbf{u}\cdot\nabla\mathbf{u}\right] = -\nabla p + \operatorname{div}\boldsymbol{\tau} . \tag{3}$$

9.3 Analysis of quasilinear systems in spherical coordinates

The analysis of quasilinear systems in orthogonal curvilinear systems is facilitated by the fact that only the terms with the highest derivatives enter into the analysis. This means that all of the extra terms, over and above Cartesian ones, which enter from geometry from the differentiation of base vectors, may be neglected. The systems of quasilinear equations that arise in this way have the same form, at highest order, as in Cartesian coordinates, except that the components are relative to the curvilinear system.

In the present analysis we consider (1)–(3) in spherical polar coordinates (r,θ,ϕ) where θ is the polar angle. In the axisymmetric problem we denote the velocity by

$$\mathbf{u} = (u,v,w) = (u,v,0) \tag{4}$$

and the extra stress by

$$\boldsymbol{\tau} = \begin{bmatrix} \sigma & \tau & 0 \\ \tau & \gamma & 0 \\ 0 & 0 & \beta \end{bmatrix} . \tag{5}$$

In addition, $\partial/\partial\phi = 0$. Equations (1)–(3) may then be written as

$$u_r + \frac{1}{r} v_\theta = -\frac{1}{r}(2u + v\cot\theta) , \tag{6}$$

$$\rho\left(uu_r + \frac{v}{r}u_\theta\right) + p_r - \sigma_r - \frac{1}{r}\tau_\theta = \frac{1}{r}\,[\rho v^2 + 2\sigma - \gamma - \beta + \tau\cot\theta] , \tag{7}$$

$$\rho\left(uv_r + \frac{v}{r}v_\theta\right) + \frac{1}{r}p_\theta - \tau_r - \frac{1}{r}\gamma_\theta$$
$$= \frac{1}{r}[-\rho uv + 3\tau + (\gamma - \beta)\cot\theta] , \tag{8}$$

$$u\sigma_r + \frac{v}{r}\sigma_\theta - \left[2(a\sigma+\mu)u_r + (a-1)\tau v_r + (a+1)\frac{\tau}{r}u_\theta\right]$$
$$= -\frac{\sigma}{\lambda} - (a-1)\frac{v\tau}{r} , \tag{9}$$

$$u\gamma_r + \frac{v}{r}\gamma_\theta - \left[2(a\gamma+G)\frac{1}{r}v_\theta + (a-1)\frac{\tau}{r}u_\theta + (a+1)\tau v_r\right]$$
$$= -\frac{\gamma}{\lambda} - \frac{1}{r}[(a-1)\tau v - 2(a\gamma+G)u] , \tag{10}$$

$$u\beta_r + \frac{v}{r}\beta_\theta = -\frac{\beta}{\lambda} + \frac{2}{r}(a\beta+G)(u + v\cot\theta) , \tag{11}$$

$$u\tau_r + \frac{v}{r}\tau_\theta - \frac{1}{2}\Big[[(a-1)\sigma + (a+1)\gamma + 2G]\frac{1}{r}u_\theta$$
$$+ [(a-1)\gamma + (a+1)\sigma + 2G]v_r\Big] \tag{12}$$
$$= -\frac{\tau}{\lambda} - \frac{1}{2r}\{[(a+1)\sigma + (a-1)\gamma + 2G]v + 2a\tau u + 2a\tau v\cot\theta\} ,$$

where the subscripts denote differentiation.

In axisymmetric flow there is only one component of vorticity, $\zeta \mathbf{e}_\phi$, and, using (1)–(3), we find that ω satisfies the following second-order equation:

$$A\frac{\partial^2\zeta}{\partial r^2} + 2B\frac{1}{r}\frac{\partial^2\zeta}{\partial\theta\partial r} + C\frac{1}{r^2}\frac{\partial^2\zeta}{\partial\theta^2} = \text{lower-order terms}\,, \tag{13}$$

where

$$\left.\begin{aligned} A &= \rho u^2 - \left[\frac{1}{2}(a+1)\sigma + \frac{1}{2}(a-1)\gamma + G\right], \\ B &= \rho uv - \tau\,, \\ C &= \rho v^2 - \left[\frac{1}{2}(a-1)\,\sigma + \frac{1}{2}(a+1)\,\gamma + G\right]. \end{aligned}\right\} \tag{14}$$

The differential equation for the characteristics of the vorticity is in the form

$$r\frac{d\theta}{dr} = \frac{-B\pm(B^2-AC)^{1/2}}{A}\,. \tag{15}$$

There are real characteristics if and only if

$$0 < B^2 - AC\,. \tag{16}$$

If $B^2-AC < 0$ the vorticity satisfies an elliptic equation. Since A,B,C depend on the solution, we might expect to see regions of the flow where the problem is hyperbolic and regions where it is elliptic.

It is also of interest to compute characteristics from the first-order system (6)–(12). We write this as

$$\boldsymbol{A}\frac{\partial \boldsymbol{q}}{\partial r} + \frac{\boldsymbol{B}}{r}\frac{\partial \boldsymbol{q}}{\partial\theta} = \boldsymbol{f}\,, \tag{17}$$

where

$$\boldsymbol{q} = (u,v,p,\sigma,\gamma,\beta,\tau)\,,$$

and $\boldsymbol{A},\boldsymbol{B},\boldsymbol{f}$ depend on $\boldsymbol{q}$, but not on its derivatives. We suppose that $\boldsymbol{q}$ is given on a curve $\phi(r,\theta)=0$. Suppose s is the arclength on the curve. Then

$$\frac{d\boldsymbol{q}}{ds} = \frac{\partial \boldsymbol{q}}{\partial r}\frac{\partial r}{\partial s} + \frac{\partial \boldsymbol{q}}{\partial \theta}\frac{\partial \theta}{\partial s} \tag{18}$$

is known on the curve. The curve $\phi(r,\theta)=0$ is characteristic if (17) and (18) cannot be uniquely solved for the derivatives of the components of $\boldsymbol{q}$.

This leads to the condition that

$$\det\left[\boldsymbol{A}\frac{d\theta}{ds} - \frac{\boldsymbol{B}}{r}\frac{dr}{ds}\right] = 0 . \tag{19}$$

So $r\, d\theta/dr$ is an eigenvalue of $\boldsymbol{B}$ relative to $\boldsymbol{A}$. Applying (19) to (6)–(12), we find that

$$(\alpha^2+1)(\alpha u+v)^3\left\{\left[\rho u^2 - \left(\frac{1}{2}(a+1)\sigma + \frac{1}{2}(a-1)\gamma + G\right)\right]\alpha^2 + 2(\rho uv-\tau)\alpha + \left[\rho v^2 - \left(\frac{1}{2}(a-1)\sigma + \frac{1}{2}(a+1)\gamma + G\right)\right]\right\} = 0 , \tag{20}$$

where

$$\alpha = -r\frac{d\theta}{dr} .$$

It follows that streamlines are triply characteristic. Streamlines always have real characteristics and do not exhibit a change of type. On the other hand, the remaining real root is the one we have already associated with the vorticity, and is given by (14) and (15). It is a real root only if (16) is satisfied.

There is a large literature on slip at walls. Slip is a discontinuity of velocity across a streamline. Slip surfaces must lie along streamlines. Slip mechanisms are said to be important in some of the phenomena collectively called melt fracture, but the appearance of slip surfaces in such phenomena seems to be controversial (for an interesting discussion and extensive list of references see Petrie and Denn [1976]). It is not known if the phenomenon called slip is a manifestation of some weak solution of the governing hyperbolic system that allows shocks or is something different, like cohesive fracture under shear.

The streamlines are characteristic but do not change type. In the models considered here the vorticity changes type and the characteristic directions for the vorticity do not lie along streamlines.

9.4 Quasilinear theory and linear theory

In §9.3 we classified the quasilinear system (1)–(3) without approximation. Certain definite results can be obtained from the quasilinear analysis under linearization. We first suppose that we have some exact solution of the governing equations. We are going to consider sink flow. We then linearize the equation around the exact solution and carry out an analysis of hyperbolicity and change of type of the linearized equations. The characteristics depend, of course, on the unperturbed flow. The analysis of the linearized equations is greatly simplified by the formulae we have already derived for the exact quasilinear system. To compute characteristic directions for the linearized systems we have only to put the values that the field variables take on in the unperturbed motion in the equations for the characteristics.

In the work to follow we treat the flow into a hole as a perturbation of sink flow. Sink flow is the axisymmetric irrotational solution in $\mathbb{R}^3$ given by

$$u = -\frac{Q}{r^2}, \quad v = 0\,. \tag{21}$$

The stresses in this flow are given by (9)–(12) as

$$\left.\begin{aligned} &\tau = 0\,, \\ &\frac{\sigma}{G} = 4e^{r^3/3\lambda Q}\, r^{-4a} \int_r^\infty s^{4a-1}\, e^{-s^3/3\lambda Q}\, ds\,, \\ &\frac{\gamma}{G} = -2e^{r^2/3\lambda Q}\, r^{2a} \int_r^\infty s^{-2a-1}\, e^{-s^3/3\lambda Q}\, ds\,. \end{aligned}\right\} \tag{22}$$

We shall also find β from (11) and then p from (7) and (8). Formulae for these quantities are not given here, because they are not needed in the formula for the characteristics of the perturbed flow.

Now we consider axisymmetric perturbations of sink flow. We linearize (6)–(12) around (21) and (22). The characteristics for the vorticity of the linearized equations are given by (15), using the expressions (21) and (22):

$$r\frac{d\theta}{dr} = \pm\left\{\frac{\frac{a-1}{2}\frac{\sigma}{G}+\frac{a+1}{2}\frac{\gamma}{G}+1}{\frac{\rho Q^2}{Gr^2}-\left[\frac{a+1}{2}\frac{\sigma}{G}+\frac{a-1}{2}\frac{\gamma}{G}+1\right]}\right\}^{\frac{1}{2}}. \tag{23}$$

It is generally true that the characteristics of the linearized problem are determined by the unperturbed problem. However, the streamlines, vorticity and stresses of the linearized problem cannot be determined by the unperturbed problem. To determine these quantities we must solve linearized boundary-value problems. In the present case we are considering the flow into a hole as representable in some sense by an axisymmetric perturbation of sink flow. Though we do not believe that such a large perturbation could be accurately represented by a linearized problem, we may hope to find results in some qualitative agreements with experiments.

The linearized problems should be solved relative to decay conditions on the solution as $r\to\infty$ and subject to the condition that $v=w=0$ on the plane containing the hole and $u=-Q/r^2$ on this plane. In fact, we shall not solve this linearized problem, and seek now only to compare the characteristics given by (23) with observations.

9.5 Characteristics for the vorticity of axisymmetric flow perturbing sink flow

Our aim now is to evaluate σ and γ for sink flow to obtain explicit forms of the characteristics (23) of the vorticity of axisymmetric flow which perturb sink flow. We first introduce dimensionless parameters

$$[R,\Sigma,\Gamma,S] = \left[\frac{r}{(\lambda Q)^{1/3}}, \frac{\sigma}{\mu}, \frac{\gamma}{\mu}, \frac{\rho Q^{2/3}}{\eta\lambda^{1/3}}\right], \tag{24}$$

where $\Sigma(R)$ and $\Gamma(R)$ are given by

$$\frac{d\Sigma}{dR} + 4a\frac{\Sigma}{R} - R^2\Sigma = -\frac{4}{R}, \tag{25}$$

$$\frac{d\Gamma}{dR} - 2a\frac{\Gamma}{R} - R^2\Gamma = \frac{2}{R}, \tag{26}$$

$$\Sigma = 4R^{-4a}\, e^{(1/3)R^3} \int_R^\infty t^{4a-1}\, e^{-(1/3)t^3}\, dt, \tag{27}$$

$$\Gamma = -2R^{2a}\, e^{(1/3)R^3} \int_R^\infty t^{-2a-1}\, e^{-(1/3)t^3}\, dt. \tag{28}$$

The characteristic equation (23) may be written in the dimensionless variables as

$$R\frac{d\theta}{dR} = \pm\left[\frac{\frac{1}{2}(a-1)\Sigma + \frac{1}{2}(a+1)\Gamma + 1}{S/R^4 - \left[\frac{1}{2}(a+1)\Sigma + \frac{1}{2}(a-1)\Gamma + 1\right]}\right]^{\frac{1}{2}}. \tag{29}$$

These are real characteristics whenever

$$B^2 - AC = \left\{S/R^4 - \left[\frac{1}{2}(a+1)\Sigma + \frac{1}{2}(a-1)\Gamma + 1\right]\right\} \left\{\frac{1}{2}(a-1)\Sigma + \frac{1}{2}(a+1)\Gamma + 1\right\} > 0. \tag{30}$$

We shall consider three cases: a=1,0, and –1. Then it can be readily shown by integration by parts from (27)–(28) that

$$(\Sigma,\Gamma) = \begin{cases} \left[\dfrac{4}{R^3} + \dfrac{4\Psi}{R^4},\ -1 + R^2\Psi\right] \ (a = 1), \\ (-2\Gamma,\Gamma) \ (a = 0), \\ \left[1 - R^3 - \dfrac{1}{2}R^4\Phi,\ \dfrac{1}{R^2}\Phi\right] \ (a = -1), \end{cases} \tag{31}$$

where

$$\left.\begin{aligned} \Psi &= e^{(1/3)R^3}\int_R^\infty e^{-(1/3)t^3}dt = \frac{1}{R^2} - \frac{2}{R^5} + \frac{5\cdot 2}{R^8} - \frac{8\cdot 5\cdot 2}{R^{11}} + \ldots, \\ \Gamma &= -2e^{(1/3)R^3}\int_R^\infty t^{-1}e^{-(1/3)t^3}dt = -2\left[\frac{1}{R^3} - \frac{3}{R^6} + \frac{6\cdot 3}{R^9} - \frac{9\cdot 6\cdot 3}{R^{12}} + \ldots\right], \\ \Phi &= -2e^{(1/3)R^3}\int_R^\infty te^{-(1/3)t^3}dt = -\frac{2}{R}\left[1 - \frac{1}{R^3} + \frac{4}{R^6} - \frac{7\cdot 4}{R^9} + \ldots\right]. \end{aligned}\right\} \tag{32}$$

It is also useful to note by applying L'Hospital's rule to (27)–(28) that as $R\to 0$

$$(\Sigma,\Gamma) \to \begin{cases} \left[\dfrac{4}{3}3^{(4/3)a}R^{-4a}\tilde{\Gamma}\left(\dfrac{4}{3}a\right),\ -\dfrac{1}{a}\right] & (0 < a \le 1), \\ (-4\ln R,\ 2\ln R) & (a = 0), \\ \left[-\dfrac{1}{a},\ -\dfrac{2}{3}3^{-(2/3)a}R^{2a}\tilde{\Gamma}\left(-\dfrac{2}{3}a\right)\right] & (-1 \le a < 0), \end{cases} \tag{33}$$

where $\tilde{\Gamma}$ is the gamma function.

The regions of hyperbolicity $B^2-AC>0$ and the characteristic curves are now given by

$$\left[B^2-AC\,,\frac{d\theta}{dR}\right] =$$

$$\begin{cases}\left[R^2\Psi\left\{\frac{S}{R^4}-\left(\frac{4\Psi}{R^4}+\frac{4}{R^3}+1\right)\right\},\pm\left\{\frac{R^4\Psi}{S-(4\Psi+4R+R^4)}\right\}^{1/2}\right] & (a=1)\,,\\[2ex] \left[\left(\frac{S}{R^4}+\frac{3}{2}\Gamma-1\right)\left(\frac{3}{2}\Gamma+1\right),\pm\left\{\frac{R^2\left(\frac{3}{2}\Gamma+1\right)}{S+R^4\left(\frac{3}{2}\Gamma-1\right)}\right\}^{1/2}\right] & (a=0)\,,\\[2ex] \left[\left(\frac{S}{R^4}+\frac{\Phi}{R^2}-1\right)\left(R^3+\frac{1}{2}R^4\Phi\right),\pm\left\{\frac{R^2\left(R^3+\frac{1}{2}R^4\Phi\right)}{S+R^2\Phi-R^4}\right\}^{1/2}\right] & (a=-1).\end{cases}$$

In the regions of hyperbolicity the characteristics are of the form $\theta = \theta_0 \pm g(R)$, where $\theta_0 \in [0,\pi)$ is a constant of integration. The functions g(R) are more or less spirals in planes containing the axis of symmetry, the exact form of these spirals are shown in Figure 9.1 and in the Figures of §9.6. When a=1 and a=–1 the vorticity is hyperbolic in a sphere around the origin. The characteristic surfaces are cones near the origin, and they are tangent to the spherical border between hyperbolic and elliptic regions. This border is an inflow boundary, so that the characteristics are tangent to a spherical inflow boundary defining change of type. A typical characteristic net for a=1 is shown in Figure 9.1. The characteristic net for the corotational model a=0 is in a spherical shell, hyperbolic in the shell, elliptic in a neighborhood of the origin and outside the shell (see Figure 9.4).

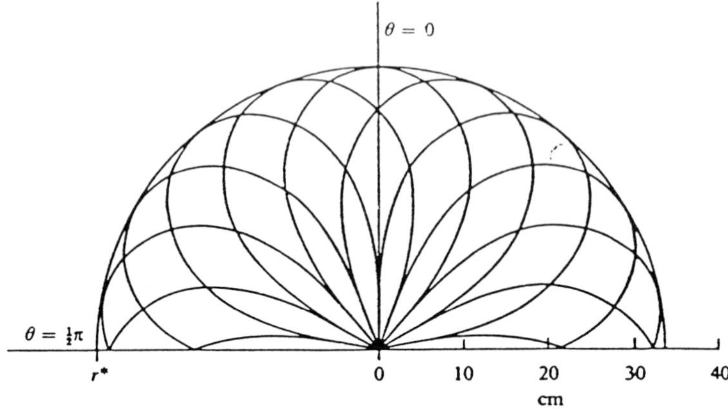

Figure 9.1. Typical characteristics net for the vorticity of an axisymmetric flow perturbing sink flow of an upper convected Maxwell model (a=1, Q=3354 cm^3/s, r^*=33.4 cm).

9.6 Discontinuities of vorticity in steady flow into a hole

In the experiments of Metzner *et al.* [1969] a fluid is sucked from a pipe of large diameter through a sudden contraction (Figure 9.2). If the hole into which the flow goes is small the problem may be thought to be a form of sink flow. Because there are boundary walls, the flow through a sudden contraction is not a sink flow in a strict sense. We shall imagine first that the flow into the hole is not strongly influenced by the walls of the large pipe. We then have a hole in the semi-infinite region above a plane. This flow is then regarded as an axisymmetric perturbation of sink flow without boundaries.

There are many reports of experiments on flow through a sudden contraction but the only one which has information about vorticity near the sink is the paper of Metzner, *et al.* [1969]. They consider high-speed flow of viscoelastic fluids into a sudden contraction (see Figure 9.2). They say that

A tentative analysis of the observed velocity field suggests the flow upstream of the small duct to be radially directed toward the origin of the spherical coordinate system. If this is so the continuity equation gives

$$ur^2 = f(\theta) . \tag{34}$$

They actually measure velocities in the cone, and they report that their measurements were accurate and that $f(\theta)$ may be taken as constant when $0 \le \theta \le 10°$. They give experimental results and say that:

In the central conical region of the velocity field in which the field shearing deformations are negligibly small we may write equation (34) as

$$u = \frac{Q}{r^2} , \tag{35}$$

where $f(\theta)$ has been replaced by the constant Q, which is proportional to the volumetric flow rate q. Under the conditions studied, equation (35) applies to about 70% of all the fluid entering the tube from upstream. For purposes of further analysis we will restrict our attention to this central region ($0 \le \theta \le 10°$); in this region the kinematics of the flow process, in spherical coordinates, are especially simple:

$$u_r = u = \frac{Q}{r^2} , \tag{36}$$

$$u_\theta = u_\phi = 0 . \tag{37}$$

These equations may be shown to give rise to a diagonal deformation rate tensor.

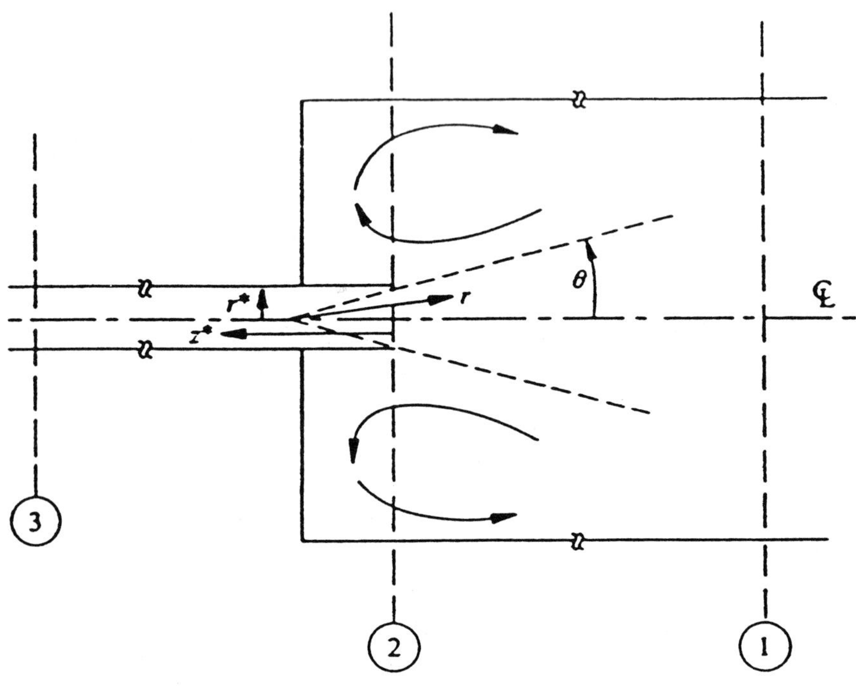

Figure 9.2. Schematic diagram of flow into a sudden contraction (Metzner *et. al.* [1969]).

Newtonian fluid: 70% Globe brand white corn syrup in water: μ = 1.96 poise

Viscoelastic fluid: 0.5% Separan AP 30, a partially hydrolysed polyacrylamide, in water

Flow behavior index n = 0.424, measured rheogonimetrically†

Flow rates:

Figure IX.3: 5.45 gal./min.
Figure IX.4: 7.25 gal./min.
Figure IX.5: 21.75 gal./min.

Geometry of equipment:

Downstream tube:

I.D. = 1.48 in.
O.D. = 2.00 in.

Upstream reservoir: square cross section 18 in. on a side. A distributor plate, to develop equal axial velocities at all radial positions, was mounted 10 in. upstream from the entrance to the small tube.

Illumination: through the vertical central plane of the apparatus. Width (depth) of illuminated plane: $\frac{1}{16}$ in.

Tracer particles: small air bubbles.

Temperature: all data reported were taken at 22.0±2 °C.

† Curves of both the shearing stress and the first normal stress differences are available for the non-Newtonian fluid used.

Table 9.1. Experimental conditions corresponding to Figures 9.3 to 9.5.

The observed flow is irrotational and has a potential $\phi = Q/r$ in the central region. There is apparently a non-zero vorticity outside this central region. Derivatives of the vorticity must be discontinuous on the border of the central region.

The reader should now examine the characteristic nets shown for sink flow in Figure 9.1. We take two members of this characteristic family separated by a cone angle of 10°. The hyperbolic character of the system allows potential flow in the cone and flow with vorticity outside of it. We wish to compare these theoretical cones of hyperbolicity with the three experiments reported by Metzner *et al.* The three experiments correspond to sinks of increasing strength. Their Figure 8 is for q=7.25 gal/min=460 cm^3/s, Figure 9 is for q=14.5 gal/min=920 cm^3/s, Figure 10 is for q=21.75 gal/min=1380 cm^3/s.

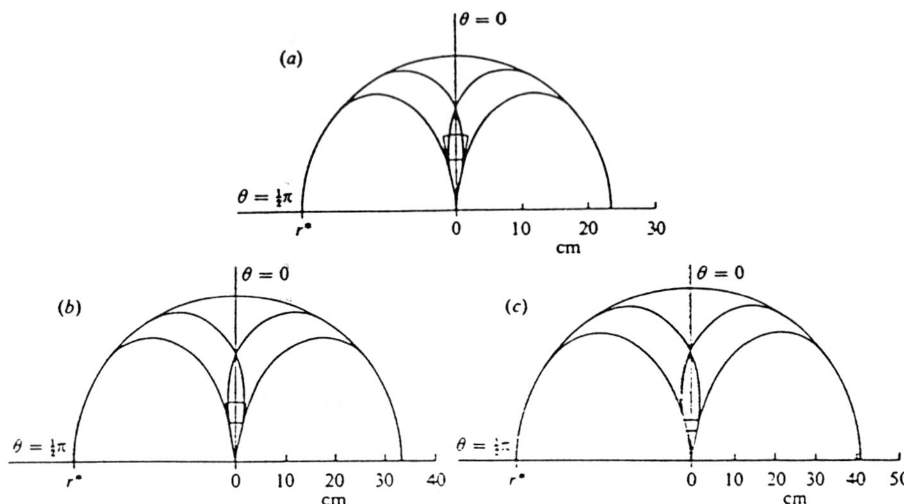

Figure 9.3. Comparison of the characteristics for the vorticity of an upper convected Maxwell model in an axisymmetric flow perturbing sink flow with experiments on flow into a hole: (a) q=460 cm^3/s, S=143.21, r^*=23.430 cm, R^*=3.3723 ($S \approx 4R^* + R^{*4}$); (b) q=920 cm^3/s, S=227.33, r^*=33.396 cm, R^*=3.8150 ($S \approx 4R^* + R^{*4}$); (c) q=1380 cm^3/s, S=297.89, r^*=41.036 cm, R^*=4.0952 ($S \approx 4R^* + R^{*4}$).

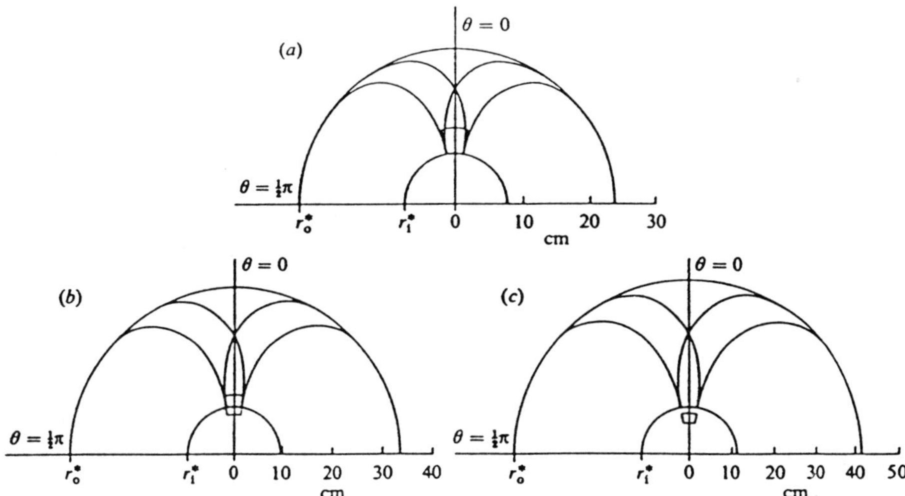

Figure 9.4. Comparison of the characteristics for the vorticity of a corotational Maxwell model in an axisymmetric flow perturbing sink flow with experiments on flow into a hole: (a) q=460 cm^3/s, S=143.21, r_i^*=7.6009 cm, R_i^*=1.094, r_o^*=23.625 cm, R_o^*=3.4003 ($S \approx R_o^{*4}$); (b) q=920 cm^3/s, S=227.33, r_i^*=9.5765 cm, R_i^*=1.094, r_o^*=33.572 cm, R_o^*=3.8351 ($S \approx R_o^{*4}$); (c) q=1380 cm^3/s, S=297.89, r_i^*=10.962 cm, R_i^*=1.094, r_o^*=41.20 cm, R_o^*=4.1123 ($S \approx R_o^{*4}$).

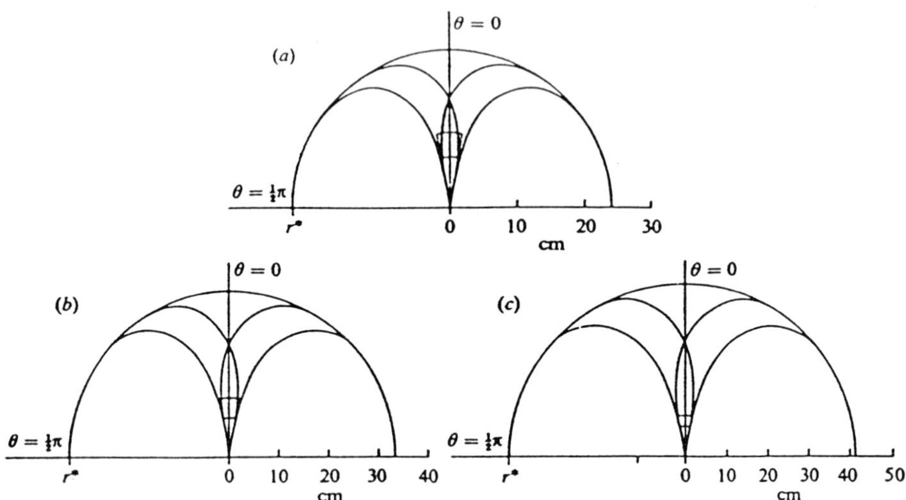

Figure 9.5. Comparison of the characteristics for the vorticity of a lower convected Maxwell model in an axisymmetric flow perturbing sink flow with experiments on flow into a hole: (a) q=460 cm^3/s, S=143.21, r^*=23.89 cm, R^*=3.4378 ($S \approx R^{*4}$); (b) q=920 cm^3/s, S=227.33, r^*=33.70 cm, R^*=3.8542 ($S \approx R^{*4}$); (c) q=1380 cm^3/s, S=297.89, r^*=41.340 cm, R^*=4.1255 ($S \approx R^{*4}$).

To compute the regions of hyperbolicity and the form of the characteristic surfaces we need values for the dimensionless flow-rate parameter S and the scale length $(\lambda Q)^{1/3}$, which is used to form the dimensionless radius. For these parameters we need values for the viscosity η, a relaxation time λ, the sink strength Q, and the density ρ. The viscoelastic fluid used in the experiments was 0.5% Separan AP30, a partially hydrolysed polymer. We take the density of this solution as ρ=0.89 g/cm^3. We used the values η=3 poise, λ=0.1 s given by Metzner [1967] for a 0.45% Separan AP30 solution. The shear-wave speed is $C = (\eta/\lambda\rho)^{1/2} = 5.63$ cm/s. To compute Q we use the observation that the sink-flow formula (35) of Metzner, *et al.* applies to 'about 70%...'. Then 0.7q gal/min is equal to the flow into the conical region (of hyperbolicity?) with apex angle (θ=10°); that is

$$0.7q = u_r\, A = \frac{Q}{r^2} A = Q2\pi(1-\cos\theta)$$

where $2\pi r^2(1-\cos\theta)$ is the area A of a spherical cap on the cone. Therefore

$$Q = 0.7q/2\pi(1-\cos 10°)\ . \tag{38}$$

With these values given we compute the values given in Table 9.1.

We shall now compare the results of our analysis with the experiment. This is done in three groups (9.3a-c; 9.4a-c; 9.5a-c) of Figures corresponding to three entries of Table 9.1. For each of the three sink strengths q we show the following.

(i) The region of hyperbolicity. These are spheres of outer radius r^* when a=1 or a=–1 and annular regions between spheres of inner radius r_i^* and outer radius r_o^* when a=0. The regions inside the circles of radius r^* in Figures 9.3 and 9.5 are hyperbolic. The annular region in Figure 9.4 between r_o^* and r_i^* is hyperbolic. The vorticity is elliptic where it is not hyperbolic.

(ii) On each of Figures 9.3–5 we draw a cone with semivertex angle 10° centered at the origin. Metzner *et al.* found that the flow in this cone was the potential flow (36)–(37). Actually they verified the relationships (36)–(37) for different values of the radius r, say $A \le r \le B$ and different θ, $0 \le \theta \le 10°$. The conical region between the spheres of radius A and B where (36)–(37) was verified in the experiments is shown in Figures 9.3-5.

(iii) We next draw on each of the Figures 9.3-5 all of the cross-sections of characteristic surfaces of revolution that are tangent to the central cone at the origin. There are two characteristic surfaces of revolution, which come in at an angle of 10° at the origin (see Figure 9.1), and the four dark lines on the Figure represent the cross-sections of the two surfaces.

We regard the comparisons of theory and experiment of this analysis as exploratory and not definitive. It is of course striking that the experiments of Metzner *et al.* [1969] do appear to involve a vorticity of changing type. It would be interesting to see if this striking type of experimental result could be repeated by other investigators using different fluids and experimental arrangements. We hasten to add that the Separan solution used in the experiment is not a Maxwell model and surely cannot be characterized by a viscosity and relaxation time. In fact only special models give the vorticity precisely as the quantity which changes type. We have already remarked that models with true viscosity, e.g. retardation times, will smooth discontinuities, with only a little smoothing if the retardation 'viscosity' parameters is small. Probably all the polymer solutions used in experiments have some small smoothing. In view of all these uncertainties in theory and experiments, it would be premature to make strong claims.

Exercises

Exercise 9.1.
Put a=0 in (25)–(28) and show that $\gamma-\sigma\to 6\mu \ln r$ as $r\to 0$. Now put a=0 and $\rho=0$ in (23) and show that sink flow of a corotational Maxwell fluid is Hadamard unstable.

Exercise 9.2.
Derive the vorticity equation (13) from (6)–(12).

Solutions for the exercises

Exercise 9.1.
Put a=0 in (25) and (26), we have

$$\frac{d\sigma}{dR} - R^2\sigma = -\frac{4}{R}G,$$

$$\frac{d\gamma}{dR} - R^2\gamma = \frac{2}{R}G$$

or

$$\frac{d(\gamma-\sigma)}{dR} - R^2(\gamma-\sigma) = \frac{6}{R}G.$$

As $R \to 0$, $R^2(\gamma-\sigma)$ can be neglected relative to the other terms. Thus

$$\frac{d}{dR}(\gamma-\sigma) = \frac{6}{R}G$$

that is $\gamma-\sigma = 6\mu \ln R$ as $R \to 0$. The characteristics for the vorticity of linearized equations are

$$r\frac{d\theta}{dr} = \pm\left\{\frac{\frac{a-1}{2}\frac{\sigma}{G}+\frac{a+1}{2}\frac{\gamma}{G}+1}{\frac{\rho Q^2}{Gr^2}-\left[\frac{a+1}{2}\frac{\sigma}{G}+\frac{a-1}{2}\frac{\gamma}{G}+1\right]}\right\}^{\frac{1}{2}}.$$

For the corotational Maxwell model (a=0), if $\rho=0$ we have

$$r\frac{d\theta}{dr} = \pm\left[\frac{\gamma-\sigma+2G}{\gamma-\sigma-2G}\right]^{1/2}$$

where $r \to 0$, $\gamma-\sigma \to 6G \ln R \to -\infty$. Therefore

$$\frac{\gamma-\sigma+2G}{\gamma-\sigma-2G} > 0 \quad \text{as} \quad r \to 0 .$$

This steady sink flow is hyperbolic when $\rho=0$. Therefore the unsteady vorticity equation is elliptic and the quasilinear system is ill-posed (Hadamard unstable).

Exercise 9.2.
Derive the vorticity equation (13) from (6)–(12).

$$\left.\begin{aligned} \sigma_r + \frac{\tau_\theta}{r} &= \rho\left(uu_r + \frac{v}{r}u_\theta\right) + p_r + \text{lot} \\ \tau_r + \frac{\gamma_\theta}{r} &= \rho\left(uv_r + \frac{v}{r}v_\theta\right) + \frac{1}{r}p_\theta + \text{lot} \end{aligned}\right\} \tag{39}$$

and

$$u\sigma_r + \frac{v}{r}\sigma_\theta - \left[2(a\sigma+G)u_r + (a-1)\tau v_r + (a+1)\frac{\tau}{r}u_\theta\right] = \text{lot}\,, \tag{40}$$

$$u\tau_r + \frac{v}{r}\tau_\theta - \frac{1}{2}\Big\{\big[(a-1)\sigma+(a+1)\gamma+2G\big]\frac{u_\theta}{r} + \big[(a-1)\gamma+(a+1)\sigma+2G\big]v_r\Big\} = \text{lot}\,, \tag{41}$$

$$u\gamma_r + \frac{v}{r}\gamma_\theta - \left[2(a\gamma+G)\frac{v_\theta}{r} + (a-1)\frac{\tau}{r}u_\theta + (a+1)\tau v_r\right] = \text{lot}\,. \tag{42}$$

We may write

$$\frac{\partial}{\partial r}(40) + \frac{\partial}{r\partial\theta}(41)$$

as

$$u\frac{\partial}{\partial r}\left(\sigma_r + \frac{\tau_\theta}{r}\right) + \frac{v}{r}\frac{\partial}{\partial\theta}\left(\sigma_r + \frac{\tau_\theta}{r}\right) - \left[2(a\sigma+G)u_{rr} + (a-1)\tau v_{rr} + (a+1)\frac{\tau}{r}u_{\theta r}\right] - \left[\frac{c_1}{r^2}u_{\theta\theta} + \frac{c_2}{r}v_{r\theta}\right] = \text{lot} \tag{43}$$

where

$$c_1 = \frac{a-1}{2}\sigma + \frac{a+1}{2}\gamma + G$$
$$c_2 = \frac{a-1}{2}\gamma + \frac{a+1}{2}\sigma + G$$

We may write

$$\frac{\partial}{\partial r}(41) + \frac{\partial}{r\partial\theta}(42)$$

as

$$u\frac{\partial}{\partial r}\left(\tau_r + \frac{\gamma_\theta}{r}\right) + \frac{v}{r}\frac{\partial}{\partial\theta}\left(\tau_r + \frac{\gamma_\theta}{r}\right) - \left[\frac{c_1}{r}u_{r\theta} + c_2 v_{rr}\right] - \left[2(a\gamma+G)\frac{v_{\theta\theta}}{r^2} + (a-1)\frac{\tau}{r^2}u_{\theta\theta} + (a+1)\tau\frac{v_{r\theta}}{r}\right] = \text{lot} \tag{44}$$

We may write

$$\frac{\partial}{\partial r}(44) - \frac{\partial}{r\partial\theta}(43)$$

as

$$u\left[\frac{\partial^2}{\partial^2 r}\left(\tau_r + \frac{\gamma\theta}{r}\right) - \frac{\partial^2}{r\partial\theta\partial r}\left(\sigma_r + \frac{\tau_\theta}{r}\right)\right]$$

$$+ \frac{v}{r}\left[\frac{\partial^2}{\partial\theta\partial r}\left(\tau_r + \frac{\gamma\theta}{r}\right) - \frac{\partial^2}{r\partial\theta^2}\left(\sigma_r + \frac{\tau_\theta}{r}\right)\right]$$

$$-\left(\frac{c_1}{r}u_{\theta rr} + c_2 v_{rrr}\right) + \left(\frac{c_1}{r^3}u_{\theta\theta\theta} + \frac{c_2}{r^2}v_{r\theta\theta}\right) \tag{45}$$

$$-\left[2(a\gamma+G)\frac{v_{\theta\theta r}}{r^2} + (a-1)\frac{\tau}{r^2}u_{\theta\theta r} + (a+1)\tau\frac{v_{r\theta r}}{r}\right]$$

$$+\left[2(a\sigma+G)\frac{u_{rr\theta}}{r} + (a-1)\frac{\tau}{r}v_{rr\theta} + (a+1)\frac{\tau}{r^2}u_{\theta r\theta}\right] = \text{lot}\,.$$

After substituting (39) into (45) and take the highest order derivatives of u and v, we get

$$\rho u\left[u v_{rrr} + \frac{v}{r}v_{\theta rr} - u\frac{u_{rr\theta}}{r} - \frac{v}{r^2}u_{\theta r\theta}\right]$$

$$+ \rho\frac{v}{r}\left[u v_{rr\theta} + \frac{v}{r}v_{\theta r\theta} - u\frac{u_{r\theta\theta}}{r} - \frac{v}{r^2}u_{\theta\theta\theta}\right]$$

$$+ (-c_1 + 2a\sigma + 2G)\frac{u_{\theta rr}}{r} - c_2 v_{rrr} + \frac{c_1}{r^3}u_{\theta\theta\theta} + (c_2 - 2a\gamma - 2G)\frac{v_{r\theta\theta}}{r^2}$$

$$+ 2\frac{\tau}{r^2}u_{\theta\theta r} - 2\frac{\tau}{r}v_{rr\theta} = \text{lot}\,.$$

The vorticity is given by $\zeta \mathbf{e}_\phi$ where

$$\zeta = \frac{1}{r}(rv)_r - \frac{u_\theta}{r} = v_r - \frac{u_\theta}{r} + \frac{v}{r}\,.$$

Recalling that

$$c_1 + c_2 = a(\sigma+\gamma) + 2G\,,$$

we find that

$$\rho u\left[u\frac{\partial^2\zeta}{\partial r^2}+\frac{v}{r}\frac{\partial^2\zeta}{\partial\theta\partial r}\right]+\rho\frac{v}{r}\left[u\frac{\partial^2\zeta}{\partial\theta\partial r}+\frac{v}{r}\frac{\partial^2\zeta}{\partial\theta^2}\right]-2\tau\frac{\partial^2\zeta}{r\partial r\partial\theta}$$
$$+\left[c_2+a(\sigma-\gamma)\right]\frac{u_{\theta rr}}{r}-c_2 v_{rrr}+\left[-c_1+a(\sigma-\gamma)\right]\frac{v_{\theta r\theta}}{r^2}+c_1\frac{u_{\theta\theta\theta}}{r^3}=\text{lot}$$

or

$$\rho u^2\frac{\partial^2\zeta}{\partial r^2}+(2\rho uv-2\tau)\frac{\partial^2\zeta}{r\partial\theta\partial r}+\rho v^2\frac{\partial^2\zeta}{r^2\partial\theta^2}-c_2\frac{\partial^2\zeta}{\partial r^2}-c_1\frac{\partial^2\zeta}{r^2\partial\theta^2}$$
$$+a(\sigma-\gamma)\left(\frac{u_{rr\theta}}{r}+\frac{v_{\theta r\theta}}{r^2}\right)=\text{lot}\,.$$

From the continuity equation

$$u_r=-\frac{v_\theta}{r}+\text{lot}\,.$$

Thus

$$u_{rr\theta}=-\frac{v_{\theta r\theta}}{r}+\text{lot}\,.$$

Hence

$$(\rho u^2-c_2)\frac{\partial^2\zeta}{\partial r^2}+2(\rho uv-\tau)\frac{\partial^2\zeta}{r\partial\theta\partial r}+(\rho v^2-c_1)\frac{\partial^2\zeta}{r^2\partial\theta^2}=\text{lot}$$

or

$$A\frac{\partial^2\zeta}{\partial r^2}+2B\frac{\partial^2\zeta}{r\partial\theta\partial r}+C\frac{\partial^2\zeta}{r^2\partial\theta^2}=\text{lot}$$

where

$$A=\rho u^2-\left(\frac{a+1}{2}\sigma+\frac{a-1}{2}\gamma+\mu\right),$$
$$B=\rho uv-\tau\,,$$
$$C=\rho u^2-\left(\frac{a-1}{2}\sigma+\frac{a+1}{2}\gamma+\mu\right).$$

CHAPTER 10

Anomalous Elongational Flows and Change of Type

Anomalous effects on elongational flows at high rates of elongation reported by Ferguson, *et al.* [1987] are here treated as a change of type following the work of Joseph and Chen [1988]. Analysis predicts that the vorticity near the drum is hyperbolic, elliptic away from the drum, under the supercritical conditions in the experiment when the shear-wave speeds have the values which we measured with our meter.

10.1 Introduction

Recent experiments of Ferguson, Hudson and Warren [1987] on elongational flows of polymer solutions at high rates of elongation have revealed anomalous effects which appear to arise from a qualitative change to more elastic responses in the neighborhood of certain critical drum speeds. They suggested that these changes are brought about by a stress-induced phase change called complexation, and they gave evidence to back up their idea. The drum speeds at which these anomalous effects were reported are of the order of critical extrusion velocities in similar solutions reported by Joseph, Matta and Chen [1987] for the delay in delayed die swell. Delayed die swell can also be connected to phase changes in some cases (Giesekus, [1968]) but another explanation based on a change of type of the vorticity equation from elliptic to hyperbolic seems to be in better agreement with the experiments. In this paper, we do what we can to establish the change of type explanation for the anomalous effects observed by Ferguson, *et al.* Our effort is two-fold: (1) On our wave-speed meter, we measured the shear-wave speeds for the liquids used by Ferguson, *et al.* The measured speeds range from about forty to eighty per cent of the post critical drum speeds reported in the experiments. This means that the Mach number M = W/c, where W is the drum speed and c the shear-wave speed, is between 1.20 and 2.5, roughly. (2) We did an analysis of an idealized problem for the experiment, using different models. The analysis shows that the vorticity near the drum in the idealized problem changes type when the wave speeds have the measured values and the idealized thread line is longer than a model dependent critical length ranging between one and twenty centimeters.

10.2 Experiments I

We shall consider the experiments of Ferguson, *et al.* [1987] and the measurements of wave speeds for the same fluids which we did on the wave-speed meter (Joseph, Riccius, and Arney, [1986]) in our laboratory.

Ferguson *et al.*, did experiments on elongation flows at high rates of extension using their extensional rheometer. With their device, they extended a thread of liquid by winding it onto a drum. They report the winder speed and the zero-shear viscosity of the liquids used, but they do not give the rate of extension or the length of the extending thread. They describe anomalous effects as follows:

> During elongational flow experiments on polymer solutions certain anomalous examples of flow behavior have been noted. These are,
>
> (1) a non-uniform velocity profile in the radial direction in strongly non-Newtonian fluids being elongated at high rates.
>
> (2) the loss of tack in the filament as rate of elongation increased.
>
> (3) the bouncing of the filament off a rotating drum at high elongation rates, and
>
> (4) the development of sustained rubber-like lateral vibrations in the fluids elongated at high rate.

Polymer solution	Drum speed ($m \cdot s^{-1}$)	Total strain at the drum	Phenomena observed
Silicone Oil (η_0 = 1.3 Pa•s)	0.0	0.8	Pure extensional flow
with Al_2O_3 particles	0.35	2.2	Pure extensional flow
0.5% PAM in ethylene glycol	0.0	0.8	Pure extensional flow
(η_0 = 1.4 Pa•s)	0.35	2.2	Anomolous flow. Particles moving at different speeds dependent on their radial position. Faster particles in the body overtaking slower ones on the outside of the filament.
5% PEO in water (η_0 = 45 Pa•s)	0.05	0.9	Tack high. Fluid adheres to syringe needle toughing filament.
	1.50	3.4	No obvious tack. Syringe needle passes through filament without fluid adhering.
4.5% PEO in water	0.05	0.9	Filament deflected by air jet, returns to equilibrium position when the disturbance is removed.
	2.00	3.5	When the deflecting air jet is removed, the filament oscillates laterally. The vibrations die out slowly.
2% PAA in ethylene glycol (η_0 = 98 Pa•s)	0.80	1.8	Fluid adheres to and is drawn by the rotating drum.
	2.00	3.3	Filament bounces off drum surface. Temporary adherence is followed by ligaments being thrown off the drum.

Table 10.1. (After Ferguson, Hudson and Warren [1987]) Some typical flow phenomena observed during elongational flow.

The anomalous behavior is described more precisely, with experimental details, in Table 10.1 which is reproduced on page 275. We are going to assume that anomalous flow occurs as a critical phenomenon, at a certain speed, though this may be masked by the action of a large effective viscosity as was true for some cases of delayed die swell.

10.3 Theory

In the theory, we assume that the flow is a pure cylindrical extension, r is the radial distance from the axis of the jet, z is the axial distance (see Figure 10.1) and the components of velocity are $\mathbf{u} = (u,v,w) = (-sr,0,2sz)$ in cylindrical coordinates (r,θ,z) and s is the extensional rate. The analysis will proceed along the lines introduced by Joseph, Renardy, and Saut [1985] and Joseph and Saut [1986]. We choose a constitutive equation and find the stresses corresponding to pure cylindrical extension. Then we perturb the system with a small but otherwise arbitrary perturbation and linearize. We can find the regions of hyperbolicity and the characteristics from the linearized equations. Universally, the same regions of hyperbolicity and characteristics apply to each and every small perturbation of uniform extension. This information is cheap. To go further, we would need to solve a linearized problem perturbing our already idealized problem and the form that such a problem should take is not clear to us. Since we do not solve the problem, or even a linearized version of it, the best we can say is that it appears that a change of type is probable in the set of parameters for which anomalous effects are observed. The physical implications of a change of type are left obscure.

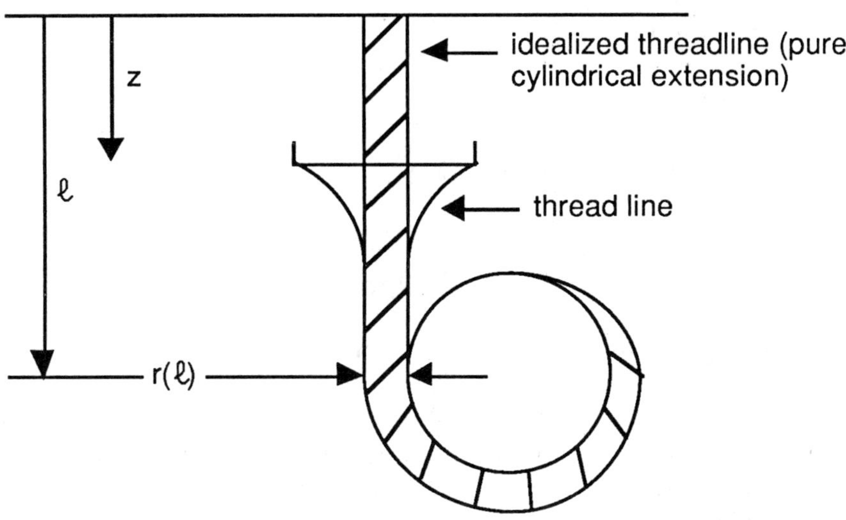

Figure 10.1. A liquid thread is elongated by winding it on a rotating drum with peripheral speed W. The idealized thread line is in pure cylindrical extension with the same speed (u,W) = (−sr(ℓ),2sℓ) as the real thread. We imagine measuring u, W, and r in an experiment, fixing ℓ and s. Only W was given by Ferguson, *et al.*

Let $\mathbf{T} = -p\mathbf{1}+\boldsymbol{\tau}$ be the stress. We assume that the evolution of τ is governed by an upper convected Maxwell model

$$\lambda \frac{\mathcal{D}\boldsymbol{\tau}}{\mathcal{D}t} + \boldsymbol{\tau} = 2\eta \mathbf{D}[\mathbf{u}] \,, \tag{1}$$

where $\mathbf{D}[\mathbf{u}]$ is the symmetric part of $\nabla\mathbf{u}$, $\mathbf{u}$ is the velocity, λ a relaxation time, η the viscosity, η/λ the rigidity and

$$\frac{\mathcal{D}\boldsymbol{\tau}}{\mathcal{D}t} = \frac{\partial\boldsymbol{\tau}}{\partial t} + (\mathbf{u}\cdot\nabla)\boldsymbol{\tau} - \mathbf{W}\boldsymbol{\tau} + \boldsymbol{\tau}\mathbf{W} - (\mathbf{D}\boldsymbol{\tau} + \boldsymbol{\tau}\mathbf{D}) \tag{2}$$

where $\mathbf{W}[\mathbf{u}]$ is the skew-symmetric part of $\nabla\mathbf{u}$ and $\mathcal{D}/\mathcal{D}t$ is an invariant derivative. Pure cylindrical extension is possible if

$$[\tau_{rr},\tau_{\theta\theta},\tau_{zz},\tau_{rz}] \overset{\text{def}}{=} [\sigma,\gamma,\beta,\tau]$$

have constant values and p is a quadratic polynomial,

$$\begin{vmatrix} \sigma \\ \gamma \\ \beta \\ \tau \\ p \end{vmatrix} = \begin{vmatrix} -2\eta s/(1+2\lambda s) \\ -2\eta s/(1+2\lambda s) \\ 4\eta s/(1-4\lambda s) \\ 0 \\ -\frac{1}{2}\rho s^2 r^2 - 2\rho\sigma^2 z^2 + \text{const} \end{vmatrix} \tag{3}$$

The stresses are bounded if

$$s < \frac{1}{4}\lambda \,. \tag{4}$$

In the idealized flow $W=2s\ell$ and (4) implies that

$$\ell > 2\lambda W \,. \tag{5}$$

Now we consider the linearized problem for flows which perturb pure cylindrical extension. We can frame our theory in terms of the vorticity. We shall assume that the disturbances of pure cylindrical extension are axisymmetric. Then the disturbance vorticity ζ is equal to the total vorticity. For the moment, let us think that ζ is the total vorticity, so we are looking at axisymmetric problems for the UCM without linearization. Then,

$$A\frac{\partial^2\zeta}{\partial r^2} + 2B\frac{\partial^2\zeta}{\partial r\partial z} + C\frac{\partial^2\zeta}{\partial z^2} = \text{lot} \tag{6}$$

where lot means lower order terms, 3rd derivatives of the velocity components at most,

$$A = \rho u^2 - \left(\frac{\eta}{\lambda} + \sigma\right),$$

$$B = \rho uw - \tau \,,$$

$$C = \rho w^2 - \left(\frac{\eta}{\lambda} + \beta\right).$$

It is well known from the theory of 2nd order PDE's that (6) is hyperbolic wherever $B^2 - AC > 0$. In those regions, the first order differential equations which determine the characteristics are

$$\frac{dz}{dr} = \frac{B \pm \sqrt{B^2 - AC}}{A} \,. \tag{7}$$

In the linearized case, the characteristics are determined a priori by (7) with A, B, and C evaluated on pure cylindrical extension. We find that

$$B^2 - AC = c^4\rho^2 \left\{ \frac{M_1^2(r)}{1-4\lambda s} + \frac{M_2^2(z)}{1+2\lambda s} - \frac{1}{(1-4\lambda s)(1+2\lambda s)} \right\}, \tag{8}$$

where

$$M_1(r) = -\frac{u(r)}{c} = \frac{sr}{c} \,,$$
$$M_2(z) = \frac{w(z)}{c} = \frac{2sz}{c} \,,$$

and

$$c = \sqrt{\eta/\rho\lambda}$$

is the speed of vorticity waves (shear waves) into rest for the Maxwell model. Then the region in the thread in which the vorticity is hyperbolic is defined by

$$(1+2\lambda s)M_1^2(r) + (1-4\lambda s)M_2^2(z) > 1 \tag{9}$$

where $s < 1/4\lambda$. The border of the hyperbolic region is defined by an equality in (9) and the inequality says that the vorticity is hyperbolic outside an ellipsoid.

In Figure 10.2 we have sketched an ellipsoid

$$\frac{M_1^2(r)}{a_1^2} + \frac{M_2^2(z)}{a_2^2} = 1 \tag{10}$$

centered on the origin of the idealized extending thread. For the UCM, the semimajor axis is $ca_2/2s$, where $a_2 = (1\text{-}4\lambda s)^{-1/2}$.

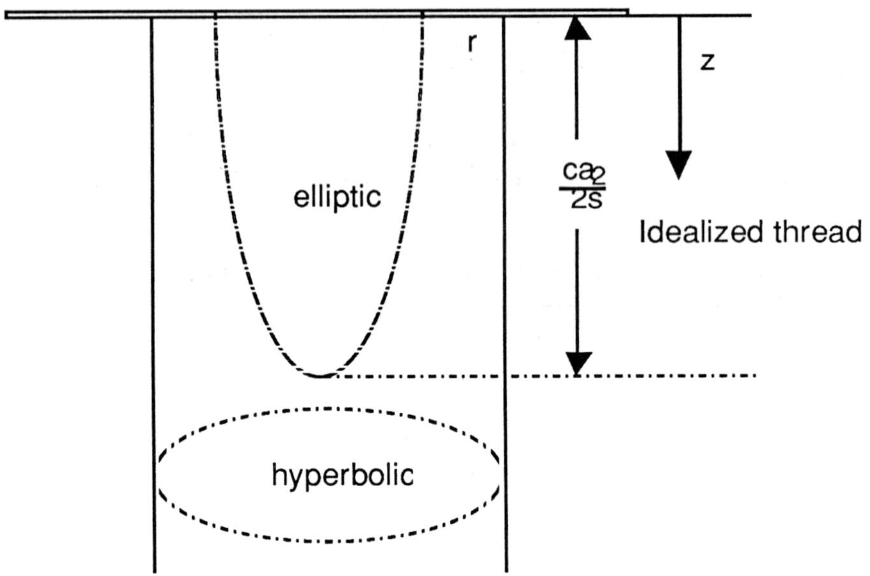

Figure 10.2. The vorticity outside the ellipsoid is hyperbolic; inside, it is elliptic.

Every point of the extending thread with $z>ca_2/2s$ is in the hyperbolic region. The portion of the thread which is hyperbolic touches the drum when

$$M_2(\ell) = \frac{W}{c} = \frac{2s\ell}{c} > a_2 \tag{11}$$

and

$$\ell > cg(\lambda,s) \tag{12}$$

where

$$g(\lambda, s) = \frac{1}{2\sqrt{s^2-4\lambda s^3}}$$

is a concave function such that $g(\lambda,0) = g(\lambda,1/4\lambda) = \infty$. There is a minimum thread length in the idealized theory

$$\ell > \ell_{min} = c \min g(\lambda,s) = cg(\lambda,\lambda/6) = 3\sqrt{3}\,\lambda c\ . \tag{13}$$

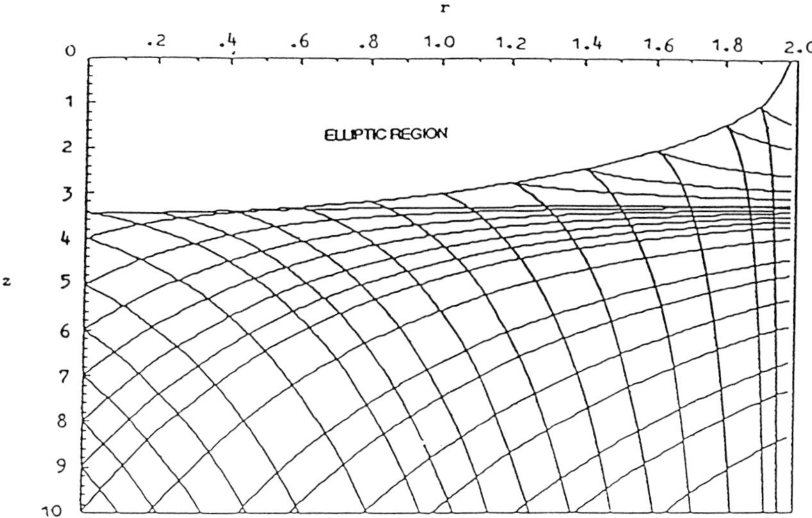

Figure 10.3. Characteristics on the cross-section of characteristic surfaces of revolution covering the portion of the thread in which the vorticity is hyperbolic. The computation is for 0.5% PAM in ethylene glycol: λ=0.022 sec, s=10.0/sec, $\tilde{\mu}$=1.4 Pa-s, c=23.9 cm/sec, ρ=1.117 g/cm^3.

The idealized theory says that we may expect a hyperbolic transition for $0<s<1/6\lambda$ and $\ell>\ell_{min} = 3\sqrt{3}\,\lambda c$.

In Figure 10.3, we have exhibited the cross-sections of the characteristic surfaces of revolution for the vorticity generated by (7).

The results just given depend on two false assumptions: that the unperturbed problem is a pure cylindrical extension and the fluid is an UCM. We tested the results by using other models. We looked at interpolated Maxwell models with a parameter a, $-1\leq a\leq 1$, multiplying the last bracket of (2). This alters the principal part of the rate equation. Some Phan Thien-Tanner (Ph.TT) models use this altered principal part and adjust the lower order terms, replacing $\boldsymbol{\tau}$ with $\boldsymbol{\tau}(1+\text{tr}\,\boldsymbol{\tau})$. When a=0, tr $\boldsymbol{\tau}$=0 in pure cylindrical extension, there is no difference between the Ph.TT models and the interpolated Maxwell models. A Giesekus model can be obtained from (2) by replacing $\boldsymbol{\tau}$ with $\boldsymbol{\tau} + \alpha\lambda/\eta\ \boldsymbol{\tau}^2$ where α=1/2 for a Leonov-like model.

The general form of the change of type results are similar for all the models but a_2, $g(\lambda,s)$ and ℓ_{min} differ from model to model, as can be seen in Table 10.2.

Joseph and Saut [1986] showed that the initial value problem for plane extensional flow of all interpolated Maxwell models with $a^2 \neq 1$ become ill-posed for sufficiently large values of s smaller than a yet larger value of s for which one of the stresses becomes unbounded. Ill-posed problems are badly unstable. In the present case, the corotational model becomes unstable when $s>1/3\lambda$, but the stresses are all finite at finite s. The value $s=1/3\lambda$ is used as a limit of validity for the corotational Maxwell model in Table 10.2, whereas the values $s=1/4\lambda$ and $s=1/2\lambda$ for $a=1,-1$ are obtained from singular stresses in the two cases in which there is no loss of well-posedness. The restriction $\ell>2\lambda W$, arising from the requirement (5) that stresses remain bounded, may also be formulated for the other entries in Table 10.2: $\ell>\lambda W$ for the LCM, $\ell>3\lambda W/2$ for the COM and $\ell>2\lambda W$ for the Leonov model. This ℓ is not larger than ℓ_{min} for the fluids used in the experiments (Table 10.4). If the idealized flow is hyperbolic at the drum, the stresses will stay bounded.

Equation (11) and the column of values of a_2 in Table 2 show that a necessary condition for a hyperbolic transition is that $M_2(\ell)>1$ or $W>c$. In the idealized problem, it is also necessary that $\ell>\ell_{min}$ where ℓ_{min} is achieved as an interior minimum for some models and at the end points of model validity in others.

Model	a_2	$g(\lambda,s)$	ℓ_{min}	Restrictions for models
UCM	$\frac{1}{\sqrt{1-4\lambda s}}$	$\frac{1}{2\sqrt{s^2-4\lambda s^3}}$	$3\sqrt{3}\,\lambda c$	$0<s<\frac{1}{4\lambda}$
LCM	$\frac{1}{\sqrt{1-2\lambda s}}$	$\frac{1}{2\sqrt{s^2-2\lambda s^3}}$	$\frac{3\sqrt{3}}{2}\lambda c$	$0<s<\frac{1}{2\lambda}$
COM	$\sqrt{1+3\lambda s}$	$\frac{\sqrt{1+3\lambda s}}{2s}$	$\frac{3}{\sqrt{2}}\lambda c$	$0<s<\frac{1}{3\lambda}$ *
Giesekus	$\sqrt{1+\frac{1}{2\alpha}\left[4\lambda s-1+\sqrt{(1-4\lambda s)^2+16\alpha\lambda s}\right]}$	$\frac{a_2}{2s}$		$0<s<\frac{1}{4\lambda}$
Leonov-like	$\sqrt{4\lambda s+\sqrt{1+16\lambda^2 s^2}}$	$\frac{\sqrt{4\lambda s+\sqrt{1+16\lambda^2 s^2}}}{2s}$	$2\sqrt{1+\sqrt{2}}\,\lambda c$	$0<s<\frac{1}{4\lambda}$

Table 10.2. Model parameters for pure cylindrical extension calculated from theory. The restrictions for model validity are associated with singular extensional stresses or with stability*. We have, in addition, the restrictions of inequality (5) for the UCM, $\ell>\lambda W$ for the LCM, $\ell>3\lambda W/2$ for the COM and $\ell>2\lambda W$ for the Leonov model.

10.4 Experiments II

We say that anomalous effects observed by Ferguson, *et al.* are consistent with an explanation based on change of type if W>c and ℓ_{min} is of the order one might expect to observe in the experiments. We have W from their experiments and we measured c in ours. Thread lengths were not reported in their paper. We may compute ℓ_{min}, however, from measured data remembering that ℓ is an idealized quantity defined on a purely cylindrical extensional flow. To compute ℓ_{min}, we need the wave speed and a relaxation time λ. A relaxation time $\lambda=\eta_0/G_c$ may be found from the zero shear viscosity η_0 given in Table 10.1 and the rigidity $G_c=\rho c^2$.

We measured wave speeds in the four solutions used in the experiments reported in Table 10.1. These are reported in the standard way (Joseph, Riccius and Arney, [1986] and Riccius, Joseph and Arney, [1987]) in Table 10.3 where $\bar{c}$ is the average wave speed over different size gaps h and $\tilde{\mu}=\eta_0$. The computation of ℓ_{min} for different models, using the data in Table 10.3, is given in Table 10.4. These values appear to be reasonable. In Table 10.5, we compute $M_2(\ell)=W/c$ where W is from Table 10.1 and c from Table 10.3. Since c<W in all the experiments exhibiting anomalous behavior, we cannot discount the possibility that this behavior is associated with a change of type.

Fluid	$\tilde{\mu}$ [Pa•s]	$\bar{c}$ [10^{-2} m/s]	G [Pa]	λ [s]	h [10^{-3} m/s]	c [10^{-2} m]	T [°C]	ρ [kg/m^3]
4.5% (Polyethelene Oxide	31	86.40	943	0.033	1.0	87.01	25	1263
WSR-301) in Water					3.0	83.03	25	
MN = 4,000,000					6.75	87.30	24	
5.0% PEO (Polyethylene Oxide	45	125.7	1841	0.024	2.0	119.62	24	1165
WSR-301) in Water					3.0	131.86	25	
MN = 4,000,000								
0.5% PAN (Polyacrylamide)	1.4	23.9	63.8	0.022	1.0	27.21	24	1117
with Al_2O_3					2.0	24.49	24	
					3.0	19.99	24	
2.0% PAA (Polyacrylic	98	167.9	3707	0.026	2.0	168.03	24	1315
Acid) MN = 4,000,000					3.0	167.92	24	

Table 10.3. Wave speeds and rigidities. The columns contain the following data: $\tilde{\mu}$ zero-shear viscosity, $\bar{c}$ average shear-wave speed, G_c effective shear modulus, λ relaxation time, h gap size, c shear-wave speed for each gap size, T temperature, ρ density.

	ℓ_{min} (cm)			
Fluid	UCM	LCM	COM	Leonov-like
4.5% PEO in Water	14.8153	7.4076	6.0483	8.8602
5% PEO in Water	15.6758	7.8379	6.3996	9.3749
0.5% PAM in Ethylene Glycol with Al_2O_3	2.7321	1.3661	1.1154	1.6339
2% PAA in Ethylene Glycol	22.6833	11.3416	9.2604	13.5657

Table 10.4. Calculated values ℓ_{min} for various models using the model parameters of Table 10.2 and the data from Table 10.3. ℓ_{min} is larger than the ℓ required for model validity (see caption to Table 10.3) for the upper and lower convected models and is of the same magnitude for the other two cases.

	W (cm/sec)	c (cm/sec)	M
4.5% PEO	200	86.4	2.31
5.0% PEO	150	125.7	1.19
0.5% PAM	35	23.9	1.46
2.0% PAA*	200	167.9	1.19

Table 10.5. Viscoelastic Mach numbers for supercritical elongational flow.

Exercises

Exercise 10.1.

Show that steady axisymmetric flow perturbing the pure cylindrical elongation of a Maxwell model with $-1 \leq a \leq 1$ satisfies equations of the form:

$$\mathbf{A}\mathbf{q}_r + \mathbf{B}\mathbf{q}_z = \mathbf{f}$$

where in coordinates (r,θ,z)

$$\mathbf{v} = (u,0,w) \ ,$$

$$\frac{\partial}{\partial\theta} = 0 \ ,$$

* The bouncing filament phenomenon was described by Ferguson, *et al.* as follows:

The drum speed was set at 200 cm/s. When the filament touched the drum, it was, at first, taken up in the usual manner. However, it very quickly changed its behaviour. It began to bounce off the drum, with a definite and sharp angle of deflection, very reminiscent of an elastic collision of an object with a wall. Its velocity on meeting the drum was approximately 160 cm/s. Occasionally, it attached itself to the drum, but was then thrown off in coherent ligaments before returning to the bouncing behaviour. The same phenomenon occurred when the drum was started from rest with the filament already falling, the bouncing effect starting a drum speed of about 180 cm/s.

$$\mathbf{T} = -p\mathbf{1} + \boldsymbol{\tau} ,$$

$$[\tau_{rr},\tau_{\theta\theta},\tau_{zz},\tau_{r\theta},\tau_{rz},\tau_{\theta z}] = [\sigma,\gamma,\beta,0,\tau,0] ,$$

$$\mathbf{q} = [u,w,p,\sigma,\gamma,\beta,\tau]^T ,$$

$$\mathbf{f} = \left[-\frac{u}{r}, \frac{\sigma-\gamma}{r}, \frac{\tau}{r}, -\frac{\sigma}{\lambda}, 2\,G\frac{u}{r} + \left(2a\frac{u}{r} - \frac{1}{\lambda}\right)\gamma, -\frac{\beta}{\lambda}, -\frac{\tau}{\lambda} \right] .$$

$$\mathbf{A} = \left| \begin{array}{ccccccc} 1 & 0 & 0 & 0 & 0 & 0 & 0 \\ \rho u & 0 & 1 & -1 & 0 & 0 & 0 \\ 0 & \rho u & 0 & 0 & 0 & 0 & -1 \\ -2(G+a\sigma) & (1-a)\tau & 0 & u & 0 & 0 & 0 \\ 0 & 0 & 0 & 0 & u & 0 & 0 \\ 0 & -(1+a)\tau & 0 & 0 & 0 & u & 0 \\ -a\tau & \left(-\frac{1+a}{2}\sigma+\frac{1-a}{2}\beta-G\right) & 0 & 0 & 0 & 0 & u \end{array} \right| ,$$

$$\mathbf{B} = \left| \begin{array}{ccccccc} 0 & 1 & 0 & 0 & 0 & 0 & 0 \\ \rho w & 0 & 0 & 0 & 0 & 0 & -1 \\ 0 & \rho w & 1 & 0 & 0 & -1 & 0 \\ -(1+a)\tau & 0 & 0 & w & 0 & 0 & 0 \\ 0 & 0 & 0 & 0 & w & 0 & 0 \\ (1-a)\tau & -2(G+a\beta) & 0 & 0 & 0 & w & 0 \\ \left(\frac{1-a}{2}\sigma - \frac{1+a}{2}\beta - G\right) & -a\tau & 0 & 0 & 0 & 0 & w \end{array} \right| .$$

Exercise 10.2.
Derive the characteristic equation

$$\frac{dz}{dr} = \frac{B \pm \sqrt{B^2 - AC}}{A} ,$$

where

$$A = \rho u^2 - \left(G + \frac{1+a}{2}\sigma + \frac{a-1}{2}\beta\right),$$

$$B = \rho uw - \tau ,$$

$$C = \rho w^2 - \left(G + \frac{a-1}{2}\sigma + \frac{a+1}{2}\beta\right).$$

Exercise 10.3.
Linearize the equations around pure cylindrical elongation and show that the initial value problem is ill-posed when a=0.

Solutions for the exercises

Exercise 10.1.
Since the equations are

$$\text{div } \mathbf{v} = 0 ,$$

$$\rho\left[\frac{\partial \mathbf{v}}{\partial t} + (\mathbf{v}\bullet\nabla)\mathbf{v}\right] - \text{div } \boldsymbol{\tau} + \nabla p = 0 ,$$

$$\frac{\partial \boldsymbol{\tau}}{\partial t} + (\mathbf{v}\bullet\nabla)\boldsymbol{\tau} + \frac{1-a}{2}(\boldsymbol{\tau}\mathbf{L} + \mathbf{L}^T\boldsymbol{\tau}) - \frac{1+a}{2}(\mathbf{L}\boldsymbol{\tau}+\boldsymbol{\tau}\mathbf{L}^T) - 2GD[\mathbf{u}] = -\frac{1}{\lambda}\boldsymbol{\tau}$$

In cylindrical coordinates $(X_1,X_2,X_3) = (r,\theta,z)$ and for steady asymmetric flow:

$$\text{div } \mathbf{v} = 0 , \tag{14}$$

$$\rho(\mathbf{v}\bullet\nabla)\mathbf{v} - \text{div } \boldsymbol{\tau} + \nabla p = 0 , \tag{15}$$

$$(\mathbf{v}\bullet\nabla)\boldsymbol{\tau} + \frac{1-a}{2}(\boldsymbol{\tau}\mathbf{L} + \mathbf{L}^T\boldsymbol{\tau}) - \frac{1+a}{2}(\mathbf{L}\boldsymbol{\tau} + \boldsymbol{\tau}\mathbf{L}^T) - 2GD[\mathbf{v}] = -\frac{\boldsymbol{\tau}}{\lambda}. \tag{16}$$

where

$$\mathbf{v} = (u,0,w) ,$$

$$\mathbf{L} = \begin{pmatrix} u_r & 0 & u_z \\ 0 & \frac{u}{r} & 0 \\ w_r & 0 & w_z \end{pmatrix},$$

and

$$\boldsymbol{\tau} = \begin{pmatrix} \sigma & 0 & \tau \\ 0 & \gamma & 0 \\ \tau & 0 & \beta \end{pmatrix}.$$

Since

$$\operatorname{div} \boldsymbol{\tau} = \frac{\partial}{\partial x_i}$$

$$\left(\sigma\hat{\mathbf{e}}_r\otimes\hat{\mathbf{e}}_r + \tau\hat{\mathbf{e}}_r\otimes\hat{\mathbf{e}}_z + \gamma\hat{\mathbf{e}}_\theta\otimes\hat{\mathbf{e}}_\theta + \tau\hat{\mathbf{e}}_z\otimes\hat{\mathbf{e}}_r + \beta\hat{\mathbf{e}}_z\otimes\hat{\mathbf{e}}_z\right)\cdot\hat{\mathbf{e}}_i$$

$$= \left(\frac{\partial\sigma}{\partial r} + \frac{\partial\tau}{\partial z} + \frac{\sigma-\gamma}{r}\right)\hat{\mathbf{e}}_r + \left(\frac{\partial\tau}{\partial r} + \frac{\partial\beta}{\partial z} + \frac{\tau}{r}\right)\hat{\mathbf{e}}_z .$$

Thus equations (14) and (15) become

$$\frac{\partial u}{\partial r} + \frac{u}{r} + \frac{\partial w}{\partial z} = 0 ,$$

$$\rho\left(u\frac{\partial u}{\partial r} + w\frac{\partial w}{\partial z}\right) + \frac{\partial p}{\partial r} - \left(\frac{\partial\sigma}{\partial r} + \frac{\partial\tau}{\partial z} + \frac{\sigma-\gamma}{r}\right) = 0 , \tag{17}$$

$$\rho\left(u\frac{\partial w}{\partial r} + w\frac{\partial w}{\partial z}\right) + \frac{\partial p}{\partial z} - \left(\frac{\partial\tau}{\partial r} + \frac{\partial\beta}{\partial z} + \frac{\tau}{r}\right) = 0 ,$$

and (16) becomes:

$$u\begin{vmatrix} \frac{\partial\sigma}{\partial r} & 0 & \frac{\partial\tau}{\partial r} \\ 0 & \frac{\partial\gamma}{\partial r} & 0 \\ \frac{\partial\tau}{\partial r} & 0 & \frac{\partial\beta}{\partial r} \end{vmatrix} + w\begin{vmatrix} \frac{\partial\sigma}{\partial z} & 0 & \frac{\partial\tau}{\partial z} \\ 0 & \frac{\partial\gamma}{\partial z} & 0 \\ \frac{\partial\tau}{\partial z} & 0 & \frac{\partial\beta}{\partial z} \end{vmatrix}$$

$$+ (1-a)\begin{vmatrix} (\sigma u_r+\tau w_r) & 0 & \frac{1}{2}\left[\sigma u_z+\tau(w_z+u_r)+\beta w_r\right] \\ 0 & \frac{u}{r}\gamma & 0 \\ \frac{1}{2}\left[\sigma u_z+\tau(w_z+u_r)+\beta w_r\right] & 0 & (\tau u_z+\beta w_z) \end{vmatrix}$$

$$- (1+a)\begin{vmatrix} (\sigma u_r+\tau u_z) & 0 & \frac{1}{2}\left[\sigma w_r+\tau(u_r+w_z)+\beta u_z\right] \\ 0 & \frac{u}{r}\gamma & 0 \\ \frac{1}{2}\left[\sigma w_r+\tau(w_z+u_r)+\beta u_z\right] & 0 & (\tau w_r+\beta w_z) \end{vmatrix}$$

$$- G\begin{vmatrix} 2u_r & 0 & u_z+w_r \\ 0 & 2\frac{u}{r} & 0 \\ u_z w_r & 0 & 2w_z \end{vmatrix} = -\frac{1}{\lambda}\begin{vmatrix} \sigma & 0 & \tau \\ 0 & \gamma & 0 \\ \tau & 0 & \beta \end{vmatrix}$$

or

$$u\frac{\partial\sigma}{\partial r} + w\frac{\partial\sigma}{\partial z} + (1-a)(\sigma u_r + \tau w_r) - (1+a)(\sigma u_r+\tau u_z) - 2Gu_r = -\frac{\sigma}{\lambda},$$

$$u\frac{\partial\gamma}{\partial r} + w\frac{\partial\gamma}{\partial z} - 2a\frac{u}{r}\gamma - 2G\frac{u}{r} = -\frac{\gamma}{\lambda},$$

$$u\frac{\partial\beta}{\partial r} + w\frac{\partial\beta}{\partial z} + (1-a)(\tau u_z+\beta w_z)-(1+a)(\tau w_r+\beta w_z)-2Gw_z=-\frac{\beta}{\lambda}, \quad (18)$$

$$u\frac{\partial\tau}{\partial r} + w\frac{\partial\tau}{\partial z} + \frac{1-a}{2}(\sigma u_z+\tau w_z+\tau u_r+\beta w_r)$$
$$-\frac{1+a}{2}(\sigma w_r+\tau u_r+\tau w_z+\beta u_z) - G(u_z + w_r) = -\frac{\tau}{\lambda}.$$

Let $\mathbf{q} = [u, w, p, \sigma, \gamma, \beta, \tau]^T$.

Then equations (17) and (18) can be put into the form

$$\mathbf{A}\,\mathbf{q}_r + \mathbf{B}\,\mathbf{q}_z = \mathbf{f}.$$

Exercise 10.2.

Let

$$c_1 = -\left(G + \frac{1+a}{2}\sigma + \frac{a-1}{2}\beta\right),$$

$$c_2 = -\left(G + \frac{a-1}{2}\sigma + \frac{a+1}{2}\beta\right).$$

Assume that across characteristic surface $\phi(r,z)=0$ there are simple jumps:

$$[\![\mathbf{q}]\!]_\phi \neq 0 \quad \text{but} \quad [\![\mathbf{q}]\!] = 0\,.$$

Thus

$$(\mathbf{A}\phi_r + \mathbf{B}\phi_z)\,[\![\mathbf{q}]\!]_\phi = 0$$

hence

$$\det\,(\mathbf{A}\phi_r + \mathbf{B}\phi_z) = 0\,.$$

that is

$$\det \begin{vmatrix} \phi_r & \phi_z & 0 & 0 & 0 & 0 & 0 \\ \rho v_0 & 0 & \phi_r & -\phi_r & 0 & 0 & -\phi_z \\ 0 & \rho v_0 & \phi_z & 0 & 0 & -\phi_z & -\phi_r \\ \begin{matrix}[-2(G+a\sigma)\phi_r \\ -(1+a)\tau\phi_z]\end{matrix} & (1-a)\tau\phi_r & 0 & v_0 & 0 & 0 & 0 \\ 0 & 0 & 0 & 0 & v_0 & 0 & 0 \\ (1-a)\tau\phi_z & \begin{matrix}[-(1+a)\tau\phi_r \\ -2(G+a\beta)\phi_z]\end{matrix} & 0 & 0 & 0 & v_0 & 0 \\ (-a\tau\phi_r + c_2\phi_z) & (-a\tau\phi_z + c_1\phi_r) & 0 & 0 & 0 & 0 & v_0 \end{vmatrix} = 0$$

where $v_0 = u\phi_r + w\phi_z$.

The determinant is reduced to:

$$v_0^3(-\phi_r^2+\phi_z^2)(c_1\phi_r^2 - c_2\phi_z^2) + v_0^3\phi_r^2\phi_z\left[(1+a)\tau\phi_r + 2(G+a\beta)\phi_z\right]$$

$$+ v_0^3\phi_z^2\phi_z\left[(1+a)\tau\phi_z + 2(G+a\sigma)\phi_r\right] + v_0^3\phi_r\left[(1-a)\tau\phi_r^2\phi_z + \rho v_0^2\phi_r\right]$$

$$- v_0^3\phi_z\left[\rho v_0^2\phi_z - (1-a)\tau\phi_z^2\phi_z\right] = 0\,.$$

This is finally evaluated as:

$$v_0^3\Big\{(-c_1\phi_r^4 - c_2\phi_z^4 - (\phi_r^2+\phi_z^2)(u^2\phi_r^2+w^2\phi_z^2+2uw\phi_r\phi_z)\rho + 2\tau\phi_r\phi_z(\phi_r^2+\phi_z^2) - \phi_r^2\phi_z^2(c_1+c_2)\Big\} = 0$$

where

$$c_1+c_2 = -\big[2G + a(\sigma+\beta)\big]$$

using

$$A = \rho u^2+c_1 = \rho u^2 - \left(G + \frac{1+a}{2}\sigma + \frac{a-1}{2}\beta\right)$$

$$C = \rho w^2+c_2 = \rho w^2 - \left(G + \frac{a-1}{2}\sigma + \frac{1+a}{2}\beta\right)$$

$$B = \rho uw - \tau$$

we get

$$v_0^3\Big[-A\phi_r^4 - C\phi_z^4 - 2B\phi_r\phi_z(\phi_r^2 + \phi_z^2) - (A+C)\phi_r^2\phi_z^2\Big] = 0$$

or

$$v_0^3(\phi_r^2+\phi_z^2)(A\phi_r^2 + 2B\phi_r\phi_z + C\phi_z^2) = 0 .$$

The characteristics for the last factor are given by

$$\frac{dz}{dr} = \frac{B\pm\sqrt{B^2-AC}}{A} .$$

Exercise 10.3.
For pure cylindrical elongation we have

$$\mathbf{v} = (-sr,0,2sz) .$$

Using the Maxwell models (18) for steady flow

$$2as\sigma + 2Gs = -\frac{\sigma}{\lambda},$$

$$2as\gamma + 2Gs = -\frac{\gamma}{\lambda},$$

$$-4as\beta - 4Gs = -\frac{\beta}{\lambda},$$

$$-as\tau = -\frac{\tau}{\lambda}.$$

The solution is

$$\begin{pmatrix} \sigma \\ \gamma \\ \beta \\ \tau \end{pmatrix} = \begin{pmatrix} -2\eta s / (1+2a\lambda s) \\ -2\eta s / (1+2a\lambda s) \\ 4\eta s / (1-4a\lambda s) \\ 0 \end{pmatrix}.$$

where

$$\eta = G\lambda .$$

When a=0, for inertialess flow (ρ=0) :

$$\begin{pmatrix} \sigma \\ \gamma \\ \beta \\ \tau \end{pmatrix} = \begin{pmatrix} -2\eta s \\ -2\eta s \\ 4\eta s \\ 0 \end{pmatrix}.$$

$$b_{11} \stackrel{\text{def}}{=} \rho u^2 - \left(G + \frac{1+a}{2}\sigma + \frac{a-1}{2}\beta \right) = -G + 3\eta s ,$$

$$b_{22} \stackrel{\text{def}}{=} \rho w^2 - \left(G + \frac{a-1}{2}\sigma + \frac{1+a}{2}\beta \right) = -G - 3\eta s ,$$

$$b_{12} \stackrel{\text{def}}{=} \rho uw - \tau = 0 .$$

Thus

$$\Delta = b_{12}^2 - b_{11}b_{12} = -(-G+3\eta s)(-G-3\eta s) = -G^2(1-3\lambda s)(1+3\lambda s)\ .$$

Therefore if $s>\frac{1}{3\lambda}$, $\Delta>0$. That is, the inertialess steady flow is hyperbolic, thus the initial value problem is ill-posed (Hadamard unstable).

CHAPTER 11

Similarity Solutions that Give Rise to Hyperbolicity and Change of Type in Steady Flows of Viscoelastic Fluids

11.1 Introduction

N. Phan-Thien [1983a,b] has found a similarity solution of an Oldroyd B fluid between parallel plates which rotate at different speeds around a common axis perpendicular to the plates. The solution is modeled according to a famous one by T. Von Karman in which the fluid variables are resolved into a quadratic polynomial in r with coefficients to be determined as functions of z. This problem is important because it may model the flow away from edges of the finite disks used in rheometers.

A different similarity solution for an upper convected Maxwell fluid has been given by R. Larson [1988b]. Larson's problem is to determine the flow when the fluid is driven by an axisymmetric accelerated surface whose axial velocity w is proportional to the axial coordinate z. This kind of boundary condition is not a common one for devices which could be constructed but it is of interest in a subject where solutions so close to exact ones are so few.

The work of Phan-Thien [1983a, b] and Larson [1988b] was used by Verdier and Joseph [1989a] to calculate transcritical steady flows of the upper convected Maxwell model, to which the Oldroyd B reduces when the Newtonian viscosity is put to zero, without linearizing and without relying completely on numerical methods.

In each case one finds an interesting transcritical flow. Unfortunately the hope that the well known neck-in instability at the free surface on a liquid sheared between rotating parallel disks might be explained as a change of type did not work out. We are therefore able to give a satisfactory analysis of transcritical steady flow, but we do not yet understand the physical phenomena to which this change of type gives rise.

11.2 Analysis of characteristics

We use the steady constitutive equations for an upper convected Maxwell model

$$\lambda\left[(\mathbf{u}\cdot\nabla)\boldsymbol{\tau} - \mathbf{L}\boldsymbol{\tau} - \boldsymbol{\tau}\mathbf{L}^T\right] + \boldsymbol{\tau} = \eta(\mathbf{L} + \mathbf{L}^T) \qquad (1)_1$$

where $\mathbf{L}=\nabla\mathbf{u}$.

To this we append the equations of motion

$$\text{div }\mathbf{u} = 0\,, \qquad (1)_2$$

$$\rho(\mathbf{u}\cdot\nabla)\mathbf{u} = -\boldsymbol{\nabla}p + \text{div }\boldsymbol{\tau} \qquad (1)_3$$

and write the whole system as a 10 x 10 matrix equation, when the flow is axisymmetric,

$$\mathbf{A}\mathbf{q}_r + \mathbf{B}\mathbf{q}_z = \mathbf{f} \qquad (2)$$

for the system vector $\mathbf{q} = (u,v,w,\tau_{rr},\tau_{r\theta},\tau_{rz},\tau_{\theta\theta},\tau_{\theta z},\tau_{zz})$ in cylindrical coordinates where $\mathbf{A}$ and $\mathbf{B}$ are given by

$$\mathbf{A} = \begin{vmatrix} -2a & 0 & 0 & 0 & u & 0 & 0 & 0 & 0 & 0 \\ -c & -a & 0 & 0 & 0 & u & 0 & 0 & 0 & 0 \\ -b & 0 & -a & 0 & 0 & 0 & u & 0 & 0 & 0 \\ 0 & -2c & 0 & 0 & 0 & 0 & 0 & u & 0 & 0 \\ 0 & -b & -c & 0 & 0 & 0 & 0 & 0 & u & 0 \\ 0 & 0 & -2b & 0 & 0 & 0 & 0 & 0 & 0 & u \\ \rho u & 0 & 0 & 1 & -1 & 0 & 0 & 0 & 0 & 0 \\ 0 & \rho u & 0 & 0 & 0 & -1 & 0 & 0 & 0 & 0 \\ 0 & 0 & \rho u & 0 & 0 & 0 & -1 & 0 & 0 & 0 \\ 1 & 0 & 0 & 0 & 0 & 0 & 0 & 0 & 0 & 0 \end{vmatrix},$$

$$\mathbf{B} = \left|\begin{array}{cccccccccc} -2b & 0 & 0 & 0 & w & 0 & 0 & 0 & 0 & 0 \\ -d & -b & 0 & 0 & 0 & w & 0 & 0 & 0 & 0 \\ -e & 0 & -b & 0 & 0 & 0 & w & 0 & 0 & 0 \\ 0 & -2d & 0 & 0 & 0 & 0 & 0 & w & 0 & 0 \\ 0 & -e & -d & 0 & 0 & 0 & 0 & 0 & w & 0 \\ 0 & 0 & -2e & 0 & 0 & 0 & 0 & 0 & 0 & w \\ \rho w & 0 & 0 & 0 & 0 & 0 & -1 & 0 & 0 & 0 \\ 0 & \rho w & 0 & 0 & 0 & 0 & 0 & 0 & -1 & 0 \\ 0 & 0 & \rho w & 1 & 0 & 0 & 0 & 0 & 0 & 0 \\ 0 & 0 & 1 & 0 & 0 & 0 & 0 & 0 & 0 & 0 \end{array}\right|$$

where a,b,c,d,e are given by

$$a = \tau_{rr} + G\ ,$$
$$b = \tau_{rz}\ ,$$
$$c = \tau_{r\theta}\ ,$$
$$d = \tau_{\theta z}\ ,$$
$$e = \tau_{zz} + G\ ,$$

and $G = \dfrac{\eta}{\lambda}$ is the rigidity.

The characteristic surfaces are generated by the real values of $r' \overset{\text{def}}{=} \dfrac{dr}{dz}$ which satisfy

$$\det\ [\mathbf{A} - r'\mathbf{B}] = 0\ . \tag{3}$$

This leads to the following expression

$$(u - r'w)^4\ (r'^2 + 1)\ (Ar'^2 - 2Br' + C)^2 = 0 \tag{4}$$

where

$$A = \rho w^2 - \tau_{zz} - G\ ,$$
$$B = \rho uw - \tau_{rz}\ ,$$
$$C = \rho u^2 - \tau_{rr} - G\ . \tag{5}$$

We solved (3) using the symbolic manipulator program “reduce2” on the Cyber. Streamlines are quartically characteristic.

Change of type is associated with the real roots of the last parenthesis of (4). These roots are associated with the vorticity $\zeta = \zeta_\theta$.

$$A\frac{\partial^2\zeta}{\partial z^2} + 2B\frac{\partial^2\zeta}{\partial r\partial z} + C\frac{\partial^2\zeta}{\partial r^2} = \text{lot} \tag{6}$$

where lot are lower order terms. The two other components ζ_r and ζ_z of the vorticity ζ = curl **u** satisfy an equation of the same kind as (6) with the same left- hand side but different right-hand sides.

It follows that the projection of the characteristic surfaces of revolution onto the (r,z) plane are determined by integration of the equations

$$\frac{dr}{dz} = \frac{B \pm \sqrt{B^2 - AC}}{A} \tag{7}$$

in hyperbolic regions where $B^2 - AC > 0$.

11.3 Flow between parallel plates which rotate at different speeds around a common axis

We are now going to use the solutions given by N. Phan-Thien [1983a, b] to determine the characteristics given by (7). The angular velocities of the lower and upper plates are Ω_1 and Ω_2. The distance between the plates is d, $Re = \dfrac{\Delta\Omega d^2}{\nu}$ is the Reynolds number where

$$\Delta\Omega \overset{\text{def}}{=} \Omega_2 - \Omega_1 \tag{8}$$

and $W = \lambda\Delta\Omega$ is the Weissenberg number. There are ten unknowns, six stresses, three velocities and the pressure. After introducing the similarity variable

$$\xi = \frac{z}{d} \tag{9}$$

he derives ten equations for the velocity functions $h(\xi)$, $g(\xi)$ and the eight stress like quantities with hats which are listed below

$$u = \Delta\Omega r h'(\xi) ,$$

$$v = \Delta\Omega r g(\xi) + \Omega_1 r , \quad w = -2\Delta\Omega d h(\xi) ,$$

$$\tau_{rr} = \eta\Delta\Omega \left(\hat{RR}_0 + \frac{r^2}{d^2} \hat{RR}_1 \right),$$

$$\tau_{r\theta} = \eta\Delta\Omega \frac{r^2}{d^2} \hat{R\theta} ,$$

$$\tau_{rz} = \eta\Delta\Omega \frac{r}{d} \hat{RZ} , \tag{10}$$

$$\tau_{\theta\theta} = \eta\Delta\Omega \left(\hat{\theta\theta}_0 + \frac{r^2}{d^2} \hat{\theta\theta}_1 \right),$$

$$\tau_{\theta z} = \eta\Delta\Omega \frac{r}{d} \hat{\theta Z} ,$$

$$\tau_{zz} = \eta\Delta\Omega \hat{ZZ} .$$

The equations show that $\hat{RR}_0 = \hat{\theta\theta}_0$ if the initial conditions for the corresponding unsteady problem are the same. We have noted that the governing equations imply that

$v, \tau_{r\theta}, \tau_{\theta z}$ are odd functions,

$u, w, \tau_{\theta\theta}, \tau_{zz}, \tau_{rr}$ are even functions of $\Delta\Omega$ in the sense that

$$(\Omega_1, \Omega_2) \rightarrow -(\Omega_1, \Omega_2).$$

Note that in deriving the symmetry W and Re changes sign.

The boundary conditions are:

$$h=0 \text{ at } \xi=0 \quad \text{and} \quad \xi=1 ,$$

$$h'=0 \text{ at } \xi=0 \quad \text{and} \quad \xi=1 ,$$

$$g=0 \text{ at } \xi=0 \quad \text{and} \quad g=1 \text{ at } \xi=1\ , \tag{11}$$

$$\hat{RR}_0=0 \text{ at } \xi=0 \quad \text{and} \quad \xi=1\ ,$$

$$\hat{ZZ}=0 \text{ at } \xi=0 \quad \text{and} \quad \xi=1\ .$$

Our system of equations is singular because we divide by $h(\xi)$ and $h(0)=0$. We used a shooting method (RK4) to solve the equations and started the numerical integration using a power series solution at the origin which eliminates the apparent singularity.

$$\begin{aligned}
\hat{RR}_0 &= 2\alpha\xi + a_{12}\xi^2\ ,\\
\hat{RR}_1 &= 2W\alpha^2 + a_{21}\xi + a_{22}\xi^2\ ,\\
\hat{R\theta} &= 2W\alpha\beta + a_{31}\xi + a_{32}\xi^2\ ,\\
\hat{RZ} &= \alpha + a_{41}\xi + a_{42}\xi^2\ ,\\
\hat{\theta\theta}_1 &= 2W\beta^2 + a_{51}\xi + a_{52}\xi^2\ ,\\
\hat{\theta Z} &= \beta + a_{61}\xi + a_{62}\xi^2\ ,\\
\hat{ZZ} &= -4\alpha\xi + a_{72}\xi^2\ ,\\
h &= \frac{\alpha}{2}\xi^2 + \frac{\gamma}{6}\xi^3 + a_{84}\xi^4\ ,\\
g &= \beta\xi + a_{92}\xi^2 + a_{93}\xi^3\ .
\end{aligned} \tag{12}$$

where $\alpha=h''(0)$, $\beta=g'(0)$, $\gamma=h'''(0)$ and the a's are functions of α,β,γ only. After each shot we get $h(1),h'(1)$, and $g(1)-1$ and formulate a Jacobian matrix to iterate for α,β,γ using Newton's method so that we finally land on $h=0,h'=0$, and $g=1$ at $\xi=1$. This is the same procedure used by Larson [1988b] except that we are shooting for three variables.

The computations of Verdier and Joseph [1989a] agree with those given by Phan-Thien [1983a, b]. He presented the graphs of h, h', g but did not give graphs of the stresses. We have computed results for various values of Re and W and we present here graphs of the velocities, stream function, and all the stresses defined by (10) when $W=1$ and $Re \leq 150$.We could not get convergent results for $W=1$ and $Re > 180$ because of extreme sensitivity to initial guesses at high Reynolds numbers.

In Figures 11.1 and 11.2 we have given the plots of $h(\xi)$ and $h'(\xi)$ giving the velocities u and w, $(10)_{1,3}$, of the secondary motion. The velocity v, $(10)_2$, is determined by $g(\xi)$ shown in Figure 11.3. The various quantities related to stresses are plotted in Figures 11.4 to 11.10. The development of a boundary layer on the rotating wall at $\xi=1$ as Re is increased at fixed W is strongly suggested by all these results. We shall see in the next section that these boundary layers appear in the elliptic regions rather than in the hyperbolic regions of flow. In thinking about stresses, given by (10) it is useful to think of $\frac{r}{d}$ being a large multiplier as would be typical at the free edge $r=r_0$ of parallel plate rheometers. For example, in the parallel plate rheometer used by Broyer and Macosko [1975] $25 < \frac{r}{d} < 250$. The shear stresses τ_{rz} and $\tau_{\theta z}$ are proportional to $\hat{RZ}$ in Figure 11.7 and $\hat{\theta Z}$ in Figure 11.9 with $\frac{r}{d}$ as the constant of proportionality. The shear stress $\tau_{r\theta}$ is proportional to $\hat{R\theta}$ in Figure 11.6 with a constant of proportionality $\frac{r^2}{d^2}$. After comparing the three stresses we find surprisingly that there is a gigantic growth in the shear stress $\tau_{r\theta}$ in the boundary layer near $\xi=1$ when Re and $\frac{r}{d}$ are large. The normal stresses τ_{rr} and $\tau_{\theta\theta}$ also scale with $\frac{r^2}{d^2}$ with amplitudes $\hat{RR}_1$ (Figure 11.5) and $\hat{\theta\theta}_1$ (Figure 11.8). The extra tension $\tau_{\theta\theta}$ near $\xi=1$ at the edge of parallel plate rheometers can be truly enormous.

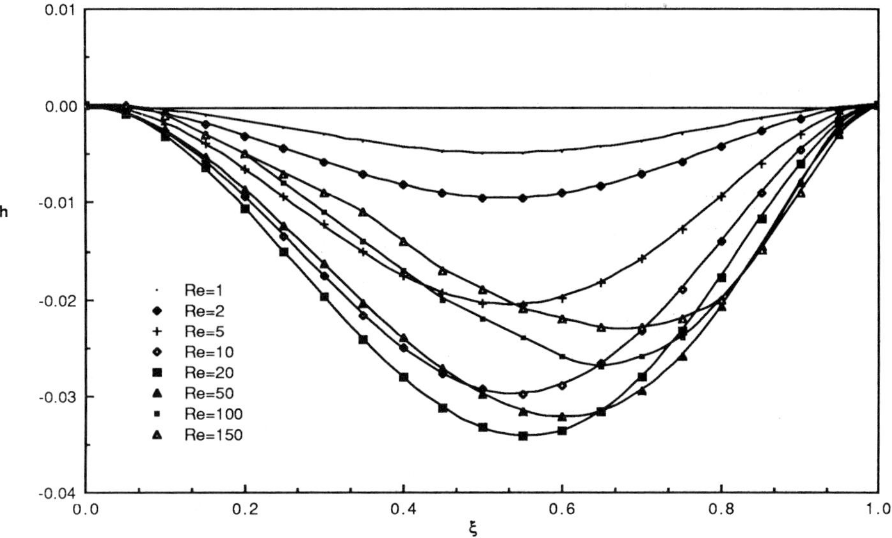

Figure 11.1 h(ξ) for Re=1, 2, 5, 10, 20, 50, 100, 150 at W=1

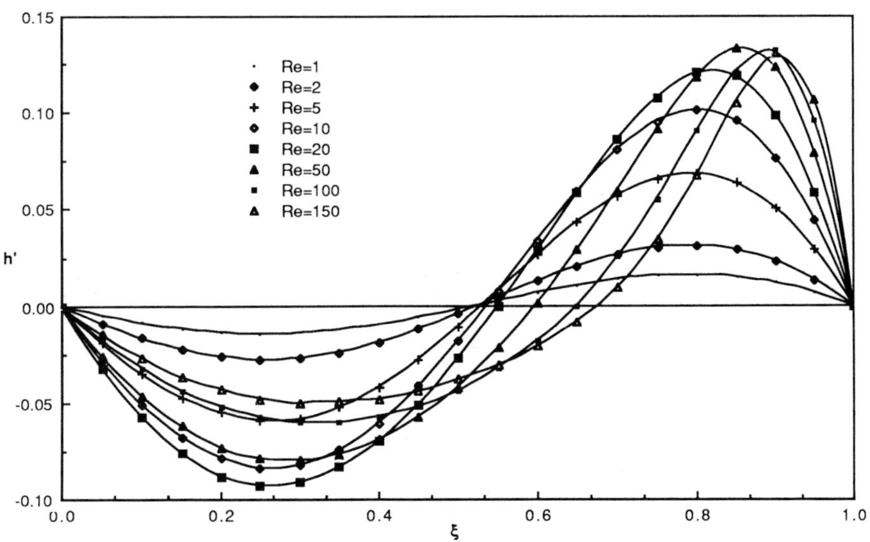

Figure 11.2 h′(ξ) for Re=1, 2, 5, 10, 20, 50, 100, 150 at W=1

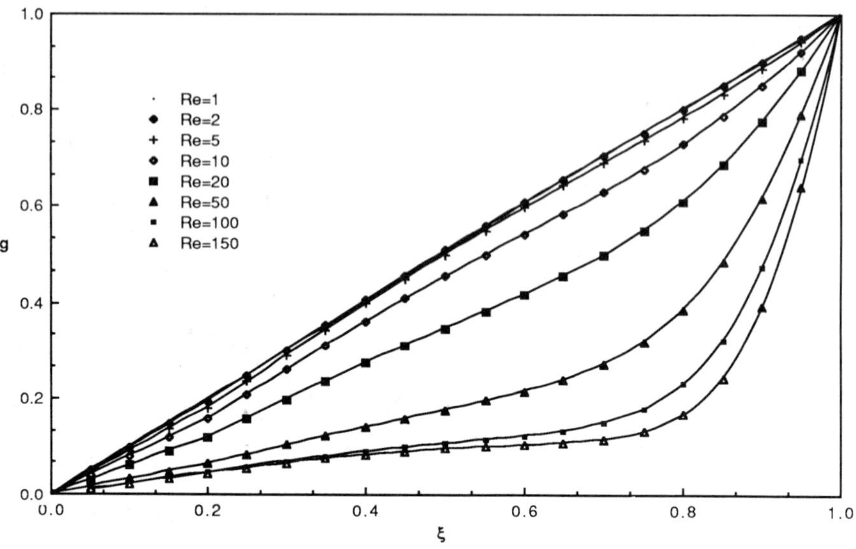

Figure 11.3 g(ξ) for Re=1, 2, 5, 10, 20, 50, 100, 150 at W=1

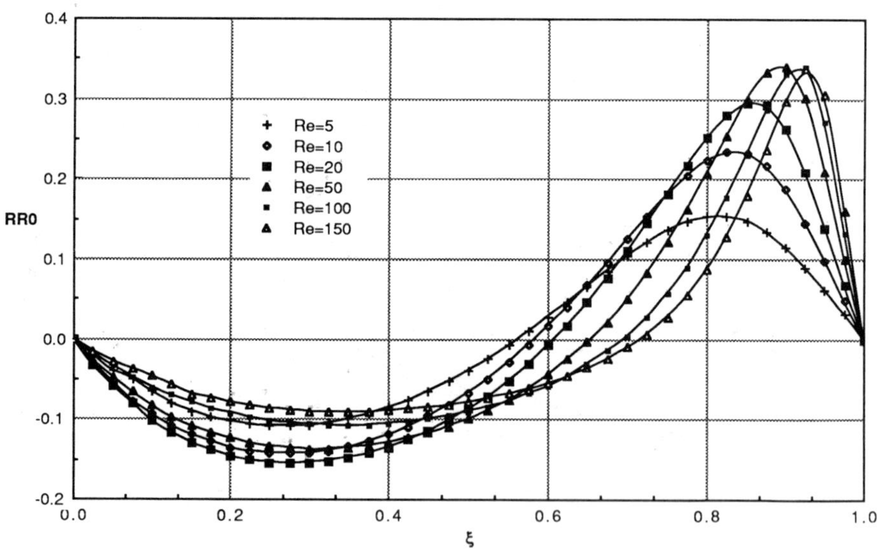

Figure 11.4 $\hat{RR}_0(\xi)$ for Re=5, 10, 20, 50, 100, 150 at W=1

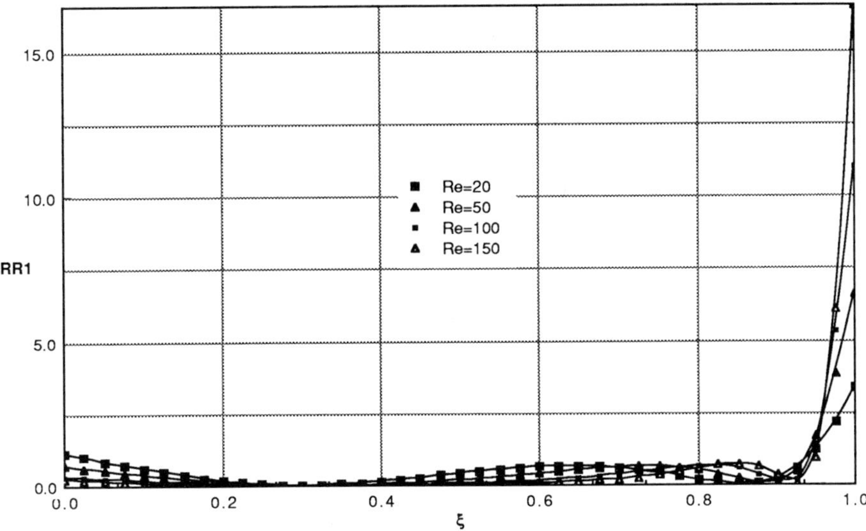

Figure 11.5 $\hat{RR}_1(\xi)$ for Re=20, 50, 100, 150 at W=1

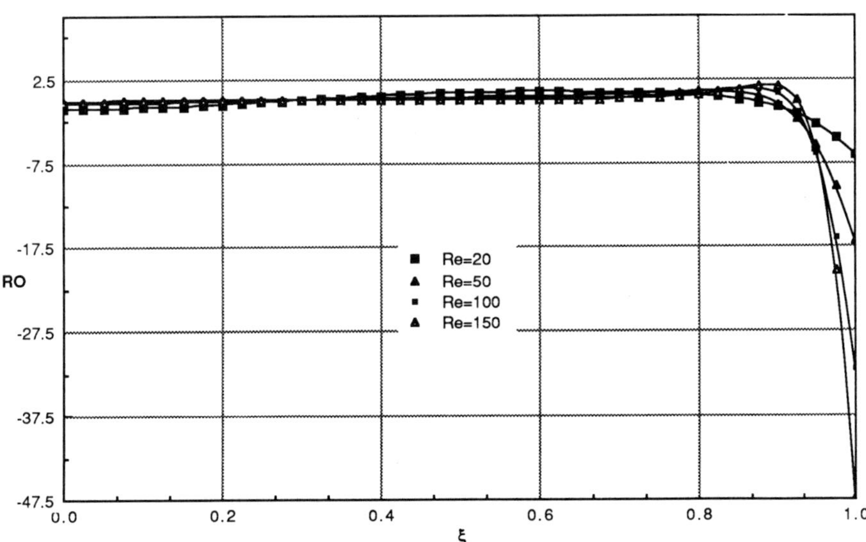

Figure 11.6 $\hat{R\theta}(\xi)$ for Re=20, 50, 100, 150 at W=1

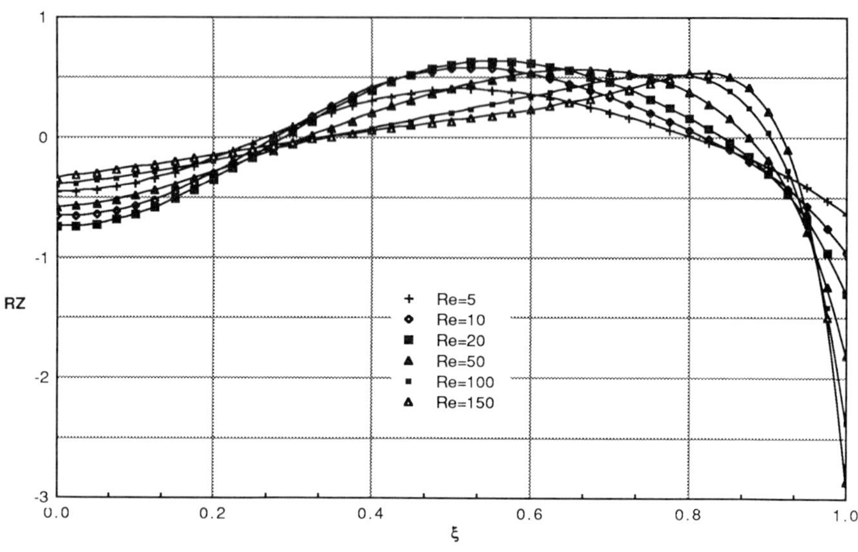

Figure 11.7 $\hat{RZ}(\xi)$ for Re=5, 10, 20, 50, 100, 150 at W=1

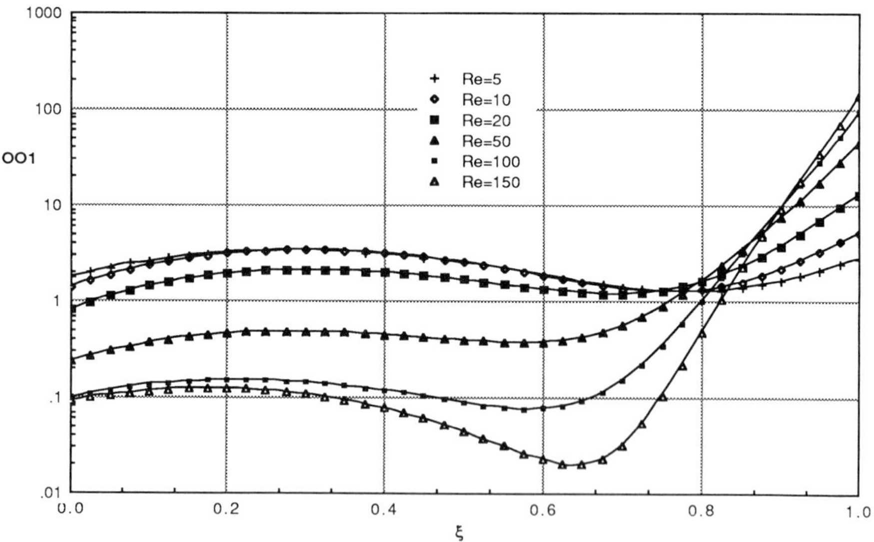

Figure 11.8 $\hat{\theta\theta}_1(\xi)$ for Re=5, 10, 20, 50, 100, 150 at W=1

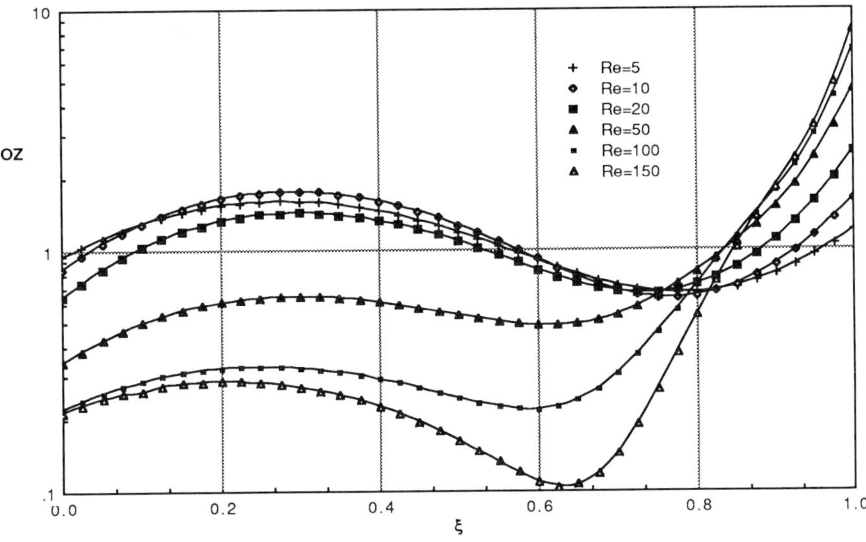

Figure 11.9 $\hat{\theta Z}(\xi)$ for Re=5, 10, 20, 50, 100, 150 at W=1

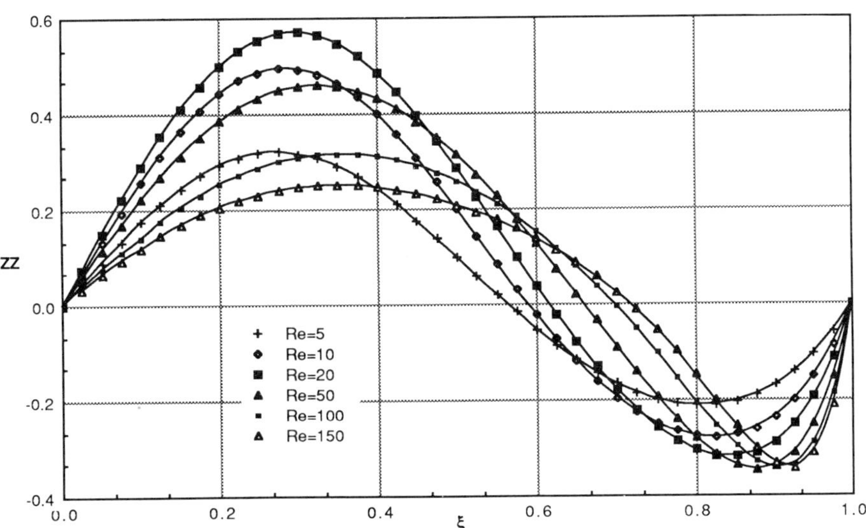

Figure 11.10 $\hat{ZZ}(\xi)$ for Re=5, 10, 20, 50, 100, 150 at W=1

For small Reynolds numbers in the operating range $10^{-5}<\mathrm{Re}<10^{-2}$ which are typical in the experiments of Broyer and Macosko [1975] on the neck-in instability, the stresses $\tau_{\theta z}$ and $\tau_{\theta\theta}$ are huge compared to other stresses. They can be 10^4 times larger. This could be important for the generation of instability to non-axisymmetric disturbances (see §11.6). All the stresses moderate in the interior and grow strongly as Re is increased.

In Figures 11.11 and 11.12 we have plotted the streamlines and isovorticity curves for Re=50 and W=1,where

$$\zeta = \frac{\partial u}{\partial z} - \frac{\partial w}{\partial r} = \Delta\Omega \frac{r}{d} h''(\xi) . \tag{13}$$

The isovorticity curves are not interesting because they do not suggest the underlying hyperbolic structure of the dynamics. To see this we should have to look at the propagation of disturbed vorticity in steady flows perturbing this one. We expect to see the perturbed vorticity propagating along the characteristics in hyperbolic regions, as in the problems studied by Yoo and Joseph [1985] and Ahrens, Yoo, and Joseph [1986]. Some features of this type of perturbation are discussed in §11.6.

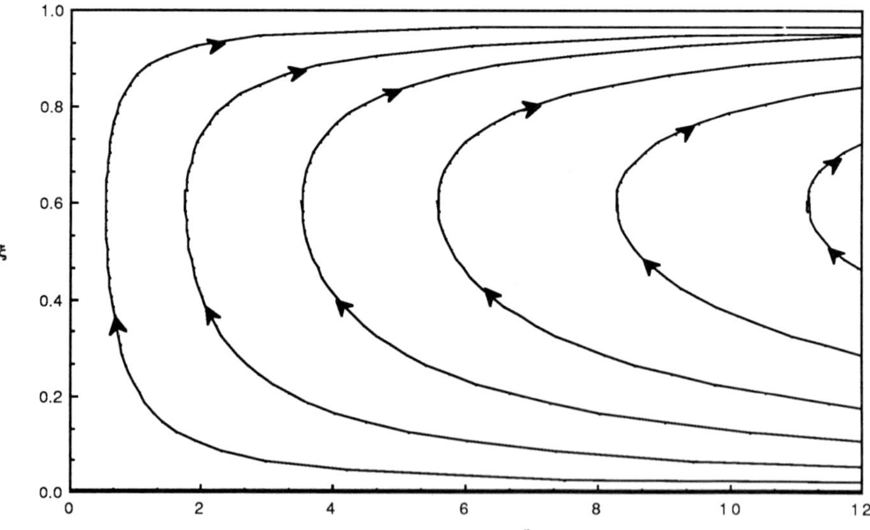

Figure 11.11 Streamlines: Re=50, W=1

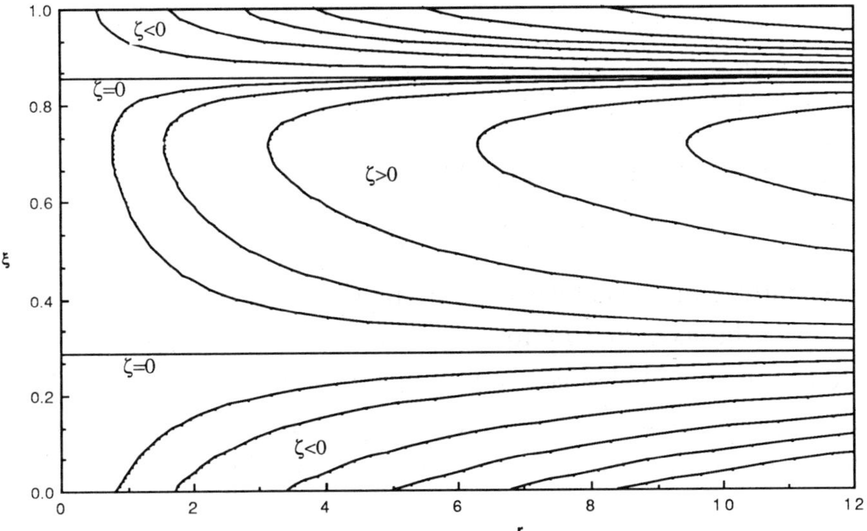

Figure 11.12 Isovorticity curves: Re=50, W=1

11.4 Change of type in the flow between rotating parallel plates

Now we look for the regions $B^2–AC>0$ defined in (7) where the vorticity is hyperbolic. We find

$$\Delta \overset{\text{def}}{=} \frac{B^2-AC}{\Delta\Omega^2\eta^2} = \alpha(\xi)\,\frac{r^2}{d^2} + \beta(\xi) \tag{14}$$

where

$$\alpha(\xi) = (2\mathrm{Re}\, hh' + \hat{R}Z)^2 - \left(4\mathrm{Re}\, h^2 - \hat{Z}Z - \frac{1}{W}\right)(\mathrm{Re}\, h'^2 - \hat{R}R_1)$$

$$\beta(\xi) = \left(4\mathrm{Re}\, h^2 - \hat{Z}Z - \frac{1}{W}\right)\left(\hat{R}R_0 + \frac{1}{W}\right) \tag{15}$$

We found that $\beta(\xi)<0$ in all cases. Using the boundary conditions (11) and the asymptotic expressions (12) we find that

$$\Delta = -\alpha^2 \frac{r^2}{d^2} - \frac{1}{W^2}, \quad z=0 ,$$
$$\Delta = -\delta^2 \frac{r^2}{d^2} - \frac{1}{W^2}, \quad z=1 \tag{16}$$

where $\alpha=h''(0)$ and $\delta=h''(1)$. Moreover

$$\Delta = \beta(\xi) < 0 \quad \text{for} \quad r = 0. \tag{17}$$

Therefore there is always an elliptic region close to the disks and close to the axis of rotation. Since $\beta(\xi)<0$, the change of type occurs when $\alpha(\xi)$ changes sign. In our computations we found that $\alpha(\xi)$ changes sign four times when ξ goes from 0 to 1. Therefore there are two hyperbolic regions whose boundaries are defined by

$$r_c(\xi) = d\sqrt{-\frac{\beta(\xi)}{\alpha(\xi)}}. \tag{18}$$

where the limiting radius $r_c(\xi)\to\infty$ at the four roots of $\alpha(\xi)$ as shown in Figure 11.13.

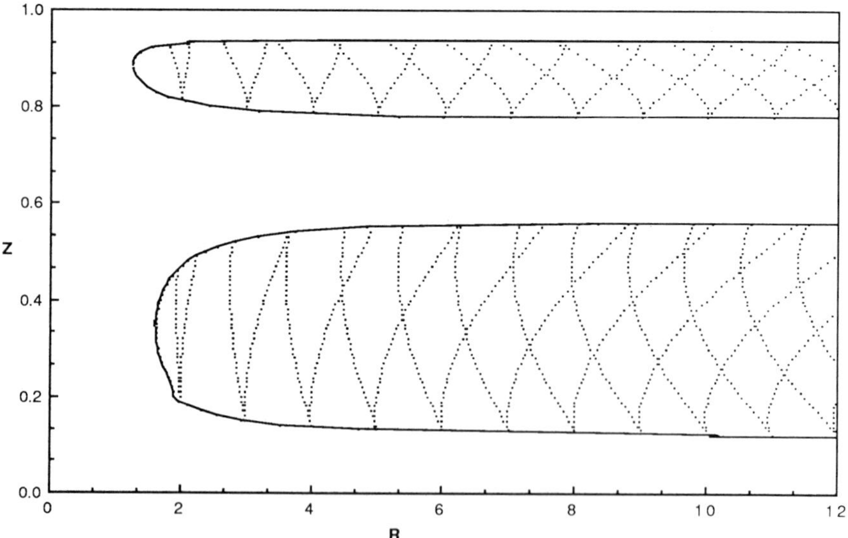

Figure 11.13 Regions of hyperbolicity and characteristics: Re=50, W=1

We used an asymptotic expansion in powers of Re,W (valid for moderately small Re and W) first used by Stewartson [1953]. When the lower disk is fixed the leading terms of this expansion are

$$
\begin{aligned}
h &= -\frac{Re}{60}\,\xi^2(\xi-1)^2(\xi+2)\,,\\
\hat{RR}_0 &= 2h'\,,\\
\hat{RR}_1 &= 2Wh''^2\,,\\
\hat{RZ} &= h''\,,\\
\hat{ZZ} &= -4h'\,.
\end{aligned}
\tag{19}
$$

Those five quantities are the only ones that are needed to see whether a change of type occurs at small Re and W. We obtain

$$
\begin{aligned}
\alpha(\xi) &= (2Re\,hh' + h'')^2 - \left(4Re\,h^2 + 4h' - \frac{1}{W}\right)(Re\,h'^2 - 2Wh''^2)\,,\\
\beta(\xi) &= \left(4Re\,h^2 + 4h' - \frac{1}{W}\right)\left(2h' + \frac{1}{W}\right)
\end{aligned}
\tag{20}
$$

where h is given in terms of ξ by $(19)_1$.

It is not hard to show , using (20) and after neglecting the smaller terms that

$$
\begin{aligned}
&\beta(\xi) < 0\,,\\
&\alpha(\xi) < 0 \quad \text{for} \quad \xi_4<\xi<1\,,\\
&\xi_2<\xi<\xi_3\,,\\
&0<\xi<\xi_1\,,\\
&\alpha(\xi) > 0 \quad \text{for} \quad \xi_1<\xi<\xi_2\,,\\
&\xi_3<\xi<\xi_4
\end{aligned}
$$

where the ξ_i (i=1,2,3,4) are roots of $\alpha(\xi)=0$ and such that

$$0 < \xi_1 < 0.237 < \xi_2 < 0.525 < \xi_3 < 0.808 < \xi_4 < 1\,.$$

This agrees with our computations.

In Figure 11.13 we plotted the hyperbolic regions and a net of characteristics for a Reynolds number of 50 and Weissenberg number of 1.

The effect of the parameters on the size and position of the regions of hyperbolicity are shown in Figures 11.14 and 11.15. In general the hyperbolic regions move in and their width decreases as the Weissenberg number W is increased at a fixed value of Re. At a fixed Weissenberg number the regions move in if the Reynolds number is increased.

Figure 11.14 Hyperbolic regions: (a) Re=50, W=0.6; (b) Re=50, W=1.8

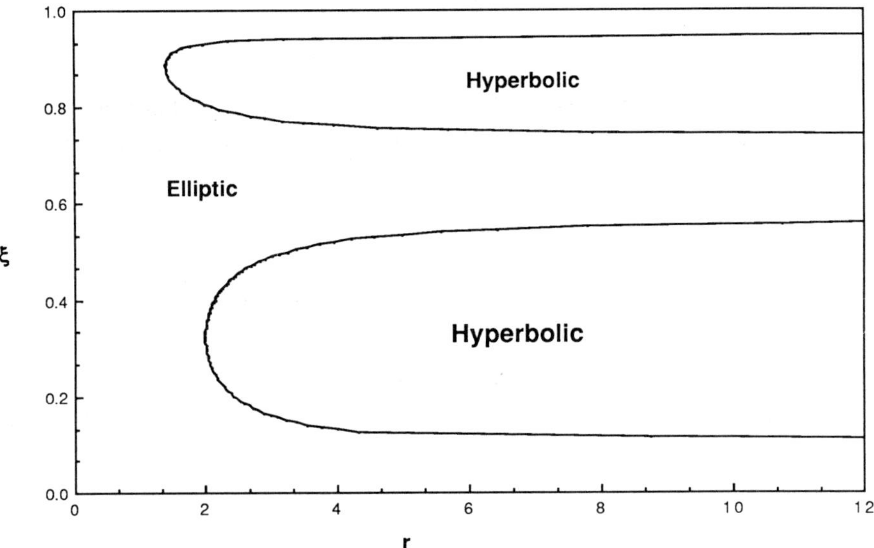

(a) Re=50, W=0.6

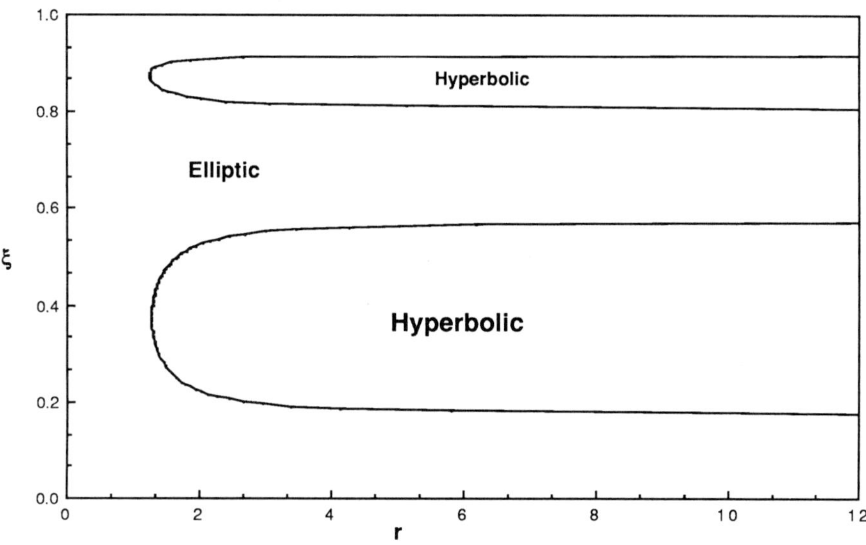
Hyperbolic
Elliptic
Hyperbolic
ξ
r
1.0
0.8
0.6
0.4
0.2
0.0
0
2
4
6
8
10
12

(b) Re=50, W=1.8

Figure 11.15 Hyperbolic regions: (a) Re=100, W=0.8; (b) Re=100, W=1.8

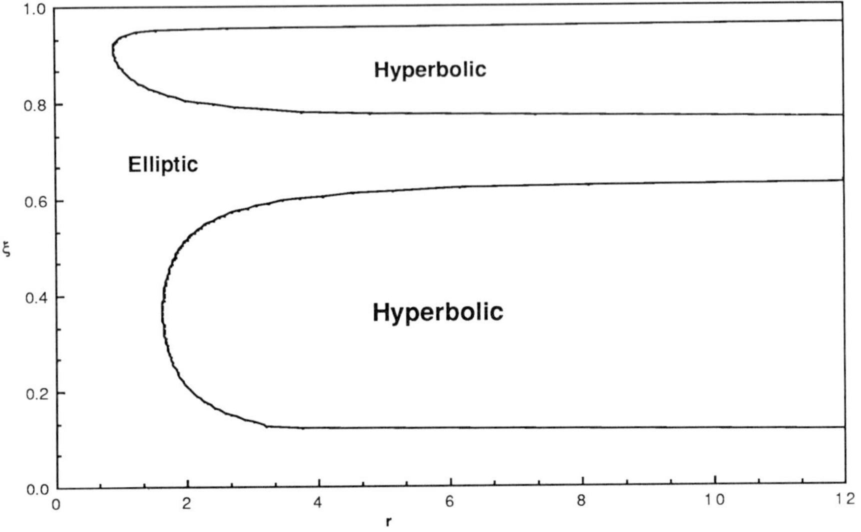

(a) Re=100, W=0.8

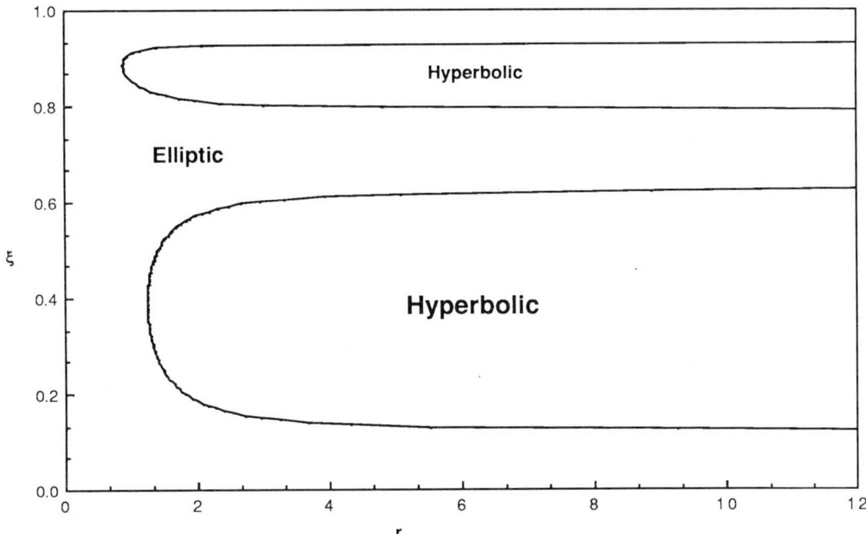

(b) Re=100, W=1.8

11.5 The viscoelastic Mach number

It may be more relevant here to introduce a local Mach number defined by

$$M = \frac{v}{c} = \frac{rG}{c} = \sqrt{WRe}\,\frac{r}{d}\,g(\xi)\ . \tag{21}$$

This Mach number increases with r and ξ. We want to check the value of the Mach number at the edge of each hyperbolic region i.e. $M=M_c$ at the point where $r_c(\xi)$ is minimum. We now introduce

$$E = \frac{W}{Re} = \frac{\lambda\nu}{d^2} \tag{22}$$

where E is the elasticity number.

We now think about one fluid, and we only vary the angular velocity Ω. Therefore E remains constant. Then we plot M_c as a function of Re (or similarly Ω). We did that in Figure 11.16 for

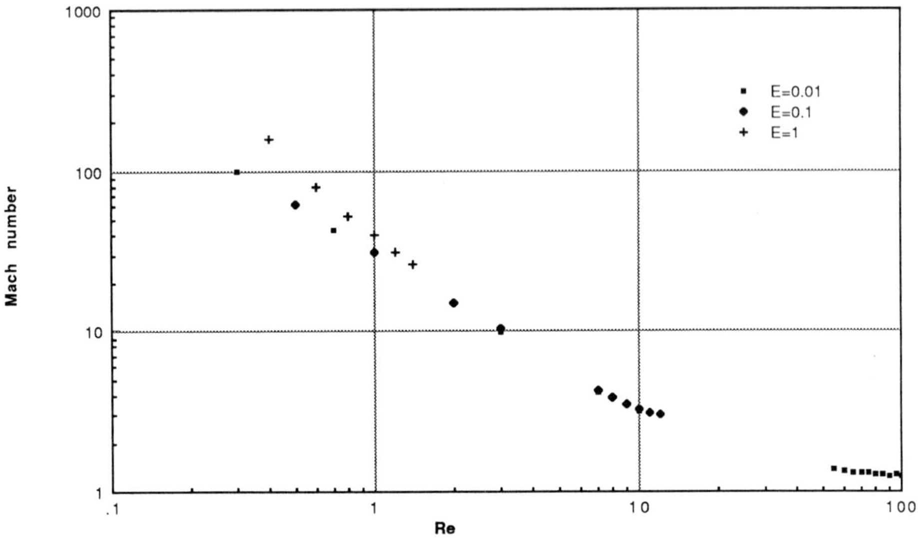

Figure 11.16 Mach number at the edge of the lower hyperbolic region vs. Re, at elasticity numbers E=0.01, 0.1, 1

different Elasticity numbers. We only plotted the smallest Mach number corresponding to the lower region. It appears that $M_c>1$. Hence the Mach number is larger than one throughout the hyperbolic region.

11.6 Three dimensional perturbations of the similarity solution

Consider a small, but otherwise arbitrary perturbation of the similarity solution. For example, one could consider a perturbation induced by a wavy top plate $z = d+\varepsilon \sin n\theta$ for small $\varepsilon \to 0$. Each and every such perturbation gives rise to a linear system of equations; in particular the linearized vorticity equation for the upper convected Maxwell model is

$$A\frac{\partial^2\zeta}{\partial r^2}+\frac{B}{r^2}\frac{\partial^2\zeta}{\partial\theta^2}+C\frac{\partial^2\zeta}{\partial z^2}+\frac{2D}{r}\frac{\partial^2\zeta}{\partial r\partial\theta}+2E\frac{\partial^2\zeta}{\partial r\partial z}+\frac{2F}{r}\frac{\partial^2\zeta}{\partial\theta\partial z}=\text{lot} \quad (23)$$

where A,B,C,D,E,F depend on the basic flow.

$$\begin{aligned}
A &= \rho u^2-\tau_{rr}-G=A_0(z)+r^2A_1(z)\,,\\
B &= \rho v^2-\tau_{\theta\theta}-G=B_0(z)+r^2B_1(z)\,,\\
C &= \rho w^2-\tau_{zz}-G=C_0(z)\,,\\
D &= \rho uv-\tau_{r\theta}=r^2D_1(z)\,,\\
E &= \rho uw-\tau_{rz}=rE_1(z)\,,\\
F &= \rho vw-\tau_{\theta z}=rF_1(z)\,,
\end{aligned} \quad (24)$$

and

$$\begin{aligned}
A_0(z) &= -\eta\Delta\Omega\,\widehat{RR}_0-G\,,\\
A_1(z) &= \rho H_z^2-\frac{\eta\Delta\Omega}{d^2}\widehat{RR}_1\,,\\
B_0(z) &= -\eta\Delta\Omega\,\widehat{\theta\theta}_0-G\,,\\
B_1(z) &= \rho G^2-\frac{\eta\Delta\Omega}{d^2}\widehat{\theta\theta}_1\,,\\
C_0(z) &= 4\rho H^2-\eta\Delta\Omega\,\widehat{ZZ}-G\,,\\
D_1(z) &= \rho GH_z-\frac{\eta\Delta\Omega}{d^2}\widehat{R\theta}\,,\\
E_1(z) &= -2\rho HH_z-\frac{\eta\Delta\Omega}{d}\widehat{RZ}\,,\\
F_1(z) &= -2\rho HG-\frac{\eta\Delta\Omega}{d}\widehat{\theta Z}\,,
\end{aligned} \quad (25)$$

are defined completely by the similarity solution.

Characteristic surfaces $\phi(r,\theta,z)=0$ are those across which second derivatives can be discontinuous with continuity for lower order derivatives. After introducing characteristic coordinates into (23) we find that

$$(A_0(z) + r^2A_1(z))\phi_r^2 + \left(\frac{B_0(z)}{r^2} + B_1(z)\right)\phi_\theta^2 + C_0(z)\phi_z^2$$

$$+ 2rD_1(z)\phi_r\phi_\theta + 2rE_1(z)\phi_r\phi_z + 2F_1(z)\phi_\theta\phi_z = 0 \qquad (26)$$

where

$$\phi_r\,dr + \phi_\theta\,d\theta + \phi_z\,dz = 0\,. \qquad (27)$$

In §11.4 we calculated the form of characteristic surfaces $\phi(r,z)=0$ which are independent of θ. The coefficients of terms with ϕ_θ can be very large. We have not been successful in solving (26) and (27).

It is perhaps of interest to consider (26) in the case when $\Omega_1=\Omega_2$, giving rise to rigid rotation. In this case (26) becomes

$$\frac{\partial^2\zeta}{\partial r^2} + \frac{\partial^2\zeta}{\partial z^2} + \left(\frac{1}{r^2} - \frac{\Omega^2}{c^2}\right)\frac{\partial^2\zeta}{\partial\theta^2} = \text{lot} \qquad (28)$$

giving rise to characteristic surfaces

$$\phi_r^2 + \phi_z^2 + \left(\frac{1}{r^2} - \frac{\Omega^2}{c^2}\right)\phi_\theta^2 = 0\,. \qquad (29)$$

A three dimensional solution of (29) is

$$\phi(r,\theta,z) = \alpha z + \beta\theta + g(r) \qquad (30)$$

where g(r) satisfies

$$g'^2 + \alpha^2 + \beta^2\left(\frac{1}{r^2} - \frac{\Omega^2}{c^2}\right) = 0\,. \qquad (31)$$

The condition $r > \frac{\Omega}{c}$ is sufficient to guarantee that the regions of flow with $r>r_c$ are hyperbolic. In Figures 11.17 and 11.18 we plotted the characteristic surfaces (30) and (31).

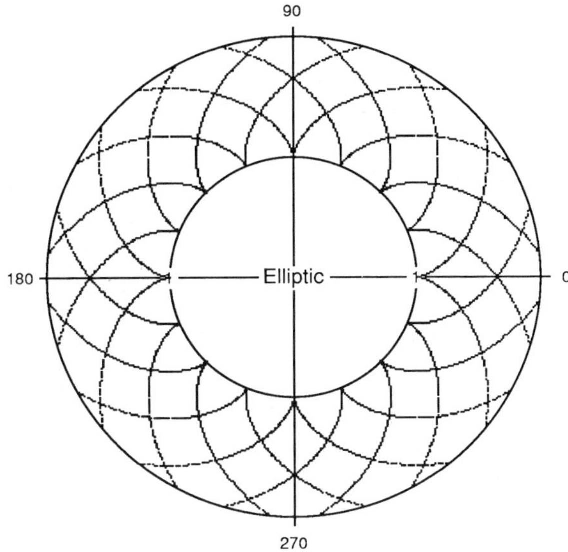

Figure 11.17 Characteristics of rigid motion $\frac{\Omega}{c} = 1$, $\alpha=0$, $\beta=1$

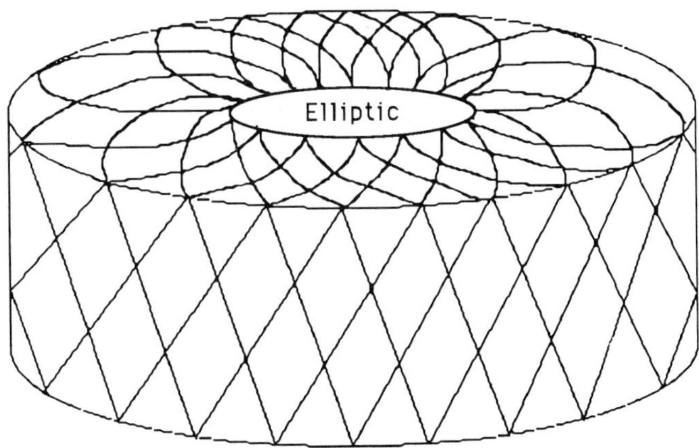

Figure 11.18 Characteristic surfaces of rigid motion

It is perhaps of interest to recall that elastic cylinders under severe twists develop instabilities with spiral form, as in Figure 11.17. The well known neck-in instability at the edge of a sample between rotating parallel plate disks develops spirals of this type before giving away. Broyer and Macosko [1975] studied this instability and they showed that it was characterized by a critical velocity rather than a critical shear rate. This finding is unusual and it suggests a change of type. However the magnitude of the velocities in the experiments are several orders of magnitude smaller than the ones needed to move the hyperbolic regions into the disk. We have not completely ruled out the possibility that this instability is connected to hyperbolicity in a way presently not understood by us, but on the face of it we cannot make the instability fit the change of type even given a wide latitude in the selection of the magnitude of the model parameters.

All this still leaves us in the dark about the physical consequences of the change of type we have found in the Phan-Thien solution.

11.7 Fluid driven by an accelerated surface

Larson [1988b] found a similarity solution for the steady flow of a Maxwell fluid in an infinitely long cylinder of dimensionless radius r=1 whose boundaries move with an axial velocity proportional to the axial coordinate z(w(1)=z, u(1)=0).

We assume axisymmetry again and note that

$$\begin{aligned} v &= 0\,, \\ \tau_{r\theta} &= 0\,, \\ \tau_{\theta z} &= 0\,. \end{aligned} \tag{32}$$

Equations (1) still apply and the similarity solution is given by

$$
\begin{aligned}
u &= \frac{f(r)}{r} , \\
w &= -z \frac{f'(r)}{r} , \\
\tau_{rr} &= - i(r) , \\
\tau_{\theta\theta} &= -j(r) , \\
\tau_{rz} &= -zk(r) , \\
\tau_{zz} &= -g(r) - \frac{1}{2} h(r) z^2 , \\
p &= p_0(r) + \frac{1}{2} p_1 z^2 .
\end{aligned}
\tag{33}
$$

where f,g,h,i,j,k,p_0 are functions of r only $(0 \leq r \leq 1)$ and p_1 is a constant pressure coefficient. The variables in (33) have been made dimensionless with a scale length based on the radius R of the cylinder and the velocity U of the accelerated surface. The problem depends on two dimensionless parameters, the Reynolds number $Re = \dfrac{UR}{\nu}$ and the Deborah number $De = \dfrac{\lambda U}{R}$ (where λ is the relaxation time of the fluid). The boundary conditions are

$$
\begin{aligned}
f'(1) &= 1, \\
f(1) &= 0 , \\
f(0) &= 0 , \\
f'(0) &= 0 , \\
i(0) &= \frac{-f''(0)}{1 - De\, f''(0)} , \\
g(0) &= \frac{2f''(0)}{1 + 2\, De\, f''(0)} , \\
k(0) &= 0 , \\
h(0) &= 0 .
\end{aligned}
\tag{34}
$$

To solve the equations, a rescaling technique is used, corresponding to the change of variables:

$$
\xi = \sqrt{De\, |p_1|}\, r ,
$$

$$
p^* = \frac{p1}{|p_1|} = \text{sign } (p_1) ,
$$

$$f(r) = \frac{F(\xi)}{De^2|p_1|},$$

$$h(r) = |p_1|\, H(\xi),$$

$$k(r) = \sqrt{\frac{|p_1|}{De}}\, K(\xi), \tag{35}$$

$$i(r) = \frac{I(\xi)}{De},$$

$$g(r) = \frac{G(\xi)}{De},$$

$$Re = Re^* De^2\, |p_1|.$$

We are left with six nonlinear first order equations for F, F', G, H, I, K depending on the parameters Re^* and p^* together with the boundary conditions

$$F(0) = 0,$$
$$F'(0) = 0,$$
$$K(0) = 0,$$
$$H(0) = 0, \tag{36}$$

$$I(0) = \frac{-F''(0)}{1-F''(0)},$$

$$G(0) = \frac{2F''(0)}{1+2\,F''(0)},$$

The solution is carried out using different values of $F''(0)$ and terminated when $F(\xi_0)=0$ (ξ_0 corresponds to the rescaled position of the wall). The integration is initiated by an expansion in powers of ξ to remove the singularity at $\xi=0$. Then determine De and p_1 from the stopping conditions.

$$1 = F'(\xi_0)\, De^{-3/2}\, |p_1|^{-1/2},$$

$$1 = De^{-1/2}\, |p_1|^{-1/2}\, \xi_0. \tag{37}$$

Our results are in good agreement with Larson's computations. Once the flow is determined we calculate B^2-AC using (5), B^2-AC has the same sign as Δ', where

$$\Delta' = \alpha(\xi)z^2 + \beta(\xi) \tag{38}$$

and $\alpha(\xi)$ and $\beta(\xi)$ are given by

$$\alpha(\xi) = \left(K-Re^*\frac{FF'}{\xi^2}\right)^2 - \left(\frac{F'^2Re^*}{\xi^2} + \frac{1}{2}H\right)\left(Re^*\frac{F^2}{\xi^2}+I-1\right),$$

$$\beta(\xi) = \frac{1-G}{De\,p_1}\left(Re^*\frac{F^2}{\xi^2}+I-1\right). \tag{39}$$

Again we look at the signs of $\alpha(\xi)$ and $\beta(\xi)$ to determine where Δ' changes sign

$$\begin{aligned}
&\beta(\xi) \le 0 && \text{for } 0 \le \xi \le \xi_0\,,\\
&\alpha(\xi) \le 0 && \text{for } \xi_1 \le \xi \le \xi_2\,,\\
&\alpha(\xi) \ge 0 && \text{for } 0 \le \xi \le \xi_1 \text{ and } \xi_2 \le \xi \le \xi_0\,,
\end{aligned}$$

where ξ_1 and ξ_2 are roots of $\alpha(\xi)=0$, i=1,2.

Therefore there are two regions that are hyperbolic, one close to the centerline, the other one close to the accelerated surface. These regions correspond to the domains of higher velocities. The regions appear when z is greater than a critical value $z_c(\xi)$ given by

$$z_c(\xi) = \sqrt{-\frac{\beta(\xi)}{\alpha(\xi)}} \tag{40}$$

In Figure 11.19 we plotted the regions of hyperbolicity and the characteristic grids on those regions. As we increase the Reynolds number, the regions of hyperbolicity clearly move in toward the origin as shown in Figures 11.20a,b,c. As the Deborah number is increased we observe the same behavior, but one region gets bigger (the one at the center line) and the other one gets smaller (the one close to the accelerated surface). This is shown in Figures 11.21a to 11.21c.

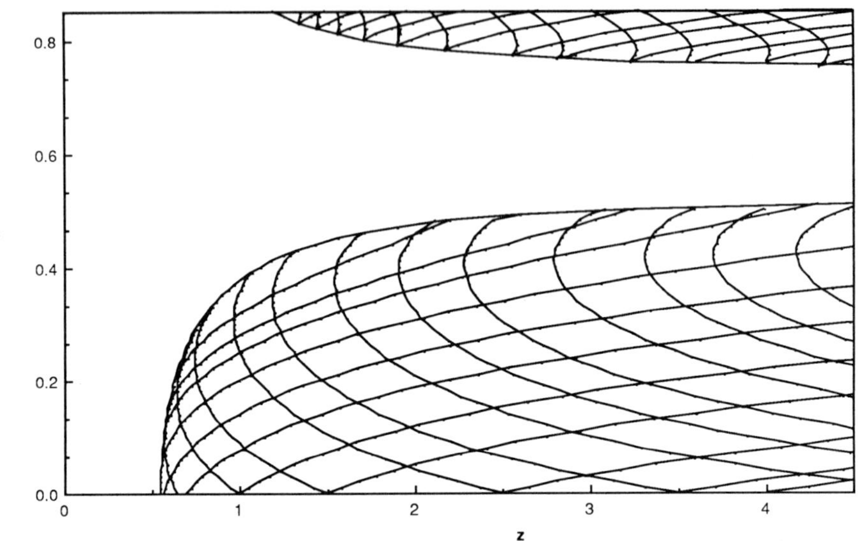

Figure 11.19 Hyperbolic regions with characteristics at De=0.2, Re=9.08, M=1.35

Figure 11.20 Hyperbolic regions: Influence of the Reynolds number: (a) De=0.1, Re=3.84; (b) De=0.1, Re=9; (c) De=0.1, Re=11.9

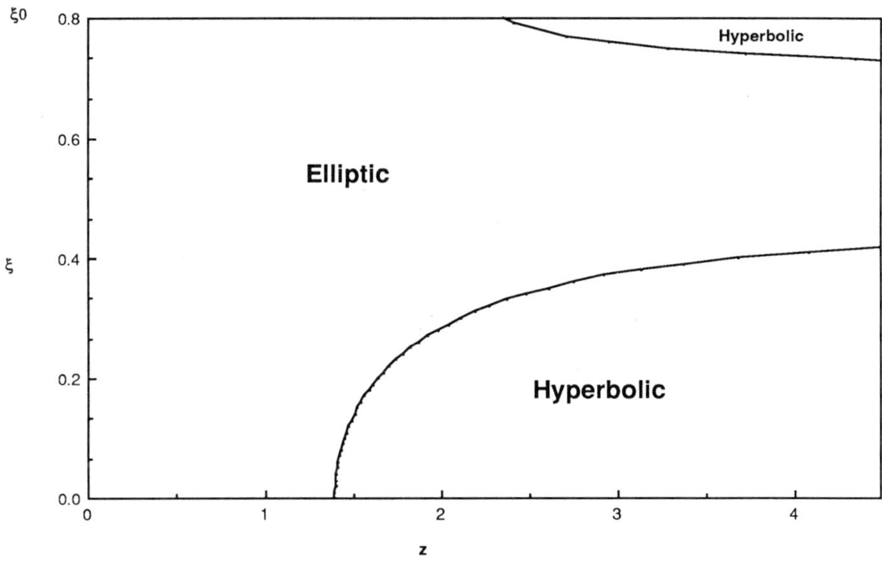

(a) De=0.1, Re=3.84

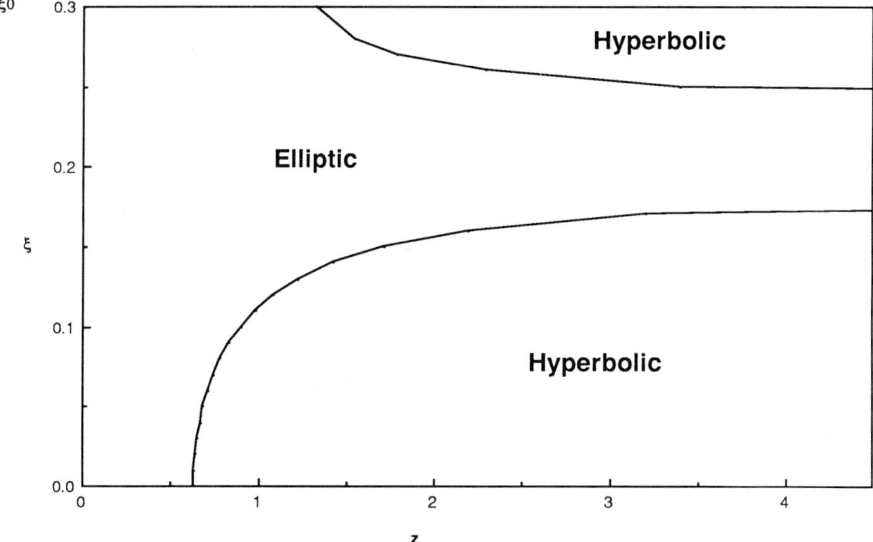

(b) De=0.1, Re=9

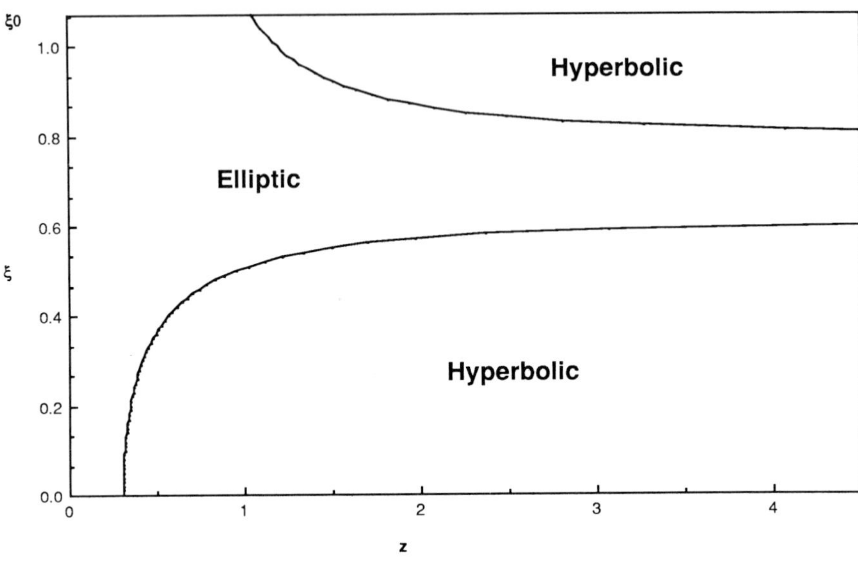

(c) De=0.1, Re=11.9

Figure 11.21 Hyperbolic regions: Influence of the Deborah number: (a) De=0.115, Re=10; (b) De=0.24, Re=10; (c) De=0.4, Re=10

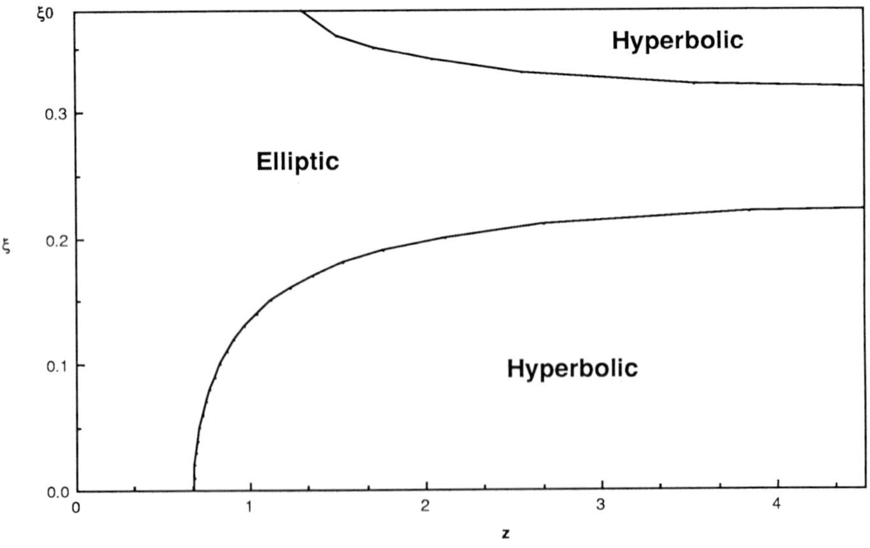

(a) De=0.115, Re=10

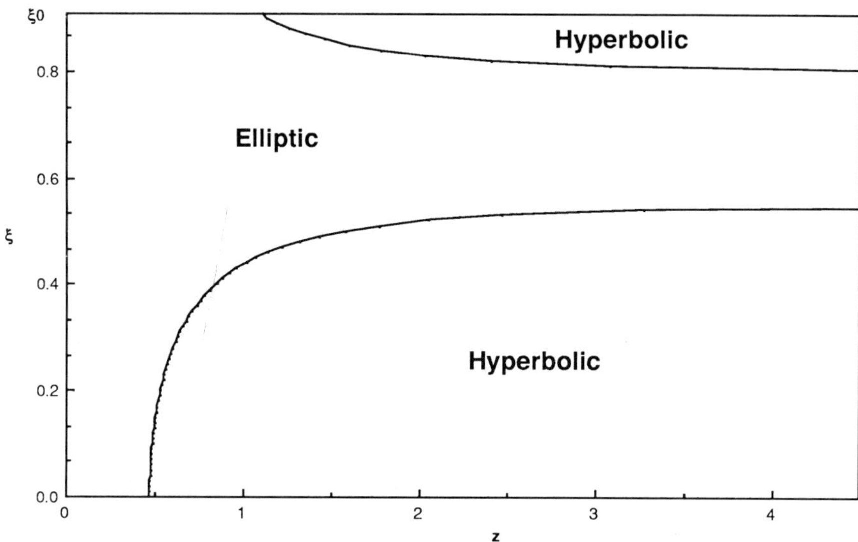

(b) De=0.24, Re=10

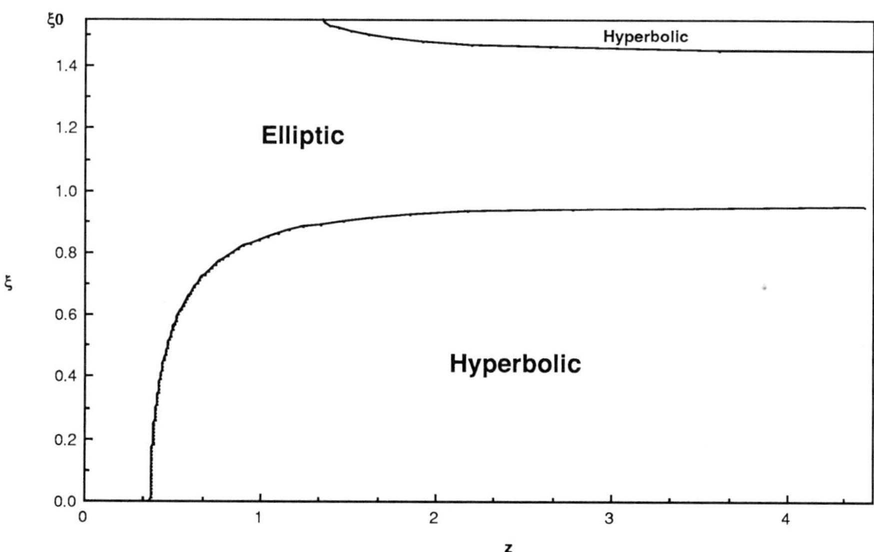

(c) De=0.4, Re=10

CHAPTER 12

Poiseuille Flows

This chapter is based on the [1985] paper of Yoo and Joseph and the [1987] paper of Ahrens, Yoo, and Joseph.

12.1 Introduction

We consider steady flow of an upper convected Maxwell fluid through a channel (or a pipe) with wavy walls. The vorticity of this flow will change type when the velocity in the center of the channel is larger than a critical value defined by the propagation of shear waves. There is then a central region of the channel in which the vorticity equation is hyperbolic and a low speed region near the walls where the vorticity equation is elliptic. We linearize the problem for small amplitude waviness and the linearized problem is solved in detail. The characteristic nets depend on the viscoelastic "Mach" number, which is the ratio (M=U/c) of the unperturbed maximum velocity U to the speed of shear waves c into the fluid at rest, and on the elasticity number E. There is a supercritical (hyperbolic) region around the center of the channel when M>1. When M»1, the thickness of this hyperbolic region is small when E is large, and large when E is small. Regions of positive and negative vorticity are swept out along forward facing characteristics in the hyperbolic region. There is a rapid damping of vorticity in the hyperbolic region away from the boundary when $M \gg 1$ and the Weissenberg number $W=M\sqrt{E} \leq O(1)$. (The Weissenberg number is proportional to the relaxation time of the fluid.)

The rate of damping of vorticity decreases as W is increased. Flows with high M appear to be more "elastic" when W is large in the sense that the damping is suppressed as the relaxation time of the fluid is increased.

12.2 Governing equations for steady flow of Maxwell fluid

We consider the steady flow of an upper convected Maxwell fluid through a plane channel with wavy walls as sketched in Figure 12.1. The governing equations for this flow are

$$\text{div}\, \mathbf{u} = 0\,, \tag{1a}$$

$$\rho\left[\frac{\partial \mathbf{u}}{\partial t} + (\mathbf{u}\cdot\nabla)\mathbf{u}\right] = -\nabla p + \nabla\cdot\boldsymbol{\tau}\,, \tag{1b}$$

$$\lambda\left[\frac{\partial \boldsymbol{\tau}}{\partial t} + (\mathbf{u}\cdot\nabla)\boldsymbol{\tau} - \nabla\mathbf{u}\cdot\boldsymbol{\tau} - \boldsymbol{\tau}\cdot\nabla\mathbf{u}^T\right] + \boldsymbol{\tau} = \eta(\nabla\mathbf{u} + \nabla\mathbf{u}^T)\,, \tag{1c}$$

where ρ is the density, λ is the stress relaxation time, η is the zero shear rate viscosity, t is the time, $\mathbf{u}=(u,v)$ is the velocity, p is the constitutively indeterminate part of the Cauchy stress tensor and $\boldsymbol{\tau}=(\tau_{xx},\tau_{yy},\tau_{xy})=(\sigma,\gamma,\tau)$ is the determinate part. The Reynolds number $R=\rho Ud/\eta$ is based on the center line velocity U.

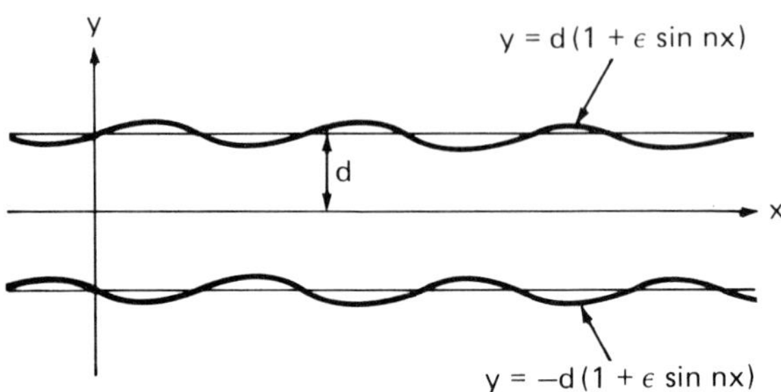

Figure 12.1. Schematic diagram of plane channel with wavy walls.

Only two of the parameters R,W,E,M defined in (7.40) appear in the analysis. We shall express our results in terms of E and M.

Flows which perturb uniform motions go supercritical when M>1. Wave speeds for water based polymer solutions are of the order 1 cm/s or 10 cm/s. The wave speeds of polymer melts are of the order of 100 cm/s. At the same time, the large viscosities of melts reduce the possibility of achieving large velocities. Hence for the flow of melts we expect that the flow is usually subcritical M<1. On the other hand, in water based polymers of the kind which are effective in drag reduction, we expect to see many flows with M>1.

The parameters M and E appear in the dimensionless equations (2) which arise from (1) and $d, U^*, \eta/\lambda$ are introduced as scales for length, velocity and stress.

$$u_x + v_y = 0\,, \tag{2a}$$

$$uu_x + vu_y + \frac{1}{M^2}(p_x - \sigma_x - \tau_y) = 0\,, \tag{2b}$$

$$uv_x + vv_y + \frac{1}{M^2}(p_y - \tau_x - \gamma_y) = 0\,, \tag{2c}$$

$$u\sigma_x + v\sigma_y - 2(\sigma+1)u_x - 2\tau u_y = -\frac{\sigma}{M\sqrt{E}}, \tag{2d}$$

$$u\tau_x + v\tau_y - (\gamma+1)u_y - (\sigma+1)v_x = -\frac{\tau}{M\sqrt{E}}, \tag{2e}$$

$$u\gamma_x + v\gamma_y - 2\tau v_x - 2(\gamma+1)v_y = -\frac{\gamma}{M\sqrt{E}}. \tag{2f}$$

We shall seek and find a solution of these equations satisfying no slip conditions at the walls

$$u = v = 0 \text{ at } y = \pm\,(1+\varepsilon \sin nx) \tag{3}$$

with symmetric streamlines

$$\frac{\partial u}{\partial y} = v = 0 \text{ at } y=0\ . \tag{4}$$

It is noteworthy that our solution is completely determined by data (3) and (4) on the velocity alone. It is not necessary, and it would be wrong to prescribe more about velocity or stresses.

12.3 The vorticity equation

Flows of incompressible fluids in the plane are conveniently represented by a streamfunction $\psi(x,y)$

$$(u, v) = \left[\frac{\partial\psi}{\partial y}, -\frac{\partial\psi}{\partial x}\right].$$

The vorticity $\boldsymbol{\zeta}$=curl $\mathbf{u}$ is related to the streamfunction

$$\boldsymbol{\zeta} = \text{curl}\ \mathbf{u} = \zeta\mathbf{e}_z = -\left[\frac{\partial^2\psi}{\partial x^2} + \frac{\partial^2\psi}{\partial y^2}\right]\mathbf{e}_z, \tag{5}$$

where $\zeta(x,y)$ satisfies

$$(M^2u^2 - \sigma - 1)\,\zeta_{xx} + 2(M^2uv-\tau)\,\zeta_{xy} + (M^2v^2 - \gamma - 1)\,\zeta_{yy}$$

$$+\left[-p_x + \frac{M}{\sqrt{E}}u\right]\zeta_x + \left[-p_y + \frac{M}{\sqrt{E}}v\right]\zeta_y \tag{6}$$

$$+ (u_x - v_y)(\sigma_{xy} + \gamma_{xy} + \tau_{xx} + \tau_{yy}) + (u_y + v_x)(\gamma_{yy} - \sigma_{xx})$$

$$+ \sigma_y u_{xx} + 2\tau_y u_{xy} + \gamma_y u_{yy} - \sigma_x v_{xx} - 2\tau_x v_{xy} - \gamma_x v_{yy} = 0 .$$

Equation (6) may be written as

$$A\zeta_{xx} + 2B\zeta_{xy} + C\zeta_{yy} + \text{l.o.t.} = 0 , \tag{7}$$

where the terms l.o.t. are of lower order for hyperbolic analysis (see Joseph, Renardy and Saut [1985]). Characteristic directions

$$\begin{aligned} \frac{dy}{dx} &= \frac{B}{A} \pm \frac{\sqrt{B^2 - AC}}{A} . \\ A &= M^2u^2 - \sigma - 1 , \\ B &= M^2uv - \tau , \\ C &= M^2v^2 - \gamma - 1 \end{aligned} \tag{8}$$

for the vorticity exist whenever

$$\Sigma \overset{\text{def}}{=} \tau^2 - 2M^2\tau uv - (1+\gamma)(1+\sigma) + M^2v^2(1+\sigma) + M^2u^2(1+\gamma) > 0 . \tag{9}$$

The expression (9) is expressed in terms of unknown velocity and stress fields. The criterion (9) for hyperbolicity can be satisfied in some regions of flow and not in others. The border $\Sigma=0$ between the elliptic and hyperbolic regions of flow is like the sonic line in gas dynamics. Across the line the equations are said to change type.

Equation (2) and all the equations of this section are general in that they apply to every plane problem, not just the channel flow problem introduced in §12.1.

12.4 Characteristics nets for problems perturbing plane Poiseuille flow

Now we shall solve the governing equations for flow in a channel with straight walls $\varepsilon=0$:

$$(u_0,v_0) = (1-y^2,0)\,, \quad (p_0,\tau_0) = -2M\sqrt{E}\,(x,y)\,,$$

$$(\sigma_0,\gamma_0) = (8M^2Ey^2,0)\,, \quad \omega_0 = 2y\,. \tag{10}$$

The basic motion depends exclusively on the Weissenberg number $W=M\sqrt{E}$ measuring the size of stresses. The solution (10) is relatively featureless and, in particular, it gives no indication of hyperbolicity.

Now we consider any plane perturbation of (10). The problem with wavy walls is one such perturbation, but there are infinitely many others. We may linearize the formula (8) for the characteristics of *any* flow slightly perturbing the Poiseuille flow (10). The characteristics for all these perturbations

$$\frac{dy}{dx} = \frac{1}{M}\left[\frac{-2\sqrt{E}y \pm \sqrt{(y^2-1)^2-4Ey^2-M^{-2}}}{-(y^2-1)^2+8Ey^2+M^{-2}}\right] \tag{11}$$

are defined in terms of quantities defined for the basic flow (10) and are given once and for all, independent of the perturbation.

Equation (11) shows that flows perturbing plane Poiseuille flow can exhibit a change of type with a "sonic" line $\Sigma=0$ given by

$$\Sigma(y^2) = (y^2-1)^2 - 4Ey^2 - M^{-2} = 0 \quad (-1 \le y \le 1)\,.$$

Since $\Sigma(y^2)$ is monotonically decreasing, it has a maximum at $y^2=0$ and

$$\Sigma(0) = 1 - M^{-2} > 0$$

if and only if the viscoelastic "Mach" number $M>1$. The "sonic" line across which the flow changes type is $y=y^*$ where $\Sigma(y^{*2})=0$,

$$y^* = \pm\left[1 + 2E - 2\left(E^2 + E + \frac{1}{4M^2}\right)^{1/2}\right]^{1/2}.$$

In thinking about the regions of hyperbolicity and characteristics nets it is useful to consider the strongly supercritical case $M \gg 1$. Then, when E is small, we obtain

$$y^* = \pm(1-2\sqrt{E})^{1/2} \tag{12}$$

and the elliptic region $y^* \leq y \leq 1$ is near the wall $y=\pm 1$. Most of the flow is hyperbolic. On the other hand, when E is larger, we get

$$y^* = \pm\left[1 + 2E - 2E\left(1 + \frac{1}{E}\right)^{1/2}\right]^{1/2} = \pm O\left[\frac{1}{E}\right]^{1/2} \tag{13}$$

and the hyperbolic region $0 \leq y \leq y^*$ is confined to a small strip centered on the center of the channel.

The two families of characteristics defined by (11) enter the region of hyperbolicity from the elliptic strip on the wall with the same slope

$$\left.\frac{dy}{dx}\right|_{y=y^*} = -\frac{1}{2M\sqrt{E}y^*} = -\frac{1}{2Wy^*} . \tag{14}$$

This common entering slope of the two families of characteristics is proportional to the reciprocal of the Weissenberg number. One of the two families of characteristics has an infinite slope at the zero of the denominator in (11). This family changes slope, has a turning point, at this zero. The other family (+) is monotone because the numerator in (11) vanishes at the same value of y as the denominator. The slopes of the characteristics tend to zero for each fixed E as $M \to \infty$ except near the turning point of the (–) family of characteristics.

In Figures 12.2–3 we have exhibited the characteristics nets for typical values of E and M. The graphs were obtained by numerical integration of (11). The computer graphics are not sufficiently fine to bring out the feature that the entering slope of families of characteristics have a common value at the "sonic" line.

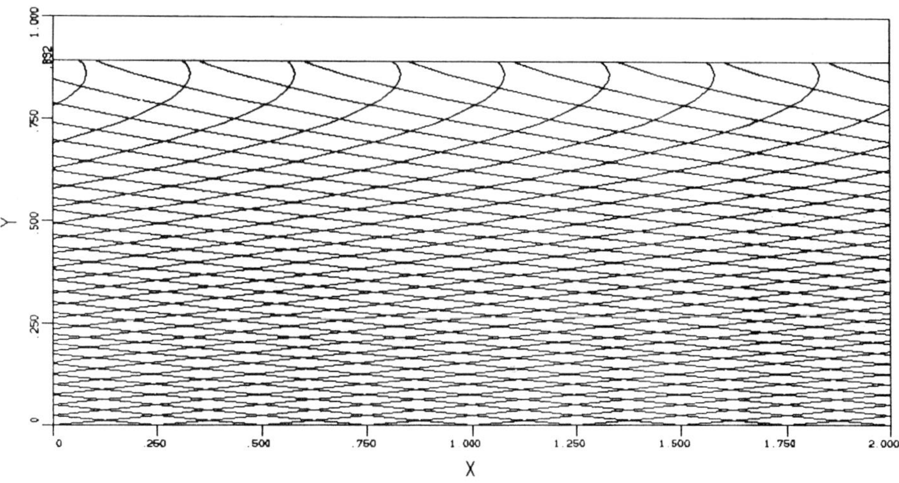

Figure 12.2. Characteristics net for E=0.01 and M=10, (R, W)=(100,1).

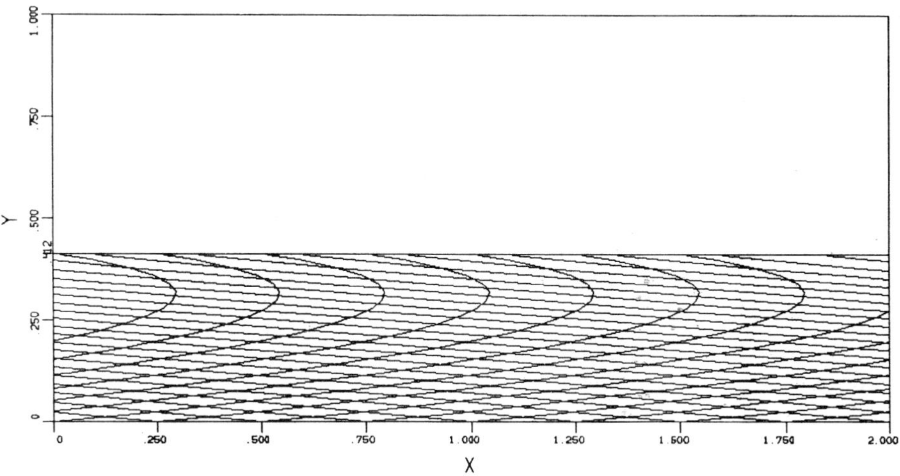

Figure 12.3. Characteristics net for E=1 and M=10, (R,W)=(10, 10).

We emphasize again that these characteristics are universal in the sense that they hold for every linearized problem perturbing plane Poiseuille flow. To go further we must choose a specific perturbation and solve a definite boundary value problem. In the next section we undertake to solve the linearized problem associated with Figure 12.1.

12.5 Perturbation equations for wavy walls

We return to the problem of flow in channels with wavy walls which was introduced in §12.1 and study it using a method of domain perturbation in the amplitude ε of the waviness of the wall (see Sturges, [1977]) for a similar application). We define a one-to-one mapping

$$x = x_0\,, \quad y = y_0\,(1 + \varepsilon \sin nx_0) \tag{15}$$

of the wavy domain

$$V_\varepsilon = \{-1-\varepsilon \sin nx \le y \le 1 + \varepsilon \sin nx\,, \quad -\infty<x<\infty\}$$

into the flat domain

$$V_0 = \{-1\le y\le 1\,, \quad -\infty<x<\infty\}$$

and we seek a solution of (2,3,4) by expanding $[u,v,p,\sigma,\gamma,\tau]$ (x,y,ε) and $y(\varepsilon)$ in powers of ε. The solution at zeroth order is the Poiseuille flow (10). For example

$$u(x,y,\varepsilon) = u_0(x_0,y_0) + \varepsilon u^{[1]}(x_0,y_0) + 0(\varepsilon^2)\,.$$

The function $u^{[1]}(x_0,y_0)$ is a derivative with respect to ε at $\varepsilon=0$ holding y_0 fixed. After applying the chain rule, we get

$$u^{[1]}(x_0,y_0) = u^{<1>}(x_0,y_0) + \left.\frac{\partial y}{\partial \varepsilon}\right|_{(\varepsilon=0)} \frac{\partial u_0}{\partial y_0}\,, \tag{16}$$

where

$$\frac{\partial y}{\partial \varepsilon} = y_0 \sin nx_0\,.$$

The field $u^{<1>}(x_0,y_0)$ is a partial derivative holding y fixed (and equal to y_0).

We are going to compute the derivatives $[u^{<1>}, v^{<1>}, p^{<1>}, \sigma^{<1>}, \gamma^{<1>}, \tau^{<1>}]$ from a linearized boundary value problem. After finding these fields we may express the solution in the deformed domain V_ε by inverting the mapping.

For simplicity in notation we now drop the superscript <1> and find that

$$u_x + v_y = 0 \,, \tag{17a}$$

$$(1-y^2)u_x - 2yv + \frac{1}{M^2}(p_x - \sigma_x - \tau_y) = 0 \,, \tag{17b}$$

$$(1-y^2)v_x + \frac{1}{M^2}(p_y - \tau_x - \gamma_y) = 0 \,, \tag{17c}$$

$$\begin{aligned}(1-y^2)\sigma_x &+ 16M^2Eyv - 2(8M^2Ey^2+1)u_x \\ &+ 4y\tau + 4M\sqrt{E}yu_y + \frac{\sigma}{M\sqrt{E}} = 0 \,,\end{aligned} \tag{17d}$$

$$(1-y^2)\gamma_x + 4M\sqrt{E}\,yv_x - 2v_y + \frac{\gamma}{M\sqrt{E}} = 0 \,, \tag{17e}$$

$$\begin{aligned}(1-y^2)\tau_x &- 2M\sqrt{E}v + 2y\gamma - (8M^2Ey^2+1)v_x \\ &- u_y + \frac{\tau}{M\sqrt{E}} = 0,\end{aligned} \tag{17f}$$

where

$$\begin{gathered} u(x,\pm 1) = 2\sin nx \,, \\ v(x,\pm 1) = 0 \,, \\ v(x,0) = u_y(x,0) = 0 \,. \end{gathered} \tag{18}$$

We can solve the equations (17) using only the velocity data (18).

The velocity can then be obtained from a stream function $\psi(x,y)$

$$(u,v) = (\psi_y, -\psi_x) \,.$$

The vorticity

$$\zeta \overset{\text{def}}{=} v_x - u_y = -\nabla^2\psi \tag{19}$$

satisfies an equation

$$\left[(1-y^2)^2 - 8Ey^2 - \frac{1}{M^2}\right]\zeta_{xx} + 4\frac{\sqrt{E}}{M}y\zeta_{xy}$$

$$-\frac{\zeta_{yy}}{M^2} + \left[\frac{2\sqrt{E}}{M} + \frac{1}{M\sqrt{E}}(1-y^2)\right]\zeta_x$$

$$= \frac{-2v}{M\sqrt{E}} - 16Eyu_{xx} + \frac{4\sqrt{E}}{M} u_{xy} - 2(1-y^2)v_x \tag{20}$$

$$+ \frac{1}{M^2}\left[2\tau_x + 2y\gamma_{yy} - 2y\sigma_{xx} + 4\gamma_y\right]$$

of changing type whose characteristics are given by (8).

12.6 Separation of variables

The column vector $\mathbf{q}$ with components $[u,v,p,\sigma,\gamma,\tau]$ satisfies a linear system

$$\mathbf{A}(y)\mathbf{q}_x + \mathbf{B}(y)\mathbf{q}_y = \mathbf{C}(y)\mathbf{q}\ . \tag{21}$$

The boundary conditions (18), linearity, and the fact that the coefficients of the linear equation are independent of x, imply that

$$\mathbf{q}(x,y) = \mathbf{q}_1(y) \cos nx + \mathbf{q}_2(y) \sin nx\ . \tag{22}$$

The existence of eigensolutions of (21) can be ruled out by a stability argument when M and W are both small. In this argument we establish that the rest state is stable. All the variables of the linearized problem have a decomposition like (22).

We insert the decomposition (22) into the equation (21), where $\mathbf{A}(y)$, $\mathbf{B}(y)$, $\mathbf{C}(y)$ are defined by (17), and we put the coefficients of cos nx and sin nx to zero. As a result, we obtain a system of 12 ordinary differential equations of the form

$$\begin{bmatrix} \mathbf{B} & \mathbf{0} \\ \mathbf{0} & \mathbf{B} \end{bmatrix}\begin{bmatrix} \mathbf{q}_1'(y) \\ \mathbf{q}_2'(y) \end{bmatrix} - \begin{bmatrix} \mathbf{C} & -n\mathbf{A} \\ n\mathbf{A} & \mathbf{C} \end{bmatrix}\begin{bmatrix} \mathbf{q}_1(y) \\ \mathbf{q}_2(y) \end{bmatrix} = \mathbf{0}\ , \tag{23}$$

which can be solved sequentially in terms of $v_1(y)$, $v_2(y)$ and their derivatives. By "sequentially" we mean that (u_1,u_2), (γ_1,γ_2), (τ_1,τ_2), (σ_1,σ_2) and (p_1,p_2) are found in that order as functions of v_1, v_2, and their derivatives. We may then obtain two equations for v_1 and v_2 (or ψ_1 and ψ_2) by differentiating p_1 and p_2, and equating these two derivatives to their expressions (p_1',p_2') as given by the equations (17c). In this way we come up with the two fourth order differential equations for ψ_1 and ψ_2 given below:

$$\psi_1^{iv} = A_{11}\psi_1''' + A_{12}\psi_2''' + A_{21}\psi_1'' + A_{22}\psi_2'' + A_{31}\psi_1' + A_{32}\psi_2' + A_{41}\psi_1 + A_{42}\psi_2 ,$$

$$\psi_2^{iv} = -A_{12}\psi_1''' + A_{11}\psi_2''' - A_{22}\psi_1'' + A_{21}\psi_2'' - A_{32}\psi_1' + A_{31}\psi_2' - A_{42}\psi_1 + A_{41}\psi_2 , \tag{24}$$

where, denoting

$$X = \frac{M^2E}{1 + M^2En^2(1-y^2)^2} ,$$

$$A_{11} = X\,[-4n^2y\,(1-y^2)] ,$$

$$A_{12} = X\,[4M\sqrt{E}\,n^3y\,(1-y^2)^2] ,$$

$$A_{21} = n^2\,[-M^2\,(1-y^2)^2 + 2 + 6M^2Ey^2] + Xn^2\,[-6 + 2M^2En^2y^2\,(1-y^2)^2] + X^2\,[-16n^4y^2\,(1-y^2)^2] ,$$

$$A_{22} = \frac{M}{\sqrt{E}}n\,(1-y^2) + 6X\,[M\sqrt{E}\,n^3(1-y^2)^2] + X^2\,[16M\sqrt{E}\,n^5y^2\,(1-y^2)^3] ,$$

$$A_{31} = Xn^4\,[4y\,(1-y^2) + 16M^2Ey\,(1-y^4)] + X^2\,[-32n^4y\,(1-y^4)] ,$$

$$A_{32} = X\,[-4M\sqrt{E}\,n^5y\,(1-y^2)^2] + X^2\,[32M\,\sqrt{E}\,n^5y\,(1-y^2)\,(1-y^4)] ,$$

$$A_{41} = n^2\,[-2M^2\,(1-y^2) + n^2M^2\,(1-y^2)^2 - n^2] + Xn^2\,[-12 + 2n^2\,(1-y^2) + 48M^2En^2y^2 - 16M^2En^2y^4 - 8M^2En^4y^2\,(1-y^2)^2] + X^2n^4\left[-96y^2 - \frac{16}{M^2E}\,y^2 + 32M^2En^2y^4\,(1-y^2)^2\right] ,$$

$$A_{42} = 2\frac{M}{\sqrt{E}}n - \frac{M}{\sqrt{E}}n^3(1-y^2)$$
$$+ Xn^3\left[-2M\sqrt{E}n^2(1-y^2)^2 - 44M\sqrt{E}y^2 + 12M\sqrt{E}\right]$$
$$+ X^2n^3\left[128M\sqrt{E}n^2y^2(1-y^2) + \frac{16}{M\sqrt{E}}n^2y^2(1-y^2)\right.$$
$$\left. + \frac{32}{M\sqrt{E}}y^2\right].$$

The boundary conditions are

$$\psi_1(0) = \psi_1''(0) = \psi_2(0) = \psi_2''(0) = \psi_1'(1) = \psi_1(1)$$
$$= \psi_2(1) = 0\ ,\ \psi_2'(1) = 2\ . \tag{25}$$

The perturbed vorticity $\zeta(x,y) = -\nabla^2\psi$ can be obtained from $\psi_1(y)$ and $\psi_2(y)$

$$\zeta(x,y) = [-\psi_1''(y) + n^2\psi_1(y)]\cos nx$$
$$+ [-\psi_2''(y) + n^2\psi_2(y)]\sin nx. \tag{26}$$

Equations (24) and (25) can be solved by a standard shooting technique employing a fourth order Runge-Kutta method.

The numerical results will be given in §12.7. Nearly all of these are for $M \geq 5$. If we restrict our considerations to the case of large M with $W = M\sqrt{E} \leq O(1)$, we find that (24) may be written

$$(1-y^2)^2n^2\psi_1^{(ii)} + \psi_1^{(iv)} - \frac{n}{M\sqrt{E}}\psi_2^{(ii)}$$
$$= \frac{1}{M}\frac{4M^2En^2(1-y^2)y}{1+M^2En^2(1-y^2)^2}\left(nM\sqrt{E}(1-y^2)\psi_2^{(iii)} - \psi_1^{(iii)}\right) + O\left[\frac{1}{M^2}\right] \tag{27}$$

along with a second equation, say (28), which is the same as (27) when the subscripts 1 and 2 are interchanged and $-n$ replaces n. The derivatives in (27) are

$$\psi^{(i)} \overset{\text{def}}{=} \frac{1}{M}\psi'\ .$$

For large M, the vorticity is given by $\zeta = -\nabla^2\psi^{(ii)}$. If E is fixed and M is sufficiently large the vorticity in the interior of hyperbolic regions oscillates as y is increased from zero with a frequency given roughly by

$$\Omega = (1-y^2)Mn = (1-y^2)M2\pi N ,$$

corresponding to half periods of oscillation given by

$$\frac{\pi}{\Omega} = \frac{1}{2(1-y^2)MN} . \tag{28}$$

This formula for successive zeros of ζ fits the numerical results well. When $E \to 0$ for fixed M we get $\psi_1'', \psi_2'' \to 0$ to leading order and the vorticity in the hyperbolic region tends to zero.

12.7 Results

We shall display our numerical results in graphical form. We have three parameters M, E, and the wave number n of the wavy wall. The characteristics for this problem are universal for all problems perturbing Poiseuille flow. Graphs of the characteristics as functions of E and M were given in §12.4.

A few preliminary results may help the reader to interpret the graphics shown in this section. It will be recalled that the boundary data which forces our problem is periodic in x with period $2\pi/n$. In fact the whole solution is periodic in x with period $2\pi/n$. Lines of zero vorticity separate regions of positive vorticity from regions of negative vorticity. The stream function is periodic in x in the same way. All the graphs cover one full period of the solution.

Our first observations are about the dependence of the solution on the wave number $n=2\pi N$. Numerical computation with ever larger wave numbers (short waves) becomes increasingly difficult, first with slow convergence, then failure. For moderate and long waves ($N<1$) we get good convergence. The solution does not seem to be very sensitive to changes in the wave number. Zero vorticity lines and characteristics are plotted on the same picture in each of the three cases, N=0.25, N=0.50, N=1.00, represented in Figures 12.4,5,6. The form of the zero vorticity line does not appear to depend on n for $0.25 \le N \le 1$.

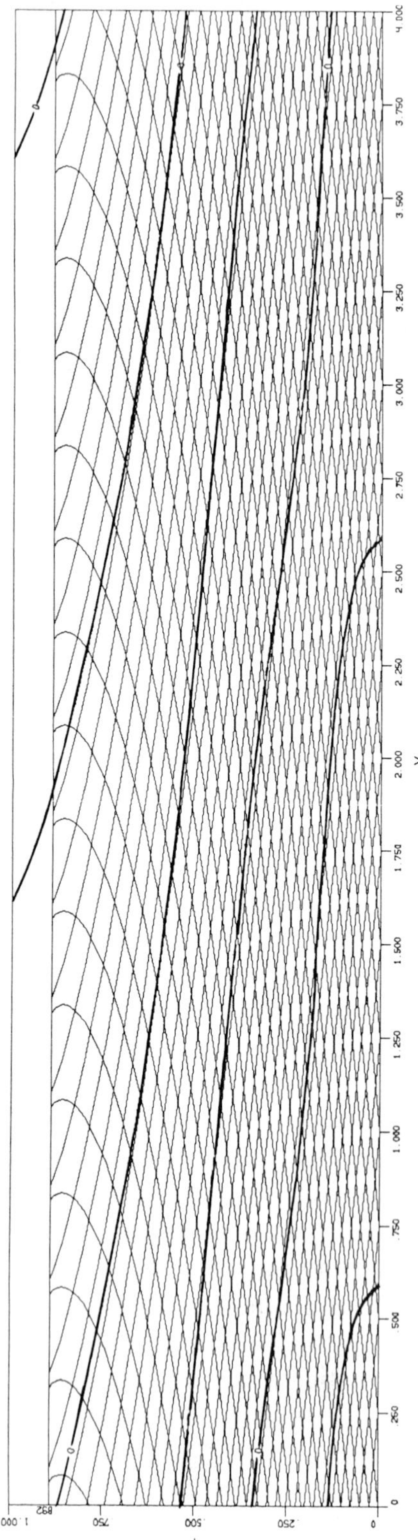

Figure 12.4. Zero-vorticity curves for E=0.01, M=10, and N=0.25. (R, W)=(100, 1).

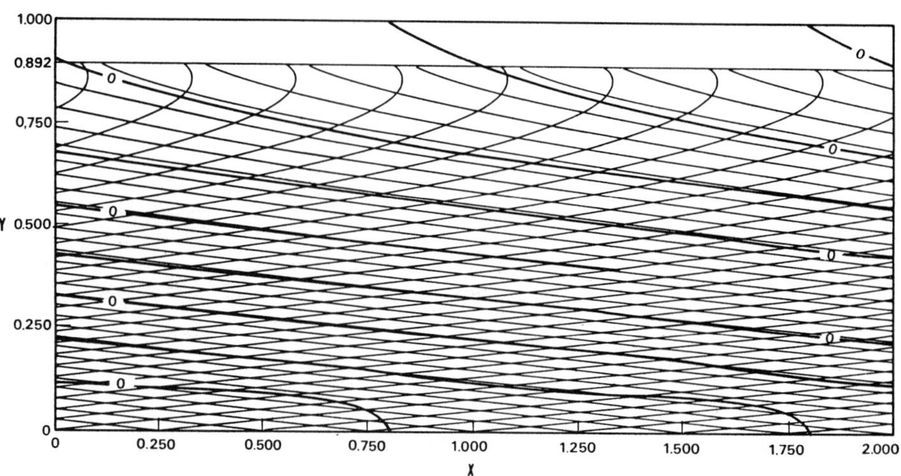

Figure 12.5. Zero-vorticity curves for E=0.01, M=10, and N=0.50. (R,W)=(100,1).

The most striking feature of Figures 12.4–6 is the way regions of positive and negative perturbed vorticity are swept out along characteristics. This sweeping property occurs in all the other supercritical cases computed by us. We have no good theoretical explanation of this property. The sweeping property, evident in the supercritical region is also evident in the subcritical region near the wall. It is apparent that in problems of change of type the supercritical flow will react with the subcritical flow so that the elliptic and hyperbolic regions of flow are compatible. We should get the same solution from a Cauchy problem set on the "sonic" line with the correct "initial conditions" taken from the compatible "true" solution. If regions of positive and negative data were then replaced with constant data having the same mean value as true data we should have formed a Cauchy problem with approximating but discontinuous data, and the discontinuities would propagate along characteristics carrying regions of positive and negative perturbed vorticity between characteristics. In fact these regions of positive and negative perturbed vorticity would be damped, as in the true solution, with fast damping for low Weissenberg number fluids (small λ, short memory).

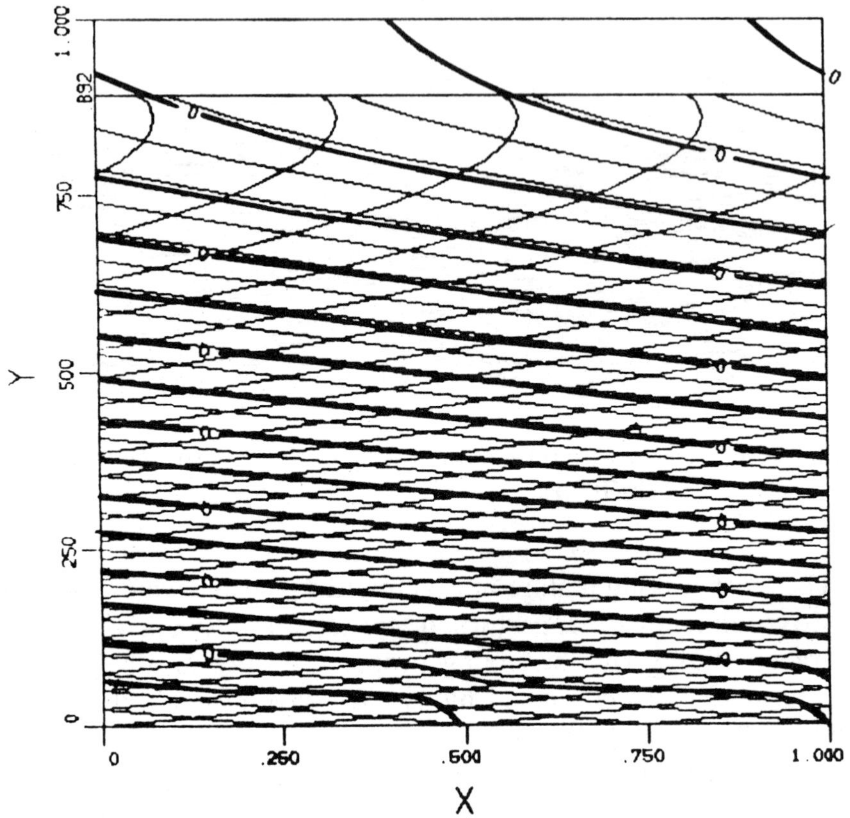

Figure 12.6. Zero-vorticity curves for E=0.01, M=10, and N=1. (R, W)=(100, 1).

The graphs shown in Figures 12.7–9 are computed for N=0.5. In each Figure there is (a) a plot of the iso-vorticity lines and (b) a plot of the streamlines. We are interested in how the solution varies with E and M. The reader should recall that the solution is periodic in x with period $2\pi/n$ and one period is on each graph.

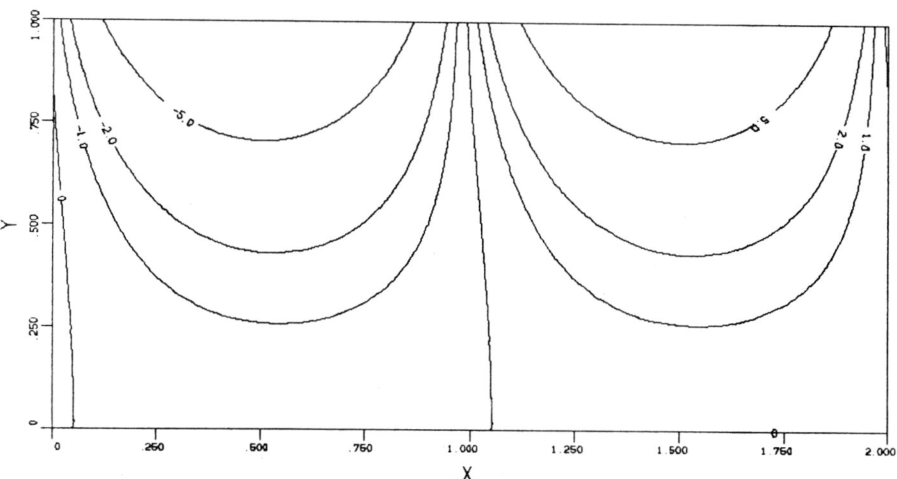

Figure 12.7(a). Iso-vorticity curves for E=0.01, M=0.1, (R, W)=(1, 0.01).

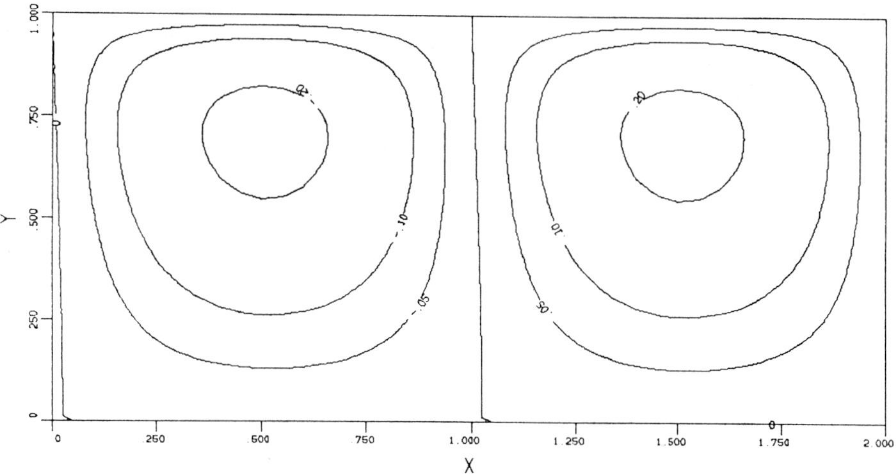

Figure 12.7(b). Streamlines for E=0.01, M=0.1, (R, W)=(1, 0.01).

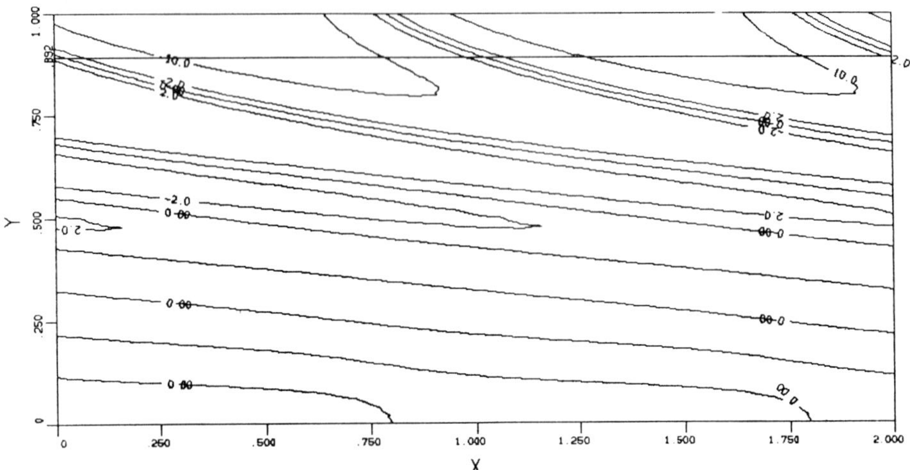

Figure 12.8(a). Iso-vorticity curves for E=0.01, M=10, (R, W)=(100, 1).

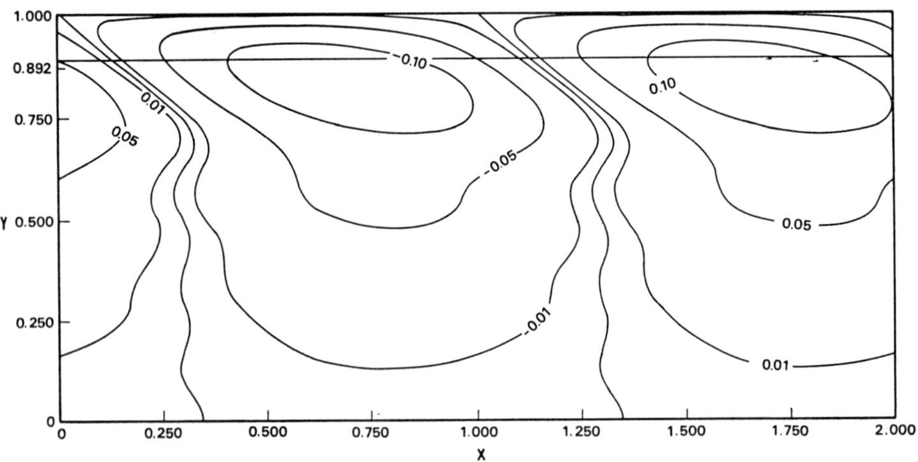

Figure 12.8(b). Streamlines for E=0.01, M=10, (R,W)=(100,1).

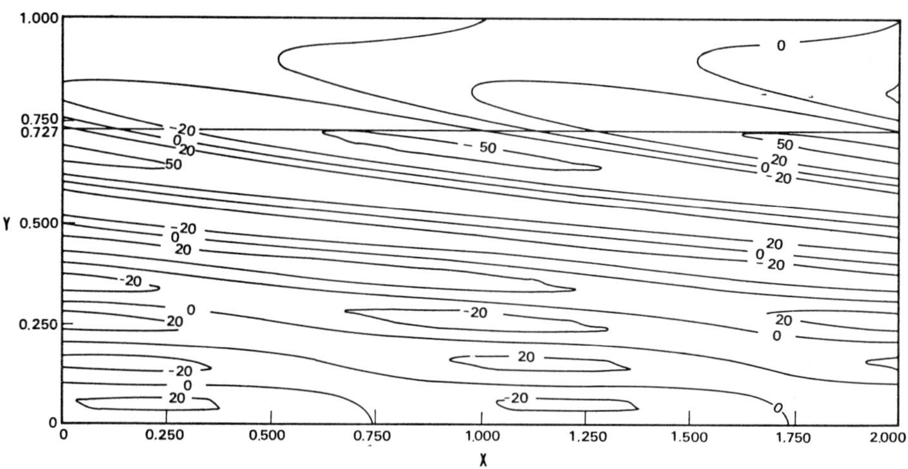

Figure 12.9(a). Iso-vorticity curves for E=0.1, M=10, (R, W)=(31.6, 3.16).

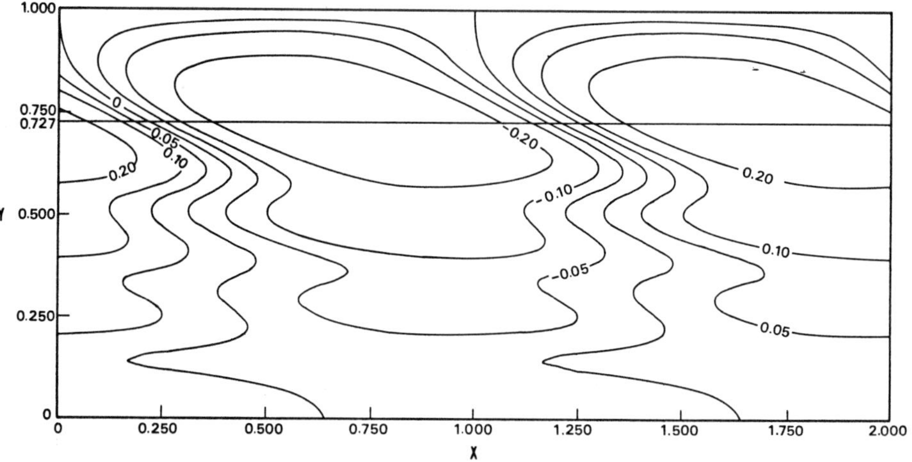

Figure 12.9(b). Streamlines for E=0.1, M=10.00, (R, W)=(31.6, 3.16).

Figure 12.7 is representative of subcritical flow (M<1). There is no strong shift in the phase of the perturbing data at the boundary. The lines of zero vorticity and the zero streamlines do not deviate strongly from the vertical. We draw attention to the difference between the iso-vorticity curves shown in Figure 12.7 and the supercritical cases which are exhibited in all the other Figures.

The iso-vorticity curves in the supercritical cases shown in Figures 12.8(a)–12.9(a) are dominated by the hyperbolic structure of the solution. The positive and negative vorticity are swept out along characteristics (shown in Figures 12.4–6) in a manner already described. The damping of the vorticity decreases as W (or M) is increased. For larger values of M an oscillatory variation of the vorticity persists all the way to the channel center. The frequency of the oscillation increases with M (or W) at a fixed E. This property of damping is most clearly brought by comparing Figures 12.9(a) and 12.8(a). In these Figures M=10 is fixed and E=0.1,0.01, respectively. The damping is more rapid for the small values of E and the channel core is essentially free of perturbation vorticity.

Crochet and Delvaux [1990] have carried out a numerical analysis of the channel problem under discussion using their numerical codes and an upper convected Maxwell model. Their results are gratifyingly close to those given here.

12.8 Pipe flow problem

We now consider an extension of the analysis for channels to the same problem for round pipes. We take note of the following differences between the problems.

We use a different scale to non-dimensionalize the stresses, which leads us to an extra factor W, the Weissenberg number, in front of every stress variable. There are of course extra lower order terms when the equations are written in polar coordinates. These extra terms give rise to terms with singular factors proportional to r^{-1} and r^{-2}. These singular factors give rise to difficulties in the numerical solution which we have avoided by computing directly from the first order quasilinear system (1) rather than the streamfunction (24).

It is necessary to call attention to the fact that the analysis of characteristics in the linear approximation does not depend on the type of perturbation of Poiseuille flow, provided it is small. Though we do analyze wavy walls, the characteristics for the vorticity would be the same for any other problem. We are obliged then to consider the relevance of the analysis of the type of a partial differential equation for the flow near the exit of a pipe from which a viscoelastic fluid is extruded. The experiments of Joseph, Matta, and Chen [1987] do suggest a strong dependence on a change of type correlated with the viscoelastic Mach number.

12.9 Governing equations

We consider the steady flow of an incompressible upper convected Maxwell fluid through a pipe whose radius varies sinusoidally along the axis. The equations governing this flow are (1).

The boundary value problem for our problem is the system (1) together with no-slip conditions at the wall of the pipe

$$u = w = 0 \quad \text{at} \quad r = b(1 + \varepsilon \sin nz)\,, \tag{29}$$

where ε is the amplitude perturbation of the sinusoidal variation of period $2\pi/n$ of the wall. We assume flow symmetry about the centerline

$$u = \frac{\partial w}{\partial r} = 0 \quad \text{at} \quad r = 0\,, \tag{30}$$

and that all flow variables are periodic in the axial direction z.

We solve this problem by considering a linear perturbation of Poiseuille flow. We first compute the solution for Poiseuille flow when $\varepsilon=0$. We then derive and solve a perturbed problem for $0<\varepsilon\ll 1$, using the technique of domain perturbation. We consider only the first order perturbation here. The perturbed problem will be linear and of mixed type; that is, if $M>1$, then in a region near the centerline of the pipe the vorticity equation will be hyperbolic, and elliptic elsewhere.

12.10 Poiseuille flow

By a Poiseuille flow we mean a solution of the form

$$u = 0\,,\; w = w(r) \tag{31a}$$

in a pipe with straight walls, $\varepsilon=0$. We also assume a constant axial pressure gradient

$$\frac{\partial p}{\partial z} = -k^2 , \tag{31b}$$

and that the stress components $\sigma_1,\sigma_2,\sigma_3$, and τ are functions of r only. A solution of system (1) and boundary conditions (29) and (30) for $\varepsilon=0$ is

$$\begin{aligned} u &= 0 , \\ w &= \frac{k^2b^2}{4\eta}\left(1 - \frac{r^2}{b^2}\right), \\ p &= -k^2z + \text{constant} , \\ \sigma_1 &= \sigma_2 = 0 , \\ \sigma_3 &= \frac{\lambda k^4}{2\eta} r^2 , \\ \tau &= -k^2r/2 . \end{aligned} \tag{32}$$

Using b, $w_0=w(0)=k^2b^2/4\eta$, and $bk^2/4$ as characteristic values for lengths, velocities, and stresses, respectively, we can write (32) in the non-dimensional form

$$\begin{aligned} u &= \sigma_1 = \sigma_2 = 0 , \\ w &= 1 - r^2 , \\ p &= -4z + \text{constant} , \\ \sigma_3 &= 8Wr^2 , \\ \tau &= -2r . \end{aligned} \tag{33}$$

We are using the same symbols for the physical variables in (32) and the non-dimensional variables in (33), and we will rely on the context to make our intention clear.

12.11 Characteristics for the vorticity

Before continuing with the perturbation problem, we remark on the change of type of system (1). Using the scales introduces after (32) we may write the dimensionless form of the system (1) for axisymmetric problems as a quasilinear system:

$$\mathbf{A}(\mathbf{q})\frac{\partial\mathbf{q}}{\partial r} + \mathbf{B}(\mathbf{q})\frac{\partial\mathbf{q}}{\partial z} = \mathbf{f}(\mathbf{q}) \ , \tag{34a}$$

where

$$\mathbf{q} = (u,w,p,\sigma_1,\sigma_2,\sigma_3,\tau) \ , \tag{34b}$$

$$\mathbf{A} = \begin{vmatrix} 1 & 0 & 0 & 0 & 0 & 0 & 0 \\ Ru & 0 & 1 & -1 & 0 & 0 & 0 \\ 0 & Ru & 0 & 0 & 0 & 0 & -1 \\ -2(W\sigma_1+1) & 0 & 0 & Wu & 0 & 0 & 0 \\ 0 & 0 & 0 & 0 & Wu & 0 & 0 \\ 0 & -2W\tau & 0 & 0 & 0 & Wu & 0 \\ 0 & -(W\sigma_1+1) & 0 & 0 & 0 & 0 & Wu \end{vmatrix} \tag{34c}$$

$$\mathbf{B} = \begin{vmatrix} 0 & 1 & 0 & 0 & 0 & 0 & 0 \\ Rw & 0 & 0 & 0 & 0 & 0 & -1 \\ 0 & Rw & 1 & 0 & 0 & -1 & 0 \\ -2W\tau & 0 & 0 & Ww & 0 & 0 & 0 \\ 0 & 0 & 0 & 0 & Ww & 0 & 0 \\ 0 & -2(W\sigma_3+1) & 0 & 0 & 0 & Ww & 0 \\ -(W\sigma_3+1) & 0 & 0 & 0 & 0 & 0 & Ww \end{vmatrix} \tag{34d}$$

and

$$\mathbf{f} = \begin{vmatrix} -u/r \\ (\sigma_1 - \sigma_2)/r \\ \tau/r \\ -\sigma_1 \\ 2(W\sigma_2 + 1)u/r - \sigma_2 \\ -\sigma_3 \\ -(Wu/r + 1)\tau \end{vmatrix} \tag{34e}$$

The characteristic curves of such a quasilinear first order system are given by

$$\det\left[\mathbf{A} - \mathbf{B}\,\frac{dr}{dz}\right] = 0\,. \tag{35}$$

That is,

$$\left[u - w\,\frac{dr}{dz}\right]^3\left[\left(\frac{dr}{dz}\right)^2 + 1\right]\left[a_2\left(\frac{dr}{dz}\right)^2 - 2a_1\,\frac{dr}{dz} + a_0\right] = 0\,, \tag{36a}$$

where

$$\begin{aligned} a_0 &= M^2w^2 - W\sigma_3 - 1\,,\\ a_1 &= M^2uw - W\tau\,,\\ a_2 &= M^2u^2 - W\sigma_1 - 1\,. \end{aligned} \tag{36b}$$

Hence there are always real characteristics

$$\frac{dr}{dz} = \frac{u}{w}\,,$$

which are streamlines, and imaginary characteristics

$$\frac{dr}{dz} = \pm i\,.$$

The streamline characteristics may be associated with the convective derivative, and the imaginary characteristics may be associated with the incompressibility condition.

The remaining pair of characteristic curves are determined by

$$\frac{dr}{dz} = \frac{a_1}{a_2} \pm \frac{\left(a_1^2 - a_2a_0\right)^{1/2}}{a_2}\,, \tag{37}$$

and these are real if and only if

$$a_1^2 - a_2a_0 = (M^2uw - W\tau)^2 - (M^2w^2 - W\sigma_3 - 1)(M^2u^2 - W\sigma_1 - 1) > 0\,. \tag{38}$$

There is therefore the possibility of a change of type from imaginary to real characteristics in crossing from one region of the flow to another, depending on the local relative magnitudes of the velocities and stresses in (38).

The characteristics given by (37) may be associated with the vorticity equation. A routine manipulation to derive the vorticity equation from system (34) gives the form

$$a_2 \frac{\partial^2 \zeta}{\partial z^2} + 2a_1 \frac{\partial^2 \zeta}{\partial r \partial z} + a_0 \frac{\partial^2 \zeta}{\partial r^2} = g\,, \tag{39}$$

where a_2, a_1, and a_0 are given in (36b), ζ is the non-zero component of vorticity ($\boldsymbol{\zeta}$=curl $\mathbf{u}=\zeta\mathbf{e}_\theta$), and g is a function of the flow variables and their first and second derivatives, which includes ζ and its first derivatives. Thus the vorticity equation is locally hyperbolic if condition (38) holds, or elliptic if the inequality in (38) is reversed. In regions of the flow where (39) is hyperbolic, there is the possibility that the derivatives of the vorticity may be discontinuous across the characteristics given by (37).

The results of this section, so far, apply to any axisymmetric problem, linear or non-linear, and not just to the problem introduced in §12.9. Since the problem of §12.9 does not have discontinuous boundary data, we do not expect to discover discontinuities of the vorticity derivatives in the hyperbolic region of the flow.

For definiteness, let us now consider a perturbation of the Poiseuille flow solution given in §12.10. That is, we may write $\mathbf{q}=\mathbf{q}_0+\mathbf{q}_1$, where $\mathbf{q}_0$ is the Poiseuille flow variables and $\mathbf{q}_1$ is any perturbation. The linearization of system (34) would take the form

$$\mathbf{A}(\mathbf{q}_0) \frac{\partial \mathbf{q}_1}{\partial r} + \mathbf{B}(\mathbf{q}_0) \frac{\partial \mathbf{q}_1}{\partial z} = \mathbf{f}_1(\mathbf{q}_1;\mathbf{q}_0)\,. \tag{40}$$

That is, $\mathbf{A},\mathbf{B}$ are given by (34c,d) with all variables given by the Poiseuille flow (33). Here $\mathbf{f}(\mathbf{q}_1;\mathbf{q}_0)$ represents the linearization of $\mathbf{f}(\mathbf{q})$. The characteristics given by (37) represent characteristics for the $\mathbf{q}_1$, and may be found by substituting the solution (33) into (37):

$$\frac{dr}{dz} = 2Wr \pm [M^2(1-r^2)^2 - 4W^2r^2 - 1]^{1/2}\,, \tag{41}$$

with the condition (38) for real characteristics becoming

$$M^2(r^2-1)^2 - 4W^2r^2 - 1 > 0\,. \tag{42}$$

This condition always fails near r=1; but is satisfied near r=0, provided $M^2-1>0$. Hence there is a region near the centerline of the pipe for which the perturbed vorticity equation is hyperbolic, if M>1; and a region of ellipticity near the wall. The radius at which the change of type occurs is found from (42) to be

$$r^* = (1+2E)^{1/2}\,[1-[1-(M^2-1)\,M^{-2}(1+2E)^{-2}]^{1/2}]^{1/2}\,. \tag{43}$$

In the next section we return to the pipe flow with wavy walls discussed in §12.9. The linearized problem derived will have characteristics given by (41), with a region $0<r<r^*$ in which the perturbed vorticity satisfies a hyperbolic equation (for $M>1$), and a region $r^*<r<1$ in which it satisfies an elliptic equation.

12.12 Linearized problem for pipe flow with wavy walls

The problem for the steady flow of an upper convected Maxwell fluid through a periodically constricted pipe is governed by system (1) (non-dimensionalized as in (34) with boundary values (29) and (30) in the domain

$$V_\varepsilon = \{(r,z)\colon z \in (-\infty,\infty),\ r \in [0,1+\varepsilon \sin nz]\}\,. \tag{44}$$

There is a one-to-one map of V_ε from the straight wall pipe

$$V_0 = \{(r_0,z_0)\colon z_0 \in (-\infty,\infty),\ r_0 \in [0,1]\} \tag{45}$$

given by

$$z = z_0\,,$$

$$r = r_0(1+\varepsilon \sin nz_0)\,. \tag{46}$$

To solve the problem we use the technique of domain perturbation and solve the first order perturbation problem. Thus, the flow variables (34b) may be written as

$$\mathbf{q}(r,z,\varepsilon) = \mathbf{q}^{[0]}(r_0,z_0) + \varepsilon\mathbf{q}^{[1]}(r_0,z_0) + O(\varepsilon^2)\,, \tag{47}$$

where

$$\mathbf{q}^{[1]}(r_0,z_0) = \frac{d}{d\varepsilon}\,\mathbf{q}(r,z,\varepsilon)\Big|_{\varepsilon=0} = \mathbf{q}^{<1>}(r_0,z_0) + \frac{\partial r}{\partial \varepsilon}\Big|_{\varepsilon=0}\frac{\partial \mathbf{q}^{[0]}}{\partial r_0}\,, \tag{48}$$

and where

$$\mathbf{q}^{<1>}(r_0,z_0) = \frac{\partial}{\partial \varepsilon}\,\mathbf{q}(r,z,\varepsilon)\Big|_{\varepsilon=0}\,, \tag{49a}$$

and

$$\left.\frac{\partial r}{\partial \varepsilon}\right|_{\varepsilon=0} = r_0 \sin nz_0 \,. \tag{49b}$$

Of course, the zeroth order variables are precisely the Poiseuille variables given by (33).

The problem for the first order perturbation is found by simply taking the derivative with respect to ε and evaluating at ε=0, of system (34) and of the boundary conditions (29) and (30). For simplicity of notation, we will drop the subscripts 0 on (r_0,z_0), and we will not use the superscript <1> for the flow variables. With these notations, the problem for the first order flow variables $q^{<1>}(r_0,z_0)$ is the system of PDE's

$$\begin{aligned}
u_r + w_z &= -\frac{u}{r} \,, \\
R(1-r^2)u_z + p_r - (\sigma_{1r}+\tau_z) &= \frac{(\sigma_1-\sigma_2)}{r} \,, \\
R[(1-r^2)w_z - 2ru] + p_z - (\tau_r+\sigma_{3z}) &= \frac{\tau}{r} \,, \\
W[(1-r^2)\sigma_1 + 4ru_z] - 2u_r &= -\sigma_1 \,, \\
W(1-r^2)\sigma_{2z} &= -\frac{2u}{r} - \sigma_2 \,, \\
W[(1-r^2)\sigma_{3z} + 4rw_r] - 2(8W^2r^2+1)w_z &= -\sigma_3 - 4Wr(\tau+4Wu) \,, \\
W(1-r^2)\tau_z - (8W^2r^2+1)u_z - w_r &= -\tau + 2W(2u-r\sigma_1) \,,
\end{aligned} \tag{50a}$$

together with the boundary conditions

$$u = 0 \,, \quad w = 2 \sin nz \text{ at } r = 1 \,, \tag{50b}$$

and

$$u = w_r = 0 \text{ at } r = 0 \,. \tag{50c}$$

In (50), subscripts of r and z denote partial derivatives. In addition, the perturbation of the vorticity equation may be written as

$$\begin{aligned}
[M^2(1-r^2)^2 - 8W^2r^2 - 1]\zeta_{zz} + 4Wr\zeta_{rz} - \zeta_{rr} &= -[R(1-r^2) + 6W]\zeta_z \\
&+2W[w_{rz} + u_{zz} + 3\sigma_{1r} - 8Wrw_{zz} - r(\sigma_{1rr} - \sigma_{3zz})]
\end{aligned} \tag{51}$$

$$-\frac{w_{rr}+w_{zz}}{r}-\frac{2u_z-w_r+4Wu}{r^2} .$$

In §12.11 we noted that the system (50) has characteristics given by (41) and the vorticity equation (51) has characteristics given by (41). There is a change of type from hyperbolic to elliptic at r^* given by (43) provided that M>1.

The linear system (50) was solved numerically, after first reducing it to a system of ODE's. The system of ODE's arose by a separation of variables of the form

$$\mathbf{q}(r,z) = \mathbf{q}_1(r) \cos nz + \mathbf{q}_2(r) \sin nz , \tag{52}$$

where $\mathbf{q}=(u,w,p,\sigma_1,\sigma_2,\sigma_3,\tau)$ is the vector of perturbation variables. The form of (52) makes explicit the assumed periodicity of all flow variables; which combined with the boundary conditions (50b,c), is sufficient to complete the determination of the first order perturbation.

12.13 Results

The characteristics of the perturbed vorticity, as given by (41) in the region $0<r<r^*$, are precisely the same as for channel flow, with (x,y) replaced with (z,r). Thus we refer the reader to the graphics of characteristics in §12.4. Here we display a graph of the characteristics in Figure 12.10 for parameter values also used in Figures 12.11 and 12.15. The value of r^* is indicated in the Figure, and both sets of characteristics of (41) are drawn.

In general, the results of the perturbed pipe problem are similar to the results of the channel problem. The isovorticity curves presented there are like the isovorticity curves of pipe flow with the same values of M,E, and n. In addition, we have plotted contours of constant perturbed stress for both the channel and the pipe problems, again with similar graphs for corresponding parameter values. We found that the stresses are slightly larger in the pipe flow case, about 10-20% larger.

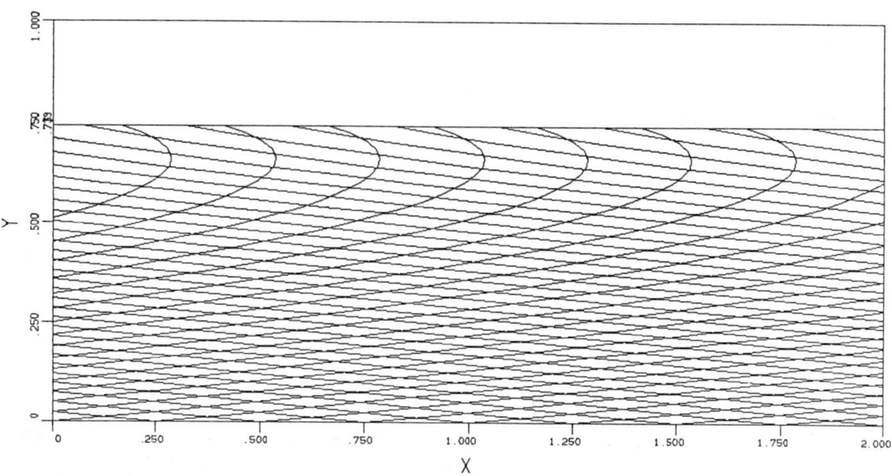

Figure 12.10. Characteristics: E=0.09, M=10, R=33.33, W=3, N=0.5

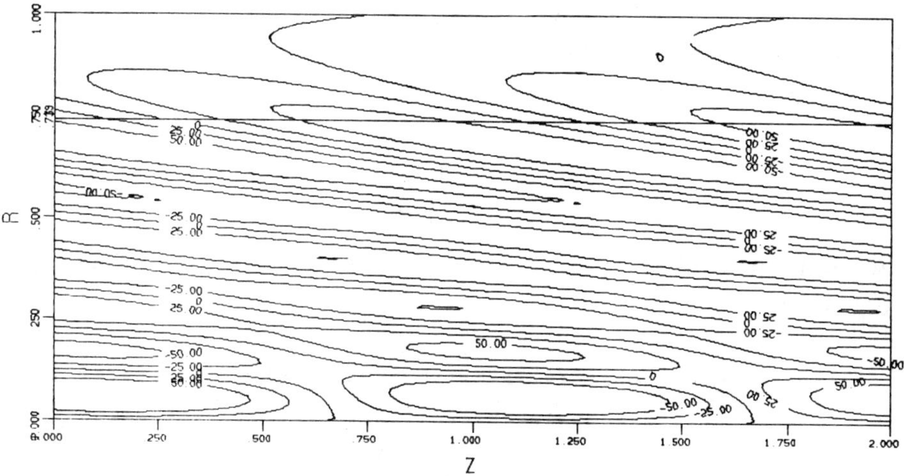

Figure 12.11. Isovorticity curves: E=0.09, M=10, R=33.33, W=3, N=0.5

A particularly striking feature of both problems is the behavior of the vorticity for different values of E. When M>1 is fixed, the vorticity in the hyperbolic region oscillates. For larger values of E, the oscillations are not damped as r decreases to zero. This feature appears in the isovorticity plots as isolated regions of positive and negative vorticity, with little change in maximum absolute value.

This is seen in Figure 12.11 (see also Figure 12.9(a)). For small E, however, the oscillations are damped severely as r decreases. In fact, the perturbed vorticity nearly vanishes in the center of the pipe. To make this more apparent, we have graphed the vorticity in the separated form

$$\zeta^{[1]}(r_0,z_0) = \zeta_1(r_0)\cos nz_0 + \zeta_2(r_0)\sin nz_0 \,. \tag{53}$$

In Figures 12.12–15 we present graphs of $\zeta_1(r_0)$ for various values of E. In those Figures FN=n/2π=1/2 is fixed, and ζ_1 is labelled VOR1. Notice the strong decay of the vorticity for small E, Figures 12.12-14, and the lack of decay for larger values of E, Figure 12.15. Also clear from these graphs is the observation that the oscillations have a period governed by M. That is, in Figure 12.14, where M=20, the interval between the zeros of the oscillations is nearly precisely half that in Figures 12.13 and 12.15, where M=10. This can be shown from the vorticity equation (51) by assuming M>1 and $W=ME^{1/2}<O(1)$, and keeping only terms of highest order in powers of 1/M. In performing the expansions the radial derivatives are scaled with M, in accordance with the observation of the period; that is,

$$\zeta_{r*} = \frac{1}{M}\zeta_r \,. \tag{54}$$

After using the separation (53), we find the following pair of equations

$$\zeta_1'' + n^2(1-r^2)\zeta_1 - \frac{n(1-r^2)^2}{W}\zeta_2 = O(1/M)\,,$$

$$\zeta_2'' + n^2(1-r^2)\zeta_2 + \frac{n(1-r^2)^2}{W}\zeta_1 = O(1/M)\,. \tag{55}$$

We take $1-r^2$ to be constant and look for exponential solutions for ζ_1 and ζ_2. This leads to

$$\zeta_1,\zeta_2 \sim \exp(\lambda r)\,, \tag{56}$$

where

$$\lambda = E^{-1/2}/2 + n\,(1-r^2)\,M\,. \tag{57}$$

Hence, the vorticity has a period of about

$$2\pi/n\ (1-r^2)\ M\ , \tag{58}$$

which for the value $n=\pi$, and for $1-r^2=1$, agrees very well with Figures 12.12-15. In addition, the perturbed vorticity decays, as r decreases, by the exponential rate governed by $E^{-1/2}/2$. This agrees with the observation that the damping is most rapid for small E. This damping factor also agrees, though only approximately so, with the measurements of the damping taken from Figures 12.12–15.

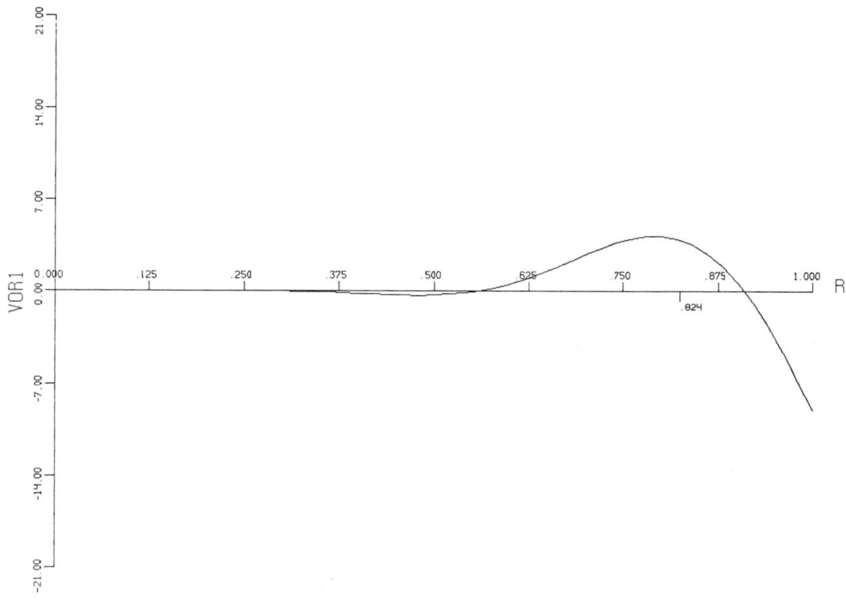

Figure 12.12. Pipe flow vorticity: E=0.001, M=3.16, R=100, W=.10, N=0.5

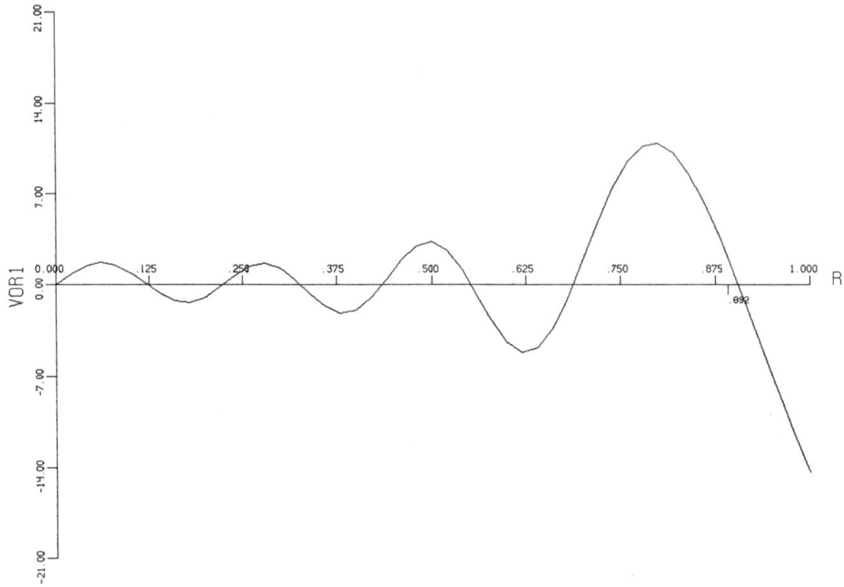

Figure 12.13. Pipe flow vorticity: E=0.001, M=10, R=100, W=1, N=0.5

The lack of decay of the vorticity for larger values of E, coupled with the fact that the values of the vorticity are larger for larger E, gives large gradients of vorticity. These large gradients prohibit computing solutions for larger values of E. In fact, the acceptable limit on computing costs limited computations to $W=ME^{1/2} < 3$. For the corresponding channel flow problem the limit was $W<7$.

An attempt to further use the idea of the change of type of the vorticity equation for Poiseuille flow was made by trying to correlate the data from experiments on delayed die swell by Joseph, Matta, and Chen [1987]. They found that when the flow rate through a die is larger than some critical rate, the initial swelling of the extruded material does not occur at the end of the die, but is delayed to a position further downstream. The appearance of the swell is much like a hydraulic jump. If one assumes that the dynamics can be related to a perturbation of Poiseuille flow at the pipe exit, then the characteristics of the vorticity equation in the hyperbolic region might also play a role. To use the analysis of the characteristics, we must also assume that the interpolated Maxwell model (1), or another model which responds elastically and not viscously to step velocity data, is sufficient to predict the behavior of actual fluids in the die swell flow conditions. Joseph, Renardy, and Saut [1985] show that models with Newtonian viscosity, like Jeffreys models, which smooth discontinuous initial data do not have a vorticity equation which changes type.

Another difficulty in applying the present analysis is the determination of the fluid constants appropriate for the interpolated Maxwell model (1). The main difficulty is the determination of a relaxation time λ. The conclusion of Joseph, Riccius, and Arney [1986] is that the wave speed of a shear wave and the measurement of a viscosity require the concept of at least two relaxation time scales (and in fact even two time scales may not be enough). The shortest relaxation times give very large shear wave speeds, c, and decay quickly. They give rise to an effective Newtonian viscosity. The elasticity is associated with slower modes. Hence we have an elastic response smoothed by an effective viscosity. The elastic response has a characteristic rigidity and wave speed which appears in the tables published by Joseph, Riccius, and Arney [1986]. We used these data.

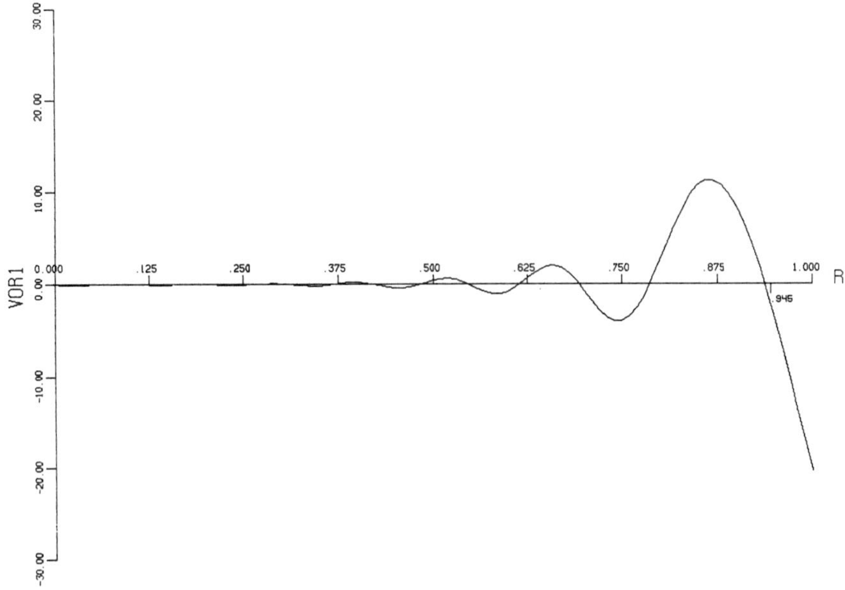

Figure 12.14. Pipe flow vorticity: E=0.0025, M=20, R=400, W=1, N=0.5

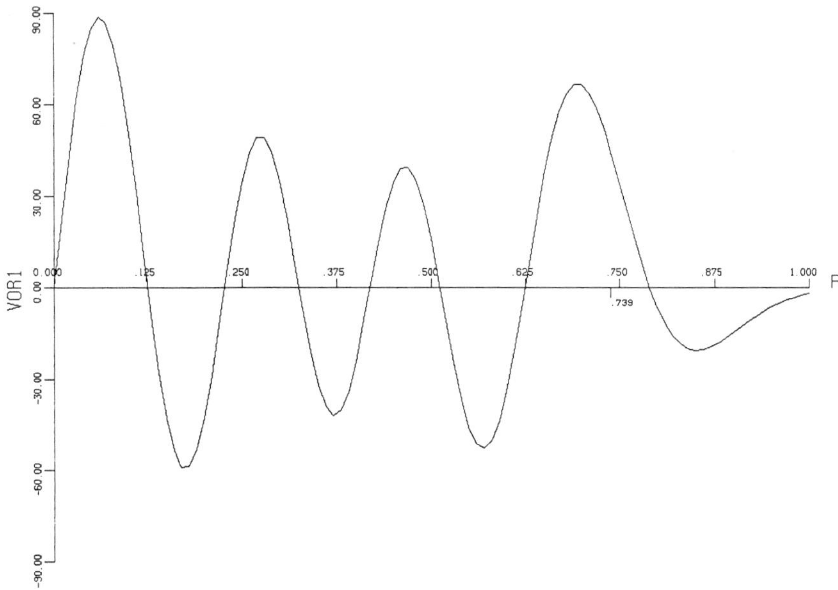

Figure 12.15. Pipe flow vorticity: E=0.09, M=10, R=33.33, W=3, N=0.5

Using the relaxation function for 1% CMC (Figure 18 of Joseph, Riccius, and Arney [1986]) we estimated the relaxation time in two ways. First, we used the simple formula

$$G(0) = \eta/\lambda \tag{59}$$

for a Maxwell model. This gives λ=0.176 s. Second we assumed an exponential decay from the shear wave modulus $G(0)=G_c$ to the peak at which the instrument began to record G(t). This gives λ=0.005 s. The first estimate is too large since η includes relaxation modes that are much slower than the initial modes. Using b=0.05 cm as a typical die radius for the experiments of Joseph, Matta, and Chen [1987], these values of λ give

$$25 < E < 32{,}000\ . \tag{60}$$

The delayed die swell was observed to occur at about M=4 for this fluid, so that (43) gives the hyperbolic region to be $0<r<r^*$ where

$$0.003 < r^* < 0.10 , \tag{61}$$

where the larger value of E gives the smaller value of r^*. For such large values of E, and of M, (43) may be expanded to give

$$r^* = E^{-1/2} + O(1/E) . \tag{62}$$

This agrees with (61), and also shows that r^* is proportional to the radius b of the pipe for any given fluid. Thus the hyperbolic region is larger in larger pipes.

The large values of E we estimate in (60) for the experiments of Joseph, Matta, and Chen [1987] should be related to our previous remarks on the analysis of the wavy pipe problem. The perturbation vorticity is very robust; it does not decay across the pipe. Thus the smallness of the hyperbolic region for large E may not be important since the vorticity is not damped even in the elliptic region. In addition, the Weissenberg number $W=ME^{1/2}$ is very large:

$$20 < W < 715 . \tag{63}$$

This is beyond the range we can compute numerically for the wavy pipe problem; and is in fact beyond the range of any reported numerical study. This indicates that any numerical study of the delayed die swell would be extremely difficult with present techniques (but see §13.10).

CHAPTER 13

Die Swell and Delayed Die Swell

A jet of viscoelastic fluid which is extruded from a pipe can swell to three or four times the diameter of the pipe from which it is extruded. The old name for this is "die swell" or more recently extrudate swell, since the extrudate rather than the die, swells.* The phenomenon of die swell is one important signature of the nonlinearity of flow in non-Newtonian fluids. It is a very important problem in industry because so many processes involve the extrusion of viscoelastic fluids, say plastic in their molten state. Die swell is not well understood. There are a number of approximate approaches and numerical simulations based on assuming different constitutive models. These approaches are well represented in the books of reference and will not be discussed here.

There is a general critical condition associated with die swell which Joseph, Matta and Chen [1987], hereinafter designated as JMC, called delayed die swell. Delayed die swell seems to occur at a critical speed not so different than the wave speed for the fluids used in the experiments; $M=U_0/c>1$ where U_0 is the centerline velocity in the pipe. We noted in the last chapter that when this inequality is satisfied, the vorticity of the flows perturbing Poiseuille

---

* A related phenomenon, extrudate distortion, occurs when the cross-section of the jet is not circular (see Bechtel, Lin, and Forest, [1988]).

flows of an upper convected Maxwell model is hyperbolic in the center of the pipe. The form of the jet reminds one of a hydraulic jump which is the shock phenomenon corresponding to shocks in gas dynamics. This type of shock phenomenon is also well known in the theory of inextensible elastic bars in which the long waves are associated with a propagating change of diameter given by $c/\sqrt{3}$, where c is the speed of a shear wave.

The phenomenon of die swell may be one driven by elastic recovery. This is suggested by the analysis of the momentum of the jet given in §13.1 and by arguments which are presented in the books of reference. Some aspects of the dynamics of low speed jets treated as a perturbation of the rest state by the method of domain perturbation, Appendix E, will be discussed in Chapter 17.

In §13.2–9 we are going to report the facts about delayed die swell as they were reported in the paper by Joseph, Matta and Chen (JMC) [1987]. Delayed die swell is an important candidate for a shock transition falling in the frame of this book. Our aim is to help to prepare the ground for future investigations into the fluid mechanics of this phenomenon.

13.1 Momentum balance

One advantage of the momentum balance is that we can avoid making a constitutive assumption. A disadvantage is that we don't solve the problem. It is good to derive the balance because one sees that mechanisms are at work. In the case of die swell we have readjustment of the jet to an abrupt change of the boundary conditions from zero velocity on the pipe wall to zero shear stress on the jet-air interface. This is bound to cause a singularity at the exit lip where this type of discontinuity is actually prescribed. We could think of how this discontinuity at the boundary makes itself known in the interior, by diffusion, wave propagation or both. Globally, the fluid flow must adjust from a fully developed one like Poiseuille flow in the pipe to an inviscid flow far downstream in the jet. In the analysis below (Joseph, [1980a]) gravity is neglected. A momentum analysis with gravity is given by Joseph, Nguyen and Matta [1983].

A liquid is forced to move from left to right (x increasing) down a round pipe of length L by high pressure imposed at the entrance x=–L of the pipe. The flow is assumed to be axisymmetric but the pressure and velocity which is prescribed at x=–L is otherwise arbitrary. At x=0 the liquid is extruded into a zero gravity field, and an axially-symmetric capillary jet of radius h(x) is formed. It is assumed that the wind shear is negligible, and that far downstream ($x \to \infty$) the flow must become rectilinear, with uniform velocity U_f and final jet diameter h_f, thus

$$U_f = \lim_{x\to\infty} \mathbf{e}_x \bullet \mathbf{u}(x,r) \,, \quad h_f = \lim_{x\to\infty} h(x) \,. \tag{1}$$

The mass flux is constant and invariant for the whole flow ($-L \le x \le \infty$).

The motion of the liquid is governed by the following system of equations,

$$\mathbf{u} = \mathbf{e}_r w + \mathbf{e}_x u \quad \text{(velocity)} \,,$$
$$\mathbf{T} = -p\mathbf{1} + \mathbf{S} \quad \text{(stress)} \,,$$

$$\frac{1}{r}\partial_r(rw) + \partial_x u = 0 \tag{2}$$

$$\rho(w\partial_r u + u\partial_x u) = \partial_x T_{xx} + \frac{1}{r}\partial_r(rT_{rx}) \tag{3}$$

and

$$\rho(w\partial_r w + u\partial_x w) = \partial_r T_{rr} + \partial_x T_{xr} + \frac{1}{r}(T_{rr} - T_{\theta\theta}) \,. \tag{4}$$

The velocity of the fluid in the pipe is bounded on the centerline and vanishes at the pipe wall

$$u(x,h_0) = w(x,h_0) = 0 \,. \tag{5}$$

The free boundary conditions, listed below, state, respectively, that the normal component of the velocity, the shear stress, the difference between the jump in the normal stress and the surface tension force (σ=coefficient of surface tension), the difference between the jet radius at x=0 and the pipe radius, and the slope of the free surface $x \to \infty$ all vanish; thus

$$w - h'u = h'(T_{rr} - h'\,T_{xr}) + (T_{xr} - h'\,T_{xx}) = 0$$

$$T_{rr} - h'T_{xr} = -\frac{\sigma}{hh'}\frac{d}{dx}\left(\frac{h}{(1+h'^2)^{1/2}}\right). \tag{6}$$

$$h(0) - h_0 = h'(\infty) = 0 \tag{7}$$

where $h' \equiv dh/dx$.

The stress and velocity are continuous throughout the flow; in particular, they are continuous across the exit plane x=0. The possibility that the stress at the exit lip x=0, r=h has a singularity is not excluded. At the entrance x=–L of the pipe there holds

$$(u,w,T_{xx}) = (\hat{u}(r),\hat{w}(r),\hat{T}_{xx}(r)) . \tag{8}$$

As $x \to \infty$ *the flow is essentially uniform and rectilinear in a cylinder of constant radius* h_f, with an isotropic stress which balances the hoop stress induced by surface tension. Thus

$$\lim_{x\to\infty} (u,w,h,\mathbf{T}) = (U_f,0,h_f,-\sigma\mathbf{1}/h_f) . \tag{9}$$

An integral invariant ((12) below) is defined in terms of the *axial thrust*

$$\mathcal{M}(x) = \int_0^{h(x)} r(\rho u^2 - T_{xx})\, dr . \tag{10}$$

It is also of interest to introduce the *radial thrust*

$$\mathcal{R}(x) = \int_0^{H(x)} r(\rho uw - T_{xr})\, dr . \tag{11}$$

The *axial thrust*

$$\mathcal{M}(x) - \frac{\sigma h(x)}{[1+h'^2(x)]^{1/2}} = \frac{1}{2}(\rho U_f^2 h_f^2 - \sigma h_f) \tag{12}$$

is conserved in the jet; that is, (12) *holds for all* $x \geqq 0$. *And in the pipe itself,* $-L \leqq x \leqq 0$, *the rate of change of* $\mathcal{M}(x)$ *is proportional to the wall shear stress*

$$\frac{d\mathcal{M}(x)}{dx} = h_0 T_{rx}(x,h_0) \,. \tag{13}$$

The radial thrust satisfies

$$\frac{d\mathcal{R}(x)}{dx} + \frac{\sigma}{h'}\frac{d}{dx}\left(\frac{h}{(1+h'^2)^{1/2}}\right) = -\int_0^{h(x)} T_{\theta\theta}\, dr \tag{14}$$

in the jet. And in the pipe, the rate of change of $\mathcal{R}(x)$ *is governed by*

$$\frac{d\mathcal{R}(x)}{dx} = h_0 T_{rr}(x, h_0) - \int_0^{h_0} T_{\theta\theta}\, dr \,. \tag{15}$$

To prove these relations, we introduce the notation $[A] = A(x,h(x))$ for any function $A(x,r)$. We then form the area average of (3) and (4), using (2), and find that

$$rw\partial_r u + ru\partial_x u = \partial_r(rwu) + r\partial_x u^2 \,,$$

$$\int_0^{h(x)} r\left(w\partial_r u + \frac{1}{2}\partial_x u^2\right) dr = \frac{d}{dx}\int_0^{h(x)} ru^2\, dr + [hu\,(w-h'u)] \,,$$

$$\int_0^{h(x)} \{\partial_r(rT_{rx}) + \partial_x(rT_{xx})\}\, dr = \frac{d}{dx}\int_0^{h(x)} rT_{xx}\, dr + [h(T_{rx}-h'\,T_{xx})] \,,$$

$$rw\partial_r w + ur\partial_x w = \partial_r(rw^2) + \partial_x(ruw) \,,$$

$$\int_0^{h(x)} (\partial_r(rw^2) + \partial_x(ruw))\, dr = \frac{d}{dx}\int_0^{h(x)} rwu\, dr + [hw(w-h'u)] \,,$$

and

$$\int_0^{h(x)} \{\partial_r\,(rT_{rr}) + \partial_x(rT_{xr})\}\, dr = \frac{d}{dx}\int_0^{h(x)} rT_{xr}\, dr + [h(T_{rr}-h'T_{xr})] \,.$$

Inserting these expressions in the averaged equations of motion gives the differential equations satisfied by $\mathcal{M}(x)$ and $\mathcal{R}(x)$. In the region $x \geq 0$, (6) and (7) hold, whence

$$\lim_{x\to\infty} \int_0^{h(x)} T_{xx}(x,r)r\,dr = \int_0^{h_f} -\frac{\sigma}{h_f} r\,dr = -\frac{\sigma h_f}{2} \tag{16}$$

and the averaged equations of motion reduce to (12) and (14).

The relations are of interest in the study of die swell. For example, it is frequently assumed in approximate studies of die swell that the main mechanism for the die swell is due to elastic recovery. Perhaps this assumption means that when the swell is large, say $h_f > 2h_0$, the average axial stress at the exit is a tension; that is,

$$\int_0^{h_0} T_{xx}(0,r)r\,dr > 0\,. \tag{17}$$

One reason to accept (17) is expressed in the inequality

$$\frac{\rho}{2} h_f^2 U_f^2 \left(\frac{h_f^2}{h_0^2} - 1 \right) + \sigma \left(\frac{h_f}{2} - \frac{h_0}{(1+h'(0)^2)^{1/2}} \right) \leq \int_0^{h_0} T_{xx}(0,r)r\,dr\,. \tag{18}$$

To prove (18), we first define the average velocity

$$\bar{u} = \frac{2}{h_0^2} \int_0^{h_0} u(0,r)r\,dr\,.$$

Conservation of mass requires that

$$\int_0^{h(x)} u(x,r)r\,dr = Q$$

be independent of x for $-L \leq x < \infty$. Hence,

$$\bar{u} = \frac{h_f^2}{h_0^2} U_f = \frac{2Q}{h_0^2}$$

and

$$0 \le \int_0^{h_0} r(u(0,r) - \bar{u})^2 \, dr = \int_0^{h_0} u^2(0,r) r \, dr - \frac{h_f^4}{2h_0^2} U_f^2 . \tag{19}$$

Evaluation of (12) on x=0 gives

$$\rho \int_0^{h_0} u^2(0,r) r \, dr = \int_0^{h_0} T_{xx}(0,r) r \, dr$$
$$+ \frac{1}{2}\rho U_f^2 h_f^2 + \sigma\left(\frac{h_0}{(1+h'(0)^2)^{1/2}} - \frac{h_f}{2}\right). \tag{20}$$

Taken together, (19) and (20) imply (18).

The inequality (18) shows that the assumptions

$$\sigma = 0 = \int_0^{h_0} T_{xx}(0,r) r \, dr , \tag{21}$$

which are sometimes invoked to explain swelling of viscoelastic jets (see Joseph, [1974], p. 126), cannot lead to swelling since (12) and (18) imply that $h_f \le h_0$.

To go further with the application of integral invariants further assumptions must be made. A popular but incorrect assumption is that the flow in the pipe is fully developed up to the exit. In this case

$$\frac{\partial T_{rr}}{\partial r} + \frac{T_{rr} - T_{\theta\theta}}{r} = 0 \tag{22}$$

everywhere in the pipe $x \le 0$. Integration of r^2 times (22) on x=0 from r=0 to $r=h_0$ gives

$$\int_0^{h_0} T_{xx}(0,r) r \, dr = \frac{h_0^2}{2} T_{rr}(0,h_0) + \int_0^{h_0} r\left\{N_1(r) + \frac{1}{2}N_2\right\} dr \tag{23}$$

where $N_1 = T_{xx} - T_{rr}$ and $N_2 = T_{rr} - T_{\theta\theta}$ are the first and second normal stress difference and $T_{rr}(0,h_0)$ is the negative of the pressure at the exit. One advantage of (23) is that the average axial stress may be expressed in terms of N_1 and N_2. We have (20), (23), and, from the conservation of mass,

$$U_f = \bar{u}(h_0/h_f)^2 . \tag{24}$$

This may be written as

$$\rho\left\{\bar{u}^2 - \frac{\bar{u}^2}{\chi}\right\} = T_{rr}(0,h_0) + \overline{N_1+N_2/2} + \frac{\sigma}{h_0}\left\{\frac{2}{\left[1+h'^2(0)\right]^{1/2}} - \chi\right\} \tag{25}$$

where $\chi=h_f/h_0$ is the swell ratio. The condition $\chi>1$ implies an inequality around the remaining quantities in (25).

13.2 Description of delayed die swell

It is easiest to understand what is meant by delayed die swell by examining the photographs exhibited. A sequential description of the phenomenon is exhibited in photographs shown as Figures 13.1-5. The working fluid is a 9.5% solution of polyisobutylene in decalin (PIB/D, 9.5%). The values of material parameters for this fluid and the other fluids used in the experiments are given in Table 13.1. The fluid is extruded vertically downward in the direction of gravity. The effects of gravity are the same in all experiments and are important in downstream regions away from the exit where data were taken. The nozzle diameter d is 1 mm and the length of the capillary tube is 6.125 cm.

The local velocity $\mathbf{u}(x,r)=(u,w)$ where u is the axial and w the radial component, $\bar{u}(x)$ is the area averaged velocity and $\bar{u}(0)$ is the average velocity at the pipe exit. The diameter of maximum swell at $x=L$ is $D=2h_f$ and the swell ratio is D/d where $d=2h_0$.

Figure 13.1 shows the jet under pre-critical conditions, before the delay. The swell ratio increases with $\bar{u}(0)>u_c$. Figure 13.2 shows the jet at criticality $\bar{u}(0)=u_c$. Figures 13.3 and 13.4 show what happens as the extrusion velocity $\bar{u}(0)$ is increased. First D increases with $\bar{u}(0)$, then near $\bar{u}(0)=1519$ cm/s, D begins to decrease. The configurations shown in Figures 13.1–3 appear to be steady. There is evidently a point of instability at $\bar{u}(0)=1012$ cm/s which leads to an oscillation in which the point of delay moves between extreme values. The maximum and minimum delay distance over a cycle is shown for $\bar{u}(0)$=1519 cm/s in Figures 13.5(a) and (b).

Figures 13.1–5 show delayed die swell in a dramatic form. In general the delay starts in a less dramatic way: a point of inflection in the swell profile suddenly appears at the pipe exit when the extrusion velocity is raised above a critical value. The strength of the delay varies widely over different fluids.

Table 13.1. Values of material parameters at 23±1°C.

Liquid	Density	Viscosity	Wave speed	Rigidity	Relaxation time
	ρ(g/cc)	$\tilde{\mu}$(Pa-s)	c(cm/s)	Gc(Pa)	$\tilde{\mu}$/Gc(s)
AP30, 1%	1.16	53.80	30.7	109.33	0.4921
AP30, 1.5%	1.17	160.0	38.4	172.52	0.9274
CMC, 1%	1.19	44.76	50.6	304.68	0.1468
CMC, 1.2%	1.20	84.48	57.0	389.88	0.2167
CMC, 1.3%	1.21	129.8	81.0	793.88	0.1628
ELVACITE, 9.8%	1.05	0.90	67.0	471.35	0.0019
K-125, 5%	1.08	1.10	45.4	222.61	0.0049
PIB/D, 6%	0.90	11.01	90.1	730.62	0.0151
PIB/D, 9.5%	0.91	139.0	162.0	2204.5	0.0631
PIBM, 1%	0.93	0.065	18.8	32.87	0.0020
PIB/T, 4%	0.87	0.87	53.5	249	0.0035
PIB/T, 5.5%	0.88	5.44	60.6	323.2	0.0168
PIB/T, 6%	0.89	41.0	115.0	1177.0	0.0348
PMMA, 1%	1.06	0.20	17.9	33.90	0.0059
PMMA, 2%	1.07	13.0	25.0	66.94	0.1194
POLYOX, 1.3%	1.00	11.9	24.7	61.01	0.1950
POLYOX, 2.5%	1.01	203.3	51.2	264.77	0.7679

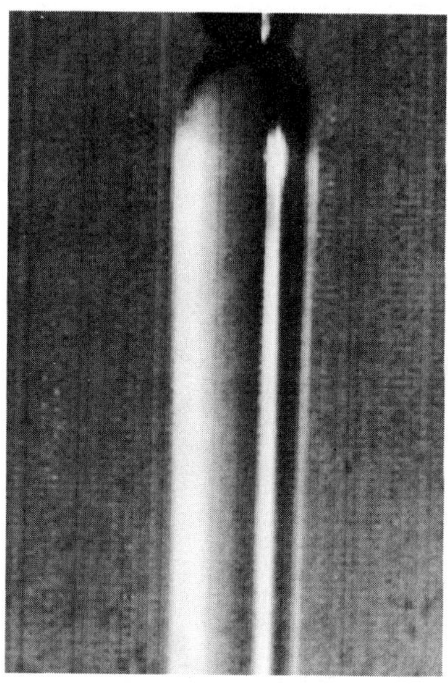

Figure 13.1. Extrusion of PIB/D, 9.5% under pre-critical conditions: $[d,\bar{u}(0),D] = [1\text{ mm}, 404\text{ cm/s}, 3.33\text{ mm}]$.

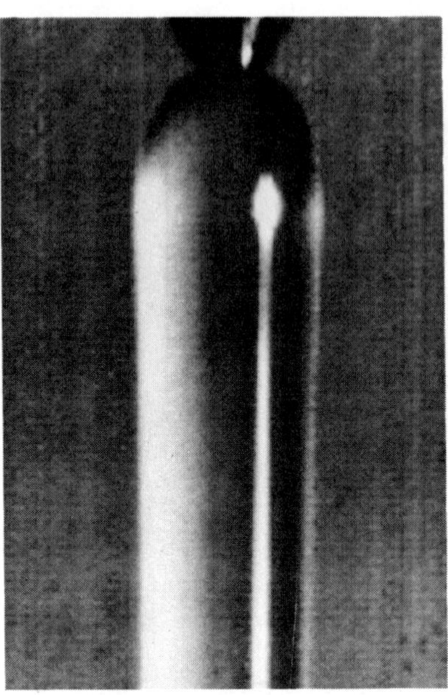

Figure 13.2. Extrusion of PIB/D, 9.5% at criticality: $[d,\bar{u}(0),D]$ = [1 mm, 785 cm/s, 3.83 mm].

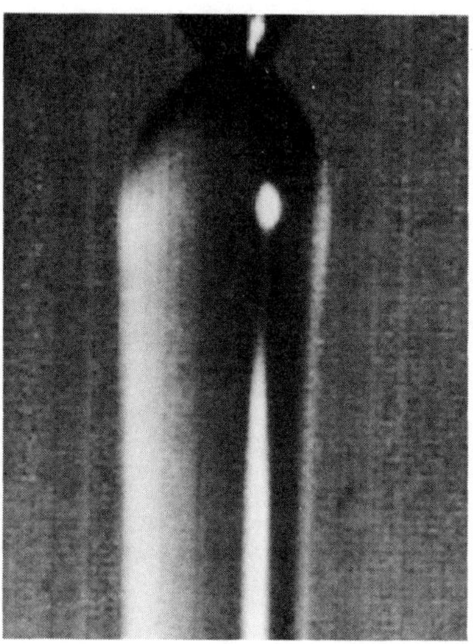

Figure 13.3. Extrusion of PIB/D, 9.5% under post-critical conditions: $[d,\bar{u}(0),D] = [1\ \text{mm}, 934\ \text{cm/s}, 4.17\ \text{mm}]$.

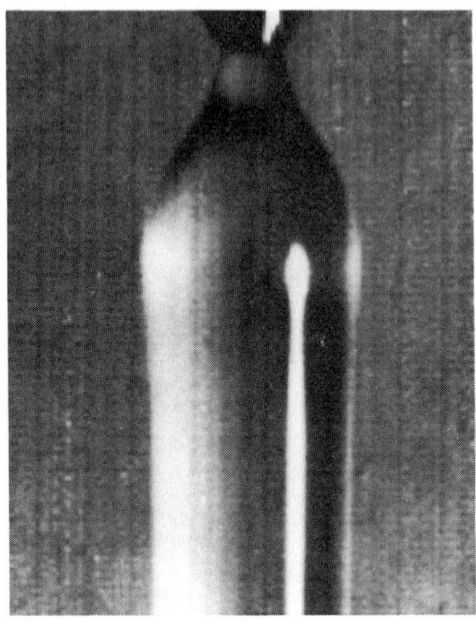

Figure 13.4. Unsteady extrusion of PIB/D, 9.5% under post-critical conditions. The oscillation has just commenced. $[d,\bar{u}(0),D]$ =[1 mm, 1214 cm/s, 4.5 mm].

Figure 13.5. Unsteady extrusion of PIB/D, 9.5% under post-critical conditions: $[d,\bar{u}(0),D]$ = [1 mm, 1519 cm/s, 3.92 mm]. (a) shows the maximum and (b) the minimum delay on one cycle.

(a) Maximum delay

(b) Minimum delay

The dynamics of die swell is not well understood and delayed die swell is perhaps less well understood. However, there is an apparent connection between various critical values of the velocities associated with the delay and the shear wave speed we measure using a wave-speed meter (see Joseph, Riccius and Arney [1986]). This might be important in die swell and in delayed die swell.

13.3 Previous work

Many casual photographs of die swell under subcritical conditions have been published in the last fifty years. Photographs of delayed die swell have appeared in the literature less frequently. We know of photographs of Merrington [1943], Giesekus [1968], Metzner, White and Denn [1966], Middleman [1968] and Brenschede and Klein [1970]. Some of these photographs are reproduced here in Figures 13.6,7,8. Among these authors only Giesekus followed by Brenschede and Klein have noted that delayed die swell is a critical phenomenon. Their explanation is a molecular one which involves the precipitation of polymer. We think this explanation cannot explain that the delay is a general property independent of the nature of solute and solvent and is not consistent with other observed properties reported.

The first paper to report systematic experiments which look at delayed die swell are by JMC. They state that the delay is a critical phenomenon, that is is a general phenomenon which occurred in all eighteen solutions tested by them and that there is a relation between die swell and relaxation through wave propagation.

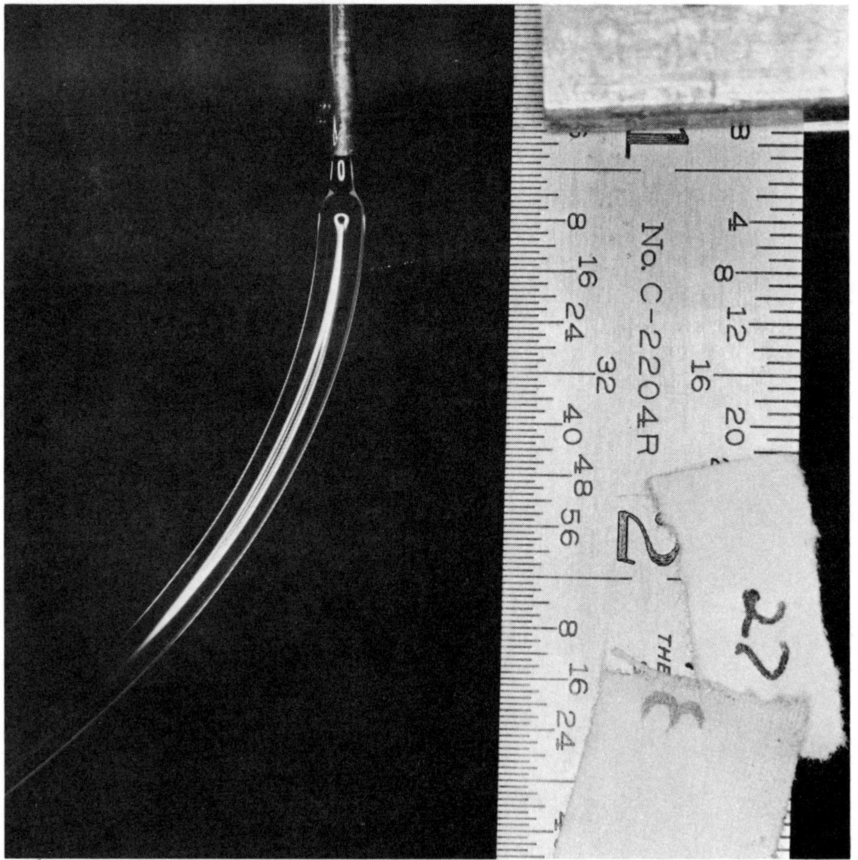

Figure 13.6. (After Metzner, White and Denn, [1966]). Delayed die swell of 5% solution of Vistanex L-100 dissolved in decalin. Wall shear rate = 7,800 sec^{-1}.

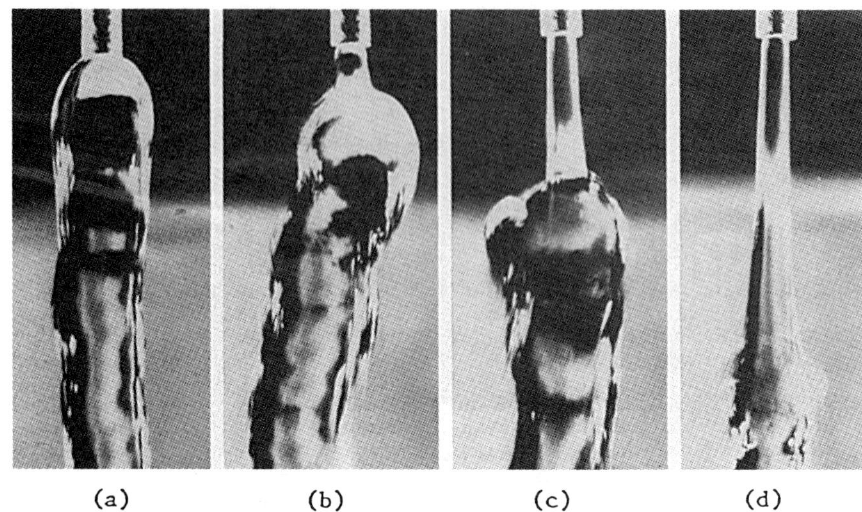

Figure 13.7 (After Brenschede and Klein, [1970]). Delayed die swell in polyisobutilene solution in toluol. $D=4\mathring{Q}/\pi R^3$ where $\mathring{Q}$ is the volume flow rate and R is the pipe radius. The critical value D is 14×10^4 reciprocal seconds. (a) $D=8\times10^4$ (b) $D=15.2\times10^4$ (c) $D=22.8\times10^4$ (d) $D=29\times10^4$.

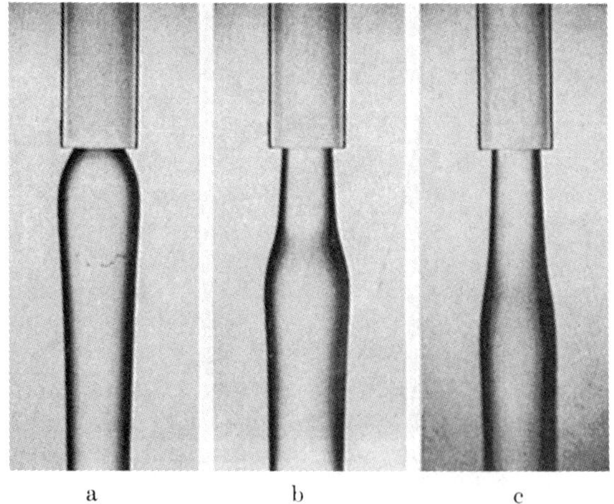

Figure 13.8 (After H. Giesekus, [1968]). Delayed die swell in a 0.5% polyacrylamide solution in water. (a) slow speed, (b) post-critical speed, (c) high speed.

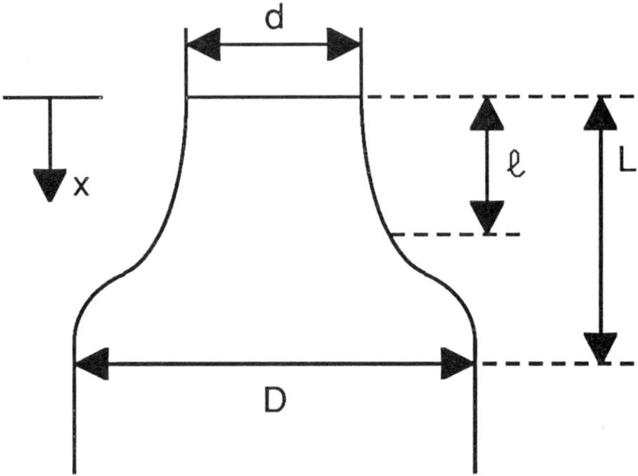

Figure 13.9. Sketch of delayed die swell. Q is the volume flow rate constant throughout the jet. $\bar{u}(x)$ is the area averaged velocity; $\bar{u}(0)=Q/\pi a^2$ where a=d/2 is the radius of the pipe. The delay distance is ℓ. The distance to the maximum swell is L.

13.4 Notations

To prepare for our discussion it is useful to define certain quantities. Many of these quantities are exhibited in Figure 13.9.

List of symbols

ρ	density (g/cm^3)
$\tilde{\mu}$	static or zero shear rate viscosity (Pa-s)
c	speed of vorticity waves (shear waves) into a fluid at rest (or rigid motion) (cm/s)
G_c	effective rigidity, $G_c=\rho c^2$ (Pa)
$\bar{\lambda}$	mean relaxation time, $\bar{\lambda}=\tilde{\mu}/G_c$ (s)
R	Reynolds number at the pipe exit based on static viscosity, $R=\rho d\bar{u}(0)/\tilde{\mu}$
ϕ	Shear rate at the pipe exit, $\phi=8\bar{u}(0)/d$
D/d	Swell ratio
L	Terminal swell distance, x=L is position where D is maximum

ℓ	delay distance
$\overline{u}_c$	$= \overline{u}_c(0) = Q_c/\pi a^2$ Mean velocity at the flow rate of incipient delay. There is a delayed die swell when $Q>Q_c$.
M_0	$= U/c$ Viscoelastic Mach number at exit $x=0$
U	$= 2\overline{u}(0)$ Definition of U, this is an approximation of the center line velocity at exit which is exact when the profile at exit is parabolic.
M_c	$= U_c/c$ $(U_c=2\overline{u}(0))$ Viscoelastic Mach number at criticality
M(L)	$= \overline{u}(L)/c$ Viscoelastic Mach number at the point of termination of the swell. It may be assumed the velocity $u(L,r)=\overline{U}(L)$ is uniform at the point of maximum swell.

13.5 Experiments

Two devices were used to extrude the test liquids, a controlled displacement device at the University of Minnesota and a controlled pressure device at the Aberdeen Army Laboratory. Results obtained from the two devices are consistent with one another.

The first device was a piston-cylinder apparatus in which the piston velocity could be accurately controlled by a microprocessor (manufactured commercially by MTS, Materials Testing System). The flow rate Q in the capillary can be determined from the rate of advance of the piston in the cylinder. The inner diameter of the capillaries were 0.81, 1, 1.57, 2.08, and 3.175 mm respectively. The length/diameter ratio for these capillaries was greater than thirty. The smallest and largest nozzles were late additions in the experiments.

The second device employs pressurized nitrogen to force the stored fluid from a 600 ml reservoir vertically downward through a 15 cm long capillary that extended 2 cm up into the reservoir. Capillaries with various inner diameters were used: 2.39, 1.81, 1.7, and 1.25 mm. When pressurized the fluid extruded vertically downward from the orifice and was collected in a sample cup placed about 10 cm below the nozzle tip. The average velocity was determined by weighing fluid extruded in a certain interval of time.

All of the delayed swell data on both devices, except the data exhibited in §13.8, were taken on a Spin Physics high speed video and motion analysis system. The critical condition for delayed swell was determined by iteration of visual observations on the video from above and below. The critical condition is very sharp and it is possible to determine it with good accuracy. The same method of iterative determination of critical conditions was made on the pressurized system, by monitoring the gauge pressure iteratively around the critical condition. The reticule feature on the video system allows for accurate measurements of the delay distance, swell ratio, and shock layer thickness. Most of the photographs shown in the chapter are taken by photographing the video screen with a polaroid camera.

The static viscosities were measured on a standard cone-and-plate rheometer and the wave speeds with the wave-speed meter. Material properties of test liquids are exhibited in Table 13.2.

JMC used the wave-speed meter to measure c for the seventeen liquids listed in Table 13.2. The wave speeds are listed in Table 13.1; they vary over a decade, from 17.9 to 162 cm/s.

List of test liquids	
β	Percent by weight of the polymer in solution
DEM	Diethylmalonate is the solvent used with PMMA, K-125, and ELVACITE
AP-30, β	Separan, Polyacrylamide in 50%–β water, 50% glycerin (β=1%, 1.5%)
CMC, β	Carboxymethycellulose in 50%–β water, 50% glycerin (β=1%, 1.2%, 1.3%)
ELVACITE, β	Polymethyl/methacrylate, 4×10^5 molecular weight, in DEM (β=9.8%)
K-125, β	A copolymer of 80% PMMA and 20% poly (ethyl/butyl acrylate) with a 1.9×10^6 viscosity molecular weight (β=5%)
PIBM, β	Polyisobutyl methacrylate, 5.5×10^6 molecular weight, in bis (2-ethyl-hexyl) hydrogen phosphite (BIS-2) (β=1%)
PIB/D, β	Polyisobutylene in Decalin (β=6%, 9.5%)
PIB/T, β	Polyisobutylene in Toluene (β=4%, 5.5%, 6%)
POLYOX, β	Polyethylene oxide (WSR-301) in water (β=1.3%, 2.5)
PMMA, β	Polymethyl methacrylate, 6×10^6 molecular weight, in DEM (β=1%)

Table 13.2. List of test liquids.

The remarkable connection between the wave speed and delayed die swell may be described as follows: Let U(x) be the maximum value of the velocity U(x,r) over the cross-section at x. Suppose that $U(0)=2\bar{u}(0)$ where $\bar{u}(0)$ is the area average of u(x,r) at the nozzle exit x=0. Because of shear thinning, $U(0)=\alpha\bar{u}(0)$, $1<\alpha\leq 2$, so we are a little high, but less high when the nozzle diameter is larger and the shear rate is smaller. Suppose further that after the swell the velocity in the jet is uniform. Then we found that the maximum velocity was always (nearly always, see Figure 13.10) supercritical

$$U(0) \geq c \tag{26}$$

before the swell and subcritical

$$\bar{u}(L) < c \tag{27}$$

after the swell, at x=L. The forward velocity of some point in the region of swelling is c.

These observations suggest that delayed die swell may be associated with a change in type. For Maxwell models the change of type is a change in the type of the second order equation governing the vorticity. The problem of change of type for the flow in the pipe is analyzed for an upper convected Maxwell model in Chapter 12.

If we assume that the velocity at exit is fully developed, then

$$u(0,r) = U(0)(1-4r^2/d^2) \tag{28}$$

and the vorticity ζ is given by

$$\zeta(0,r) = \frac{-\partial u(0,r)}{\partial r} = \frac{8rU(0)}{d^2}\ . \tag{29}$$

It is perhaps useful to associate die swell with the relaxation of the developed distribution (29) at exit to zero downstream where the flow is uniform. This relaxation is induced by a discontinuous prescription of the data, no slip on the wall (r = d/2), no shear at the surface of the jet. This prescribed discontinuity gives rise to singular behavior when the fluid is non-Newtonian. In any event the discontinuous prescription of the boundary data requires that the flow readjust from pipe flow with a linear vorticity profile (29) at exit to a uniform flow with zero vorticity downstream. This readjustment is powered by waves which carry the boundary data into the jet. In general the speed of such waves at each point depends on the state of stress there, but $c = \sqrt{G(0)/\rho}$ in all motions perturbing rigid ones and in shearing motions of all amplitudes in one space dimension of upper convected Maxwell fluids. The vorticity in the pipe is greatest at the wall and is zero at the center of the pipe and downstream where the flow is uniform. The vorticity at the boundary is impulsively reduced to zero at the exit lip and the region of zero vorticity consumes more and more of the jet as x increases.

We could imagine the relaxation to zero of all the vorticity in the jet is accomplished in a distance L at a time τ and that there is a velocity $\hat{u}<\bar{u}(0)$ such that $L\approx\hat{u}\tau$. At this time and place the wave of relaxation has penetrated all the way to r=0, a distance D/2. The speed of penetration might be estimated as $w=c-\hat{v}$ where $\hat{v}$ is the outward velocity of particles which could itself be estimated $\hat{v}\approx(D-d)/\tau$. This gives $D/2\approx w\tau$ and

$$\frac{2L}{D} \approx \frac{\hat{u}}{w} = \frac{\hat{u}}{c-\hat{v}} \,. \tag{30}$$

The usual way to estimate relaxation times uses data for the first normal stress. This might be good for the die swell application, since the swelling is associated with normal stresses, especially with excess tension. We do not have the rheological data for this estimate. The mean relaxation time $\tilde{\mu}/G_C$ could also be used for τ, but it seems to be smaller than the time which is actually effective in the dynamics of swelling. We always get $\hat{u}<\bar{u}(0)$ when $\bar{\lambda}=\tau$, and the estimate (30) is also in rough agreement with measured data when $\bar{u}(L)\leq\hat{u}\leq\bar{u}(0)$.

The considerations of the previous paragraph apply to pre-critical and post-critical flows and they do not explain the delay.

The question is: can there be an important hyperbolic region in the jet outside the pipe? An affirmative, though speculative, answer can be framed along the following line. The equation for flows perturbing uniform flows are hyperbolic everywhere that U>c, and in three-dimensional flows they give rise to the Mach cones described in the appendix of the paper by Ahrens, Yoo and Joseph

[1987]. When a fluid is extruded at high speed from a pipe, the center line velocity U>c, but u(x,r) vanishes at the exit lip (x,r)=(0,d/2). The velocity at the surface r=R(x) of the jet accelerates rapidly and if U were large enough the surface velocity would exceed c already for small values of x>0. The velocity profiles for these values of x may not be too different than uniform ones, locally at each x. At the very least, we might expect that each and every value of the velocity on the cross-section of the delay would be greater than c, with an associated hyperbolic vorticity equation.

13.6 Values of parameters at criticality

Delayed die swell seems to be a general phenomenon in elastic liquids. We found a robust and reproducible critical condition for delay in each and every one of the seventeen test fluids measured at the University of Minnesota. The existence of delay is less apparent in fluids of small viscosity and time of relaxation. The 1% solutions of PMMA and PIBM are in this category. These fluids are close to a Newtonian limit and considerable smoothing, probably viscous smoothing, is evident. We could not determine a die swell or a delay in soybean oil.

The flow at criticality can be steady or unsteady. This property of the critical flow evidently depends only on the fluid and not on the experiment. Flows which were unsteady in the constant flow rate apparatus were also unsteady in the constant pressure gradient apparatus. The flows which were unsteady at criticality are listed in Table 13.3. The unsteady flows sometimes emerge at an angle from the axis of the nozzle. The sidewise deviation from the vertical rotates around the axis of the pipe. We could not eliminate unsteadiness from the flows in which it occurred. We do not know why some fluids give rise to unsteady flow at criticality and some do not. Fluids with the longest mean times of relaxation (Table 13.1) were unsteady at criticality.

The Tables 13.3 and 13.4 we have listed the measured values of swell parameters at criticality. The values of c for different fluids are given in Table 13.1. Critical Reynolds number (based on the static viscosity) and critical shear rates can be computed from these tables. The critical swell ratio D/d, terminal swell distance ratio L/d, and the terminal value $\bar{u}(L)/c$ of the viscoelastic Mach number are not well defined when the flow is unsteady. The values given in Table 13.3 are average values over one cycle. Since the flow rate is accurately prescribed, $\bar{u}(0)$ is steady. The unsteady flow appears to rise as an instability.

Polymer	d(mm)	$\bar{u}(0)/c$	L/d	D/d	$\bar{u}(L)/c$
AP30, 1%	1	3.989	2.3333	2.8333	0.4970
AP30, 1.5%	1	4.477	4.3333	3.3333	0.4030
POLYOX, 1.3%	1	3.975	3.6667	3.4167	0.0292
POLYOX, 2.5%	0.81	2.899	2.7692	3.4615	0.2419
	1	2.283	2.6154	3.3846	0.1993
	1.57	1.342	1.9600	2.5200	0.2144
	2.08	0.879	1.4848	2.0000	0.2198
	3.175	0.5491	1.4359	1.9231	0.1485
PMMA, 2%	2.39	7.6000	2.5000	3.2222	0.8055

Table 13.3. Critical parameters for delayed die swell when the flow at criticality is unsteady.

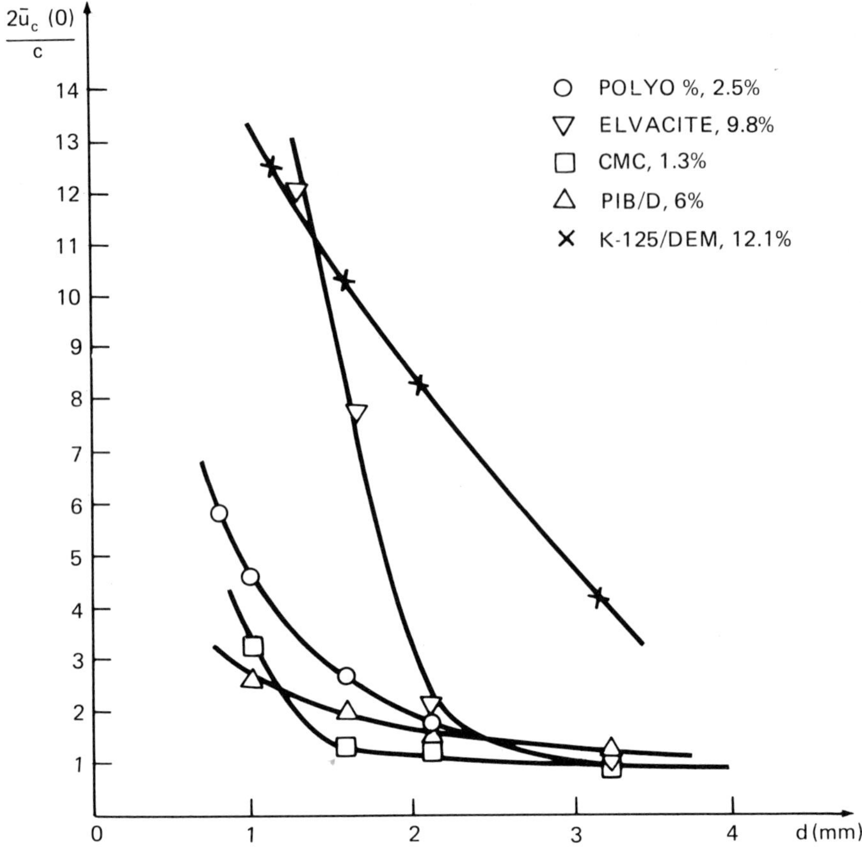

Figure 13.10. Critical Viscoelastic Mach number based on the approximation $2\bar{u}_C$ of the center line velocity at exit as a function of nozzle diameter.

Polymer	d(mm)	$\bar{u}(0)/c$	L/d	D/d	$\bar{u}(L)/c$
PIB/T, 4%	1	0.8042	0.6429	1.6429	0.2979
PIB/T, 5.5%	1	1.1341	0.9231	2.0769	0.2629
PIB/T, 6%	1	2.2121	1.0000	2.3330	0.4045
CMC, 1%	1	2.1400	1.0000	1.9170	0.5823
K-125, 5%	1	2.2444	2.8333	3.7500	0.1596
PIB/D, 6%	1	1.2869	2.0000	3.3636	0.1138
	1.57	0.9790	1.7950	2.8952	0.1168
	2.08	0.7437	1.3986	2.4039	0.1287
	3.175	0.7147	1.2312	1.8898	0.2002
PIB/D, 9%	1	4.6795	1.8333	3.8333	0.3185
CMC, 1.2%	1.25	2.0702	0.8750	1.6250	0.7831
	2.39	1.7544	0.5926	1.5000	0.7789
ELVACITE, 9.8%	1.25	5.9701	1.5000	2.4500	0.9728
	1.76	3.9254	1.2500	2.1875	0.8024
	2.08	0.9798	1.5769	2.3077	0.1840
	3.175	0.5046	1.0000	1.7500	0.1648
CMC, 1.3%	1	1.5970	0.9167	2.0833	0.3680
	1.57	0.6273	1.0616	1.9639	0.1626
	2.08	0.5875	0.8413	1.8429	0.1730
	3.175	0.4772	0.7087	1.5748	0.1924

Table 13.4. Critical parameters for delayed die swell when the flow at criticality is steady.

Several conclusions follow from inspection of Tables 13.3 and 13.4.

1. The critical velocity $\bar{u}_c$, the swell ratio D/d, and the terminal swell distance ratio L/d are decreasing functions of d, the pipe diameter.

2. The critical Mach number at exit, $M_c=U_c/c$, $U_c=2\bar{u}_c$ were nearly always larger than one. (The value $M_c=0.954$ for 1.3% CMC was the only exception.)

3. There seems to be a limiting value of M_c for large values of d. The evidence for this conclusion is exhibited in Figure 13.10. This Figure is not inconsistent with a limiting value $M_c=1$, but there are too few data to make a certain statement.

4. Whenever the die swell is delayed, there are points on the exit plane where $u(0,r)=c$.

5. The critical terminal Mach number, after the swell M(L) $=\overline{u}(L)/c$, is always definitely less than one.

The data here, and in §13.7, are not consistent with a brutal analogy between delayed die swell and hydraulic jumps. The velocity $\overline{u}(0)$ at the exit can be much larger than c in pipes of small diameter even under pre-critical conditions. It is likely that the speed c of vorticity waves in regions of uniform or rigid motion is not the relevant wave speed in the very stressed region at the exit of a pipe of small diameter. We also did not rule out the possibility that there is a fluid and a pipe of large diameter such that delayed die swell can occur with $U<c$. (The exception mentioned under (2) above may or may not realize this possibility.)

13.7 Post-critical values of the flow parameters

More energy is put into the flow as the extrusion velocity $\overline{u}(0)$ is increased. The delay distance increases with $\overline{u}(0)$ and the swell ratio first increases, then decreases with $\overline{u}(0)$. The shape of the jets with delayed swells differs from fluid to fluid. More dilute solutions, with smaller times of relaxation have rather smooth transitions as might be expected of fluids with a large Newtonian viscosity (see Figures 13.17,18). The Separan and Polyox solutions have a slight swell at exit followed by a large delayed swell. In general the small diameter of the jet downstream from the delay increases slightly with distance and abruptly at the point $x=l$ of maximum delay.

There is no possibility for the delay to persist. It must eventually be destroyed by the viscous action of decayed elastic modes or by the stretching effects associated with the acceleration of gravity.

We took data for increasing values $\overline{u}(0)>\overline{u}_c$ in the post-critical region. The delay distance increases monotonically from zero at $\overline{u}(0)=\overline{u}_c$. The swell first increases, or does not increase, then decreases as $\overline{u}(0)$ is increased. The taking of data was terminated at a certain $\overline{u}(0)$ for three different reasons.

1. We did not wish to extend the rated capacity of the transducer on the constant displacement extruder.

2. A flow instability intervened and we could not measure unique values of the delay distance or swell ratio.

3. The swell ratio showed marked decreases and the swell region was smoothed. Examples of this type of smoothing are shown in strong form in Figures 13.16, 17, 18.

In those cases in which we can identify a definite ℓ and a definite L we may speak of a shock layer transition. This is the region of thickness $L-\ell$ in which there is a sharp change in the radius of the jet. There are many cases in which a shock layer may be easily identified, there are marginal cases, and cases in which the identification of such a layer is not possible. This definition of a shock layer holds under pre-critical conditions, but we do not look for shock-like transitions in the pre-critical case. We are introducing the notion of shock layers for the post-critical case, in which there is a delay, $\ell>0$. Here we only list the measurements for PIB/D, 6% for different pipes (see Figures 13.11–14).

The post-critical case is defined by $\bar{u}(0)>\bar{u}_c$. The following conclusions are evident from the entries in the tables.

1. $\ell>0$ increases with $\bar{u}(0)$.

2. $M_0=2\bar{u}(0)/c>1$.

3. $M(L)=\bar{u}(L)/c$ is an increasing function of $\bar{u}(0)$.

4. $M(L)<1$ for all shock layer transitions.

5. $M(L)$, D/d, $(L-\ell)/d$ decrease with d for each fixed ℓ. The strength of the shock is greater in pipes of small diameter.

$\bar{u}(0)/c$	ℓ/d	L/d	D/d	$\bar{u}(L)/c$
0.8533	0.0000	1.9091	3.0909	0.0898
0.9656	0.0000	1.9091	3.0909	0.1011
1.1036	0.0000	1.9091	3.1818	0.1090
1.2875	0.0000	2.0000	3.3636	0.1138
1.4856	0.3636	2.1818	3.5455	0.1182
1.7557	1.1818	3.2727	3.8182	0.1204
2.1458	3.0909	5.0909	3.8182	0.1472
2.5750	5.4554	7.2727	3.4545	0.2158
2.7589	6.9091	8.2727	3.3636	0.2438
2.9711			2.4167	0.5087

Table 13.5. Measured values of flow parameters under post-critical conditions for PIB/D, 6%, d=1 mm.

$\bar{u}(0)/c$	ℓ/d	L/d	D/d	$\bar{u}(L)/c$
0.3918	0.0000	1.1582	2.1424	0.0855
0.4477	0.0000	1.1582	2.1424	0.0975
0.5223	0.0000	1.2160	2.3162	0.0974
0.6268	0.0000	1.4476	2.4899	0.1011
0.7835	0.0000	1.5634	2.4899	0.1264
0.9794	0.0000	1.7950	2.8952	0.1168
1.1193	0.4054	2.0845	3.0110	0.1235
1.3058	1.2160	2.7215	2.9531	0.1497
1.5670	2.4899	4.0532	2.9531	0.1797
1.7411	3.5901	5.1534	2.8373	0.2163
1.9587	5.4430	6.6589	2.6636	0.2761
2.2385			2.3162	0.4172

Table 13.6. Measured values of flow parameters under post-critical conditions for PIB/D, 6%, d=1.57 mm.

When the extrusion rates are high we get unsteady flow of fluids with large mean times of relaxation (Table 13.1) and smooth steady flow for fluids with small mean times of relaxation. When the degree of unsteadiness becomes too great, it is impossible to measure unique values of ℓ/d, L/d, and D/d. In certain periodic cases we can identify maximum and minimum values of these ratios over one cycle; in other cases, when the amplitude of oscillations is small, we record average values. Some photographs in which the effects of unsteadiness is important are shown in Figure 13.5(a) and (b).

$\bar{u}(0)/c$	ℓ/d	L/d	D/d	$\bar{u}(L)/c$
0.2232	0.0000	0.7430	1.8357	0.0662
0.2551	0.0000	1.0926	1.8794	0.0722
0.2976	0.0000	1.0926	1.9231	0.0805
0.3571	0.0000	1.0926	1.9231	0.0966
0.4058	0.0000	1.0926	2.0105	0.1004
0.4960	0.0000	1.6213	2.1416	0.1081
0.6377	0.0000	1.6213	2.2727	0.1235
0.7440	0.0000	1.3986	2.4039	0.1287
0.8928	0.1311	1.5734	2.5787	0.1343
0.9920	0.3934	1.8357	2.6224	0.1442
1.1160	1.1581	2.3601	2.6224	0.1623
1.2754	1.6171	3.0157	2.6224	0.1855
1.4879	3.0157	4.4580	2.4913	0.2397
1.7855			2.0105	0.4417

Table 13.7. Measured values of flow parameters under post-critical conditions for PIB/D, 6%, d=2.08 mm.

$\bar{u}(0)/c$	ℓ/d	L/d	D/d	$\bar{u}(L)/c$
0.2554	0.0000	0.7731	1.4889	0.1152
0.2979	0.0000	0.9162	1.4889	0.1344
0.3575	0.0000	0.9449	1.5748	0.1442
0.4469	0.0000	0.9735	1.6607	0.1620
0.5959	0.0000	1.1739	1.7752	0.1891
0.7150	0.0000	1.2312	1.8898	0.2002
0.8938	0.2577	1.4603	2.0329	0.2163
1.0215	0.8590	1.9184	2.0329	0.2472
1.1172	0.9162	2.1761	2.0902	0.2557
1.1917	1.5462	2.8633	2.0043	0.2966
1.2768	1.5748	2.9492	1.9757	0.3271

Table 13.8. Measured values of flow parameters under post-critical condition for PIB/D, 6%, d=3.175 mm.

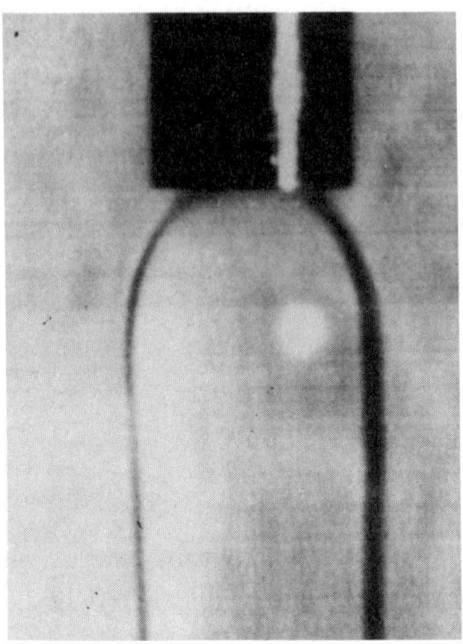

Figure 13.11. Extrusion of PIB/D, 6% at criticality: $[d,\bar{u}(0)/c,M(L)] = [3.175\ \text{mm},0.72,0.20]$.

Figure 13.12. Extrusion of PIB/D, 6% under post-critical conditions: $[d,\bar{u}(0)/c,M(L)] = [3.175\ \text{mm},1.02,0.247]$.

Figure 13.13. Extrusion of PIB/D, 6% under post-critical conditions: $[d, \bar{u}(0)/c, M(L)] = [3.175\text{ mm}, 1.28, 0.327]$.

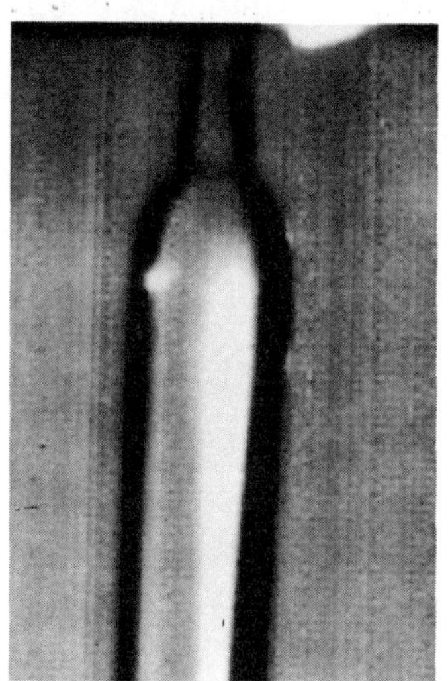

Figure 13.14. Extrusion of PIB/D, 6% under post-critical conditions: $[d, \bar{u}(0)/c, M(L)] = [1\text{ mm}, 1.29, 0.114]$.

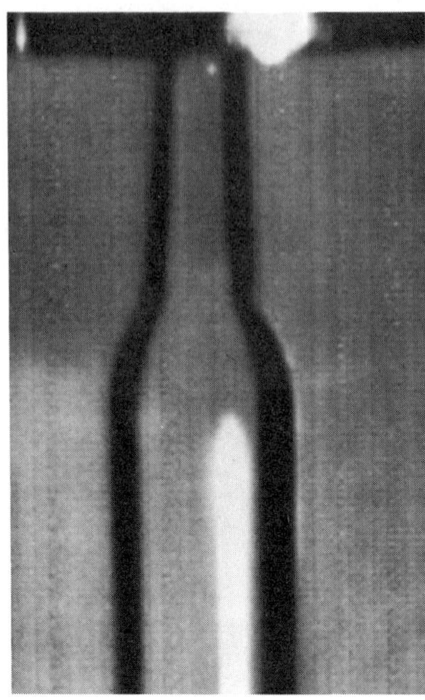

Figure 13.15. Steady response in extrusion of PIB/.D, 6% under post-critical conditions: $[d, \bar{u}(0)/c, M(L)] = [2.08\ \text{mm}, 1.49, 0.240]$.

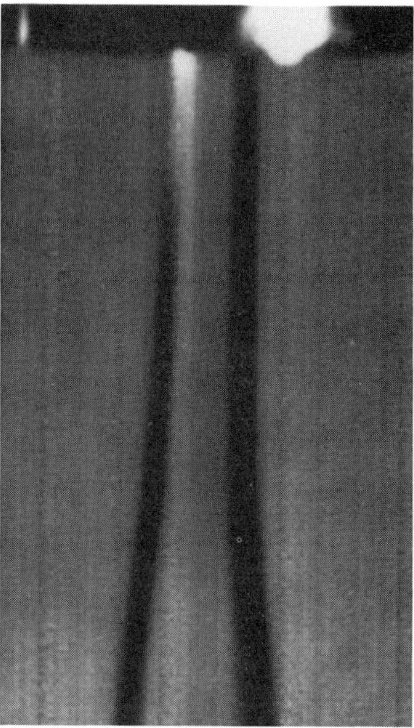

Figure 13.16. Steady response in extrusion of PIB/.D, 6% under post-critical conditions: $[d,\bar{u}(0)/c,M(L)]$ =[2.08 mm,1.79,0.442].

The case of steady flow at very high rates of extrusion in our experiments are characterized by smoothing the shock layer. In the more extreme cases the degree of smoothing is so great that the idea of a delay distance ℓ is not useful. Examples of such smoothing can be seen in photographs shown in §13.8. Figures 13.16,17,18 represent extreme cases of high speed extrusion without shock layers. The viscoelastic Mach number M(L) is greater than one in all of these cases. The type of complete smoothing which is exhibited in these Figures may be general for high speed extrusion in the steady case.

The empty rows at the bottom of the columns in Tables 13.5–8 signal difficulties in defining and measuring the missing entry.

It is of definite interest to consider delayed die swell when the extrusion rates are high. The argument given at the end of §13.5 shows how high speed flow might be used to study hyperbolicity and change of type for delayed die swell. Unfortunately our extruding devices have a restricted speed capacity which set an upper limit on what we could study.

13.8 Post-critical dependence of the swell ratio on the shear rate and Reynolds number in fluids with a small mean time of relaxation

The four liquids with the shortest mean time of relaxation listed in Table 13.2 were studied in the Aberdeen laboratory of Dr. J. Matta before the Minnesota experiments were undertaken. These four liquids were the most Newtonian and the most stable of all those tested. The shape of the jets of these fluids is smoother and the flow is steadier than in more elastic fluids. The smooth shape may be the result of the action of an effective Newtonian viscosity associated with the rapid decay of the elastic response. The Aberdeen experiments were designed to determine the swell ratio as a function of the shear rate in the pipe and did not attempt to determine the critical condition for delay. Two of these fluids, Elvacite/9.8% and K-125/5% were also studied in Minnesota and critical conditions were determined.

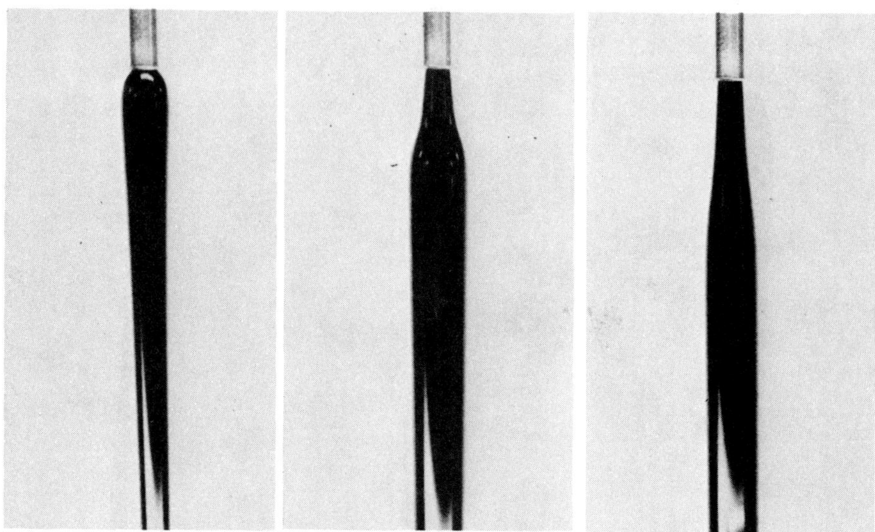

Figure 13.17 Steady response of Elvacite, 9.8% under post-critical conditions: [d,c] = [2.38 mm, 67 cm/s], [Q(cm^3/s), ϕ(1/s), $\bar{u}$(0)/c, D/d, M(L)] = (a) [6.244, 4418, 1.97, 1.99, 0.50], (b) [20.1, 14283, 4.27, 2.43, 1.08], (c) [24.2, 17194, 5.14, 2.13, 1.69]. The Mach number M(L)=$\bar{u}$(ℓ)/c at the maximum swell is inferior to one when the maximum is achieved abruptly. Mach numbers M(L) larger than one appear only after the shock has smoothed. This fluid has a short time of relaxation, but a large viscosity.

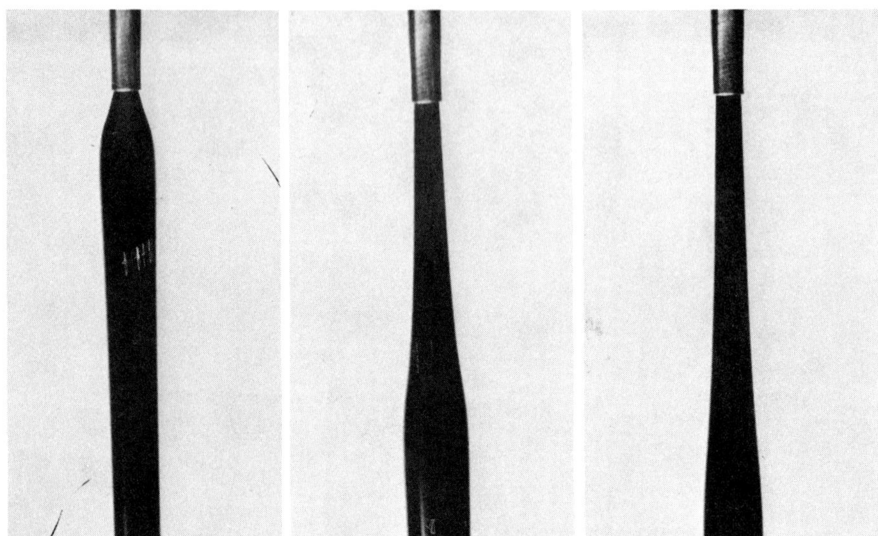

Figure 13.18 Steady response of K-125, 5% under post-critical conditions: [d,c] = [2.38 mm, 45.5 cm/s], [Q(cm^3/s), ϕ(1/s), $\bar{u}(0)/c$, D/d, M(L)] = (a) [8.13, 5780, 3.8, 2.3, 0.72], (b) [21.5, 15360, 10, 2.6, 1.5], (c) [23.8, 16930, 11.1, 2.41, 1.92]. The response of K-125, 5% is similar to Elvacite, 9.8%. The swell profile is "smooth" when M(L)>1.

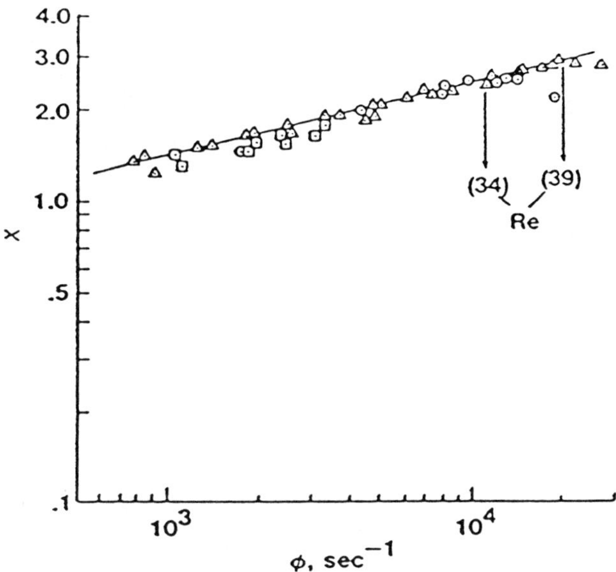

Figure 13.19 Swell ratio χ=D/d as a function of the apparent rate of shear $\phi=8\bar{u}(0)/d$ for Elvacite/9.8% at T=24±1°C, ⊙=2.39 mm, ▲=1.76 mm, and ⊡=1.25 mm. Figures 13.17(a), (b), and (c) correspond, respectively, to d=2.39 mm, ϕ=4418, ϕ=14823, and ϕ=17194.

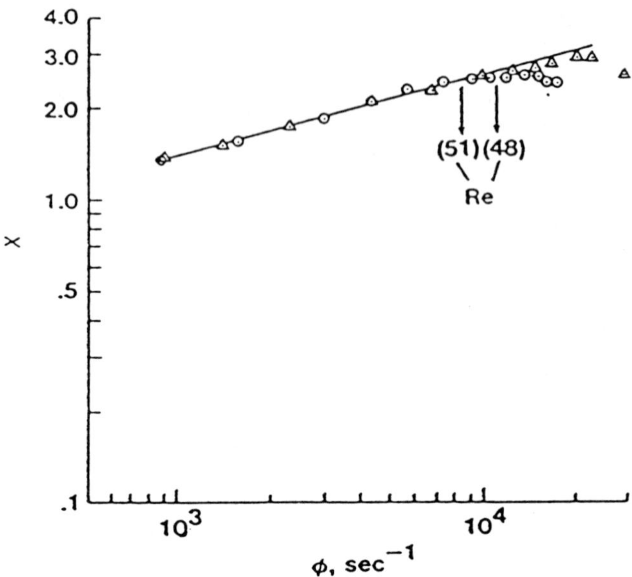

Figure 13.20 Swell ratio χ=D/d as a function of the apparent rate of shear $\phi=8\bar{u}(0)/d$ for K-125/5% as T=24±1°C, ʘ=2.39 mm, and ▲=1.86 mm. Figure 13.18(a), (b), and (c) correspond, respectively, to d=2.39, ϕ=5780, ϕ=15360, and ϕ=16930.

Some results of the Aberdeen experiments are summarized in Figures 13.17-20 where we have plotted the swell ratio D/d as a function of the shear rate ϕ with the nozzle diameter as a parameter. The straight lines in the Figures suggests that a power law expression exists between the swell ratio and the apparent shear rate $\phi=8\bar{u}(0)/d$ that is independent of nozzle diameter. This may be expressed as $D/d=a\phi^b$ where a and b are constants dependent on material parameters of the fluid. At high shear rates, however, D/d begins to level out and eventually decreases with a further increase in the shear rate. The point of onset of this deviation is nozzle diameter dependent and occurs at lower shear rates for larger diameters. The Reynolds number, $R=\rho\bar{u}(0)d/\eta(\phi)$, using the shear dependent viscosity $\eta(\phi)$ at the current rate of shearing, at which the deviation becomes obvious, was determined within limits, and indicated on each Figure. The viscosities were deduced using standard capillary procedures and are consistent with lower shear rate cone and plate values. For a given fluid, the onset of deviation occurs roughly at the same Reynolds number for different nozzle diameters.

The effects of gravity and inertia may be important in the eventual decrease of the swell ratio. It is well known that inertia will decrease the swell ratio. This effect is more important at high than at low speeds. Gravity alone would eventually accelerate the jet to speeds greater than c and beyond with a constant decrease of the jet radius.

13.9 Conclusions

When a viscoelastic fluid is extruded from a pipe the diameter of the extruded jet usually increases. At low rates of extrusion the swell appears to start at the pipe exit. At high rates of extrusion the tendency of the jet to swell at the exit is suppressed and there is a delay in the swell. We found that:

1. Delayed die swell appears to be a general phenomenon. It occurred in each and every polymeric solution of the seventeen tested by us.

2. Delayed die swell is a critical phenomenon. There is a definite critical extrusion velocity for the delay. When the extrusion velocity is greater than critical, there is a delay, a finite stand off before the major jet expansion.

3. Delayed die swell occurred as steady or unsteady response to steady extrusion, depending on the fluid. The response was unsteady in fluids with large mean times of relaxation and was steady for fluids with short mean times of relaxation.

4. The critical extrusion velocity $\bar{u}_c$, the swell ratio D/d, and the terminal swell distance L/d were decreasing functions of d, the pipe diameter.

5. The critical Mach number at exit, $M_c=U_c/c$, $U_c=2\bar{u}_c$, where c is the speed of vorticity waves into rest, was nearly always

larger than one. (The value M_c=0.954 for 1.3% CMC in a large pipe was the only exception.)

6. There seems to be a limiting value of M_c for large values of d and the limit of M_c might be one.

7. Whenever the die swell is delayed there are points r on the exit plane where u(0,r)=c.

8. The critical terminal Mach number, after the swell M(L) $=\bar{u}(L)/c$, is definitely less than one. The post-critical case is defined by $\bar{u}(0)>\bar{u}_c$, $\ell>0$.

The following facts hold in the post-critical case:

9. $\ell>0$ increases with $\bar{u}(0)$.

10. $M_0=2\bar{u}(0)/c>1$.

11. $M(L)=\bar{u}(L)/c$ is an increasing function of $\bar{u}(0)$.

12. M(L)<1 for all shock layer functions.

13. M(L), D/d, $(L-\ell)/d$ decrease with d for each fixed ℓ. The strength of the shock is greater in pipes of small diameter.

14. The response of fluids with moderate to long times of relaxation, $\lambda\geq 0.01$, became unsteady, perhaps time periodic, at some post-critical value of velocity less than terminal.

The following conclusions hold for large post-critical values of the extrusion velocity (the data taken here were restricted by limitations of the capacity of the extruder):

15. The delayed die swell terminated in a large amplitude oscillation or by smoothing.

16. Whenever the termination was stable, by smoothing, the shock layer of radius change was drastically smoothed and large values M(L) could be achieved for downstream.

Conclusion (12) did not apply when the shock layer was suppressed.

17. The smoothing occurred only in the four solutions with the smallest mean time of relaxation.

13.10 A numerical simulation of the nonlinear problem (this section is a last-minute addition)

Delvaux and Crochet [1989] have done a numerical study of delayed die swell in a plane jet using an Oldroyd B model, the upper convected Maxwell model plus a very small perturbing Newtonian viscosity μ such that $\mu/(\mu+\eta)=0.05$ (see Exercise 13.1). The results of their calculation are very interesting. They confirm the conjectures of Joseph, Matta and Chen [1987] which have been expounded, and introduce some new understandings. The main new result can be described as "the breakout of the region of the hyperbolic vorticity." At small supercritical values of the velocity (Mach numbers not too greatly in excess of one) the hyperbolic region extends slightly downstream into the jet but does not touch the jet boundary, as can be seen in panel (a) and (b) of Figure 13.21.

As the velocity increases, M, R, and W all increase and more and more of the jet is consumed by the hyperbolic region. At a certain velocity, between panel (b) and (c) of Figure 13.21, the hyperbolic region first touches the jet boundary, then consumes more and more of the jet boundary. Evidently the change in the curvature of the jet is associated with the breakout of the hyperbolic region. This explains why the delay is not observed at small supercritical values of the velocity ($M>1$) but only at larger postcritical breakout values. It would be good if we could find a way to explain the way the delay depends on the jet diameter.

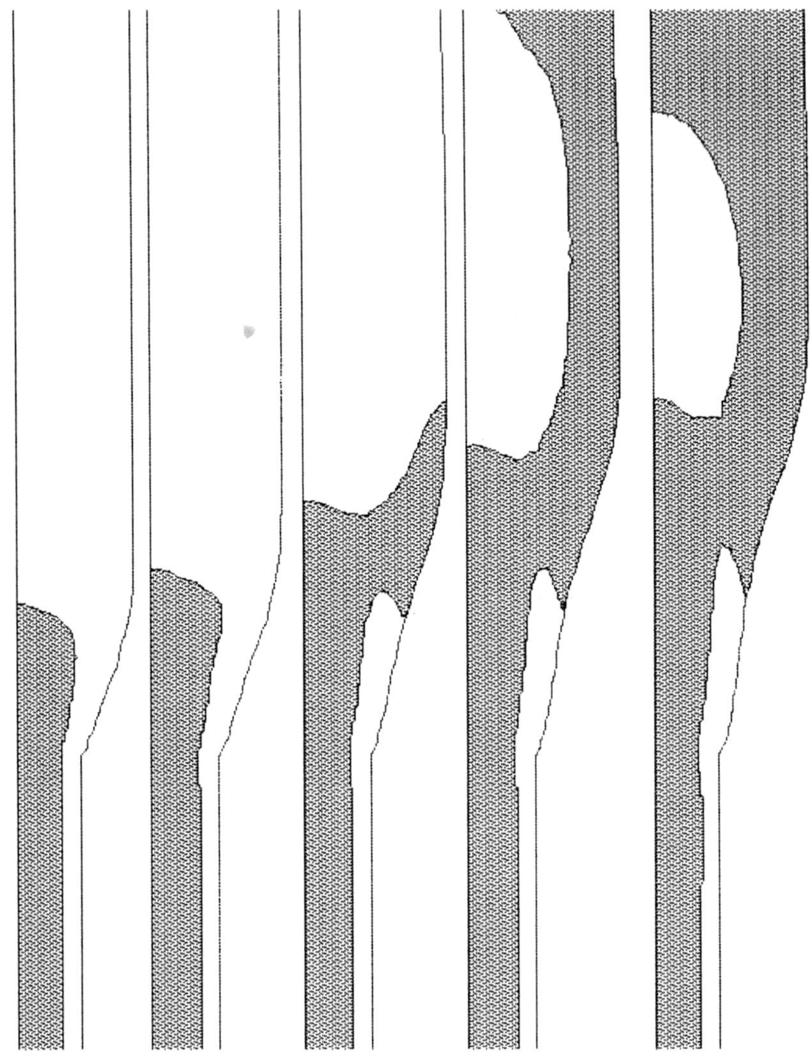

Figure 13.21. (after Delvaux and Crochet, 1989). Jet profile and hyperbolic regions of vorticity under different conditions in a plane jet of an Oldroyd B fluid with a very small Newtonian viscosity. (a) (M, R,W)=(2.3, 13.5, 0.39); (b) (2.9, 17, 0.49); (c) (4.1, 29.9, 0.87), (d) (5.1, 34.6, 1)

Exercise 13.1

Consider the decomposition used in Oldroyd B $\boldsymbol{\tau}=\boldsymbol{\tau}_N+\boldsymbol{\tau}_E$ into a Newtonian part $\boldsymbol{\tau}_N=2\mu\mathbf{D}(\mathbf{u})$ and elastic part as in Exercise 2.1. Show that $\boldsymbol{\tau}_E+\rho\mathbf{uu}$ is a positive tensor when $\sqrt{B^2-AC}>0$ where

$B^2\text{–}AC$ is defined by (6.2) and that $\boldsymbol{\tau}_E+\rho\mathbf{uu}$ lose positivity when $B^2\text{–}AC$ changes sign. If μ is small, we may therefore expect a perturbed change of type when $\boldsymbol{\tau}_E+\rho\mathbf{uu}$ changes sign. (You may write a little mathematical paper providing the mathematical details for this categorical statement.)

CHAPTER 14

Hyperbolicity and Change of Type in the Flow Between Rotating Cylinders When the Inner Cylinder Is Corrugated*

We may think of this problem as the flow in a gear box. We get this flow by perturbing Couette flow between two concentric rotating cylinders, putting a wave on the inner cylinder. As in the previous examples, the vorticity equation changes type from elliptic to hyperbolic whenever the velocity at some point in the domain exceeds the speed of shear waves into rest. There are two cases: the inner cylinder rotates and the outer cylinder rotates. The isovorticity lines even in weakly subcritical flow lie in directions swept out by characteristics. This direction is uniquely determined in the regions where the two families lie in the same direction. In the strongly supercritical case, high Mach numbers, one finds that hyperbolicity dominates the flow field. The isovorticity lines are oriented along forward facing characteristics and the vorticity of the perturbed motion is then swept out along these characteristics.

The mean radius of the inner cylinder is r_0. The outer cylinder is of radius r_2.

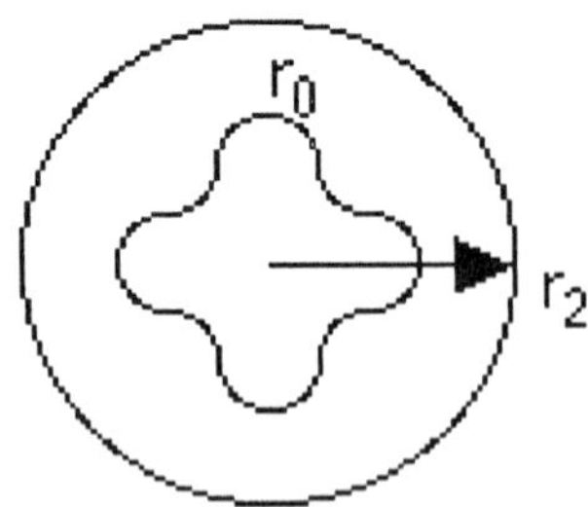

Figure 14.1. The mean radius of the inner cylinder is r_0.

---

* This chapter is based on the thesis of Riccius [1989b].

We use r_0 as a length scale, $r_0\Omega$ as a velocity scale where Ω is the angular speed of the rotating member, and $\eta\Omega$ as a stress scale. The parameters of the problem are

$$R = \frac{\rho r_0^2 \Omega}{\eta}, \qquad \text{(Reynolds number)},$$

$$W = \lambda\Omega, \qquad \text{(Weissenberg number)},$$

$$E = \frac{\lambda\eta}{\rho_0^2}, \qquad \text{(elasticity number)},$$

$$M^2 = \frac{\rho\lambda r_0^2 \Omega^2}{\eta}, \qquad \text{(Mach number)},$$

We choose M and W as parameters.

The resulting set of equations is a quasilinear first order system in r and θ. The inner cylinder is corrugated

$$r_0 = 1 + \varepsilon \sin n\theta, \tag{1}$$

where n is a wave number and ε an amplitude, supposed small. The coordinates in the perturbed (r) and unperturbed (r^*) domain are related through

$$r = r^*\left(1 + \varepsilon \frac{r_2 - r^*}{r_2 - 1} \sin(n\theta)\right), \quad \theta = \theta^*. \tag{2}$$

The expansion for the field variables, $\mathbf{q}=(u,v,\sigma,\tau,\gamma,p)$, follows along the lines of Chapter 12 and Appendix E. The zeroth order solution for the case when the inner cylinder rotates is

$$u_0 = 0,$$

$$v_0 = \left(\frac{1}{r} - \frac{r}{r_2^2}\right)\frac{1}{a},$$

$$\sigma_0 = 0,$$

$$\tau_0 = -\frac{2}{ar^2}, \tag{3}$$

$$\gamma_0 = 2\tau_0^2 ,$$

$$p_0 = \left\{\frac{R}{r_2^2}\left(\frac{r^2}{2r_2^2} - 2\ln r - \frac{r_2^2}{2r^2}\right) + \frac{2}{r^4}\right\}\frac{1}{a^2} ,$$

where $a = 1 - \frac{1}{r_2^2}$. When the outer cylinder rotates v_0, τ_0, and p_0 are given by

$$v_0 = \left(r - \frac{1}{r}\right)\frac{1}{a} ,$$

$$\tau_0 = \frac{2}{ar^2} , \tag{4}$$

$$p_0 = \left\{R\left(\frac{r^2}{2} - 2\ln r - \frac{1}{2r^2}\right) + \frac{2}{r^4}\right\}\frac{1}{a^2} .$$

In order to determine the characteristics we bring the vorticity equation again into the form

$$A\frac{\partial^2\zeta}{\partial r^2} + 2B\frac{\partial^2\zeta}{r\partial r\partial\theta} + C\frac{\partial^2\zeta}{r^2\partial\theta^2} = \text{lot} , \tag{5}$$

where "lot" means terms of lower order derivatives in ζ and with $A = M^2u^2 - (W\sigma + 1)$, $B = M^2uv - W\tau$ and $C = M^2v^2 - (W\gamma + 1)$. In the linearized case it is evident from (5) that the vorticity of the perturbed flow is associated with the zeroth-order solution alone such that the characteristics are given through

$$\frac{d\theta}{dr} = \frac{B}{A} \pm \frac{\sqrt{B^2 - AC}}{A} = \frac{\tau_0}{r} \pm \frac{1}{r}\sqrt{M^2v_0^2 + W(W-2)\tau_0^2 - 1} . \tag{6}$$

In order to calculate the characteristics from (6) one has to know the critical radii, r_c, which separate elliptic and hyperbolic regimes. The critical value r_c are roots of $P(r^2) = M^2v_0^2 + W(W-2)\tau_0^2 - 1 = 0$, a cubic polynomial in r^2. There can be three real roots and at most two fall in the gap. Let us consider $M^2 > 1$ and define $\beta = 4W^3(W-2)/a^2$ where $a = 1 - 1/r_2^2$.

When the inner cylinder rotates we have exactly one root inside the gap whenever

$$M^2 > 1-\beta \quad \text{and}\ \beta < r_2^4 . \tag{7}$$

On the other hand, if $M^2<1-\beta$ and $M>1$, then $W<2$ and if $P>0$, then we have $W<2$ for $M^2>1$, and if also $P>0$ in $(1,r_2)$, then there are two roots r_c which bound a hyperbolic region inside the gap.

When the outer cylinder rotates there is exactly one root whenever $\beta<1$ and

$$M^2 > \frac{1}{r_2^2}\left(1-\frac{\beta}{r_2^4}\right). \tag{8}$$

If (8) holds and $\beta>1$, $M^2>1$ and $P<0$ in $(1,r_2)$ then there are two roots which bound an elliptic regime inside the gap. When there is only one root in $(1,r_2)$ it is given by

$$r_{x1} = \sqrt{h\left(\cos\frac{\phi}{3} - \sqrt{3}\sin\frac{\phi}{3}\right) - \frac{a_2}{3a_1}} ,$$

with

$$\phi = \operatorname{acos}\left(\frac{q}{h^3}\right),$$

$$q = \frac{a_2^3}{(3a_1)^3} - \frac{a_2}{6a_1^2} + \frac{a_4}{2a_1},$$

$$a_4 = 4W^2(W-2)/R,$$

$$(a_1, a_2) = \left(\frac{1}{r_2^4}\,,\ -\frac{2}{r_2^2} - \frac{a^2}{M^2}\right) \text{inner cylinder rotates,}$$

$$(a_1, a_2) = \left(1,\ -2 - \frac{a^2}{M}\right) \text{outer cylinder rotates,}$$

$$h = \operatorname{sign}(q)\sqrt{|p|} ,$$

$$p = \frac{1}{3a_1} - \frac{a_2^2}{(3a_1)^2} .$$

When there are two roots, one root is r_{x1} and the other is given by

$$r_x = \sqrt{h\left(\cos\frac{\phi}{3} + \sqrt{3}\sin\frac{\phi}{3}\right) - \frac{a_2}{3a_1}} , \qquad \text{inner cylinder rotates}$$

or

$$r_x = \sqrt{-2h\cos\frac{\phi}{3} - \frac{a_2}{3a_1}} . \qquad \text{outer cylinder rotates}$$

For the first order solution the boundary conditions are as follows. When the inner cylinder rotates, we have

$$\begin{aligned} u(r_2,\theta) &= v(r_2,\theta) = 0 , \\ u(1,\theta) = 0, &\quad v(1,\theta) = \sin n\theta. \end{aligned} \tag{9}$$

When the outer cylinder rotates we have

$$\begin{aligned} u(r_2,\theta) &= v(r_2,\theta) = 0 , \\ u(1,\theta) &= v(1,\theta) = 0. \end{aligned} \tag{10}$$

Some results for the first order problem are given in the Figures where the gap is mapped into a rectangle. The horizontal axis represents $\theta \in (0,\pi/2)$ and the vertical axis the radius between $r_0=1$ and $r_2=2$. The direction of flow is from left to right. The plots show characteristic nets and the iso-curves for the perturbed vorticity. The coordinates are those of the unperturbed domain. The wave number for the perturbation is n=5 throughout. Different outer radii, r_2, as well as different wave numbers, n, do not change the characteristic features in the graphs.The length scale is the same for both boundary conditions. Thus the flow speed at r_2 is twice that at r_0 for identical parameters.

Let us look at the case when the inner cylinder is rotating. For M=1.4 and W=2 (Figure 14.2) the two families of characteristics lie in the same backward direction. The isovorticity contours are also in this direction even in the subcritical case shown in Figures 14.3

and 14.5. The Mach numbers in Figures 14.4 and 14.6 are much greater than unity and we find characteristic nets with forward and backward facing characteristics. The alignment of the vorticity lines has changed here. The orientation at the moving wall and through large parts of the hyperbolic regime follows forward facing characteristics. The plot "Vorticity vs. r at $\pi/4$" for M=14.1 and W=2 illustrates the oscillations of the vorticity in the hyperbolic region. The damping is evidently associated with the elliptic region.

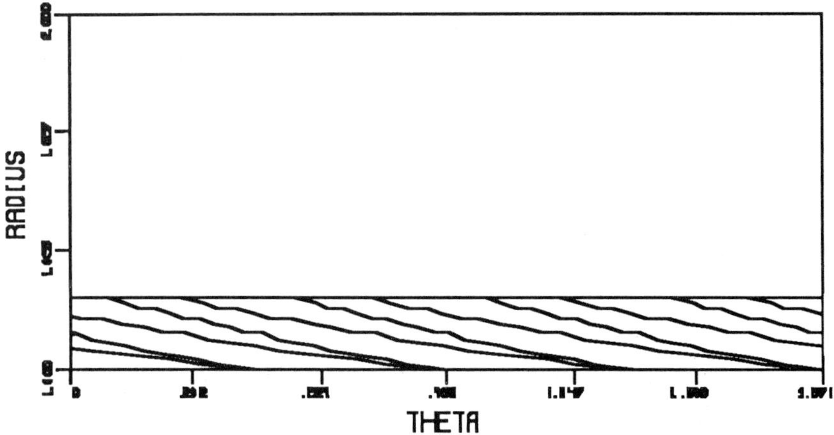

Figure 14.2. Characteristics. Inner cylinder rotates. R=1, W=2, M=1.4, r_x=1.203. Both sets of characteristics face backwards.

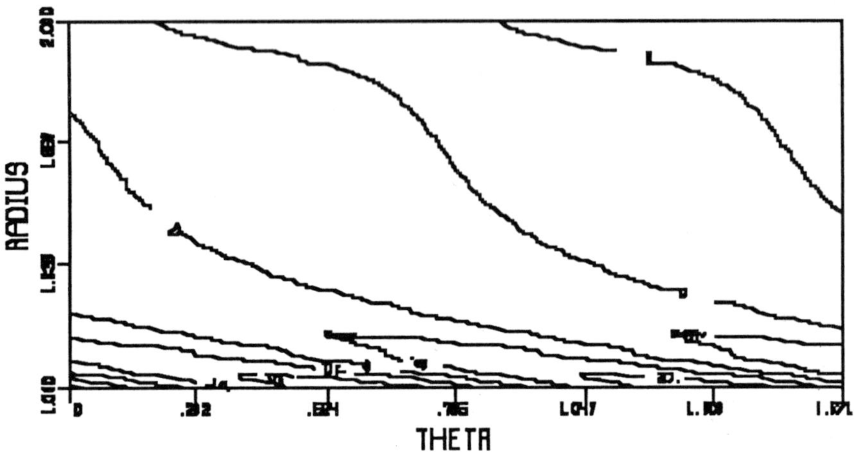

Figure 14.3: Isovorticity lines for a subcritical case. Inner cylinder rotates. R=0.32, W=2, M=0.8.

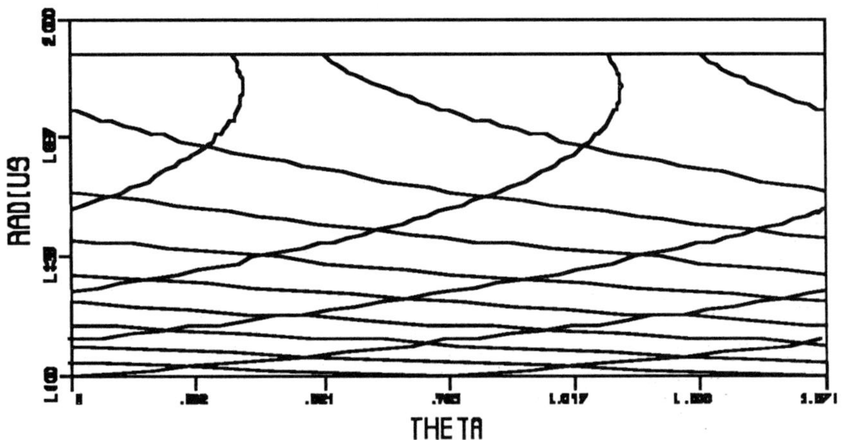

Figure 14.4: Characteristics. Inner cylinder rotates. R=100, W=2, M=14.1, r_x=1.027.

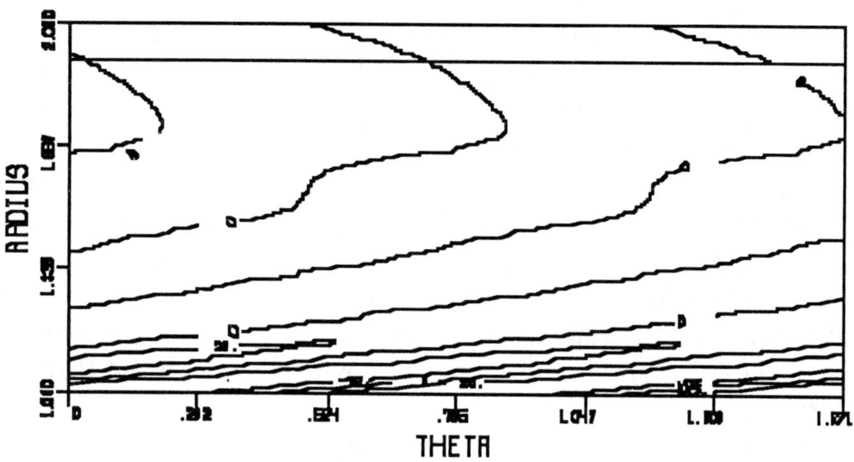

Figure 14.5: Isovorticity lines. Inner cylinder rotates. R=100, W=2, M=14.1, r_x=1.897.

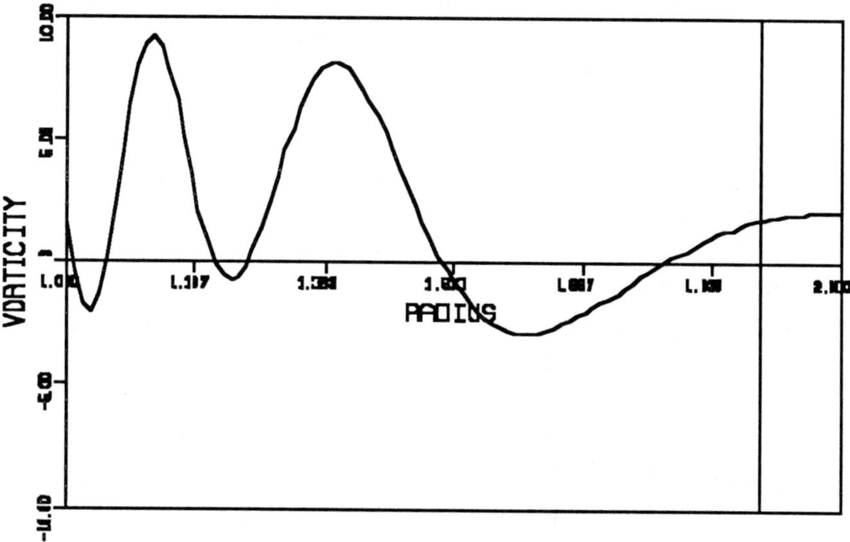

Figure 14.6: Vorticity versus radius for q=π/4. Inner cylinder rotates. R=100, W=2, M=14.1, r_x=1.897.

When the outer cylinder rotates it is again the inner wall which causes the perturbation. The characteristic net for M=1.4 and W=2 shown in Figure 14.5 can be compared with the isovorticity lines in the subcritical case M=0.8 and W=2 shown in Figure 14.8. The isovorticity lines for M=1.4 and W=2 are nearly the same We note that the zero vorticity lines lie along the characteristics in the region near the center where the slopes are in the same direction. In the high Mach number case shown in Figure 14.9 and 14.10 we see strong alignment along the forward facing characteristics. The streamline plot for M=14.1 and W=2 shows the typical waves due to the radial oscillations of the vorticity. In general, damping is large when W is small and vice versa, whether or not the flow is hyperbolic.

Hyperbolicity has a particularly strong effect on the flow of an upper convected Maxwell fluid in a Couette cell.

We summarize the results:

1. The fluid can be partitioned into elliptic and hyperbolic regime depending on the choice of M and W. The hyperbolic part is dominated through elastic effects like wave propagation along characteristics and undamped oscillations while one finds diffusive properties like smoothing of discontinuities and damping of oscillations in the elliptic part.

2. The perturbed vorticity is swept out along forward facing characteristics through large parts of the hyperbolic regime at Mach numbers much greater than unity.

3. The isovorticity lines go through a change of orientation with increasing Mach number until they are aligned along forward facing characteristics.

4. The vorticity shows strong radial oscillations in the hyperbolic regime when M»1 and W is large. They are damped when the elliptic part is not small and dominate the whole flow field otherwise. For modest and even large M the oscillations are damped if W is small.

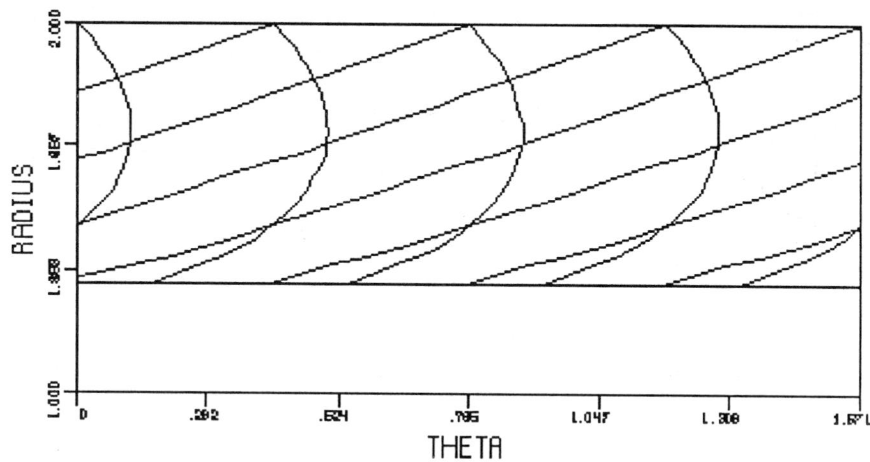

Figure 14.7: Characteristics. Outer cylinder rotates. R=1, W=2, M=1.4, r_X=1.300.

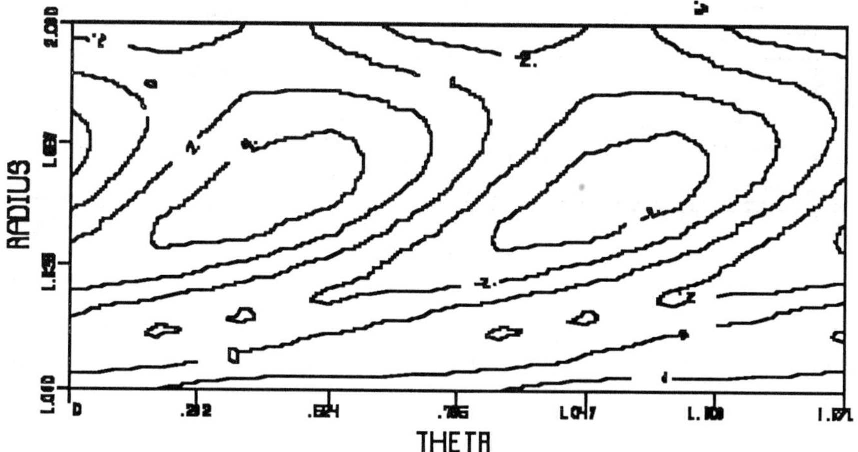

Figure 14.8: Isovorticity lines for a subcritical case. The outer cylinder rotates. R=0.32, W=2, M=0.8.

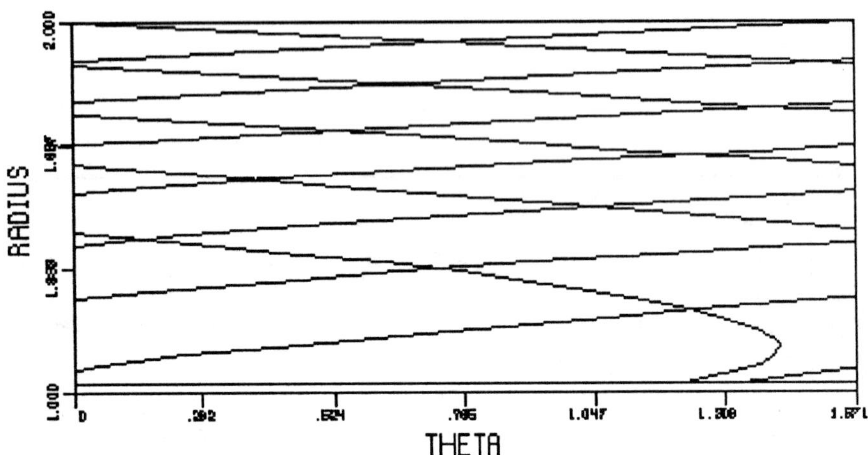

Figure 14.9: Characteristics. Outer cylinder rotates. R=100, W=2, M=14.1, r_x=1.027.

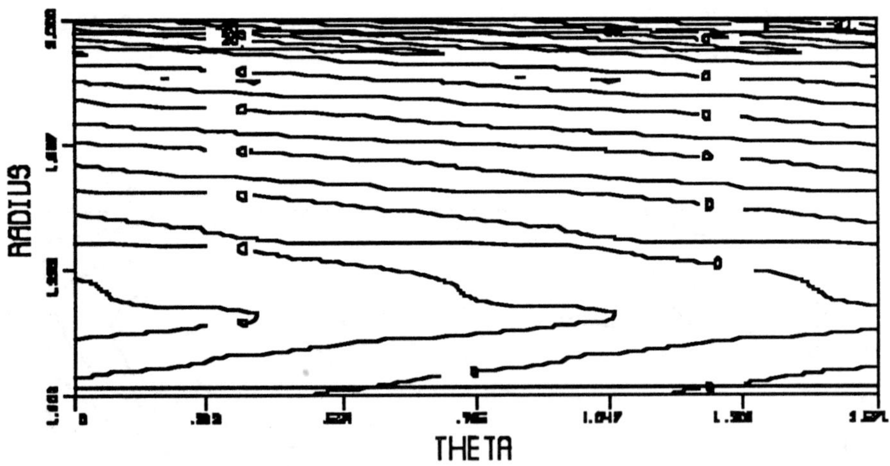

Figure 14.10: Isovorticity lines. The outer cylinder rotates. R=100, W=2, M=14.1, r_x=1.027.

CHAPTER 15

Simple Fluids and Fading Memory

Incompressible simple fluids are fluids which depend only on the history of the first spatial gradient of the deformation, on $\nabla\boldsymbol{\chi}_t(\mathbf{x},\tau)=\mathbf{F}_t(\tau)$. Nearly all of the constitutive equations used for amorphous liquids are of this type. On the face of it, this is a very restrictive assumption. Certainly, we would not want to restrict attention to fluids whose histories are represented only by first derivatives with respect to time. Fluids whose constitutive equation depends on second or higher gradients, or the objects to which series of such higher gradients might approximate, are not described as simple fluids, but they could be important.

We are now going to follow Noll's [1958] method about how one can use invariance under a change of frame to get some reduction in the form of the constitutive equation. This comes down to a guarantee that a functional on the history of $\mathbf{F}_t(\tau)$ can be replaced, without loss of generality, by another functional of, say, the Cauchy-Green tensor $\mathbf{C}_t(\tau)=\mathbf{F}_t^T(\tau)\mathbf{F}_t(\tau)$ or many other symmetric tensors functionally related to this one, like $\mathbf{C}_t^n(\tau)$, integer n, or the Finger tensor, $\mathbf{C}_t^{-1}(\tau)$. The second ingredient in the theory of simple fluids is a theory of fading memory. The physical idea of fading memory is that more recent deformations of a material have a greater effect on the present value of the stress. The mathematical theory of fading memory is about how to study the dependence of constitutive equations on the history of deformation, using functional analysis. This comes down to a statement about the domain of the functional giving the stress. It is necessary to have one or another of such theories, because the notion of a functional is meaningless without a specification of the domain on which it is defined.

The type of theory generally understood by "simple fluids with fading memory" is good for persons with a mathematical turn of mind. It is an excellent framework for asymptotic theories because one needs at least some vague notion of what is approximated by approximate theories. The theory is rather too general for direct application, and it is in no way general enough to take in all we might wish to consider.

15.1 Noll's representation

It cannot be established that the stress $\mathbf{T}$ is indifferent without some further assumption. The assumption made by Noll is that the traction vector, expressed as hypothese H.1 below, is indifferent. The theory of *simple materials* rests on the following three assumptions.

H.1 The traction vector is indifferent $\mathbf{t}^*=\mathbf{Qt}$.

H.2 The stress at a particle depends on the history of first spatial gradients $\mathbf{F}(\tau)$ (or $\mathbf{F}_t(\tau) = \mathbf{F}(\tau)\mathbf{F}^{-1}(t)$).

H.3 The constitutive equation is form invariant (indifferent).

H.1 and H.3 say that observers in two different frames will compute the same stress from the constitutive equation if the deformation which each measures in his own frame is the same.
H.1 implies that the stress is indifferent.

$$\mathbf{t}^* = \mathbf{Qt} = \mathbf{QT^Tn}\ .$$

Since $\mathbf{n}$ is a point difference, hence indifferent, we have

$$\mathbf{t}^* = \mathbf{T}^{*\mathrm{T}}\mathbf{n}^* = \mathbf{Q}^{\mathrm{T}}\mathbf{T}^{\mathrm{T}}\mathbf{Q}\mathbf{n}^*\ .$$

Hence,

$$\mathbf{T}^{*\mathrm{T}} = \mathbf{Q}^{\mathrm{T}}\mathbf{T}^{\mathrm{T}}\mathbf{Q}\ , \quad \mathbf{T}^* = \mathbf{QTQ}^{\mathrm{T}}\ .$$

H.4 (Fluids) $\mathbf{x} = \boldsymbol{\chi}_t(\mathbf{x},t)$ is given as a reference particle and kinematics in terms of the history of $\mathbf{F}_t(t)$.

Noll's result is that the stress at a particle $\mathbf{x}$ is given by a symmetric tensor-valued operator whose components relative to any basis are functionals defined on the history of the right relative Cauchy-Green tensor

$$\mathbf{T}(\mathbf{x},t) = \underset{\tau=-\infty}{\overset{t}{\mathbf{T}}}[\mathbf{C}_t(\mathbf{x},\tau)] \; . \tag{1}$$

Moreover, $\mathbf{T}[\;\;]$ is an isotropic functional of $\mathbf{C}_t(\tau)$

$$\mathbf{QTQ}^T = \mathbf{T}[\mathbf{QC}_t(\tau)\mathbf{Q}^T] \tag{2}$$

for all orthogonal tensors $\mathbf{Q}$.

Proof: For simple writing, I will omit the range of τ. First, I will derive a functional equation

$$\left.\begin{array}{l} \mathbf{T}^* = \mathbf{T}^*[\mathbf{F}_t^*(\tau)] \\ \\ \mathbf{T} = \mathbf{T}[\mathbf{F}_t(\tau)] \end{array}\right| \quad \text{(H.2) and (H.4)} \; .$$

$$\mathbf{T}^*[\;\;] \underset{\text{(H.3)}}{=} \mathbf{T}[\;\;] \; .$$

$$\mathbf{T}^* = \mathbf{QTQ}^T \, , \quad \mathbf{Q} = \mathbf{Q}(t) \; .$$

According to (1.85),

$$\mathbf{F}^*(\tau) = \mathbf{Q}(\tau)\mathbf{F}_t(\tau)\mathbf{Q}^T(t) \, ,$$

hence,

$$\begin{aligned} \mathbf{T}^* = \mathbf{QTQ}^T = \mathbf{QT}[\mathbf{F}_t(\tau)]\mathbf{Q}^T = \mathbf{T}^*[\mathbf{F}^*(\tau)] \\ = \mathbf{T}[\mathbf{Q}(\tau)\mathbf{F}_t(\tau)\mathbf{Q}^T(t)] \; . \end{aligned}$$

The functional equation which we have to solve is

$$\mathbf{T}[\mathbf{Q}(\tau)\mathbf{F}_t(\tau)\mathbf{Q}^T(t)] = \mathbf{Q}(t)\mathbf{T}[\mathbf{F}_t(\tau)]\,\mathbf{Q}^T(t) \; . \tag{3}$$

We find that

$$\mathbf{T}[\mathbf{F}_t(\tau)] = \mathbf{T}[\mathbf{U}_t(\tau)] \tag{4}$$

where $\mathbf{U}_t(\tau)$ is the positive-definite, symmetric tensor which arises after applying the polar decomposition (Appendix A, §12) to $\mathbf{F}_t(\tau)$

$$\mathbf{F}_t(\tau) = \mathbf{Q}(\tau)\mathbf{U}_t(\tau) \tag{5}$$

where, for a given $\mathbf{F}_t(\tau)$ there is a *unique* $\mathbf{Q}(\tau)$ and a *unique* stretch $\mathbf{U}_t(\tau)$.

Equation (4) says that no matter what $\mathbf{T}[\quad]$ may be, if it is indifferent, it does not depend on the rotation $\mathbf{Q}(\tau)$, but only on the stretch. (A simple example of this is given in Exercise 15.1.) Since $\mathbf{F}_t(t)=\mathbf{1}$, we have $\mathbf{Q}(t)=\mathbf{1}$, by uniqueness, so that $\mathbf{U}_t(t)=\mathbf{1}$. The $\mathbf{Q}(\tau)$ in (5) must be corotational.

To prove (4), consider the corotational $\mathbf{Q}(\tau)=\mathbf{R}^T(\tau)$ which appears in the polar decomposition

$$\begin{gathered}\mathbf{F}_t(\tau) = \mathbf{R}^T(\tau)\mathbf{U}_t(\tau)\ , \\ \mathbf{F}_t(t) = \mathbf{R}^T(t) = \mathbf{U}_t(t) = \mathbf{1}\ .\end{gathered} \tag{6}$$

Now we put $\mathbf{Q}(\tau)=\mathbf{R}^T(\tau)$ and (6) into (3) and find that

$$\mathbf{T}[\mathbf{U}_t(\tau)] = \mathbf{T}[\mathbf{F}_t(\tau)]\ .$$

Since $\mathbf{C}_t(\tau) = \mathbf{F}_t^T(\tau)\mathbf{F}_t(\tau)$, we may define a new functional

$$\mathop{\mathbf{T}}_{\tau=-\infty}^{t}[\mathbf{F}_t(\tau)] = \mathop{\mathbf{S}}_{\tau=-\infty}^{t}[\mathbf{C}_t(\tau)]\ .$$

A simple fluid is automatically isotropic. Since $\mathbf{T}$ and $\mathbf{C}_t(\tau)$ are indifferent,

$$\begin{gathered}\mathbf{T}^* = \mathbf{QTQ}^T = \mathbf{S}[\mathbf{C}_t^*(\tau)]\ , \\ \mathbf{C}_t^*(\tau) = \mathbf{QC}_t\mathbf{Q}^T\ .\end{gathered}$$

Hence, for all $\mathbf{Q}$,

$$\mathbf{QTQ}^T = \mathbf{S}[\mathbf{QC}_t(\tau)\mathbf{Q}^T]\ . \tag{7}$$

In fact, for incompressible fluids, we should think of $\mathbf{S}$ as the extra stress. For compressible fluids,

$$\mathbf{S}[\rho(t), \mathop{\mathbf{C}_t(\tau)}_{\tau=-\infty}^{t}](\tau) \text{ depends on the density.}$$

Isotropic simple solids

$$\mathbf{T}[\mathbf{x}(\mathbf{X},t),t] = \mathop{\mathbf{S}}_{\tau=-\infty}^{t}[\mathbf{B}(t),\mathbf{C}_t(\tau)] \,,$$

$$\mathbf{Q}\mathbf{S}[\mathbf{B}(t),\mathbf{C}_t(\tau)]\mathbf{Q}^T = \mathbf{S}[\mathbf{Q}\mathbf{B}\mathbf{Q}^T,\mathbf{Q}\mathbf{C}_t(\tau)\mathbf{Q}^T] \tag{8}$$

for all orthogonal tensors $\mathbf{Q}$, where

$\mathbf{B}=\mathbf{F}\mathbf{F}^T$ is defined relative to $\mathbf{X}$, $\mathbf{F} = \dfrac{\partial \boldsymbol{\xi}}{\partial \mathbf{X}}$,

$\mathbf{C}_t(\tau)=\mathbf{F}_t^T\mathbf{F}_t$ is defined relative to $\mathbf{x}=\boldsymbol{\xi}(\mathbf{X},t)$.

Static deformations: $\boldsymbol{\xi}(\mathbf{X},\tau)$ is independent of τ and

$$\boldsymbol{\xi}(\mathbf{X},\tau) = \mathbf{x} \quad \text{for all } \tau \le t \,.$$

Hence,

$$\boldsymbol{\chi}_t(\mathbf{x},\tau) = \mathbf{x} \quad \text{for all } \tau \le t$$

in static deformations and

$$\mathbf{F}_t(\tau) = \mathbf{1} = \mathbf{C}_t(\tau) \,, \quad \tau \le t \,.$$

Then, in static deformations

$$\mathbf{T} = \mathbf{S}[\mathbf{B};\mathbf{1}] \,. \tag{9}$$

In the undeformed state, $\boldsymbol{\xi}(\mathbf{X},\tau)=\mathbf{x}=\mathbf{X}$ and $\mathbf{B}=\mathbf{1}$. Then (8) implies that

$$\mathbf{Q}\mathbf{S}\mathbf{Q}^T = \mathbf{S}[\mathbf{1},\mathbf{1}] = \alpha\mathbf{1}$$

so that the stress in the undeformed state is isotropic.

Elastic bodies: An elastic body satisfies (3) with an isotropic $\mathbf{S}$. The undeformed state is called the natural state of the body. Elasticity is the special case of viscoelasticity which arises under static deformations.

15.2 Fading memory

What I have in mind here is some sort of representation founded in functional analysis theory for the functional giving the extra stress as a functional of the history of the relative strain.

$$\boldsymbol{\tau} = \underset{0}{\overset{\infty}{\boldsymbol{F}}}[\mathbf{G}(s)] = \underset{\tau=-\infty}{\overset{t}{\mathbf{S}}}[\mathrm{C}_t(\tau)] \tag{10}$$

where $s=t-\tau$ is the lapse time, $\mathbf{G}=\mathrm{C}_t(\tau)-\mathbf{1}$ and arranged so that the stress of a body at rest vanishes

$$\boldsymbol{F}[\mathbf{0}] = 0\,. \tag{11}$$

The basic idea is to choose some topology for the history $\mathbf{G}(s)$ and then to carry out an analysis of all the linearized forms of stresses. The basic tool is Riesz's representation theorem.

To start this, let us review the topological choice of the domain of $\boldsymbol{F}$ which was used by Coleman and Noll [1960]. They let

$$\mathrm{dom}\,\boldsymbol{F} = L_h^2(0,\infty) \tag{12}$$

where $L_h^2(0,\infty)$ is a weighted Hilbert space.

$$<\mathbf{u},\mathbf{v}>_h \overset{\mathrm{def}}{=} \int_0^\infty h^2(s)\ \mathrm{trace}\ [\mathbf{u}(s)\mathbf{v}(s)]\ ds \tag{13}$$

h is smooth, $h(0)=1$, $h'(s)<0$, $h(s)\to 0$ as $s\to\infty$

$$\|\mathbf{u}\|^2_{L_h^2(0,\infty)} = <\mathbf{u},\mathbf{u}>_h\,. \tag{14}$$

The Riesz theorem says that each and every linear functional on $L_h^2(0,\infty)$ may be represented by an inner product with a fixed element, a 4th order tensor, call it $\mathbf{U}\in L_h^2(0,\infty)$. So every linear functional of $\mathbf{G}(s)\in L_h^2$ may be represented as

$$<\mathbf{U}(s),\mathbf{G}(s)>_h = \int_0^\infty h^2 \operatorname{tr}[\mathbf{U}(s)\mathbf{G}(s)]\, ds. \tag{15}$$

The first derivative of $\boldsymbol{F}\left[\mathbf{G}(s)\right]_{s=0}^{\infty}$ on $\mathbf{G}(s)=0$ is a tensor valued linear functional on L_h^2

$$\boldsymbol{F}_1[\mathbf{0}|\mathbf{G}(s)] \overset{\text{Riesz theorem}}{=} <\mathbf{U}(s),\mathbf{G}(s)>_h$$

$$(= \quad <U_{ijkl}G_{kl}> \text{ symmetric in ij}) , \tag{16}$$

$$\boldsymbol{F}_1[\mathbf{0}|\mathbf{G}(s)] = \int_0^\infty h^2 \operatorname{tr}[\mathbf{U}(s)\mathbf{G}(s)]\, ds . \tag{17}$$

Each component of $\boldsymbol{F}_1$ is a functional

$$\boldsymbol{F}_{1ij} = \int_0^\infty h^2(s)U_{ijkl}(s)G_{kl}(s)\, ds$$

$$\overset{\text{def}}{=} \int_0^\infty K_{ijkl}(s)G_{kl}(s)\, ds \tag{18}$$

Now we already know that the isotropy of $\boldsymbol{F}_1$ implies that

$$K_{ijkl} = Q_{ia}Q_{jb}Q_{kc}Q_{ld}K_{abcd} \tag{19}$$

is an isotropic tensor (see Appendix A, §5). Hence, we have

$$\boldsymbol{F}_1[0|\mathbf{G}(s)] = \int_0^\infty K(s)\mathbf{G}(s)\, ds \tag{20}$$

with $\dfrac{K(s)}{h^2(s)} \in L^2(0,\infty)$, from the Schwartz inequality.

Consider next a perturbation of an arbitrary motion.

$$\mathop{\boldsymbol{F}}_{s=0}^{\infty}[G(s)] = \boldsymbol{F}[\mathbf{G}_0(s) + \delta\mathbf{G}(s) + \ldots]$$
$$= \boldsymbol{F}_1[\mathbf{G}_0(s)|\delta\mathbf{G}] + 0_{\|\delta G\|^2 L_h^2} \,. \tag{21}$$

$\boldsymbol{F}$ is given by a scalar product with a fixed fourth order tensor $\mathbf{U} \in L_h^2$

$$\boldsymbol{F}_1[\mathbf{G}_0(s)|\delta\mathbf{G}] = \int_0^{\infty} h^2(s)\ \mathbf{U}(\mathbf{G}_0,s)\ \delta\mathbf{G}(s)\ ds\,. \tag{22}$$

The fixed $\mathbf{U}[\mathbf{G}_0(s),s]$ means there is one of these which will do it for all the $\delta\mathbf{G}$.

The symmetries of $\mathbf{K}[\mathbf{G}_0(s),s]$ are determined by the symmetry of $\mathbf{G}_0$.

$$\boldsymbol{F}_{ij}[\mathbf{G}_0|\delta\mathbf{G}] = \int_0^{\infty} K_{ijkl}(\mathbf{G}_0,s)\ \delta G_{kl}(s)\ ds\,. \tag{23}$$

Finding symmetries is a hard problem. If $\mathbf{G}_0=0$, as is the case when perturbing rest, K_{ijkl} is an isotropic tensor and (23) reduces to (24), below.

The main things about L_h^2:

The kernels are smooth,

$$\frac{K(s)}{h^2(s)} \in L^2(0,\infty)\,.$$

This means that delta function $\delta(s-s_0)$ are not allowed. δ functions are integrable, but not quadratically. Using Figure 15.1, we compute

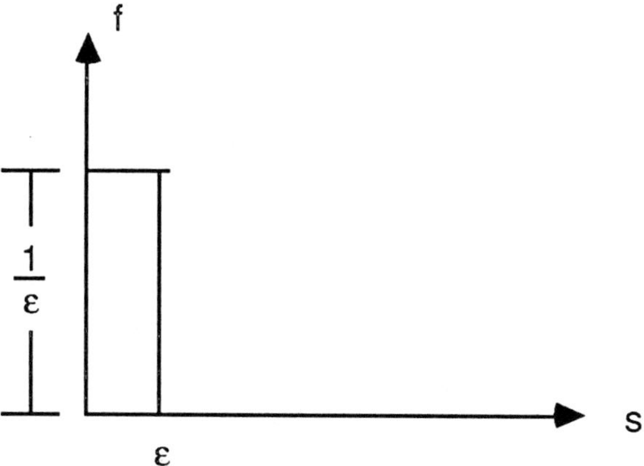

Figure 15.1. Sequence function for a one-sided delta function.

$$\int_0^\infty f^2(s,\varepsilon)\, ds = \int_0^\varepsilon \frac{1}{\varepsilon^2} ds = \frac{1}{\varepsilon}$$

which tends to infinity as $\varepsilon \to 0$. Piecewise continuous functions are in L_h^2 but not delta functions.

Physical limitations:

$$\boldsymbol{F}_1[0|\mathbf{G}(s)] = \int_0^\infty K(s)\mathbf{G}(s)\; ds \tag{24}$$

does not contain the Newtonian fluid as a special case. Suppose

$$K(s) = \mu \frac{d\delta(s)}{ds}.$$

Then, using (C.16),

$$\mu \int_0^\infty \frac{d\delta}{ds} \mathbf{G}(s) ds = \mu \int_0^\infty \frac{d\delta}{ds}\left[-s\mathbf{A}_1(t) + \frac{s^2}{2}\mathbf{A}_2(t) + \ldots \right] ds =$$

$\mu\mathbf{A}_1[t]$.

Quadratic integrals are defined in $L_h^2(0,\infty)$

$$\int_0^\infty m(s)\mathbf{G}(s)\mathbf{G}(s)\,ds = \int_0^\infty \frac{m(s)}{h(s)}\mathbf{G}(s)h(s)\mathbf{G}(s)\,ds$$

$$\leq \left\{\int_0^\infty \frac{m^2(s)}{h^2(s)}\mathbf{G}^2(s)\,ds\right\}^{1/2} \|\mathbf{G}(s)\|_{L_h^2}$$

$$\leq \max_{[0,\infty]} \frac{m(s)}{h^2(s)} \|\mathbf{G}(s)\|^2_{L_h^2}\,.$$

Cubic integrals are not defined in $L_h^2(0,\infty)$

$$\int_0^\infty \frac{m(s)}{h(s)}\mathbf{G}^2 h\mathbf{G}\,ds = \leq \left[\int_0^\infty \frac{m^2}{h^2}\mathbf{G}^4\,ds\right]^{1/2} \|\mathbf{G}(s)\|_L$$

$$\leq \left[\int_0^\infty \frac{m^4}{h^6}\mathbf{G}^6\,ds\right]^{1/4} \|\mathbf{G}(s)\|^{3/4}$$

$$\leq \left[\int_0^\infty \frac{m^8}{h^{10}}\mathbf{G}^{10}\,ds\right]^{1/8} \|\mathbf{G}(s)\|^{7/4}\,,\quad \text{etc.}$$

Polynomial integrals are likewise not defined. Neither are integrals of BKZ type.

$$\boldsymbol{F} = \int_0^\infty W(\mathrm{II},\mathrm{III})\ \mathbf{G}(s)\,ds\,,$$

$$\mathrm{II} = \operatorname{tr}\mathbf{G}^2\,,$$
$$\mathrm{III} = \det\mathbf{G}$$

in which W(,) is a polynomial.

Shortcomings of fading memory space $L_h^2(0,\infty)$ of Coleman and Noll are

1. Newtonian fluids aren't in it.

2. Many nonlinear single integral models like BKZ with polynomial kernels of II and III are not in it.

These shortcomings can be avoided in different ways. Saut and Joseph [1983] addressed this by introducing other domain spaces for $\boldsymbol{F}$. For example, certain Sobolev spaces will do the job. By restricting the domain of $\boldsymbol{F}$ we enrich the topological dual.

The Sobolev Space. $H_h^k(0,\infty)$: Suppose

$$<\mathbf{u},\mathbf{v}>_{L_h^2} = \int_0^\infty h^2 \operatorname{tr}(\mathbf{u}(s)\mathbf{v}(s))\, ds \, .$$

Then the Sobolev space $H_h^k(0,\infty)$ is defined by

$$H_h^k(0,\infty) = \{\mathbf{u} \in L_h^2(0,\infty), \frac{d^j\mathbf{u}}{ds^j} \in L_h^2(0,\infty), \quad 1 \le j \le k\}$$

$$[\mathbf{u},\mathbf{v}] = \sum_{j=0}^{k} < \frac{d^j\mathbf{u}}{ds^j}, \frac{d^j\mathbf{v}}{ds^j} >_{L_h^2} .$$

Take the case k=1, for scalar functions

$$[u,v] = \int_0^\infty h_1^2(s)u(s)v(s)\, ds + \int_0^\infty h_2^2(s)\, u'(s)v'(s)\, ds \, .$$

For simplicity, we could take $h_1=h_2$. The Riesz theorem says that any linear functional f(v) on $H_h^k(0,\infty)$ may be expressed as a scalar product against some fixed function u, that is

$$f(v) = [u,v]$$

where $u \in L_h^2$

$$[u,v] = \int_0^\infty h^2(s)u(s)v(s)\,ds + \int_0^\infty h^2(s)u'(s)v'(s)\,ds\ ,$$

where u and v are elements like $\dot{\mathbf{G}}$, $\mathbf{G} \in H_h^2(0,\infty)$. If we think of how this looks for $\boldsymbol{F}_1[0,\mathbf{G}(s)]$, we get, after using isotropy,

$$\boldsymbol{F}_1[0,\mathbf{G}(s)] = \int_0^\infty \mu(s)\mathbf{G}(s)\,ds + \int_0^\infty \upsilon(s)\dot{\mathbf{G}}(s)\,ds.$$

The kernel υ can contain delta functions. It gives rise to Newtonian contributions in the constitutive equation. To see this consider

$$\int_0^\infty h^2(s)u'(s)v'(s)ds =$$

$$\int_0^\infty [h^2(s)u'(s)v(s)]'\,ds - \int_0^\infty v(s)[h^2(s)u'(s)]'\,ds\ .$$

The first term on the right vanishes and $[h^2u'(s)]'$ is the kernel in what remains. Now we note that $u''(s)$ appears in our kernel. If $u'(s)$ is a step function, then

$$u'(s) \in L_h^2\ .$$

But derivatives of step functions give rise to δ functions.

Consider next

$$h(s)g(s) = \int_0^\infty (h'g+hg')\,ds \le 2\sqrt{s}\left\{\int_0^\infty (h^2g^2 + h'^2g'^2)ds\right\}^{1/2}$$

$$\le 2\sqrt{s}\,\|g\|_{H'}\ .$$

g(s) is bounded in the sup norm on any finite interval.

$$\int_0^\infty m(s)g^3(s)ds = \int_0^\infty \frac{m(s)}{h^3(s)} h^3g^3ds \le 8\|g\|_{H'}^{3/2} \int_0^\infty \frac{s^{3/2}m(s)}{h^3(s)} ds .$$

Multiple integrals in powers of g are well defined in H′.

15.3 Rate equations for fluids with instantaneous elasticity

It is perhaps useful to try to understand hyperbolicity in fluids with instantaneous elasticity without choosing special models. We want to know which phenomena are intrinsic and which are accidental, arising only for one or the other of the special models.

A general theory of rate equations for fluids with instantaneous elasticity was given by Joseph, Renardy and Saut [1985]. Here we give a slightly different derivation leading to the general form of quasilinear problems associated with motions of fluids with instantaneous elasticity.

Let $\mathbf{C}_t(\mathbf{x},\tau) = \nabla\boldsymbol{\chi}_t^T\nabla\boldsymbol{\chi}_t$ be the relative Cauchy strain where $\boldsymbol{\chi}_t(\mathbf{x},t)$ is the position of $\mathbf{x}=\boldsymbol{\chi}_t(\mathbf{x},t)$ at τ=t. The extra stress $\boldsymbol{\tau}$ may be expressed by a functional of the history of the Cauchy strain.

$$\boldsymbol{\tau} = \mathbf{S}[t, \mathbf{C}_t(\mathbf{x},\tau)]_{\tau=-\infty}^{t} . \tag{25}$$

For the moment, I state only that the domain of **S** is equipped with the topology of a Hilbert space. **S** depends on the history of $\mathbf{C}_t(\mathbf{x},\tau)$ and explicitly on the present time, perhaps through a kernel function and in other ways.

It follows from differentiating (25) that

$$\frac{d\boldsymbol{\tau}}{dt} = \frac{\partial}{\partial t}\mathbf{S} + \mathbf{S}_1[t,\mathbf{C}_t \mid \frac{d\mathbf{C}_t}{dt}] \tag{26}$$

where $\mathbf{S}_1[t,\mathbf{C}_t \mid]$ is a Frechet derivative, evaluated on $\mathbf{C}_t(\mathbf{x},\tau)$. The representation theorem for a linear functional on a Hilbert space leads us to

$$\mathbf{S}_1[t,\mathbf{C}_t \mid \frac{d\mathbf{C}_t}{dt}] = \int_{-\infty}^{t} \mathbf{K}(t-\tau,\mathbf{C}_t(\mathbf{x},\tau)) \frac{d\mathbf{C}_t}{dt}(\mathbf{x},\tau)\, d\tau \tag{27}$$

where $\mathbf{K}(t-\tau,\mathbf{C}_t(\mathbf{x},\tau))$ is a fourth order, tensor valued kernel function, not too singular, satisfying some invariance conditions arising from isotropy (see JRS). The term $\partial\mathbf{S}/\partial t$ is a derivative holding the history of $\mathbf{C}_t$ fixed. This term is again a functional of the form (25); in the case of single integral models with a kernel proportional to $e^{-(t-\tau)/\lambda}$, we have $\partial\mathbf{S}/\partial t = -\boldsymbol{\tau}/\lambda$. Motivated by this last observation, we introduce the suggestive notation

$$\boldsymbol{\tau} \overset{\text{def}}{=} -\lambda \frac{\partial \mathbf{S}}{\partial t} \tag{28}$$

for a constant λ; $\boldsymbol{\tau}$ is stress-like and λ is a "relaxation time."

Saut and Joseph [1983] have shown that the smoothness of the kernel $\mathbf{K}$ in (27) depends on the choice of the topology of convergence for the history $\mathbf{C}_t(\mathbf{x},\tau)$. If $d\mathbf{C}_t/dt\ (\mathbf{x},\tau)$ lies in the weighted $L_h^2(0,\infty)$ (fading memory) spaces of Coleman and Noll, then the kernel $\mathbf{K}$ must have at least the smoothness which is required for the validity of the application of the Schwartz inequality to the integral in (27). If $d\mathbf{C}_t/dt$ lies in the (fading memory) Sobolev spaces used by Saut and Joseph [1983], then various derivatives of Dirac measures can appear in the kernel $\mathbf{K}$. These give rise to "viscosity" terms like the retardation term in (2.8). The fading memory spaces of Coleman and Noll [1960] allow only those constitutive equations which exhibit instantaneous elasticity. The $L_h^2(0,\infty)$ fading memory spaces of Coleman and Noll, therefore, do not form a general basis for discussion of problems of rheology. They do form a general basis for discussion of problems of instantaneous elasticity. In the work below, dom $\mathbf{S}=L_h^2(0,\infty)$.

$$\frac{dC_t(x,\tau)}{dt} = \frac{dF_t^T}{dt} F_t + F_t^T \frac{dF_t}{dt} = -L^T(\mathbf{x},t)C_t(\mathbf{x},\tau) - C_t(\mathbf{x},\tau)L(\mathbf{x},t) \quad (29)$$

and

$$\left(\int_{\infty}^{t} \mathbf{K}(t-\tau,\mathbf{C}_t) \frac{d\mathbf{C}_t}{dt} d\tau \right)_{ij} = \int_{-\infty}^{t} K_{ijkl} \frac{dC_{kl}}{dt} d\tau = M_{ijkp}(\mathbf{x},t)L_{pk}(\mathbf{x},t) \quad (30)$$

where

$$M_{ijkp}(\mathbf{x},t) = -\int_{-\infty}^{t} (K_{ijkl} + K_{ijlk})C_{pl}(\mathbf{x},t)\, dt \quad (31)$$

is symmetric in the first pair of indices and $L_{pk} = \partial u_p/\partial x_k$. We may write (31) as

$$\frac{d\boldsymbol{\tau}}{dt} = \mathbf{M}(\mathbf{x},t)\mathbf{L}(\mathbf{x},t) - \frac{1}{\lambda}\boldsymbol{\tau} \quad (32)$$

where **M** is the fourth order tensor valued functional defined by (31).

The equations governing the evolution of fluids satisfying (32) were derived in Exercises 3.1 and 3.2. We may always decompose the fourth order tensor **M**=**S**+**A** into symmetric and skew symmetric parts as in (3.66). We can ask for the most general **M**, giving rise to vorticity equation of hyperbolic type in unsteady flow and of changing type in steady flow. One general class of models which gives rise to a hyperbolic vorticity are those for which the symmetric fourth order tensor **S** may be written as a sum of all possible combinations of Kronecker's delta and an arbitrary second order tensor **P**. I call these JRS models (after Joseph, Renardy, and Saut, [1985]). They are discussed in Exercise 3.2.

15.4 Rate equations for single integral models

Single integral models may be motivated in different ways. The K-BKZ model (Kaye, [1962]; Bernstein, Kearsley and Zapas, [1963]) is a single integral model motivated by the theory of elasticity and supported by thermodynamic arguments. Others see the general single integral models as a logical outcome of the model-building process using the rubber-like model of Lodge [1956] as the starting point. Rivlin and Sawyers [1971] associate such models with the requirement that the effects on the stress at time t of the deformations existing in a fluid element at different times are independent of each other. Narain [1986] has given a general interpretation of single integral models in terms of concepts of relaxing elasticity.

The development of "rate equations" from single integral models will be illustrated using the K-BKZ model. Let $\mathbf{C}(\tau)$ and $\mathbf{C}^{-1}(\tau)$ be the right relative Cauchy strain tensor with invariants

$$(\mathrm{I},\mathrm{II}) = \left(\mathrm{tr}\,\mathbf{C}(\tau),\ \mathrm{tr}\,\mathbf{C}^{-1}(\tau)\right)$$

and "strain energy"

$$W(t-\tau,\mathrm{I},\mathrm{II})\ .$$

The extra stress is given by

$$\boldsymbol{\tau} = \int_{-\infty}^{t} \{-W_{,\mathrm{II}}(t-\tau,\mathrm{I},\mathrm{II})\mathbf{C}^{-1}(\tau) + W_{,\mathrm{I}}(t-\tau,\mathrm{I},\mathrm{II})\mathbf{C}(\tau)\}\, d\tau \qquad (33)$$

where $W_{,\mathrm{I}}=\partial W/\partial \mathrm{I}$, etc. The extra stress vanishes on the rest state ($\mathbf{C}=\mathbf{C}^{-1}=\mathbf{1}$); hence,

$$W_{,\mathrm{II}}(t-\tau,3,3) = W_{,\mathrm{I}}(t-\tau,3,3)\ . \qquad (34)$$

The rate equation satisfied by $\boldsymbol{\tau}$ may be obtained from (33) by differentiating with respect to t, using

$$\frac{d\mathbf{C}}{dt} = -\mathbf{L}^{T}(t)\mathbf{C}(\tau) - \mathbf{C}(\tau)\mathbf{L}(t)\ ,$$

$$\frac{d\mathbf{C}^{-1}}{dt} = \mathbf{L}(t)\mathbf{C}^{-1}(\tau) + \mathbf{C}^{-1}(\tau)\mathbf{L}^{T}(t)\ , \qquad (35)$$

$$\frac{dI}{dt} = \mathrm{tr}\,\frac{d\mathbf{C}}{dt} = -2\,\mathrm{tr}\,[\mathbf{C}(\tau)\mathbf{L}(t)]\,,$$

$$\frac{dII}{dt} = \mathrm{tr}\,\frac{d\mathbf{C}^{-1}}{dt} = 2\mathrm{tr}\,[\mathbf{C}^{-1}(\tau)\mathbf{L}^{T}(t)]\,.$$

We find that

$$\frac{d\boldsymbol{\tau}}{dt} + \frac{1}{\lambda}\mathbf{t} = \mathbf{L}\mathbf{P}_1 + \mathbf{P}_1\mathbf{L}^T + \mathbf{P}_2\mathbf{L} + \mathbf{L}^T\mathbf{P}_2$$

$$+ \int_{-\infty}^{t} \{\mathbf{F}_1(\tau)\,\mathrm{tr}\,[\mathbf{C}(\tau)\mathbf{L}(t)] + \mathbf{F}_2(\tau)\,\mathrm{tr}\,[\mathbf{C}^{-1}(\tau)\mathbf{L}^{T}(t)]\}\,d\tau \quad (36)$$

where

$$\mathbf{t} = \int_{-\infty}^{t} \left[\frac{\partial W_{,II}}{\partial t}\mathbf{C}^{-1} - \frac{\partial W_{,I}}{\partial t}\mathbf{C}\right] d\tau\,,$$

$$\mathbf{P}_1 = -\int_{-\infty}^{t} W_{,II}(t-\tau,I,II)\mathbf{C}^{-1}(\tau)\,d\tau\,,$$

$$\mathbf{P}_2 = -\int_{-\infty}^{t} W_{,I}(t-\tau,I,II)\mathbf{C}(\tau)\,d\tau\,,$$

$$\mathbf{F}_1(\tau) = 2[W_{,I\,II}\mathbf{C}^{-1}(\tau) - W_{,I\,I}\mathbf{C}(\tau)]\,,$$

$$\mathbf{F}_2(\tau) = 2[-W_{,II\,II}\mathbf{C}^{-1}(\tau) + W_{,I\,II}\mathbf{C}(\tau)]\,.$$

The last terms of (36) (under the integral) are of leading order in hyperbolic analysis, because they contain first derivatives $L_{ij}=\partial_j u_i(\mathbf{x},t)$ of $\mathbf{u}(\mathbf{x},t)$ at the present time. These terms are spoilers, because they lead to third derivatives of $\mathbf{u}$ which do not reduce to second derivatives of $\boldsymbol{\tau}$. This shows that not every popular fluid model has a vorticity of changing type.

It may be helpful to draw a distinction between two types of rate equations. The first type among our special examples depends only on the velocity and the stress. We may regard the stress components as unknown, to-be-determined, field variables. The other type has coefficients which are functionals on the history of the Cauchy tensor. One could say that these equations are not "true" rate equations, because the coefficients do not depend directly on the field variables and problems based on these equations must be posed as initial history rather than initial value problems. Rate equations of the first type are clearly convenient, because the balance equations of mechanics are also relations among the velocity and stresses and the whole system of balance and constitutive equations is then closed.

Exercise 15.1

The stress in any simple fluid linear in $\mathbf{L}=\nabla\mathbf{u}$ is necessarily independent of the vorticity.

Solution

Simple fluids are isotropic (there are anisotropic fluids). Hence,

$$T_{ij} = -p\delta_{ij} + \mu_{ijkl}\partial_k u_l$$

where

$$\mu_{ijkl} = \mu[\delta_{ik}\delta_{jl}+\delta_{il}\delta_{jk}] + \gamma\delta_{ij}\delta_{kl}$$

is the only isotropic tensor symmetric in i,j. Hence,

$$T_{ij} = (-p + \gamma \operatorname{div} \mathbf{u})\delta_{ij} + \mu(L_{ij}+L_{ji}) .$$

This shows that one need not assume that the stress in a Newtonian fluid depends on the symmetric part of $\mathbf{L}$; it comes out of isotropy.

CHAPTER 16

Asymptotic Theories for Simple Fluids

Notations. From now on we deal only with relative quantities. To simplify notation we drop the subscript t, and write

$$(\boldsymbol{\chi},\mathbf{F},\mathbf{C}) \overset{\text{def}}{=} (\boldsymbol{\chi}_t,\mathbf{F}_t,\mathbf{C}_t) .$$

In perturbation studies we write

$$f(\varepsilon) = f_o + \varepsilon f_1 + \varepsilon^2 f_2 + \ldots$$

where factorials are suppressed,

$$f_n = \frac{1}{n!}\frac{d^n f}{d\varepsilon^n} .$$

In previous chapters we used $\boldsymbol{\tau}$ to denote the extra stress,

$$\mathbf{T} = -p\mathbf{1} + \boldsymbol{\tau} .$$

In the next two chapters we will use $\mathbf{S}$ for the extra stress writing

$$\mathbf{T} = -p\mathbf{1} + \mathbf{S} .$$

This was the notation used in the original papers from which the materials in the next two chapters was taken. This is to say that

$$\boldsymbol{\tau} \overset{\text{def}}{=} \mathbf{S} .$$

Asymptotic theories can be generated by restricting the domain of deformation to special sets on which the stress functional reduces to a manageable form. The simplest of these special deformations are rigid motions. For these $\mathbf{F}(\tau)$ is a corotational orthogonal tensor for all past times $\tau \leq t$. Hence $\mathbf{C}(\tau)=\mathbf{1}$ and $\mathbf{G}(s)=\mathbf{C}(\tau)-\mathbf{1}=\mathbf{0}$ for all lapse times $s=t-\tau \geq 0$. We can always chose a stress functional for an incompressible fluid

$$\boldsymbol{F}\,[0] = \mathbf{0}\,. \tag{1}$$

Rigid motion of a fluid does not excite a viscoelastic response. However a very rich theory of motions perturbing rigid ones can be developed.

16.1 Functional expansion perturbing rigid motion

We form approximations for the stress perturbing rigid motions in terms of a "Taylor series" of functional derivatives evaluated on the zero history. An n term approximation of $\boldsymbol{F}$, with an error going to zero faster than the nth power of some appropriate norm of $\mathbf{G}(s)$ may be written as

$$\boldsymbol{F}\,[\mathop{\mathbf{G}}_{s=0}^{\infty}(s)] \cong \boldsymbol{F}_1[\mathbf{0}|\mathop{\mathbf{G}}_{s=0}^{\infty}(s)] + \boldsymbol{F}_2[\mathbf{0}|\mathop{\mathbf{G}}_{s_1=0}^{\infty}(s_1)|\mathop{\mathbf{G}}_{s_2=0}^{\infty}(s_2)] + \ldots$$

$$+\boldsymbol{F}_n[\mathbf{0}|\mathop{\mathbf{G}}_{s_1=0}^{\infty}(s_1)|\mathop{\mathbf{G}}_{s_2=0}^{\infty}(s_2)|\ldots|\mathop{\mathbf{G}}_{s_n=0}^{\infty}(s_n)] \tag{2}$$

where $\boldsymbol{F}_\ell$ is an ℓ-linear, continuous tensor-valued form in $\mathbf{G}(\bullet)$, a functional derivative of order ℓ whose linear arguments are histories of $\mathbf{G}(s)$.

The possibility that functional derivatives do not exist at the origin seems not to have been considered in the books of reference. Such nonanalytic behavior, with square root dependence at the origin rather than integral powers does arise in computer simulations of certain idealized liquids. For example, the molecular dynamic simulations of liquid behavior using certain gas potentials at the triple point gives rise to nonanalytic behavior (see Evans, Hanley, and Hess, [1984] and the discussion following (85)).

There are two types of functional derivatives, Gateaux derivatives and Fréchet derivatives. Fréchet derivatives may be defined when the topology of dom $\boldsymbol{F}$ (=domain of $\boldsymbol{F}$) is given by a **norm**.

Gateaux derivatives may be defined for elements belonging to linear topological vector spaces, even when these spaces are not normable. We are going to say a few words about the mathematical properties of functional derivatives. This type of theory, however, is not too useful in practice because it is never possible to verify if some real fluid you like, say aqueous polyox, has one or another kind of functional derivative.

To compute Gateaux and Fréchet derivatives it is necessary that the domain of $\boldsymbol{F}$ be a linear space. The tensor $\mathbf{G}(s)$ cannot be in a linear space because the eigenvalues of $\mathbf{G}-\mathbf{1}=\mathbf{C}(\tau)$ are positive (det $\mathbf{C}(\tau)=1$ for incompressible materials). It is possible to embed these tensors in a space of symmetric-tensor valued functions which can form the basis of a linear space. The functional analysis is then worked in this extended space. We shall suppose now that $\mathbf{G}(s)$ lies in the extended space of symmetric tensors endowed with a topology suitable for a topological vector space. We call this space dom $\boldsymbol{F}$. The same algorithms are used to compute Fréchet and Gateaux derivatives

$$\frac{\partial}{\partial\lambda}\underset{s=0}{\overset{\infty}{\boldsymbol{F}}}\,[\mathbf{G}(s)+\lambda\mathbf{J}(s)]_{\lambda=0}\overset{\text{def}}{=}\boldsymbol{F}_1[\underset{s=0}{\overset{\infty}{\mathbf{G}(s)}}|\underset{s_1=0}{\overset{\infty}{\mathbf{J}(s_1)}}]\,,\quad \mathbf{G},\mathbf{J}\in \text{dom}\,\boldsymbol{F} \tag{3}$$

is linear and continuous in $\mathbf{J}(\bullet)$. The second derivative may be defined by

$$\frac{\partial^2}{\partial\lambda_1\partial\lambda_2}\underset{0}{\overset{s=\infty}{\boldsymbol{F}}}[\mathbf{G}(s)+\lambda_1\mathbf{J}(s)+\lambda_2\mathbf{K}(s)]_{\lambda_1=\lambda_2=0}\overset{\text{def}}{=}\boldsymbol{F}_2[\underset{s=0}{\overset{\infty}{\mathbf{G}(s)}}|\underset{s_1=0}{\overset{\infty}{\mathbf{J}(s_1)}}|\underset{s_2=0}{\overset{\infty}{\mathbf{K}(s_2)}}] \tag{4}$$

where $\boldsymbol{F}_2$ is linear in $\mathbf{J}(\bullet)$, $\mathbf{K}(\bullet)\in$ dom $\boldsymbol{F}$. Higher derivatives are computed in the same way. We get derivates on the zero history by putting $\mathbf{G}=\mathbf{0}$.

It is obvious that the functional derivatives are symmetric to all transpositions of their linear arguments.

Notation. We frequently omit reference to the zero history on which the $\boldsymbol{F}_n$ are evaluated. And we also omit explicit reference to the range of lapse times. For example

$$\boldsymbol{F}_2[\mathbf{0}\,\overset{\infty}{\underset{s_1=0}{|}}\mathbf{G}(s_1)\overset{\infty}{\underset{s_2=0}{|}}\mathbf{G}(s_2)] \overset{\text{def}}{=} \boldsymbol{F}_2[\mathbf{G}|\mathbf{G}] \ . \tag{5}$$

16.2 Multiple integral expansions

It is clear that (2) is an expansion for small $\mathbf{G}(s)$ and different choices for the topologies defining smallness are possible. Green and Rivlin [1957] introduce expansions of the form (2) in which the $\boldsymbol{F}_n$ were expressed by multiple integrals. Coleman and Noll [1961] carried out an analysis of multiple integrals in their weighted fading memory space. Their work is for fluids with instantaneous elasticity, i.e. delta functions and their derivatives are not allowed in the kernel. Pipkin [1964] simplified the second order theory for incompressible fluids and he gave the isotropic forms through terms of order three. A multiple integral expression for the stress with products of four $\mathbf{G}(s)$ was given by Joseph and Beavers [1977a, b]. The reduction of the integrands in the multiple integral expansion to isotropic form uses the fact that isotropic tensors of even order may be expressed as products of Kronecker's delta and the symmetry properties of functional derivatives.

A three term multiple integral approximation to $\boldsymbol{F}$, for incompressible fluids is given by

$$\boldsymbol{F}^{(3)} = \boldsymbol{F}_1 + \boldsymbol{F}_2 + \boldsymbol{F}_3 \ , \tag{6}$$

$$\boldsymbol{F}_1 = \int_0^\infty \zeta(s)\mathbf{G}(s)\, ds \ , \tag{7}$$

$$\boldsymbol{F}_2 = \int_0^\infty \int_0^\infty \{\beta(s_1, s_2)\, \mathbf{G}(s_1)\, \mathbf{G}(s_2) + \alpha(s_1,s_2)\, [\operatorname{tr} \mathbf{G}(s_1)]\, \mathbf{G}(s_2)\}\, ds_1\, ds_2 \ , \tag{8}$$

$$F_3 = \int_0^\infty \int_0^\infty \int_0^\infty \Big\{ \psi_1(s_1,s_2,s_3)\, \mathbf{G}(s_1)\mathbf{G}(s_2)\, \mathbf{G}(s_3)$$
$$+ \psi_2(s_1,s_2,s_3)\, [\mathrm{tr}\, \mathbf{G}\,(s_1)]\, \mathbf{G}(s_2)\, \mathbf{G}(s_3) \tag{9}$$
$$+ \psi_3(s_1,s_2,s_3)\, [\mathrm{tr}\, \mathbf{G}\,(s_1)]\, [\mathrm{tr}\, \mathbf{G}\,(s_2)]\, \mathbf{G}(s_3)$$
$$+ \psi_4(s_1,s_2,s_3)\, \mathrm{tr}\, [\mathbf{G}\,(s_1)\, \mathbf{G}(s_2)]\, \mathbf{G}(s_3) \Big\}\, ds_1\, ds_2\, ds_3$$

where $\beta(s_1,s_2)$, $\psi_1(s_1,s_2,s_3)$ are unchanged under transposition of arguments.

These integral expressions cannot be used to obtain useful expressions about dynamics without giving more precise details about the kernels. For example if $\zeta(s) = \mu\frac{d\delta(s)}{ds}$ where $\delta(s)$ is the Dirac delta function, then $F_1=\mu\mathbf{D}[u]$ is appropriate for Newtonian fluids. Green and Rivlin [1957] allowed singularities in their kernel.

For fluids with instantaneous elasticity we should assume that the kernels are smooth and bounded. This kind of assumption was made by Coleman and Noll [1961].

The analytical nature of the kernels is restricted by certain considerations, like those mentioned at the end of Chapter 15. By assuming a topology for $\mathbf{G}(s)$ we imply a certain mathematical structure for the kernels.

In the end we don't know what to assume about constitutive equations until we see what works in practice, say in experiments. It is possible to waste a lot of time with foolish assuming of this and that.

16.3 Nonuniqueness of multiple integrals

The integrals given in the multiple integral expansion are not unique. We could base a theory of multiple integrals on the Finger tensor $\mathbf{C}^{-1}(\tau)$ and the corresponding history $\mathbf{H}(s)=\mathbf{1}-\mathbf{C}^{-1}(\tau)$. Then we would again get (6) through (9) with $\mathbf{H}(s)$ instead of $\mathbf{G}(s)$. This expansion would be different because $\mathbf{C}^{-1}\neq\mathbf{C}$, but just as good, maybe better.

We could as well use some appropriate tensor-valued functional of **G** or **H** which could be small when **G** or **H** is not small. There is no uniqueness.

The reason for lack of uniqueness is that the expansion (6) through (9), and beyond is not based on a systematic expansion of the nonlinear measures of relative strain (for example, **G**(s) or **H**(s)) which arise under perturbation of pathlines.

16.4 Canonical forms of the stress for perturbation of the rest state

For perturbations of the rest state the pathlines are given by

$$\boldsymbol{\chi}(\mathbf{x},\tau,\varepsilon) = \mathbf{x} + \sum_{n=1} \boldsymbol{\chi}_n(\mathbf{x},\tau)\varepsilon^n \tag{10}$$

where

$$\mathbf{x} = \boldsymbol{\chi}(\mathbf{x},\tau,0) \tag{11}$$

for all τ because the fluid is at rest. Ultimately these expansions are generated by the solutions of the equations of motion where ε is a small parameter in the data. Since

$$\boldsymbol{\chi}(\mathbf{x},t,\varepsilon) = \mathbf{x}$$

it follows that

$$\boldsymbol{\chi}_n(\mathbf{x},t) = 0\ , \quad n \geq 1\ . \tag{12}$$

The velocity of the particle at time τ which at time t is at place $\mathbf{x}$ (the particle at $\mathbf{x}$ when ε=0) is given by

$$\mathbf{U}(\boldsymbol{\chi}(\mathbf{x},\tau,\varepsilon),\tau,\varepsilon) \equiv \tilde{\mathbf{U}}(\mathbf{x},\tau,\varepsilon) = \frac{\partial \boldsymbol{\chi}}{\partial \tau}(\mathbf{x},\tau,\varepsilon)$$

$$= \sum_{n=1} \tilde{\mathbf{U}}_n(\mathbf{x},\tau)\varepsilon^n = \sum_{n=1} \varepsilon^n \frac{\partial \boldsymbol{\chi}_n}{\partial \tau}(\mathbf{x},\tau)\ . \tag{13}$$

Hence

$$\chi_n(\mathbf{x},\tau) = \int_{\tau}^{t} \tilde{\mathbf{U}}_n(\mathbf{x},t') \, dt' \tag{14}$$

where, by differentiating (13), with respect to ε at ε=0, we find that

$$\tilde{\mathbf{U}}_1(\mathbf{x},\tau) = \mathbf{U}_1(\mathbf{x},\tau) \,,$$
$$\tilde{\mathbf{U}}_2(\mathbf{x},\tau) = \mathbf{U}_2(\mathbf{x},\tau) + (\chi_1 \cdot \nabla)\mathbf{U}_1(\mathbf{x},\tau)$$

etc., where $\mathbf{U}_n(\mathbf{x},\tau)$ is n! times the nth partial derivative of U with respect to ε at ε=0 holding $\boldsymbol{\chi}=\mathbf{x}$ fixed. We note that since

$$\nabla_{\xi} \cdot \mathbf{U}(\boldsymbol{\xi},\tau,\varepsilon) = 0 \,, \quad \boldsymbol{\xi} = \boldsymbol{\chi}(\mathbf{x},\tau,\varepsilon)$$

is an identity in $\boldsymbol{V}$

$$\nabla_{\xi} \cdot \mathbf{U}_n(\boldsymbol{\xi},\tau) = 0$$

in $\boldsymbol{V}$ and, taking ε=0, $\boldsymbol{\xi}=\mathbf{x}$ we have

$$\nabla \cdot \mathbf{U}_n(\mathbf{x},\tau) = 0$$

for n≥0.

With these kinematic preliminaries aside we turn next to the stresses. We confine our attention to the second order theory and seek to purge $\boldsymbol{F}^{(2)}$ of redundant terms. We first rewrite the two term approximation of $\boldsymbol{F}$ as

$$\boldsymbol{F}^{(2)} = \int_0^{\infty} G(s)\boldsymbol{\Pi}(t-s,\varepsilon)\, ds + \int_0^{\infty}\int_0^{\infty} \gamma(s_1,s_2)\boldsymbol{\Pi}(t-s_1,\varepsilon)$$
$$\boldsymbol{\Pi}(t-s_2,\varepsilon)\, ds_1\, ds_2 + O(\varepsilon^3) \,, \tag{15}$$

where

$$\boldsymbol{\Pi}(t-s,\varepsilon) = \mathbf{F}^T(t-s,\varepsilon)\, \mathbf{A}_1[\mathbf{U}(\boldsymbol{\xi},t-s,\varepsilon)]\mathbf{F}(t-s,\varepsilon) \,, \tag{16}$$

$$\Pi_{ij} = \frac{\partial \xi_\ell}{\partial x_i}\left[\frac{\partial U_m}{\partial \xi_\ell} + \frac{\partial U_\ell}{\partial \xi_m}\right]\frac{\partial \xi_m}{\partial x_j} = \frac{\partial U_m}{\partial x_i}\frac{\partial \xi_m}{\partial x_j} + \frac{\partial \xi_\ell}{\partial x_i}\frac{\partial U_\ell}{\partial x_j} \,. \tag{17}$$

The material constant $\alpha(s_1,s_2)$ does not appear in the second order theory because $\mathbf{G}(s,\varepsilon)=\mathbf{O}(\varepsilon)$, tr $\mathbf{G}(s,0)=0$ (for incompressible materials) and tr $\mathbf{G}(s,\varepsilon)=O(\varepsilon^2)$. The material functions G(s) and $\gamma(s_1,s_2)$ are, respectively, the linear and quadratic shear relaxation moduli,

$$\frac{dG}{ds}=\zeta(s)\ ,\quad \gamma(s_1,s_2)=\frac{\partial^2\beta(s_1,s_2)}{\partial s_1\,\partial s_2}\ .$$

G(s) and $\gamma(s_1,s_2)$ vanish with sufficient rapidity to justify the integration by parts leading to (15). To expand $\boldsymbol{F}$, at constant $\mathbf{x}$ we write

$$\pi_{ij}=\frac{\partial\tilde{\mathbf{U}}}{\partial x_i}(\mathbf{x},\tau,\varepsilon)\bullet\frac{\partial\boldsymbol{\chi}}{\partial x_j}(\mathbf{x},\tau,\varepsilon)+\frac{\partial\tilde{\mathbf{U}}}{\partial x_j}(\mathbf{x},\tau,\varepsilon)\bullet\frac{\partial\boldsymbol{\chi}}{\partial x_i}(\mathbf{x},\tau,\varepsilon) \qquad (18)$$

and

$$\frac{\partial\tilde{\mathbf{U}}}{\partial x_i}(\mathbf{x},\tau,\varepsilon)\bullet\frac{\partial\boldsymbol{\chi}}{\partial x_j}(\mathbf{x},\tau,\varepsilon)$$

$$=\sum_{\ell=1}\varepsilon^{\ell}\frac{\partial\tilde{\mathbf{U}}_{\ell}}{\partial x_i}+\sum_{\ell=1}\sum_{m=1}\varepsilon^{m+\ell}\frac{\partial\mathbf{U}_{\ell}}{\partial x_i}\bullet\frac{\partial\boldsymbol{\chi}_m}{\partial x_j}\ .$$

After inserting these expansions into (15) we find that to second order in ε

$$\begin{aligned}\boldsymbol{F}\sim\boldsymbol{F}^{(2)}&=\varepsilon\int_0^\infty G(s)\,\mathbf{A}_1[\mathbf{U}_1(s)]\,ds+\varepsilon^2\int_0^\infty G(s)\mathbf{A}_1[\mathbf{U}_2(s)]\,ds\\&\quad+\varepsilon^2\int_0^\infty G(s)\,\Big\{(\boldsymbol{\chi}_1\bullet\nabla)\,\mathbf{A}_1\,[\mathbf{U}_1(s)]\\&\quad+\mathbf{A}_1\,[\mathbf{U}_1(s)]\,\mathbf{F}_1+\mathbf{F}_1^T\mathbf{A}_1\,[\mathbf{U}_1(s)]\Big\}\,ds \qquad (19)\\&\quad+\varepsilon^2\int_0^\infty\int_0^\infty\gamma(s_1,\,s_2)\,\mathbf{A}_1\,[\mathbf{U}_1(s_1)]\,\mathbf{A}_1\,[\mathbf{U}_1(s_2)]\,ds_1\,ds_2\end{aligned}$$

where

$$\mathbf{U}_n(s) \stackrel{\text{def}}{=} \mathbf{U}_n(\mathbf{x},\ t-s)\ ,$$
$$\mathbf{F}_1 = \nabla\boldsymbol{\chi}_1(\mathbf{x},t-s)\ ,\quad \nabla = \frac{\partial}{\partial\mathbf{x}}\ . \tag{20}$$

The terms neglected are of order greater than two in ε. The position $\mathbf{x}$ is the position of a particle in the rest state, so $\mathbf{x}$ is independent of s under the integral

$$\rho\frac{\partial\mathbf{U}_1}{\partial t}(\mathbf{x},t) + \nabla p_1 - \int_0^\infty G(s)\ \nabla^2\mathbf{U}_1\ (\mathbf{x},t-s)\ ds = 0 \tag{21}$$

and

$$\rho\left[\frac{\partial\mathbf{U}_2}{\partial t} + \mathbf{U}_1\boldsymbol{\cdot}\nabla\mathbf{U}_1\right] + \nabla p_2 - \int_0^\infty G(s)\nabla^2\mathbf{U}_2(\mathbf{x},t-s)$$
$$-\nabla\boldsymbol{\cdot}\int_0^\infty G(s)\ \left\{(\boldsymbol{\chi}_1\boldsymbol{\cdot}\nabla)\mathbf{A}_1(s) + \mathbf{A}_1(s)\nabla\boldsymbol{\chi}_1 + \nabla\boldsymbol{\chi}_1^T\mathbf{A}_1(s)\right\}ds$$
$$-\nabla\boldsymbol{\cdot}\int_0^\infty\int_0^\infty \gamma(s_1,\ s_2)\ \mathbf{A}_1(s_1)\mathbf{A}_1(s_2)\ ds_1\ ds_2 = 0\ , \tag{22}$$

where $\boldsymbol{\chi}_1=\boldsymbol{\chi}_1(\mathbf{x},t-s)$, $\mathbf{A}_1(s)=\mathbf{A}_1[\mathbf{U}_1(\mathbf{x},t-s)]$ and $\nabla\boldsymbol{\cdot}\mathbf{U}_1=\nabla\boldsymbol{\cdot}\mathbf{U}_2=0$. Eqs. (21) and (22) are to be solved subject to boundary conditions arising from the perturbation of given data. When this data is steady we find that $\boldsymbol{\chi}_1(\mathbf{x},t-s)=-s\mathbf{U}_1(\mathbf{x})$ and

$$\nabla p_1 - \mu\nabla^2\mathbf{U}_1(\mathbf{x}) = 0 \tag{23}$$

and

$$\rho\mathbf{U}_1\boldsymbol{\cdot}\nabla\mathbf{U}_1 + \nabla p_2 - \mu\nabla^2\mathbf{U}_2 - \nabla\boldsymbol{\cdot}\left\{\alpha_1\mathbf{A}_2 + \alpha_2\mathbf{A}_1^2\right\} = 0\ . \tag{24}$$

where μ is the viscosity, α_1 and α_2 are quadratic constants defined by (42) and (45) and $\mathbf{A}_2$ is the second Rivlin-Ericksen tensor defined in Appendix C.10.

The perturbation problems may be solved sequentially. First we find $\mathbf{U}_1$ from (21). Then we compute $\boldsymbol{\chi}_1(\mathbf{x},\tau)$ from (14). Then we compute $\mathbf{U}_2$ from (22). We may then compute $\boldsymbol{\chi}_2(\mathbf{x},\tau)$ from (14) and proceed to higher orders.

16.5 Canonical forms for the stress perturbing rigid motion

The theory of rotating simple fluids was given by Joseph [1977]. The calculation shows clearly that perturbations expressed in multiple integrals can be regarded as arising uniquely from a perturbation of the path lines. The history of the motion, the integrand under the integral, is computed relative to the pathline of the unperturbed motion. This endows the originally isotropic forms for the stress with the symmetries of the unperturbed motion. This property holds in all perturbation theories, perturbation of the rest state, rigid motions, viscometric motions, elongational motions in which particle paths may be expanded relative to unperturbed particle paths. If the perturbed motion is unsteady, the integral forms of the stress should be used.

The perturbation of particle paths induces a perturbation of the velocity in the following manner:

$$\boldsymbol{\chi}(\mathbf{x},\tau,\varepsilon) = \boldsymbol{\chi}_0(\mathbf{x},\tau) + \varepsilon\boldsymbol{\chi}_1(\mathbf{x},\tau) + \varepsilon^2\boldsymbol{\chi}_2(\mathbf{x},\tau) + \ldots \tag{25}$$

and

$$\begin{aligned} \mathbf{U}(\boldsymbol{\chi}(\mathbf{x},\tau,\varepsilon),\tau,\varepsilon) &\equiv \widetilde{\mathbf{U}}(\mathbf{x},\tau,\varepsilon) \\ &= \widetilde{\mathbf{U}}_0(\mathbf{x},\tau) + \varepsilon\widetilde{\mathbf{U}}_1(\mathbf{x},\tau) + \varepsilon^2\widetilde{\mathbf{U}}_2(\mathbf{x},\tau) + \ldots . \end{aligned} \tag{26}$$

Since

$$\frac{\partial\boldsymbol{\chi}(\mathbf{x},\tau,\varepsilon)}{\partial\tau} = \widetilde{\mathbf{U}}(\mathbf{x},\tau,\varepsilon) = \mathbf{U}(\boldsymbol{\chi}(\mathbf{x},\tau,\varepsilon),\tau,\varepsilon) ,$$

we have, assuming the particle label

$$\mathbf{x} = \boldsymbol{\chi}(\mathbf{x},t,\varepsilon) \tag{27}$$

is independent of ε, so that for n=0,1,2,...,

$$\frac{\partial \chi_n(\mathbf{x},\tau)}{\partial \tau} = \tilde{\mathbf{U}}_n(\mathbf{x},\tau) \; . \tag{28}$$

Moreover, using (25), (27), and (26), we find that

$$\boldsymbol{\chi}_0(\mathbf{x},t) = \mathbf{x} \; , \tag{29}$$

$$\boldsymbol{\chi}_n(\mathbf{x},t) = 0 \; , \quad n \geq 1 \; , \tag{30}$$

and

$$\tilde{\mathbf{U}}_n(\mathbf{x},t) = \mathbf{U}_n(\mathbf{x},t) \; , \quad n \geq 1 \; .$$

The function $\tilde{\mathbf{U}}(\mathbf{x},\tau)$ is an auxiliary function used to facilitate our computation of particle paths. The perturbation of the particle paths and velocities induces a perturbation of the stress. These perturbations may be arranged into an ordered sequence of initial-boundary value problems which determines the perturbation fields sequentially.

The velocity derivatives which enter into the hereditary integrals giving the stress are in the form $\mathbf{U}_n(\boldsymbol{\chi}_0(\mathbf{x},\tau),\tau)$ where $-\infty<\tau\leq t$, $\mathbf{x}$ is the particle which coincides with the field point $\mathbf{x}$ in the present configuration and $\boldsymbol{\chi}_0$ is determined by integration of $d\boldsymbol{\chi}_0/d\tau = \boldsymbol{\Omega}\wedge\boldsymbol{\chi}_0$ subject to the condition that $\boldsymbol{\chi}_0(\mathbf{x},t)=\mathbf{x}$. The history of the velocity derivative $\mathbf{U}_n$ belonging to $\mathbf{x}$ is therefore computed for $\boldsymbol{\chi}_0(\mathbf{x},\tau)$ on circles of fixed radius and elevation around the fixed axis $\mathbf{e}_z = \boldsymbol{\Omega}/|\boldsymbol{\Omega}|$ of rotation. For such motions the only coordinate which changes with $s=t-\tau$ is the angular displacement $\theta_0=\theta-\Omega s$. The perturbation equations are therefore naturally framed in coordinates having axial symmetry; for example, in cylindrical coordinates (r,θ,z). In these coordinates,

$$\mathbf{U}_0 = \mathbf{e}_\theta r\Omega \; , \quad p_0 + \rho\Omega^2 r = \text{const} \; , \tag{31}$$

and we introduce the vectors

$$\begin{aligned}\tilde{\mathbf{U}}_n(s) &= \mathbf{e}_r(\theta)U_n^{\langle r\rangle}(r,\theta-\Omega s,z,t-s) + \mathbf{e}_\theta(\theta)U_n^{\langle\theta\rangle}(r,\theta-\Omega s,z,t-s) \\ &\quad + \mathbf{e}_z U_n^{\langle z\rangle}(r,\theta-\Omega s,z,t-s) \equiv \mathbf{e}_\ell(\theta)U_n^{\langle\ell\rangle}(r,\theta-\Omega s,z,t-s) \; ,\end{aligned} \tag{32}$$

and

$$\mathbf{U}_n = \tilde{\mathbf{U}}_n(0) , \tag{33}$$

for n>1. At each order ε^n there are four equations in four unknowns

$$\operatorname{div} \mathbf{U}_n = 0 \tag{34}$$

and

$$\rho(\partial_r + \Omega\partial_\theta + \boldsymbol{\Omega}\wedge)\mathbf{U}_n + \nabla p_n - \int_0^\infty G(s)\nabla^2\tilde{\mathbf{U}}_n(s)\, ds = \mathbf{f}_n . \tag{35}$$

Here G(s) is the shear relaxation modulus, $\mathbf{f}_n$ depends on lower-order fields $\ell<n$; for example $\mathbf{f}_1=0$,

$$\mathbf{f}_2 = \operatorname{div} \Big\{ \int_0^\infty G(s)\mathbf{A}_2(s)\, ds$$

$$+ \int_0^\infty \int_0^\infty \gamma(s_1, s_2)\mathbf{A}_1(s_1)\mathbf{A}_1(s_2)\, ds_1\, ds_2 \Big\}$$

where

$$\mathbf{A}_1(s) = \nabla\tilde{\mathbf{U}}_1(s) + \nabla^T\tilde{\mathbf{U}}_1(s) ,$$
$$\mathbf{A}_2(s) = (\boldsymbol{\chi}_1\boldsymbol{\cdot}\nabla)\mathbf{A}_1(s) + \mathbf{A}_1(s)\nabla\boldsymbol{\chi}_1 + (\mathbf{A}_1(s)\nabla\boldsymbol{\chi}_1)^T$$

and

$$\frac{d\boldsymbol{\chi}_1(\mathbf{x},\tau)}{d\tau} = \mathbf{e}_\ell(\theta-\Omega s)U_1^{<\ell>}(r,\theta-\Omega s,z,t-s) , \quad \boldsymbol{\chi}_1(\mathbf{x},t) = 0 . \tag{36}$$

$\mathbf{U}_1$ and p_1 are determined by inverting (34) and (35) for prescribed boundary values and initial-history. Then $\boldsymbol{\chi}_1$ follows from (36) and we may compute $\mathbf{f}_2$. $\mathbf{U}_2$ and p_2 are then determined by inverting (34) and (35) (when n=2). Then we get $\boldsymbol{\chi}_2$ from an equation like (36) and compute $\mathbf{f}_3$, and so on.

The perturbation equations have many interesting properties:

1. The principal (linear) parts of the equations (34) and the left side of (35) do not depend on any constitutive assumption less general than that the stress be Fréchet differentiable on the zero history in Hilbert space topologies defined by integrals. An integral representation for the stress is then implied by the Riesz theorem.

2. The theory of perturbations of the rest state (Joseph, [1976]) is a special case of the present theory with $\Omega=0$. When $\Omega=0$ and G(s) is physically reasonable the rest state is linearly stable and solution of the perturbation problems exist and are unique (Slemrod, [1976]).

3. Rigid rotation is linearly stable if G(s) is physically reasonable. The solutions of the perturbation problems exist and are unique (Laklalech, [1985]).

4. If the velocity $\mathbf{U}_n(r,\theta,z)$ at a fixed point (r,θ,z) is steady, the history of the particle which is presently at that fixed point is unsteady because it was at a different place $(r,\theta-\Omega s,z)$ earlier on, and its velocity $\tilde{\mathbf{U}}(r,\theta-\Omega s,z)$ at that place was different. The same remark goes for steady flow in a rotating basis $(r,\theta(t),z)$ but, of course, the inertial terms in (35) take a different form in the rotating bases.

5. If the flow is axisymmetric in a fixed or rotating basis, the perturbation fields are independent of the angular variable θ and $\theta-\Omega s$. The terms in the equations of motion are then identical to those which arise in the theory of perturbations of the rest state. If, in addition, the flow is steady in some fixed or rotating basis, the perturbation equations for particle paths may be integrated in terms of the steady, but unknown, velocity, the time variables t and τ appear only in the combination $s = t-\tau$, and the stresses may be expressed in terms of the Rivlin-Ericksen tensors and constants familiar in the theory of slow steady motions. The constants are defined

by various moments of the kernel functions in the integrals giving the stress (see (55), (56)).

6. Various simplifications of the equations result from decomposing solutions of (35) and (36) as a Fourier series in θ. For time-periodic data, the equations may be further simplified by decomposing solutions into a Fourier series in time. These decompositions lead to an infinite number of different "complex viscosities" of the form

$$\eta^*(\omega m+\Omega\ell) = \int_0^\infty G(s)e^{-i(\omega m+\Omega\ell)s}\, ds$$

where ℓ is the index of azimuthal periodicity and $m\omega$ the mth harmonic of the fundamental frequency. The "complex viscosity" is not a material function because Ω, ω, m, and ℓ depend on the prescribed data.

7. I call (35) with n=1 ($\mathbf{f}_1=0$) the linearized equation for rotating simple fluids. The well known linearized theory of rotation Newtonian fluids for large Ω and small ε applies to (34) and (35) when the boundary data is steady. This theory was applied to the orthogonal rheometer by Joseph [1977]. When Ω is large the stress in the interior of the fluids may be neglected and inertia dominates. At the boundary there is a layer in which the stress is important. The thickness of this layer is much smaller in a viscoelastic fluid than in a Newtonian fluid; the ratio of thickness tends to zero with $1/\sqrt{\Omega}$ as $\Omega\to\infty$.

16.6 Nearly steady slow motion

The asymptotic expansion appropriate for these motions was introduced by Coleman and Noll [1960]. They called these slow nearly steady motions retarded. They take a given history $\mathbf{G}(s)$ and define a retarded history $\mathbf{G}_\varepsilon(s)=\mathbf{G}(\varepsilon s)$, $0\leq\varepsilon<1$. Since $\mathbf{G}(0)=\mathbf{0}$, $\mathbf{G}(\varepsilon s)$ is small if ε is sufficiently small. The $L^2_h(0,\infty)$ norm (Chapter 15) is used as the measure of smallness. Conceptually, and certainly as a matter of practice, it may be best not to retard a given history. Instead the given history should be regarded as the $\varepsilon=1$ value of a real history determined by a nearly steady slow motion generated from a perturbation of rest with a small parameter ε.

We start our analysis with some general consideration recalling that

$$\mathbf{A}_n(t) = (-1)^n \frac{d^n}{ds^n}\mathbf{G}(s)\bigg|_{s=0} .$$

Consider all motions which can be approximated by a (possibly finite) Taylor series

$$\mathbf{G}(s) \sim -s\mathbf{A}_1(t) + \frac{s^2}{2}\mathbf{A}_2(t) + \ldots + (-1)^n \frac{s^n}{n!}\mathbf{A}_n(t) + \mathbf{o}(s^n) . \quad (37)$$

The expansion we seek requires that we introduce retarded motions or nearly steady slow motions. We shall proceed as far as possible without introducing these special motions.

In all situations for which (37) holds we find, using the property of linearity, that

$$\begin{aligned}\boldsymbol{F}_1[\mathbf{G}] &= \boldsymbol{F}_1[-s\mathbf{A}_1(t) + \frac{s^2}{2}\mathbf{A}_2(t) + \ldots + \mathbf{o}(s^n)]\\ &= -\mathbf{f}_1[s]\mathbf{A}_1 + \frac{1}{2}\mathbf{f}_1[s^2]\mathbf{A}_2 + \ldots + \frac{(-1)^n}{n!}\mathbf{f}_1[s^n]\mathbf{A}_n\\ &\quad + \mathbf{o}(s^n) \qquad (38)\end{aligned}$$

where

$$\boldsymbol{F}_1[\mathbf{B}(t), s^n\mathbf{A}_n(t)] = \mathbf{f}_1[\mathbf{B}, s^n]\mathbf{A}_n(t) \quad \text{and}$$

$$\mathbf{f}_1[q(s)] \overset{\text{def}}{=} \mathbf{f}_1\left[\overset{\infty}{\underset{s=0}{q(s)}}\right]$$

is a fourth-order tensor valued functional of the function q(•). For isotropic materials $\mathbf{f}_1$ is a fourth order isotropic tensor expressible as dyadic products of the unit tensor with scalar coefficients. Using the fact that $\boldsymbol{F}_1$ and $\mathbf{A}_n(t)$ are symmetric, we find modulo terms proportional to **1** that

$$\mathbf{f}_1[s^n]\mathbf{A}_n = F_1[s^n]\mathbf{A}_n \tag{39}$$

where

$$F_1[q(s)] \overset{\text{def}}{=} F_1\overset{\infty}{\underset{s=0}{[q(s)]}} \tag{40}$$

is a linear functional from $[0,\infty] \rightarrow \mathbb{R}$. Omitting terms proportional to **1** in the expansion for $\boldsymbol{F}_1$, we find that

$$\boldsymbol{F}_1[\mathbf{G}] = \mu\mathbf{A}_1(t) + \alpha_1\mathbf{A}_2(t) + \beta_1\mathbf{A}_3(t) + \gamma_1\mathbf{A}_4(t) + \ldots + \mathbf{o}(s^n) \tag{41}$$

where

$$\mu = -F_1[s]\,,$$

$$\alpha_1 = \frac{1}{2}F_1[s^2]\,,$$

$$\beta_1 = \frac{1}{3!}F_1[s^3]\,,$$

$$\gamma_1 = \frac{1}{4!}F_1[s^4]\,,\quad \text{etc.} \tag{42}$$

Obviously, $\mu, \alpha_1, \beta_1, \gamma_1$ are not independent.

The same type of consideration applies to higher order functional derivatives. Using bilinearity

$$\boldsymbol{F}_2[\mathbf{G}(s_1)|\mathbf{G}(s_2)] = \mathbf{f}_2[s_1|s_2]\,\mathbf{A}_1^2 - \mathbf{f}_2\left[s_1\left|\frac{s_2^2}{2}\right.\right](\mathbf{A}_1\mathbf{A}_2 + \mathbf{A}_2\mathbf{A}_1)$$
$$+ \mathbf{f}_2\left[\frac{s_1^2}{2}\left|\frac{s_2^2}{2}\right.\right]\mathbf{A}_2^2 + \ldots \qquad (43)$$

where

$$\mathbf{f}_2[q(s_1)|m(s_2)] \overset{\text{def}}{=} \mathbf{f}_2\left[\begin{array}{c|c} \infty & \infty \\ q(s_1) & m(s_2) \\ s_1=0 & s_2=0 \end{array}\right]$$

is a sixth order isotropic tensor, symmetric in successive pairs of indices and to interchanges of q(•) and m(•). These tensors are expressible by dyadic products of the unit with the required symmetries. Omitting terms proportional to **1**, we find

$$\boldsymbol{F}_2[\mathbf{G}|\mathbf{G}] = \alpha_2\mathbf{A}_1^2 + \beta_2[\mathbf{A}_2\mathbf{A}_1 + \mathbf{A}_1\mathbf{A}_2] + \gamma_3\mathbf{A}_2^2 + \ldots \qquad (44)$$

where

$$\alpha_2 = \mathrm{F}_2[s_1|s_2]\,,$$
$$\beta_2 = -\frac{1}{2}\,\mathrm{F}_2[s_1|s_2^2]\,, \qquad (45)$$
$$\gamma_3 = \frac{1}{4}\,\mathrm{F}_2[s_1^2|s_2^2]$$

are defined by the bilinear functional

$$\mathrm{F}_2[q(s_1)|m(s_2)] = \mathrm{F}_2\left[\begin{array}{c|c} \infty & \infty \\ q(s_1) & m(s_2) \\ s_1=0 & s_2=0 \end{array}\right]\,.$$

Obviously α_2, β_2, γ_3, etc. need not be independent.

Finally we find that (44) induces the following expansion

$$\boldsymbol{F}\,[\mathbf{G}] = \mu\mathbf{A}_1+\alpha_1\mathbf{A}_2+\beta_1\mathbf{A}_3+\ldots$$
$$+\alpha_2\mathbf{A}_1^2+\beta_2(\mathbf{A}_2\mathbf{A}_1+\mathbf{A}_1\mathbf{A}_2)+\gamma_3\mathbf{A}_2^2+\ldots\,. \qquad (46)$$

Such a representation is useful only for motions in which the higher order $\mathbf{A}_\ell$, or their coefficients are sufficiently small.

We are now ready to consider nearly steady slow motions. By a slow nearly steady motion we mean

$$\mathbf{U}(\mathbf{x},t) = \varepsilon \mathbf{v}(\mathbf{x},\tau) \tag{47}$$

where τ time derivatives are order one and t time derivatives are slow

$$\frac{\partial}{\partial t} = \varepsilon \frac{\partial}{\partial \tau} . \tag{48}$$

Steady motions are those for $\mathbf{U}$ is independent of τ. It is easy to show, using (48), that

$$\mathbf{A}_m[\mathbf{U}(\mathbf{X},t)] = \varepsilon^m \mathbf{A}_m[\mathbf{v}(\mathbf{x},\tau)] . \tag{49}$$

Hence, from (37), we get

$$\mathbf{G}(s) = \sum_1^N \frac{(-\varepsilon s)^m}{m!} \mathbf{A}_m[\mathbf{v}(\mathbf{x},\tau)] + \mathbf{O}([\varepsilon s]^{N+1}) \tag{50}$$

which is a retardation of the history of $\mathbf{U}(\mathbf{x},t)$.

16.7 Fluids of grade N, stability of the rest state

Fluids of grade N can be regarded as the Nth order approximation to $\boldsymbol{F}$ which arises from (49) and (46)

$$\mathbf{S}^{(1)} = \varepsilon\mu\mathbf{A}_1 ,$$

$$\mathbf{S}^{(2)} = \mathbf{S}^{(1)} + \varepsilon^2(\alpha_1\mathbf{A}_2 + \alpha_2\mathbf{A}_1^2) ,$$

$$\mathbf{S}^{(3)} = \mathbf{S}^{(2)} + \varepsilon^3\{\beta_1\mathbf{A}_3 + \beta_2(\mathbf{A}_1\mathbf{A}_2+\mathbf{A}_2\mathbf{A}_1) + \beta_3\mathbf{A}_1 \operatorname{tr} \mathbf{A}_2\} , \quad \text{etc.} \tag{51}$$

where

$$\mathbf{A}_n \overset{\text{def}}{=} \mathbf{A}_n[\mathbf{v}] .$$

If we now reverse the transformation we find that when $\mathbf{A}_n=\mathbf{A}_n[\mathbf{U}]$

$$\mathbf{S}^{(1)} = \mu \mathbf{A}_1 ,$$

$$\mathbf{S}^{(2)} = \mathbf{S}^{(1)} + \alpha_1 \mathbf{A}_2 + \alpha_2 \mathbf{A}_1^2 , \tag{52}$$

$$\mathbf{S}^{(3)} = \mathbf{S}^{(2)} + \beta_1 \mathbf{A}_3 + \beta_2(\mathbf{A}_1\mathbf{A}_2 + \mathbf{A}_2\mathbf{A}_1) + \beta_3 \mathbf{A}_1 \operatorname{tr} \mathbf{A}_2 .$$

It is usual to define

$$\mathbf{S}_{(1)} = \mu \mathbf{A}_1 ,$$

$$\mathbf{S}_{(2)} = \alpha_1 \mathbf{A}_2 + \alpha_2 \mathbf{A}_1^2 , \tag{53}$$

$$\mathbf{S}_{(3)} = \beta_1 \mathbf{A}_3 + \beta_2(\mathbf{A}_1\mathbf{A}_2 + \mathbf{A}_2\mathbf{A}_1) + \beta_3 \mathbf{A}_1 \operatorname{tr} \mathbf{A}_2 ,$$

$$\mathbf{S}_{(n)} = \phi_n \mathbf{A}_n + \text{nonlinear terms} .$$

Then, for example,

$$\mathbf{S}^{(3)} = \mathbf{S}_{(1)} + \mathbf{S}_{(2)} + \mathbf{S}_{(3)} \tag{54}$$

and $\mathbf{S}^{(n)}$ is known as a fluid grade N.

We can express the coefficients of fluids of grade N in terms of the expansions (6) of $\boldsymbol{F}$ in terms of multiple integrals. To do this we replace $\mathbf{G}(s)$ in (6) with its expanded form (37). Since the $\mathbf{A}_n(t)$ are independent of s, they come out of the integrals. For example $\boldsymbol{F}_1$, given by (7), becomes

$$\boldsymbol{F}_1 = \sum_{n=1} \frac{(-1)^{n-1}}{(n-1)!} \phi_n \mathbf{A}_n(t) \tag{55}$$

where

$$\phi_n = \int_0^\infty s^{n-1} \, G(s) \, ds . \tag{56}$$

This procedure leads to the expressions (42) and (45) where the different functionals F_1, F_2, etc. are expressed in terms of the kernels. We get

$$\phi_1 = \mu = -F_1[s] = -\int_0^\infty s\zeta(s)\, ds = \int_0^\infty G(s)\, ds\ ,$$

$$\phi_2 = -\alpha_1 = -\frac{1}{2} F_1[s^2] = -\int_0^\infty \frac{s^2}{2}\zeta(s)\, ds = \int_0^\infty sG(s)\, ds\ , \quad (57)$$

$$\frac{1}{2}\phi_3 = \beta_1 = -\frac{1}{3!} F_1[s^3] = -\int_0^\infty \frac{s^3}{3!}\zeta(s)\, ds = \frac{1}{2}\int_0^\infty s^2 G(s)\, ds$$

and

$$\alpha_2 = F_2[s_1,s_2] = \int_0^\infty \int_0^\infty \beta(s_1,s_2)\, ds_1\, ds_2\ ,$$

$$\beta_2 = -\frac{1}{2} F_2[s_1,s_2^2] = -\int_0^\infty \int_0^\infty \beta(s_1,s_2)\, s_2\, ds_1\, ds_2\ , \quad (58)$$

Very little is known about the kernel $\beta(s_1,s_2)$. The only attempt to use these multiple integrals in a computation known to me is the study of the free surface and flow induced by the torsional oscillation of a rod in a viscoelastic fluid by Joseph and Beavers [1977a, b]. Generally, when people compute with integrals, they use single integral models like the BKZ model.

Coleman and Markovitz [1964] were the first to note that the coefficients of the fluid of grade N could be expressed in terms of integrals of kernels in the multiple integral expansion.

Sometimes the fluids of grade N are introduced as a constitutive equation which is supposed to apply exactly to some fluid in all of its motions. This point of view suppresses the fact that the constitutive equations are asymptotic expressions designed only for nearly steady slow motion. There are a huge number of examples of incorrect application of these asymptotic expansions to flows which are neither slow nor nearly steady. The results of analysis using fluids of grade N outside their domain of applicability usually disagree with experiment and intuition. Many such examples are given in the books of reference; the one by Bird, Armstrong and Hassager [1977a, b] has a good collection. It is frequently true that asymptotic expressions hold well beyond the range in which they are supposed to be valid. It does not seem that we have this kind of good luck with fluids of grade N.

Perhaps the most striking example of the incorrect use of fluids of grade N is to the problem of the stability of rest. Clearly, only bad or incorrectly applied theory will tell you that the state of rest of an isolated chemically passive fluid is unstable. This type of unacceptable result was given by Joseph [1980c] who showed that the rest state of fluids of grade N, any $N \neq 1$, is unstable in the spectral sense of linearized theory when the ratio of the coefficients of $\mathbf{A}_n$ and $\mathbf{A}_{n-1}$ in the constitutive equation is negative. Negative ratios, and only negative ratios, are implied by integral expansions of the stress. Moreover, in the only case (N=2) which has been checked in experiments with polymeric liquids, the ratio is negative.*

To start the proof of the unacceptable stability result we note that if the shear relaxation function G(s) is given then the ϕ_n are all known. Since G(s) is everywhere positive, integral inequalities among the different ϕ_n are implied (Joseph, [1980b]). For example

$$-\alpha_1 \leq \mu^{1/2}(2\beta_1)^{1/2} . \tag{59}$$

For the demonstration here, it is enough that all the moments ϕ_n of G(s) are positive. Then

$$\phi_1 = \mu>0 , \quad \phi_2 = -\alpha_1>0 , \quad \phi_3 = 2\beta_1>0 \tag{60}$$

and so on. All the experimental determinations of material parameters of fluids which are known to me are consistent with (60).

In the linearized theory we judge stability, as in Lyapounov's theory, by the sign of the real part of the principal spectral value

---

* The first normal stress N_1 (see (87)) has been measured in many polymeric liquids. In all cases so far reported $N_1 \geq 0$ with inequality when polymers are present. At low rates of shearing, the ratio of the coefficients of $\mathbf{A}_2$ on $\mathbf{A}_1$ is α_1/μ which is negative.

associated with the linearized equations of dynamics. First we linearize the stress. From (52), using the recursion rotation (10), App. C, for the $\mathbf{A}_n$, we get

$$\mathbf{S}^{(1)} = \mu \mathbf{A}_1 ,$$

$$\mathbf{S}^{(2)} = \mathbf{S}^{(1)} + \alpha_1 \frac{\partial \mathbf{A}_1}{\partial t} , \quad \alpha_1 = -\phi_2 ,$$

$$\mathbf{S}^{(3)} = \mathbf{S}^{(2)} + \beta_1 \frac{\partial^2 \mathbf{A}_1}{\partial t^2} , \quad \beta_1 = \frac{\phi_3}{2} ,$$

$$\mathbf{S}^{(n)} = \mathbf{S}^{(n-1)} + \frac{(-1)^{n-1}}{(n-1)!} \phi_n \frac{\partial^{n-1} \mathbf{A}_1}{\partial t^{n-1}} . \tag{61}$$

Rivlin-Ericksen fluids of complexity n have the same linearized stresses. Thus, in the work here, there is no difference between complexity and grade. On the other hand, under linearization $\dot{\mathbf{G}}(s) \rightarrow -\mathbf{A}_1(s)$ and the integral representations give

$$\boldsymbol{F}^{(n)} \rightarrow \int_0^\infty G(s) \mathbf{A}_1(s) \, ds \tag{62}$$

which have the *same linearization independent of n.* The only tensor that comes up in the linearized stresses is

$$\mathbf{A}_1(s) = \mathbf{L}(s) + \mathbf{L}^T(s) , \quad \mathbf{L}(s) = \nabla \mathbf{U}(t-s) .$$

For each and every linearization we get

$$\mathbf{S} = \boldsymbol{L}(\mathbf{A}_1) \tag{63}$$

where $\boldsymbol{L}$ is linear and

$$\nabla \cdot \mathbf{S} = \boldsymbol{L}(\nabla^2 \mathbf{U}) .$$

To study the stability of the rest state, we consider a container of fluid V with closed boundary ∂V on which zero velocity is prescribed. A state of rest with $\mathbf{U}=0$ solves the equations of dynamics and we perturb this rest state keeping only linear terms:

$$\rho \frac{\partial \mathbf{U}}{\partial t} = -\nabla P + \boldsymbol{L}(\nabla^2 \mathbf{U}) , \quad \text{div}\, \mathbf{U} = 0 , \quad \mathbf{U}|_{\partial V} = 0 . \tag{64}$$

The spectral problem for (64) is obtained by putting

$$(\mathbf{U},P) = e^{-\sigma t}(\boldsymbol{\zeta},\pi) . \tag{65}$$

For $\boldsymbol{\zeta}$ and π we get

$$-\rho\sigma\boldsymbol{\zeta} = -\nabla\pi + k(\sigma)\nabla^2\boldsymbol{\zeta} , \quad \operatorname{div}\boldsymbol{\zeta} = 0 , \quad \boldsymbol{\zeta}|_{\partial V} = 0 \tag{66}$$

where the $k(\sigma)$ associated with (64) are

$$k(\sigma) = \mu - \alpha_1\sigma = \phi_1 + \sigma\phi_2 ,$$

$$k(\sigma) = \mu - \alpha_1\sigma + \beta_1\sigma^2 = \phi_1 + \sigma\phi_2 + \frac{1}{2}\sigma^2\phi_3 , \tag{67}$$

$$k(\sigma) = \sum_{\ell=1}^{n} \frac{\phi_\ell \sigma^{\ell-1}}{(\ell-1)!} , \quad n > 1 ,$$

and the $k(\sigma)$ associated with (21) is given by

$$k(\sigma) = \int_0^\infty G(s)\, e^{\sigma s}\, ds . \tag{68}$$

It is well known and easy to show that (66) are Euler equations for the Rayleigh quotient

$$\Lambda[\mathbf{u}] = \frac{\int_V (\nabla \mathbf{u})^2 \, dV}{\int_V |\mathbf{u}|^2 \, dV} \tag{69}$$

where

$$\mathbf{u} \in H = (\mathbf{u}:\operatorname{div}\mathbf{u}=0, \mathbf{u}|_{\partial V}=0, \int (\nabla\mathbf{u})^2 \, dV<\infty) .$$

Moreover the critical values Λ_n of the functional (69) are positive, bounded below and cluster only at infinity:

$$0<\Lambda_n , \quad \Lambda_n \geq \Lambda_{n-1} , \quad \Lambda_n \to \infty \quad \text{as } n\to\infty. \tag{70}$$

These critical values are related to the spectral values σ by the equation

$$k(\sigma) = \sigma/\Lambda_n . \tag{71}$$

Any σ solving (71) for any integer n is an acceptable spectral value for (66).

Consider the case $n\to\infty$; then if $k(\sigma)=0$ determines bounded spectral values σ, (71) will imply that $k(\sigma)=0$ as $n\to\infty$. Equation (67) shows that when $n>1$, $k(\sigma)=0$ is a polynomial of degree $n-1$ in σ with strictly positive coefficients. The coefficient of σ^{n-2} in this polynomial is minus the sum of all eigenvalues and it is positive because the coefficient of σ^{n-2} is positive; that is, the roots σ_ℓ of $(n-1)!k(\sigma)/\phi_n = (\sigma-\sigma_1)(\sigma-\sigma_2)\ldots(\sigma-\sigma_{n-1})=0$ satisfy

$$-(\sigma_1 + \sigma_2 + \ldots + \sigma_{n-1}) = (n-1)\frac{\phi_{n-1}}{\phi_n} > 0 . \tag{72}$$

Hence at least one root has a negative real part, leading to *instability* (see (65)).

The instability of the rest state of a layer of "fluid" of grade two against two dimensional disturbances was demonstrated in Craik [1968]. Putting the members of $(67)_1$ equal to zero, we get Craik's result:

$$\sigma = \frac{\mu}{\alpha_1} = \frac{\phi_1}{\phi_2} < 0 .$$

When n=3, we put $(67)_2$ equal to zero and find that

$$\sigma = \frac{1}{2}\frac{\alpha_1}{\beta_1} \pm \frac{1}{2\beta_1}\sqrt{\alpha_1^2-4\mu\beta_1} = -\frac{\phi_2}{\phi_3} \pm \frac{1}{\phi_3}\sqrt{\phi_2^2-\phi_1^2\phi_3} .$$

The inequality (59) shows that $\alpha_1^2-4\mu\beta_1<0$ so that

$$\text{re}\,\sigma = -\frac{\phi_2}{\phi_3} < 0 ,$$

and the rest state of the fluid of grade 3 is unstable. Equation (72) shows that we get this same instability for each and every n.

Equation (72) does not require that one accepts the view that the parameters of the fluid of grade n be expressible as moments of the kernel functions. The general result implied by (72) is stated in the following theorem

The rest state of each and every "fluid" of grade n≠1 is unstable in the sense of the spectral problem of linearized theory if the ratio of the coefficients of A_n *to* A_{n-1} *is negative.*

In contrast, Craik [1968], Joseph [1976], and Slemrod [1976] have shown that the rest state of fluids obeying (62) is stable in the spectral sense of linearized theory provided only that G(s) is a positive, monotone, decreasing function.

I say that it is wrong to study stability using the constitutive expressions for "fluids" of grade n. These expressions arise in response to slow deformations. In fact, there is no such thing as a constitutive equation without prior specification of the domain of deformations in which the constitutive equation lives. However good rigid body mechanics is for some problems, it is obviously no good for studying the deformation of strained bodies. It is equally no good to study stability with constitutive expressions which do not allow you to consider small disturbances with large frequencies. Such disturbances can never be classed as retarded and they do not lead to fluids of grade n.

16.8 Dynamics of slow steady motion

We want to solve a boundary value problem

$$\rho(\mathbf{U}\cdot\nabla)\mathbf{U} = -\nabla P + \operatorname{div}\mathbf{S},\ \operatorname{div}\mathbf{U} = 0 \text{ in } V\,, \qquad \mathbf{U} = \varepsilon\mathbf{U}_0(\mathbf{x}) \text{ on } \partial V \tag{73}$$

with a power series

$$\begin{bmatrix} \mathbf{U}(\mathbf{x},\varepsilon) \\ P(\mathbf{x},\varepsilon) \end{bmatrix} = \sum_{1} \varepsilon^n \begin{bmatrix} \mathbf{U}_n(\mathbf{x}) \\ P_n(\mathbf{x}) \end{bmatrix}. \tag{74}$$

This solution will generate a solution based on approximating the extra stress with fluids of grade n; that is

$$\mathbf{S} = \mathbf{S}_{(1)} + \mathbf{S}_{(2)} + \ldots = \varepsilon\mathbf{S}_1 + \varepsilon^2\mathbf{S}_2 + \varepsilon^3\mathbf{S}_3 + \ldots \tag{75}$$

where

$$S_1 = \mu A_1[U_1] ,$$

$$S_2 = \mu A_1[U_2] + \alpha_1 A_2[U_1,U_1] + \alpha_2 A_1^2[U_1] ,$$

$$\begin{aligned} S_3 = \mu A_1[U_3] &+ 2\alpha_1 A_2[U_1,U_2] + 2\alpha_2 A_1[U_1]A_1[U_2] \\ &+ \beta_1 A_3[U_1,U_1,U_1] + \beta_3(\mathrm{tr}\, A_2[U_1,U_1])A_1[U_1]) \\ &+ \beta_2\{A_2[U_1,U_1]A_1[U_1] + \text{transpose}\} . \end{aligned} \tag{76}$$

The reader should note that $\mathbf{A}_n$ is n-linear form in $\mathbf{U}$, through the recursion relation

$$A_{m+1} = \overset{\bullet}{A}_m + A_m \nabla U + \nabla U^T A_m$$

giving rise through terms like $A_2[U_1,U_2]$, etc.

The first problem $O(\varepsilon)$ is

$$\begin{aligned} -\nabla P_1 + \mu\nabla^2 U_1 = 0 , \quad \operatorname{div} U_1 = 0 \text{ in } V , \\ U_1(x) = U_0(x) \text{ on } \partial V . \end{aligned} \tag{77}$$

The second problem is given by (24)

The problems with n>1 are in the form

$$\begin{aligned} -\nabla P_n + \mu\nabla^2 U_n = f_n(U_\ell) , \quad \operatorname{div} U_n = 0 \text{ in } V, \\ U_n(x) = 0 \text{ on } \partial V \end{aligned} \tag{78}$$

where $\mathbf{f}_n(\mathbf{U}_\ell)$ is known from earlier $\ell<n$ solutions

$$f_n(U_\ell) = \rho \sum_{m+\ell=n} (U_m \cdot \nabla)U_\ell - \operatorname{div}\{S_n - \mu A_1[U_n]\} . \tag{79}$$

Theorem. Problem (78) is a Stokes flow problem and is uniquely invertible in the same space as in the classical theory. Problems (78) can be solved sequentially. For each fixed n, (78) is uniquely invertible.

The proof of this theorem is by the methods used to study the Navier-Stokes equations, under certain technical conditions.

Convergence. For each and every n we may compute $\mathbf{U}_n(\mathbf{x})$, $P_n(\mathbf{x})$ and these functions (1) are unique. We don't know if, and in what sense the series "solution" (74) converges to a solution of (73); if it does represent a solution, is it the only small solution of (73)?

The problem is that we have no a-priori estimates of $\mathbf{f}_n(\mathbf{U}_\ell)$, independent of $n>\ell$ because $\mathbf{f}_n(\mathbf{U}_\ell)$ is a differential operator of order higher than two. The high order of differentiation enters through the terms

$$\begin{aligned}(\mathbf{U}\cdot\nabla)\mathbf{A}_n &= (\mathbf{U}\cdot\nabla)(\mathbf{U}\cdot\nabla\mathbf{A}_{n-1}+\ldots)\\ &= (\mathbf{U}\cdot\nabla)[(\mathbf{U}\cdot\nabla)(\mathbf{U}\cdot\nabla\mathbf{A}_{n-2}+\ldots)]\end{aligned} \tag{80}$$

which enter into the last member of (79).

The convergence problem generated by the slow motion expansions is a deep one from the mathematical point of view. We play smallness in ε against growth due to differentiation. This problem is still unsolved. In (1979) I formulated an easier problem with the same difficulty.

$$u'' = \alpha_0 uu' + \alpha_1 uu'u'' + \alpha_2 uu'u''u''' + \ldots$$

where we could terminate the right with differentiation at order $n>2$. The "smallness" ε is introduced at the boundary

$$u(0) = 0\,, \quad u(1) = \varepsilon\,. \tag{81}$$

These are just the right number of conditions for a second order, but not higher order differential equation.

We can "solve" this problem with a "slow motion" power series

$$u(x,\varepsilon) = \varepsilon u_0 + \varepsilon^2 u_1 + \varepsilon^3 u_2 + \ldots\,, \tag{82}$$

$$u_0(x) = x\,,$$

$$u_1(x) = \alpha_0 \frac{x}{3}(x^2-1)\,, \tag{83}$$

etc.

The question here is convergence. We want to know if and in what sense this formal power series solution converges to a solution. Perhaps it converges and is the unique "small amplitude" solution. Uniqueness in general could not be expected in such a highly nonlinear and vaguely specified problem. Existence and uniqueness of small solutions is all that is at issue.

Niggeman [1982], in the group of K. Kirchgässner, partially solved this problem posed by me, but it is generally still unsolved.

16.9 Functional expansions perturbing viscometric flows

Viscometric flows are discussed in every book of reference and they will not be treated in detail in this book. Coleman, Markovitz, and Noll [1966] define a viscometric flow at a particle as differing from simple shear by a time dependent change of frame. Yin and Pipkin [1970] say that the history of deformation of a fluid element is viscometric if its motion, viewed from a system of possibly rotating axes, is a steady simple shearing motion. The shear rate $\kappa(\mathbf{x},t)$ of a particle in viscometric flow is an invariant following the motion; that is

$$\frac{d\kappa}{dt} = \frac{\partial\kappa}{\partial t} + (\mathbf{u}\bullet\nabla)\kappa = 0 \ . \tag{84}$$

Basic to these definitions is the concept of a simple shear flow $\mathbf{U}=\mathbf{e}_x u(y)$, $u(y)=\kappa y$ where κ is the rate of shear.* This type of flow is dynamically admissible, satisfies the equations of motion, no matter what the constitutive equation. Some of the properties of simple shear flows are discussed in Appendix C.

From the definition of simple shear it is clear that the stress components will depend only on the shear rate κ. Let (x,y) be the plane of shearing motion. Then, using symmetry arguments one shows that $\tau_{xz}=0$. This leaves four independent components of the stress. We would like to express the remaining shear stress and the normal stress in terms of κ. This is impossible for the normal stresses because of the pressure. However, the normal stress differences are independent of pressure and can be constitutively determined. The choice of the two normal stresses is not unique but there is a good choice: $N_1=T_{11}-T_{22}$, the first normal stress difference and $N_2=T_{11}-T_{33}$, the second normal stress difference where $x_1=x$, $x_2=y$, $x_3=z$. The first normal stress is an extra tension in the direction of flow and is usually large in the shear flow of polymeric liquids while N_2 is usually negative and smaller than N_1,

---

* Rheologists call the shear rate $\overset{\circ}{\gamma}$. I prefer here to use the simple notation with $\kappa=\overset{\circ}{\gamma}$.

say a tenth of N_1. $N_1(\kappa)$, $N_2(\kappa)$, and the viscosity $\mu(\kappa)$ are the three viscometric functions.

Symmetry arguments are given to show that the shear stress is an odd function and the normal stresses even functions of κ.

It is customary to assume that the viscometric functions can be developed in powers of κ, although counter-examples have been presented. Nonanalytic viscosity functions of the form

$$\mu(\kappa) = \mu_0 - A\sqrt{\kappa} \tag{85}$$

have been computed in molecular dynamic simulations of simple shear for fluids obeying a Lennard Jones potential by Evans, Hanley, and Hess [1984]. The square root behavior occurs in three dimensions, log κ in two, and linearity in κ in four dimensions. This shows a certain artificiality since results ought to rely only on the fluid and not on dimensionality.

Assuming now that the viscometric functions may be developed in powers of κ we may evaluate the fluids of grade N on simple shear to find that

$$\mu(\kappa) = \mu_0 + 2(\beta_1+\beta_3)\kappa^2 + 0(\kappa^4) \tag{86}$$

where $\tau_{12}=\tau_{xy}=\mu(\kappa)\kappa$,

$$\begin{aligned} N_1(\kappa) &= T_{11}-T_{22} = T_{xx}-T_{yy} = \\ &-2\alpha_1\kappa^2 - 4(\gamma_3+\gamma_4+\gamma_5)\kappa^4 + 0(\kappa^6)\,, \end{aligned} \tag{87}$$

$$\begin{aligned} N_2(\kappa) &= T_{11}-T_{33} = T_{xx}-T_{zz} = \\ &(2\alpha_1+\alpha_2)\kappa^2 + \{4(\gamma_3+\gamma_4+\gamma_5) + 2\gamma_6\}\kappa^4 + 0(\kappa^6)\,. \end{aligned} \tag{88}$$

The property of shear thinning or thickening are associated with nonlinear terms in (86). In the limit of small κ

$$\begin{aligned} \tau_{12} &= \kappa\mu_0 + 0(\kappa^3)\,, \\ N_1(\kappa) &= -2\alpha_1\kappa^2 + 0(\kappa^4)\,, \\ N_2(\kappa) &= (2\alpha_1+\alpha_2)\kappa^2 + 0(\kappa^4)\,. \end{aligned} \tag{89}$$

The viscometric flows are very important because they are used as the theory for many of the rheometrical instruments used in industrial and academic laboratories. The common flows which fit the definition of viscometric flow are plane Couette flow, Couette flow between rotating and sliding cylinders, Poiseuille flow in channels and between concentric cylinders, combined Poiseuille-Couette flow in channels and between rotating and sliding cylinders, flow between rotating disks which rotate with different speeds around a common axis, flow between rotating cones which rotate with different speeds around a common axis, flow between a rotating cone and disk. Yin and Pipkin [1970] have exhibited some other viscometric flows.

The concept of a viscometric flow is kinematic. Dynamics is ignored. Some viscometric flow can satisfy the equations of motions and some cannot (unless artificial body forces are assumed to balance the equations). For example, the flow between rotating disks and cones cannot satisfy the equations of motion unless inertia is neglected.

A huge number of flows might be described as nearly viscometric. A general theory for perturbed viscometric flow was considered by Pipkin and Owen [1976] . They derived the lowest order theory in the perturbation sequence and showed that thirteen independent components of a kernel, functions of the shear rate and lapse time, are required to describe any fluid in a nearly viscometric flow. The nearly viscometric theory of Pipkin and Owen is discussed in some of the books of reference. However, I like the presentation derived by Dunwoody and Joseph [1984] as presented by Ahrens, et. al [1984], given below.

Consider a viscoelastic fluid in which the extra stress is given by a functional of the deformation gradients

$$\boldsymbol{S} = \boldsymbol{S}^{\mathrm{T}} = \mathbf{T} + \mathrm{P}\mathbf{1} = \boldsymbol{F}\underset{s=0}{\overset{\infty}{[}}\mathbf{G}(s)] \,. \tag{90}$$

Here $\mathbf{T}$ is the Cauchy stress, P the pressure, and $\mathbf{G}$ the relative Cauchy strain: $\mathbf{G}(s)=\nabla\boldsymbol{\chi}^{\mathrm{T}}(s)\nabla\boldsymbol{\chi}(s)-\mathbf{1}$, where $\boldsymbol{\chi}(s)=\boldsymbol{\chi}(\mathbf{x},t-s)$ is the position at time t–s of the particle presently (i.e. at time t) at $\mathbf{x}$. We represent the basic flow (as yet unspecified) by a "$_0$" subscript, and write the perturbation as follows:

$$
\begin{aligned}
&\text{(a) } \boldsymbol{\chi}(s) = \boldsymbol{\chi}_0(s) + \boldsymbol{\xi}(s)\ ,\\
&\text{(b) } \mathbf{U}(s) = \mathbf{U}_0(s) + \mathbf{u}(s)\ ,\\
&\text{(c) } P = P_0 + p\ ,\\
&\text{(d) } \boldsymbol{S} = \boldsymbol{S}_0 + \boldsymbol{S}\ .
\end{aligned}
\tag{91}
$$

Here the solenoidal velocity fields $\mathbf{U}_0$ and $\mathbf{u}$ satisfy

$$
\begin{aligned}
&\text{(a) } \mathbf{U}_0(\boldsymbol{\chi}_0,t-s) = -\frac{\partial \boldsymbol{\chi}_0}{\partial s},\quad \boldsymbol{\chi}_0(0) = \mathbf{x}\ ,\\
&\text{(b) } \mathbf{u}(\boldsymbol{\chi}_0,t-s) = -\frac{\partial \boldsymbol{\xi}}{\partial s} - (\boldsymbol{\xi}\cdot\nabla)\mathbf{U}_0\ ,\quad \boldsymbol{\xi}(0) = \mathbf{0}\ .
\end{aligned}
\tag{92}
$$

The linearized perturbation of the extra stress is given by

$$
\mathbf{S}_0 = \underset{s=0}{\overset{\infty}{\boldsymbol{F}}}\,[\mathbf{G}_0(s)]\ ,\quad \mathbf{G}_0 = \nabla\boldsymbol{\chi}_0^T\nabla\boldsymbol{\chi}_0 - \mathbf{1}\ , \tag{93}
$$

$$
\mathbf{S} = \underset{s=0}{\overset{\infty}{\boldsymbol{F}_1}}[\mathbf{G}_0(s)|\delta\mathbf{G}(s)]\ ,\quad \delta\mathbf{G} = \nabla\boldsymbol{\xi}^T\nabla\boldsymbol{\chi}_0 + \nabla\boldsymbol{\chi}_0^T\nabla\boldsymbol{\xi} \tag{94}
$$

where $\boldsymbol{F}_1$ is the first functional derivative of $\boldsymbol{F}$, evaluated on the basic flow; that is, $\boldsymbol{F}_1$ is a linear operator acting on $\delta\mathbf{G}$. We assume an integral representation for $\boldsymbol{F}_1$:

$$
\underset{s=0}{\overset{\infty}{\boldsymbol{F}_1}}[\mathbf{G}_0|\delta\mathbf{G}] = \int_0^\infty \mathrm{K}(s,\mathbf{G}_0)\delta\mathbf{G}(s)\ ds\ . \tag{95}
$$

The isotropy of $\boldsymbol{F}$ implies that

$$
\mathbf{Q}\mathbf{S}_0\mathbf{Q}^T = \underset{s=0}{\overset{\infty}{\boldsymbol{F}}}\,[\mathbf{Q}\mathbf{G}_0\mathbf{Q}^T]\ , \tag{96}
$$

$$
\mathbf{Q}\underset{s=0}{\overset{\infty}{\boldsymbol{F}_1}}[\mathbf{G}_0|\delta\mathbf{G}]\mathbf{Q}^T = \underset{s=0}{\overset{\infty}{\boldsymbol{F}_1}}[\mathbf{Q}\mathbf{G}_0\mathbf{Q}^T|\mathbf{Q}\delta\mathbf{G}(s)\mathbf{Q}^T] \tag{97}
$$

which holds for all orthogonal $\mathbf{Q}$ and symmetric $\delta\mathbf{G}$.

For simple shear, with shear rate κ=U/h, we have

$$
\boldsymbol{\chi}_0 = (\mathbf{1} - s\mathbf{N})\mathbf{x}\ , \tag{98}
$$

$$G_0 = -s(N+N^T) + s^2 N^T N\ , \tag{99}$$

$$\delta G = \nabla\xi + \nabla\xi^T - s(\nabla\xi^T N + N^T \nabla\xi) \tag{100}$$

where

$$N = \kappa \begin{pmatrix} 0 & 1 & 0 \\ 0 & 0 & 0 \\ 0 & 0 & 0 \end{pmatrix}. \tag{101}$$

The extra stress of the basic shear flow is given by the viscometric functions

$$S_0 = T_0 - \left(\frac{1}{3}\operatorname{tr} T_0\right)\mathbf{1} = \frac{1}{3}\begin{pmatrix} 2N_1+N_2 & 3\tau & 0 \\ 3\tau & -N_1+N_2 & 0 \\ 0 & 0 & -N_1-2N_2 \end{pmatrix} \tag{102}$$

where τ is the shear stress and $N_1=T_{11}-T_{22}$, $N_2=T_{22}-T_{33}$ are the first and second normal-stress differences.

Consistency relations between the kernel functions **K** and the viscometric functions can be obtained as follows. First, we consider families $Q(\lambda)$ of orthogonal matrices such that $Q(0)=\mathbf{1}$ and then differentiate (96) with respect to λ, at $\lambda=0$. Since $\dot{Q}(0)$ is an arbitrary skew-symmetric matrix, this will yield three matrix equations. A fourth matrix equation is obtained by differentiating (96) with respect to κ, i.e. by perturbing the shear rate. Using the notation $(K)_{ijlm}=K_{lm}$, we find the following four equations

$$\begin{pmatrix} \frac{2N_1'+N_2'}{3} & \tau' & 0 \\ \tau' & \frac{-N_1'+N_2'}{3} & 0 \\ 0 & 0 & \frac{-N_1'-2N_2'}{3} \end{pmatrix} \tag{103}$$

$$= \int_0^\infty -2s[K_{12}-\kappa s K_{22}]\, ds\ ,$$

$$\begin{pmatrix} -2\tau & N_1 & 0 \\ N_1 & 2\tau & 0 \\ 0 & 0 & 0 \end{pmatrix} = \int_0^\infty 2\kappa s[\mathbf{K}_{11} - \mathbf{K}_{22} - \kappa s \mathbf{K}_{12}]\, ds\,, \tag{104}$$

$$\begin{pmatrix} 0 & 0 & N_1+N_2 \\ 0 & 0 & \tau \\ N_1+N_2 & \tau & 0 \end{pmatrix} = \int_0^\infty -2\kappa s \mathbf{K}_{23}\, ds\,, \tag{105}$$

$$\begin{pmatrix} 0 & 0 & \tau \\ 0 & 0 & N_2 \\ \tau & N_2 & 0 \end{pmatrix} = \int_0^\infty 2\kappa s(\kappa s \mathbf{K}_{23} - \mathbf{K}_{13})\, ds\,. \tag{106}$$

Clearly, only (103) and (104) are relevant for two-dimensional flows.

Equation (97) for the isotropy of $\boldsymbol{F}_1$ can be used to show, by choosing special forms for $\mathbf{Q}$, that

$$K_{ijlm} = 0 \tag{107}$$

whenever there are an odd number of 3's in the set {i,j,l,m}. The symmetry of $\mathbf{S}$ implies that

$$K_{ijlm} = K_{jilm}\,, \tag{108}$$

Since the fluid is incompressible

$$\det \mathbf{C} = \det(\mathbf{C}_0 + \delta\mathbf{G}) = \det \mathbf{C}_0^{-1} \det(\mathbf{C}_0 + \delta\mathbf{G})$$

$$= \det(\mathbf{1} + \mathbf{C}_0^{-1}\delta\mathbf{G}) = 1 \tag{109}$$

where $\mathbf{G}=\mathbf{C}-\mathbf{1}$, $\mathbf{C}=\mathbf{C}_t(\mathbf{x},\tau)$ and

$$[\mathbf{C}_0^{-1}] = \begin{bmatrix} 1+\kappa^2 s^2 & \kappa s & 0 \\ \kappa s & 1 & 0 \\ 0 & 0 & 1 \end{bmatrix}. \tag{110}$$

Equation (109) is now expanded for small $\mathbf{C}_0^{-1}\delta\mathbf{G}$ and one finds that to leading order

$$\mathrm{tr}\,[\mathbf{C}_0^{-1}\delta\mathbf{G}] = 0\,. \tag{111}$$

If we choose

$$\sum_{\ell=1}^{3} \mathbf{K}_{\ell\ell} + 2\kappa s\mathbf{K}_{12} + \kappa^2 s^2 b\mathbf{K}_{11} = \mathbf{0} \tag{112}$$

then the stress $\boldsymbol{F}_1$ will vanish for each and every $\delta\mathbf{G}$ satisfying the incompressibility condition (111). From (95) and (100), we find the following expression for the perturbed extra stress:

$$\mathbf{S} = \int_0^\infty \sum_{\alpha,\beta=1}^{3} \mathbf{K}_{\alpha\beta}\delta\mathbf{G}_{\alpha\beta}\,ds$$

$$= 2\int_0^\infty \{\mathbf{K}_2\xi_{1,1} + \mathbf{K}_1(\xi_{1,2}+\xi_{2,1}) + \mathbf{K}_{13}(\xi_{1,3} + \xi_{3,1}) \tag{113}$$

$$+ \mathbf{K}_{23}(\xi_{2,3} + \xi_{3,2} - \kappa s\xi_{1,3}) + (\mathbf{K}_{33}-\mathbf{K}_{22})\xi_{3,3}\}\,ds$$

where

$$\mathbf{K}_1 = -\kappa s\mathbf{K}_{22} + \mathbf{K}_{12}\,, \tag{114}$$

$$\mathbf{K}_2 = \mathbf{K}_{11} - \mathbf{K}_{22} - \kappa s\mathbf{K}_{12}\,. \tag{115}$$

Note that $\mathbf{K}_1$, $\mathbf{K}_2$, $\mathbf{K}_{13}$, and $\mathbf{K}_{23}$ are all related to viscometric functions by the consistency relations (103)-(106), but that $\mathbf{K}_{22}-\mathbf{K}_{33}$ does not occur in any of the consistency relations.

Many of the ideas put forward by Pipkin and Owen were developed independently by Goddard and Miller [1967] for application to the problem of stability of Couette flow between rotating cylinders. In fact it is not possible to give a general discussion of the linearized theory of stability without linearizing the stress on a viscometric flow.

The application of the theory of perturbed viscometric flows has been restricted because of the difficulty of dealing with thirteen unknown kernels. Some applications are discussed in the book by Huilgol [1975]. The perturbation problem is simpler in two dimensions. Only four kernels need to be determined and these are related by the consistency conditions. A further simplification is based on a drastic simplification of the perturbed strain where

$$\delta\mathbf{G} \cong -s\mathbf{A}_1[\mathbf{u}(\mathbf{x},t)] \tag{116}$$

and $\mathbf{u}(\mathbf{x},t)$ is the perturbed velocity. (In fact, Dunwoody and Joseph [1984] formulated the Pipkin-Owen theory in terms of $\mathbf{u}(\mathbf{x},t-s)$). Kramer and Johnson [1972] studied a flow in a cylinder driven at the top by a rotating disk using (116). They regard (116) as a "slow flow" so that they are assuming the motion is slow rather than the memory is short. Becker, *et al.* [1980] tried to justify the simplifications implied by (116) with an idea based on the fluid having a memory so short that perturbation terms proportional to higher orders in s don't enter into the stress. This theory was applied to boundary layer and lubrication theory and to stability of simple shear. A good discussion of these applications is given by Böhme [1987] and in the original references. The short memory idea was applied to the study of stability of simple shear by Akbay, Becker, Krozer, and Sponagel [1980a, b]. Their work is reviewed below.

Assuming the memory so short that only the leading term of $\delta\mathbf{G}$, given by (116) enters into the stress, we can write (113) as

$$\begin{aligned}
\mathbf{S} &= -2\int_0^\infty s\left\{\sum_{\alpha,\beta=1}^{3}\mathbf{K}_{\alpha\beta}D_{\alpha\beta}\right\}ds \\
&= -2\int_0^\infty s(\mathbf{K}_{11}-\mathbf{K}_{22})\,ds\,\frac{\partial u_1}{\partial x_1} - 2\int_0^\infty s\mathbf{K}_{12}\,ds\left(\frac{\partial u_1}{\partial x_2}+\frac{\partial u_2}{\partial x_1}\right) \\
&\quad - 2\int_0^\infty s\mathbf{K}_{13}\,ds\left(\frac{\partial u_1}{\partial x_3}+\frac{\partial u_3}{\partial x_1}\right) - 2\int_0^\infty s\mathbf{K}_{23}\,ds\left(\frac{\partial u_2}{\partial x_3}+\frac{\partial u_3}{\partial x_2}\right) \\
&\quad - 2\int_0^\infty s(\mathbf{K}_{33}-\mathbf{K}_{22})\,ds\,\frac{\partial u_3}{\partial x_3}\,.
\end{aligned} \tag{117}$$

where $u_\ell = u_\ell(\mathbf{x},t)$ is independent of s. The first four integrals are precisely those occurring in the consistency relations, and if terms of order κs can be neglected compared to others, they are approximately equal to expression (103)–(106) involving viscometric functions. With this approximation, we arrive at the following expression for the perturbed stress:

$$\mathbf{S} = -\frac{1}{\kappa}\begin{pmatrix} -2\tau & N_1 & 0 \\ N_1 & 2\tau & 0 \\ 0 & 0 & 0 \end{pmatrix}\frac{\partial u_1}{\partial x_1}$$
$$+\begin{pmatrix} \frac{2N_1'+N_2'}{3} & \tau' & 0 \\ \tau' & \frac{-N_1'+N_2'}{3} & 0 \\ 0 & 0 & \frac{-2N_2'-N_1'}{3} \end{pmatrix}\left(\frac{\partial u_1}{\partial x_2}+\frac{\partial u_2}{\partial x_1}\right)$$
$$+\frac{1}{\kappa}\begin{pmatrix} 0 & 0 & \tau \\ 0 & 0 & N_2 \\ \tau & N_2 & 0 \end{pmatrix}\left(\frac{\partial u_1}{\partial x_3}+\frac{\partial u_3}{\partial x_1}\right) \tag{118}$$
$$+\frac{1}{\kappa}\begin{pmatrix} 0 & 0 & N_1+N_2 \\ 0 & 0 & \tau \\ N_1+N_2 & \tau & 0 \end{pmatrix}\left(\frac{\partial u_2}{\partial x_3}+\frac{\partial u_3}{\partial x_2}\right)$$
$$-2\int_0^\infty s(\mathbf{K}_{33}-\mathbf{K}_{22})\,ds\,\frac{\partial u_3}{\partial x_3}\,.$$

We see that even in this approximation there remains a term not related to viscometric functions. More such terms would occur, if higher order terms were included. In two-dimensional flows, Equation (118) reduces to

$$\mathbf{S} = -\frac{1}{\kappa}\begin{pmatrix} -2\tau & N_1 \\ N_1 & 2\tau \end{pmatrix}\frac{\partial u_1}{\partial x_1}$$
$$+\begin{pmatrix} \frac{2N_1'+N_2'}{3} & \tau' \\ \tau' & \frac{-N_1'+N_2'}{3} \end{pmatrix}\left(\frac{\partial u_1}{\partial x_2} + \frac{\partial u_2}{\partial x_1}\right). \tag{119}$$

The equation of motion, after linearizing at plane Couette flow and eliminating the pressure, becomes

$$\rho\left(\frac{\partial}{\partial t} + \kappa x_2\frac{\partial}{\partial x_1}\right)\left(\frac{\partial u_1}{\partial x_2} - \frac{\partial u_2}{\partial x_1}\right)$$
$$= \left(\frac{\partial^2}{\partial x_2^2} - \frac{\partial^2}{\partial x_1^2}\right)S_{12} + \frac{\partial^2(S_{11}-S_{22})}{\partial x_1\partial x_2}. \tag{120}$$

We introduce a stream function Ψ such that

$$(u_1,u_2) = \left(-\frac{\partial\Psi}{\partial x_2}, \frac{\partial\Psi}{\partial x_1}\right).$$

Then we obtain

$$\rho\left(\frac{\partial}{\partial t} + \kappa x_2\frac{\partial}{\partial x_1}\right)\left(\frac{\partial^2\Psi}{\partial x_2^2} + \frac{\partial^2\Psi}{\partial x_1^2}\right)$$
$$= \left(N_1' - \frac{N_1}{\kappa}\right)\frac{\partial^2}{\partial x_1\partial x_2}L\Psi + \tau' L^2\Psi + \frac{4\tau}{\kappa}\frac{\partial^4\Psi}{\partial x_1^2\partial x_2^2} \tag{121}$$

where

$$L = \frac{\partial^2}{\partial x_2^2} - \frac{\partial^2}{\partial x_1^2}.$$

The boundary conditions are

$$\Psi = \frac{\partial\Psi}{\partial x_2} = 0 \quad \text{at} \quad x_2 = 0 \quad \text{and } x_2 = h. \tag{122}$$

Equations (120, 121) give rise to an instability if a certain "Weissenberg number" exceeds a critical value. This instability will be associated with a change of type. We first recall that to obtain the symbol of an operator, one simply replaces the derivatives $\partial/\partial x_i$ by numbers D_i. Thus we can assign to any differential operator a polynomial. For example, the differential operator on the right-hand side of (121) would have the symbol

$$\left(N_1{}' - \frac{N_1}{\kappa}\right) D_1 D_2 \,(D_2^2 - D_1^2) + \tau'(D_2^2 - D_1^2)^2 + \frac{4\tau}{\kappa} D_1^2 D_2^2 \,. \quad (123)$$

Two differential operators are said to be of the same type, if—up to a transformation of the independent variables—their symbols have the same asymptotic behavior at infinity. If the asymptotic behavior of the symbol changes, we say that a change of type occurs. For example, the operator $\frac{\partial^2}{\partial x^2} + y\frac{\partial^2}{\partial y^2}$ changes type if y goes through zero: For $y>0$ the symbol $D_1^2 + yD_2^2$ is a positive definite quadratic form (the operator is then called "elliptic"), for $y<0$ the symbol is an indefinite quadratic form (the operator is then called "hyperbolic"). The type describes how the operator acts on rapidly varying functions. For such cases, one expects that locally the variations of the coefficient can be neglected, and that, in the interior of the domain where the problem is posed, the influence of boundary conditions can also be neglected. For constant coefficient problems without boundaries, the symbol is just the Fourier transform of the differential operator. This then again leads to "frozen coefficients" and normal modes.

How does the foregoing apply to (121)? If we formally set $\partial/\partial t = \sigma$, $\partial/\partial x_1 = D_1 = i\alpha$, $\partial/\partial x_2 = D_2 = i\beta$, then the left side of the equation becomes

$$\rho(\sigma + \kappa x_2 i\alpha)(-\beta^2 - \alpha^2)\,, \quad (124)$$

and the right side becomes

$$-\left(N_1{}' - \frac{N_1}{\kappa}\right)(\alpha^2 - \beta^2)\,\alpha\beta + \tau'(\alpha^2 - \beta^2)^2 + \frac{4\tau}{\kappa}\,\alpha^2\beta^2\,. \quad (125)$$

The expression (125) is a homogeneous fourth-degree polynomial. It turns out that this polynomial is positive definite if and only if

$$\mathrm{We}^2 = \left[\frac{d}{d\kappa}\frac{N_1}{\kappa}\right]^2 \kappa^2/(\eta(\kappa)\,\tau'(\kappa)) < 16\,. \tag{126}$$

If $\mathrm{We}^2>16$, the polynomial is indefinite. If we now formally put (124) equal to (125), then we find that σ is negative if α,β are large and $\mathrm{We}^2<16$. If $\mathrm{We}^2>16$, then there is a sector in the α,β-plane for which σ becomes arbitrarily large in this case as $\alpha,\beta\to\infty$.

This heuristic argument can be made rigorous (see Figure 16.1). If $\mathrm{We}^2<16$, then at least highly oscillatory modes will be stable. If instabilities exist, then they will only affect a finite number of modes. If, on the other hand, $\mathrm{We}^2>16$, then we expect to find an infinite number of unstable modes. In addition, the growth constants of these modes can be arbitrarily large. Thus the change of type leads to an instability which is in a sense much stronger than those studied in bifurcation theory. One cannot expect a secondary flow whose dynamics would be governed by the evolution of one or several modes. Rather, the initial value problem becomes ill-posed, i.e., there are flow fields for $\mathrm{We}^2>16$, which would not occur even as transient states, since "random" disturbances containing all modes would blow up instantly. In situations where this "forbidden" unstable region lies between two stable ones, a phase transition can occur. That is, there can be discontinuous solutions whose values lie in both of the stable regions. The two regions are then referred to as different "phases." If the forbidden region stretches all the way to infinity and a second phase does not exist, the result of a change of type may be something like "fracture."

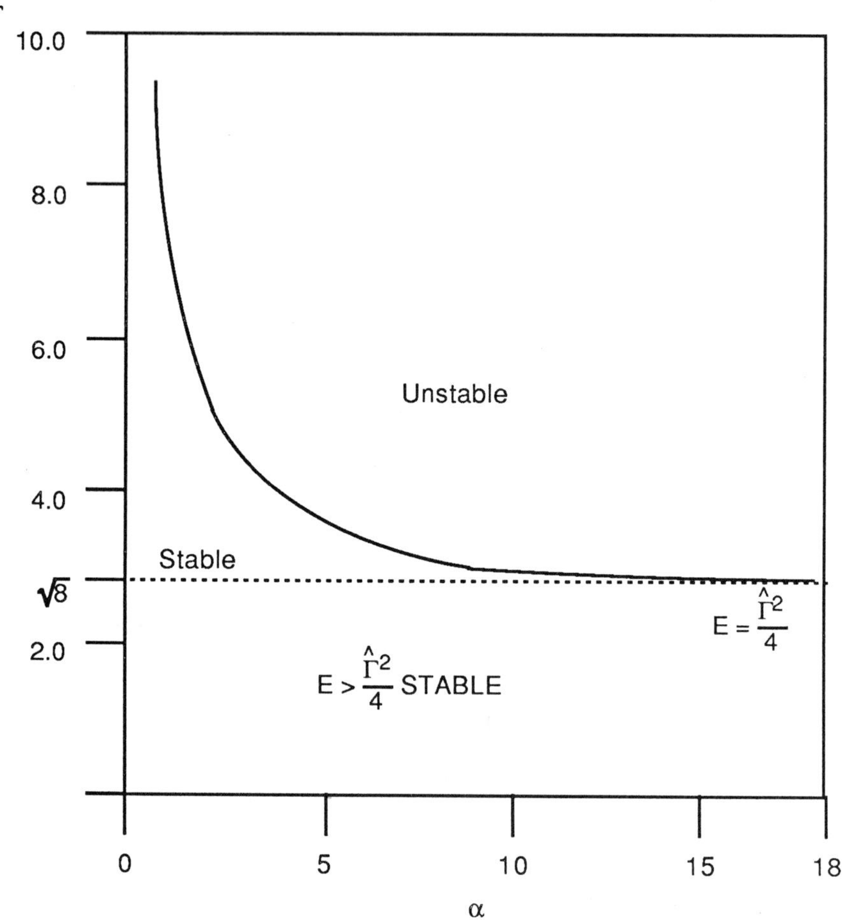

Figure 16.1. Stability region for We and the wave number α, for zero Reynolds number. The neutral stability curve is asymptotic to the line We=4 so that an infinite number of highly oscillating modes become unstable as We approaches 4. This curve is given by the exact solution to the eigenvalue problem obtained from (121) by using normal modes in the t and x_1 variables, and for zero Reynolds number. Similar curves are given for non-zero Reynolds number in Ahrens, et. al [1984], obtained from numerical approximation of the eigenvalue problem. Here $\Gamma = (\kappa/\tau'(\kappa))\ (N_1(\kappa)/\kappa)'$ and $E = 4\mu(\kappa)/\tau'(\kappa)$, so that $We = 2\Gamma/\sqrt{E}$. In this Figure E=2.κ

Ahrens, *et al.* [1984] gave a critical assessment of the theory of Akbay et. al. They point out that there is a problem with the short memory assumption. It turns out that if this assumption is applied to the basic shear flow, then in some models the stability criterion (126) is always satisfied. It appears that in some cases where (126) need not be satisfied, for example, for the upper and lower convected Maxwell model, no change of type can occur (see (1.66) and (1.67)) and direct studies of stability without first invoking short memory do not give instability. On the other hand the criterion (126) is in good agreement with experiments, particularly those of Gleissle [1982].

Exercises

Exercise 16.1.
(Example of functional differentiation). Define

$$\begin{cases} \underset{s=0}{\overset{\infty}{P}}[u(s)] = \int_0^\infty F(s,u(s))\, ds, \quad P[0] \overset{def}{=} \beta \\ \underset{s=0}{\overset{\infty}{\boldsymbol{F}}}[u(s)] = \exp P[u(s)]\, ds, \quad \alpha \overset{def}{=} \exp \beta . \end{cases} \tag{127}$$

Show that

$$\begin{cases} \boldsymbol{F}_1\underset{s=0}{\overset{\infty}{[0|u_1(s)]}} = \alpha \int_0^\infty F_u(s,0)\, u_1(s)\, ds \\ \boldsymbol{F}_2\underset{s_1=0}{\overset{\infty}{[0|u_1(s_1)}}\underset{s_2=0}{\overset{\infty}{|u_2(s_2)]}} = \int_0^\infty \int_0^\infty \alpha^2\{F_u(s_1)F_u(s_2)+ \\ \qquad \alpha\delta(s_1-s_2)F_{uu}(s_1)\}\, u_1(s_1)\, u_2(s_2)\, ds_1\, ds_2 \end{cases} \tag{128}$$

where $\delta(\bullet)$ is the Dirac delta function and $u, u_1, u_2 \in \text{dom}\ \boldsymbol{F}$. Consider

$$\boldsymbol{F}\,[u(s)+\lambda_1 u_1(s)+\lambda_2 u_2(s)] = \exp \int_0^\infty F(s,u(s)+\lambda_1 u_1(s)+\lambda_2 u_2(s))\, ds$$

$$\frac{\partial \boldsymbol{F}}{\partial \lambda_1} = \boldsymbol{F}\,[u(s)+\lambda_1 u_1(s)+\lambda_1 u_2(s)]$$

$$\int_0^\infty \{u_1(s)F'(s,u(s)+\lambda_1 u_1(s)+\lambda_2 u_2(s))\}\, ds\ ,$$

$$\frac{\partial^2 \boldsymbol{F}}{\partial \lambda_1 \partial \lambda_2} = \boldsymbol{F}\,[\ ]\Big\{ \int_0^\infty u_1(s)F'(\quad)\, ds \int_0^\infty u_2(s)F'(\quad)\, ds \Big\}$$

$$+\,\boldsymbol{F}\,[\ ] \int_0^\infty F''(\quad) u_1(s)u_2(s)\, ds\ ,$$

$$\left.\frac{\partial \boldsymbol{F}}{\partial \lambda_1}\right|_{\lambda_1=0} = \boldsymbol{F}\,[u(s)] \int_0^\infty F'(s,u(s))\ u_1(s)\, ds\ ,$$

$$\left.\frac{\partial^2 \boldsymbol{F}}{\partial \lambda_1 \partial \lambda_2}\right|_{\lambda_1=\lambda_2=0} = \boldsymbol{F}\,[u(s)] \int_0^\infty u_1(s)F'(s,u(s))\, ds$$

$$\int_0^\infty u_2(s)F'(s,u(s))\, ds + \boldsymbol{F}\,[u(s)] \int_0^\infty F''(s,u(s))u_1(s)\ u_2(s)\, ds\ .$$

If we evaluate these functional derivatives at u(s)=0 we find (128).

Exercise 16.2.
Show that

$$\mathbf{T}+p\mathbf{1} = \mu(\kappa)\mathbf{A}_1 + \frac{N_1+N_2}{\kappa^2}\,\mathbf{A}_1^2 - \frac{1}{2}\frac{N_1}{\kappa^2}\,\mathbf{A}_2$$

whenever $\mathbf{A}_1$ and $\mathbf{A}_2$ are evaluated on a viscometric flow.

CHAPTER 17

Second Order Fluids

Notations. We simplify the notation by writing $\mathbf{A}_1=\mathbf{A}$ and $\mathbf{A}_2=\mathbf{B}$. In steady flow the recursion (10), Appendix C, for $\mathbf{A}_n$ shows that $\mathbf{A}_n$ is an n-linear functional of $\mathbf{u}$. The notation $\mathbf{B}[\mathbf{u}_1,\mathbf{u}_2] = \mathbf{B}[\mathbf{u}_2,\mathbf{u}_1]$ is in recognition that $\mathbf{B}$ is a symmetric bilinear functional of $\mathbf{u}$.

We shall do some domain deformation problems in this chapter, following Appendix E. We replace the notation $(\bullet)^{<\ell>}$ for ℓth derivative of the domain, holding $\mathbf{x}$ fixed with $(\bullet)_\ell$.

A second order fluid is an asymptotic approximation to the stress for nearly steady slow flow. The constitutive equation for a second order fluid is given by $(16.52)_1$ and $(16.52)_2$. The second order fluid is characterized by material constants, functions of the temperature and possibly other parameters: the zero shear viscosity μ

$$\mu = \int_0^\infty G(s)ds = \lim_{\kappa\to 0} \frac{T_{12}(\kappa)}{\kappa}, \tag{1}$$

the first quadratic constant

$$\alpha_1 = -\int_0^\infty sG(s)ds = -\lim_{\kappa \to 0} \frac{N_1(\kappa^2)}{2\kappa^2} \stackrel{\text{def}}{=} -n_1/2 \tag{2}$$

and the second quadratic constant (see (16.58))

$$\alpha_2 = \int_0^\infty \int_0^\infty \gamma(s_1, s_2)\, ds_1\, ds_2 = \lim_{\kappa \to 0} \frac{N_1(\kappa^2) + N_2(\kappa^2)}{\kappa^2} \stackrel{\text{def}}{=} -n_1 + n_2\,. \tag{3}$$

These relations imply that μ is positive, α_1 is negative and the sign of α_2 is indeterminant.

The parameter α_1 is associated with the extra tension $N_1 = T_{11} - T_{33}$ along streamlines in shear flow. At the same time $\frac{-\alpha_1}{\mu}$ may be thought to be a kind of relaxation time for the fluid, associated with the memory or elasticity. It could be said that N_1 is associated with elasticity. On the other hand, $\alpha_2 = \phi(0,0)$ is the only coefficient to arise in the expansion around rest of a Reiner-Rivlin fluid. Though no real fluid could approximate a Reiner-Rivlin fluid in general, all simple fluids reduce to Reiner-Rivlin fluids in pure extension.

Many problems of flow using the constitutive equation for second order fluids were solved in the 1960s and 1970s. Frequently the theory was applied to flows which were not slow, or to slow flows which were not nearly steady. The results of these applications did not agree with observation, giving even foolish results like the instability of the rest state. On the other hand, when the theory was applied in appropriate situations, the results were very good.

We are going to discuss three features of flow in the range of validity of second order theory which are not well represented in standard books of reference (Joseph, [1989]). The first is the characteristic signature of dynamics of flow of second order fluids, the balance of inertia and normal stress at the quadratic level. The second is the persistence of normal stress in regions of vanishing dimension which I call "something for nothing." The third is the rheometrical problem of determining values of the quadratic constants from rheometrical measurements.

17.1 Balance of normal stresses and inertia

The perturbation equations for slow steady flow are given by (16.23) and (16.24). At first order we get Stokes flow of Newtonian fluid. In the limit $\varepsilon \to 0$, defined in §16.6, there is no difference between Newtonian and non-Newtonian fluids. At second order we get

$$\mu\nabla^2\mathbf{U}_2 - \nabla p_2 = \rho(\mathbf{U}_1\cdot\nabla)\mathbf{U}_1 - \operatorname{div}(\alpha_1\mathbf{B}+\alpha_2\mathbf{A}^2)\,. \tag{4}$$

The inhomogeneous terms are the force of inertia

$$\rho(\mathbf{U}_1\cdot\nabla)\mathbf{U}_1 \tag{5}$$

and the divergence of the normal stress

$$\operatorname{div}(\alpha_1\mathbf{B}+\alpha_2\mathbf{A}^2) \tag{6}$$

where $\mathbf{A}[\mathbf{U}_1]$ and $\mathbf{B}[\mathbf{U}_1,\mathbf{U}_1]$ are evaluated on $\mathbf{U}_1$. Suppose now that $\mathbf{U}_1$ scales with U and that $\mathbf{U}_1$ has an order one variation in a distance L. Then inertia scales like $\rho U^2/L$ and normal stresses like $\alpha U^2/L^3$ where α is representative of an appropriate linear combination of α_1 and α_2. The two effects are equally important at distances of the order

$$L=\sqrt{\alpha/\rho}\,. \tag{7}$$

In many problems there is a critical distance, of order L, with inertia dominant for larger distances and normal stresses dominant for smaller distances. We could call the critical distance an inertial radius. This type of balance appears in many problems. There is an inertial radius for secondary motions which in rotating flows takes form as a dividing surface for eddies with a different direction of circulation, as in Figures (17.1, 3, 4, 6). There is also an inertial radius for the pressure. Suppose that $U=\omega r$, $L=r$, where ω is the angular velocity of rotation in the first order Newtonian problem. Inertia

$$\rho\omega^2 r^2 \tag{8}$$

goes to zero quadratically with r, but the normal stresses

$$\alpha \frac{U^2}{r^2} \sim \alpha\omega^2 \tag{9}$$

do not go to zero, no matter how small is the distance r. This last astonishing result (something for nothing) appears to be a valid description of normal stress effects in rod climbing, at "pressure holes" (where $\omega=\kappa d$, the shear rate times the hole diameter) and elsewhere, though its effects at the smallest distance are probably masked by other small dynamical effects like surface tension (see (77)), which become relatively large in this limit.

The concept of an inertial radius or critical distance dividing fields into regions dominated by inertia and regions dominated by normal stress requires a competition between normal stresses and inertia. Sometimes people refer to this as a competition between elasticity and inertia, but since α_2 is involved, this choice of words may be misleading. Examples of competing effects are abundant. In steady flows around rotating bodies, it appears that inertia drives eddies with one sense, and normal stresses drive them in the opposite sense, as in Figures 17.1, 3, 4, 6. The inertia of fluid will cause it to sink near a rotating rod and normal stresses will make it rise. The effect of normal stresses is to make the extrudate from a pipe swell, while inertia makes it contract. Inertia produces an error of one sign at pressure hole and the normal stresses produce an error of the other sign.

It is not obligatory that normal stresses and inertia must compete. Counter examples can be found. For example, when a rod is rotated in a two-layer system, with the viscoelastic liquid above, this liquid will climb down the rod. Analysis shows that the effect of gravity on the rotating heavy fluid below is to depress the free surface into a "bathtub vortex" with greater depression near the rod. The normal stresses push the viscoelastic fluid up into the air and down into the heavy fluid. Hence, inertia and normal stresses are in conflict at the air-liquid surface and in concert at the two-fluid interface. In the flow through straight pipes with non-circular cross-section, there are no inertial effects at low orders, but the normal stress effect produces secondary motions. A casual inspection of the equations of motion, say (28), might suggest that normal stresses and inertia do indeed act in opposition in general, in spite of the counter-examples in which they are not opposed.

17.2 The vorticity equations at first and second order

At first order we find that the vorticity $\boldsymbol{\zeta}_1$=curl $\mathbf{u}_1$ is a harmonic vector

$$\nabla^2\boldsymbol{\zeta}_1 = 0 .$$

At second order we get

$$\rho\,\frac{\partial\boldsymbol{\zeta}_2}{\partial t} + \rho\,\text{curl}(\boldsymbol{\zeta}_1\wedge\mathbf{u}_1) = \mu\nabla^2\boldsymbol{\zeta}_2 + (\alpha_1+\alpha_2)\,\text{curl div}\,\mathbf{A}^2 \quad (10)$$

where $\mathbf{A}=\mathbf{A}[\mathbf{u}_1]$ and the Giesekus [1963] relation div $\mathbf{B}$=div $\mathbf{A}^2$+grad ϕ (Exercise 17.1) has been used. Equation (10) shows that no couples arise from normal stresses at second order when

$$\alpha_1+\alpha_2 = \lim_{\kappa\to 0}\frac{\left(\frac{1}{2}N_1 + N_2\right)}{\kappa^2} \quad (11)$$

vanishes. In polymeric liquids so far studied N_2 is small and negative. For these liquids $\alpha_1+\alpha_2$ is positive.

After scaling the $\mathbf{u}_1$ terms as in §17.1, we conclude that the inertial radius for secondary flows is estimated by

$$L = 0\left(\frac{\alpha_1+\alpha_2}{\rho}\right)^{1/2}$$

At distances greater than the inertial radius for secondary motions, inertia will drive secondary motions; at lesser distances secondary motions are driven by normal stresses.

We can show that fluids which have large extensional viscosities at low stretching are likely to give rise to intense secondary motions.

The motion known as simple extension in the direction $\mathbf{e}_1$ is given by

$$[\mathbf{A}_1] = \begin{pmatrix} 2\dot{s} & 0 & 0 \\ 0 & -\dot{s} & 0 \\ 0 & 0 & -\dot{s} \end{pmatrix},$$

$$[\mathbf{A}_1^2] = \begin{pmatrix} 4\dot{s}^2 & 0 & 0 \\ 0 & \dot{s}^2 & 0 \\ 0 & 0 & \dot{s}^2 \end{pmatrix},$$

$$\mathbf{A}_2 = \mathbf{A}_1^2 , \tag{12}$$

where $\dot{s}$ is the rate of stretching, $d\mathbf{U}_1/dx_1 = 2\dot{s}$.

The stress differences in extension

$$\mathbf{T}_{11}-\mathbf{T}_{22} = \mathbf{T}_{11}-\mathbf{T}_{33} = \mathbf{S}_{11}-\mathbf{S}_{22} = \mathbf{S}_{11}-\mathbf{S}_{33}$$

are evaluated on $\mathbf{S}_{(2)} = \mu\mathbf{A}_1 + (\alpha_1+\alpha_2)\,\mathbf{A}_1^2$ as

$$\mathbf{T}_{11}-\mathbf{T}_{22} = 3[\dot{s}\mu + (\alpha_1+\alpha_2)\,\dot{s}^2] = 3\dot{s}\,[\mu + (\alpha_1+\alpha_2)\,\dot{s}] \tag{13}$$

where $\mu + (\alpha_1+\alpha_2)\dot{s}$ is the extensional viscosity at low stretching (Roscoe, [1965]).

Second order fluids with $\alpha_1=0$ and $\alpha_2=\phi(0,0)$ are second order Reiner-Rivlin fluids. For these the extensional viscosity is larger than for any second order fluid with the same α_2, because α_1 is negative. Many intense secondary flows, like those with strong vortices in the flow through contractions, should then arise from Reiner-Rivlin fluids. The larger the α_2, the more intense the motion.

17.3 Axisymmetric flow induced by rotating bodies

We are considering the motions which are induced by the steady rotation of an axisymmetric body which rotates with a steady angular velocity ε

$$\mathbf{U} = \mathbf{e}_\theta r(z)\varepsilon. \tag{14}$$

There is no motion when $\varepsilon=0$. The equations of motion for such axisymmetric motions may be expressed in terms of a stream function $\Psi(r,z)$ and angular velocity function $\Omega(r,z)$. The equations satisfied by these two functions are derived in Appendix D. For steady flow these equations may be written as

$$\frac{\rho}{r^2}\left(\frac{\partial\Omega}{\partial z}\frac{\partial\Psi}{\partial r}-\frac{\partial\Omega}{\partial r}\frac{\partial\Psi}{\partial z}\right)=\frac{\mu}{r}D^2\Psi+\mathbf{e}_\theta\cdot\text{div}\,\boldsymbol{\tau} \tag{15}$$

and

$$\frac{\rho}{r^2}\frac{\partial\Psi}{\partial r}\frac{\partial D^2\Psi}{\partial z}-\frac{\rho\partial\Psi}{\partial z}\frac{\partial}{\partial r}\left(\frac{1}{r^2}D^2\Psi\right)+\frac{\rho\Omega}{2r^3}\frac{\partial\Omega}{\partial z}$$
$$=\frac{\mu}{r}D^4\Psi-\mathbf{e}_\theta\cdot\text{curl div}\,\boldsymbol{\sigma} \tag{16}$$

where the nonlinear stresses $\boldsymbol{\sigma}$ and $\boldsymbol{\tau}$ are defined in (D.18) and $D^2\Psi=\frac{\partial^2\Psi}{\partial r^2}-\frac{1}{r}\frac{\partial\Psi}{\partial r}+\frac{\partial^2\Psi}{\partial z^2}$. To get the piezometric pressure or head $\Phi=p+\rho gz$ we must solve

$$-\mathbf{e}_r\rho\left(\frac{\Omega}{r^2}\frac{\partial\Omega}{\partial r}+\frac{\partial\Psi}{\partial r}\frac{D^2\Psi}{r^2}\right)-\mathbf{e}_z\rho\left(\frac{\Omega}{r^2}\frac{\partial\Omega}{\partial z}+\frac{\partial\Psi}{\partial z}\frac{D^2\Psi}{r^2}\right)$$
$$=-\nabla_2(\Phi+\rho|\mathbf{U}|^2)-\mathbf{e}_r\left(\frac{\mu}{r}\frac{\partial D^2\Psi}{\partial z}-\mathbf{e}_r\cdot\text{div}\,\boldsymbol{\sigma}\right) \tag{17}$$
$$+\mathbf{e}_z\left(\frac{\mu}{r}\frac{\partial D^2\Psi}{\partial r}+\mathbf{e}_z\cdot\text{div}\,\boldsymbol{\sigma}\right)$$

The problems which fall in the frame of (14) through (17) are invariant with respect to changes in the sign of the angular velocity ε in the following sense

$$\Omega,\boldsymbol{\tau}(S_{r\theta},S_{z\theta}) \text{ change sign with } \varepsilon,$$
$$\Psi,\Phi,\boldsymbol{\sigma}(S_{rr},S_{rz},S_{zz},S_{\theta\theta}) \text{ are even functions of } \varepsilon. \tag{18}$$

We may use the theory of slow steady motions to solve these problems. Hence

$$\begin{pmatrix}\Psi\\ \Phi\\ \mathfrak{S}\end{pmatrix}=\sum_{n=1}\varepsilon^{2n}\begin{pmatrix}\Psi_{2n}\\ \Phi_{2n}\\ \mathfrak{S}_{2n}\end{pmatrix}, \tag{19}$$

$$\begin{pmatrix}\Omega\\ \mathfrak{T}\end{pmatrix}=\sum_{n=1}^{p}\varepsilon^{2n-1}\begin{pmatrix}\Omega_{2n-1}\\ \mathfrak{T}_{2n-1}\end{pmatrix} \tag{20}$$

where $\mathfrak{T}_1$ is automatically zero since $\mathfrak{T}=\mathbf{0}(\varepsilon^3)$.
Since $\Omega=r^2\omega$ (see D.5), we get

$$D^2\Omega_1 = D^2r^2\omega_1 = 0 \tag{21}$$

from (15) and, from (14)

$$\omega_1=1 \text{ on } r=r(z) \tag{22}$$

on the body surface. At second order, we find that

$$\frac{\mu}{r}D^4\Psi_2-\frac{\rho}{r^3}\frac{\partial}{\partial z}\Omega_1^2-\mathbf{e}_\theta\cdot\text{curl div }\mathfrak{S}_2=0 \tag{23}$$

where

$$\mathfrak{S}_2=\alpha_1\,\mathbf{B}[\mathbf{e}_\theta r\omega_1]+\alpha_2\mathbf{A}^2[\mathbf{e}_\theta r\omega_1] \tag{24}$$

and the normal derivative of Ψ_2 vanishes on the body surface and $\mathbf{B}[\mathbf{e}_\theta r\omega]$ may be evaluated using the recursion formulas (9) and (10) of Appendix C. We find that

$$\text{div }\mathfrak{S}_2=(\alpha_1+\alpha_2)\text{ div }\mathbf{A}^2+\alpha_1\nabla_2\text{ tr }\mathbf{A}^2 \tag{25}$$

where

$$\mathbf{A}=r[\mathbf{e}_\theta\nabla_2\omega_1+\nabla_2\omega_1\mathbf{e}_\theta] \tag{26}$$

After computing the last term in (23) using (25), replacing (ω_1,Ψ_2,Φ_2) with (ω,Ψ,Φ), we get

$$D^2r^2\omega=0\text{ , }\omega(r,z)=\varepsilon\text{ on }r=r(z) \tag{27}$$

and

$$\mu D^4\Psi = r^2 \frac{\partial}{\partial z} \{\rho\omega^2 - 2(\alpha_1+\alpha_2) |\nabla_2\omega|^2\} \tag{28}$$

where $\Psi=\partial\Psi/\partial n=0$ on $r=r(z)$. Since $\alpha_1+\alpha_2$ is positive, the term in braces suggests a competition between inertia and normal stresses in driving secondary motions.

Equations (27) and (28) were derived by Böhme [1974]. They show that there will always be secondary motions at second order when $\partial\omega/\partial z\neq 0$. We see clearly from (28) that secondary motions induced by normal stresses scale with $\alpha_1+\alpha_2$.

The piezometric pressure Φ may be determined from ω and Ψ,

$$\frac{\partial}{\partial r}\left[\Phi - \frac{\alpha_1}{2} r^2 |\nabla_2\omega|^2\right] = -\frac{\mu}{r}\frac{\partial}{\partial z} D^2\Psi + \rho\omega^2 r \tag{29}$$

$$+ (\alpha_1+\alpha_2)\left\{\frac{\partial}{\partial r}\left(r^2\left[\frac{\partial\omega}{\partial r}\right]^2\right) + \frac{\partial}{\partial z}\left(r^2\frac{\partial\omega}{\partial r}\frac{\partial\omega}{\partial z}\right) - r\left(\frac{\partial\omega}{\partial z}\right)^2\right\},$$

$$\frac{\partial}{\partial z}\left[\Phi - \frac{\alpha_1}{2} r^2 |\nabla_2\omega|^2\right] = \frac{\mu}{r}\frac{\partial}{\partial r} D^2\Psi \tag{30}$$

$$+ (\alpha_1+\alpha_2)\left\{\frac{\partial}{\partial r}\left[r^2\frac{\partial\omega}{\partial r}\frac{\partial\omega}{\partial z}\right] + \frac{\partial}{\partial z}\left(r^2\left[\frac{\partial\omega}{\partial z}\right]^2\right)\frac{\partial\omega}{\partial r} + r\,\frac{\partial\omega}{\partial r}\frac{\partial\omega}{\partial z}\right\}.$$

17.4 Rotating rod

A rod of radius a rotates with constant angular velocity ε in an infinite sea of liquid. The solution of (27)

$$\omega = \frac{a^2\varepsilon}{r^2} \tag{31}$$

is independent of z and the right side of (28) vanishes. We find that $\Psi=0$ identically. There is no secondary motion at second order, in fact at any order. For large ε, this is a viscometric flow, the flow between rotating cylinders, the outer cylinder far away.

The piezometric pressure Φ may be determined from (29). We find that

$$\Phi = p + \rho gz = p_0 + \left(2\,\hat{\beta}\,\frac{a^4}{r^4} - \frac{\rho a^4}{2r^2}\right)\varepsilon^2 \tag{32}$$

where p_0 is the pressure at z=0 as $r\to\infty$ and

$$\hat{\beta} = 3\alpha_1 + 2\alpha_2 = \lim_{\kappa\to 0}\frac{N_1/2 + 2N_2}{\kappa^2} \tag{33}$$

is the climbing constant. When $\hat{\beta}>0$ we see an inertial radius

$$r_c = \sqrt{\frac{4\hat{\beta}}{\rho}} \tag{34}$$

for the pressure. Inertia dominates when $r<r_c$, normal stress dominates when $r>r_c$. When $r=r_c$, the pressure is purely hydrostatic. If $a>r_c$, the entire flow will be dominated by inertia. When r=a, (32) reduces to

$$\Phi - p_0 = (2\hat{\beta} - \rho a^2/2)\,\varepsilon^2 \tag{35}$$

This shows that inertial effects vanish and normal stress effects persist as $a\to 0$.

A formula for the shape of a free surface z=h(r) generated by a rotating rod may be obtained from (32) by requiring that $p=p_0$ when z=h, interpreting p_0 as the atmospheric pressure. This leads to the theory of rod climbing which will be discussed more completely in §17.10.

17.5 Rotating wavy rod

The problem of the flow induced by the rotation of a wavy rod $r=a(1+\delta\sin\lambda z)$ was solved by Sturges [1977]. He used the perturbation in ε to reduce his problem to (27) through (31). He solved this problem by the method of domain perturbations (see Appendix E). In accordance with this method, we define an invertible, one to one mapping

$$r = r_0[1+\delta \sin \lambda z] , \quad z = z_0 \tag{36}$$

which takes the region outside the wavy rod onto the region outside the straight rod of radius a. We then expand the variable in a power series in δ

$$\begin{aligned} \omega(r,z,\delta) &= \omega_0(r_0,z_0) + \delta\omega^{[1]}(r_0,z_0) + \dots , \\ \Psi(r,z,\delta) &= \Psi_0(r_0,z_0) + \delta\Psi^{[1]}(r_0,z_0) + \dots \end{aligned} \tag{37}$$

where, for example

$$\Psi^{[1]}(r_0,z_0) = \frac{d}{d\delta}\Psi[r(\delta),z,\delta] \Big|_{\delta=0} .$$

Using the chain rule we get

$$\begin{aligned} \Psi^{[1]}(r_0,z_0) &= \frac{\partial\Psi}{\partial\delta} + \frac{\partial\Psi}{\partial r}\frac{dr}{d\delta}\Big|_{\delta=0} \\ &= \Psi_1(r_0,z_0) + r_0 \sin \lambda z_0 \frac{\partial\Psi_0}{\partial r_0}(r_0,z_0) . \end{aligned} \tag{38}$$

Substituting these series into equations (27) to (31), we identify powers of δ and obtain the equations satisfied by the partial derivative ω_1,Ψ_1. At the lowest order (δ=0), we recover the results

$$\omega_0(r_0,z_0) = (a/r_0)^2 ,$$

$$\Psi_0(r_0,z_0) = 0 . \tag{39}$$

At first order in δ we find that

$$D^2 r_0^2 \omega_1 = 0 \tag{40}$$

where D^2 is unchanged except that r and z are replaced with r_0 and z_0. At the rod boundary in the reference domain

$$\omega^{[1]}(a,z_0) = 0 . \tag{41}$$

After relating $\omega^{[1]}(a,z_0)$ to $\omega_1(a,z_0)$ through the chain rule, we find that

$$\omega_1(a,z_0) = 2 \sin \lambda z_0 . \tag{42}$$

The solution of (40) and (42) is expressible in terms of modified Bessel functions

$$\omega_1(r_0,z_0) = \frac{2a}{r_0}\,\frac{K_1(\lambda r_0)}{K_1(\lambda a)}\sin\lambda z_0\,. \tag{43}$$

As observed earlier, the z_0 variation in this rotation field will drive a secondary motion. The equation for the secondary motion is formulated for Ψ_1 rather than $\Psi^{[1]}$ by methods in Appendix E. We find that

$$\mu D^2\Psi_1 = 2a^2\left(\rho\,\frac{\partial\omega_1}{\partial z_0} + \frac{4(\alpha_1+\alpha_2)}{r_0}\,\frac{\partial^2\omega_1}{\partial r_0\partial z_0}\right) \tag{44}$$

with $\Psi_1=\partial\Psi_1/\partial r_0=0$ on $r_0=a$ and $\Psi_1/r_0^2\to 0$ as $r_0\to\infty$. The solution of this problem is also given in terms of modified Bessel functions.

$$\Psi_1(r_0,z_0) = r_0\chi(\lambda r_0)\cos\lambda z_0 \tag{45}$$

where

$$\chi(\lambda r_0) = AK_1(\lambda r_0) + Br_0K_0(\lambda r_0) - \frac{\rho}{\mu\lambda^4}\,\frac{(a\lambda^3)}{K_1(\lambda a)}$$
$$\left[\frac{K_0(\lambda r_0)}{\lambda r_0} + \ln(\lambda r_0)K_1(\lambda r_0)\right] - \frac{2(\alpha_1+\alpha_2)}{\mu\lambda^2}\,\frac{(a\lambda)^3}{K_1(\lambda a)}\,\frac{K_0(\lambda r_0)}{\lambda r_0}$$

$$A = \frac{\rho(a\lambda)^2}{\mu\lambda^4K_1(\lambda a)}$$
$$\left\{\lambda a\ln(\lambda a) + \frac{2K_0^2(\lambda a) - \lambda aK_0(\lambda a)\,K_1(\lambda a)}{\lambda aK_0^2(\lambda a) - \lambda aK_1^2(\lambda a) + 2K_0(\lambda a)K_1(\lambda a)}\right\}$$
$$+\ \frac{4(\alpha_1+\alpha_2)}{\mu\lambda^2}\,\frac{(a\lambda)^2}{K_1(\lambda a)}\bullet\frac{K_0^2(\lambda a)}{\lambda aK_0^2(\lambda a) - \lambda aK_1^2(\lambda a) + 2K_0(\lambda a)K_1(\lambda a)}$$

$$B = \frac{\rho\lambda a}{\mu\lambda^4 K_1(\lambda a)} \left\{ \frac{\lambda K_0^2(\lambda a)}{\lambda a K_0^2(\lambda a) - \lambda a K_1^2(\lambda a) + 2K_0(\lambda a)K_1(\lambda a)} \right\}$$

$$+ \frac{2(\alpha_1+\alpha_2)(a\lambda)^2}{\mu\lambda^2 K_1(\lambda a)} \bullet \frac{\lambda[K_0^2(\lambda a) - K_1^2(\lambda a)]}{\lambda a K_0^2(\lambda a) - \lambda a K_1^2(\lambda a) + 2K_0(\lambda a)K_1(\lambda a)}$$

Some properties of the secondary flows are determined by the numbers

$$R_1 = \frac{K_0(\lambda a)}{4\lambda a K_1(\lambda a)},$$

$$R_2 = \frac{K_0^2(\lambda a)}{2(\lambda a)^2[K_1^2(\lambda a) - K_0^2(\lambda a)]}. \tag{46}$$

There are three cases:

(i) $\frac{\alpha_1+\alpha_2}{a^2\rho} < R_1$ inertia dominates, (47)

(ii) $\frac{\alpha_1+\alpha_2}{a^2\rho} > R_2$ normal stresses dominate, (48)

(iii) $R_1 < \frac{\alpha_1+\alpha_2}{a^2\rho} < R_2$ Flow field divides into a part dominated by inertia and a part dominated by normal stresses. (49)

The variation of these regions with λa is shown in Figure 17.2.

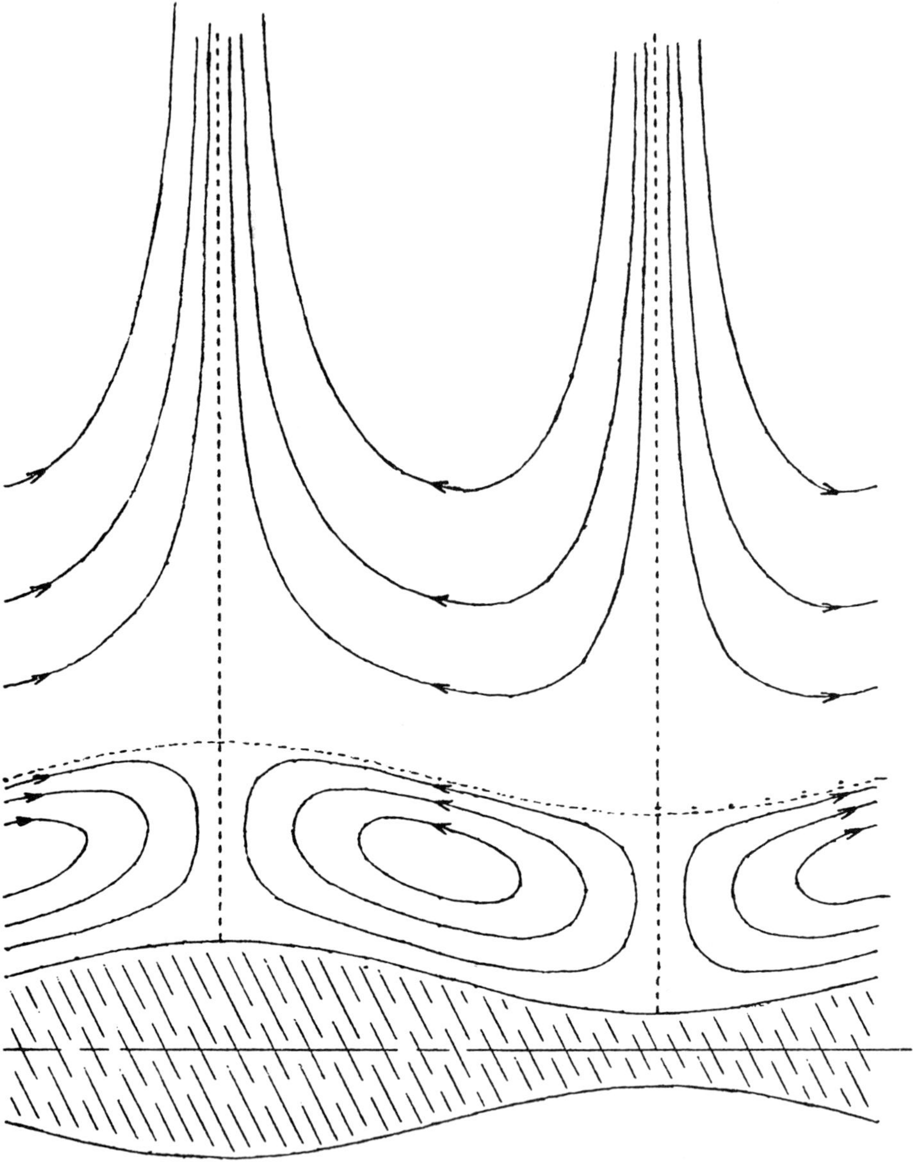

Figure 17.1 (Sturges, [1977]). Streamlines of the secondary flow for the wavy rod. The parameters used in the computation are: α_1=1.475 g/cm, α_2=2.70 g/cm, ρ=0.89 g/cm^3, a=1 cm, λ=0.07 cm^{-1}. For these values the critical radius for pressure is 2.1 cm and the critical radius for secondary flow is 3.7 cm.

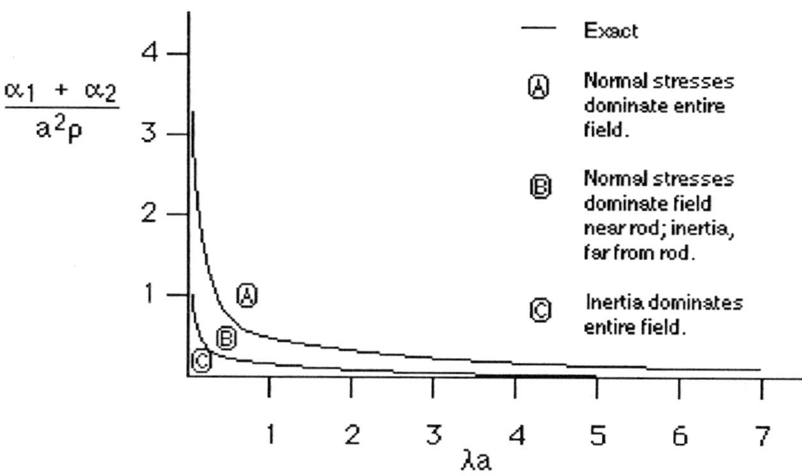

Figure 17.2 (Sturges, [1977]). Dependence of regions dominance of normal stresses and inertia on the wave length λa of the wavy rod.

17.6 Rotating sphere, cone, and plate

The rotating sphere was first treated by Giesekus [1968] and the secondary motion between a cone and plate was discussed by Giesekus [1967]. These problems were treated by perturbations, and solutions through second order were given. These problems can be conveniently treated with Böhme's equations (27) through (31), after changing variables from cylindrical to spherical polar coordinates. We will not repeat the mathematical development, but will just summarize some relevant results.

In the case of the rotating sphere, the flow field can divide into eddies driven by normal stresses against eddies driven by inertia, as in Figure 17.3. If

$$a^2 < \frac{4(\alpha_1+\alpha_2)}{\rho} \tag{50}$$

the normal stresses dominate, there is only one set of eddies and the normal stresses determine the direction of flow. If, on the other hand,

$$a^2 > \frac{12(\alpha_1+\alpha_2)}{\rho} \tag{51}$$

then there is again only one set of eddies, here driven by inertia. The situation when

$$\frac{4(\alpha_1+\alpha_2)}{\rho} < a^2 < \frac{12(\alpha_1+\alpha_2)}{\rho} \tag{52}$$

is the one shown in Figure 17.3 with eddies driven by normal stresses inside and inertial eddies outside. A good discussion of this problem in the quadratic approximation can be found in Böhme [1987]. Higher order approximations have been given by Fosdick and Kao [1980].

The problem of secondary flow between a cone and plate is like the sphere with counter eddies, the inner eddies being driven by normal stresses and the outer ones by inertia.

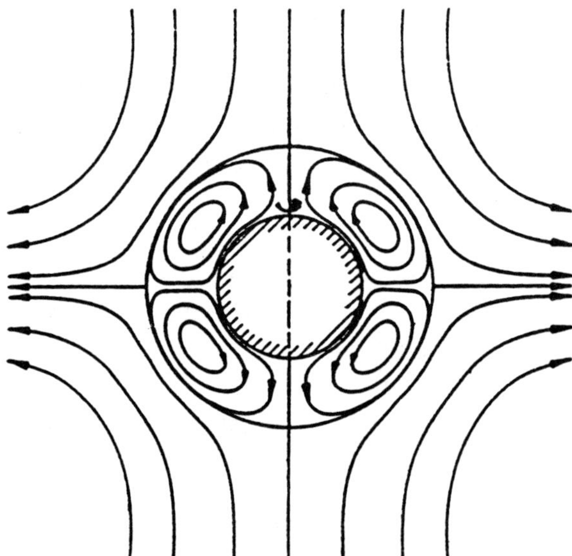

Figure 17.3 Eddies around a rotating sphere when (52) holds. The outside eddies are driven by inertia and the fluid is thrown out at the equator.

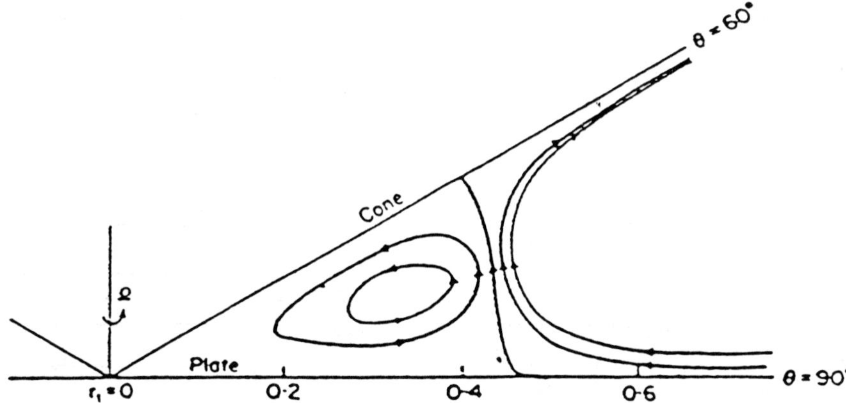

Figure 17.4 (Walters and Waters, [1968]). Typical streamlines of the secondary flow of a viscoelastic liquid $(\alpha_1+\alpha_2)=0.5$. The inner eddies are driven by normal stresses, the outer eddies by inertia.

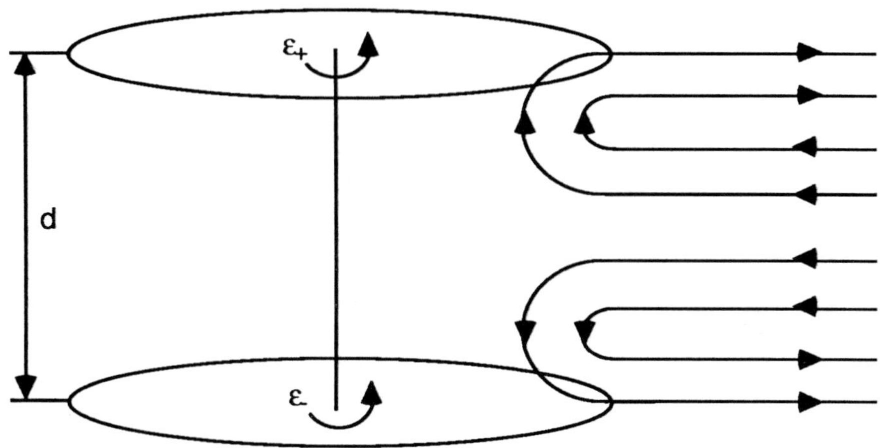

Figure 17.5 Parallel rotating disk for the case $\varepsilon_- = \varepsilon_+$.

17.7 Flow between rotating disks

This problem was solved by Joseph [1974] and others. In this treatment we neglect edge effects treated there. We consider parallel disks, $0 \le r < \infty$, separated by a distance d, $-d/2 \le z \le d/2$ as in Figure 17.5.

We must solve $D^2(r^2\omega)=0$ for $\omega=\varepsilon_+$ at $z=d/2$ and $\omega=\varepsilon_-$. For $z=-d/2$:

$$\omega = Az+B ,$$
$$A = \frac{\varepsilon_+ - \varepsilon_-}{d} , \tag{53}$$
$$B = \frac{\varepsilon_+ + \varepsilon_-}{2} .$$

The stream function is determined from (28) which, after accounting for (53), becomes

$$\mu D^4 \Psi = 2r^2 \rho A(Az+B) \tag{54}$$

and

$$\Psi(r,\pm d/2) = \frac{\partial \Psi}{\partial z}(r,\pm d/2) = 0\ . \tag{55}$$

We may find a function

$$\hat{\Psi}(z) = \frac{\rho}{\mu}\left[\frac{A^2 z}{60}\left(z^2 - \frac{d^2}{4}\right) + \frac{AB}{12}\left(z^2 - \frac{d^2}{4}\right)^2\right] \tag{56}$$

of z alone such that

$$\Psi(r,z) = r^2\hat{\Psi}(z)\ . \tag{57}$$

There is no inertial radius for secondary flow in this problem, the eddies areas shown in Figure 17.5, with as much fluid going in as coming out.*

The fact that ω is independent of r rules out an inertial radius. However, even in this problem, there is a competition between inertia and the normal stresses. To see this, we need the pressure. We have $D^2\Psi = r^2\hat{\Psi}''$ and

$$\begin{aligned}\frac{\partial}{\partial r}\left[\Phi - \frac{\alpha_1}{2}r^2A^2\right] &= -r\ \left\{\mu\hat{\Psi}''' + \rho(Az+B)^2 - A^2(\alpha_1+\alpha_2)\ \right\}\\ &= r\left\{\rho\frac{A^2d^2}{20} + \rho B^2 - (\alpha_1+\alpha_2)A^2\right\},\end{aligned} \tag{58}$$

and

$$\frac{\partial}{\partial z}\left[\Phi - \frac{\alpha_1}{2}r^2A^2\right] = 2\mu\hat{\Psi}''\ . \tag{59}$$

Hence

$$\Phi = 2\mu\hat{\Psi}' + \frac{r^2}{2}\left[\frac{\rho A^2d^2}{20} + \rho B^2 - \alpha_2A^2\right]. \tag{60}$$

The normal stress is

$$T_{zz} = -p + \mathbf{e}_z\cdot(\alpha_1\mathbf{A}_2+\alpha_2\mathbf{A}_1{}^2)\cdot\mathbf{e}_z$$

---

* There is a secondary motion (Figure 17.7) showing a competition between inertia and normal stresses in the rotating disk in a cylinder system studied in the experiments of Hill [1972] and theoretically by Kramer and Johnson [1972].

where $\mathbf{A}_1$ and $\mathbf{A}_2$ are evaluated on $\mathbf{u}=\mathbf{e}_\theta r(Az+B)$ and

$$T_{zz} = -\Phi + \rho gz + (2\alpha_1+\alpha_2)r^2A^2 . \tag{61}$$

Finally, after combining (60) and (61) we get an explicit formula for T_{zz}:

$$T_{zz} = C(z) + \frac{1}{2}r^2A^2\left[3\alpha_2 + 4\alpha_1 - \rho\left(\frac{d^2}{20} + \frac{B^2}{A^2}\right)\right] . \tag{62}$$

Using (53) we may write $A=\Delta\varepsilon/d$

$$T_{zz} - C(z) = \frac{1}{2}\Delta\varepsilon^2\frac{r^2}{d^2}\left[3\alpha_2 + 4\alpha_1 - \rho d^2\left(\frac{1}{20} + \frac{B^2}{\Delta\varepsilon^2}\right)\right] . \tag{63}$$

If $3\alpha_2+4\alpha_1>0$, there is a critical gap size

$$d_c^2 = \frac{3\alpha_2+4\alpha_1}{\rho}\left(\frac{1}{20} + \frac{B^2}{\Delta\varepsilon^2}\right)^{-1} \tag{64}$$

with normal stress dominant when $d<d_c$ and inertia dominant when $d>d_c$.

We may imagine a manometer board mounted on a stationary upper plate, $\varepsilon_+=0$. Then

$$d_c^2 = 10\,\frac{3\alpha_2+4\alpha_1}{3\rho} , \tag{65}$$

the manometer readings will take form as in Figure 17.6 and (63) reduces to

$$T_{zz} - C(z) = \frac{\varepsilon^2}{2}\frac{r^2}{d^2}\left[3\alpha_2 + 4\alpha_1 - \frac{3\rho d^2}{10}\right] .$$

We may use (2) and (3) to write

$$4\alpha_1 + 3\alpha_2 = \lim_{\kappa\to 0} \frac{N_1(\kappa^2) + 3\,N_2(\kappa^2)}{\kappa^2} \tag{66}$$

This shows that $4\alpha_1+3\alpha_2>0$ for all fluids with $N_1+3N_2>0$, a common situation for polymeric liquids.

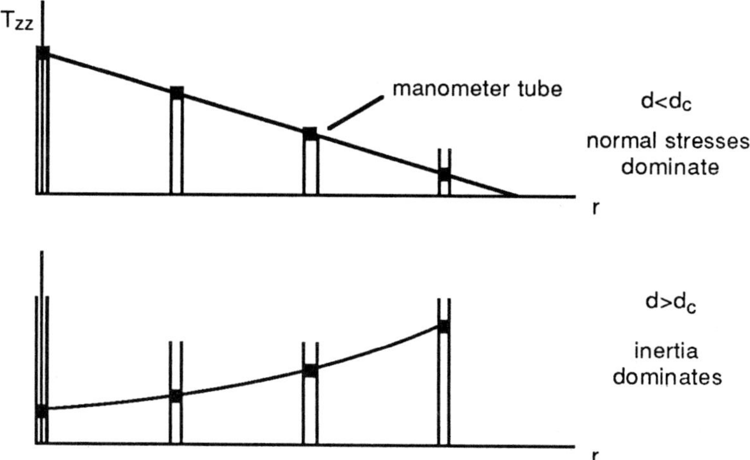

Figure 17.6. Pressure readings induced by normal thrust due to nonlinear stresses in the flow between rotating parallel disks.

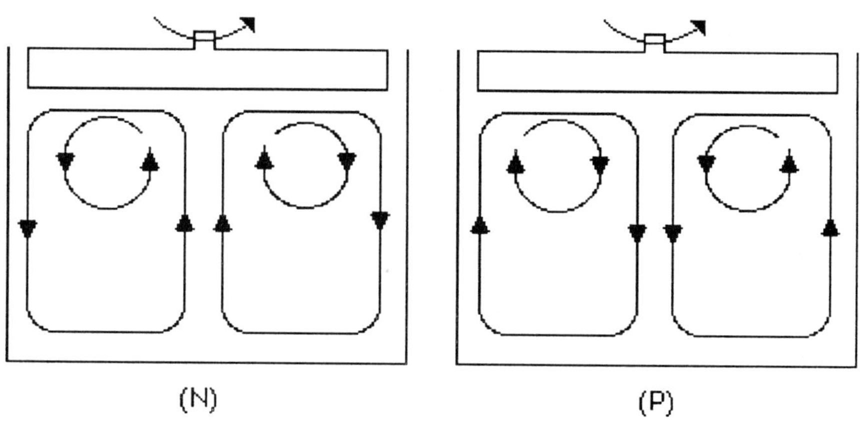

Figure 17.7. Secondary flows in the disk-cylinder system. **N** The Newtonian fluid moves up at the center whereas **P**, the viscoelastic fluid, polyacrylamide (Separan 30)/glycerol/water, moves down at the center. (C. T. Hill, *Trans. Soc. Rheol.* **16**, 213–245 [1972])

17.8 Die swell in a low speed jet

The problem of die-swell was treated as a perturbation from rest by Tieu and Joseph [1983] following an earlier analysis of the plane jet by Sturges [1981] in which nearly identical methods and results were employed. The rest state here is taken in the form of a motionless jet held together by surface tension in a perfect right circular cylinder. The problem is analyzed using the theory of domain perturbations and is solved by Sturges with biorthogonal series and by Tieu and Joseph with finite elements. The cylinder in the rest state is surely unstable, but is perhaps stabilized by the small ε, which is the volume flow rate. If the surface tension is very large, the perfect cylinder could be a good choice for the rest configuration. A different case, small surface tension, was treated by a different method by Trogdon and Joseph [1981]. In these perturbation studies, the rheology of the fluids may be expressed by three constants, the viscosity and the two constants α_1 and α_2 appearing at order two in the expansion of the extra stress around zero shear. The analysis predicts that no matter what the constitutive equation may be, the jet swell will persist even as the diameter of the pipe is shrunk to zero. The analysis leads to the following expression for the shape of the jet.

$$\frac{h(\hat{x})}{a} = 1 + \frac{\mu\bar{u}}{\sigma}\left(\frac{\hat{h}_1(\hat{x})}{2}\right) + \frac{\rho a\bar{u}^2}{\sigma}\left[\frac{\hat{h}_{20}(\hat{x})}{4}\right] + \frac{\alpha_1\bar{u}^2}{a\sigma}\left[\frac{\hat{h}_{21}(\hat{x})}{4}\right]$$
$$+ \frac{\alpha_2\bar{u}^2}{a\sigma}\left[\frac{\hat{h}_{22}(\hat{x})}{4}\right] + \left(\frac{\mu\bar{u}}{\sigma}\right)^2\left[\frac{\hat{h}_{23}(\hat{x})}{4}\right] + O(|\bar{u}|^3)\,, \tag{67}$$

$$\hat{h}_{NN}(\hat{x}) = \frac{\hat{h}_{22}(\hat{x})}{2c} - \frac{\hat{h}_{21}(\hat{x})}{4}\,, \quad \text{where } c \in [1, \tfrac{4}{3}]\,.$$

where $a=h_0$ is the radius of the pipe, $\hat{x}=x/a$ is the radius of the jet, $\bar{u}$ is the average velocity flow and σ is the coefficient of surface tension. When $\varepsilon=0$, $h(x)=a$ for all $x\geq 0$. The average velocity $\bar{u}$ is proportional to ε and $\hat{h}_1$ is a first order contribution which is determined by a Stokes flow. The coefficients $\hat{h}_{2\ell}$, $\ell=0,1,2,3$ are determined at second order. These functions $\hat{h}_1$ and $\hat{h}_{2\ell}$ are determined from a finite element calculation, but once and for all since they do not depend on the parameters. The graphs of these universal functions are exhibited in Figure 17.8.

Figure 17.8 (After Tieu and Joseph [1983]). Universal shape functions appearing in (67).

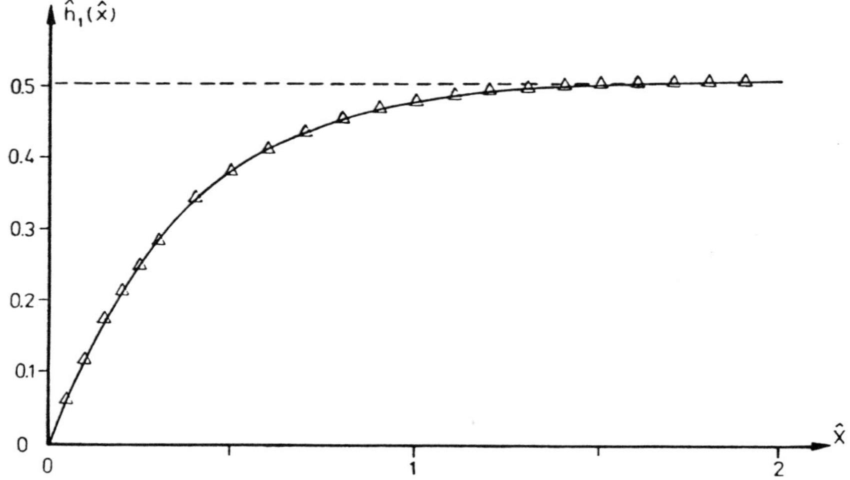

(a) Surface shape correction at first order. Results computed on Meshes I–IV are essentially the same. Mesh I □; Mesh II ×; Mesh III o; Mesh IV Δ.

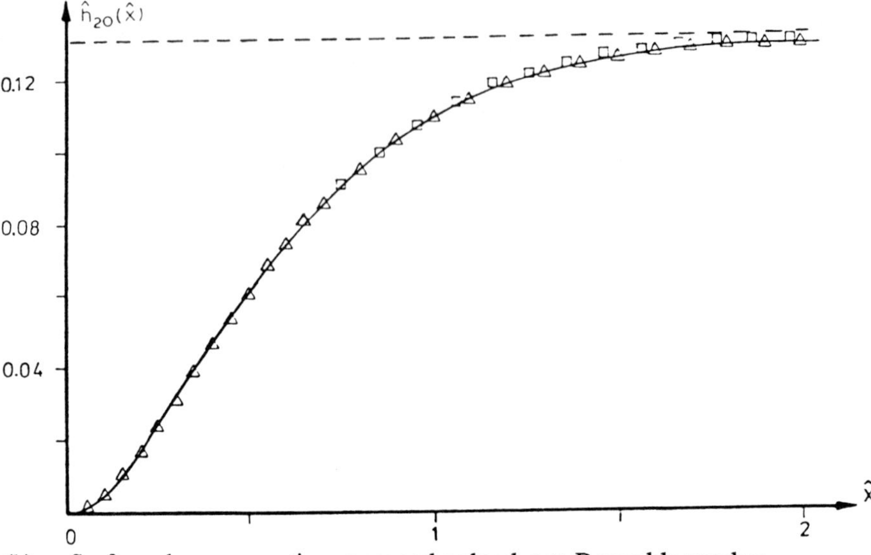

(b) Surface shape correction at second order due to Reynolds number.

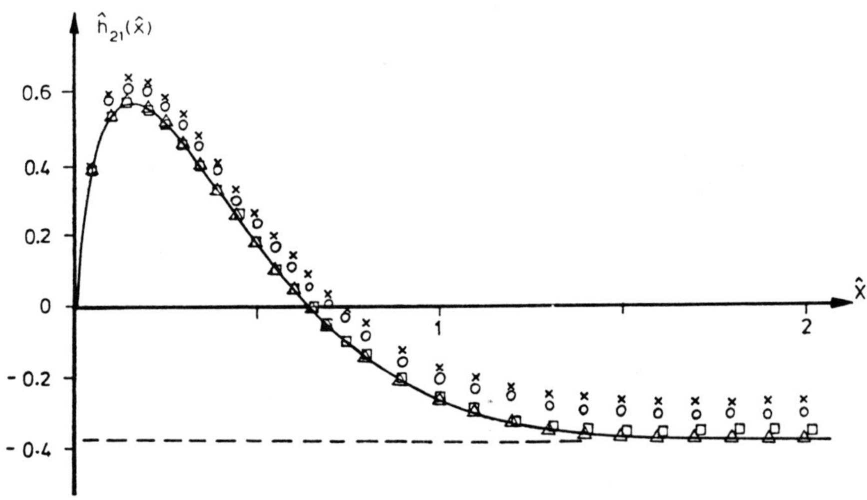

(c) Surface shape correction at second order due to α_1.

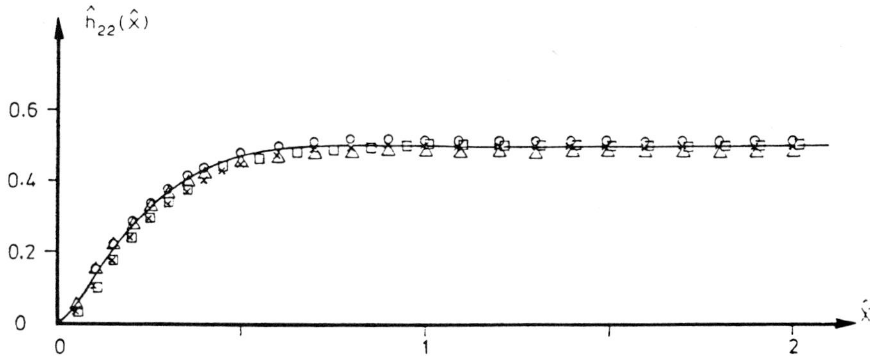

(d) Surface shape correction at second order due to α_2.

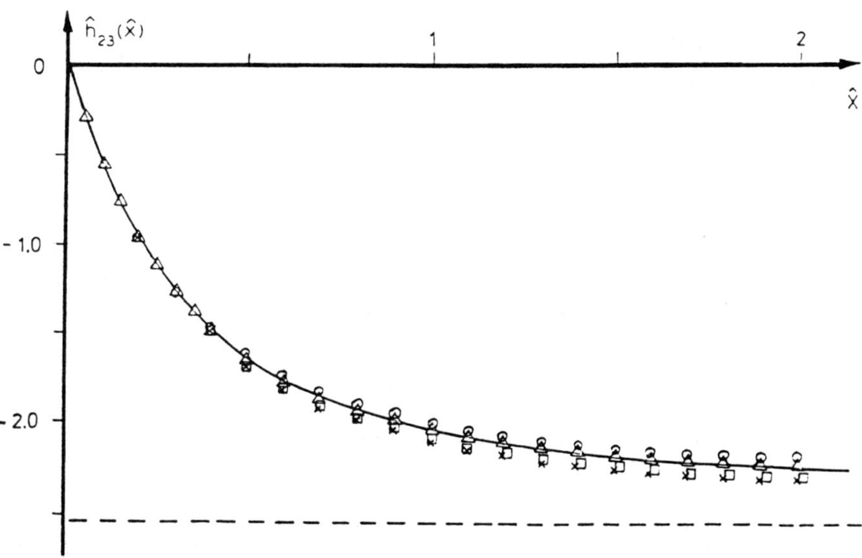

(e) Surface shape correction at second order.

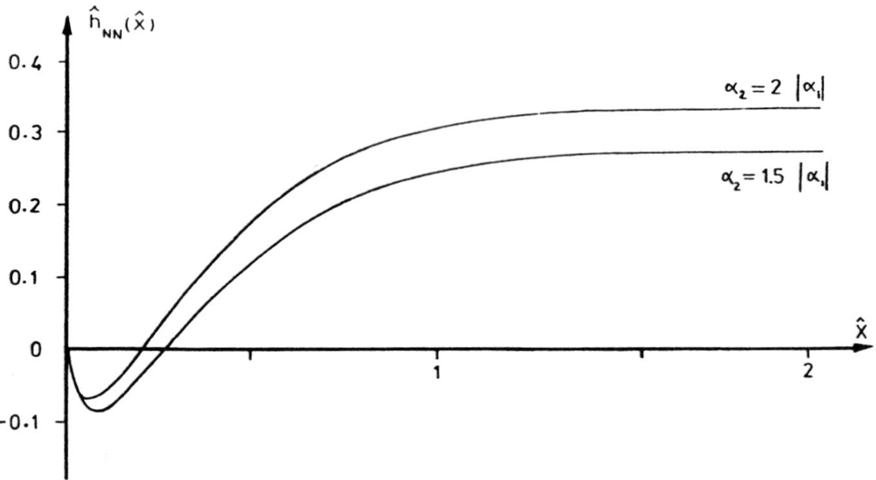

(f) Surface shape correction at second order due to viscoelasticity in a fluid.

One of the most interesting features about (67) is that it predicts a finite swell as $\hat{x}\to\infty$ even as $a\to 0$, then

$$h \to a + \left[\alpha_1 \hat{h}_{21}\left(\frac{x}{a}\right) + \alpha_2\, \hat{h}_{22}\left(\frac{x}{a}\right)\right]\frac{\bar{u}^2}{\sigma} \tag{68}$$

and $x/a\to\infty$. Of course the ordering of the series is lost in this kind of limit, since the second order terms are becoming larger than the first. It is generated by squared gradients which do not vanish as $L\to 0$ (see §17.1). Though it is unlikely that this type of phenomenon will be observed in the limit, the physical effect in which the normal stress effects are longest lived as the characteristic length goes to zero appears to be an accurate description of what is in fact observed.

It is of interest to examine the combined behavior of the two functions expressing normal stress effects in the square bracket of (68). If the second normal stress is negative, then $\alpha_2=2|\alpha_1|$; if the fluid climbs a rod, then $\alpha_2=1.5|u_1|$ with the consequences shown in Figure 17.8. It might be thought that an initial decrease in jet is an artifact of the perturbation solution. This feature does not appear in direct numerical simulations. These simulations do not isolate the different effects or go to the limit of small a, and they are all based on special rheological models.

The perturbation solutions given by Trogdon and Joseph [1980], Sturges [1981], and Tieu and Joseph [1983] are probably not valid near the exit lip where there are singular stresses. In this case the expansion of the stress into fluids of grade one, two, etc. is likely to break down locally. It is not known whether a local error in the neighborhood of the lip would vitiate the solution over the entire field of flow.

17.9 Inertial and normal stress effects on pressure readings in pressure holes

The determination of the pressure, or more generally the normal stress at points on a wall over which a fluid flows, is frequently accomplished by attaching a pressure device at the base of a small hole in the wall. Manometer tubes could be offered as an example. The presence of the hole disturbs the flow, introducing errors. For examples, burrs at the opening of pressure holes used to measure pressure gradients are particularly onerous. But even if the holes are perfectly made, there are disturbing errors which need to be taken into account.

Consider flow over a hole of diameter d and depth D, as in Figures 17.9 and 17.10. Let κ be the undisturbed shear rate as if the hole were not present; d and D are the only natural lengths, and in deep holes D is unimportant. In these circumstance the characteristic velocity of the eddies in the hole is $U=\kappa d$ and the Reynolds number $R=(\kappa d)d\,\rho/\mu$ is small for small holes. The thrust p is positive when pointing in the direction of the inward normal, like the pressure; p is the thrust at the hole, say in the center, or the average thrust and p_g is the gauge pressure, say the average pressure at a level in the tube at which the pressure is to be read. In the best of all possible worlds $p_g=p$. In the flow of a Newtonian fluid over a slot, shown in Figure 17.9, the difference in the thrust is balanced by the shear stresses on the wall of the slot and dimensional analysis shows that

$$p_g - p = \tau_w f(R;d/D) > 0\,. \tag{69}$$

For a deep slot $d/D \to 0$ and the thrust difference depends on the Reynolds number alone. The pressure p_g pushing out of the hole is greater that the thrust pushing it in. This pressure difference must be balanced by flow induced shear stresses at the wall. To have the imbalance it is necessary to have a faster flow, or larger stress on the wall of the hole on the upstream side, as in Figure 17.9. An asymmetry in the flow is required.

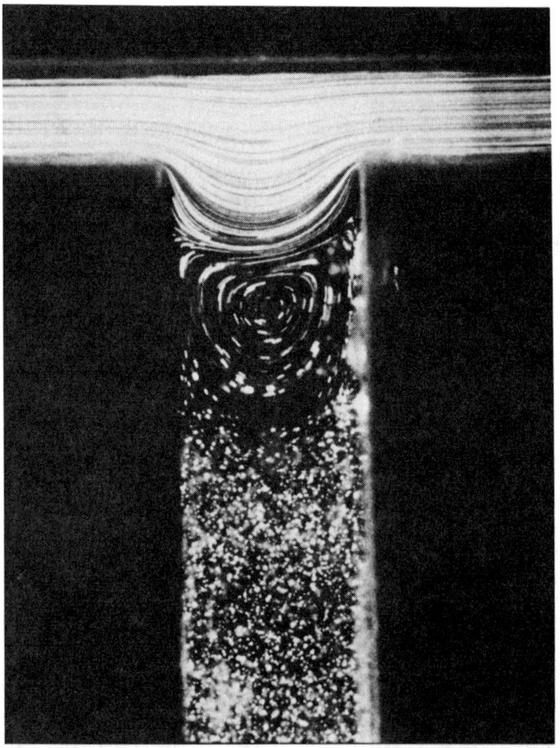

Figure 17.9. (Cochrane, Walters, Webster, [1981]). The flow is from the right to the left. The fluid is Newtonian with a Reynolds based on slot diameter of six. The streamlines which descend into the hole are crowded on the upstream side, producing a downward shear force which is balanced by the pressure discrepancy.

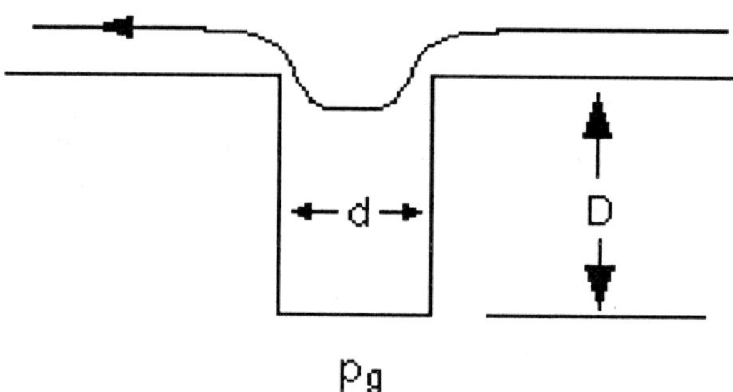

Figure 17.10. Flow of a viscoelastic fluid over a deep slot. It is assumed that the flow is symmetric as in Stokes flow. The extra tension produces an upward force which must be balanced by the pressure discrepancy.

This asymmetry disappears as the hole size and Reynolds number is reduced to zero and symmetrical streamlines result from the prescription of symmetric flow at the top of the hole.

Experimenters make holes of small size to avoid big errors, but in 1968, Broadbent, Kaye, Lodge and Vale suggested that small holes would not be enough to put disturbances from viscoelasticity to rest. The explanation of this is generally framed in terms of the extra tension, N_1, along streamlines. In this explanation, it is best to remove inertia by considering slow flow over a slot with the streamlines symmetric. The effect of the extra tension in the streamlines which occur in a viscoelastic fluid is to "lift" the fluid out of the hole, as a weight on a sagging chain. Of course, the fluid is not lifted out of the hole, the tension on the chain is balanced by the difference in thrust

$$p - p_g > 0 . \tag{70}$$

This argument suggests that the difference in thrust should be determined by the first normal stress difference giving the tension on the chain. A simple and elegant analysis using this idea and the exploiting consequences of symmetry was given by Tanner and Pipkin [1969]. Their result is for flow across two-dimensional slots where N_2 cannot enter (see Exercise 17.5)

$$p - p_g = -\frac{\alpha_1 \kappa^2}{4} \; . \tag{71}$$

This formula is in good agreement with experiments at low rates of shear and not bad even at higher rates of shear. The symmetric streamlines arise only as perturbation of the flow in the slot from rest. The small parameter could be taken as the shear force at the mouth of the hole, proportional to the area of the mouth.

One of the more interesting features of the thrust difference due to normal stresses is that it is independent of hole size for small d. This is a consequence of self similarity of streamlines and the general argument about the persistence of normal stresses which was presented in (9). In this case, the scale velocity is κd and (9) is replaced by

$$\alpha \frac{(\kappa d)^2}{d^2} \sim a\kappa^2 \; . \tag{72}$$

Simple direct measurements of Tanner and Pipkin in [1969] have shown that for sufficiently small holes, the thrust difference is independent of hole size.

Since normal stresses and inertia have opposing effects, and since the effects of inertia increase with Reynolds number, it seems probable that there is a critical inertial diameter $d_c=\sqrt{\alpha/\rho}$ where α is a suitable combination of the two quadratic constants such that inertia will dominate when $d>d_c$ and normal stresses will dominate when $d<d_c$.

17.10 Rod climbing

It is well known that many non-Newtonian fluids will climb a rotating rod, whereas Newtonian fluids will not climb. The climb is associated with nonlinear effects, normal stresses, which cannot occur in fluids like Newtonian fluids, in which the stress is linearly related to the gradient of velocity. The climbing property of non-Newtonian fluids can be used to characterize important rheological parameters in those fluids. The most important of these is the climbing constant $\hat{\beta}$. This constant appears in some perturbation studies of Kaye [1973], Böhme [1974], and Joseph and Fosdick [1973], and implicitly in a study of Hoffman and Gottenberg [1973]. Joseph, Beavers, and Fosdick [1973] showed that it is necessary to include surface tension effects if $\hat{\beta}$ is to be computed from measured values of the climb. Values of $\hat{\beta}$ for some liquids were given by Beavers and Joseph [1975, 1977], by Yoo, Joseph, and Beavers [1979], by Beavers, Yoo, and Joseph [1980], by Kolpin, Beavers, and Joseph [1980], and Joseph and Beavers [1976]. These values and others were collected into table of climbing constants for various liquids by Joseph, Beavers, Cers, Dewald, Hoger, and Than [1984]. Spera, Borgia and Strimple [1988] measured a huge value $\hat{\beta}=2.53\text{x}10^4$ g/cm for a rhyolitic magma melt at 1220°C.

Joseph and Fosdick [1973] and Beavers and Joseph [1975] studied rod climbing using the domain perturbation method of Appendix E. We refer readers to the two aforementioned papers for details. It is necessary to calculate normal stresses at second order. An approximate method for doing a special problem which involves this type of computation is given in exercise 17.6. The expansion parameter is ε, the angular velocity of the rod. The first order problem is exactly the rotating rod of §17.4. The velocity is $\mathbf{U}=\mathbf{e}_\theta a^2\varepsilon/r$. The angular velocity is $\omega(r)=\varepsilon a^2/r^2$. This goes to zero with a, but not uniformly. If r=a and then $a\to 0$, we get a finite $\omega(0)=\varepsilon$, something for nothing. They found that there was no secondary motion at order two and that the shape of the free surface can be expressed as

$$z = h(r,\varepsilon) = h_0 + h_2(r)\varepsilon^2 + O(|\varepsilon|^4) \tag{73}$$

where $h_0(r)$ is the static climb. They found that when surface tension is neglected,

$$h_2(r) = \frac{1}{\rho g}\left(\frac{2a^4}{r^4}\hat{\beta} - \frac{\rho a^4}{2r^2}\right), \tag{74}$$

where a is the radius of the rotating rod, ρ is the density of the liquid, and g is gravity.

Equations (73) and (74) suggest the climbing is not possible when ε is small and

$$r^2 > \frac{4\hat{\beta}}{\rho} \tag{75}$$

the fluid inside the critical radius $r=2(\hat{\beta}/\rho)^{1/2}$ climbs, the fluid outside sinks as in a Newtonian fluid. This implies that it is not possible to get climbing on a rod whose radius is greater than the critical one.

If we put r=a and let $a \rightarrow 0$, we get a finite climb $h_2(a)=2\hat{\beta}/\rho g$, again something for nothing.

Equation (74) could be used to compute $\hat{\beta}$ from measurements of $h(r,\varepsilon)$ for small ε. Joseph, Beavers, and Fosdick [1973] and Beavers and Joseph [1975] showed that to get the shapes of $h(r,\varepsilon)$ to agree with measured ones, it is necessary to retain the effects of surface tension. In this case, the height rise function $h_2(r)$ is governed by the following problem:

$$\begin{gathered} \frac{T}{r}(rh'_2)' - \rho g h_2 = -\frac{2a^4}{r^4}\hat{\beta} + \frac{\rho a^4}{2r^2}, \\ h'_2(a) = 0, \quad h_2(r) \rightarrow 0 \text{ as } r \rightarrow \infty, \end{gathered} \tag{76}$$

where T is the surface tension. A very accurate approximate solution of Equation (76) was derived by Joseph, Beavers, and Fosdick [1973]. When evaluated at the rod this solution gives

$$h_2(a) \sim \frac{a}{2T\sqrt{S}}\left(\frac{4\hat{\beta}}{4+\lambda} - \frac{\rho a^2}{2+\lambda}\right) \tag{77}$$

where $\lambda^2=a^2S$ and $S=\rho g/T$. Computations based on (76) were in very good agreement with experiments summarized in the [1984] paper of Joseph, *et al.* (see Joseph and Beavers, [1977b]) for a description of experimental procedures).

Equation (77) shows that surface tension prevents one from getting a finite climb at r=a as $a \to 0$. The climb now scales with a, vanishing as $a \to 0$. As in the previous case, the climb at small a is increasingly dominated by normal stresses. This formula is in wonderful agreement with experiments.

Beavers and Joseph [1975] proposed (76) and (77) as a basis for a rotating rod viscometer. A possible defect of this kind of rheometer is that it requires values for the surface tension. Fortunately, the values of T which have been measured in the different liquids used in experiments are nearly the same (see Joseph, *et al.*, [1984]). The value computed for $\hat{\beta}$ from the graph of the measured height of climb at the rod is not strongly affected by small changes in surface tension as the following argument shows.

Under the conditions which hold in most of the experiments, the second term on the right of (77) is small compared with the first. Then

$$\hat{\beta} \sim \frac{2T\sqrt{S}\,h_2(a)}{a}\left(\frac{4+\lambda}{4}\right)$$

Thus, for fixed values of $h_2(a)$, a, and ρ, we have

$$\frac{d\hat{\beta}}{\hat{\beta}} \sim \left(\frac{2}{4+\lambda}\right)\frac{dT}{T}\,. \tag{78}$$

For the operating conditions of most of the experiments ($a \sim 0.3$ cm, $\rho \sim 1$ g/cm^3, $T \sim 30$ dyn/cm), expression (78) gives

$$\frac{d\hat{\beta}}{\hat{\beta}} \sim 0.35\,\frac{dT}{T}$$

So an error in the value of T of 1% leads to an error in the computed value for $\hat{\beta}$ of about 0.35%.

The value of $\hat{\beta}$ is determined from the measured values of the height of climb on the rod. The rod is rotated slowly, at a speed for which a measurable height of climb can just be distinguished. The climb is measured as a function of increasing rotational speed, and the slope of the plot of $h(a,\varepsilon)$ versus ε^2 as $\varepsilon\to 0$ is read off the graph. This slope is compared

$$h_2(a) = dh/d\varepsilon^2 \text{ as } \varepsilon\to 0 \tag{79}$$

from (74) to determine $\hat{\beta}$.

17.11 Rheological measurements of α_1 and α_2

We have seen that the motions of viscoelastic fluids at low rates of shearing are determined through quadratic order when the material constants ρ, μ, α_1, and α_2 are known. We can determine α_1 and α_2 using free surface rheometers, the rotating rod giving $\hat{\beta}=3\alpha_1+2\alpha_2$ (Joseph, Beavers, Cers, Dewald, Hoger, and Than, [1984]) and the tilted trough giving $2\alpha_1+\alpha_2$ (Keentok, Georgescu, Sherwood, and Tanner, [1980]). Probably many other devices could be found which would yield α_1 and α_2, but the rheometers have not been developed.

The quadratic constants α_1 and α_2 are directly related to limiting values of the two normal stress-difference functions by (2) and (3). In principle, separate measurements of α_1 and α_2 would appear to be redundant, but this is not true in practice.

The rheometers used to measure $N_1(\kappa)$ are nearly useless when κ is very small (less than about 1.0 s^{-1}) and they do not give accurate values of $N_2(\kappa)$ for small or even large values of κ. The graphs of $N_1(\kappa)$ which are common in the rheological literature usually have blank places where κ is small. The values of $N_1(\kappa)$ for small κ are determined by (2) when α_1 is known. It is well known that $\alpha_1<0$ in polymeric liquids, and there are theoretical reasons for this. One could hope that it would be possible to back extrapolate N_1/κ^2 to $\kappa=0$.

The ratio $-N_2(\kappa^2)/N_1(\kappa^2)=1+\alpha_2/2\alpha_1+O(\kappa^4)$ is slowly varying for small κ and it appears to be possible to back extrapolate the ratio to $\kappa=0$ when the values of $N_2(\kappa^2)$ at small κ can be measured. This property appears to be evident in the data of Keentok, *et al.* [1980] and in data presented in Figure 14 of the paper by Alvarez, Lodge and Cantow [1985]. The measurements so far reported give $N_2<0$. However, there is no theoretical justification from continuum mechanics for the belief that $N_2<0$.

Weissenberg conjectured that $N_2(\kappa)=0$, a conjecture that is realized in some old molecular models. The climbing constant is

$$\hat{\beta} = 3\alpha_1+2\alpha_2 > 0 . \tag{80}$$

Let

$$n_2 = \lim_{\kappa\to 0}\frac{N_2(\kappa)}{\kappa^2} = 2\alpha_1+\alpha_2 . \tag{81}$$

Since $\alpha_1<0$ we have $\alpha_2>0$, and if $n_2<0$, then $\alpha_2<-2\alpha_1$. The parameters α_1 and α_2 of $S_{(2)}$ are therefore determined by n_2 and $\hat{\beta}$ as

$$\begin{aligned} \alpha_1 &= 2n_2 - \hat{\beta} , \\ \alpha_2 &= 2\hat{\beta} - 3n_2 . \end{aligned} \tag{82}$$

When $n_2=0$, $\alpha_1=-\hat{\beta}$ and $\alpha_2=2\hat{\beta}$ are determined by rod climbing. If we suppose that a fluid climbs a rod ($\hat{\beta}>0$) and bulges out when falling down a trough ($n_2<0$), we may conclude that there is a number k, $1\leq k\leq 4/3$, such that

$$2\alpha_1 + k\alpha_2 = 0 . \tag{83}$$

It is even more useful to put

$$\lim_{\kappa\to 0} (-N_2/N_1) = 1/m ,$$

where m is a real number which is larger than one in liquids for which N_2 and N_1 are known. Then

$$k = \frac{m}{m-1}$$

and

$$n_2 = -\frac{2}{m-4}\hat{\beta} .$$

Thus, for $\hat{\beta}>0$ and $n_2<0$, we have $m>4$. Also,

$$\alpha_1 = -\left(\frac{m}{m-4}\right)\hat{\beta}$$

and

$$\alpha_2 = \left(\frac{2(m-1)}{m-4}\right)\hat{\beta} . \tag{84}$$

For many polymeric liquids, m is about 10. In this case

$$\alpha_1 = -(5/3)\,\hat{\beta}$$

and

$$\alpha_2 = 3\hat{\beta} .$$

As an example of the application of this method, we note that according to data of Alvarez, *et al.* [1985] and Walters [1983] the value of m for polyisobutylene/decalin solution "D1" varies roughly between 10 to 14.3. We measured the climbing constant $\hat{\beta}$=62.2 g/cm and the viscosity μ=95 g/cm s at 25° C. We find the values α_1 and α_2 using (84) and the results:

m for D1	10	14.3
α_1 (g/cm)	-103	-86
α_2 (g/cm)	187	161
$\alpha_1+\alpha_2$	84	75

Some other values of m are given in Table 17.1.

Polymer fluid	Polymer concentration by weight	$-\frac{N_2}{N_1} = \frac{1}{m}$
Polyox WSR 30 / water	1.0% 0.7%	0.075 ± .01 0.065 ± .01
Separan AP-30 solutions water	0.8% 0.5% 0.2%	0.100 ± .01 0.085 ± .015 0.065 ± .01
Opponal B200/ Dekalin	1.1%	0.11 ± .01
NBS Nonlinear Fluid No. 1	7.1%	0.30 ± .04

Table 17.1. (After Keentok, *et al.*, [1980]) Normal stress ratios for different fluids

The second normal stress in the NBS Nonlinear Fluid #1 is very large. With this value we may compute

$$\hat{\beta} = \lim_{\kappa \to 0} \frac{N_1}{2} + 2N_2 = \lim_{\kappa \to 0} -\frac{1}{10} N_1 \, . \tag{85}$$

It follows that the NBS Nonlinear Fluid No. 1 should not climb a rotating rod at low rates of shear. Lodge, Schieber and Bird [1987] have considered the problem of rod climbing at finite rotation speeds. They obtained (81) and its consequences and report that unpublished experiments of E. Kearsly have shown that NBS Nonlinear Fluid No. 1 does climb a rotating rod. We do not know whether the speed of rod was sufficiently low.

A very complete study of normal stresses was done on the fluid M1 (0.244% Polyisobutylene in a solvent of 7% kerosene and 93° polybutene (HYVIS)) in many different laboratories world wide. The paper by Hu, et al [1990] reports measured values of $\hat{\beta}$ for M1 over a wide range of temperatures. The measurements are summarized in Figure 17.12 and photographs of rod climbing are shown in Figure 17.13. Measured values for $N_1/2\kappa^2$ at 20° C are shown in Figure 17.14 and at 27.2° C in Figure 17.15. Table 17.2

gives the values of the second order constants for M1 obtained from the climbing constant and backward extrapolation of $N_1/2\kappa^2$. This table shows that N_2/N_1 is very insensitive to changes in the temperature even over a range of temperature in which the values of N_1 and N_2 depend strongly on the temperature.

Using the values in Table 17.2, and the values of the viscosity μ=[29.5, 11] poise at 20° and 27.2° C, we may give the coefficient in the expression (13) for the extensional viscosity. At 20° C

$$T_{11}-T_{22}\sim 3[29.5\overset{\circ}{s}+2.34\overset{\circ}{s}{}^2],$$

and at 27.2° C

$$T_{11}-T_{22}\sim 3[11\overset{\circ}{s}+0.77\overset{\circ}{s}{}^2],$$

Comparing now the value $0.77\leq\alpha_1+\alpha_2<2.34$ with the values in Table 17.1, we can conclude that the increase of extensional viscosity at second order is more than fifty times greater in D1 than in M1.

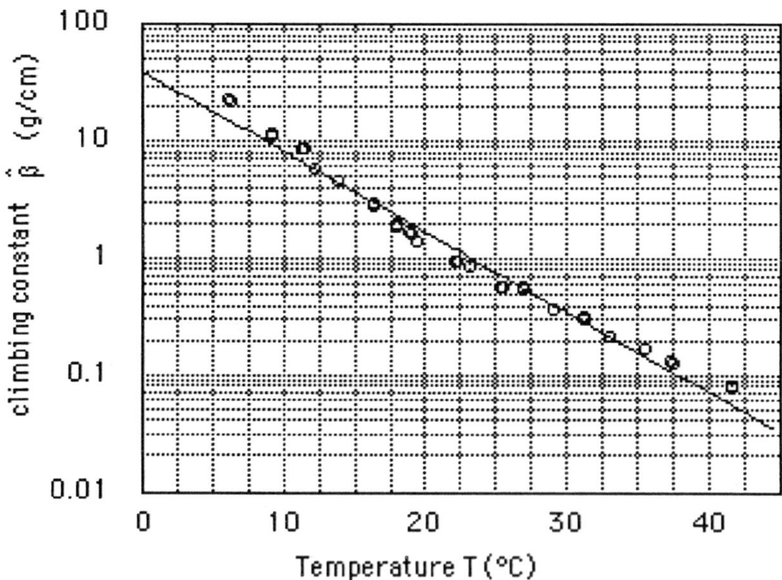

Figure 17.12 (Hu, et al, [1990]) Summary of the dependence of climbing constant $\hat{\beta}$ for M1 on temperature T (6.2 – 41.5 °C). The equation of the line is $\hat{\beta} = 39.7 * \exp(-0.158 * T)$ (g/cm). At T=20 °C, $\hat{\beta} = 1.68$ g/cm.

Figures 17.13a–f Photographs of rod climbing in M1 at 21.5° C. (a) At rest, there is a small static climb due to capillary; (b) Stable steady climb ω=3rev/sec; (c) and (d), Slightly post-critical time-periodic motion at ω=4.94 rev/sec; (e) and (f), Large amplitude time-periodic motion at ω=6.55rev/sec.

Figure 17.13a

Figure 17.13b

Figure 17.13c

Figure 17.13d

Figure 17.13e

Figure 17.13f

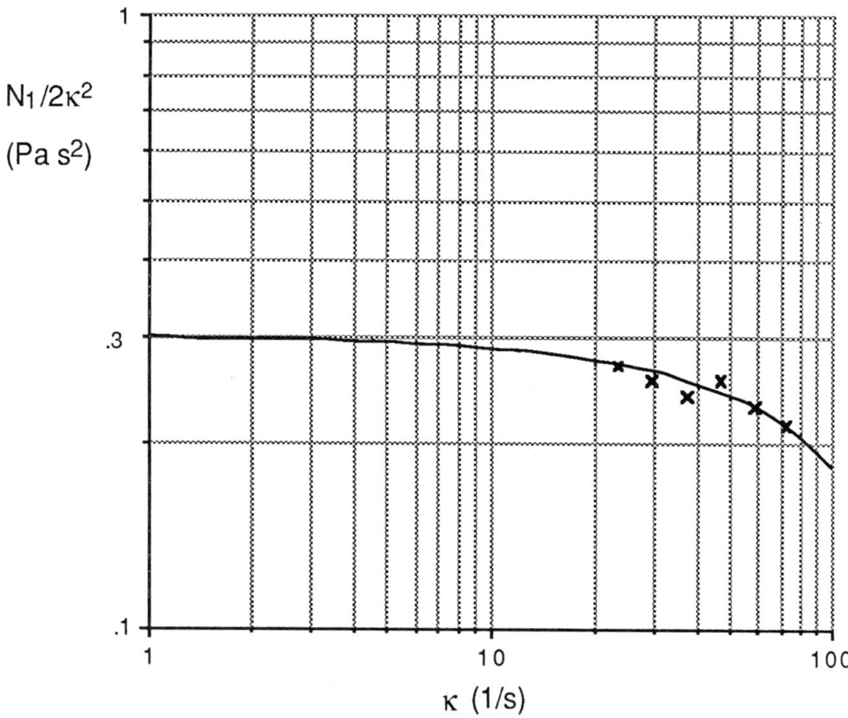

Figure 17.14 (After Binding, Jones and Walters, [1989]) Measured values of $N_1/2\kappa^2$ for M1 at 20°C. The backward extrapolation gives that $N_1/2\kappa^2 \to -\alpha_1$ is about 0.3 Pa s^2 or 3 g/cm.

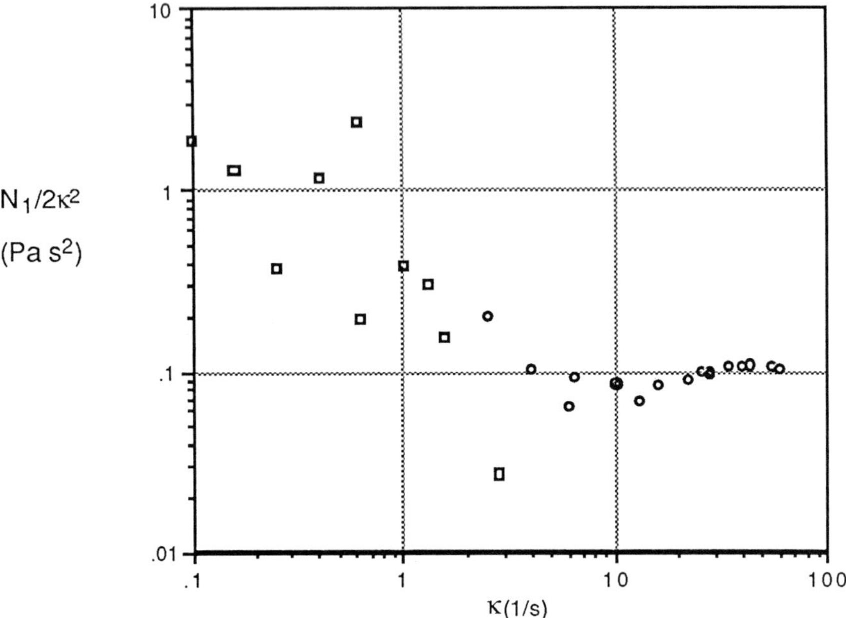

Figure 17.15 (After R.K. Prud'homme, [1989], personal communication) Measured values of $N_1/2\kappa^2$ for M1 at 27.2°C. The measured values of the stress for □ are below the allowed transducer limits and are discarded. We took the limiting value to be $-\alpha_1 = 0.1$ Pa·s^2 or 1 g/cm.

	20° C	27.2° C
$-\frac{n_1}{2} = \alpha_1$ (g/cm)	–3.0	–1.0
$\hat{\beta}$ (g/cm)	1.68	0.54
α_2 (g/cm)	5.34	1.77
$\alpha_1 + \alpha_2$	2.34	0.77
N_2/N_1	–0.11	–0.12

Table 17.2 Material parameters for M1 at 20° C and 27.2° C using (17.13) and (17.14)

17.12 Some other free surface problems

The paper "Free surface problems in rheological fluid mechanics" by Joseph and Beavers [1977a] is still one of the better sources of reference for work in the area, including the unsteady problems not mentioned here. This type of work has not developed much for more than a decade. Perhaps it is more fair to say that attention in this field has shifted to direct numerical simulation using special models. One problem with direct numerical simulation is that the results depend on the model selected. You fit the model parameters in some limited set of experiments and hope for the best. No model works for all flows; maybe they all work only for limited sets. Perhaps delimiting sets of deformations is good science, but in any event is not treated in this book.

17.12.1 Tilted trough

One of the most important, and earliest, analyses of a free surface flow is the study of the shape of the free surface on a viscoelastic liquid flowing down a tilted open channel by Wineman and Pipkin [1966]. They analyzed the problem as a domain perturbation of the state of rest. They found that the rise of the free surface was proportional to the hard to measure second normal stress at lowest $O(\beta^2)$ where β is the angle of tilt. They found that the free surface bulges outward if the second normal stress is negative. Kuo and Tanner [1974] reported experiments showing an outward bulge in the fluids studied by them. More measurements were made by Keentok, Georgescu, Sherwood and Tanner [1980] and values of the ratio of N_2/N_1 as a function of the shear rate with $N_2<0$ always were presented.

Tanner [1970] presented a simplified analysis of the tilted trough with a semi-circular cross-section framed only in terms of the viscometric functions N_1 and N_2. His solution is valid under restrictive assumptions, which disappear in the limit of small β in which the Wineman and Pipkin solution is valid. Sturges and Joseph [1975] did a systematic domain perturbation through terms of $O(\beta^4)$ and showed that secondary motion would not develop until $O(\beta^6)$ when the channel is infinite or has a semicircular cross section. The shear rates which can be achieved in a trough tilted through a small angle are rather small, so that the deflections which develop on the free surface would not be easy to measure in many fluids like M1.

17.12.2 Density difference singularities and normal stress amplifiers

The density which one fluid feels in the presence of another is a relative density, another recognition of Archimedes' principle. Suppose we look at the problem of a rotating rod when the fluid on the top is not air, but some other liquid. Then the term ρgh would be replaced by $(\rho_1-\rho_2)$ where ρ_1 is the density of the heavy fluid. If interfacial tension is neglected, we would find an equation of the form

$$(\rho_1-\rho_2)gh = f$$

where f depends on the motion and not on the densities, or not in an important way on the densities. This gives rise to a density difference singularity in the the height rise

$$h = f/g(\rho_1-\rho_2) \tag{86}$$

can be raised indefinitely by matching the densities. This density difference singularity can be realized. It amplifies the effects of normal stresses by reducing the opposing force of gravity.

The perturbation analysis of rod climbing can be carried through second order without complication when a layer of liquid is floated on another immiscible liquid. Beavers and Joseph [1977] have given an analysis and reported an experiment testing the analysis for the problem in which a non-Newtonian liquid (STP) is floated on water. The STP climbs up the rotating rod into the air and down the rod into the water. The down-climb is much larger than the up-climb, roughly in the ratio

$$\frac{h_B^{[2]}(a)}{h_T^{[2]}(a)} \sim \frac{\sqrt{T_T(\rho-\rho_a)}}{\sqrt{T_B(\rho_w-\rho)}} \, \frac{4+a\sqrt{(\rho-\rho_a)g/T_T}}{4+a\sqrt{(\rho_w-\rho)g/T_B}} \tag{87}$$

where the quantity $h_B^{[2]}(a)$ is the down-climb at the bottom STP-water interface on the rod surface at r=a, $h_T^{[2]}(a)$ is the up-climb at the air-STP interface, and ρ, ρ_w, and ρ_a are densities of STP, water

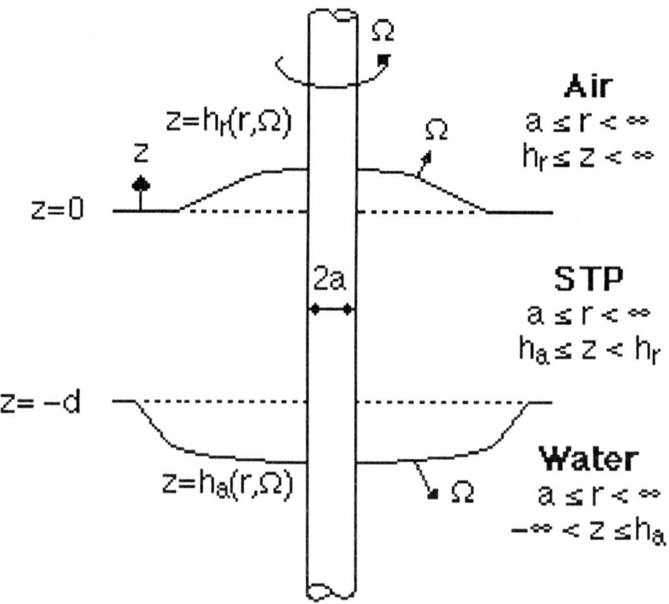

Figure 17.16 Schematic sketch of the free surfaces on STP floating on water

and air respectively, and T_T and T_B are surface tension coefficients. The dynamical problem is solved in all three layers. Correct coupling conditions for the continuity of stresses at each of the two interfaces are prescribed, and it is not assumed that the interface is free of tangential tractions.

The analysis shows that the effect of inertia and gravity is to push the heavy fluid down near the rod. The normal stresses push the STP up into the air and down into the water. Hence, inertia and normal stresses are in conflict at the STP-air interface and in concert at the water-STP interface (see Figure 17.16).

In the experiments reported by Beavers and Joseph [1977], the values of the ratio (87) are, respectively, [3.7, 3.9, 4.3] for three different rods of radius a=(0.479, 0.627, 0.951 cm). The down-climb/up-climb ratio (87) may be further approximated by $h_B^{[2]}(a)/h_T^{[2]}(a) \sim [(\rho-\rho_a)/(\rho_w-\rho)^{1/2} \sim 2.7$. The value of $\hat{\beta}$ for SPT was computed from the observed up-climb and the observed down-climb. There is presumably one and only one value of $\hat{\beta}$ for STP at a given temperature. The values of $\hat{\beta}$ computed from the up-climb, from the down-climb, and from the independent measurements are in good agreement with one another.

Some years ago, by accident we noticed that STP floats on TL227, and the two fluids do not mix. We thought that the immiscibility was unusual since both fluids have petroleum oil as solvents, but the polymers are different. The densities of these two fluids are almost the same; the density difference is

$$\rho_{227} - \rho_{STP} = 0.90 - 0.86 = 0.04 \text{ g/cm}^3$$

We rotated a rod in the composite layered fluid (air, STP, TL227). At very small speeds of rotation the TL227 climbed right through about four inches of STP. The magnitude of the amplification of the climb due to the small density difference is spectacular.

A photograph of the spectacular magnification of the climb is shown as Figure 6 in the [1977a] paper of Joseph and Beavers. Afterwards, we constructed a demonstration device to illustrate the phenomena. The device is a bottle whose cavity dimensions are: inner radius 3 cm, height 27 cm. A rod fixed in the center of the cylinder is free to turn when driven manually by the top disk which is turned by a human finger in the eccentric hole in the disk shown in Figure 17.7. The bottle is filled with TL277 and STP as in (a). Then the rod is turned very slowly and one gets the spectacular climb shown in (b). The normal stresses are about twenty times larger in TL227 than in STP; at 26.7°C, $\hat{\beta}_{STP}=0.97\frac{g}{cm}$, $\hat{\beta}_{227}=18\frac{g}{cm}$. The normal stresses of TL227 overwhelm those in STP and push the fluid up against a small effective gravity $g(\rho_{227}-\rho_{STP})$ where $\rho_{227}-\rho_{STP}=0.04\text{g/cm}^3$. It takes many hours for the climb to fall after the rotation has ceased. It is astonishing that there is an interface between STP and TL227 since they have the same type of petroleum oil solvent. The interfacial tension is very small, estimated crudely as 0.4 dyne/cm. In Figure 17.17(b) you see a huge climb of TL227 in STP, where the effective gravity is small, and a small climb of TL227 into air, where the effective gravity is large. In treating the problem of the demonstration device approximately according to the method of (86) it is necessary to take into account the shearing action of the STP on the climbing cylinder of TL227 in the manner indicated in exercise 17.6.

Tieu, Joseph and Beavers [1984] studied the motion and evolution of the interface between two polymeric liquids in a rotating cylinder, without a rod. This gives rise to rigid rotation of both fluids with a convex interface when the rotation is steady and to other shapes when the motion is oscillatory. They solved the problem through second order using domain perturbation and integral expression for the stresses. They did an experiment using STP and TL227 and showed that the mean motion at second order totally dominates the part which oscillates at twice the frequency of the oscillating cylinder. You can centrifuge by oscillating a cylinder, as well as in steady rotation. In this case both the centripetal acceleration differences and gravity are reduced by multiplication by the density difference, so that density matching does not produce a big effect.

Figure 17.17 Demonstration device illustrating the amplification of the climb of TL227 into STP by density material.

The use of density matching to enhance the bulge on a free surface in a tilted trough was considered by Joseph and Sturges [1980]. They noted that the magnitude of height rise usefulness of the device can be enhanced by using two different fluids in the trough. They noted that for some fluids flowing down a trough, the actual bulging, without amplification, would be very small (only about 0.013 mm for STP in a 2cm trough at a slope of 30 deg.) and it would be difficult to get accurate values of N_2 from the bulge. As in the rod climbing problem, the magnitude of the height rise is roughly proportional to the reciprocal of the density difference when surface tension is neglected. The analysis of the problem of two fluid flow down a trough is more complicated than the rod because the shear stresses are not matched and must be taken into account, even at low orders. No experiments using the two fluid device have been reported.

Exercises

In exercises 17.1 and 2

$$\operatorname{div}\mathbf{u}=0\,,\quad \mathbf{A}[\mathbf{u}]=\mathbf{A}_1[\mathbf{u}]\,,\quad \mathbf{B}=\mathbf{A}_2[\mathbf{u}]$$

and

$$\operatorname{div}\mathbf{A}[\mathbf{u}]=\nabla Q$$

Exercise 17.1.
(Giesekus, [1963])

$$\operatorname{div}\mathbf{B}=\operatorname{div}\mathbf{A}^2+\operatorname{grad}\left\{\frac{\partial Q}{\partial t}+(\mathbf{u}\cdot\nabla)Q+\frac{1}{4}\operatorname{tr}\mathbf{A}^2\right\}.$$

The proof is given in nearly all standard books of reference.

Exercise 17.2.
Suppose that

$$\operatorname{div}\mathbf{A}^2=\nabla\Gamma\,.$$

Then

$$\operatorname{div} \mathbf{B} = \operatorname{grad}\left\{\Gamma + \frac{\partial Q}{\partial t} + (\mathbf{u}\cdot\nabla)Q + \frac{1}{4}\operatorname{tr}\mathbf{A}^2\right\}$$

and the divergence of the stress

$$\mathbf{T} = -p\mathbf{1}+\mathbf{S}\ ,$$

in a second order fluid

$$\mathbf{S} = \mu\mathbf{A} + \alpha_1\mathbf{B} + \alpha_2\mathbf{A}^2\ , \tag{88}$$

vanishes

$$\operatorname{div} \mathbf{T} = -\nabla p + \operatorname{div} \mathbf{S} = 0$$

if

$$p = \mu Q + (\alpha_1+\alpha_2)\Gamma + \alpha_1\left(\frac{\partial Q}{\partial t} + (\mathbf{u}\cdot\nabla)Q + \frac{1}{4}\operatorname{tr}\mathbf{A}^2\right). \tag{89}$$

It is convenient to express this when $Q=p_N/\mu$ is the potential (the "pressure") for a Newtonian flow

$$p = (\alpha_1+\alpha_2)\Gamma + \left(p_N + \frac{\alpha_1}{\mu}\frac{dp_N}{dt}\right) + \alpha_1\frac{1}{4}\operatorname{tr}\mathbf{A}^2\ . \tag{90}$$

Exercise 17.3.
Show that

$$\Gamma = 2\left(\frac{\partial^2\phi}{\partial x_j\,\partial x_j}\right)^2$$

when $\mathbf{u}=\nabla\phi$ has a velocity potential.

Exercise 17.4.
In plane flows $\mathbf{u}=(u,v)$ where u and v depend on x,y but not z

$$\Gamma = 4\left(\frac{\partial u}{\partial x}\right)^2 + \left(\frac{\partial u}{\partial y} + \frac{\partial v}{\partial x}\right)^2$$

and

$$\mathbf{A}^2 = \begin{bmatrix} \Gamma & 0 & 0 \\ 0 & \Gamma & 0 \\ 0 & 0 & 0 \end{bmatrix} .$$

Hence

$$\text{div}\ \mathbf{A}^2 = 2\nabla\Gamma .$$

Tanner's [1966] theorem is that the velocity of plane Stokes flow also satisfies div **T**=0 for a second order fluid (88) with the pressure (90).

Exercise 17.5.
The plane part of the stress **T** given by (88) for plane Stokes flow is independent of α_2. To prove this we need to note only that

$$-p\mathbf{1}+\alpha_2\mathbf{A}^2 = \begin{bmatrix} -p+\alpha_2\Gamma & 0 & 0 \\ 0 & -p+\alpha_2\Gamma & 0 \\ 0 & 0 & -p \end{bmatrix}$$

and

$$-p + \alpha_2\Gamma = -\frac{3}{2}\alpha_1 + p_N + \frac{\alpha_1}{\mu}\frac{dp_N}{dt} .$$

Exercise 17.6*
Suppose the interface between two polymeric liquids which has been amplified by density matching is nearly vertical, as in Figure 17.17. Compute an approximate value of f in (86), imagining a perturbation from a perfectly flat interface and a perfect cylindrical vertical interface of radius R, neglecting the effects of a small interfacial tension.

Solution
We know that for each liquid the equations of motion through order two are

---

* This calculation was carried out by H. Hu.

$$\nabla \bullet \mathbf{u} = 0 \,, \tag{91}$$

$$\rho(\mathbf{u}\bullet\nabla)\mathbf{u} = -\nabla\Phi + \nabla\bullet\mathbf{S} \tag{92}$$

where

$$\Phi = p + \rho g z \,,$$
$$\mathbf{S} = \mu\mathbf{A}_1\mathbf{u} + (\alpha_1\mathbf{A}_2 + \alpha_2\mathbf{A}_1^2) \,,$$
$$\mathbf{A}_1 = \nabla\mathbf{u} + (\nabla\mathbf{u})^T \,,$$
$$\mathbf{A}_2 = (\mathbf{u}\bullet\nabla)\mathbf{A}_1 + \nabla\mathbf{u}\bullet\mathbf{A}_1 + (\nabla\mathbf{u}\bullet\mathbf{A}_1)^T \,,$$

and for each liquid ρ, μ, α_1 and α_2 are different.

Flat interface. The boundary conditions are

$\mathbf{u} = a\varepsilon\mathbf{e}_\theta$ on $r = a$ (radius of the rod),
$\mathbf{u} = 0$ on $r = b$ (radius of the container). (93)

At the interface between the two liquids $z=h(r,\varepsilon)$ we have

$$\begin{gathered} w - h'u = 0 \,, \\ [\![\mathbf{u}]\!] = 0 \,, \\ [\![S_{z\theta} - h'S_{r\theta}]\!] = 0 \,, \\ [\![h'(S_{zz} - S_{rr}) + (1-h'^2)\,S_{rz}]\!] = 0 \,, \\ [\![-\Phi + \rho g h + S_{zz} - h'S_{rz}]\!] = 0 \,. \end{gathered} \tag{94}$$

At the zeroth order of approximation the solution is

$$\mathbf{u}^{<0>} = \mathbf{S}^{<0>} = h^{<0>} = 0 \,. \tag{95}$$

The reference domain is

$$V_0 = \{r, z \mid r \in [a, b] \text{ and } z \in (-\infty, 0) \cup (0, \infty)\} .$$

The problem at the first order approximation is as follows:

$$\begin{cases} \nabla \bullet \mathbf{u}^{<1>} = 0 \,, \\ -\nabla\Phi^{<1>} + \nabla \bullet \mathbf{S}^{<1>} = 0 & \text{in } V_0, \end{cases}$$

where

$$\mathbf{S}^{<1>} = \mu\left(\nabla\mathbf{u}^{<1>} + (\nabla\mathbf{u}^{<1>})^T\right) .$$

The boundary and interface conditions for this problem are:

on $r = a$ $\mathbf{u}^{<1>} = a\mathbf{e}_\theta$,

on $r = b$ $\mathbf{u}^{<1>} = 0$,

on $z = 0$
$$\begin{cases} w^{<1>} = 0 , \\ [\![\mathbf{u}^{<1>}]\!] = 0 , \\ [\![S^{<1>}_{z\theta}]\!] = 0 , \\ [\![S^{<1>}_{rz}]\!] = 0 , \\ [\![-\Phi^{<1>} + S^{<1>}_{zz}]\!] = -(\rho_1-\rho_2)gh^{<1>} . \end{cases}$$

The solution of this problem is

$$\begin{cases} \mathbf{u}^{<1>} = v\mathbf{e}_\theta , \\ h^{<1>} = 0 , \\ \Phi^{<1>} = 0 , \end{cases} \tag{96}$$

where

$$v = Ar + \frac{B}{r} , \tag{97}$$

and

$$A = -\frac{a^2}{b^2-a^2} , \quad B = \frac{a^2b^2}{b^2-a^2} . \tag{98}$$

At second order, the problem to be solved is

$$\begin{cases} \nabla\cdot\mathbf{u}^{<2>} = 0 , \\ \rho(\mathbf{u}^{<1>}\cdot\nabla)\mathbf{u}^{<1>} = -\nabla\Phi^{<2>} + \nabla\cdot\mathbf{S}^{<2>} \qquad \text{in } V_0, \end{cases} \tag{99}$$

where

$$\mathbf{S}^{<2>} = \mu\mathbf{A}_1[\mathbf{u}^{<2>}] + \alpha_1\mathbf{A}_2[\mathbf{u}^{<1>}] + \alpha_2\mathbf{A}_1^2[\mathbf{u}^{<1>}] .$$

The boundary and interfacial conditions are

$\mathbf{u}^{<2>} = 0$ on r=a and r=b,

on $z = 0$
$$\begin{cases} w^{<2>} = 0 \ , \\ [\![\mathbf{u}^{<2>}]\!] = 0 \ , \\ [\![S^{<2>}_{z\theta}]\!] = 0 \ , \\ [\![S^{<2>}_{rz}]\!] = 0 \ , \\ [\![-\Phi^{<2>} + S^{<2>}_{zz}]\!] = -(\rho_1 - \rho_2) g h^{<2>} \ . \end{cases}$$

Using (96) we calculate

$$\mathbf{u}^{<1>} \bullet \nabla \mathbf{u}^{<1>} = -\frac{v^2}{r} \mathbf{e}_r \ ,$$
$$\nabla \mathbf{u}^{<1>} = \frac{dv}{dr} \mathbf{e}_r \mathbf{e}_\theta - \frac{v}{r} \mathbf{e}_\theta \mathbf{e}_r \ ,$$
$$\mathbf{A}_1[\mathbf{u}^{<1>}] = \left(\frac{dv}{dr} - \frac{v}{r} \right) (\mathbf{e}_r \mathbf{e}_\theta + \mathbf{e}_\theta \mathbf{e}_r) \ ,$$
$$\mathbf{A}_1^2[\mathbf{u}^{<1>}] = \left(\frac{dv}{dr} - \frac{v}{r} \right)^2 (\mathbf{e}_r \mathbf{e}_r + \mathbf{e}_\theta \mathbf{e}_\theta) \ ,$$
$$\mathbf{A}_2[\mathbf{u}^{<1>}] = 2\left(\frac{dv}{dr} - \frac{v}{r} \right)^2 \mathbf{e}_r \mathbf{e}_r \ .$$

Therefore

$$\mathbf{S}^{<1>}_{2} \overset{\text{def}}{=} \alpha_1 \mathbf{A}_2[\mathbf{u}^{<1>}] + \alpha_2 \mathbf{A}_1^2[\mathbf{u}^{<1>}]$$
$$= \left(\frac{dv}{dr} - \frac{v}{r} \right)^2 [(2\alpha_1 + \alpha_2)\, \mathbf{e}_r \mathbf{e}_r + \alpha_2\, \mathbf{e}_\theta \mathbf{e}_\theta] \ .$$

Using (97) we find that

$$\mathbf{S}^{<1>}_{2} = \frac{4B^2}{r^4} [(2\alpha_1 + \alpha_2)\, \mathbf{e}_r \mathbf{e}_r + \alpha_2\, \mathbf{e}_\theta \mathbf{e}_\theta] \ ,$$

and

$$\nabla \bullet \mathbf{S}^{<1>}_{2} = -\frac{8B^2}{r^5} (3\alpha_1 + 2\alpha_2)\, \mathbf{e}_r = -\frac{8B^2}{r^5} \hat{\beta} \mathbf{e}_r \ .$$

Substituting these expressions into (99), we see that

$$\nabla \cdot \mathbf{S}^{<1>}_{2} - \rho \mathbf{u}^{<1>} \cdot \nabla \mathbf{u}^{<1>} = \mathbf{e}_r \frac{d}{dr}$$
$$\left[\frac{2B^2}{r^4} \hat{\beta} + \rho \left(\frac{A^2}{2} r^2 + 2AB \ln r - \frac{B^2}{2r^2}\right)\right]$$

is a gradient which can be balanced by a pressure without motion, thus the solution of the second order is

$$\mathbf{u}^{<2>} = 0 ,$$
$$\Phi^{<2>} = \frac{2B^2}{r^4} \hat{\beta} + \rho\left(\frac{A^2}{2} r^2 + 2AB \ln r - \frac{B^2}{2r^2}\right), \tag{100}$$
$$h^{<2>} = \frac{[\![\Phi^{<2>}]\!]}{[\![\rho]\!] g} = \frac{2B^2}{g r^4} \frac{\hat{\beta}_1 - \hat{\beta}_2}{\rho_1 - \rho_2} + \frac{1}{g}\left(\frac{A^2}{2} r^2 + 2AB \ln r - \frac{B^2}{2r^2}\right).$$

In the limit of $b \to \infty$, we have $A \to 0$ and $B \to a^2$, therefore

$$h^{<2>} = \frac{2a^4}{g r^4} \frac{\hat{\beta}_1 - \hat{\beta}_2}{\rho_1 - \rho_2} - \frac{a^4}{2 g r^2} . \tag{101}$$

Cylindrical interface. The boundary conditions in this case are unaltered, we must satisfy (93); but the interfacial conditions are replaced by the following conditions at $r = \eta(z;\varepsilon)$:

$$\begin{gathered} u - \eta' w = 0 , \\ [\![\mathbf{u}]\!] = 0 , \\ [\![S_{\theta r} - \eta' s_{\theta z}]\!] = 0 , \\ [\![(1-\eta'^2) S_{zr} + \eta' (S_{rr} - S_{zz})]\!] = 0 , \\ [\![-\Phi + \rho g z + S_{rr} - \eta' S_{rz}]\!] = 0 . \end{gathered} \tag{102}$$

The zeroth order solution is

$$\mathbf{u}^{<0>} = \mathbf{S}^{<0>} = 0 , \ \eta^{<0>} = R . \tag{103}$$

The reference domain is

$$V_0 = \{ r,z \mid r \in [a,R] \cup [R,b],\ z \in (0,\infty) \} .$$

The normal stress condition is not satisfied at zeroth order, the fluid at rest will fall slowly if the density is nearly matched. We will enforce the unbalanced normal stress in an approximate manner at second order. The first order solution takes the form:

$$\mathbf{u}^{<1>} = v\mathbf{e}_\theta \,, \quad \eta^{<1>} = 0 \,, \quad \Phi^{<1>} = 0 \,, \tag{104}$$

where

$$v = \begin{cases} A_1 r + \dfrac{B_1}{r} & a \le r \le R \,, \\ A_2 r + \dfrac{B_2}{r} & R \le r \le b \,, \end{cases} \tag{105}$$

$$\begin{aligned} A_1 &= 1 - mR^2b^2/K \,, \\ B_1 &= mR^2a^2b^2/K \,, \\ A_2 &= -R^2a^2/K \,, \\ B_2 &= R^2a^2b^2/K \,, \end{aligned} \tag{106}$$

and

$$m = \mu_2/\mu_1 \,, \; K = R^2(mb^2 - a^2) + (1-m)\, a^2b^2 \,.$$

Following the same steps as in the case of flat interface, we find the solution in the form

$$\mathbf{u}^{<2>} = 0 \,, \quad \Phi^{<2>} = \frac{2B^2}{r^4}\hat{\beta} + \rho\left(\frac{A^2}{2} r^2 + 2AB \ln r - \frac{B^2}{2r^2}\right) \tag{107}$$

where A, B, $\hat{\beta}$, ρ take different values depending on whether $r \in [a,R]$ or $r \in [R,b]$.

The normal stress balance determines the shape of the interface, thus

$$\frac{z}{\varepsilon^2} = \frac{[\![\Phi^{<2>}]\!]}{g[\![\rho]\!]} = \frac{2}{gR^4}\frac{[\![B^2\hat{\beta}]\!]}{[\![\rho]\!]} + \frac{R^2}{2g}\frac{[\![\rho A^2]\!]}{[\![\rho]\!]} + \frac{2 \ln r}{g}\frac{[\![\rho AB]\!]}{[\![\rho]\!]} - \frac{1}{2R^2 g}\frac{[\![\rho B^2]\!]}{[\![\rho]\!]} \,. \tag{108}$$

If $b \to \infty$, (106) becomes

$$A_1 = \frac{(1-m)a^2}{mR^2 + (1-m)a^2} \,, \qquad B_1 = \frac{mR^2a^2}{mR^2 + (1-m)a^2} \,,$$

$$A_2 = 0\,, \qquad B_2 = \frac{R^2 a^2}{mR^2+(1-m)a^2}\,.$$

Substituting back into (108), we have

$$\frac{z}{\varepsilon^2} = \frac{a^4}{[mR^2+(1-m)a^2]^2 g(\rho_1-\rho_2)}\Big[2(m^2\hat{\beta}_1-\hat{\beta}_2)$$
$$+\frac{R^2}{2}\rho_1(1-m)^2 + 2\rho_1 \ln R\bullet R^2 m(1-m) - \frac{R^2}{2}(\rho_1 m^2-\rho_2)\Big]\,. \quad (109)$$

The maximum height to which the liquid 1 climbs is

$$h = \frac{1}{g(\rho_1-\rho_2)}$$
$$\Big[2(m^2\hat{\beta}_1-\hat{\beta}_2) + \frac{a^2}{2}\rho_1(1-2m) + \frac{a^2}{2}\rho_2 + 2\rho_1 \ln a\bullet a^2 m(1-m)\Big]\ \varepsilon^2\,.$$

This formula may be simplified for small values of a

$$h = \frac{2(m^2\hat{\beta}_1-\hat{\beta}_2)}{g(\rho_1-\rho_2)}\ \varepsilon^2\,. \qquad (110)$$

Comparing (110) with the result of (101) for the flat interface, we see that the viscosity ratio m enters into the analysis when the unperturbed interface is cylindrical.

If liquid 1 is TLA227 and liquid 2 is STP, we have

$\rho_1 = 0.9$ g/cm^3, $\rho_2 = 0.86$ g/cm^3 ,
$\mu_1 = 200$ p. , $\mu_2 = 120$ p. ,
$\hat{\beta}_1 = 18$ g/cm and $\hat{\beta}_2 = 0.97$ g/cm at 26.7°C.

The maximum height to which the liquid 1 climbs is, respectively,

$h = 0.281\ \varepsilon^2$ (cm) when the unperturbed interface is cylindrical,
$h = 0.869\ \varepsilon^2$ (cm) when the unperturbed interface is flat.

A typical slow angular velocity of ten radian seconds leads to amplified climbs of 28 and 87 centimeters respectively.

CHAPTER 18

Elasticity and Viscosity of Liquids

18.1 Constitutive equations

There is a great simplification in the problem of constitutive modeling when the deformations are a small perturbation of states of rest. These deformations depend on a Newtonian viscosity μ and a smooth relaxation function G(s), where G(s)>0, G′(s)<0, and G′′(s)>0, for $0 \le s=t-\tau < \infty$, and τ is the past time. The stress $\boldsymbol{\tau}$ is given by

$$\boldsymbol{\tau} = 2\mu\mathbf{D}[\mathbf{u}] + 2\int_0^\infty G(s)\mathbf{D}[\mathbf{u}(\mathbf{x},t-s)]\, ds \tag{1}$$

where μ is the Newtonian viscosity, $\mathbf{u}(\mathbf{x},t)$ is the velocity, and $\mathbf{D}[\mathbf{u}]$ is the symmetric part of the velocity gradient. Equation (1) is a Jeffreys' type of generalization of Boltzmann's equation of linear viscoelasticity in which the presence of a Newtonian viscosity is acknowledged. Equation (1) also holds in the class of small perturbations of rigid motions with $\mathbf{x}$ replaced by $\boldsymbol{\chi}_0(\tau)$, the pathline on the unperturbed motion.

A constitutive equation of the rate type may be obtained as the time derivative of (1).

$$\frac{\partial \boldsymbol{\tau}}{\partial t} = 2\mu\frac{\partial \mathbf{D}}{\partial t} + 2G(0)\mathbf{D} + 2\int_0^\infty G'(s)\mathbf{D}[\mathbf{u}(\mathbf{x},t-s)]\ ds. \tag{2}$$

Jeffreys' model is a special case of (2) in which

$$G(s) = \frac{\eta}{\lambda}e^{-s/\lambda} \tag{3}$$

where λ is the relaxation time and η is the elastic viscosity. Combining (2) and (3), we get

$$\lambda\frac{\partial \boldsymbol{\tau}}{\partial t} = 2\mu\lambda\frac{\partial \mathbf{D}}{\partial t} + 2\tilde{\mu}\mathbf{D} - \boldsymbol{\tau} \tag{4}$$

where $\tilde{\mu} = \mu+\eta$. A retardation time

$$\Lambda = \mu\lambda/\eta \tag{5}$$

is usually defined for (4). When $\mu=0$, (4) gives rise to a Maxwell model

$$\lambda\frac{\partial \boldsymbol{\tau}}{\partial t} = 2\eta\mathbf{D} - \boldsymbol{\tau}\,. \tag{6}$$

Fluids with $\mu=0$ are like relaxing elastic solids. They propagate shock waves. Fluids with $\mu\neq 0$ are diffusive; they smooth shocks.

Equations (1) and (2) are perturbation equations and naturally not invariant under changes of frame which do not satisfy the same restrictions of linearization. Various invariant theories which are said to be linear have been proposed (see Chapter 15). In these theories, a functional is linearized for small values of the norm of some nonlinear measure of the history of the strain; for example, for motions whose Finger tensor is small in some norm, this linearization gives

$$\boldsymbol{\tau} = \int_{-\infty}^{t} \frac{\partial G(t-\tau)}{\partial \tau} [\mathbf{C}_t^{-1}(\tau) - \mathbf{1}]\, d\tau$$

$$= -\int_{-\infty}^{t} G(t-\tau) \frac{\partial \mathbf{C}_t^{-1}(\tau)}{\partial \tau} d\tau \,. \tag{7}$$

This is a linear expression, linear in this nonlinear tensor. None of these so-called linear equations are completely linearized. When they are completely linearized, they reduce universally to (1) and (2). These two equations are model independent. They apply to all viscoelastic fluids in motions which perturb rest. This shows that the Newtonian viscosity μ and the relaxation function G(s) are genuine material parameters which are also model independent.

It will not be possible to understand wave propagation or the elasticity of viscoelastic fluids without first understanding well the linearized response to small pulses governed by (1).

18.1.1 Stress relaxation and fading memory

The first question we need to consider is whether μ is zero or not zero, if it is zero, in some liquids and not in others. This question would be meaningless if we allowed G(s) to have a delta function contribution. We shall disallow this by requiring from outset that

$$G(0) \text{ is finite.} \tag{8}$$

After applying a step in strain at $\tau=t-s=0$ in (1), which is equivalent to

$$\mathbf{D}[\mathbf{u}(\mathbf{x},t-s)] = \mathbf{E}(\mathbf{x})\delta(t-s) = \mathbf{E}(\mathbf{x})\delta(\tau)$$

we find that when $t>\tau>0$, $\mathbf{D}[\mathbf{u}(\mathbf{x},t)] = 0$ and (1) reduces to

$$\boldsymbol{\tau}(\mathbf{x},t) = 2\mathbf{E}(\mathbf{x}) \int_{-\infty}^{t} G(t-\tau)\delta(\tau)\, d\tau = 2G(t)\, \mathbf{E}(\mathbf{x}) \,. \tag{9}$$

The stress should decay continuously after

$$G'(t) < 0 \tag{10}$$

moreover, since we are restricting our attention to fluids, we must require that the stress vanishes when the fluid is at rest

$$G(t) \to 0 \text{ as } t \to \infty . \qquad (11)$$

We may also, following Boltzmann, look at fading memory in terms of the strain, rather than strain rate. The linearization of (7) is

$$\boldsymbol{\tau} = -\int_{-\infty}^{t} \frac{\partial G(t-\tau)}{\partial \tau} [\mathbf{E}(\mathbf{x},t) - \mathbf{E}(\mathbf{x},\tau)]\, d\tau . \qquad (12)$$

Hence, the effect of the strain in the distant past is smaller than in the near past if

$$\frac{\partial^2 G(t-\tau)}{\partial \tau^2} = G''(s) > 0 . \qquad (13)$$

The conditions $G(t)>0$, $G'(t)<0$, and $G''(t)>0$ are sufficient conditions for stability of the rest state and the first two conditions are necessary.

Summarizing, unless otherwise stated, we shall assume that

$$\begin{gathered} G(t) > 0, \\ G'(t) < 0, \\ G''(t) > 0, \\ G(t) \to 0 \text{ as } t \to \infty, \\ G(0) \text{ is finite.} \end{gathered} \qquad (14)$$

18.1.2 Static viscosity, Newtonian viscosity, elastic viscosity

It is desirable to characterize the static or zero shear rate viscosity for (1) in the following way. Suppose that we are in the case of steady shearing with one component of velocity $u(x)$ depending on one variable x. The shear stress $\tau(\kappa)=\tau_{12}$ of $\boldsymbol{\tau}$ depends then on the rate of shear $\kappa(x)=D_{12}$ of $\mathbf{D}$ and (1) reduces to $\tau = (\mu+\eta)\kappa$, where

$$\tilde{\mu} = \mu + \eta \qquad (15)$$

is the zero shear viscosity and

$$\eta = \int_0^\infty G(s)\ ds \tag{16}$$

is the elastic viscosity. In the next section, we shall review important historical ideas about the origin of viscosity in elasticity. For each steady flow

$$\boldsymbol{\tau} = 2\tilde{\mu}\mathbf{D}[\mathbf{u}(\mathbf{x})]$$

as for a Newtonian fluid. In general, Newtonian fluids are those for which $\tilde{\mu}=\mu$, $\eta=0$. Elastic fluids have $\mu=0$, $\tilde{\mu}=\eta$. Clearly,

$$\tilde{\mu} \geq \eta \tag{17}$$

with equality for elastic fluids. It is easy to measure the zero shear viscosity $\tilde{\mu}$, but the measurement leaves μ and η undetermined. If it were possible to measure the relaxation function G(t) accurately, we could find

$$\mu = \tilde{\mu} - \eta \tag{18}$$

from measurements. In Figures 18.1-4, we have presented the graph of relaxation functions taken directly off a commercial rheometer by a microprocessor. The zero shear viscosity $\tilde{\mu}$ is represented as the area of a square with side length $\sqrt{\tilde{\mu}}$.

The comparison in Figures 18.1-4 cannot be used to determine μ in practice. The problem is that G(t) is not measured accurately by the rheometers. Simplified theories governing the use of the rheometer for step in strain experiments are given to standard books (for example, Bird, Armstrong and Hassager, [1977a]). The working formula is

$$M = \frac{2}{3}\pi R^3 \gamma_0 G(t) \tag{19}$$

where M is the torque on the plate of a cone and plate rheometer, R is the radius of the sample, γ_0 is the shear strain and G(t) the relaxation function. Inertia is neglected in the simplified theory but it is not negligible near t=0. A theory that accounts for inertia is given in the paper by Narain and Joseph [1983a, b]. The neglect of inertia in the simplified theory may be justified by the fact that these rheometers do not work for $t<0.02$ seconds where inertia is most important. Moreover G(t) is not measured for large times when the stress level drops below a critical value for the transducer. So we can't get μ this way.

To learn about elasticity and viscosity we could consider ever more dilute solutions of polymer chains of large molecules in solvents which are thought to be Newtonian. What happens when we reduce the amount of polymer? There are two good ideas about this which are in collision. The first (Jeffreys') idea says that there is always a viscosity and some elasticity with an ever greater viscous contribution as the amount of polymer is reduced. The popular molecular models of polymer solution, those of Rouse, Bueche, and Zimm (see, for example, Bird, *et al.* [1977b] or Larson, [1988a]) are biased on this idea. They think of polymers in a Newtonian solvent of viscosity η_S. When the polymers are taken away we are left with a Newtonian fluid of viscosity η_S. In fact, these molecular theories all give rise to a relaxation time which does not go to zero with concentration. At zero concentration the relaxation time is the time of relaxation of a single polymer in a sea of solvent. We already remarked in §7.9 that this time of relaxation is not relevant to the dynamics of polymeric liquids under extreme dilution. There is an effective viscosity in these molecular models, $\mu=\eta_S$ but η_S could not be the effective viscosity in oils or amorphous polymers in which there is no solvent and I do not think that $\mu=\eta_S$ even in solutions.

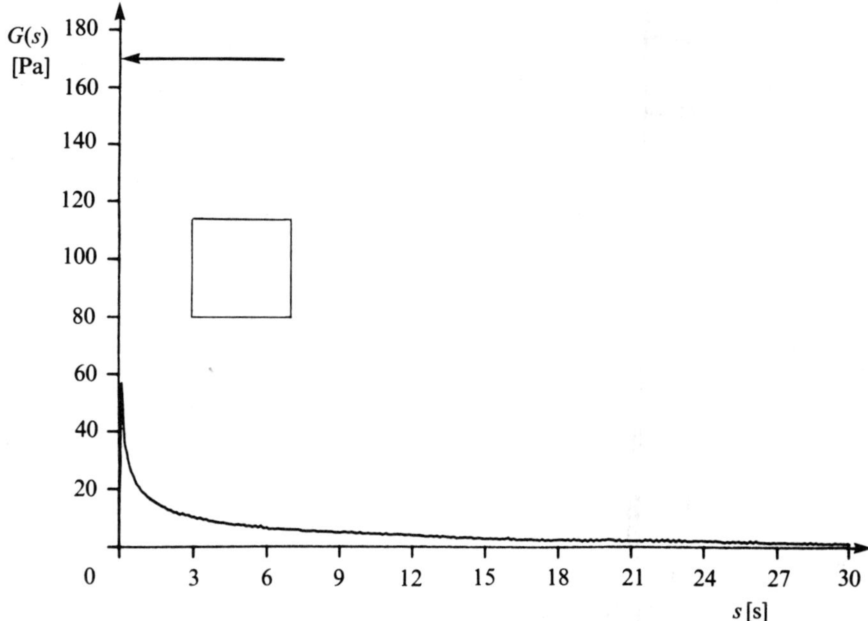

Figure 18.1. Shear modulus G_C=172 Pa and relaxation G(s) for an observed wave speed c=38.4 cm/s in a 1.5% Poly(acrylamide) Separan AP30 in 50% glycerin and 48.5% water. The zero shear viscosity, $\tilde{\mu}$=160 Pa-s is the area of the square. Bird, *et al.* [1977a] exhibit frequency data (ω<1000 rad/s) of Huppler, *et al.* [1967] for this solution. They say that the storage modulus $G'(\omega)$ is nearly at its limiting value with $G'(\omega)$=140 Pa at frequencies of ω=100 Hz.

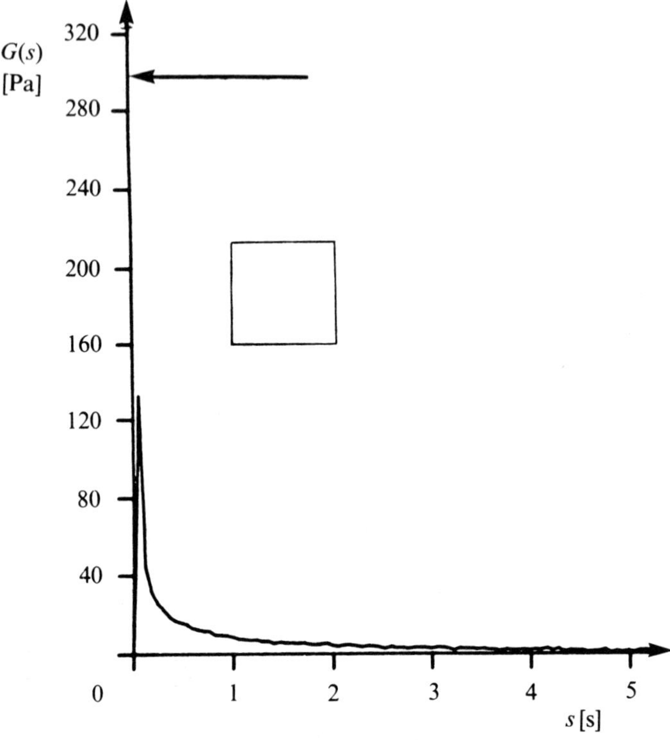

Figure 18.2. Shear modulus G_C=304 Pa and relaxation function G(s) for an observed wave speed c=50.6 cm/s in a 1% carboxymethyl cellulose (CMC) solution in 49% water and 50% glycerin.

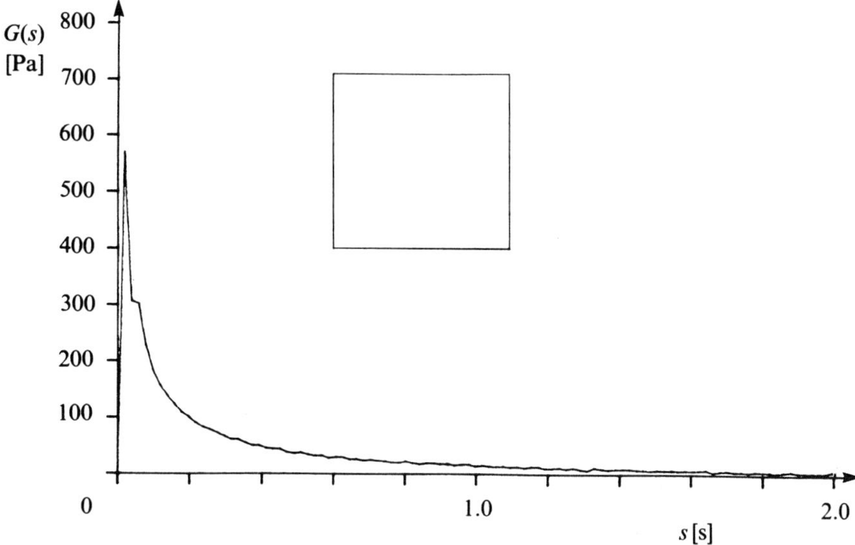

Figure 18.3. Shear modulus G_C=2210 Pa and relaxation function G(s) for 9.5% of poly(isobutylene)(Vistanex L-100) in decalin, MN=1,000,000, (T=25°C, a=1). The zero-shear viscosity $\tilde{\mu}$=139 Pa s (T=25°C) is the area of the square.

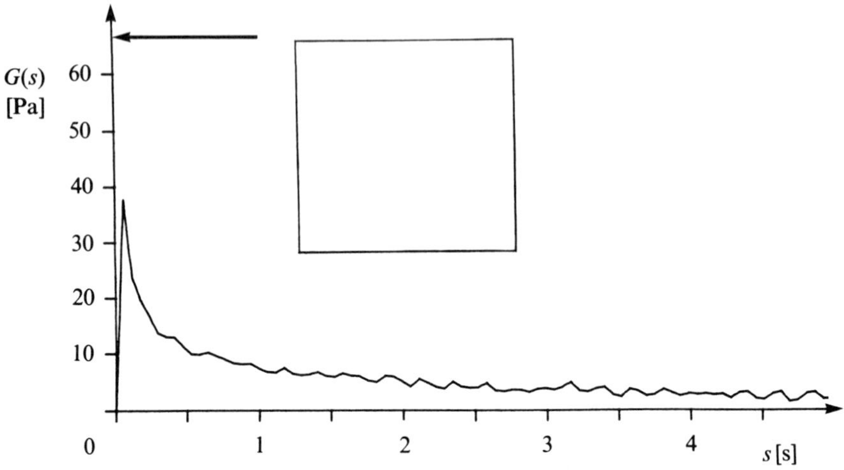

Figure 18.4. Shear modulus G_C=68.0 Pa and relaxation function G(s) for 1.7% polyox coagulant, MN=4,000,000, (T=28°C, g=1). The zero-shear viscosity $\tilde{\mu}$=58.4 Pa s (T=27°C) is the area of the square.

18.1.3 Effective relaxation function, rigidity and viscosity

On the other hand we may suppose that the liquid is elastic so that $\mu=0$ and the static viscosity $\tilde{\mu}=\eta$ is the area under the graph of the relaxation function. Since η is finite in all liquids, we have always a mean time of relaxation

$$\bar{\lambda} = \eta/G(0) \ . \tag{20}$$

This mean time is the base of the rectangle of height G(0) whose area is η and it is not Maxwell's relaxation time which is the one and only one relaxation time of the Maxwell fluid. If we accept Maxwell's idea, then as $\lambda\to 0$, $G(0)\to\infty$ in such a way that

$$\eta_s = \lambda G(0) \ .$$

The more dilute solutions must have every higher rigidities as the amount of polymer is reduced. There is no distinction, no decomposition into a Newtonian plus elastic part in Maxwell's theory.

Maxwell's and Jeffreys' ideas can be reconciled by replacing the notion of one relaxation time with a distribution of relaxation times. This notion is consistent with structural theories of liquids in which different times of relaxation correspond to different modes of molecular relaxation. It is convenient again to think of polymers in a solvent, but now we can imagine that the solvent is elastic, but with an enormously high rigidity. In fact many of the so called Newtonian solvents have a rigidity of the order 10^9 Pascals, which is characteristic of glass, independent of variations of the chemical characteristics among the different liquids (for example, see Harrison, [1976]). To find this glassy modulus it is necessary to use high frequency devices operating in the range of 10^9 Hertz. These devices were first introduced by Mason [1947] and Mason, Baker, McSkimin and Hess [1949] at Bell labs. To measure the glassy modulus in low molecular weight liquids it is also necessary to supercool them to temperatures near the glassy state. In these circumstances the liquid acts like a glassy solid, the molecular configurations cannot follow the rapid oscillations of stress, the liquid cannot flow. For slower processes it is possible for the liquid to flow and if the relaxation is sufficiently fast the liquid will appear to be Newtonian in more normal flows. For practical purposes there is no difference between Newtonian liquids and liquids with rigidities of order 10^9 and mean relaxation times of 10^{-10} seconds or so. In fact it is convenient to regard such liquids as Newtonian, even though $\mu=0$ and $\tilde{\mu}=\eta$.

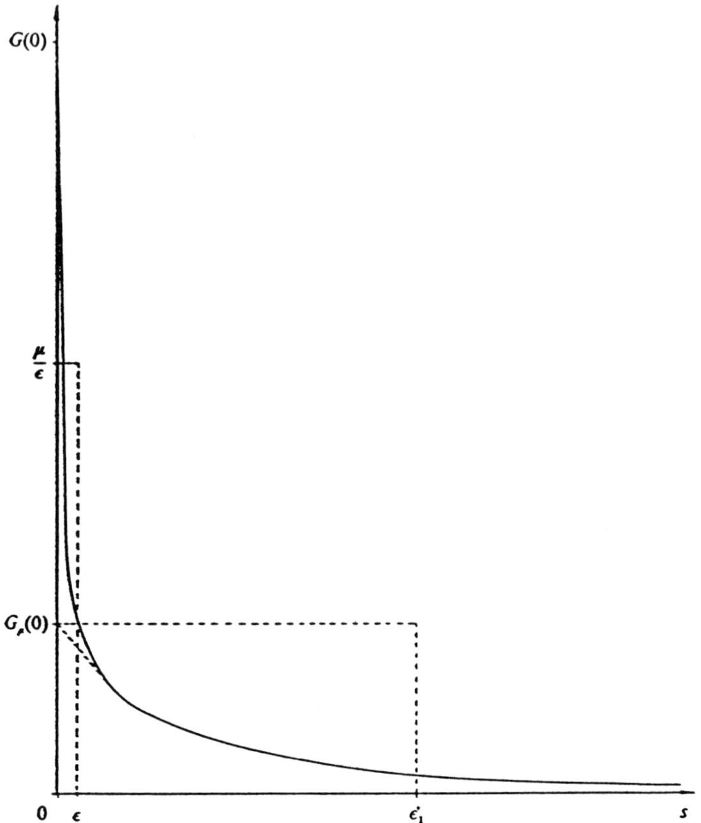

Figure 18.5. Typical relaxation function for a polymer solution. At very short times the molecular order is frozen and the liquid does not flow. The wave instantaneously sees a material that looks like an organic glass. Then the stress associated with the molecular forces between small molecules relaxes. The relaxation of large structures, like those associated with macromolecules, is much slower with an effective shear modulus more typical of soft rubbers. Eventually all the elastic response decays in a diffusive limit with a viscosity given by the area under the graph of G(s). We define an effective relaxation function $G_\mu(s)$ with the tangent construction. First put $G_\mu(0) = G_c$; then drop the tangent to G(s). This relaxation function may be modelled by a double step. The decay of the glassy modes under the first step gives rise to a "Newtonian" viscosity μ. The viscosity associated with the second step is $G_\mu(\varepsilon_1-\varepsilon)$, much larger than μ. The total viscosity is $\tilde{\mu}=\mu+G_\mu(\varepsilon_1-\varepsilon)$. The usual type of estimate for the time of relaxation $\bar{\lambda}G(0)=\tilde{\mu}$ would lead one to a grossly misleading value of $\bar{\lambda}$ of the true time of relaxation.

The presence of polymers would not allow the liquid to enter the region of viscous relaxation at such early times. Instead much slower relaxation processes associated with the interaction of polymers and solvent would be induced (*cf.* Mason, *et al., op. cit.*). The second slower regime of relaxation starts in a neighborhood of small times which terminate fast relaxation from the glassy state. A value $G_\mu(0)$ representative of G(t) at these early times can be said to loosely define a second smaller modulus, an effective modulus (see Figure 18.5) It can be said that *the effective modulus is associated with times so long that fast modes have relaxed and so short that slower modes have not.*

It should be clear from Figure 18.5 that the effective viscosity μ and effective rigidity $G_\mu(0)$ are not uniquely defined by our construction. It is apparent from the picture that the construction would depend sensitively on the choice of the small time ε. After this time the constitutive equation would behave as if there were an effective viscosity, even if $\mu=0$.

To develop the thought just introduced, we define an approximate kernel

$$G_M(s) = \begin{cases} G'(\varepsilon) + G_M(0)\,, & s\leq\varepsilon \\ G(s)\,. & s\geq\varepsilon \end{cases} \tag{21}$$

This approximate kernel G_M (s) is a little different than $G_\mu(s)$ in Figure 18.5. In $G_\mu(s)$ we actually choose $G_\mu(0)$ and then find ε by projecting to G(s) at $s=\varepsilon$ in a thought experiment. Actually we get $G_\mu(0)$ from real experiments by measuring the wave speed $c = \sqrt{G_\mu(0)/\rho}$ with the wave speed meter (Appendix F). In the construction (21) we actually choose ε and then we must compare G(0) and $G_M(0)$. We may write

$$\int_0^\infty G(s)\ \mathbf{D}[\mathbf{u}(\mathbf{x},t-s)]ds$$

$$= \int_0^\infty G_M(s)\ \mathbf{D}[\mathbf{u}(\mathbf{x},t-s)]ds \tag{22}$$

$$+ \int_0^\varepsilon [G(s)-G_M(s)]\ \mathbf{D}[\mathbf{u}(\mathbf{x},t-s)]\ ds.$$

Using the mean value theorem, the last integral may be written

$$\varepsilon[G(\bar{s}) - G_M(\bar{s})]\ \mathbf{D}[\mathbf{u}(\mathbf{x},t-\bar{s})]\ ,\quad 0 < \bar{s} < \varepsilon\ .$$

Then with $\varepsilon \to 0$ we get $\bar{s}(\varepsilon) \to 0$ and approximate the perturbing term with

$$\varepsilon[G(0)-G_M(0)]\quad \mathbf{D}[\mathbf{u}(x,t)]\ .$$

The approximating problem is like one perturbed by a small viscosity

$$\mu = \varepsilon[G(0)-G_M(0)]\ . \tag{23}$$

The heuristic argument of Narain and Joseph [1983b] just given applies to any two kernels which coincide for $s>\varepsilon$. The implication is that an approximation corresponding to one kernel may be obtained by solving a problem with the other kernel, perturbed by a viscous term with a suitably selected viscosity coefficient.

The analytic structure of G(s) near $s=0$, $\tau=t$, is important for wave propagation. Renardy [1982] studied singular kernels

$$G(0)\ \text{finite}\ ,\quad G'(0) = -\infty\ ,\quad \mu = 0\ . \tag{24}$$

He showed that shock waves of velocity in Stokes' first problem propagate with a speed $c=\sqrt{G(0)/\rho}$ but that the amplitude of the wave vanishes. This means that the support of the shock wave propagates but there is no discontinuity of velocity. In fact, in one example, the velocity is C^∞ smooth at the support, but not analytic. This singular kernel is intermediate between a regular kernel, in which the discontinuous velocity propagates, and a regular kernel plus viscosity in which the shock front is smoothed in an analytic way; the Renardy smoothing is intermediate between viscous smoothing and shocks; it is well approximated by a regular kernel plus a viscosity. In either case, we get propagation of discontinuities with smoothing at the front.

There is a mystery about the nonuniqueness apparent in the construction given in Figure 18.5. We already mentioned that apparently repeatable values of $G_\mu(0)$ are obtained from an instrument which we use to measure the speed of shear waves (Appendix F). This suggests that whatever the smoothing of the wave front due to the action of an effective viscosity, the dynamics are governed by a wave with a more or less unique speed. We could get $G_\mu(0)$ such that $G_\mu(0)=G(\varepsilon)$ was insensitive to the effective time ε of initiation of the wave, as would be true for the kernel shown in Figure 18.6.

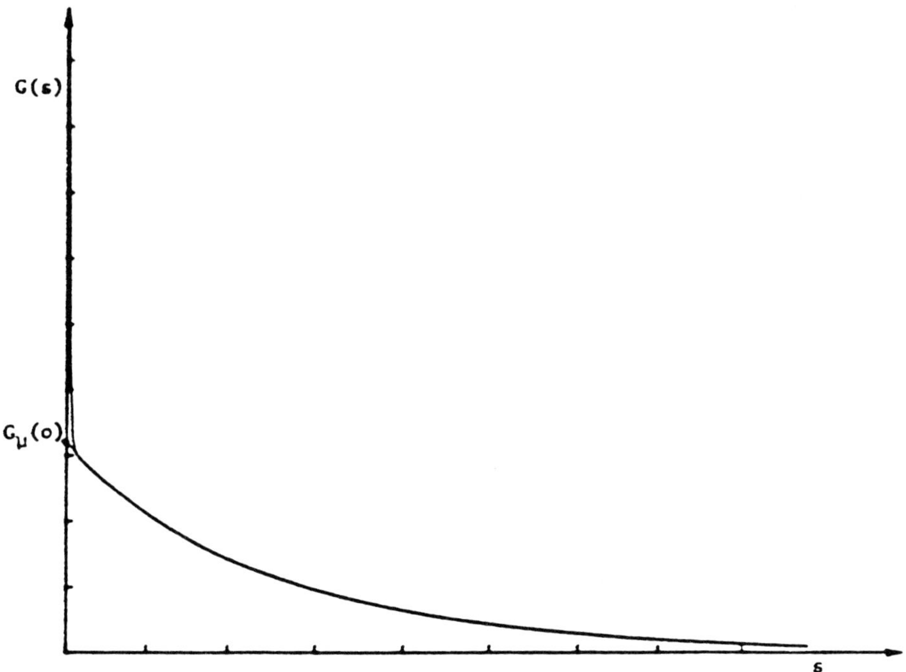

Figure 18.6. Typical relaxation function for a polymer solution on a longer time scale than shown in Figure 18.5. Glassy modes have decayed at times of observations and rubbery ones are decaying. The observer sees the effective rigidity as the more or less unique starting $G_\mu(0)$ which is actually G(s) in a time window large compared to the relaxation of glassy modes and small (say 10^{-6}–10^{-3} sec) on the observer's watch.

18.2 The origins of viscosity in elasticity

There are recurrent arguments starting with Poisson [1829], stimulated by Maxwell [1873], which suggests that every liquid should have some instantaneous elasticity. Indeed, important constitutive theories, like those of Maxwell [1873], or Boltzmann [1874] are founded on the idea that liquids will accommodate short-lived elastic responses. This idea appears to run counter to the notion that liquids like water and glycerin have a purely viscous response and that other liquids, like extremely dilute solutions, have some elasticity and some viscosity. The differences are more apparent than real. It is all a question of time scales with purely elastic responses to impulsive forces of short duration or high frequency. As in the Maxwell model, the viscosity η arises after all the elasticity has decayed. One might say that Newtonian liquids have a bad rather than fading memory; they remember their viscosity forever, as long as they flow, and not much else.

The idea that diffusion arises in the rapid decay of elastic modes is of general application. In the case of heat transport, it may be argued that an effective thermal conductivity arises out of fast decay of electronic modes at time scales at which phonons have just begun to relax (see Joseph and Preziosi, [1989], [1990]).

18.2.1 Maxwell

In the mathematical description of Newtonian fluids important differences between liquids and gases do not arise provided that the Mach number is small. Maxwell's constitutive theory, derived in his 1867 treatise on gases, implies that formula relating the relaxation function to the elastic viscosity with $G(s) = \frac{\eta}{\lambda} e^{-s/\lambda}$. Despite the fact that Maxwell thought his equation should apply to gases and liquids because "the phenomenon of viscosity is independent of hypothesis," gases should not be used as a model for the elastic response of liquids. It is simply wrong to think of a molecule of liquid with a mean free path, bumping into other molecules. You don't move a liquid molecule without moving the neighbors by short range forces; the motions are undoubtedly highly cooperative. In the words of J.M. Burgers [1939]:

> The formation of an adequate picture of the internal structure of a liquid is a problem of great difficulty. It has been pointed out by many physicists that the structure must have much more in common with that of solid matter, than with that of a gas. The following exposition of this view may be taken over from Poynting (37): "A liquid may be imagined to be, in the main, a solid structure in as much as the molecules cohere and resist strain of any kind. But the molecules have so much energy, potential or kinetic or both, that they are not very far from instability. In a mass of connected molecules irregularly vibrating, concentrations of energy must occur, and at the points of concentration individual molecules may receive so much energy that they are able to do the work needed to free them from their immediate surroundings. Thus the solid structure is

continually breaking down and renewing itself." Poynting then continues with an explanation of the phenomenon of viscosity on the general lines already exposed before: "If we impose a shear strain on the structure, the strain will of course disappear with the structure in which it is produced. But the breaking down will always lag slightly behind the imposition of the shear, and the still surviving shear strain will be accompanied by a resistance the same in kind as the resistance to shear in a solid, though in a liquid it is only recognized as viscosity. This is the view first set forth by Poisson and developed by Maxwell." By introducing a factor expressing the fraction of the molecules which are set free (or, "mobilized") per second, the coefficient of viscosity is then obtained as the quotient of the modulus of rigidity and this factor.

Maxwell's ideas about the elasticity were strongly motivated by the earlier ideas of Poisson. In his [1873] paper, "On double refraction in a viscous fluid in motion," he says,

According to Poisson's theory of the internal friction of fluids, as viscous fluid behaves as an elastic solid would do if it were periodically liquified for an instant and solidified again, so that each fresh start it becomes for the moment like an elastic solid free from strain.

... I have enumerated these instances of the application of polarized light to the study of the structure of solid bodies as suggestions with respect to the application of the same method to liquids so as to determine whether a given liquid differs from a solid in having a very small "Rigidity", but a sensible "Time of relaxation", or in both ways. Those which, like Canada balsam, act strongly on polarized light, have probably a small "Rigidity," but a sensible "Time of relaxation." Those which do not show this action are probably much more "rigid," and owe their fluidity to the smallness of their "time of relaxation."

A definite distinction between solids, with permanent elasticity, and fluids with relaxing elasticity is explicit in Maxwell's equations and in his thinking

If, therefore, we define a fluid as a substance which cannot remain in permanent equilibrium under a stress not equal in all directions, we must call these substances (cold pitch and asphalt) fluids, though they are so viscous that we can walk on them without leaving any footprints. What is required to alter the form of a soft solid is a sufficient force and this, when applied, produces its effect at once. In the case of a viscous fluid it is "time" which is required, and if enough time is given, the very smallest force will produce a sensible effect.

Maxwell's idea that fluids may be described as relaxing elastic solids is in conflict with the instantaneous changes of configuration required by the Newtonian fluids embraced by Maxwell's friend, Stokes. Maxwell evidently felt a need to show how his theory could be reconciled with Stokes. He writes (Larmor, [1907])

GLENLAIR, DALBEATTIE

25 Sept. 1875

MY DEAR STOKES,

I quite concur with you that Mr. Gore's paper should be printed *in extenso* in the proceedings.

Mrs. Maxwell has been keeping pretty well and has been able to get out on her pony.

Would you agree with the following statements about elasticity and viscosity, as related to a molecular theory?

When after being strained the groups of molecules in a body tend to return to the same stable configurations as when unstrained, the body is elastic. If at corresponding stages of the straining and restitution the stresses are the same, the body is perfectly elastic. If the stress during restitution is less than that during deformation the elasticity is imperfect.

If, when the strain exceeds a certain value, complete restitution does not occur, this value is called the limiting strain and the stress the limiting stress of elasticity.

If, when the stress is removed, the body does not completely return to its original form the body is said to be plastic or viscous. A viscous body, if kept strained long enough, loses all tendency to change its form. A plastic body does not permanently change its form unless the stress exceeds a certain value, and if kept strained it never loses all stress.

THEORY

In an elastic solid the thermal agitation of the molecules does not carry them beyond the limits of oscillating about stable configurations.

But as the thermal agitation increases so many molecules per second are thrown out and oscillate about a new configuration, the nature of which is determined by the present form of the body and not by its unstrained form. The greater the strain it is probable that more molecules will be so thrown out; but the number is not proportional to the strain, but varies very little for small strains, and suddenly rises enormously for breaking strains.

If $1/\ell$ of the molecules are thrown out in unit of time from a state of strain represented by *e,f,g,a,b,c,* see Thomson and Tait, and if these molecules enter into the state θ where

$$\theta = e+f+g$$

then*

$$\frac{de}{dt} = \frac{1}{\ell}\left[e - \frac{1}{3}\theta\right] + \frac{du}{dx},$$

$$\frac{da}{dt} = -\frac{1}{\ell}a + \frac{dv}{dz} + \frac{dw}{dy}$$

and if *k* and *n* are the coefficients of elasticity so that

$$P = \left(k - \frac{2}{3}n\right)\theta + 2ne \text{ and } S = na\,,$$

the equations of motion will be of the form

$$\frac{1}{\frac{1}{\ell} + \frac{d}{dt}}\left[\left[\frac{k}{\ell} + \left[k + \frac{n}{3}\right]\frac{d}{dt}\right]\frac{d\theta}{dx} - n\nabla^2 u\right] + X - \rho\frac{Du}{Dt} = 0\,,$$

with 2 others and

$$\frac{d\theta}{dt} = \frac{du}{dx} + \frac{dv}{dy} + \frac{dw}{dz},$$

If ℓ is small we may put

$$P = C - k\theta \text{ and } n\ell = \mu$$

and the equation becomes

$$-\frac{dp}{dx} - \mu\nabla^2 u + X - \rho\frac{Du}{Dt} = 0$$

---

* Here (u,v,w) is velocity. The signs of the terms involving $\nabla^2 u$ below should be changed, unless $\nabla^2 u$ has itself the quaternionic sign.

Yours very truly,

J. CLARK MAXWELL

An extended discussion of Maxwell's ideas was given by his student Butcher [1876]. Readers will find the tensorial interpretation of Maxwell's derivation which was prepared by Prof. M. Braun (1987, personal communication) interesting.

Maxwell, in his 1875 letter to Stokes, suggests a constitutive law which related stress rate to deformation velocity.

The quantities e,f,g,a,b,c in Maxwell's notation are generalized strains $\varepsilon_{11}, \varepsilon_{22}, \varepsilon_{33}, 2\varepsilon_{23}, 2\varepsilon_{13}, 2\varepsilon_{12}$. The relaxation equations, as they stand, do not constitute a tensorial equation. If, however, a sign is reversed in the first of these equations, such that it reads

$$\frac{de}{dt} = -\frac{1}{\ell}\left(e - \frac{1}{3}\theta\right) + \frac{du}{dx},$$

then the relaxation equations can be combined to one tensorial equation

$$\dot{\varepsilon}_{ik} + \frac{1}{\ell}\left(\varepsilon_{ik} - \frac{1}{3}\varepsilon_{jj}\delta_{ik}\right) = V_{(i,k)}. \tag{1}$$

where V_i denotes the velocity components and $V_{(i,k)}$ is the symmetric part of $V_{i,k}$.

In Maxwell's notation P and S denote the normal stress σ_{11} and the shear stress σ_{23}, respectively. They are related to the generalized strains by Hooke's law

$$\sigma_{ik} = \left(k - \frac{2n}{3}\right)\varepsilon_{jj}\delta_{ik} + 2n\varepsilon_{ik} \tag{2}$$

with k and n denoting bulk modulus and shear modulus, respectively. The constitutive law of the material is provided by the combination of equations (1) and (2) with k,n, and ℓ as material parameters. The meaning of the constitutive equations becomes clearer, after the auxiliary variables ε_{ik} are eliminated to this end each of the tensorial quantities is decomposed in its isotropic and deviatoric parts, e.g.

$$\sigma_{ik} = \frac{1}{3}\sigma_{jj}\delta_{ik} + \sigma_{i,k}^{D}. \tag{3}$$

Thus equations (1) and (2) are decomposed in

$$\dot{\varepsilon}_{jj} = V_{j,j} \tag{4}$$

$$\dot{\varepsilon}^D_{ik} + \frac{1}{\ell}\varepsilon^D_{ik} = V^D_{(i,k)} \tag{5}$$

and

$$\dot{\sigma}_{jj} = 3k\varepsilon_{jj}\ , \tag{6}$$

$$\dot{\sigma}^D_{ik} = 2n\varepsilon^D_{i,k}\ , \tag{7}$$

respectively. Elimination of the generalized strains from the parts of equations (4), (6), and (5), (7) yields the equations

$$\dot{\sigma}_{jj} = 3kV_{j,j} \tag{8}$$

$$\dot{\sigma}^D_{ik} + \frac{1}{\ell}\sigma^D_{ik} = 2nV^D_{(i,k)} \tag{9}$$

relating the isotropic and deviatoric parts of stress and deformation velocity tensors known, respectively. For these equations to be admissible constitutive equations the time derivative has to be chosen as Oldroyd or Jaumann or any other objective time derivative.

If the proper interpretation of the time derivatives is ignored, equations (8) and (9) can be integrated, resulting in

$$\sigma_{jj}(t) = 3k \int_{\tau=0}^{\infty} V_{j,j}(t-\tau)\, d\tau\ , \tag{10}$$

$$\sigma^D_{ik}(t) = 2n \int_0^{\infty} \exp\left(-\frac{\tau}{\ell}\right) V^D_{(i,k)}(t-\tau)\, d\tau\ . \tag{11}$$

These equations can be recombined to a single integral representation of the stress tensor

$$\sigma_{ik}(t) = \int_0^\infty \left[\left(k - \frac{2n}{3} V_{j,j}(t-\tau)\delta_{ik} + 2ne^{-\tau/\ell} V_{(i,k)}(t-\tau) \right) \right] d\tau . \quad (12)$$

But these formulae are valid only if the time derivatives in (8), (9) are interpreted as partial derivatives, and this interpretation is not consistent with the principle of material frame indifference.

Maxwell himself gives only a formal resolution of the differential equation (9):

$$\sigma^D_{ik} = 2n \left(\frac{i}{\ell} + \frac{d}{dt} \right)^{-1} V^D_{(i,k)} . \quad (13)$$

Using the abbreviation

$$\theta = \varepsilon_{jj} = \frac{1}{3k} \sigma_{jj} \quad (14)$$

the stresses tensor can be written as

$$\sigma_{ik} = k\theta\delta_{ik} + \sigma^D_{ik} = \left(\frac{1}{\ell} + \frac{d}{dt} - 1 \right) \left\{ \frac{k}{\ell} + \left(k - \frac{2n}{3} \frac{d}{dt} \right) \theta\delta_{ik} + 2nV_{(i,k)} \right\} \text{q.}(15)$$

while θ satisfies the differential equation

$$\frac{d\theta}{dt} = V_{j,j} . \quad (16)$$

The divergence of the stress tensor (15) reads

$$\sigma_{ik,k} = \left(\frac{1}{\ell} + \frac{d}{dt} \right)^{-1} \left\{ \left[\frac{k}{\ell} + \left(k + \frac{n}{3} \right) \frac{d}{dt} \theta_i \right] + nV_{i,kk} \right\} . \quad (17)$$

Thus the equations of motion are obtained in the form

$$\rho \frac{Du_i}{Dt} = \left(\frac{1}{\ell} + \frac{d}{dt} \right)^{-1} \left\{ \left[\frac{k}{\ell} + \left(k + \frac{n}{3} \right) \frac{d}{dt} \right] \theta_i + nV_{i,kk} \right\} + f_i \quad (18)$$

where f_i are the volume forces. Equation (18) coincides with Maxwell's result, except for the sign of the Laplacian term $V_{i,kk}$.

Maxwell made an experiment to show that fluids, in spite of the great mobility of their particles, possess a characteristic property of solid bodies, namely rigidity (elasticity). Maxwell showed that just as glass under pressure acquires the polarizing property, so also a fluid layer of Canada balsam, by suitable deformation becomes optically active and doubly refractive. Maxwell's method was applied to other liquids by A. Kundt [1881] and G. de Metz [1888] (olive oil, mixtures of olive oil and petroleum or paraffin, collodium, castor oil, Canada balsam in bencol, glue in water, almond oil, whale liver oil). They tried and failed to verify the idea suggested by Maxwell's constitutive theory that the rigidity (elasticity) would be greater in more viscous liquids. Maxwell's theory gives the instantaneous elasticity of a liquid as η/λ. It appears that "relatively" mobile liquids like Canada balsam, gelatin solution, castor oil, and olive oil possess the polarizing power to a high degree but much more viscous liquids, glycerin, treacle, sugar solutions, and gum arabic, even under the greatest deformation, remain inactive. In fact, Maxwell himself seems to have been persuaded that the polarizing power is great in liquids with slow relaxation, rather than high viscosity.

18.2.2 Experiments following Maxwell

Schwedoff [1889], Colin [1893], Trouton and Andrews [1904], Tammann and Lauer [1908] and Tammann and Faust [1909] use direct methods involving forces and deformations.

Theodore Schwedoff was a professor of physics at the University of Odessa. He was inspired by Maxwell's ideas and sought a direct means to verify them. He says that

> On connait bien les méthodes expérimentales qui permettent de mesurer la cohésion des corps solides. Quant aux liquides, on n'en a pu mesurer que la viscosité, et, pour cette raison, leur rigidité est reputée nulle. Pourtant, la généralité des propriétés de la matière avait depuis longtemps inspiré aux physiciens l'idée que les fluides aussi, malgré l'extrême mobilité de leurs particules , doivent receler des vestiges de la rigidité. Maxwell conçut l'idée de relier la rigidité des liquides à leur viscosité par l'hypothèse suivante. La rigidité E d'un liquide serait liée au temps "t" écoulé depuis le moment de sa déformation, par la relation

$$E = E_0 e^{-(t/\tau)} \tag{19}$$

où E_0 est la rigidité du liquide au premier instant de déformation, e la base des logarithmes népériens, τ le module de relaxation, c'est-à-dire la période au bout de laquelle E tombe à $(1/e)^{\text{ième}}$ de sa valeur initiale, Cette hypothèse conduit à la loi

$$\eta = E_0\tau\ , \tag{20}$$

où η est le coefficient de frottement du liquide.

Experiments made by Schwedoff with a solution of 5 grams of gelatin in one liter of water were performed between cylinders. The outer cylinder was stationary. The inner cylinder was set into rotation by twisting a torsion wire fixed at the center of the inner cylinder. After a certain instant, a fixed angle of twist of the inner cylinder was maintained by continuously reducing the torque, detwisting the torsion wire. The function giving the detwist as a function of time, for constant angular displacement of the inner cylinder, is the relaxation function. Schwedoff found that the apparent rigidity followed a law

$$R = E\left[\frac{\lambda}{\varepsilon} + \left(1 - \frac{\lambda}{\varepsilon}\right)e^{-t/\tau}\right], \tag{25}$$

which resembles Maxwell's but has a residual rigidity as $t\to\infty$. Schwedoff concluded that

1. The modulus of stiffness (rigidity) is 0.535 dynes/cm^2, consequently the fluid is 1840×10^9 times less stiff than steel.

2. The elasticity of the fluid is perfect if the deformation does not exceed a certain amount of λ and only lasts for a moment.

3. If the deformation exceeds λ for any length of time, a deformation ρ remains over and the fluid no longer returns to its old position. Colin [1893] rediscovered the effects described by Schwedoff for a solution of gelatin of strength five grams per liter except that the rigidity was not found to vanish when the duration of the deformation was infinitely great. In Colin's experiments the angle of twist of the torsion wire is fixed, and the angle of twist of the inner cylinder is measured. Colin says that the gelatin is not a true liquid

> because its properties change with time. He rejects Schwedoff's law of rigidity but accepts Maxwell's, reporting some data for castor oil and claims for olive oil.

There are a number of papers on the measurement of rigidity of colloids, gels, and sols. Some references for these are given in §9.2 of the article by Bateman *et al* [1932] and in Chapter XI of the book by Barr [1931]. The physiologist W.R. Hess [1920] did free oscillation experiments in a Couette apparatus using gelatin solutions which did not solidify. His demonstration of elasticity in these liquids was attributed to cohesive forces. Michaud [1922] found that as the concentration of jelly in water is progressively decreased the rigidity becomes zero before the concentration. He concludes that water has no rigidity.

A different approach to the problem of rigidity of liquids was taken in England by Trouton and Andrews [1904] and in Germany by Tammann and Lauer [1908] and Tammann and Faust [1909]. H. Markovitz [1968] noted that Trouton and Andrews

> ... reported on their studies on hot glass, pitch and shoemaker's wax. In their apparatus they mounted a cylindrical sample of the material and performed torsional experiments. On applying a load they were surprised to note first that the coefficient of viscosity of bodies such as pitch is a function of the time and "second, that on removing the stress there is a flow back in the opposite direction". Thus, for the first time, we have a creep and recovery experiment on a fluid. They noted further that when they plotted the applied torque against the steady-state angular velocity they did not obtain a straight line through the origin; this was the first evidence of a shear-dependent viscosity in a fluid. Then they proceeded to perform a stress-relaxation experiment by removing weights to keep the angular deformation constant. Even now few rheological investigations are that thorough.

The German experiments of Tammann, Lauer and Faust were equally ingenious. Their apparatus is composed of concentric glass spheres, with a (large) gap of 0.65 cm. The inner cylinder moves impulsively by a magnetic device and its motions and recoil are measured. In the 1908 paper they find that glycerin and a fluid called Betol have rigidity (elasticity) when the temperature is below a certain critical value. When the temperature is above the critical one (the viscosity is below a critical viscosity) they are unable to detect the signal which indicates the passage of the wave. They found no rigidity for water. In the 1909 paper they found that twelve out of twenty-five liquids studied by them displayed rigidity when the viscosity was larger than a critical value.

The old literature on the elasticity of liquids leads to the literature of rheology. In the words of Marcus Reiner [1949]

> Rheology came into being as a separate branch of science as a result of the following development: Until quite recently, the technologist took his materials from nature, either as they were found, e.g. timber and stone, or refined according to methods gradually developed through thousands of years, e.g. iron and other metals. Only in a very few cases were the materials artificial products and then the method of their production had been invented ages ago and probably by chance as was the case with glass. The mechanical properties of these materials were taken as nature or chance had turned them out. Then the modern development of such industries as deal with paints, plastics, ceramic, lubrications, cellulose and rubber products, etc. necessitated the large scale production of artificial materials "which were to possess definite mechanical properties". The processes by which these materials were produced were developed by chemists. When these chemists had to investigate the mechanical properties of their materials, it became apparent that the scientific terminology which mechanics had to offer them in its theories of elasticity and viscosity was much too primitive.

18.2.3 Boltzmann

Ludwig Boltzmann [1874] introduced a linear integral equation for the stress in viscoelastic solids

$$\tau(x,t) = \beta\varepsilon(x,t) + \int_{-\infty}^{t} m(t-\tau)\,[\varepsilon(x,t) - \varepsilon(x,\tau)]\,d\tau \,. \qquad (26)$$

This is a one-dimensional expression of (12) in which the permanent elastic response proportional to β is not suppressed. He was motivated by experiments on creep of silk fiber by Weber [1835] and on glass fibers by F. Kohlrausch [1863,1866]. Boltzmann did not apply his equations to liquids but he thought his equation might apply to soft or viscous materials. The idea of fading memory was expressed by Boltzmann as the requirement that m is a positive monotone-decreasing function. He says that

> ... however the deformation is of lesser influence the longer ago it occurred; that is the force which is necessary for a certain amount of

> deformation is smaller if a deformation in the same direction had occurred earlier.

Boltzmann remarks that a Newtonian viscosity, a term $\eta\varepsilon_t$, can be obtained as a limiting case of (19), by choosing kernels which are large near zero and small elsewhere, equivalent to $m=\eta\delta'$ where δ is a Dirac's delta function.

An excellent review of Boltzmann's work was been written by H. Markovitz [1977]:

> In Maxwell's 1878 article on the "Constitution of Bodies" he quoted Boltzmann's integral representation and ignored his own. Perhaps because the article was published in the famous 9th edition of the *Encyclopedia Britannica*, Maxwell's first-order differential equation continued to exert a much greater influence on the future development of rheology than did Boltzmann's more encompassing theory.

One of the reasons that Boltzmann's theory is better than Maxwell's is that Maxwell's model allows only a single time constant, the relaxation time in the exponential. It is impossible to fit relaxation data for stress relaxation to an exponential curve with a single time constant. The idea of a single time constant is also incompatible with structural theories which acknowledge that each material has a distribution of sizes, structures, components, etc., which is associated with a distribution of relaxation times. For example, we could say that in a polymer solution there are at least two times of relaxation, one for the polymer and one for the solvent. This type of thinking is greatly oversimplified because the polymer and solvent do not form separate entities but they interact in a highly cooperative way and even in extreme dilution, the solution is a material in its own right, forming a very unstable lattice in which polymers can interact with one another through the solvent.

The notion of a distribution of relaxation times is in no way tied to solutions. Even in pure amorphous polymer liquids there is a molecular weight distribution, some small molecules and some large ones, which give rise to a distribution of relaxation times. An effective rigidity might be expected from a spectrum centred around two very different relaxation times. Such effective moduli have already been identified in pioneering experiments using high frequency dynamic measurements on amorphous polymers. Mason, *et al.* [1949] note that

> Both the torsional crystal and high frequency shear wave technique applied to polyisobutylene and poly-α-methylstyrene liquids, show that there are two main relaxation frequencies in liquids. At frequencies under 100 kc [khz], the shear stiffness is in the order of 3×10^7 dynes/cm^2, while in the high megacycle range it has increased to 5×10^9 dynes/cm^2. The low shear elasticity appears to be associated with a composite motion of molecular rotation and translation that allows a configurational change to occur from the most probable chain shape. When the shear stress is removed, the molecule quickly returns to its most probable shape. This results in a low shear stiffness. At high frequencies this motion cannot take place, and the shear stiffness is determined by motions within single potential wells, and the value approaches that for a crystal.

Another good example of a viscoelastic liquid with a well-defined and robust wave speed is a soybean oil (Joseph, Riccius, and Arney, [1986]). Soybean oil is a chemical compound made of fatty acids and glycerin. The fatty acids are monomer chains of carbon with hydrogen and oxygen molecules. The monomer units of carbon are held together by mostly single and some double bonds in chains of eighteen units, with a total molecular weight of approximately 850 g/mol. Actually two such chains can be hooked together on a glycerin molecule forming a chain of 36 units of modest molecular weight relative to polymer solutions but large relative to organic liquids. There are many ways such a substance might vibrate. Molecular models using springs, beads, dumbbells, etc., would seem inappropriate for substances like soybean oil. It might be interesting to try for a molecular model based on the chemical formula for this substance.

The complicated structure and elastic response of organic liquids is well documented by G. Harrison [1976]. Even chemically pure substances, with a single molecule, like water, are believed to be microscopically heterogeneous with clusters and structures of different type carrying different relaxation processes and times.

18.3 Relaxation spectrum

The idea that different stressed structures in a fluid relax at different rates is embodied in the idea of relaxation spectrum. A relaxation spectrum with N relaxation times may be defined by

$$G(s) = \sum_{1}^{N} G_k(s) \ ,$$

$$G_k(s) = \frac{\eta_k}{\lambda_k} e^{-s/\lambda_k} \ , \quad \lambda_k > \lambda_{k-1} \ , \tag{27}$$

$$\eta = \sum_{1}^{N} \eta_k \ .$$

If we suppose that there are N elastic structures each of which may be modeled by a spring and dashpot in series, as in Figure 1.1, then the total stress in the linearized approximation is given by

$$\boldsymbol{\tau}(\mathbf{x},t) = \sum_{1}^{N} \frac{2\eta_k}{\lambda_k} \int_0^{\infty} e^{-s/\lambda_k} \, \mathbf{D}[\mathbf{u}(\mathbf{x},t-s)] \ ds \ . \tag{28}$$

This corresponds to a generalized Maxwell model, N Maxwell elements in parallel. The rate equations for the generalized Maxwell models are

$$\lambda_k \frac{\partial \boldsymbol{\tau}_k}{\partial t} + \boldsymbol{\tau}_k = 2\eta_k \mathbf{D}[\mathbf{u}] \ ,$$

$$\boldsymbol{\tau} = \sum_{1}^{N} \boldsymbol{\tau}_k \ . \tag{29}$$

A generalized Jeffreys' model could be formed by adding $\boldsymbol{\tau}_N = 2\mu\mathbf{D}[\mathbf{u}]$ to the sum of $\boldsymbol{\tau}_k$ as in Exercise 18.3. It is also possible to define generalized upper convected, lower convected and interpolated models as is done in Exercises 18.1 and 18.2.

Molecular models of polymers which generate stresses in terms of a sum over discrete model elements, like spring, beads, and dumbbells arrive at some special form of (27), for example, in a model by Doi-Edwards for polymer melts (see 53.8–1 in Bird, et al [1977a]).

$$G = \frac{96\eta}{\pi^2\lambda} \sum_{k=1,3,5,\ldots}^{\infty} e^{-\pi k^2 s/\lambda} . \tag{30}$$

In the actual practice the molecule models are not used to compute the η_k and λ_k. The method actually used by rheologists is to measure some functional of G(s), like the storage and loss modulus (see §19.1) or G(s) itself and then to fit the parameters η_k, λ_k to the measured data. This is because people don't actually trust the molecular models. It's like the better developed kinetic theory of gases which leads to various mathematical expressions for the material parameters of gases which are virtually ignored in favor of measured values in the construction of the handbook values of the viscosity and specific heat.

The discrete spectral representation of G(s) is associated with the problem of approximation of a given real time function over $(0,\infty)$ by a linear combinatinon of a given number n of exponentials. This problem seems to have been studied first by Prony [1795] and later by electrical engineers. For example, McDonough and Huggins [1968] show how to choose the 2N constants to minimize the integrated squared error. There is at present no discussion in the rheology of this approximation procedure in a mathematical setting. One would like to know something about convergence for $N\to\infty$ sensitivity to errors in choice η_k,λ_k, comparison of different methods, say collocation against least squares and other mathematical properties.

Some features of the physical behavior of liquids may be conveniently described by the limiting behavior of the slowest time λ_{max} of relaxation. Newtonian fluids emerge in the limit $\lambda_{max}\to 0$. The diverging relaxation time $\lambda_{max}\to\infty$ is found in crosslinking at the gel point, a point of phase transition. The gel problem has been extensively studied by H. Winter [1987]. The important parameter in the gel study is the extent of the crosslinking p(t), which changes with time. There the critical gel point, a critical p for network formation which can be predicted from dynamic measurements of $G'(\omega, p)$ and $G''(\omega, p)$. The most surprising discovery of these experiments was that critical gels exhibit stress relaxation in a power law; if the glassy response to high frequencies is suppressed then $G(t)=St^{-n}$ where n depends on network structure.

The cooperative nature of the elastic responses of the unstable and weakly stable lattices formed from short range forces in liquids are probably better approximated by a continuum of Maxwell models

$$G(t) = \int_0^\infty \frac{\eta(\lambda)}{\lambda} \exp\left(-\frac{t}{\lambda}\right) d\lambda . \tag{31}$$

The approximation problem here is to find $\eta(\lambda)$ for G(t). Putting $\lambda=1/p$ we find that

$$G(t) = -\int_0^\infty \eta\left(\frac{1}{p}\right) \exp(-pt) \frac{dp}{p} . \tag{32}$$

The integral is the Laplace transform of

$$\frac{\eta\left(\frac{1}{p}\right)}{p} = \frac{1}{2\pi i} \int_{\gamma-i\infty}^{\gamma+i\infty} G(t) e^{pt}\, dt . \tag{33}$$

Ferry [1980] gives a fairly extensive discussion of spectra, but the spectral representations have not been used extensively by rheologists. In general, for relaxation functions that admit a spectral representation

$$\tilde{G}'\left(\frac{1}{t}\right) \geq G(t) \tag{34}$$

with equality at t=0 and $t=\Lambda$ (Ferry, [1980], pages 41, 42, and 69).

The discussions of spectra are framed in terms of the dynamic and storage modulus, which are defined on plane harmonic waves. The dynamic and storage modulus can be inverted for G(t) in different ranges of frequency using integral transform and numerical methods suggested by Kamath and Mackley [1989].

Exercises

Exercise 18.1.
Consider the following generalization of the integral form of the upper convected Maxwell model.

$$\boldsymbol{\tau} = \Sigma \boldsymbol{\tau}_k \tag{35}$$

where

$$\boldsymbol{\tau}_k = \int_{-\infty}^{t} \frac{\eta_k}{\lambda_k^2} \exp\left[\frac{-(t-\tau)}{\lambda_k}\right] (\mathbf{C}_t^{-1}(\tau) - \mathbf{1})\, dt\,. \tag{36}$$

Show that

$$\lambda_k \overset{\Delta}{\boldsymbol{\tau}}_k + \boldsymbol{\tau}_k = 2\eta_k \mathbf{D}(\mathbf{u}) \tag{37}$$

and

$$\Sigma \lambda_k \overset{\Delta}{\boldsymbol{\tau}}_k + \boldsymbol{\tau} = 2\eta \mathbf{D}[\mathbf{u}] \tag{38}$$

or

$$\overset{\Delta}{\boldsymbol{\tau}} = 2G(0)\mathbf{D}[\mathbf{u}] - \Sigma \frac{\boldsymbol{\tau}_k}{\lambda_k} \tag{39}$$

where

$$G(0) = \Sigma \frac{\eta_k}{\lambda_k}, \quad \eta = \Sigma \eta_k\,. \tag{40}$$

Show that this model gives rise to vorticity of changing type.

Exercise 18.2.
Repeat exercise 18.1 for the case in which

$$\boldsymbol{\tau}_k = \int_{-\infty}^{t} \frac{\eta_k}{\lambda_k^2} \exp\left[\frac{-(t-\tau)}{\lambda_k}\right] (\mathbf{1} - \mathbf{C}_t(\tau))\, d\tau \tag{41}$$

Exercise 18.3.
Suppose that

$$\boldsymbol{\tau} = \sum \boldsymbol{\tau}_k + \boldsymbol{\tau}_N\,, \quad \boldsymbol{\tau}_N = 2\eta \mathbf{D}[\mathbf{u}] \tag{42}$$

and

$$\lambda_k \frac{D\boldsymbol{\tau}_k}{Dt} + \boldsymbol{\tau}_k = 2\eta_k \mathbf{D}[\mathbf{u}] \tag{43}$$

where D/Dt is any one of the interpolated Maxwell models.

Show that

$$\sum_{i=1}^{r} \left(1 + \lambda_i \frac{D}{Dt}\right)\boldsymbol{\tau} = 2\eta \sum_{i=0}^{s} \left(1 + \Lambda_i \frac{D}{Dt}\right)\mathbf{D}[\mathbf{u}] \tag{44}$$

where Λ_i is a retardation time. Show r=s+1 is a generalized Maxwell model and r=s is a generalized Jeffreys' model.

CHAPTER 19

Wave Propagation in Linear Viscoelastic Fluids

There are two canonical problems:

1. External excitation of plane harmonic waves, and

2. Impulsive excitation of shear waves.

19.1 Plane harmonic waves

Polymer chemists use devices which generate small amplitude harmonic waves to study the structure of liquids. The methods which are used are called the method of dynamic measurements. Inertia is not neglected in the theory. Since the governing equations are linear, it is possible to superpose plane waves

$$u(y,t) = Ue^{i(\omega t-\beta y)}\,, \tag{1}$$

where ω is the frequency of the superimposed sinusoidal oscillation and $\beta=\gamma+i\alpha$ is to be determined as a solvability condition for the reduced equations of motion governing the amplitude. (The condition is that $\beta^2/\omega^2=\rho/\tilde{G}^*(\omega)$.) The viscosity μ and relaxation function G(s) enter these equations in a unique composition called a complex viscosity

$$\eta^*(\omega) = \mu+\int_0^\infty G(s)e^{-i\omega s}\,ds \overset{\text{def}}{=} \eta'-i\eta''\,. \tag{2}$$

The equations are satisfied when the phase speed $V(\omega) = \gamma/\omega$ is given by $V(\omega)=\sqrt{2\omega q^2/\rho(\eta''+q)}$ and the attenuation by

$\alpha(\omega)=-[\omega\rho\eta'^2/2q^2(\eta''+q)]^{1/2}$, where $q^2=\eta'^2+\eta''^2$. It is also useful to define the complex rigidity $\tilde{G}^*(\omega) \overset{\text{def}}{=} i\omega\eta^*=\tilde{G}'+i\tilde{G}''$, where $\tilde{G}''=\omega\eta'(\omega)$ is the loss modulus, η' is the dynamic viscosity, and $\tilde{G}'=\omega\eta''(\omega)$ is the storage modulus. The storage modulus is an imperfect measure of elasticity and vanishes identically for Newtonian liquids. Asymptotic forms of $\eta^*(\omega)$ can be obtained by repeated integration by parts. Thus

$$\eta' = \mu - \frac{G'(0)}{\omega^2} + 0\left(\frac{1}{\omega^4}\right), \quad \omega\eta'' = G(0) + 0\left(\frac{1}{\omega^2}\right). \tag{3}$$

The limiting value of the storage modulus is the glassy modulus G(0). When $\mu\neq 0$, the limit $\omega\to\infty$ leads to unbounded values of the phase speed and attenuation (15). On the other hand, when $\mu=0$,

$$\lim_{\omega\to\infty} [V(\omega),\alpha(\omega)] = \left[c\,,\, \frac{G'(0)}{2cG(0)}\right], \tag{4}$$

where $c=(G(0)/\rho)^{1/2}$ is the speed and $G'(0)/2cG(0)$ is the attenuation for shock waves of vorticity (shear) into liquids at rest.

The relations between the storage modulus $\tilde{G}'(\omega)$ and the relaxation function G(t) are complicated. In principle one can get G(t) from $\tilde{G}'(\omega)$ by transform methods, using numerical inversion as in the recent work of Kamath and Mackley [1989]. The relations (14) suggests that these two functions are close for large ω or small $t=1/\omega$. However, the derivative of $\tilde{G}'(\omega)$ at $\omega=\infty$ is not in general equal to the derivative of G(t) at t=0. In fact, though they have the common value G(0) at t=0, the values separate at a "maximum" rate. To see this, we differentiate the relation $\tilde{G}'(1/t)=G(t)$ which presumably holds for small t. We find that

$$\frac{d}{dt}\,\tilde{G}'\left(\frac{1}{t}\right) = -2G''(0)\,t + O(t^3) \to 0 \tag{5}$$

is stationary, whereas

$$\left|\frac{dG(t)}{dt}\right|_{t=0} \tag{6}$$

is huge, because the mobile molecules associated with the glassy rigidity relax very rapidly. In organic liquids, it is believed that $G(0)=0(10^9 \text{ Pa})$ and $G'(0)=0(G(0)/\lambda)$ where λ is the time of relaxation of the glassy modes, say, $\lambda=0(10^{-9}\text{sec})$ (see Harrison [1976]), and $G'(0)=0(10^{18} \text{ Pa/sec})$. The relation

$$G(t) \sim \tilde{G}'(1/t)\,, \quad G(1/\omega) \sim \tilde{G}'(\omega)\,, \quad t = \frac{1}{\omega} \tag{7}$$

as $\omega \to \infty$, $t \to 0$ has to be massively wrong when the decay of G(t) to G(0) is not asymptotically proportional to t^2 as required by (3). For example, the ratio of the storage modulus to the relaxation function for a Maxwell model with a single relaxation time λ

$$\frac{\tilde{G}'(1/t)}{G(t)} = \frac{\lambda^2/t^2}{1+\lambda^2/t^2} \exp\left(\frac{t}{\lambda}\right) \tag{8}$$

can be arbitrarily large, even for small t, if the relaxation time λ is sufficiently small. The reason for this is that the relaxation function decays exponentially while the storage modulus decays algebraically.

In Figures 19.1 and 19.2, we have compared G(t) and $\tilde{G}'(1/t)$ for

$$G(t) = \sum_{n=0}^{25} G_n \exp\left(\frac{t}{-\lambda_n}\right), \quad \tilde{G}'(1/t) = \sum_{n=0}^{25} G_n \frac{\lambda_n^2/t^2}{1+\lambda_n^2/t^2}\,,$$

$$(G_0, G_1, G_n) = (10^9, 10^4, 10^4 / \sum_{n=1}^{25} n^{-3})\,, \tag{9}$$

$$(\lambda_0, \lambda_1, \lambda_n) = (10^{-8}, 3\times10^{-3}, 3\times10^{-3} / n^3)\,.$$

This relaxation function has one fast and 25 slow modes. It is representative for relaxation functions with widely separated spectra. The purpose of this comparison is to show that an effective modulus $G_\mu(0)$ could be masked in the storage modulus by the slow algebraic decay of the fast mode. The dramatic differences shown in Figure 19.1 are reduced in regions in which G(t) is slowly varying.

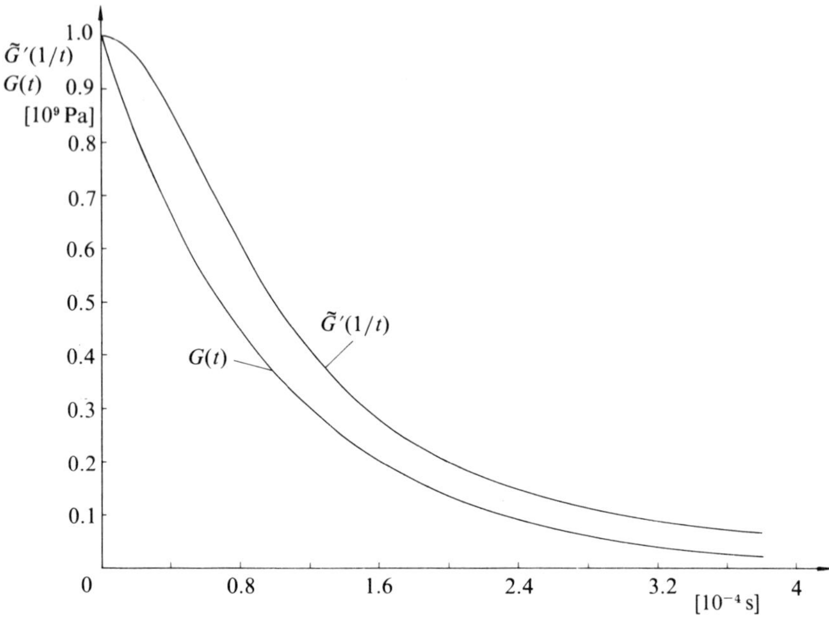

Figure 19.1. Comparison of relaxation function and storage modulus given by (9). This comparison is for short times on a linear scale. The comparison shows huge differences at early times.

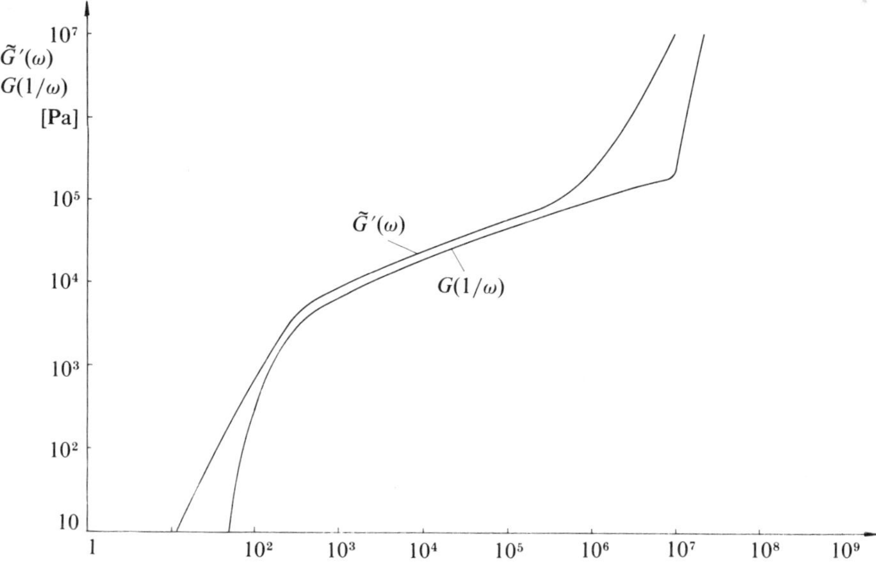

Figure 19.2. Comparison of relaxation function and storage modulus given by (9). This comparison is set in the logarithmic coordinates used by polymer chemists. The existence of an effective modulus $G_\mu(0) \approx 0(10^5)$ is not so clearly evident in the storage modulus.

The graph of $\tilde{G}'(\omega)$ does suggest an effective modulus, though it is masked by algebraic decay. The graph is like measured ones for amorphous polymers, where an effective modulus can be associated with entanglement couplings. We do not mean to identify effective moduli, which we have also associated even with not particularly dilute polymeric solutions, with plateau moduli. I do not find support for the idea of a robust effective modulus, a value $G_\mu = G(\varepsilon)$ (relative to ε) at small ε in the literature on dynamic measurements. So the study of effective moduli obliges us to look at dynamic measurements at high ω more carefully. The measured values of $\tilde{G}'(\omega)$ at large ω which appear in the literature are what we want to study. The data in this part of the graph correspond actually to smaller ω, but at lower temperatures, and are master curves constructed by the method of reduced variables (§19.2).

19.2 Reduced variables

In general, it is not possible to measure the storage modulus over a large range of frequencies using one instrument. Different instruments are required and each one has limitations of frequency and viscosity. The viscoelastic behavior of a liquid also depends on the temperature. The infinite frequency limit G(0) of the storage modulus is called the glassy modulus. Another way to reach the glassy state is to cool the material to near the glass transition temperature. In bulk polymers, there is a distinct temperature T_G at which the liquid solidifies; above this temperature, the molecules are flexible chains; below it, they are rigid. The melting temperature T_M of the crystalline solid corresponding to our polymer is larger than T_G, $T_M>T_G$ and a material in the supercooled region $T_G<T<T_M$ is a composite with both crystalline and amorphous structure. Crystallization in supercooled liquids can be very slow, the liquid structure essentially frozen, as in a glass.

The idea that the glassy modulus may be reached either by high frequency at any temperature or at low temperatures in glassy regions at any frequency is an aspect of the "method of reduced variables" or time-temperature shifting. As the temperature rises, so does the amount of molecular motion occurring in one unit of the observer's time; the material's time scale shortens so that the relaxation proceeds faster. The same response can be obtained at high frequency and high temperature as for low frequency and low temperature. The process of reducing data obtained over a range of temperatures to the behavior which would be observed at a single temperature, but over a much wider frequency, is called the method of reduced variables. The method basically involves shifting of patches of data in limited frequency ranges along the frequency axis to get a single "master curve" at one reference temperature.

The method of reduced variables is used to get a master curve for the storage modulus (see Ferry [1980], chap. 11). To illustrate this method we consider the storage modulus

$$\tilde{G}'(\omega) = \frac{G(0)\lambda^2\omega^2}{1+\lambda^2\omega^2}, \quad G(0) = \frac{\eta}{\lambda},$$

corresponding to a Maxwell model $G(s) = \frac{\eta}{\lambda} e^{-s/\lambda}$. We define two temperature-dependent functions $\alpha=\alpha(T)$, $\beta=\beta(T)$ and define $G=G(0)$ at temperature T and G_0 at temperature T_0 with the same convention for λ

$$G = \beta G_0, \quad \lambda = \alpha\lambda_0. \tag{10}$$

Solving for α, we get

$$\alpha = \frac{\lambda}{\lambda_0} = \frac{\eta}{\beta\eta_0}.$$

Moreover

$$\tilde{G}'(\omega) = \beta\tilde{G}_0'(\alpha\omega).$$

If $\alpha(T)$ and $\beta(T)$ are known, measurements of $\tilde{G}'(\omega)$ at different temperatures could be used to get $\tilde{G}_0'(\Omega)$ at the reference temperature T_0 and a different frequency $\Omega=\alpha\omega$, reducing the data to a master curve. One of the problems of the method of reduced variables is that (10) must also hold for every mode of relaxation, with only one pair of functions (α,β), when there are N modes of relaxation, no matter what the value of N. Fluids for which all modes of relaxation scale in the same way with temperature are called thermorheologically simple.

To see how reduced variables might be used to get a master relaxation function, write

$$G(s,T) = G(T)e^{-s/\lambda(T)}.$$

Then, using (10), we get

$$G(s,T) = \beta(T)G(T_0)e^{-s/\alpha\lambda(T_0)} = \beta G\left[\frac{s}{\alpha}, T_0\right]. \tag{11}$$

The reduced time s/α for supercooled liquids may actually enter into the region of small time necessary to study glassy response.

The method of reduced variables is more like a rule of thumb, rather than an exact procedure. It cannot work unless all of the relaxation times scale in the same way, which is certainly impossible in general. It does not apply to solutions. In melts it appears to work away from T_G but even then only in small T intervals. One might have expected an analogy with concentration in concentrated solutions, but the method of reduced variables does not work for changes in concentration. The experiments claim to show that the longest time of relaxation changes strongly with polymer concentration, while the other parts of the spectra are not so strongly affected.

19.3 Spectral decomposition of G(s) and effective moduli

We may understand glassy moduli, effective moduli, viscosity, fast and slow modes through a spectral decomposition of G(s) in the instantaneously elastic case, $\mu=0$. It suffices here to consider a finite basis of N modes of relaxation

$$G(s) = \sum_{1}^{N} G_k(s) \ ,$$

$$G_k(s) = \frac{\eta_k}{\lambda_k} e^{-s/\lambda_k} \ , \quad \lambda_k > \lambda_{k-1} \ ,$$

$$\tilde{\mu} = \eta = \sum_{1}^{N} \eta_k \ . \tag{12}$$

Suppose that ε is a small time defined by a physical process of interest and $\varepsilon=\lambda_M$. Then, for short times $t \ll \lambda_M$ all the modes $k \geq M$ are glassy. For $t > \lambda_M$ all the modes with $k<M$ have decayed. The effective (Newtonian) viscosity μ and rigidity $G_\mu(s)$ may then be defined as follows:

$$\tilde{\mu} = \mu + \eta ,$$

$$\mu = \sum_{1}^{M-1} \eta_k , \quad \eta = \sum_{M}^{N} \eta_k , \tag{13}$$

$$G_\mu(s) = \sum_{M}^{N} G_k(s) .$$

For many polymeric solutions $\mu/\tilde{\mu}$ is very small and the contributions of the fast modes may be expressed by a relatively small Newtonian viscosity.

Let us now suppose that we have modelled the relaxation function so that the fast modes are expressed by a small viscosity and the slow ones by a single relaxation time. Then

$$\left.\begin{aligned} \eta^*(\omega) &= \mu + \frac{\eta}{1+\lambda^2\omega^2} - \frac{i\lambda\omega\eta}{1+\lambda^2\omega^2}, \\ V^2(\omega) &= \frac{2\omega}{\rho}\,\frac{\tilde{\mu}^2 + \mu^2\lambda^2\omega^2}{\lambda\omega\eta+(1+\lambda^2\omega^2)^{1/2}(\tilde{\mu}^2+\mu^2\lambda^2\omega^2)^{1/2}}, \\ \alpha^2(\omega) &= \frac{\omega\rho[\tilde{\mu}+\mu\lambda^2\omega^2]^2}{2[\tilde{\mu}^2+\mu^2\lambda^2\omega^2]\,[\lambda\omega\eta+(1+\lambda^2\omega^2)^{1/2}(\tilde{\mu}^2+\mu^2\lambda^2\omega^2)^{1/2}]} . \end{aligned}\right\} \tag{14}$$

Large frequencies ω correspond to small times t. The asymptotic values for large ω are

$$(\eta^*, V^2, \alpha^2) = \left(\mu \, , \frac{2\omega\mu}{\rho} , \frac{\omega\rho}{2\mu}\right), \tag{15}$$

corresponding to a Newtonian fluid. If, on the other hand, $\mu/\tilde{\mu}=\varepsilon$ is small, ω large, and

$$1 \ll \lambda^2\omega^2 \ll \frac{1}{\varepsilon^2} ,$$

then

$$V^2 = c^2 + 0(\varepsilon)\,, \quad c^2 = \frac{\eta}{\lambda\rho} \tag{16}$$

is the limiting wave speed for a Maxwell model. If, in addition $\lambda^2\omega^2 \ll 1/\varepsilon$, then the attenuation $-(\alpha^2)^{1/2}$ for large ω is also that for a Maxwell model. In this case the effective modulus is η/λ and it can be obtained by measurements at large frequencies, which are sufficiently small so as not to excite the fast (here, viscous) modes.

The existence of slow and fast modes, of different times of relaxation, corresponding to different mechanisms of relaxation, is suppressed by the notion of a mean relaxation time. The equation $\eta = G(0)\bar{\lambda}$ cannot be a very useful description of stress relaxation when there are very different times of relaxation. The relaxation from the huge glassy modulus G(0) (say 10^9 Pa) is very fast and the area under this fast mode gives rise only to a very small fraction of the total viscosity. Typically the total viscosity is associated with the slow relaxation $\lambda_\mu \gg \bar{\lambda}$ of modes associated with a much smaller modulus $G_\mu(0)$ (see Figure 18.5).

19.4 Stokes' first problem for viscoelastic fluids

The theory of the wave-speed meter (Appendix F) requires that one first understands some ideas based on the analysis of propagation of waves of shear into viscoelastic fluids at rest. A planar shear flow is a unidirectional velocity field $\mathbf{u} = \mathbf{e}_x u(y,t)$ which depends only on the coordinate y perpendicular to the direction of motion. Rheological properties of liquids can have a big effect on the way that shear discontinuities propagate. In Stokes' first problem we prescribe a shock in the velocity for boundary data at y=0. We then determine how this data will propagate into the fluid above (y>0) the plate. In addition,

$$u(y,t) = 0 \quad \text{for} \quad t \le 0 \tag{17}$$

is prescribed for all $y \ge 0$. The stress and velocity in $y \ge 0$ satisfy

$$\left.\begin{aligned} &\rho \frac{\partial u}{\partial t} = \frac{\partial \tau}{\partial y}, \\ &\frac{\partial \tau}{\partial t} = \mu \frac{\partial^2 u}{\partial y \partial t} + G(0) \frac{\partial u}{\partial y} + \int_0^t G'(s) \frac{\partial u}{\partial y}(y,t-s)\, ds \end{aligned}\right\} \tag{18}$$

where

$$u(y,t) \to 0 \quad \text{as} \quad y \to \infty.$$

These equations may be combined and written as

$$\rho \frac{\partial^2 u}{\partial t^2} = G(0) \frac{\partial^2 u}{\partial y^2} + \mu \frac{\partial^3 u}{\partial y^2 \partial t} + \int_0^t G'(s) \frac{\partial^2 u(y,t-s)}{\partial y^2}\, ds\,. \tag{19}$$

At the plate we have

$$u(0,t) = UH(t)\,, \tag{20}$$

where $H(t)$ is Heaviside's step function.

When $G(\bullet)=0$ there is no elasticity and the step input is transmitted to the interior by diffusion with

$$u(y,t) = U \operatorname{erfc}\left[\frac{y}{(4\mu t/\rho)^{1/2}}\right].$$

For elastic waves $\mu=0$ and (19) is like a telegraphers equation. When

$$G(s) = \frac{\eta}{\lambda} e^{-s/\lambda}$$

then (19) reduces to

$$\frac{\partial^2 u}{\partial t^2} + \frac{1}{\lambda} \frac{\partial u}{\partial t} = \frac{\eta}{\rho\lambda} \frac{\partial^2 u}{\partial y^2}\,. \tag{21}$$

This telegraphers equation is hyperbolic and transmits waves, but the waves are damped.

It would not be useful to base a rheometrical theory on a special case. It is necessary to understand how shear-wave propagation depends on the relaxation function G(s) and the viscosity μ.

There are many papers on Stokes' first problem, or problems equivalent to it, which are based on Maxwell models. The first group of papers is for elastic fluids $\mu=0$ with $G(s)=(\eta/\lambda)e^{-s/\lambda}$. The solution of (17), (20), and (21) is given by Carslaw and Jaeger [1963]. Step jumps of velocity between two plates have been studied by Böhme [1987], Kazakia and Rivlin [1981], Rivlin [1982], [1983] and Christensen [1982]. Stokes' first problem with $\mu \neq 0$ and $G(s) = (\eta/\lambda)\, e^{-s/\lambda}$ has been studied by Morrison [1956], Tanner [1962] and Saut and Joseph [1983]. Stokes' first problem for fluids with instantaneous elasticity, $\mu = 0$ and a general relaxation function G(s) was studied by Chu [1962], Narain and Joseph [1982], [1983a, b] and Renardy [1982] and an allied problem was studied by Coleman, Gurtin and Herrara [1965]. All but the authors last mentioned use the method of Laplace transforms and all find that the discontinuity propagates into the fluid at rest with a speed and attenuation given by

$$\{c, u(y, y/c)\} = \left\{\left(\frac{G(0)}{\rho}\right)^{1/2}, U \exp\left[\frac{yG'(0)}{2cG(0)}\right]\right\}. \tag{22}$$

Narain and Joseph [1982] also gave a boundary-layer analysis for small values of μ. The solution of Stokes' first problem with $\mu\neq 0$ and a general relaxation function has been obtained and computed by Preziosi and Joseph [1987]; see Figure 19.3.

We shall begin with the general problem when $\mu = 0$ and show how an effective μ arises from the decay of the fast modes. The solution of (17)–(20) with $\mu=0$ is given by the Laplace inversion integral (see Narain and Joseph [1982], equations (5.1) and (5.10)):

$$u(y,t;G) = \frac{U}{2\pi i}\int_{\gamma-i\infty}^{\gamma+i\infty} \frac{d\omega}{\omega} \exp\left[\omega t - y\left(\frac{\rho\omega}{\bar{G}(\omega)}\right)^{1/2}\right] \tag{23}$$

where

$$\bar{G}(\omega) = \int_0^\infty G(s)e^{-\omega s}\, ds \tag{24}$$

is the Laplace transform of G(s). We assume that G(s) is positive, smooth and decreasing, and such that $\bar{G}(\omega)$ is analytic in the positive part of the complex-ω plane with the origin excised. All these hypotheses hold, for example, when G(s) is expressed by a discrete relaxation spectrum, as in (12), or by an infinite number, or even a continuum of such modes.

The notation used in (23) makes explicit that the velocity at point (y,t) is for a fluid with relaxation function G(s). The solution is a functional of G(•).

The solution of Stokes' first problem satisfies a certain type of group invariance, which we call scaling under radial shifts of y and t. The radial shift is given by

$$(y,t) = (\phi\zeta,\phi\tau)\,, \quad 0<\phi<\infty\,. \tag{25}$$

The scaling invariance is then described by the following relationship:

$$u(y,t;G) = U(\zeta,\tau;G_\phi)\,, \tag{26}$$

where U is a functional of

$$G_\phi(s) = G(\phi s)\,.$$

The scaling invariance says that different observers at different points on a radius in the (y,t)-plane see different scaled relaxation functions.

To prove (26) we change variables as in (25) and write $\omega=\Omega/\phi$. We find easily that

$$u(y,t;G) = \frac{U}{2\pi i}\int_{\gamma-i\infty}^{\gamma+i\infty} \frac{d\,\Omega}{\Omega}\exp\left[\Omega\tau-\zeta\left(\frac{\rho\Omega}{\bar{G}_\phi(\Omega)}\right)^{1/2}\right], \tag{27}$$

where

$$\bar{G}_\phi(\Omega) = \int_0^\infty G(\phi s)\, e^{-\Omega s}\, ds\, .$$

Various asymptotic results may be derived from (26). The strategy is to take (ζ,τ) as $0(1)$ and to derive asymptotic results by taking various limits for ϕ. The first result is for small t and y, $\phi\to 0$:

$$u(y,t;G) = U\left\{\exp \frac{yG'(0)}{2cG(0)} + 0(\phi y)\right\} H\left(t - \frac{y}{c}\right). \tag{28}$$

To prove this we write

$$\bar{G}_\phi(\Omega) = \int_0^\infty e^{-\Omega s}\, [G(0)+\phi sG'(0)+0(\phi^2 s^2)]\, ds$$

$$= \frac{G(0)}{\Omega} + \frac{\phi G'(0)}{\Omega^2} + 0\left(\frac{\phi^2}{\Omega^3}\right).$$

The integrand for the inversion integral may be written as

$$\frac{d\Omega}{\Omega} \exp\left[\Omega\left(\tau - \frac{\zeta}{c}\right) + \frac{\phi\zeta}{c}\frac{G'(0)}{2G(0)} + 0\left(\frac{\phi^2}{\Omega}\right)\right]$$

$$= \frac{d\Omega}{\Omega} \exp\left[\frac{\phi\zeta G'(0)}{2cG(0)}\right] \exp \Omega\left(\tau - \frac{\zeta}{c}\right)\left[1 + 0\left(\frac{\phi^2\zeta}{\Omega}\right)\right].$$

After integrating, using the contour appropriate for Heaviside functions (e.g. Figure 5.1 in Narain and Joseph [1982]), we get (28).

We recall that the exact solution of (17)–(20) with $\mu=0$ is of the form

$$u(y,t;G) = Uf(y,t)\, H\left(t - \frac{y}{c}\right), \quad f(y,y/c) = \exp \frac{yG'(0)}{2cG(0)}\, .$$

The result (28) is hardly surprising; when y and t are small we may replace f(y,t) with f(y,y/c).

The second result is for large t and y, $t-y/c>0$. If $\phi=1/\varepsilon$, $\varepsilon\to 0$, we find that

$$u(y,t;G) = U\,\mathrm{erfc}\left[\frac{y}{2\left(\frac{\eta}{\rho}t\right)^{1/2}}\right] + 0(\varepsilon)\ . \tag{29}$$

To derive this we note that

$$\bar{G}_\phi(\Omega) = \int_0^\infty G\left(\frac{s}{\varepsilon}\right)e^{-\Omega s}\,ds\ .$$

As $\varepsilon \to 0$ the relaxation function collapses on the origin, like a delta function, and

$$\bar{G}_\phi(\Omega) \to \int_0^\infty G\left(\frac{s}{\varepsilon}\right)ds = \varepsilon\eta\ ,$$

provided that $|\Omega|$ is not too large. In fact, the integrand in the inversion integral tends rapidly to zero for large Ω on the contour of integration. Then, with a small error, we get

$$\begin{aligned} u(y,t;G) &= \frac{U}{2\pi i}\int_{\gamma-i\infty}^{\gamma+i\infty}\frac{d\,\Omega}{\Omega}\exp\left[\Omega\tau-\zeta\left(\frac{\rho\Omega}{\varepsilon\eta}\right)^{1/2}\right] \\ &= \frac{U}{2\pi i}\int_{\gamma-i\infty}^{\gamma+i\infty}\frac{d\omega}{\omega}\exp\left[\omega t-y\left(\frac{\omega}{\eta/\rho}\right)^{1/2}\right] \\ &= U\,\mathrm{erfc}\left[\frac{y}{2\left(\frac{\eta}{\rho}t\right)^{1/2}}\right]. \end{aligned}$$

When t and y are large the solution looks Newtonian with viscosity η. The same type of result holds when $\mu \neq 0$ with η replaced by $\tilde{\mu}$.

The third result to come out of scaling invariance shows how an effective viscosity and modulus can arise when different molecular substructures have different times of relaxation. We shall suppose that there is an additive decomposition for G(s), as in (13), with

$$G(s) = G_0(s) + G_\mu(s) \,, \tag{30}$$

where λ_0 is a short mean time of relaxation and

$$\mu = \int_0^\infty G_0(s)\, ds \,. \tag{31}$$

We find that when $t \gg \lambda_0$, $t - y/c > 0$, then with a small error

$$u(y,t;G) = \frac{U}{2\pi i} \int_{\gamma - i\infty}^{\gamma + i\infty} \frac{d\omega}{\omega} \exp\left[\omega t - y\left(\frac{\rho\omega}{\mu + \bar{G}_\mu(\omega)}\right)^{1/2}\right] \tag{32}$$

where $\bar{G}_\mu(\omega)$ is the Laplace transform of $G_\mu(s)$. This is the solution of (17)–(20) when $\mu \neq 0$ and $G(s) = G_\mu(s)$. The elastic response of unrelaxed modes is smoothed by the viscosity μ induced by modes already relaxed.

A heuristic argument for (32), which can be made precise, starts from

$$u(\zeta,\tau;G_\phi) = \frac{U}{2\pi i} \int_{\gamma - i\infty}^{\gamma + i\infty} \frac{d\Omega}{\Omega} \exp\left[\Omega\tau - \zeta\left(\frac{\rho\Omega}{\bar{G}_0^\phi(\Omega) + \bar{G}_\mu^\phi(\Omega)}\right)^{1/2}\right], \tag{33}$$

where

$$\bar{G}_0{}^\phi(\Omega) = \int_0^\infty G_0(\phi s)\, e^{-\Omega s}\, ds$$

and

$$\bar{G}_\mu{}^\phi(\Omega) = \int_0^\infty G_\mu(\phi s)\, e^{-\Omega s}\, ds \,.$$

The function $G_0(s)$ has a mean relaxation time λ_0 and $\phi>\lambda_0$. Hence, $G_0(\phi s)$ is crowded at the origin like a δ-function, nearly zero for very small values of s for which $e^{\Omega s}\sim 1$. Then, with a small error

$$\bar{G}_0{}^{\phi}(\Omega) = \frac{\mu}{\phi}$$

and (32) follows after rescaling. Small errors are associated with large values of $|\Omega|$ on the line of integration Re $\Omega=\gamma$. The integrand is very small, and also oscillates rapidly when $|\Omega|$ is large. Of course, $G_\mu(s)$ could still be glassy at times large enough for $G_0(s)$ to have relaxed completely.

Having started the study of the problem with $\mu=0$, we turn now to the study of Stokes' first problem when the viscosity ratio

$$J \overset{\text{def}}{=} \frac{\mu}{\tilde{\mu}} \quad (0\leq J\leq 1)$$

is small. Our aim is to establish the shock-layer approximation

$$u(y,t,\mu;G_\mu) = f(y,t) \int_{-\infty}^{\xi} e^{-\psi^2}\, d\xi + 0(J) \tag{34}$$

where

$$\xi = \frac{t-y/\hat{c}}{(Jyc)^{1/2}}, \tag{35}$$

$$\hat{c} = \left(\frac{G_\mu(0)}{\rho[1-J]}\right)^{1/2} = \frac{c_\mu}{(1-J)^{1/2}} \tag{36}$$

and

$$u(y,t,0;G_\mu) = Uf(y,t)\, H\left(\frac{t-y}{c_\mu}\right), \tag{37}$$

$$f(y, y/c_\mu) = \exp\left[\frac{yG'_\mu(0)}{2c_\mu G(0)}\right].$$

This result holds for sufficiently small values of $\rho^2 Jyc_\mu/\bar{\mu}\eta$ and J.

The viscosity ratio arises from the following change to dimensionless variables:

$$[v,\tau,x,g(\tau)] = \left[\frac{u}{U}, \frac{tG_\mu(0)}{\eta}, y\left(\frac{G_\mu(0)\rho}{\eta\mu}\right)^{1/2}, \frac{G_\mu(t)}{G_\mu(0)}\right]. \quad (38)$$

Clearly, g(0)=1. After introducing this change of variables into (17), (19), and (20), using the effective modulus we find that $v(x,\tau,J)$ satisfies

$$\left.\begin{aligned} &v_{\tau\tau} = (1-J)v_{xx} + Jv_{xx\tau} + (1-J)\int_0^\tau g'(s)v_{xx}(x,\tau-s)\,ds, \\ &v(0,\tau) = H(\tau), \\ &v(x,\tau) = 0 \quad \text{for } \tau \le 0. \end{aligned}\right\} \quad (39)$$

We note that $1-J=\eta/\bar{\mu}$. When J=0,

$$\left.\begin{aligned} &v(x,\tau,0) = f(x,\tau)H(\tau-x), \\ &f(x,x) = e^{(x/2)g'(0)}, \\ &f(0,\tau) = 1. \end{aligned}\right\} \quad (40)$$

The function $f(x,\tau)$ is defined, up to the change (38) to dimensionless variables, by the expression for f(x,t) given by Equation 5.10 of Narain and Joseph [1982].

For the Jeffreys' model $g(s)=e^{-s}$ and

$$v_{\tau\tau} + v_\tau - v_{xx} = Jv_{xx\tau}.$$

In this case the viscosity ratio is the only parameter in the problem. The viscosity ratio $J=\lambda_2/\lambda_1$ is the ratio of relaxation upon retardation time.

The effect of a small viscosity is to smooth the shock. The amplitude of the shock decays rapidly and eventually the solution looks diffusive, with viscosity $\tilde{\mu}$. We look for smoothing at small values of (x,ε) where the unperturbed problem has a large $f(x,\tau)$ and a large shock. We follow Narain and Joseph [1982] and change variables

$$x = JX\,, \quad t = JT\,, \quad V(X,T,J) = U(x,t,J)\,,$$

where

$$\left.\begin{aligned} &V_{TT} = V_{XX} + V_{XXT} + JF(X,T,J)\,,\\ &V(0,T) = H(T)\,,\\ &V(X,T) = 0\,;\ T \le 0 \end{aligned}\right\} \tag{41}$$

and

$$F(X,T,J) = V_{xx}(X,T,J) + (1-J)\int_0^T g'(Js)V_{XX}(X,T-s)\,ds\,.$$

We look at small values of x and t in the limit $J \to 0$. Assuming that $g'(0)$ is finite, we find that $V(X,T,0)$ satisfies

$$\left.\begin{aligned} &V_{TT} = V_{XX} + V_{XXT}\,,\\ &V(0,T) = H(T)\,,\\ &V(X,T) = 0\,;\ T \le 0\,. \end{aligned}\right\} \tag{42}$$

This problem governs the perturbation of Stokes' first problem for pure, rather than relaxing, elasticity. An exact solution of (42) was given by Morrison ([1956], Equation A 14). This solution smooths the Heaviside function. Renardy, Hrusa, and Nohel [1987] have studied the problem

$$\left.\begin{aligned} &U_{\tau\tau} = U_{xx} + \varepsilon U_{xx\tau}\,,\\ &U(0,\tau,\varepsilon) = H(t)\,,\\ &U(x,\tau,\varepsilon) = 0\,;\ \tau \le 0\,. \end{aligned}\right\} \tag{43}$$

which arises from changing variables $[X,T,V] = [(x/\varepsilon),(\tau/\varepsilon),U]$. They estimated the Laplace inversion integral for (43) to obtain a small-ε approximation for (43) under the following change of variables:

$$x = x\ , \quad \xi = \frac{\tau - x}{(\varepsilon x)^{1/2}}\ , \tag{44}$$

and they show that with a small error

$$U(x,\ \tau,\ \varepsilon) = \int_{-\infty}^{\xi} e^{-\frac{1}{2}\psi^2} d\psi = w(\xi) \tag{45}$$

depends on ξ alone, provided that $(\varepsilon x)^{1/2}$ is small. Note that

$$w(-\infty) = 0\ , \quad w\ (\infty) = 1\ . \tag{46}$$

Clearly

$$\lim_{\varepsilon \to 0} w(\xi) = H(t-x)\ .$$

This result may also be obtained directly from (43) using the change of variables (44). We find that when εx is small and x-derivatives are relatively small

$$w_{\xi\xi\xi} + \xi w_{\xi\xi} + w_{\xi} = 0\ , \tag{47}$$

subject to (46). Equation (45) represents the effect of diffusion in smoothing the propagating step function. This analysis shows that the size of the shock layer perturbing the propagating step scales with $(\varepsilon x)^{1/2}$.

We may now identify ε=J; note that

$$U(x,\tau,J) \approx f(x,\tau)w(\xi)\ , \tag{48}$$

with the $f(x,\tau)$ as in (40) also satisfies (47), the boundary conditions and initial conditions to lowest order in J. We note next that (48) is not uniformly valid; it is not correct, for example, at x=0 when J≠0. A composite shock-layer solution, given by Narain and Joseph

([1982], Figure 18.1) with αx replaced by $(Jx)^{1/2}$, eliminates this non-uniformity.

Figure 19.3 (Preziosi and Joseph, [1987]) Flow development for Stokes' first problem with different kernels and different values for the effective viscosity.

(a) Flow development for Stokes' first problem as a function of x for different times t and when $\mu/\bar{\mu} = 0.1$, $g(s) = e^{-s}$. The dashed line through x = 0.4 and x = 1.6 are shocks. The shock strength is the value of v for $\mu = 0$ on the line x = t. The shock layer thickness for $\mu/\bar{\mu} \neq 0$ is the interval $\triangle x$ centered on x in which the change of v is 95% of its shock value. The thickness is $\triangle x = 0.710$ for x = 0.4 and $\triangle x = 0.141$ for x = 1.6, satisfying the relation $\triangle x \approx (\mu x/\bar{\mu})^{1/2}$. ■ → t = 0.0797, ● → t = 0.4, ▲ → t = 0.8, ▼ → t = 1.2, ◆ → t = 1.6.

(b) Flow development for Stokes' first problem as a function of x for different times t and when $\mu/\bar{\mu} = 0.001$, $g(s) = e^{-s}$. The dashed line through x = 0.4 and x = 1.6 are shocks. The shock strength is the value of v for $\mu = 0$ on the line x = t. The shock layer thickness for $\mu/\bar{\mu} \neq 0$ is the interval $\triangle x$ centered on x in which the change of v is 95% of its shock value. The thickness is $\triangle x = 0.0712$ for x = 0.4 and $\triangle x = 0.142$ for x = 1.6, satisfying the relation $\triangle x \approx (\mu/\mu)^{1/2}$. ■ → t = 0.0718, ● → t = 0.4, ▲ → t = 0.8, ▼ → t = 1.2, ◆ → t = 1.6, ⬟ → t = 2.0.

(c) Comparison of the solution of Stokes' first problem for the double Maxwell kernel (solid symbols) $g(s) = e^{-s} + 15e^{-15s}$ and the equivalent Jeffrey's model (open symbols) $g(s) = e^{-s}$, $\mu/\bar{\mu} = 0.5$.

(d) Comparison of the solution of Stokes' first problem for the double Maxwell kernel (solid symbols) $g(s) = e^{-s} + 3e^{-27s}$ and the equivalent Jeffrey's model (open symbols) $g(s) = e^{-s}$, $\mu/\bar{\mu} = 0.1$.

(e) Flow development for Stokes' first problem as a function of x for different times t and when $\mu/\bar{\mu} = 0.001$, $g(s) = (2/\sqrt{\pi})e^{-s^2}$. ● → t = 0.4, ▲ → t = 0.8, ▼ → t = 1.2, ◆ → t = 1.6, ⬠ → t = 2.0.

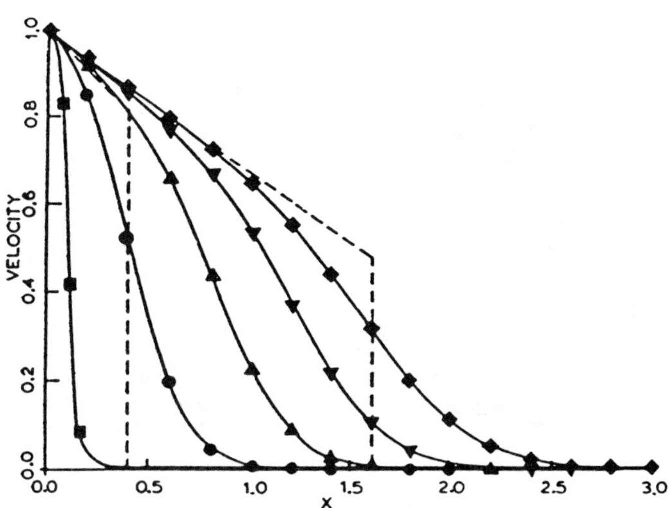

19.3(a)

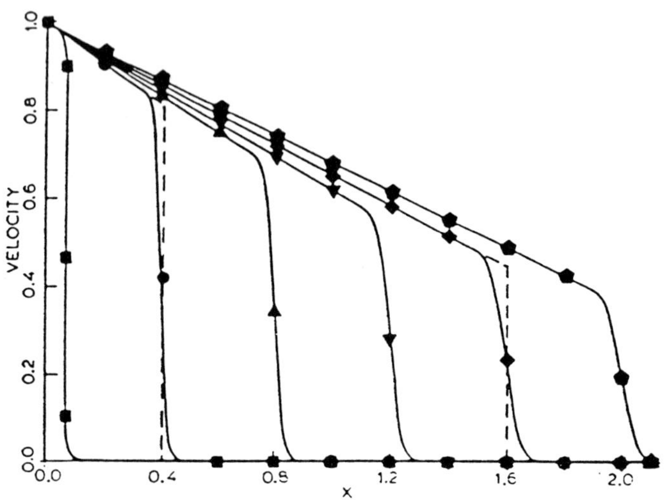
VELOCITY
0.0
0.2
0.4
0.6
0.8
1.0
0.0
0.4
0.8
1.2
1.6
2.0
X

19.3(b)

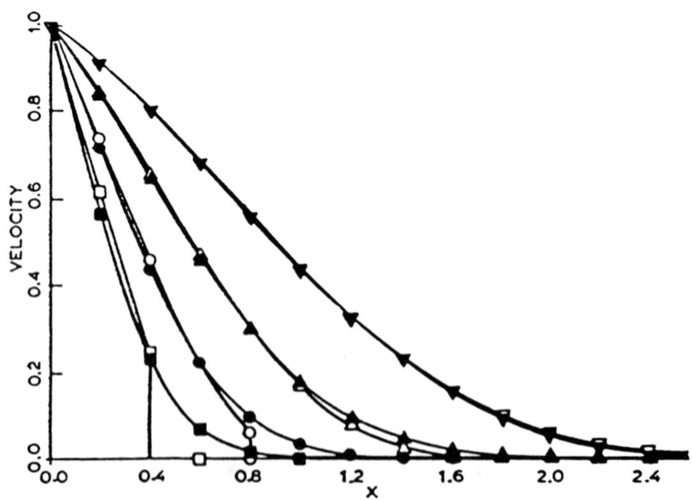
VELOCITY
0.0
0.2
0.4
0.6
0.8
1.0
0.0
0.4
0.8
1.2
1.6
2.0
2.4
X

19.3(c)

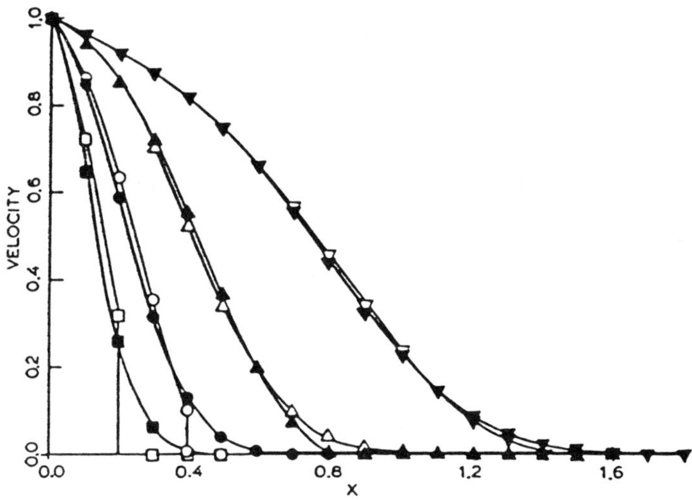
VELOCITY
0.0
0.2
0.4
0.6
0.8
1.0
0.0
0.4
0.8
1.2
1.6
X

19.3(d)

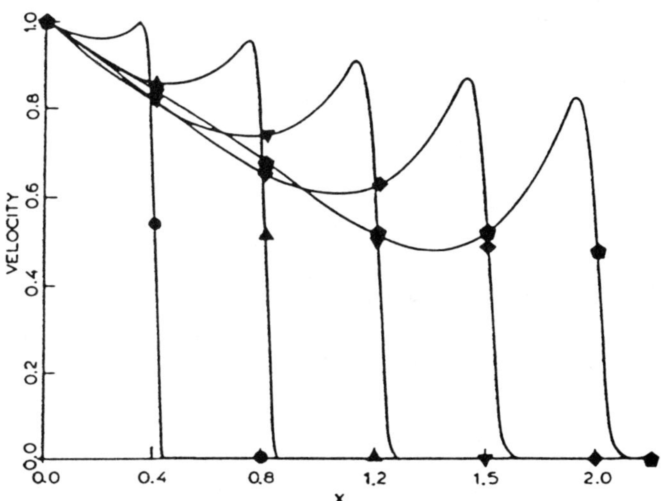
VELOCITY
0.0
0.2
0.4
0.6
0.8
1.0
0.0
0.4
0.8
1.2
1.5
2.0
X

19.3(e)

Exercises

Exercise 19.1.
Extended scale invariance for Stokes' first problem (Preziosi, 1989). Show that

$$u(y,t,\mu;G(s)) = u\left(\frac{y}{\phi_y}, \frac{t}{\phi_t}, \frac{\phi_t}{\phi_y^2}\mu, \frac{\phi_t^2}{\phi_y^2}G(\phi_t s)\right) \tag{49}$$

where

$$u(y,t,\mu;G(s)) = \frac{U}{2\pi i}\int_{\gamma-i\infty}^{\gamma+i\infty}\frac{d\omega}{\omega}\exp\left[\omega t + \left(\frac{\rho\omega y^2}{\mu+\bar{G}(\omega)}\right)\right], \tag{50}$$

$\bar{G}(\omega)$ is given by (24) and ϕ_y and ϕ_t are any given positive real numbers. Putting $(y/\phi_y, t/\phi_t) = (\zeta,\tau)$ we get a generalization of (26)

$$u(y,t,\mu;G(s)) = u\left(\zeta, \tau, \frac{t/\tau}{(y/\zeta)^2}\mu ; \left(\frac{t/\tau}{y/\zeta}\right)^2 G\left(\frac{t}{\tau}s\right)\right) \tag{51}$$

in which y and t can be independently rescaled.

Proof. Replace (y,t) in (50) with $(y/\phi_y, t/\phi_t)$. Then write $\omega=\omega'\phi_t$ in (50) and (24). The required result will follow. Select $\zeta=y$, $\tau=1$. Then

$$u(y,t,\mu;G(s)) = u\,(y,1,t\mu;t^2G(ts)) .$$

For large t, $G(ts)$ will be rapidly decreasing. The solution at any t is the same as the solution at $t=1$ for a fluid with a much larger viscosity $t\mu$, rigidity $t^2G(0)$, but with a much smaller relaxation time λ/t.

Exercise 19.2.
Stokes' first problem for a generalized Bingham fluid (H. Pascal, [1989]). This is a case of degenerate parabolic propagation.

We consider Stokes' first problem for a generalized Bingham fluid in one space dimension. The generalized Bingham fluid is a power law fluid with a yield stress τ_0. The shear stress τ is given by

$$\tau = \tau_0 + \left(\frac{\partial u}{\partial y}\right)^n C, \; n \geq 1 \tag{52}$$

where n and C are power law parameters and

$$\left(\frac{\partial u}{\partial y}\right) = 0 \text{ when } \tau \leq \tau_0 \, . \tag{53}$$

The fluid will not move unless a stress greater than τ_0 is applied. The problem is to find a function u(y,t) such that

$$\rho \left(\frac{\partial u}{\partial t}\right) = \left(\frac{\partial \tau}{\partial y}\right) . \tag{54}$$

with u(0,t)=U when t>0 and u(y,t)=0 for all y≥0 when t<0.

Solution. From (52) and (54) we have

$$\left(\frac{\partial u}{\partial y}\right)^{n-1} \left(\frac{\partial^2 u}{\partial y^2}\right) = a^2 \left(\frac{\partial u}{\partial t}\right) \tag{55}$$

where $a^2=\rho n/C$. This is a degenerate parabolic equation which reduces to the heat equation when n=1. We can give a solution of this problem in terms of the similarity variable

$$\eta = yt^{-\frac{1}{1+n}} \, . \tag{56}$$

We find that when $\eta \leq \eta_1$

$$\frac{du}{d\eta} = -B \, \eta_1^{\frac{2}{n-1}} \left(1 - \frac{\eta^2}{\eta_1^2}\right)^{\frac{1}{n-1}} , \tag{57}$$

where

$$B = \left[\frac{(n-1)a^2}{2(n+1)}\right]^{\frac{1}{n-1}}$$

and

$$\tau - \tau_0 = Ct^{-\frac{n}{n+1}} B^n \eta^{\frac{2n}{n-1}} \left(1 - \frac{\eta^3}{\eta_1^2}\right)^{\frac{n}{n-1}} . \tag{58}$$

When $\eta \geq \eta_1$, $\tau = \tau_0$ and $\partial u/\partial y = 0$. Equation (58) shows that there is an $\eta = \eta_1$, such that $\tau = \tau_0$ if and only if $n > 1$. The position $y = \ell(t)$ of the front at $\eta = \eta_1$ is given by (56) as

$$\ell(t) = \eta_1 t^{\frac{1}{1+n}} , \frac{\eta}{\eta_1} = \frac{y}{\ell(t)} . \tag{59}$$

The velocity of this front is

$$\frac{d\ell}{dt} = \frac{\eta_1}{1+n} t^{\frac{-n}{1+n}} . \tag{60}$$

The velocity of the fluid is obtained by integrating (57) for $u(0,t) = U$, $u(y,t) = 0$ for $y \geq \ell$. We get

$$u(\eta) = U\left(1 - \frac{\Psi(\eta/\eta_1)}{\Psi(1)}\right), \eta \leq \eta_1 \tag{61}$$

where

$$\Psi(\eta, \eta_1) = \int_0^{\eta/\eta_1} \left[1 - \zeta^2\right]^{-\frac{1}{n-1}} d\zeta \tag{62}$$

and, using (57)

$$\eta_1 = \left[\frac{U}{B\Psi(1)}\right]^{\frac{n-1}{n+1}} . \tag{63}$$

The shear stress at the plate $y = 0$ is given by

$$\tau(0) = C B^n \eta_1^{\frac{2n}{n-1}} t^{-\frac{n}{1+n}} + \tau_0 \tag{64}$$

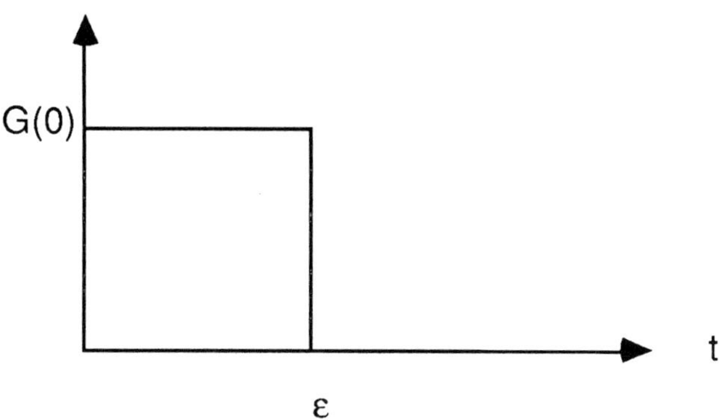

Figure 19.4 Step kernel

and is always larger than the yield stress.

Exercise 19.3
Stokes' first problem for a step kernel (Joseph and Renardy, [1990])

Consider Stokes' first problem for the step kernel (Figure 19.4).

$$G(t) = G(0)\ \{H(t) - H(t-\varepsilon)\}\ , \tag{65}$$

where H(t) is the unit step function, zero when $t<0$ and one when $t>0$.

The problem is to solve

$$\rho u_{tt} = G(0)\ u_{yy} + \int_{-\infty}^{t} G'(t-\tau)\ u_{yy}(y,\tau)d\tau \tag{66}$$

for

$$u(y,t) = 0 \text{ for } t<0 \text{ and } y\geq 0 \tag{67}$$

and

$$u(0,t) = H(t)\ . \tag{68}$$

It is not a good idea to judge *a priori* how solutions ought to behave at infinity. Because we are dealing with a hyperbolic problem, we should not be surprised to see the propagation of discontinuous derivatives in the kernel.

Solution. It is convenient to change variables, writing

$$y^2 = G(0)\, x^2/\rho \ , \ g(s) = G(s)/G(0)\ . \tag{69}$$

Then, using (67), we get

$$u_{tt} = u_{xx} + \int_0^t g'\ (t-\tau)\ u_{xx}\ (x,\tau)\ d\tau\ . \tag{70}$$

For the normalized step kernel corresponding to (65), we have

$$g'(t-\tau) = -\delta(t-\tau-\varepsilon)\ . \tag{71}$$

After combining (71) and (70), we find that

$$u_{tt}(x,t) = u_{xx}(x,t) - u_{xx}\ (x,\ t-\varepsilon)\ . \tag{72}$$

To solve (72), we introduce the characteristic coordinate

$$\begin{aligned} \alpha &= t-x, \\ \beta &= t+x\ . \end{aligned} \tag{73}$$

and find that $u_{\alpha\beta} = 0$ for $t \le \varepsilon$ and

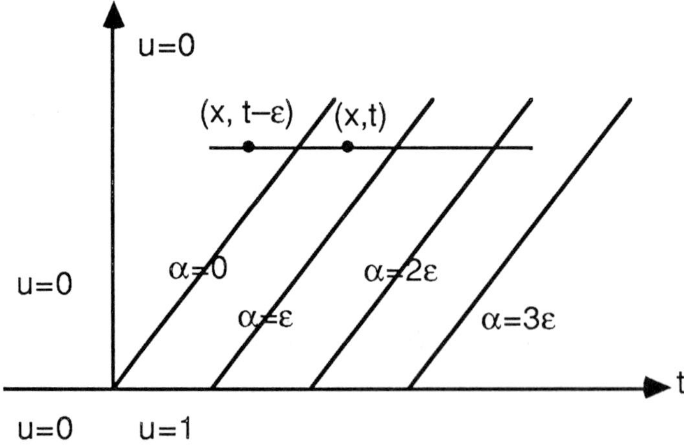

Figure 19.5 Characteristic coordinates for $\alpha = t-x$. The cross-characteristics for $\beta=t+x$ are not shown.

$$4u_{\alpha\beta} = -u_{xx}(x, t-\varepsilon) \tag{74}$$

when t>ε. This implies a stepping property:

$$\text{If } u_{xx}(x, t-\varepsilon) = 0, \text{ then } u_{\alpha\beta}\ (\alpha,\beta) = 0 \text{ at } (x,t)\ . \tag{75}$$

Using (75), we establish that $u_{\alpha\beta} = 0$ for $\alpha<0$. Therefore $u=f(\alpha)+g(\beta)$ for $\alpha<0$ and since u=0 on the line t<0 x>0, we get u=0 for $\alpha<0$. Of course, the region next to constant state is a simple wave, where only the α characteristics are important (see Figure 19.5).

We next prove that

$$u(\alpha,\beta) = H(\alpha),\ a>0,\ \alpha=n\varepsilon,\ n=1,2,\ldots \tag{76}$$

First we prove (76) when $0<\alpha<\varepsilon$, then for $\varepsilon<\alpha<2\varepsilon$, and so on. If $0<\alpha<\varepsilon$ then, by (76), $u_{\alpha\beta}=0$ and $u=f(\alpha)+g(\beta)$ and $u_\beta = g'(\beta)$. On x=0, we have 1=f(t)+g(t) and $g'(\beta)=0$. Hence $g(\beta)=0$ and $u=f(\alpha)=H(\alpha)$. Now, again using (75), we find that $u_{\alpha\beta}=0$ when $\varepsilon<\alpha<2\varepsilon$ and the previous arguments repeat. In this way we go on, interval after interval, to prove that u=1 when $\alpha>0$, except when $\alpha=n\varepsilon$.

Now we get the solution on the lines $\alpha=n\varepsilon$. It's a very singular solution. In an interval around $\alpha=\varepsilon$, we compute (74) in the form

$$4u_{\alpha\beta} = -\frac{\partial}{\partial x^2} H(\alpha-\varepsilon) = -\delta'(\alpha-\varepsilon) \tag{77}$$

where δ is Dirac's delta function. The solution of (77) which satisfies the condition u(0,t)=H(t) is

$$u(\alpha,\beta) = H(\alpha) - \frac{\beta-\varepsilon}{4}\ \delta(\alpha-\varepsilon)\ . \tag{78}$$

The term $(\beta-\varepsilon)\delta(\alpha-\varepsilon) = (t-\varepsilon)\delta(t-\varepsilon) = 0$, even at $t=\varepsilon$ when x=0. Given (78), we take the next step, writing (74) in an interval around 2ε. We get

$$4u_{\alpha\beta} = -u_{xx}(x, t-\varepsilon) = \frac{\partial^2}{\partial x^2}\left\{\frac{\beta-2\varepsilon}{4}\,\delta(\alpha-2\varepsilon)\right\}$$

$$= \frac{\beta-2\varepsilon}{4}\,\delta''\,(\alpha-2\varepsilon) - \frac{1}{2}\,\delta'\,(\alpha-2\varepsilon)\,. \tag{79}$$

The solution

$$u=H(\alpha) - \frac{\beta-\varepsilon}{4}\,\delta(\alpha-\varepsilon) - \frac{\beta-2\varepsilon}{8}\,\delta(\alpha-2\varepsilon) + \frac{(\beta-2\varepsilon)^2}{32}\,\delta'(\alpha-2\varepsilon) \tag{80}$$

is valid for $\alpha<3\varepsilon$, giving $u(0,t)=1$ when $x=0$.

At the next step, in $0<\alpha<4\varepsilon$, we would get $(\beta-3\varepsilon)^3\delta''(\alpha-3\varepsilon)$ and so on. The solution contains a countable set of terrible singularities.

Exercise 19.4

Stokes' first problem for the ramp kernel.

Repeat exercise 19.3 when the normalized kernel is the ramp function (see Figure 19.6).

$$g(s) = \begin{cases} 1-s/\varepsilon, & 0<s\leq\varepsilon, \\ 0, & s>\varepsilon. \end{cases} \tag{81}$$

The point of this problem is to show that the non-analytic point $s=\varepsilon$ of the kernel propagates in the solution.

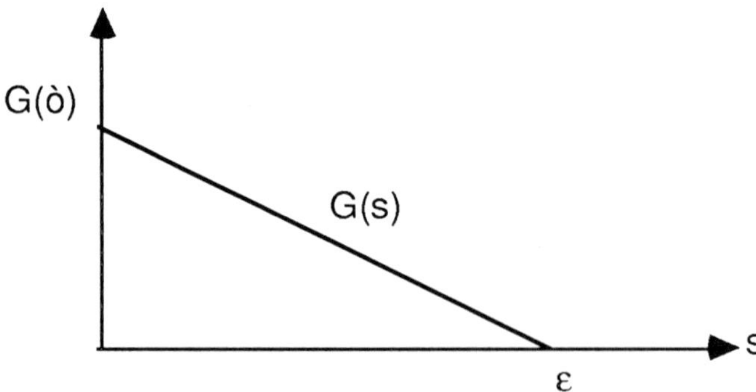

Figure 19.6 Ramp kernel.

Solution. First we compute

$$g'(t-\tau) = \frac{1}{\varepsilon}\begin{cases} -1, & 0 < t-\tau \le \varepsilon, \\ 0, & t-\tau > 0. \end{cases} \tag{82}$$

and

$$\int_0^t g'(t-\tau)\, u_{xx}(x,\tau)d\tau = -\frac{1}{\varepsilon} \int_{t-\varepsilon}^{t} u_{xx}(x,\sigma d\tau)\ . \tag{83}$$

Our governing equation then is (70) using (83). We want to solve this equation subject to (67) and (68). The solution of this problem is the same as in the classical case in the time region $\alpha<\varepsilon$ before the point of discontinuous slope $t-\tau=\varepsilon$ begins to propagate. This was also the case for the step kernel studied in exercise 19.4 where the classical solution for $\varepsilon\to\infty$ is $u=1$. We can see that the delayed propagation is associated with the delay term in the third order equation

$$u_{ttt}(x,t) = u_{xxx}(x,t) - \frac{1}{\varepsilon}\{u_{xx}(x,t) - u_{xx}(x, t-\varepsilon)\} \tag{84}$$

which governs for the ramp kernel. Of course, the solution for the ramp kernel will be less singular than for the step kernel.
The solution of (84) subject to (67) and (68) is of the form

$$u(x,t) = f(x,t)\, H(\alpha),\ \alpha = t-x \tag{85}$$

where on the shock

$$f(x,x) = \exp\,[xg'(0)/2] = \exp\,[-x/2\varepsilon]\ . \tag{86}$$

The function $f(x,t)$ is analytic and exponentially damped in the region $\alpha<\varepsilon$.

To determine the solution in the region $\alpha<2\varepsilon$, we compute

$$u_{xx}(x, t-\varepsilon) = f(x, t-\varepsilon)\, H''\,(\alpha-\varepsilon) + \text{LST} \tag{87}$$

where LST are "less singular terms." We may write (84) near $\alpha=\varepsilon$ as

$$4u_{t\alpha\beta} = f(x, t-\varepsilon)\, H''(\alpha-\varepsilon) + \text{LST}\,. \tag{88}$$

In the analysis of the most singular term, we treat the smoothly varying function f(x, t−ε) as a constant, then

$$u_{t\beta} = \tfrac{1}{4} f(x, t-\varepsilon)\, \delta'(t-\varepsilon-x) + \text{LST} \tag{89}$$

and

$$u_t = \tfrac{1}{4}\,(t-\varepsilon+x)\, \delta\,(t-\varepsilon-x)\, f(x, t-\varepsilon) + \text{LST}\,. \tag{90}$$

The condition $u_t(0, t-\varepsilon)=0$ which is implied by (67) is satisfied. After integrating, we find that

$$u(x,t) = \tfrac{1}{2} x H\,(t-\varepsilon-x)\, f(x, t-\varepsilon) + \text{LST}\,. \tag{91}$$

The condition u(0,t)=H(t) is absorbed by the LST. This solution is one derivative less singular than the one for the step kernel when $\alpha<2\varepsilon$. The problem of increasing singularities as α is increased to 2ε and beyond does not occur for the ramp function because nothing worse than $H(\alpha)$ can appear at any stage. (The analysis just given shows that the generation of step function singularities is a closed operation.)

CHAPTER 20

Nonlinear Waves

In this chapter we shall study some problems of wave propagation in the nonlinear case. We shall consider models without Newtonian viscosity which give rise to hyperbolic waves. In general we deal with first order quasilinear systems and first order nonlinear systems which are not quasilinear. In §5.7 we showed how to reduce a nonlinear system of order N in independent variables (x,t) or (x,y) into a quasilinear system of order 3N by differentiating the nonlinear system once. A nonlinear system of order N in three independent variables can be reduced to a quasilinear system of order 4N, etc. An important point is that the principal part in such differentiated systems is only of order N. N characteristic directions are possible. We expect shocks to occur in quasilinear problems when characteristics intersect. In genuinely nonlinear problems the characteristics are not parallel, they intersect. At the point of intersection of the characteristics the solution may become multiple valued. This can be interpreted as the breaking or overturn of the wave (see Figure 20.2), a point of view developed in the fine book of Whitham [1974] on nonlinear waves.

Another way to describe the formation of a shock of u(x,t) is as a blow-up in a finite time of the x derivative u_x of the quasilinear variable u. This corresponds nicely to thinking of a shock in terms of wave breaking. The wave front is about to overturn when its vertical tangent u_x is infinite. The blow-up time is just the time it takes different initial values which are carried along characteristics to intersect.

The important point here is that the x derivative of the quasilinear variable is what blows up. If we start out with a nonlinear rather than quasilinear problem; say $f(u)u_x^2$ rather than $f(u)u_x$ then we must differentiate to get our quasilinear variable. Then u_{tx} or u_{xx} will blow up. So the nonlinear systems are "more regular," they lead to blow-up of derivatives one order higher (Joseph and Saut, [1986]). This then is the explanation for the remarkable property first pointed out in the work of Slemrod [1985] in which he exhibited two models, apparently benign, one of which (quasilinear) leads to a velocity shock, and the other (nonlinear) to a vorticity shock. Other models, which follow the same rule, are analyzed in the book of Renardy, Hrusa, and Nohel [1987].

The plan of this chapter is as follows. First, in §20.1, we consider some simple kinematic wave problems, a single first order quasilinear equation. This will help to fix ideas about the formation of shocks, characteristics blow-up and the effects of damping in the simplest context. Then, in §20.2, we set up the quasilinear system which governs unsteady shearing problems in one space dimension and analyze the evolution of simple jumps in the vorticity, following the work of Coleman and Gurtin [1968]. They worked with simple jumps of acceleration, called acceleration waves. Jumps in acceleration are proportional to jumps in vorticity, as (32) shows. Coleman and Gurtin [1968] work at the level of generality of simple fluids with $L_h^2(0,\infty)$ fading memory (see Chapter 15 for a discussion of different theories of fading memory). They get blow-up of the vorticity jump when the initial amplitude of the jump is greater than critical; a typical result for blow-up for problems with damping. They do not say how the vorticity jump might be formed. The formation of jumps (called "shocks") from smooth data for special models was discussed first by Slemrod [1978]. Renardy, *et al.* [1987] extended Slemrod's results, introducing new mathematical methods which are required to treat more general models. Some parts of Slemrod's work are discussed in §20.3. In §20.4 we consider shearing flows of the interpolated Maxwell models, showing that the upper and lower convected models ($a=\pm 1$) are not genuinely nonlinear. The interpolated models are quasilinear, rather than nonlinear and therefore lead to shocks of velocity rather than shocks of vorticity. In §20.5 we derive shock relations for three-dimensional flow of an upper convected Maxwell model exploiting ideas introduced by Rutkevich [1969] and Hulsen [1986].

20.1 Blow-up, intersecting characteristics, breaking waves and threshold amplitudes

Michael Renardy pointed out to me that a good model equation governing blow-up is

$$\dot{y} = y^2 , \quad y(0) = \alpha .$$

The solution of this is

$$y(t) = \frac{\alpha}{1-\alpha t}$$

which blows up at $t=1/\alpha$. If $\alpha>0$, we have blow-up in a finite time. Renardy remarked that many problems reduce to the form

$$\dot{y} = y^2 + \text{“junk terms”}$$

and that the problem is to find the right y.

For Burger's equation

$$u_t + uu_x = 0 , \quad u(x,0) = u_0(x) , \quad x \in \mathbb{R} ,$$

the right “y” is u_x. The characteristic curves for Burger's equation are $x=\tilde{x}(\xi,t)$ where

$$\frac{d\tilde{x}}{dt} = u(\tilde{x},t) , \quad \tilde{x}(\xi,0) = \xi .$$

Actually $u(\tilde{x}(\xi,t),t)$ is constant on characteristics because

$$\frac{du}{dt}(\tilde{x}(\xi,t),t) = u_t + uu_x = 0 .$$

Hence, the characteristic curves are straight lines

$$x = \xi + u(\xi,0)t . \tag{1}$$

We may differentiate Burger's equation with respect to x

$$u_{xt} + uu_{xx} + u_x^2 = 0$$

which on characteristics can be written as

$$\frac{d}{dt}u_x(\tilde{x},t) + u_x^2(\tilde{x},t) = 0 .$$

So if $u_x(\xi,0)$ is ever negative we will get a blow-up in finite time. The blow-up of u_x may be interpreted as the breaking of the wave u(x,t), which gets steeper and then breaks. The characteristics with the larger wave speed $d\tilde{x}/dt = u(\xi,0)$ travel faster; large u's overtake small ones as in Figure 20.2.

Another way to do the blow-up is to use $x=\xi+u(\xi,0)t$ as an implicit equation for $\xi(x,t)$; $u(\xi,0)=u(\xi(x,t),0)$ and

$$u_t = u_\xi \xi_t , \quad u_x = u_\xi \xi_x .$$

The total differential on a characteristic line is

$$dx = d\xi(1+u_\xi t) + u\, dt .$$

We find ξ_t from this when dx=0 and ξ_x when dt=0 . Hence

$$u_t = \frac{-uu_\xi}{1+u_\xi t} , \quad u_x = \frac{u_\xi}{1+u_\xi t}$$

which implies Burger's equation. The blow-up time is

$$t_B = -\frac{1}{u_\xi}(\xi_B,0)$$

where ξ_B is the value ξ for u_ξ is most negative, $u_{\xi\xi}(\xi_B)=0$. The point (x_B,t_B) where $x_B = \xi_B+u_\xi(\xi_B)t_B$ is the tip of cusp formed by intersecting tangent vectors as in Figure 20.2. The cusp is envelope of the characteristics (1).

Obviously we get blow-up if $u_\xi(\xi,0)$ is ever negative. This is not true if there is damping. Consider Burger's equation with damping proportional to velocity

$$u_t + uu_x + au = 0 \tag{2}$$

where a is a positive constant. Differentiating this with respect to x we get

$$u_{xt} + uu_{xx} + u_x^2 + au_x = 0$$

which may be written as an ordinary differential equation

$$\frac{du_x}{dt} + u_x^2 + au_x = 0$$

on characteristics $x = \tilde{x}(\xi,t)$ where again $d\tilde{x}/dt = u$. It follows that blow-up occurs if and only if

$$u_x < -\frac{1}{a} \,.$$

There is a threshold amplitude $-1/a$; when u_x is larger than this a blow-up cannot occur.

Burger's equation (2) with damping may be rewritten as

$$\frac{du}{dt} = -au \,.$$

The solution of the initial value problem for $u(\tilde{x}(\xi,t),t)$ is

$$u = e^{-at}u(\xi,0)$$

and

$$\frac{d\tilde{x}}{dt} = e^{-at}u(\xi,0) \,.$$

We get

$$\tilde{x}(\xi,t) = \xi + \frac{1-e^{-at}}{a}\, u(\xi,0) \,. \tag{3}$$

The nonlinearity gives the typical distortion of the wave profile, but simultaneously the wave is damped due to the presence of the undifferentiated term in the equation.

Consider the next question of breaking. This may be investigated by seeing whether the characteristic curves (3) have an envelope. An envelope of these curves satisfies the derivative of (3) with respect to the parameter ξ:

$$0 = 1 + \frac{1-e^{-at}}{a}\, u_\xi(\xi,0) \,.$$

Since a and t are positive, this is possible if and only if

$$u_\xi(\xi,0) < -a \,.$$

Thus breaking occurs if and only if the initial curve has a large enough negative slope: the damping may prevent breaking.

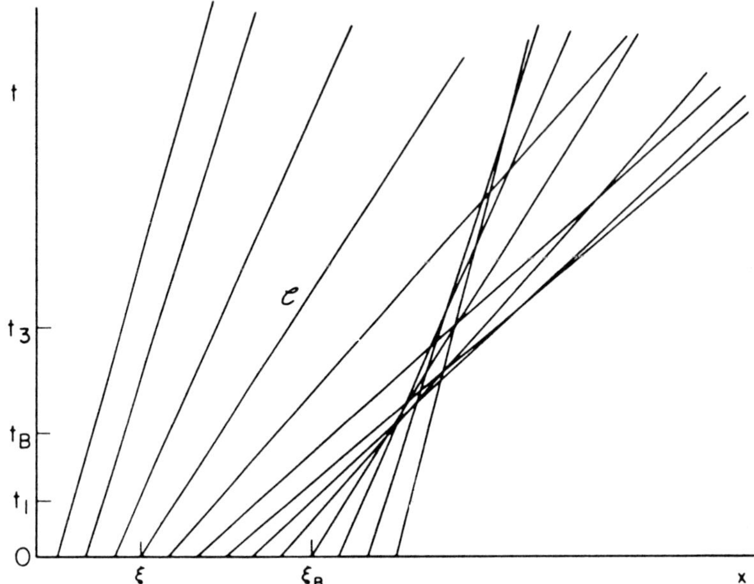

Figure 20.1. Characteristic curves $x=\xi+u(\xi,0)t$ for Burger's equation. The cusp centered at blow-up (x_B,t_B) is an envelope of neighboring characteristics. (After Whitham, [1974])

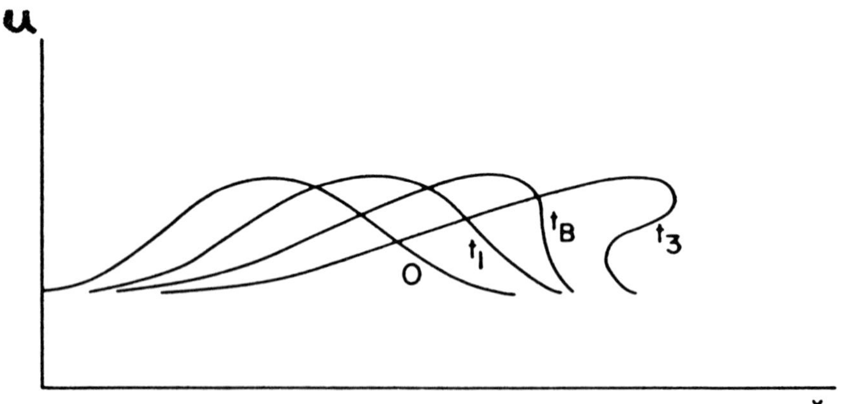

Figure 20.2. Breaking wave at blow-up of Burger's equation. (After Whitham, [1974])

20.2 One-dimensional unsteady shearing problems

We are considering one-dimensional shearing motions in the plane

$$\mathbf{u} = \mathbf{e}_x u(y,t),$$

$$[\tau] = \begin{pmatrix} \tau_{xx} & \tau_{xy} & 0 \\ \tau_{xy} & \tau_{yy} & 0 \\ 0 & 0 & \tau_{zz} \end{pmatrix} = \begin{pmatrix} \sigma & \tau & 0 \\ \tau & \gamma & 0 \\ 0 & 0 & \delta \end{pmatrix}.$$

The equation of motion for u(y,t) is

$$\rho u_t = \tau_y + p \tag{4}$$

where p is a constant.

Coleman and Noll [1961] showed that this flow is determined by three functionals on the history of the relative Cauchy tensor, the stress functional T, the extra tension normal stress functional N_1 and the second normal stress functional N_2, where

$$\tau(t) = \mathop{T}_{s=0}^{\infty} [\Lambda(t,s)] ,$$

$$\sigma(t) - \delta(t) = \mathop{N_1}_{s=0}^{\infty} [\Lambda(t,s)] , \tag{5}$$

$$\gamma(t) - \delta(t) = \mathop{N_2}_{s=0}^{\infty} [\Lambda(t,s)] .$$

where T is an odd functional and N_1 and N_2 are even functionals of

$$\Lambda(t,s) = -\int_{t-s}^{t} u_y(y,\tau)d\tau , \tag{6}$$

the time dependent part of the gradient of the backward path line

$$\boldsymbol{\chi}_t(\mathbf{x},t-s) = \mathbf{e}_x[x - \int_{t-s}^{t} u(y,\tau')d\tau'] + \mathbf{e}_y y + \mathbf{e}z \tag{7}$$

which determines $\mathbf{G}(s) = \nabla\boldsymbol{\chi}_t^T\nabla\boldsymbol{\chi}_t - \mathbf{1}$.

Basically the stress for shearing problems can depend only on four stress components. The trace of $\boldsymbol{\tau}$ can be placed in the isotropic part of the stress with the pressure, hence only three functions can be determined as constitutive relations. The other component, like the pressure, is determined by the problem and its relation to deformations cannot be specified a-priori.

Notation. By $\tau(t)$, $u_y(t)$, etc. we mean $\tau(y,t)$, $u_y(y,t)$, etc. the dependence on y is not made explicit. To simplify notations, we will drop the notation indicating the dependence of the functional on the history; for example, $T[\Lambda] = T\,[\Lambda(t-s)]_{s=0}^{\infty}$. The time derivative of this

$$\begin{aligned}\partial_t T &= T_1[\Lambda(t,s')|\partial_t\Lambda(t,s)] \\ &= T_1[\Lambda|\partial_t\Lambda(t-s)]\end{aligned} \tag{8}$$

is a linear functional on $L_h^2(0,\infty)$ of

$$\partial_t\Lambda = u_y(t-s) - u_y(t) \tag{9}$$

evaluated at the history of $\Lambda(t,s)$. The second time derivative is

$$\partial_{tt}T = T_1[\Lambda|\partial_{tt}\Lambda(t,s)] + T_2[\Lambda|\partial_t\Lambda(t,s)|\partial_t\Lambda(t,s')] \tag{10}$$

which is the sum of a linear functional of

$$\partial_{tt}\Lambda = u_{yt}(t-s) - u_{yt}(t) \tag{11}$$

on $L_h^2(0,\infty)$ and a bilinear functional T_2. T_1 is the first functional derivative of T

$$T_1[\Lambda|\phi(s)] = \left.\frac{d}{d\varepsilon}\right|_{\varepsilon=0} T\,[\Lambda(s) + \varepsilon\phi(s)]$$

and T_2 is the second functional derivative of T

$$T_2[\Lambda|\phi(s_1)|\psi(s_2)] = \left.\frac{\partial^2 T}{\partial\varepsilon_2\partial\varepsilon_1}[\Lambda(s) + \varepsilon_1\phi(s) + \varepsilon_2\psi(s)]\right|_{\varepsilon_1=0,\varepsilon_2=0}.$$

After combining (8) and (9) we get

$$\partial_t T = T_1[\Lambda|u_y(t-s)] - u_y(t)\, T_1[\Lambda|1]\,. \tag{12}$$

After combining (10) and (11) we get

$$\partial_{tt} T = T_1[\Lambda|u_{yt}(t-s)] - u_{yt} T_1[\Lambda|1] + u_y^2 T_2[\Lambda|1|1]$$

$$- 2u_y T_2[\Lambda|u_y(t-s)|1] + T_2[\Lambda|u_y(t-s)|u_y(t-s)]\,. \tag{13}$$

Now we may apply the Riesz theorem for representing linear functionals on $L^2_h(0,\infty)$

$$T_1[\Lambda|u_y(t-s)] = \int_0^\infty \partial_s G\big(\overset{\infty}{\underset{s'=0}{\Lambda}}(t,s'),s\big)\, u_y(t-s)\, ds \tag{14}$$

and

$$T_1[\Lambda|1] = -\,G\big(\overset{\infty}{\underset{0}{\Lambda}}(t,s'),0\big) \tag{15}$$

depends in general on t and y. We may write, in the same way

$$T_2[\Lambda|u_y(t-s)|1] = -\int_0^\infty \partial_s \gamma\big(\overset{\infty}{\underset{s'=0}{\Lambda}}(t,s'),s\big)\, u_y(t-s)\, ds \tag{16}$$

and

$$T_2[\Lambda|1|1] = \gamma\big(\overset{\infty}{\underset{s'=0}{\Lambda}}(t,s'),0\big)\,.$$

Under certain special conditions, say T_2 is a Hilbert-Schmidt operator on $L^2_h(0,\infty)$ we may write T_2 as a double integral

$$T_2[\Lambda|u_y(t-s)|u_y(t-s)]$$

$$= \int_0^\infty \int_0^\infty \frac{\partial^2 \underset{s'=0}{\overset{\infty}{\nu}}\left(\Lambda(t,s'),s_1,s_2\right)}{\partial s_1 \partial s_2} u_y(t-s_1)u_y(t-s_2)\, ds_1\, ds_2 \qquad (17)$$

where

$$T_2[\Lambda|\phi(s_1)|\psi(s_2)] = T_2[\Lambda|\psi(s_1)|\phi(s_2)]$$

is a symmetric operator. Of course

$$T_2[\Lambda|u_y(t-s)|1] = -\int_0^\infty \frac{\partial \underset{s'=0}{\overset{\infty}{\nu}}\left(\Lambda(t,s'),s,0\right)}{\partial s} u_y(t-s)\, ds\ . \qquad (18)$$

Hence

$$\underset{s=0}{\overset{\infty}{\gamma}}\left(\Lambda(t,s),0\right) = \underset{s=0}{\overset{\infty}{\nu}}\left(\Lambda(t,s),0,0\right)\ . \qquad (19)$$

Collecting all the results we get

$$\tau_t(t) = u_y(t)\, G(\Lambda,0) + \int_0^\infty \partial_s G(\Lambda,s)\, u_y(t-s)\, ds \qquad (20)$$

and

$$\tau_{tt}(t) = u_{yt}(t) G(\Lambda,0) + \int_0^\infty \partial_s G(\Lambda,s) u_{yt}(t-s)\, ds$$

$$+ \gamma(\Lambda,0) u_y^2(t) + 2u_y(t) \int_0^\infty \partial_s \gamma(\Lambda,s)\, u_y(t-s)\, ds$$

$$+ \int_0^\infty \int_0^\infty \frac{\partial^2 \nu(\Lambda,s_1,s_2)}{\partial s_1 \partial s_2} u_y(t-s_1)\, u_y(t-s_2)\, ds_1\, ds_2\ . \qquad (21)$$

One further reduction of the second term on the right of (21) is useful. We note that

$$u_{yt}(t-s) = -u_{ys}(t-s)$$

and integrate by parts

$$\int_0^\infty \partial_s G(\Lambda,s)\, u_{ys}(t-s)\, ds = -G'(\Lambda,s)u_y(t) - \int_0^\infty G''(\Lambda,s)u_y(t-s)\, ds$$

where primes designate derivatives with respect to s, the variable in the second place. Hence

$$\begin{aligned} \tau_{tt}(t) &= G(\Lambda,0)u_{yt}(t) + G'(\Lambda,0)u_y(t) + \int_0^\infty G''(\Lambda,s)u_y(t-s)\, ds \\ &\quad + \gamma(\Lambda,0)u_y^2(t) + 2u_y(t)\int_0^\infty \partial_s\gamma(\Lambda,s)u_y(t-s)\, ds \qquad (22) \\ &\quad + \int_0^\infty\int_0^\infty \frac{\partial^2 \nu(\Lambda,s_1,s_2)}{\partial s_1 \partial s_2}\, u_y(t-s_1)\, u_y(t-s_2)\, ds_1\, ds_2 \,. \end{aligned}$$

In the special case of steady flow the deformation gradient $\Lambda(t,s)=-\kappa s$, where $\kappa(y)=u_y(y)$ is independent of t. Then we may simplify by replacing the functional dependence of kernel on $\Lambda(t,s)$ with functions of κ and s. For example, we may write in a loose notation

$$G(\kappa,s) = G(\Lambda(t,s'),s)\Big|_{s'=0}^{\infty} .$$

The equation

$$\partial_t\tau = G(\kappa,0)\, u_y(t) + \int_0^\infty G'(\kappa,s)\, u_y(t-s)\, ds \qquad (23)$$

governs the stress perturbing steady shear. Together with (4), this implies that waves propagate into a sheared region with a speed given by

$$c = \sqrt{G(\kappa,0)/\rho}\ . \tag{24}$$

This speed varies with y if $\kappa=\kappa(y)$.

20.3 Acceleration waves (evolution of jumps in the vorticity)

The velocity is $\mathbf{u}=\mathbf{e}_x u(y,t)$. The jump of acceleration across $y=\phi(t)$ is finite

$$a(t) \overset{\text{def}}{=} u_t(y_+,t) - u_t(y_-,t) \overset{\text{def}}{=} [\![u_t]\!](\phi(t),t) \neq 0 \tag{25}$$

where y_+ is on the plus side and y_- is on the minus side of $y=\phi(t)$ at the time t. The region ahead of the wave is steady, $u(y,t) = u(y)$ for $y>\phi$ as in Figure 20.3. Of course a(t) could not be finite across $y=\phi$ if u(y,t) is discontinuous. Hence

$$[\![u]\!] = 0\ . \tag{26}$$

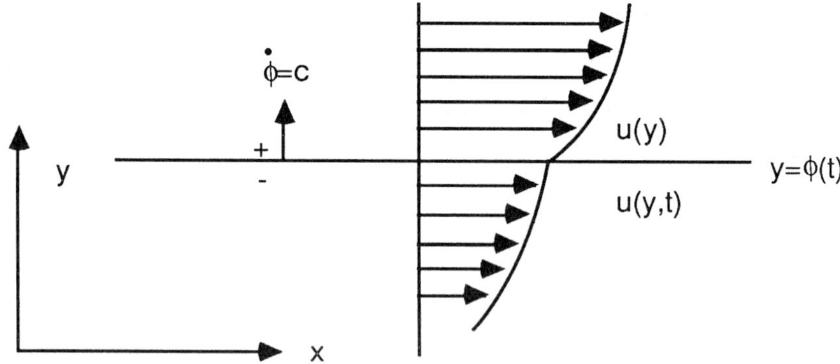

Figure 20.3 Acceleration wave.

The conservation of momentum (3.22) gives

$$[\![\tau]\!] = 0\,. \tag{27}$$

The only nonzero jumps occur across $y=\phi(t)$ at time t. Otherwise all the variables and their derivatives are continuous. Hence, for any f,

$$[\![f]\!](\phi(t),\tau) = 0 \text{ if } \tau \neq t\,. \tag{28}$$

Moreover

$$\int_{t-s}^{t} \big(f(y_+,\tau) - f(y_-,\tau)\big)\, d\tau = 0 \tag{29}$$

because even though this difference is non-vanishing at the end point $\tau=t$, it makes no contribution to the integral.

The equations

$$[\![u]\!](\phi(t),t) = 0\,, \tag{30}$$

$$[\![\tau]\!](\phi(t),t) = 0 \tag{31}$$

are identities in t. Hence, each and every time derivative of these identities vanishes. The first derivative gives

$$[\![u_t]\!] + c[\![u_y]\!] = 0 \tag{32}$$

and

$$[\![\tau_t]\!] + c[\![\tau_y]\!] = 0 \tag{33}$$

where $c = \dot{\phi}$. Equation (32) shows that an acceleration wave is a wave of vorticity.

It follows now from (32) that

$$[\![u_y]\!] = -\,a/c \neq 0\,. \tag{34}$$

Despite this

$$[\![\Lambda]\!] = 0 \tag{35}$$

where

$$\Lambda = \int_t^{t-s} u_y(y,\tau)\, d\tau$$

because of (29).

We now form the jumps of τ_t and τ_{tt} given by (20) and (22). The jump of each and every integral vanishes because the integrands of the jumps are non-zero, except at s=0, where (29) applies. Then

$$[\![\tau_t]\!] = G(\kappa)\, [\![u_y]\!] \tag{36}$$

and

$$[\![\tau_{tt}]\!] = G(\kappa)\, [\![u_{yt}]\!] + G'(\kappa)[\![u_y]\!] + \gamma(\kappa)[\![u_y^2]\!] - 2[\![u_y]\!]\kappa\gamma(\kappa) \tag{37}$$

We replaced the kernels in (20) and (22), which depend functionally on the history of $\Lambda(t,s)$, with new kernels which are ordinary functions of κ because by (35) Λ is continuous across $y = \phi$ (cf. (23)). We also used

$$[\![u_y(t) \int_0^\infty \partial_s\gamma(\Lambda,s)\, u_y(t-s)\, ds]\!] = -\kappa[\![u_y]\!]\gamma(\Lambda,0) = -\,\kappa[\![u_y]\!]\gamma(\kappa). \tag{38}$$

Equation (37) may be simplified further

$$\begin{aligned}[\![u_y^2]\!] &= \big(u_y(y_+,t)-u_y(y_-,t)\big)\big(u_y(y_+,t)+u_y(y_-,t)\big) \\ &= [\![u_y]\!]\big(\kappa+u_y(y_-,t)\big) = [\![u_y]\!](2\kappa-[\![u_y]\!])\,.\end{aligned}$$

Hence

$$[\![\tau_{tt}]\!] - G(\kappa)[\![u_{yt}]\!] = G'(\kappa)[\![u_y]\!] - \gamma(\kappa)[\![u_y]\!]^2\,. \tag{39}$$

Now we shall find an expression for the wave speed. From (33) and (36) we get

$$G(\kappa)[\![u_y]\!] + c[\![\tau_y]\!] = 0\,. \tag{40}$$

Equations (40) and (32) imply that

$$G(\kappa)[\![u_t]\!] = c^2[\![\tau_y]\!]\,. \tag{41}$$

The equation of motion (4) implies that

$$\rho[\![u_t]\!] = [\![\tau_y]\!] \, . \tag{42}$$

Equations (41) and (42) show that

$$c^2 = G(\kappa)/\rho \, . \tag{43}$$

The velocity c is the speed of a wave of acceleration, or alternatively, a wave of vorticity into a region under shear.

We turn next to the determination of the equation which governs the amplitude a(t) of this wave. Since by (32), $[\![u_y]\!] = -a/c$ we can get an amplitude equation from (39) provided that $[\![\tau_{tt}]\!]$ and $[\![u_{yt}]\!]$ can be expressed in terms of a. The strategy is to get an independent expression for the left side of (39), depending on a and c alone from the other equations. The ones we need are from the first time derivative of (4).

$$\rho[\![u_{tt}]\!] = [\![\tau_{yt}]\!] \, . \tag{44}$$

From (33) and (42) we get $\rho ac = -[\![\tau_t]\!]$ and the time derivative of this is

$$-\rho \frac{d(ac)}{dt} = [\![\tau_{tt}]\!] + c[\![\tau_{yt}]\!] \tag{45}$$

and from the time derivative of $a = [\![u_t]\!]$ we have

$$\frac{da}{dt} = [\![u_{tt}]\!] + c[\![u_{ty}]\!] \, . \tag{46}$$

After eliminating $[\![\tau_{yt}]\!]$ and $[\![u_{tt}]\!]$ from (44), (45), and (46) we get

$$\begin{aligned} -[\![\tau_{tt}]\!] + \rho c^2 [\![u_{ty}]\!] &= -[\![\tau_{tt}]\!] + G(\kappa)[\![u_{ty}]\!] \\ &= \rho \frac{d(ac)}{dt} + \rho c \frac{da}{dt} = 2\rho\sqrt{c}\,\frac{d(a\sqrt{c})}{dt} \, . \end{aligned} \tag{47}$$

Combining now (47) and (39) we get

$$2\rho\sqrt{c}\,\frac{d(a\sqrt{c})}{dt} = G'(\kappa)\,\frac{a}{c} + \gamma(\kappa)\left(\frac{a}{c}\right)^2 . \tag{48}$$

Equation (48) is the amplitude equation for the acceleration (or vorticity) wave.

We next consider the mathematical implications of (48) for the growth or decay of shocks of vorticity into rest. We first note that $\kappa(y) = \kappa(\phi(t))$ at the front is independent of s but depends on t. We may write (48) as

$$\frac{db}{dt} + \alpha(t)b = \beta(t)\, b^2 , \tag{49}$$

$$b = a\sqrt{c} ,$$

$$\alpha = -\, G'(\kappa)/2\rho c^2 ,$$

$$\beta = \gamma(\kappa)/2c^{3/2}G(\kappa) .$$

Equation (49) is a Ricatti equation, the solution of which is

$$b = \frac{b_0 e^{-\psi(t)}}{1 - b_0 \int_0^t \beta(\tau) e^{-\psi(\tau)}\, d\tau} \tag{50}$$

where $b_0 = b(t)$ at $t=0$ and

$$\psi(t) = \int_0^t \alpha(\tau)\, d\tau \geq 0 . \tag{51}$$

Equation (50) shows that when $\beta(\tau)>0$ the vorticity shock blows up in finite time.

To investigate the properties of (50) we consider the special case of propagation into a region of uniform shear, in which $\kappa(y)$ is independent of y and $\kappa(\phi(t))$ independent of ϕ. Then

$$\psi(t) = \alpha t$$

and

$$b_0 \int_0^t \beta e^{-\alpha\tau}\, d\tau = \frac{-b_0\beta}{\alpha}(e^{-\alpha t} - 1) = -\frac{a_0}{\lambda}(e^{-\alpha t} - 1)$$

where $\lambda = 2c^2/\gamma$. Then

$$a(t) = \frac{\lambda}{\left[\left(\frac{\lambda}{a_0} + 1\right) e^{\alpha t} - 1\right]} \quad . \tag{52}$$

Clearly, the behavior of a(t) is determined by the sign of λ; that is, the sign of γ, the kernel for the second functional derivative (19) of the shear stress function T.

Equation (17) shows that the second derivative $T_2[-\kappa s|\kappa|\kappa]$ is positive when $\gamma>0$; that is, when the fluid shear thickens, as is in Figure 20.4.

The properties of (52) are as follows:

- The vorticity shock a(t) either decays monotonically to zero or blows up in a finite time t_1 such that

$$\left(\frac{\lambda}{a_0} + 1\right) e^{\alpha t_1} - 1 = 0 \, .$$

- If $0 < \frac{\lambda}{a_0} + 1 < 1$, then we get a blow-up in finite time.

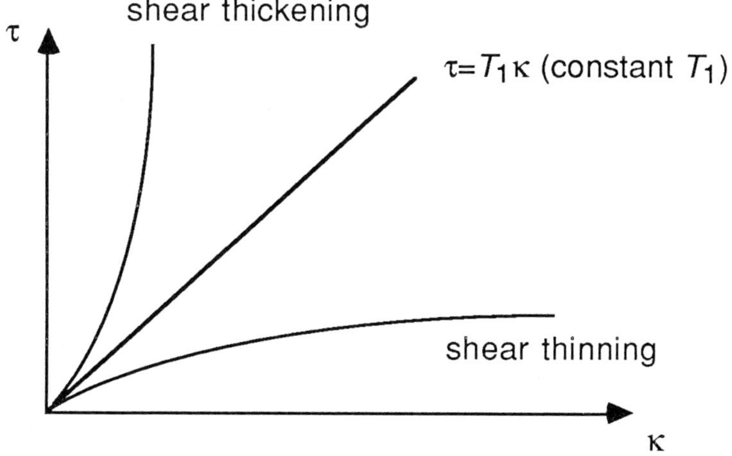

Figure 20.4. The shear stress increases faster than linear $\tau = T_1\kappa$ when it is shear thickening, less when it is shear thinning.

- Small data $|a_0|<|\lambda|$ decays monotonically to zero. Large data $|a_0|>|\lambda|$ with a suitable sign will blow up in finite time.

The interested reader may verify that a vorticity shock initiated at the wall of a pipe will blow up if the fluid is shear thickening. A vorticity shock initiated in the interior and moving toward the wall of a pipe will blow up if the fluid is shear thinning. It has been suggested that the blow-up of vorticity shocks leads to forms of extrudate distortion of melts which are known collectively as melt fracture (Tordella, [1969]), but the evidence is not clear. One problem is how the vorticity shock gets into the fluid in the first place. A second problem is that unlike the x independent solution, the forms of melt fracture depend strongly on x, with fissures and fractures all up and down the axis. A third problem is that if the glassy modulus is used without an effective viscosity, as in the theory, the numbers computed from the theory will not fit the experiment.

20.4 Breakdown of smooth shearing flow in viscoelastic fluids for two constitutive relations: The velocity shock versus the vorticity shock

M. Slemrod [1978] was the first to give an example of a constitutive equation with instantaneous elasticity which gives rise to a blow-up of the vorticity in a finite time. This result suggests the formation of a shock in the velocity as a result of the blow-up of the vorticity. In a later work, Slemrod [1985] showed that if one used a different constitutive equation with instantaneous elasticity, first derivatives of the vorticity will blow up in finite time, suggesting the formation of a vortex shock. These results show how strongly dynamics are tied to the selection of a constitutive equation even when there is no effective viscosity.

The first problem of Slemrod was for the constitutive equation

$$\tau = f(\zeta) \tag{53}$$

where $f(\bullet)$ is a nonlinear odd function of the linear functional

$$\zeta = \int_0^\infty \exp\left(-\frac{s}{\lambda}\right) u_y(y,t-s)\, ds\ . \tag{54}$$

The second problem of Slemrod was for a constitutive equation of the form

$$\tau = \int_0^\infty \exp\left(-\frac{s}{\lambda}\right) f(u_y(y,t-s))\, ds \tag{55}$$

in which τ is a linear function of a nonlinear function.

Slemrod [1985] remarks that sometime after the appearance of his [1978] paper

> ...C.J.S. Petrie remarked that a similar result (similar to blow-up for (53) could (be obtained for (55)).
>
> Upon working out Petrie's idea I noticed the results are similar but *not* the same. Specifically the above mentioned result holds with C^1 replaced by C^2. D.D. Joseph observed that this may have important physical consequences as now the appearance of jump discontinuity in $\hat{v}_x$ may be expected, i.e. formation of a *vortex shock.*

Equation (55) is the first integral of the rate equation

$$\tau_t + \frac{\tau}{\lambda} = f(u_y(y,t))\ . \tag{56}$$

On the other hand (54) leads to

$$\tau_t = f'(\zeta)\,\{u_y(y,t) - \frac{1}{\lambda}\zeta\}\ . \tag{57}$$

Simple jumps of derivatives of τ and u satisfying (56) are governed by

$$[\![\tau_t]\!] = [\![f(u_y,t)]\!] \tag{58}$$

or if τ and u satisfy (57) then

$$[\![\tau_t]\!] = f'(\zeta)[\![u_y]\!]\ . \tag{59}$$

The characteristics for the system arising from (58) and $\rho[\![u_t]\!]=[\![\tau_y]\!]$ are determined by the instantaneous values of $\tau(y,t)$ and $u(y,t)$. On the other hand the characteristic form of the system based on (59) leads to

$$\det\begin{pmatrix} c & f' \\ 1 & \rho c \end{pmatrix} = \rho c^2 - f'(\zeta) = 0 .$$

Hence the wave speed

$$c = \frac{dy}{dt} = \sqrt{\frac{f'(\zeta)}{\rho}} \tag{60}$$

depends on the history of the motion through the integral ζ defined in (54).

Now we shall follow Slemrod's analysis. He considers shearing flow in a channel of height h and requires the no-slip boundary at the stationary wall $u(0,t)=0$ and $u(h,t)=U$ is the prescribed velocity of the upper wall. A steady solution of this problem is simple shear

$$u(y) = Uy/h . \tag{61}$$

Let

$$v(y,t) = u(y,t) - \frac{Uy}{h} \tag{62}$$

be a disturbance of simple shear satisfying the boundary conditions

$$v(0,t) = v(h,t) = 0 . \tag{63}$$

A smooth velocity history consistent with the governing equations and boundary conditions is prescribed

$$v(y,t) = v_0(y,t), \ -\infty \le t \le 0 . \tag{64}$$

Slemrod's first problem leading to blow-up of vorticity is based on (53) and (54). The equation of motion

$$\rho\partial_t u = \partial_y f(\zeta) \tag{65}$$

is transformed under the change of variables (62) to

$$\rho\partial_t u = \partial_y f\left[\lambda\frac{U}{h} + \int_0^\infty \exp\left(\frac{-s}{\lambda}\right)\partial_y v(y,(t-s))\, ds\right]. \tag{66}$$

Now we convert (66) into a first order system depending on present values

$$q(y,t) \overset{\text{def}}{=} \int_0^\infty \exp\left(-\frac{s}{\lambda}\right)\partial_t v(y,t-s)\, ds\ , \tag{67}$$

$$p(y,t) \overset{\text{def}}{=} \int_0^\infty \exp\left(-\frac{s}{\lambda}\right)\partial_y v(y,t-s)\, ds\ . \tag{68}$$

We see that $p=\zeta$, defined by (54). After integrating by parts we get

$$q(y,t) = v(y,t) - \frac{1}{\lambda}\int_0^\infty \exp\left(-\frac{s}{\lambda}\right)v(y,t-s)\, ds \tag{69}$$

Hence

$$\partial_t q = \partial_t v - \frac{q}{\lambda}\ . \tag{70}$$

We can replace the integral argument of f(•) with p and write

$$\partial_t v = \partial_y F(p) \tag{71}$$

where

$$F(p) \overset{\text{def}}{=} \frac{1}{\rho} f\left[\lambda\frac{U}{h} + p\right] - \frac{1}{\rho} f\left[\lambda\frac{U}{h}\right].$$

The first order system we need is obtained from (70) and (71)

$$\partial_t q = \partial_y F(p) - \frac{q}{\lambda} \tag{72}$$

and the equation

$$\partial_y q = \partial_t p \tag{73}$$

which arises from cross differentiation of (67) and (68). Equations (72) and (73) are to be solved subject to the conditions

$$q(0,t) = q(h,t) = 0 \tag{74}$$

and the initial conditions

$$\begin{pmatrix} q(y,0) \\ p(y,0) \end{pmatrix} = \begin{pmatrix} q_0(y) \\ p_0(y) \end{pmatrix} . \tag{75}$$

The initial conditions can be computed from the prescribed initial history. Equations (72) through (75) are in canonical form for the study of blow-up. For the moment we set aside these equations and show how the same system of equations arises out of Slemrod's second problem.

Slemrod's second problem is associated with the constitutive equation (56) which using (62) may be written as

$$\tau_t + \frac{\tau}{\lambda} = \rho g[v_y(y,t)] \tag{76}$$

where

$$\rho g[v_y(y,t)] = f\left[v_y(y,t) + \frac{U}{h}\right] - f\left[\frac{U}{h}\right], \; g[0] = 0 . \tag{77}$$

After differentiating (76) with respect to y, using $\rho v_t = \tau_y$, $\rho u_{tt} = \tau_{yt}$ we get

$$u_{tt} + \frac{1}{\lambda} v_t = g_y[v_y(y,t)] = g'[v_y(y,t)]\; v_{yy}(y,t) . \tag{78}$$

Now define

$$\begin{aligned} q &= v_t , \\ p &= u_y . \end{aligned} \tag{79}$$

Then (78) becomes

$$\partial_t q = \partial_y g(p) - \frac{1}{\lambda} q \tag{80}$$

and, of course,

$$\partial_y q = \partial_t p \tag{81}$$

where the new q and p also satisfy (74) and (75).

Now we shall introduce Riemann invariants for systems of the form (72) to (75). We have reduced Slemrod's two problems to this form. We have to guarantee that the system is hyperbolic. This means that

$$F'(p) > 0 \quad (\text{and } g'(p)>0) .$$

The system has to be genuinely nonlinear with $f''(p_0)=0$ for some real p_0. We shall see in §20.4 that the upper and lower convected models are not genuinely nonlinear, but the interpolated models are genuinely nonlinear. If we choose

$$U = \frac{h}{\lambda} p_0 .$$

Then

$$F''(p) = \frac{1}{\rho} f''(p_0+p) .$$

Hence

$$F(0) = 0$$

and

$$F''(0) \neq 0 \tag{82}$$

(also $g(0)=0$ and $g''(0)\neq 0$).

Hyperbolic systems of the form (72) through (75) and (82) admit Riemann invariants

$$\binom{r}{s} = q \pm \Phi(p) \tag{83}$$

where

$$\Phi(p) = \int_0^{\infty} \sqrt{F'(s)}\, ds. \tag{84}$$

The transformation between (r,s) and (q,p) given by (83) is one to one.

Slemrod's [1978] result is expressed in terms of the Riemann invariants r and s whose initial values are $r_0(x)$ and $s_0(x)$. He proves that if $|r_0|$ and $|s_0|$ are small, $F(0)>0$, $F''(0)>0$ and if either $r_{0,x}$ or $s_{0,x}$ is positive and large, then (80) and (81) have a solution (p,q) in $C'[0,h]\times C'[0,h]$ for only a finite time. A similar result holds if $F''(0)<0$ and $r_{0,x}$ or $s_{0,x}$ is sufficiently negative at any point x. This implies that $|p_t|+|p_x|+|q_t|+|q_x|\to\infty$ in a finite time and suggests but does not prove the occurrence of a jump discontinuity in p and q.

It follows then that the quasilinear model (53) and (54) leads to blow-up of vorticity (velocity shock), whilst the nonlinear model (55) leads to blow-up of a derivative of vorticity (vorticity shock). Similar results for other models were proved by Renardy *et al.* [1987]. In general, the nonlinear first order systems lead to weaker shocks than quasilinear first order systems (Joseph and Saut, [1986]).

20.5 Equations of motion for interpolated Maxwell models

$$\sigma_t - \tau u_y(1+a) + \frac{\sigma}{\lambda} = 0\,, \tag{85}$$

$$\tau_t + \left\{\frac{1-a}{2}\sigma - \frac{1+a}{2}\gamma - G\right\} u_y + \frac{\tau}{\lambda} = 0\,, \tag{86}$$

$$\gamma_t + \tau u_y(1-a) + \frac{\gamma}{\lambda} = 0\,. \tag{87}$$

Equations (4) and (85), (86), and (87) are four equations for (u,τ,σ,γ). The second equation of momentum may be satisfied by taking

$$p_y = \gamma_y \tag{88}$$

If $a\neq\pm1$, then

$$\tau u_y = \frac{\gamma_t+\gamma/\lambda}{a-1} = \frac{\sigma_t+\sigma/\lambda}{1+a}\,. \tag{89}$$

We may write the governing system of quasilinear equations for (u,τ,σ) as

$$\begin{bmatrix} 1 & 0 & 0 \\ 0 & 1 & 0 \\ 0 & 0 & \rho \end{bmatrix}\begin{bmatrix} \sigma_t \\ \tau_t \\ u_t \end{bmatrix} + \begin{bmatrix} 0 & 0 & -\tau(1+a) \\ 0 & 0 & \sigma(1-a)-G \\ 0 & -1 & 0 \end{bmatrix}\begin{bmatrix} \sigma_y \\ \tau_y \\ u_y \end{bmatrix} = -\frac{1}{\lambda}\begin{bmatrix} \sigma \\ \tau \\ 0 \end{bmatrix} \tag{90}$$

where γ is given by (87). Analysis of the characteristics for (90) reveals that

$$\frac{dy}{dt}\left[\rho\left(\frac{dy}{dt}\right)^2 - G + (1-a)\sigma\right] = 0\,. \tag{91}$$

The solutions of (91) are y=const, streamlines are characteristic and

$$c = \frac{dy}{dt} = \pm\sqrt{\frac{G}{\rho} + (a-1)\frac{\sigma}{\rho}}\,. \tag{92}$$

Since the speed depends on $\sigma(y)$, we may consider the possibility that it is possible to get a shock up from smooth data. If a=1 that speed of waves into the sheared medium is the same as the speed of waves into rest; shocks are impossible. Moreover, since $\gamma_t+\gamma/\lambda=0$, γ is zero if it is zero initially and (86) is then linear, identical to the problem perturbing rest. If a=–1, then $\sigma_t + \sigma/\lambda = 0$; if $\sigma=0$ initially it is always zero and again (86) is linear and $c=\sqrt{G/\rho}$, $G=\eta/\lambda$.

Let us suppose now that $a^2\neq 1$. Then we use (89) to justify writing

$$\gamma = \frac{a-1}{a+1}\,\sigma\,. \tag{93}$$

Using (93) in (86) we find that

$$\tau_t + (1-a)\sigma u_y - G u_y + \frac{\tau}{\lambda} = 0\,. \tag{94}$$

Johnson and Segalman [1977] have shown (85) and (94) may be integrated once.

$$\tau = G \int_{-\infty}^{t} e^{\frac{-(t-t')}{\lambda}} \kappa(t') \cos\left\{\sqrt{1-a^2} \int_{t'}^{t} \kappa(s')\, ds'\right\} dt' \qquad (95)$$

and

$$\sigma = G\sqrt{\frac{1+a}{1-a}} \int_{-\infty}^{t} e^{\frac{-(t-t')}{\lambda}} \kappa(t')\sin\left\{\sqrt{1-a^2} \int_{t'}^{t} \kappa(s')\, ds'\right\} dt' \quad (96)$$

where $\kappa(t) = u_y(y,t)$. We may write (95) and (96) as functionals of $\Lambda(t,s)$. For example, in (95) we put $t'=t-s$ and write

$$\kappa(t') = \kappa(t-s) = u_y(t-s) = -\partial_s\Lambda(t,s)\,. \qquad (97)$$

Then (95) becomes

$$\tau = -G \int_{0}^{\infty} e^{-s/\lambda} \partial_s\Lambda(t,s) \cos\left\{\Lambda(t,s)\sqrt{1-a^2}\right\} ds\,. \qquad (98)$$

The rate equation corresponding to (98) may be obtained by differentiating:

$$\tau_t + \frac{\tau}{\lambda} = Gu_y(y,t) + G\sqrt{1-a^2} \int_{0}^{\infty} e^{-s/\lambda} u_y(y,t-s) \qquad (99)$$

$$\sin\left\{\sqrt{1-a^2} \int_{t}^{t-s} u_y(y,s')\, ds'\right\} ds\,.$$

Equation (99) is nonlinear in u_y, but the stress σ in (94) has been eliminated.

20.6 Shock relations for Maxwell models (Rankine-Hugoniot conditions)

We are going to study shock relations which we shall assume to be bounded jumps in the velocities and stresses. We derived the consequences of the basic laws of conservation in Chapter 3. Here we make use of the relation

$$m = \rho_1 v_1 = \rho_2 v_2 \,.$$

Since we are assuming that the fluids are incompressible $\rho_1 = \rho_2$ and

$$v_1 = v_2 \overset{\text{def}}{=} v = (\mathbf{u} - \mathbf{u}_\Sigma) \cdot \mathbf{n} \tag{100}$$

is the velocity of propagation relative to the discontinuity surface Σ, $\mathbf{u}_\Sigma$ is the shock velocity and $\mathbf{n}$ is the normal to shock surface Σ. If this surface is $\Phi(\mathbf{x},t)=0$, then $\mathbf{n}=\nabla\Phi/|\nabla\Phi|$.

The shock relations derived in Chapter 3 are of the following form. Suppose that every term of the equation

$$\rho \frac{d\mathbf{Q}}{dt} + \operatorname{div} \mathbf{P} = \mathbf{g} \tag{101}$$

is smooth in V except $\mathbf{Q},\mathbf{P},\mathbf{g}$ suffer simple jumps across Σ. Then

$$m[\![\mathbf{Q}]\!] + [\![\mathbf{P}]\!]n = 0 \,. \tag{102}$$

Suppose τ is governed by the equation for an upper convected Maxwell model

$$\frac{d}{dt}\tau_{ij} - u_{i,k}\tau_{kj} - u_{j,k}\tau_{ik} - \frac{\eta}{\lambda}(u_{i,j} + u_{j,i}) = -\frac{1}{\lambda}\tau_{ij} \,. \tag{103}$$

The velocity is solenoidal and satisfies the equation of motion

$$\frac{d}{dt}\rho u_i + \left(p\delta_{ik} - \tau_{ik}\right)_{,k} = \rho f_i \,. \tag{104}$$

Now multiply (104) by u_j and add the transposed equation to get

$$\frac{d}{dt}(\rho u_i u_j) - u_j\tau_{ik,k} - u_i\tau_{jk,k} = -p_{,i}u_j - p_{,j}u_i + \rho(f_i u_j + f_j u_i) \tag{105}$$

The interesting fact about the UCM is that when (103) and (105) are added the stresses may all be tucked behind derivatives

$$\frac{dQ_{ij}}{dt} + P_{ijk,k} = -p_{,i}u_j - p_{,j}u_i - \frac{1}{\lambda}\tau_{ij} + \rho(f_iu_j + f_ju_i) \tag{106}$$

where

$$Q_{ij} = \rho u_iu_j + \tau_{ij} ,$$

$$P_{ijk} = -u_i\tau_{jk} - u_j\tau_{ik} - \frac{\eta}{\lambda}(u_i\delta_{kj} + u_j\delta_{ki}) .$$

We are now ready to do the identifications that lead to shock relations. First we compare (104) and (101) and use (102) to find that

$$\rho v[\![u_i]\!] = -[\![p]\!]n_i + [\![\tau_{ik}]\!]n_k . \tag{107}$$

In treating (106), we must assume that the right side suffers simple jumps across Σ, at most. Therefore, at worst $p_{,i}$ has a simple jump across Σ; that is

$$[\![p]\!] = 0 \tag{108}$$

and

$$\rho v[\![\mathbf{u}]\!] = [\![\boldsymbol{\tau}]\!]\mathbf{n} . \tag{109}$$

After making the required identification between (105) and (101), we use (102) to find that

$$v[\![\rho u_iu_j + \tau_{ij}]\!] = [\![u_i\tau_{jk}]\!]\, n_k + [\![u_j\tau_{ik}]\!]n_k + \frac{\eta}{\lambda}[\![u_in_j + u_jn_i]\!] . \tag{110}$$

After contracting (110) with n_j, we get

$$v\left\{\rho\mathbf{u}\bullet\mathbf{n}[\![u_i]\!] + [\![\tau_{ij}]\!]n_j\right\} = \mathbf{u}\bullet\mathbf{n}[\![\tau_{ij}]\!]\, n_j + \left(\tau_{nn} + \frac{\eta}{\lambda}\right)[\![u_i]\!]$$

which together with (109) implies that

$$[\![\tau_{ij}n_j]\!]^2 = \left(\tau_{nn} + \frac{\eta}{\lambda}\right)\rho\, [\![u_i]\!]^2 . \tag{111}$$

According to inequalities proved in §1.16, with $\boldsymbol{\xi}=\mathbf{n}$, $\tau_{nn} + \dfrac{\eta}{\lambda}$ is always positive for the UCM, as required by (111).

If $\Phi(\mathbf{x},t) = 0$ is the equation of the shock surface, then

$$\frac{d\Phi}{dt} = \frac{\partial\Phi}{\partial t} + \mathbf{u}_\Sigma\boldsymbol{\cdot}\nabla\Phi = 0$$

and

$$\mathbf{u}_\Sigma\boldsymbol{\cdot}\mathbf{n} = -\frac{1}{|\nabla\Phi|}\frac{\partial\Phi}{\partial t}\,.$$

Hence

$$v = (\mathbf{u} - \mathbf{u}_\Sigma)\boldsymbol{\cdot}\mathbf{n} = \frac{\mathbf{u}\boldsymbol{\cdot}\nabla\Phi}{|\nabla\Phi|} + \frac{1}{|\nabla\Phi|}\frac{\partial\Phi}{\partial t}$$

and v is given by (109) and (111) as

$$v = \pm\left(\frac{1}{\rho}\tau_{nn} + c^2\right)^{\frac{1}{2}} \tag{112}$$

where $c = (\eta/\lambda\rho)^{1/2}$ is the speed of vorticity waves into rest. After squaring v in (109) and (112) writing $\tau_{nn} = \tau_{ij}\Phi_{,i}\Phi_{,j}/|\nabla\Phi|^2$, we find that

$$(\partial_t\Phi)^2 + 2\mathbf{u}\boldsymbol{\cdot}\nabla\Phi\,\partial_t\Phi + (\mathbf{u}\boldsymbol{\cdot}\nabla\Phi)^2 = \frac{1}{\rho}\Phi_{,i}\tau_{ij}\Phi_{,j} + |\nabla\Phi|^2c^2\,. \tag{113}$$

Comparing now (113) with (18) in §5.5 with a=1, we may conclude that the shock surfaces $\Phi=0$ are characteristic surfaces for the vorticity and the relative speed v of shock surfaces coincides with the relative normal speed

$$(\mathbf{u}-\hat{\mathbf{c}})\boldsymbol{\cdot}\mathbf{n} = v \tag{114}$$

of characteristic surfaces defined by (21) of §5.6.

We may now summarize the results which we have derived about propagation of shock discontinuities for the UCM

1. Shock surfaces are characteristic for the vorticity, $\Phi=\phi$, and they propagate with the characteristic speed of vorticity waves

$$v^2 = \frac{1}{\rho}\tau_{nn} + c^2 \tag{115}$$

where $c^2=\eta/\rho\lambda$ and $\tau_{nn}=\mathbf{n}\cdot\boldsymbol{\tau}\cdot\mathbf{n}$, $\mathbf{n}=\nabla\Phi/|\nabla\Phi|$. Usually shock speeds differ from characteristic speeds. This might suggest that shocks cannot develop in an UCM from smooth data as is true in the one-dimensional problem discussed under (92).

2. $$[\![\mathbf{u}]\!]\cdot\mathbf{n} = 0\,, \tag{116}$$

3. $$[\![\tau_{nn}]\!] = 0\,, \tag{117}$$

4. $$[\![p]\!] = 0\,, \tag{118}$$

5. $$\rho v[\![\mathbf{u}]\!] = [\![\boldsymbol{\tau}]\!]\mathbf{n}\,, \tag{119}$$

6. $$v[\![\rho\mathbf{u}\otimes\mathbf{u} + \boldsymbol{\tau}]\!] = [\![\mathbf{u}\otimes\boldsymbol{\tau}\mathbf{n} + \boldsymbol{\tau}\mathbf{n}\otimes\mathbf{u}]\!] + \frac{\eta}{\lambda}[\![\mathbf{u}\otimes\mathbf{n} + \mathbf{n}\otimes\mathbf{u}]\!]\,. \tag{120}$$

Rutkevich [1969] also considers the case of continuous velocities $[\![\mathbf{u}]\!]=\mathbf{0}$ in which case the discontinuity does not propagate relative to the medium, v=0. Then (107) implies

$$[\![p]\!]\mathbf{n} = [\![\boldsymbol{\tau}]\!]\mathbf{n}\,. \tag{121}$$

Hulsen [1986] argues that terms like

$$p_{,i}u_j = (pu_j)_{,i} - pu_{j,i}$$

in (106) will give rise to a new

$$P'_{ijk} = P_{ijk} + p\delta_{ki}u_j + p\delta_{kj}u_i\,.$$

In this case, we must assume that $u_{j,i}$ suffers simple jumps across Σ at most. Hence $[\![\mathbf{u}]\!]=0$ and $[\![p]\!]$ is not necessarily zero. Equation

(115) implies that $[\![p]\!]=[\![\tau_{nn}]\!]$ and $[\![\tau_{nt}]\!]=0$ where τ_{nt} are the shear stresses on Σ. Other components of stress are not constrained by (121). For example, let $\mathbf{n}=\mathbf{e}_3$ at a point and $\mathbf{t}=\mathbf{e}_1$ or $\mathbf{e}_2$. Then $[\![\tau_{13}]\!]=[\![\tau_{23}]\!]=0$, $[\![\tau_{33}]\!]=[\![p]\!]$ with no restrictions of $[\![\tau_{11}]\!]$, $[\![\tau_{12}]\!]$, $[\![\tau_{22}]\!]$.

The conditions

$$[\![\mathbf{u}]\!] = 0\,, \quad v = 0\,, \tag{122}$$

$$[\![p]\!]\mathbf{n} = [\![\boldsymbol{\tau}]\!]\mathbf{n} \tag{123}$$

with $[\![\boldsymbol{\tau}]\!]$ otherwise nonrestricted, holds widely for all constitutive models and not only for the UCM.

M. Renardy [1989c] has derived shock relations for the lower convected Maxwell model in three dimensions and for the interpolated models ($a^2<1$) in two space dimensions. His method of derivation is different than the method of Rutkevich in which the equations are first put into conservation form. He notes that there may be many ways of writing a given set of equations in conservation form and they do not always lead to the same jump conditions. The correct jump conditions are determined by physical rather than mathematical criteria; for example, we should like the jump conditions which arise from the shock-up of smooth data. Whitham [1974] addressed this problem in the following terms

> At first sight, the weak solution consept appears to bypass the more involved and less precise discussion of the real physical processes. But it is not really so. Corresponding to the differential equation
>
> $$\frac{\partial \rho}{\partial t} + c(\rho)\,\frac{\partial \rho}{\partial x} = 0$$
>
> there are an infinite number of conservation equations
>
> $$\frac{\partial f(\rho)}{\partial t} + \frac{\partial g(\rho)}{\partial x} = 0\,. \tag{124}$$
>
> Any choice which satisfies
>
> $$g'(\rho) = f'(\rho)c(\rho)$$

will do. For differentiable functions $\varphi(x,t)$, these are all equivalent. However, their integrated forms are *not* equivalent and lead to different jump conditions. The weak solution of (124) will require the shock condition

$$-U[\![f(\rho)]\!] + [\![g(\rho)]\!] = 0 ;$$

different choices of f and g lead to different relations between ρ_1, ρ_2, and U. Therefore a discussion of the physical processes is still necessary in order to pick out which weak solution is relevant to the particular physical problem at hand.

The method used by Renardy is to look at the constitutive equation in a smooth layer centered at the shock, as would be apparent in a structured shock smoothed, say by a very small Newtonian viscosity. We may suppose, without losing generality, that the normal to the shock is in the direction of positive x_1. Then $\mathbf{n}=\mathbf{e}_1$, $\tau_{nn}=\tau_{11}$, (116) may be written as

$$[\![u_1]\!] = 0 \tag{125}$$

and (107) becomes

$$[\![-p + \tau_{11}]\!] = 0 , \tag{126}$$

$$\rho v[\![u_2]\!] = [\![\tau_{21}]\!] , \tag{127}$$

$$\rho v[\![u_3]\!] = [\![\tau_{31}]\!] , \tag{128}$$

where $v=u_1-C_\Sigma$, $C_\Sigma=\mathbf{u}_\Sigma\cdot\mathbf{e}_1$ is the shock speed. These relations are universal; they do not depend on the constitutive equation. The constitutive equation is simplified by neglecting derivatives in the directions x_2 and x_3 which are parallel to the plane of the shock. The simplified equations for the LCM with the lower order terms $\boldsymbol{\tau}$ and tangential derivatives neglected are

$$\frac{\partial\tau_{11}}{\partial t} + u_1\frac{\partial\tau_{11}}{\partial x_1} + 2\tau_{12}\frac{\partial v_2}{\partial x_1} + 2\tau_{13}\frac{\partial v_3}{\partial x_1} = 0 , \tag{129}$$

$$\frac{\partial\tau_{12}}{\partial t} + u_1\frac{\partial\tau_{12}}{\partial x_1} + \tau_{22}\frac{\partial u_2}{\partial x_1} + \tau_{23}\frac{\partial u_3}{\partial x_1} = \frac{\eta}{\lambda}\frac{\partial u_2}{\partial x_1} , \tag{130}$$

$$\frac{\partial \tau_{13}}{\partial t} + u_1 \frac{\partial \tau_{13}}{\partial x_1} + \tau_{33} \frac{\partial u_3}{\partial x_1} + \tau_{23} \frac{\partial u_2}{\partial x_1} = \frac{\eta}{\lambda} \frac{\partial u_3}{\partial x_1} , \tag{131}$$

$$\frac{\partial \tau_{22}}{\partial t} + u_1 \frac{\partial \tau_{22}}{\partial x_1} = 0 , \tag{132}$$

$$\frac{\partial \tau_{23}}{\partial t} + u_1 \frac{\partial \tau_{23}}{\partial x_1} = 0 , \tag{133}$$

$$\frac{\partial \tau_{33}}{\partial t} + u_1 \frac{\partial \tau_{33}}{\partial x_1} = 0 . \tag{134}$$

The last three equations are in conservation form and if $m=\rho v \neq 0$ in (102), we get

$$[\![\tau_{22}]\!] = [\![\tau_{23}]\!] = [\![\tau_{33}]\!] = 0 . \tag{135}$$

It follows now that the coefficients in (130) and (131) are continuous across the shock and these equations are in conservation form. Hence

$$v[\![\tau_{12}]\!] = \left(\frac{\eta}{\lambda} - \tau_{22}\right)[\![u_2]\!] - \tau_{23}[\![u_3]\!] ,$$

$$v[\![\tau_{13}]\!] = \left(\frac{\eta}{\lambda} - \tau_{33}\right)[\![u_3]\!] - \tau_{23}[\![u_2]\!] . \tag{136}$$

After replacing $[\![\tau_{12}]\!]$ and $[\![\tau_{13}]\!]$ in these equations with the expressions (127) and (128) we find the relative shock speed v as one of the eigenvalues of the matrix

$$\begin{bmatrix} \frac{\eta}{\lambda} - \tau_{22} & -\tau_{23} \\ -\tau_{23} & \frac{\eta}{\lambda} - \tau_{33} \end{bmatrix} \tag{137}$$

and $[\![u_2]\!]$ and $[\![u_3]\!]$ are the corresponding eigenvectors. Integration of (130) and (131) in the shock layer gives rise to

$$v\tau_{12} = \left(\frac{\eta}{\lambda} - \tau_{22}\right)u_2 - \tau_{23}u_3 + \gamma_2\,, \tag{138}$$

$$v\tau_{13} = \left(\frac{\eta}{\lambda} - \tau_{33}\right)u_3 - \tau_{23}u_2 + \gamma_3\,. \tag{139}$$

where γ_2 and γ_3 are constants of integration which can be evaluated on either side of the shock layer and

$$\frac{\partial(\bullet)}{\partial t} \sim -c_\Sigma \frac{\partial(\bullet)}{\partial x_1}\,. \tag{140}$$

$$v^2[\![\tau_{11}]\!] + \left(\frac{\eta}{\lambda} - \tau_{22}\right)[\![u_2^2]\!] + \left(\frac{\eta}{\lambda} - \tau_{33}\right)[\![u_3^2]\!] - 2\tau_{23}[\![u_2u_3]\!]$$
$$+ 2\gamma_2[\![u_2]\!] + 2\gamma_3[\![u_3]\!] = 0\,. \tag{141}$$

The reader can verify that the shock wave speeds C_Σ given by the eigenvalues of (137) are the same as speeds of vorticity waves generated using frozen coefficients.

The interpolated models ($a^2<1$) are different. They give rise to a genuinely nonlinear shock speed. Renardy [1989a] gave these models for the two dimensional case. He says that the interpolated models are Hypoelastic. For these he says "only if the kinematics is essentially one-dimensional can one expect to formulate Rankine-Hugoniot conditions. This is the case in two space dimensions, where only one component of the velocity gradient is infinite at the shock.

The leading order terms of the constitutive equation for the interpolated Maxwell models in two dimensions are

$$\frac{\partial\tau_{11}}{\partial t} + u_1\frac{\partial\tau_{11}}{\partial x_1} + (1-a)\tau_{12}\frac{\partial u_2}{\partial x_1} = 0\,, \tag{142}$$

$$\frac{\partial\tau_{12}}{\partial t} + u_1\frac{\partial\tau_{12}}{\partial x_1} + \left(\frac{1-a}{2}\tau_{22} - \frac{1+a}{2}\tau_{11}\right)\frac{\partial u_2}{\partial x_1} = \frac{\eta}{\lambda}\frac{\partial u_2}{\partial x_1}\,, \tag{143}$$

$$\frac{\partial\tau_{22}}{\partial t} + u_1\frac{\partial\tau_{22}}{\partial x_1} - (1+a)\tau_{12}\frac{\partial u_2}{\partial x_1} = 0\,. \tag{144}$$

We multiply (144) by 1+a and (149) by 1–a and add. This leads to

$$\left(\frac{\partial}{\partial t}+u_1\frac{\partial}{\partial x_1}\right)((1+a)\tau_{11}+(1-a)\tau_{22})=0\,,$$

i.e.

$$v((1+a)[\![\tau_{11}]\!]+(1-a)[\![\tau_{22}]\!])=0\,.$$

Next, we set $S=\frac{a-1}{2}\tau_{22}+\frac{a+1}{2}\tau_{11}$. We find

$$\frac{\partial S}{\partial t}+u_1\frac{\partial S}{\partial x_1}=(a^2-1)\tau_{12}\frac{\partial u_2}{\partial x_1}\,, \tag{145}$$

$$\frac{\partial \tau_{12}}{\partial t}+u_1\frac{\partial \tau_{12}}{\partial x_1}-S\frac{\partial u_2}{\partial x_1}=\frac{\eta}{\lambda}\frac{\partial u_2}{\partial x_1}\,. \tag{146}$$

We multiply (145) by $\eta/\lambda+S$ and (146) by $(1-a^2)\tau_{12}$ and add them (this is permissible as long as u varies rapidly without actually being discontinuous). This yields

$$\left(\frac{\partial}{\partial t}+u_1\frac{\partial}{\partial x_1}\right)\left(\frac{S\eta}{\lambda}+\frac{1}{2}S^2+\frac{1}{2}(1-a^2)\tau_{12}^2\right)=0\,. \tag{147}$$

This leads to the shock condition

$$\left[\!\!\left[\frac{S\eta}{\lambda}+\frac{1}{2}S^2+\frac{1}{2}(1-a^2)\tau_{12}^2\right]\!\!\right]=0\,.$$

Moreover, (147) leads to a relation between S and τ_{12}:

$$S=-\frac{\eta}{\lambda}\pm\sqrt{\frac{\eta^2}{\lambda^2}-(1-a^2)\tau_{12}^2+\gamma}\,, \tag{148}$$

where γ can be determined from the knowledge of S and τ_{12} on one side of the shock. We divide (146) by $\eta/\lambda-S$, where we think of S as a function of τ_{12} given by (148). This yields

$$\frac{1}{\frac{\eta}{\lambda}+S(\tau_{12})}\left(\frac{\partial}{\partial t}+u_1\frac{\partial}{\partial x_1}\right)\tau_{12}=\frac{\partial u_2}{\partial x_1}\,.$$

With $\phi(\tau_{12})$ denoting a primitive of the function $1/((\eta/\lambda)+S(\tau_{12}))$ we obtain the shock condition

$$[\![\phi(\tau_{12})]\!] = [\![u_2]\!]\,. \tag{149}$$

We note that (149) in conjunction with (127) now leads to a shock speed which is in general different from the linear wave speed. We note that the function ϕ is monotone as long as $S>-(\eta/\lambda)$. This is indeed the case as long as the equations remain well-posed.

To solve a problem with Rankine-Hugoniot relations it is necessary to piece the shocks together with a smooth field. This is sometimes called Riemann's problem. So far we have not got firm evidence from experiments or observations that there are some flows of viscoelastic fluids which support shock waves.

APPENDIX A

Tensor Algebra

In this appendix the reader will find a list of formulas which are used in this book. No proofs are given. One source for some proofs and more formulas is Truesdell and Noll [1965].

A.1 Notations

- F scalar,
- $\mathbf{F}$ vector, tensor,
- $\mathbf{F}$ tensor of order two = a matrix valued linear operator mapping vectors into vectors $\mathbf{a}'=\mathbf{Fa}$, $\mathbb{R}^n \rightarrow \mathbb{R}^n$,
- $\{\mathbf{e}_i\}$ orthonormal base vectors $\mathbf{e}_i \cdot \mathbf{e}_m = \delta_{im}$,
- $F_{i\ell}$ physical components of $\mathbf{F}$ with respect to $\mathbf{e}_i$,

$$\mathbf{F} = F_{\ell m}\mathbf{e}_\ell \otimes \mathbf{e}_m \overset{\text{def}}{=} F_{\ell m}\mathbf{e}_\ell \mathbf{e}_m \quad (\mathbf{e}_\ell \mathbf{e}_m = \mathbf{e}_\ell \otimes \mathbf{e}_m \text{ dyad}) ,$$

$(\mathbf{e}_\ell \cdot \mathbf{e}_m = \delta_{\ell m}$ scalar product$)$.

A.2 Gradient of a vector $\mathbf{F}(x)$

$$d\mathbf{F} = \nabla\mathbf{F}\, d\mathbf{x} = (\nabla\mathbf{F})_{ij}\mathbf{e}_i\mathbf{e}_j\bullet(dx_\ell\mathbf{e}_\ell) = \frac{\partial F_i}{\partial x_\ell}\, dx_\ell\, \mathbf{e}_i,$$

$$(\nabla\mathbf{F})_{ij} = \frac{\partial F_i}{\partial x_j}\,. \tag{1}$$

A.3 Orthogonal tensors: $\mathbf{Q}$ is orthogonal if $\mathbf{Q} = \mathbf{Q}^T$

$$\mathbf{Q}\mathbf{Q}^T = \mathbf{Q}^T\mathbf{Q} = \mathbf{1}\,, \tag{2}$$

$$1 = \det \mathbf{Q}\mathbf{Q}^T = \det \mathbf{Q} \det \mathbf{Q}^T = (\det \mathbf{Q})^2\,. \tag{3}$$

$\mathbf{Q}$ is *proper orthogonal* if $\det \mathbf{Q} = 1$,
$\mathbf{Q}$ is *improper orthogonal* if $\det \mathbf{Q} = -1$.

An orthogonal tensor preserves scalar products

$$(\mathbf{Q}\mathbf{u})\bullet(\mathbf{Q}\mathbf{v}) = [\mathbf{Q}^T\mathbf{Q}\mathbf{u}]\bullet\mathbf{v} = \mathbf{u}\bullet\mathbf{v}\,. \tag{4}$$

Physical components of $\mathbf{Q}$. Let $\mathbf{e}_i$ be orthonormal base vectors. Then

$$\mathbf{e}'_i = \mathbf{Q}\bullet\mathbf{e}_i = Q_{\ell m}\mathbf{e}_\ell\mathbf{e}_m\bullet\mathbf{e}_i = Q_{\ell i}\mathbf{e}_\ell\,, \tag{5}$$

$$\mathbf{e}_j\bullet\mathbf{e}'_i = Q_{ji}\,, \quad \mathbf{e}_i = \mathbf{Q}^T\bullet\mathbf{e}'_i\,. \tag{6}$$

A.4 Determinant formulas

$$\det \mathbf{F} = \varepsilon_{ijk}F_{1i}F_{2j}F_{3k} \quad \text{expansion in rows}\,, \tag{7}$$

$$\det \mathbf{F} = \varepsilon_{ijk}F_{i1}F_{j2}F_{k3} \quad \text{expansion in columns}\,, \tag{8}$$

$$\varepsilon_{\ell mn} \det \mathbf{F} = \varepsilon_{ijk}F_{\ell i}F_{mj}F_{nk} = \varepsilon_{ijk}F_{i\ell}F_{jm}F_{kn}\,, \tag{9}$$

$$\det \mathbf{F} = \frac{1}{6}\,\varepsilon_{ijk}\varepsilon_{\ell mn}F_{i\ell}F_{jm}F_{kn}\,, \tag{10}$$

where

$$\varepsilon_{123} = \varepsilon_{231} = \varepsilon_{312} = 1 ,$$

$$\varepsilon_{213} = \varepsilon_{132} = \varepsilon_{321} = -1 ,$$

otherwise, zero.

$$\varepsilon_{ijk}\varepsilon_{i\ell m} = \delta_{j\ell}\delta_{km} - \delta_{jm}\delta_{k\ell} . \quad (11)$$

A.5 Isotropic tensors, tensor functions and functionals

A tensor function (functional) is isotropic if and only if the forms of the component functions (functionals) are the same for all orthonormal bases (Truesdell and Noll [1965], p. 23):

$$\boldsymbol{f} = \mathbf{f}(\mathbf{A},\mathbf{B},\ldots,\mathbf{C}) , \quad f_{ij} = f_{ij}(A_{\ell k},B_{mn},\ldots,C_{pq}) \quad (12)$$

is isotropic if and only if

$$\mathbf{Q}\boldsymbol{f}\mathbf{Q}^T = \mathbf{f}(\mathbf{Q}\mathbf{A}\mathbf{Q}^T,\mathbf{Q}\mathbf{B}\mathbf{Q}^T,\ldots,\mathbf{Q}\mathbf{C}\mathbf{Q}^T) , \quad (13)$$

$$f^*_{ij} = f_{ij}(A^*_{\ell k},B^*_{mn},\ldots,C^*_{pq}) .$$

A.6 Invariant multinomial forms

An invariant multinomial form is formed from the product of n different vectors with a tensor of order n, it is a scalar and is unchanged.

$$\phi = A_{ij\ldots k\ell}\, a_i b_j \ldots c_k d_\ell = A^*_{ij\ldots k\ell}\, a^*_i b^*_j \ldots c^*_k d^*_\ell = \phi^* \quad (14)$$

under orthogonal transformation, $\mathbf{a}^*=\mathbf{Q}\mathbf{a}$, etc. An invariant multinomial is form invariant if and only if

$$A_{ij\ldots k\ell} = A^*_{ij\ldots kl} , \quad (15)$$

that is, if and only if $A_{ij\ldots k\ell}$ is an isotropic tensor. To find the form of isotropic tensors, we use Cauchy's theorem: an invariant multilinear form can be expressed in terms of scalar products and triple scalar products among the n vectors.

A.7 Cauchy theorem for isotropic tensors

1. $$\phi = A_{ij}a_ib_j = \mathbf{Aa \cdot b} = A\delta_{ij}a_ib_j\,.$$
$$\text{Hence, } A_{ij} = A\delta_{ij} \quad \text{(isotropic tensor of order 2)} \tag{16}$$

2. $$\phi = A_{ijk}a_ib_jc_k = \mathbf{Ba \cdot (b \wedge c)} = B\varepsilon_{ijk}a_ib_jc_k\,.$$
$$\text{Hence, } A_{ijk} = B\varepsilon_{ijk} \quad \text{(isotropic tensor of order 3)} \tag{17}$$

3. $$\phi = A_{ijk\ell}a_ib_jc_kd_\ell = \mathbf{Aa \cdot bc \cdot d + Ba \cdot cb \cdot d + Ca \cdot db \cdot c}$$
$$= \{A\delta_{ij}\delta_{k\ell} + B\delta_{ik}\delta_{j\ell} + C\delta_{i\ell}\delta_{jk}\}a_ib_j\,c_kd_\ell\,.$$
$$\text{Hence, } A_{ijk\ell} = A\delta_{ij}\delta_{k\ell} + B\delta_{ik}\delta_{j\ell} + C\delta_{i\ell}\delta_{jk}$$
$$\text{(isotropic of order 4)} \tag{18}$$

Corollary of Cauchy's theorem: *Every even order isotropic tensor may be expressed in terms of Kronecker's delta.*

$$\varepsilon_{ijk}\varepsilon_{\ell mn} = \det \begin{vmatrix} \delta_{i\ell} & \delta_{im} & \delta_{in} \\ \delta_{j\ell} & \delta_{jm} & \delta_{jn} \\ \delta_{k\ell} & \delta_{km} & \delta_{kn} \end{vmatrix}\,. \tag{19}$$

It is not hard to find the form of an isotropic tensor of order 6. But the application of Cauchy's theorem will give you terms of the form $\alpha\varepsilon_{ijk}\varepsilon_{\ell mn}$, as well as scalar product terms. Equation (19) shows that such terms are also expressible in powers of the Kronecker delta. Terms like (19) arise at even orders ≥ 6.

A.8 Isotropic tensor polynomials multilinear in n symmetric tensors

Such polynomials are in the form

$$\underset{\substack{2n+2 \\ \text{indices}}}{K_{ijmn\ldots k\ell}\;A_{mn}D_{k\ell}} \;\underset{n \text{ tensors}}{\overset{\text{def}}{=}}\; \mu_{ij} = \mu_{ji} \tag{20}$$

with an isotropic $K_{ijmn\ldots k\ell}$ symmetric in successive pairs of indices. Applying Cauchy's theorem to K, it is easy to show that $\boldsymbol{\mu}$ can be expressed as multilinear products of tensors, trace of tensors, trace of products of tensors. For example,

$$\begin{aligned} K_{ijk\ell mn}A_{k\ell}B_{mn} &= \alpha A_{ik}B_{kj} + \beta B_{ij}\,\mathrm{tr}\,\mathbf{A} + \gamma A_{ij}\,\mathrm{tr}\,\mathbf{B} \\ &+ \mu\delta_{ij}\,\mathrm{tr}\,(\mathbf{AB}) + \nu\delta_{ij}\,\mathrm{tr}\,\mathbf{A}\,\mathrm{tr}\,\mathbf{B}\,. \end{aligned} \tag{21}$$

A.9 Invariants of a second-order tensor

The characteristic equation for the eigenvalues of a tensor **A**

$$\mathbf{Ax} = \lambda\mathbf{x} \tag{22}$$

is

$$p(\lambda) = \det\,[\mathbf{A}-\lambda\mathbf{1}] = -\lambda^3 + \mathrm{I}\lambda^2 - \mathrm{II}\lambda + \mathrm{III} = 0 \tag{23}$$

where

$$\mathrm{I} = -\mathrm{tr}\,\mathbf{A} = -(\lambda_1+\lambda_2+\lambda_3)\,, \tag{24}$$

$$\mathrm{II} = \frac{1}{2}\left[(\mathrm{tr}\,\mathbf{A})^2 - \mathrm{tr}\,\mathbf{A}^2\right] = \lambda_1\lambda_2 + \lambda_2\lambda_3 + \lambda_1\lambda_3\,, \tag{25}$$

$$\mathrm{III} = \lambda_1\lambda_2\lambda_3 = \det\,\mathbf{A}\,. \tag{26}$$

The principal invariants of **A** are I, II, and III. They are invariant under proper orthogonal transformation.

A.10 Cayley-Hamilton theorem and tensor functions

$$p(\mathbf{A}) = -\mathbf{A}^3 + \mathrm{I}\mathbf{A}^2 - \mathrm{II}\mathbf{A} + \mathrm{III}\mathbf{1} = 0\,. \tag{27}$$

Fundamental representation theorem for tensor functions of a single symmetric tensor (Truesdell and Noll [1965], p. 32). A symmetric tensor function is isotropic if and only if it has a representation,

$$\mathbf{f}(\mathbf{A}) = \phi_o(\mathrm{I},\mathrm{II},\mathrm{III})\mathbf{1} + \phi_1(\mathrm{I},\mathrm{II},\mathrm{III})\mathbf{A} + \phi_2(\mathrm{I},\mathrm{II},\mathrm{III})\mathbf{A}^2 .$$

The oldest proof of the Cayley-Hamilton theorem requires $\mathbf{f}(\bullet)$ to be analytic near zero. Then $\mathbf{f}(\mathbf{A})$ is expanded in a Taylor series and the power $\mathbf{A}^n$, $n \geq 3$ are eliminated using the Cayley-Hamilton theorem. The newer proofs do not require $\mathbf{f}$ to be analytic.

A.11 Representation theorem for tensor functions of two symmetric tensors (Rivlin, [1955])

If $\mathbf{g}(\mathbf{A},\mathbf{B})$ is a symmetric tensor-valued function of the two symmetric tensors $\mathbf{A}=\mathbf{A}^T$ and $\mathbf{B}=\mathbf{B}^T$, and $\mathbf{g}$ is also isotropic

$$\mathbf{Q}\mathbf{g}(\mathbf{A},\mathbf{B})\mathbf{Q}^T = \mathbf{g}(\mathbf{Q}\mathbf{A}\mathbf{Q}^T,\mathbf{Q}\mathbf{B}\mathbf{Q}^T) ,$$

then, for all proper orthogonal tensors,

$$\begin{aligned} \mathbf{g}(\mathbf{A},\mathbf{B}) = {} & a_o\mathbf{1} + a_1\mathbf{A} + a_2\mathbf{A}^2 + a_3\mathbf{B} + a_4\mathbf{B}^2 \\ & + a_5(\mathbf{AB}+\mathbf{BA}) + a_6(\mathbf{AB}^2+\mathbf{B}^2\mathbf{A}) \\ & + a_7(\mathbf{A}^2\mathbf{B}+\mathbf{B}\mathbf{A}^2) + a_8(\mathbf{A}^2\mathbf{B}^2+\mathbf{B}^2\mathbf{A}^2) \end{aligned}$$

where

$$\begin{aligned} a_\Gamma = \hat{a}_\Gamma(&\mathrm{tr}\,\mathbf{A},\mathrm{tr}\,\mathbf{A}^2,\mathrm{tr}\,\mathbf{A}^3,\mathrm{tr}\,\mathbf{B},\mathrm{tr}\,\mathbf{B}^2,\mathrm{tr}\,\mathbf{B}^3, \\ &\mathrm{tr}\,\mathbf{AB},\mathrm{tr}\,\mathbf{A}^2\mathbf{B},\mathrm{tr}\,\mathbf{B}^2\mathbf{A},\mathrm{tr}\,\mathbf{A}^2\mathbf{B}^2) \end{aligned}$$

are functions of invariants.

A.12 Polar decomposition theorem

Let $\mathbf{T}$ be any invertible tensor. Then $\mathbf{T}$ has two unique decompositions

$$\mathbf{T} = \mathbf{RU} \,,$$
$$\mathbf{T} = \mathbf{VR} \,,$$
$$\mathbf{R} \text{ is orthogonal} \,,$$
$$\mathbf{U} \text{ and } \mathbf{V} \text{ are symmetric and positive definite} \,,$$
$$\mathbf{V} = \mathbf{RUR}^T \,.$$

Let **T**=**F**. Then

$$\mathbf{F} = \mathbf{RU} = \mathbf{VR}$$
$$\mathbf{F}^T\mathbf{F} = \mathbf{C} = \mathbf{U}^2$$
$$\mathbf{FF}^T = \mathbf{B} = \mathbf{V}^2 \,.$$

The polar decomposition theorem is used in Noll's theory to derive the relation between stress and deformation, the constitutive equation.

APPENDIX B

Reciprocal Base Vectors, Metric Tensors, Components

B.1 Gradient of a scalar

$$\nabla\phi = \mathbf{e}_i \frac{\partial\phi}{\partial x_i} = \mathbf{e}_i \frac{\partial\eta_j}{\partial x_i}\frac{\partial\phi}{\partial\eta_j} = \nabla\eta_j \frac{\partial\phi}{\partial\eta_j} = \mathbf{g}^j \frac{\partial\phi}{\partial\eta_j} .$$

B.2 Contravariant and covariant components of vectors

$\mathbf{A} = \mathbf{g}_i A^i$ The numbers A^i are the *contravariant* components of $\mathbf{A}$.

$\mathbf{A} = \mathbf{g}^i A_i$ The numbers A_i are the *covariant* components of $\mathbf{A}$. These are projections of the vector onto the coordinate lines.

B.3 Metric tensors

$$\mathbf{g}_i \cdot \mathbf{g}_j = g_{ij} ,$$

$$\mathbf{g}^i \cdot \mathbf{g}^j = g^{ij} .$$

The arc length is expressed as $d\mathbf{x} \cdot d\mathbf{x} = \mathbf{g}_i d\eta_i \cdot \mathbf{g}_j d\eta_j = g_{ij} d\eta_i d\eta_j$. Metric tensors give the transformation between contravariant and covariant components of a vector (raising and lowering scripts).

$$\mathbf{A} = \mathrm{g}_i A^i = \mathrm{g}^i A_i \,,$$

$$\mathrm{g}^\ell \cdot \mathbf{A} = \mathrm{g}^\ell \cdot \mathrm{g}_i A^i = \mathrm{g}^\ell \cdot \mathrm{g}^i A_i$$
$$= \delta_{\ell i} A^i = \mathrm{g}^{\ell i} A_i \,, \quad A^\ell = \mathrm{g}^{\ell i} A_i \,,$$

$$\mathrm{g}_\ell \cdot \mathbf{A} = \mathrm{g}_\ell \cdot \mathrm{g}_i A^i = \mathrm{g}_\ell \cdot \mathrm{g}^i A_i$$
$$= \mathrm{g}_{\ell i} A^i = \delta_{\ell i} A_i \,, \quad \mathrm{g}_{\ell i} A^i = A_\ell \,.$$

B.4 Orthonormal bases and Cartesian bases

Orthonormal base vectors are orthogonal and have unit length $\mathbf{e}_i \cdot \mathbf{e}_j = \delta_{ij}$. Examples, Cartesian bases ($\mathbf{e}_1, \mathbf{e}_2, \mathbf{e}_3$) are independent of coordinates. Spherical polar bases ($\mathbf{e}_r, \mathbf{e}_\theta, \mathbf{e}_\phi$) depend on θ and ϕ. The *physical components* of a vector are its components relative to an orthonormal basis.

$$\mathbf{A} = A_i \mathbf{e}_i \,.$$

B.5 Components of a second-order tensor

A second-order tensor $\mathbf{T}$ is a transformation which assigns to each given vector $\mathbf{u}$ another vector $\mathbf{Tu}$

$$\mathbf{v} = \mathbf{Tu}$$

and satisfies the rules

$$\mathbf{T}(\mathbf{u}+\mathbf{w}) = \mathbf{Tu}+\mathbf{Tw} \quad \text{(additivity)}$$

and

$$\mathbf{T}(\alpha\mathbf{u}) = \alpha\mathbf{Tu} \quad \text{(homogeneity)} \,.$$

Dyad of two vectors:

$$\mathbf{ab} \ (\text{or } \mathbf{a}\otimes\mathbf{b}) \,,$$
$$\mathbf{c}\cdot\mathbf{ab} = (\mathbf{c}\cdot\mathbf{a})\mathbf{b} \,.$$

Components of a tensor

$\mathbf{g}^j \cdot \mathbf{T} \mathbf{g}^i = T^{ji}$ contravariant ,

$\mathbf{g}_j \cdot \mathbf{T} \mathbf{g}_i = T_{ji}$ covariant ,

$\mathbf{g}_j \cdot \mathbf{T} \mathbf{g}^i = T^{i}_{j}$ mixed ,

$\mathbf{e}_i \cdot \mathbf{T} \mathbf{e}_j = T_{ij}$ physical ,

$$\mathbf{T} = \mathbf{g}_j T^{ji} \mathbf{g}_i = \mathbf{g}^j T_{ji} \mathbf{g}^i = \mathbf{g}^j T^{i}_{j} \mathbf{g}_i = \mathbf{e}_j T_{ji} \mathbf{e}_i \ .$$

The contravariant and covariant components of the unit tensor **T**=**1** are the components of metric tensor.

APPENDIX C

Kinematics

(Referential and spatial descriptions; path lines for simple flows)

Points in a body (a continuum) are identified by their position; **X** is the position in some reference configuration V(**X**)

$$\xi(\mathbf{X},\tau) \quad \tau\leq t \ , \qquad \xi(\mathbf{X},t) = \mathbf{x} \ ,$$ position of a point at time τ.

$$\mathbf{u}(\xi,\tau) \overset{def}{=} \frac{d\xi}{d\tau} \overset{def}{=} \left(\frac{\partial\xi}{\partial\tau}\right)_{\mathbf{X}} \ ,$$ velocity of the point **X** .

$$\mathbf{a}(\xi,\tau) \overset{def}{=} \frac{d^2\xi}{d\tau^2} \ ,$$ acceleration .

$$\overset{(n)}{\xi} \overset{def}{=} \frac{d^n\xi}{d\tau^n} \ ,$$ higher order “acceleration” if n>2 .

$$\mathbf{F} \overset{def}{=} \nabla\xi(\mathbf{X},\tau) \ ,$$ deformation gradient (Jacobian matrix).

$$J \overset{def}{=} \det \mathbf{F} \ ,$$ Jacobian .

Transformation of volumes

$$dV(\xi) = \det \mathbf{F} \, dV(\mathbf{X}) \ , \tag{1}$$ the Jacobian is a scale for the transformation of volume.

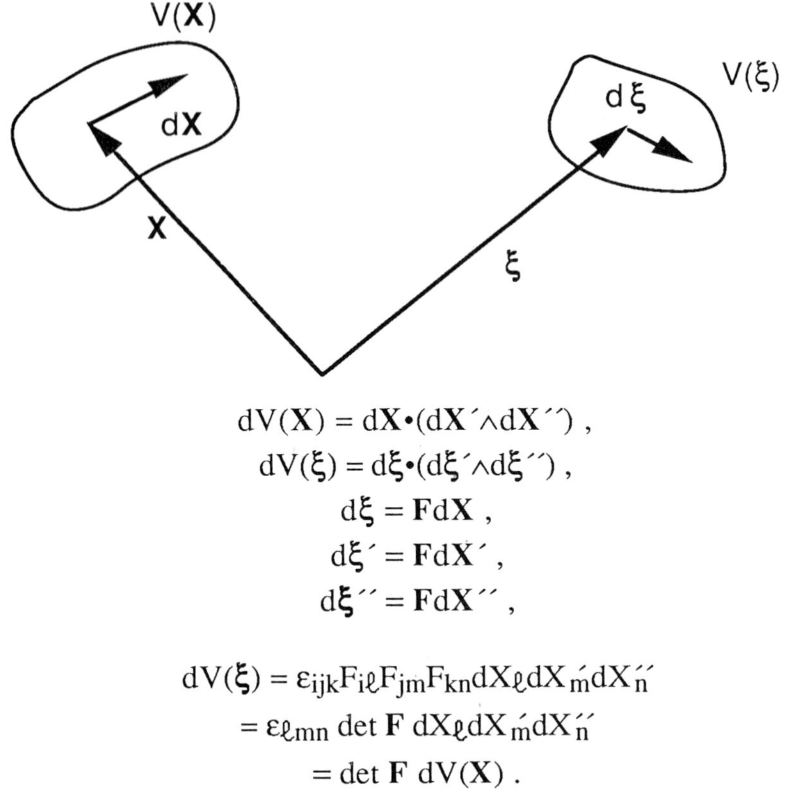

$$dV(\mathbf{X}) = d\mathbf{X}\cdot(d\mathbf{X}'\wedge d\mathbf{X}'') ,$$
$$dV(\xi) = d\xi\cdot(d\xi'\wedge d\xi'') ,$$
$$d\xi = \mathbf{F}d\mathbf{X} ,$$
$$d\xi' = \mathbf{F}d\mathbf{X}' ,$$
$$d\xi'' = \mathbf{F}d\mathbf{X}'' ,$$

$$\begin{aligned} dV(\xi) &= \varepsilon_{ijk}F_{i\ell}F_{jm}F_{kn}dX_\ell dX'_m dX''_n \\ &= \varepsilon_{\ell mn} \det \mathbf{F}\, dX_\ell dX'_m dX''_n \\ &= \det \mathbf{F}\, dV(\mathbf{X}) . \end{aligned}$$

Conservation of mass. Let $\rho[\xi(\mathbf{X},\tau),\tau]$ be the mass density of the particle which was at $\mathbf{X}$ at time τ_o. This is a special choice of $\mathbf{X}$ such that $\xi(\mathbf{X},\tau_o)=\mathbf{X}$. Let V be any material volume; that is, no mass crosses the boundary ∂V of V. Then the equation expressing the conservation of mass is

$$\int_{V(\tau)} \rho[\xi,\tau]\, dV = M = \int_{V_0} \rho_0\, dV$$

where $\rho>0$, $\rho_0=\rho[\mathbf{X},\tau_o]$, $M>0$ is constant (independent of τ). Transforming $V(\tau)$, we have

$$\int_{V_0} \{\rho[\xi(\mathbf{X},\tau),\tau]J - \rho_0\}\, dV = 0 .$$

Since V_0 is arbitrary, the integrand vanishes, and

$$\rho J = \rho_0 \,. \tag{2}$$

Incompressible material. $\rho[X,\tau]$ is independent of τ. Then $\rho=\rho_0$,

$$J = 1 \,. \tag{3}$$

Isochoric motion. $J = 1$.

Material mapping. If the map $\mathbf{X}{:}\to\xi$ maps material particles, then the ratios of volumes

$$J = \frac{\partial V(\xi)}{\partial V(\mathbf{X})} \text{ is finite and not zero } \left(\text{also } J = \frac{\rho}{\rho_0}\right)$$

$$\det \mathbf{F} = J \neq {}^{0}_{\infty} \,,$$

$$\mathbf{F}^{-1} \text{ exists} \,,$$

$$\left.\begin{aligned} d\xi &= \mathbf{F}d\mathbf{X} \\ d\mathbf{X} &= \mathbf{F}^{-1}\, d\xi \end{aligned}\right| \,. \tag{4}$$

Measures of the local change of angle and length.

$$\begin{aligned} d\xi\cdot d\xi' &= \mathbf{F}d\mathbf{X}\cdot\mathbf{F}d\mathbf{X}' = F_{i\ell}dX_\ell F_{ij}dX_j' \\ &= F^{T}_{\ell i}\, F_{ij}dX_j' dX_\ell = [\mathbf{F}^T\mathbf{F}\cdot d\mathbf{X}']\cdot d\mathbf{X} \,, \end{aligned}$$

$$\mathbf{F}^T\mathbf{F} \overset{\text{def}}{=} \mathbf{C} = \mathbf{C}^T \quad \text{(right Cauchy-Green strain tensor)} \,, \tag{5}$$

$$d\xi\cdot d\xi' = \mathbf{C}d\mathbf{X}'\cdot d\mathbf{X} \,,$$

$$\frac{|d\xi|^2}{|d\mathbf{X}|^2} = (\mathbf{C}\mathbf{e}_X)\cdot\mathbf{e}_X \,,$$

$$\mathbf{F}\mathbf{F}^T = \mathbf{B} = \mathbf{B}^T \quad \text{(left Cauchy-Green strain tensor)} \,. \tag{6}$$

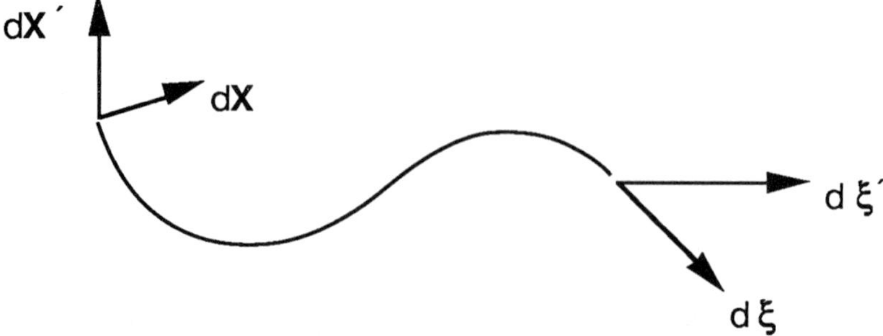

Rivlin-Ericksen tensors $\mathbf{A}_n$. These arise as a tensor expressing the time derivative of the length of a material line.

$$\overset{(n)}{|d\boldsymbol{\xi}|^2} = (\mathbf{A}_n d\boldsymbol{\xi})\cdot d\boldsymbol{\xi} ,$$

$$\overset{(n)}{|d\boldsymbol{\xi}|} = (\overset{(n)}{\mathbf{C}}\, d\mathbf{X})\cdot d\mathbf{X} = (\overset{(n)}{\mathbf{C}}\, \mathbf{F}^{-1} d\boldsymbol{\xi})\cdot \mathbf{F}^{-1} d\boldsymbol{\xi}$$
$$= [\mathbf{F}^{T^{-1}} \overset{(n)}{\mathbf{C}}\, \mathbf{F}^{-1} d\boldsymbol{\xi}]\cdot d\boldsymbol{\xi} ,$$

$$\mathbf{A}_n \overset{\text{def}}{=} \mathbf{F}^{T^{-1}} \overset{(n)}{\mathbf{C}}\, \mathbf{F}^{-1} , \quad \mathbf{A}_n = \mathbf{A}_n^T . \tag{7}$$

Recursion formula for RE tensors:

$$\mathbf{L}[\mathbf{u}(\boldsymbol{\xi},\tau)] \overset{\text{def}}{=} \nabla \mathbf{u}(\boldsymbol{\xi},\tau) = \left(\frac{\partial u_i}{\partial \xi_j}\right) \text{ (velocity gradient)} , \tag{8}$$

$$\mathbf{A}_1 = \mathbf{L} + \mathbf{L}^T . \tag{9}$$

Proof:

$$\dot{\mathbf{C}} = \dot{\mathbf{F}}^T \mathbf{F} + \mathbf{F}^T \dot{\mathbf{F}}$$

and

$$\mathbf{F}^{T^{-1}} \dot{\mathbf{C}} \mathbf{F}^{-1} = \mathbf{F}^{T^{-1}} \dot{\mathbf{F}}^T + \dot{\mathbf{F}} \mathbf{F}^{-1} = \mathbf{L}^T + \mathbf{L}$$

because

$$(\dot{\mathbf{F}}\mathbf{F}^{-1})_{ij} = \dot{F}_{i\ell}F^{-1}_{\ell j} = \frac{\partial u_i}{\partial X_\ell}\frac{\partial X_\ell}{\partial \xi_j} = \frac{\partial u_i}{\partial \xi_j} = L_{ij}\ .$$

$$\mathbf{A}_{n+1} = \dot{\mathbf{A}}_n + \mathbf{A}_n\mathbf{L} + (\mathbf{A}_n\mathbf{L})^T\ , \quad \dot{\mathbf{A}}_n \overset{\text{def}}{=} \frac{\partial \mathbf{A}_n}{\partial t} + (\mathbf{u}\cdot\nabla)\mathbf{A}_n\ . \quad (10)$$

Notation: $\mathbf{L_n}$ and $\mathbf{A}_n$ are determined by $\mathbf{u}$. The gradients are with respect to the place where $\mathbf{u}$ is evaluated. At time $t=\tau$, $\boldsymbol{\xi}=\mathbf{x}$ and the gradients are with respect to $\mathbf{x}$. We use the notation

$$\begin{aligned} \mathbf{L}(\tau) &= \mathbf{L}[\mathbf{u}(\boldsymbol{\xi}(\mathbf{X},\tau),\tau]\ , \\ \mathbf{L}(t) &= \mathbf{L}[\mathbf{u}(x),t]\ , \\ \mathbf{A}_1(\tau) &= \mathbf{L}(\tau) + \mathbf{L}^T(\tau)\ , \quad \text{etc.} \end{aligned} \quad (11)$$

Referential and spatial description:

$\mathbf{X}$	$\mathbf{x} = \boldsymbol{\xi}(\mathbf{X},t)$.
$\boldsymbol{\xi}(\mathbf{X},\tau)$	$\boldsymbol{\chi} = \chi_t(\mathbf{x},\tau)$.
particle **X**	particle which is presently at **x** .
elasticity	fluids, motion of viscoelastic solid.

Relative motion. Relative to the present $\tau=t$

$\boldsymbol{\xi}(\mathbf{X},t)=\mathbf{x}$	is the position of the particle which is presently at point **x**
$\boldsymbol{\chi} = \chi_t(\mathbf{x},\tau)$	is the position of the particle which is presently at **x** at an earlier instant τ
$\mathbf{x} = \boldsymbol{\xi}(\mathbf{X},t)$,	$\mathbf{X} = \xi^{-1}(\mathbf{x},t)$,
$\chi_t(\mathbf{x},\tau) = \boldsymbol{\xi}(\boldsymbol{\xi}^{-1}(\mathbf{x},t),\tau)$	relative position vector ,
$\mathbf{F}_t(\tau) = \nabla\boldsymbol{\chi}_t(\mathbf{x},\tau) = \left(\frac{\partial \chi_i}{\partial x_j}\right)\mathbf{e}_i\mathbf{e}_j$	relative deformation gradient, (12)

$$\mathbf{C}_t(\tau) = \mathbf{F}_t^T(\tau)\mathbf{F}_t(\tau) = \left(\frac{\partial\chi_\ell}{\partial x_i}\frac{\partial\chi_\ell}{\partial x_j}\right)\mathbf{e}_i\mathbf{e}_j \text{ right relative Cauchy-Green tensor .}$$

When τ=t

$$\begin{aligned} \boldsymbol{\chi}_t(\mathbf{x}, t) &= \mathbf{x}, \\ \mathbf{F}_t(t) &= \mathbf{1}, \\ \mathbf{C}_t(t) &= \mathbf{1}. \end{aligned} \tag{13}$$

Analysis of relative strain:

$$d\boldsymbol{\xi}\bullet d\boldsymbol{\xi}' = [\mathbf{C}_t d\mathbf{x}']\bullet d\mathbf{x} ,$$

$$\overset{(n)}{|d\xi|} = (\mathbf{A}_n[\mathbf{u}(\chi,\tau)]d\xi)\bullet d\xi , \quad (\mathbf{A}_1)_{ij} = \frac{\partial u_i}{\partial \chi_j} + \frac{\partial u_j}{\partial \chi_i}$$

where

$$\overset{(n)}{f}(t,\tau) \overset{\text{def}}{=} \frac{\partial^n f(t,\tau)}{\partial \tau^n} ,$$

and

$$\overset{(n)}{\mathbf{C}}_t(\tau) = \mathbf{F}_t^T(\tau)\mathbf{A}_n[\mathbf{u}]\mathbf{F}_t^T(\tau) . \tag{14}$$

When τ=t, $\mathbf{F}_t(t)=\mathbf{1}$ and

$$\overset{(n)}{\mathbf{C}}_t(t) \overset{\text{def}}{=} \left.\frac{\partial^n \mathbf{C}_t(\tau)}{\partial \tau^n}\right|_{\tau=t} = \mathbf{A}_n(t) \quad (= \mathbf{A}_n[\mathbf{u}(\mathbf{x},t)]) . \tag{15}$$

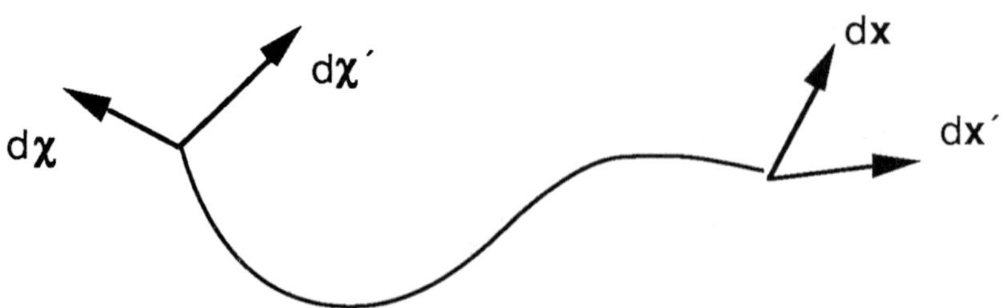

For analytic histories,

$$\mathbf{C}_t(\tau) = \mathbf{1} + \sum_{n=1}^{\infty} \frac{(\tau-t)^n}{n!} \mathbf{A}_n(t) \tag{16}$$

Path lines and velocity. We are given the velocity $\mathbf{u}(\chi_t(\mathbf{x},\tau),\tau)$ and must find the path line line $\boldsymbol{\chi}_t(\mathbf{x},\tau)$ by integrating

$$\frac{d\boldsymbol{\chi}_t}{d\tau} = \mathbf{u}(\boldsymbol{\chi}_t(\mathbf{x},\tau),\tau) \tag{17}$$

where $\mathbf{x}$ is fixed and

$$\boldsymbol{\chi}_t(\mathbf{x},t) = \mathbf{x} \ . \tag{18}$$

We will solve some problems in Cartesian coordinates and in polar cylindrical coordinates $(\tilde{r},\tilde{\theta},\tilde{z})$

$$d\boldsymbol{\chi}_t(\mathbf{x},\tau) = d\tilde{r}\,\mathbf{e}_r(\tau) + \tilde{r}\,d\tilde{\theta}\,\mathbf{e}_\theta(\tau) + d\tilde{z}\,\mathbf{e}_z \ .$$

Hence,

$$\frac{d\tilde{r}}{d\tau} = u(\tilde{r},\tilde{\theta},\tilde{z},\tau) \ ,$$

$$\tilde{r}\frac{d\tilde{\theta}}{d\tau} = v(\tilde{r},\tilde{\theta},\tilde{z},\tau) \ ,$$

$$\frac{d\tilde{z}}{d\tau} = w(\tilde{r},\tilde{\theta},\tilde{z},\tau)$$

where $[\tilde{r}(r,\theta,z,\tau), \tilde{\theta}(r,\theta,z,\tau), \tilde{z}(r,\theta,z,\tau)] = [r,\theta,z]$ when $\tau = t$.

Steady flows. Theorem: $\boldsymbol{\chi}_t(\mathbf{x},\tau) = \tilde{\boldsymbol{\chi}}(\mathbf{x},t-\tau)$; that is, τ and t enter only through $t-\tau$.

Proof (for the special case when $\boldsymbol{\chi}_t(\mathbf{x},\tau)$ is analytic in τ):

$$\begin{aligned}\boldsymbol{\chi}_t(\mathbf{x},\tau) &= \mathbf{x} + \dot{\boldsymbol{\chi}}_t\big|_{\tau=t}(\tau-t) + \frac{1}{2}\ddot{\boldsymbol{\chi}}_t\big|_{\tau=t}(\tau-t)^2 + \ldots \\ &= \mathbf{x} + \mathbf{u}(\mathbf{x},t)(\tau-t) + \frac{1}{2}\dot{\mathbf{u}}(\mathbf{x},t)(\tau-t)^2 + \ldots \ .\end{aligned}$$

In steady flow, $\mathbf{u}(\mathbf{x},t)$ is independent of t and

$$\dot{\mathbf{u}} = (\mathbf{u}\cdot\nabla)\mathbf{u} \,, \quad \text{etc.}$$

Rigid motions. A rigid motion is a motion in which no material line is stretched.

$$|d\boldsymbol{\chi}_t(\mathbf{x},\tau)|^2 = |d\mathbf{x}|^2 \quad \text{for } \tau\le t \text{ where } \mathbf{x} = \boldsymbol{\chi}_t(\mathbf{x},\tau) \,.$$

In a rigid motion,

$$\mathbf{C}_t(\tau) = \mathbf{1} \,,$$

$$\mathbf{F}_t(\tau) = \mathbf{F}_\tau^T(\tau); \quad \text{that is, } \mathbf{F}_t(\tau) = \mathbf{Q}(\tau) \text{ is orthogonal and } \mathbf{Q}(t) = \mathbf{1} \,,$$

$$\boldsymbol{\chi}_t(\mathbf{x},\tau) = \mathbf{Q}(\tau)\mathbf{x} \,, \tag{19}$$

$$\mathbf{L} = \frac{d}{d\tau}\,\mathbf{Q}(\tau) \,.$$

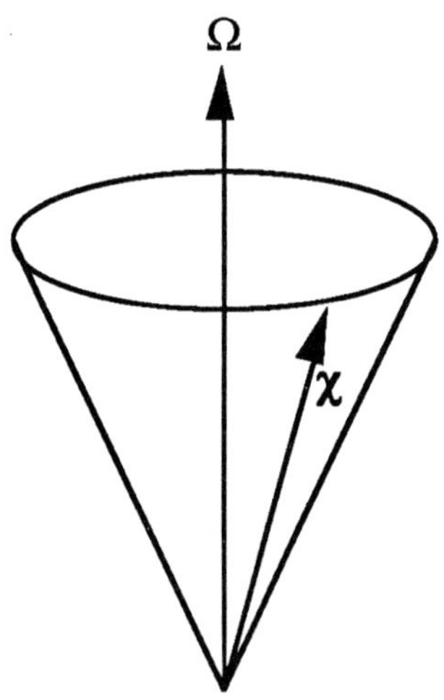

In a two-dimensional steady rigid motion,

$$\mathbf{u}(\boldsymbol{\chi},\tau) = \boldsymbol{\Omega}\wedge\boldsymbol{\chi}\ , \quad \boldsymbol{\Omega} \text{ is a constant vector,}$$
$$\boldsymbol{\chi}_t(\mathbf{x},\tau) = \mathbf{Q}(\Omega s)\mathbf{x} \quad (\Omega = |\boldsymbol{\Omega}|\ , \quad s = t-\tau)\ ,$$
$$\mathbf{Q}(0) = \mathbf{1}\ ,$$

$$[\mathbf{Q}(\Omega s)] = \begin{vmatrix} \cos\Omega s & \sin\Omega s & 0 \\ -\sin\Omega s & \cos\Omega s & 0 \\ 0 & 0 & 1 \end{vmatrix}\ , \tag{20}$$

$$\mathbf{e}_i(0) = \mathbf{Q}(\Omega s)\mathbf{e}_i(\Omega s)$$

Special flow:

i. Simple shear

$$\boldsymbol{\chi}_t(\mathbf{x},\tau) = (\chi_1,\chi_2,\chi_3)\ ,$$
$$\boldsymbol{\chi}_t(\mathbf{x},t) = (x_1,x_2,x_3)\ ,$$
$$\mathbf{u} = (u_1,u_2,u_3) = \left(\frac{d\chi_1}{d\tau}\frac{d\chi_2}{d\tau}\frac{d\chi_3}{d\tau}\right) = (\dot{\gamma}x_2,0,0)\ . \tag{21}$$

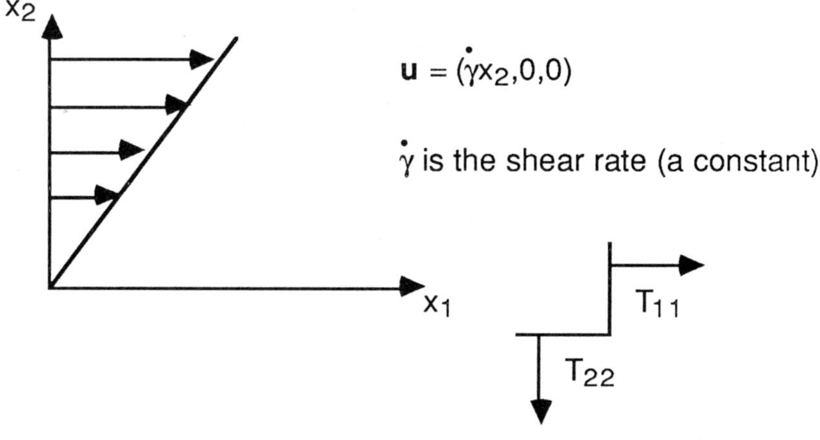

Hence,

$$\chi_2 = x_2 \ ,$$
$$\chi_3 = x_3 \ ,$$

$$\frac{d\chi_1}{d\tau} = \dot{\gamma}\chi_2 \ , \quad \text{integrate with } \chi_2 = x_2 \text{ at } \tau = t \text{ to get} \tag{22}$$

$$\chi_1(x_1,x_2,x_3,\tau,t) = x_1 - \dot{\gamma}x_2(t-\tau) \ ,$$

$$[\mathbf{L}(\tau)] = [\mathbf{L}(t)] = \begin{vmatrix} 0 & \dot{\gamma} & 0 \\ 0 & 0 & 0 \\ 0 & 0 & 0 \end{vmatrix} \ ,$$

$$[\mathbf{F}_t(\tau)] = \begin{vmatrix} 1 & -\dot{\gamma}(t-\tau) & 0 \\ 0 & 1 & 0 \\ 0 & 0 & 0 \end{vmatrix} \ ,$$

$$\mathbf{F}_t(\tau) = \mathbf{1} - s\mathbf{L} \ , \quad s = t - \tau \ ,$$

$$\mathbf{A}_1 = \begin{vmatrix} 0 & \dot{\gamma} & 0 \\ \dot{\gamma} & 0 & 0 \\ 0 & 0 & 0 \end{vmatrix} \ , \tag{23}$$

$$\mathbf{A}_2 = (\mathbf{u}\cdot\nabla)\mathbf{A}_1 + \mathbf{A}_1\mathbf{L} + \mathbf{L}^T\mathbf{A}_1 \ ,$$

$$(\mathbf{u}\cdot\nabla)\mathbf{A}_1 = 0 \ ,$$

$$[\mathbf{A}_1\mathbf{L}] = \begin{vmatrix} 0 & 0 & 0 \\ 0 & \dot{\gamma}^2 & 0 \\ 0 & 0 & 0 \end{vmatrix} \ , \quad [\mathbf{A}_2] = \begin{vmatrix} 0 & 0 & 0 \\ \dot{\gamma}^2 & 0 & 0 \\ 0 & 0 & 0 \end{vmatrix} \ , \tag{24}$$

$$\mathbf{A}_n = \mathbf{0} \quad \text{for} \quad n > 2 \ . \tag{25}$$

ii. Simple extension

$$(u_1,u_2,u_3) = (a_1\chi_1,a_2\chi_2,a_3\chi_3) = (\dot{\chi}_1,\dot{\chi}_2,\dot{\chi}_3)\ , \quad \dot{\boldsymbol{\chi}} = \frac{d\boldsymbol{\chi}}{d\tau}$$

$$\frac{d\chi_1}{d\tau} = a_1\chi_1\ , \quad \chi_1 = x_1 \quad \text{at} \quad \tau = t\ , \quad a_1 \text{ const}$$

$$\begin{aligned} \chi_1 &= x_1 e^{-a_1 s}\ , \\ \chi_2 &= x_2 e^{-a_2 s}\ , \\ \chi_3 &= x_2 e^{-a_3 s}\ . \\ s &= t-\tau\ . \end{aligned} \qquad (26)$$

$$[\mathbf{F}_t] = \begin{vmatrix} e^{-a_1 s} & 0 & 0 \\ 0 & e^{-a_2 s} & 0 \\ 0 & 0 & e^{-a_3 s} \end{vmatrix} = [\mathbf{F}_t^T]\ , \qquad (27)$$

$$[\mathbf{L}] = \begin{vmatrix} a_1 & 0 & 0 \\ 0 & a_2 & 0 \\ 0 & 0 & a_3 \end{vmatrix} = [\mathbf{L}^T]\ , \qquad (28)$$

$$\mathbf{L} = \frac{1}{2}\mathbf{A}_1\ .$$

In simple extension, we have

$$\mathbf{A}_n = \mathbf{A}^n \quad (n \geq 1)$$

$$\mathbf{C}_t(\tau) = \exp(-2\mathbf{A}_1 s)\ ,$$

$$[\mathbf{C}_t(\tau)] = \begin{vmatrix} \exp(-2a_1 s) & 0 & 0 \\ 0 & \exp(-2a_2 s) & 0 \\ 0 & 0 & \exp(-2a_3 s) \end{vmatrix}\ , \qquad (29)$$

where $a_1+a_2+a_3=0$ for isochoric flow.

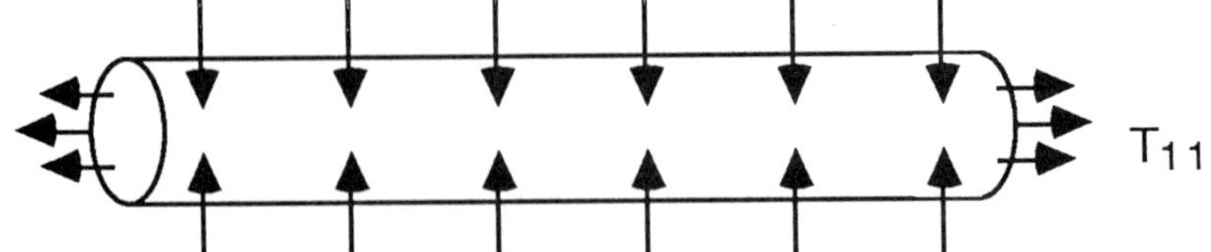

iii. Simple cylindrical extension as an isochoric motion in which two of the a_I's are equal

$a_1 = \alpha$, $a_2 = a_3 = -\alpha/2$, α is the extension rate.

Radial symmetry

$$\dot{x} = \alpha x \ ,$$

$$\dot{\theta} = 0 \ ,$$

$$\dot{r} = -\alpha r \ .$$

iv. Stagnation flow of an ideal fluid

$$[\mathbf{L}] = \begin{vmatrix} \alpha & 0 & 0 \\ 0 & 0 & 0 \\ 0 & 0 & -\alpha \end{vmatrix} = [\mathbf{L}^T] \ ,$$

Irrotationality. The antisymmetric part of **L** is zero.

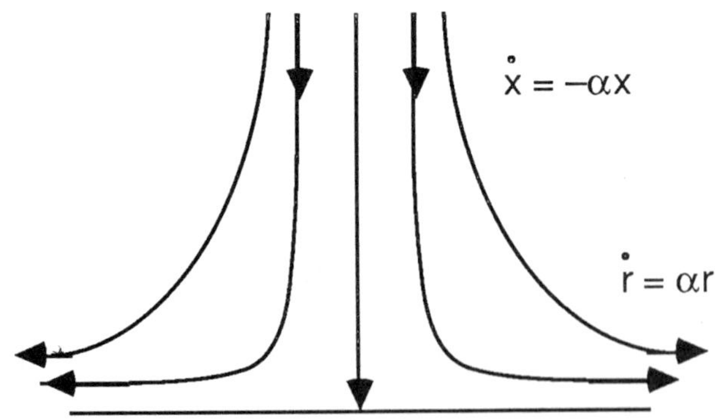

Exercises

Exercise C.1. Prove (10).

Exercise C.2. Prove that $\mathbf{F}_t(\tau) = \mathbf{F}_t(\tau)\mathbf{F}^T(t)$.

Exercise C.3. Prove (14).

Exercise C.4. Prove (19).

Exercise C.5. Prove (20).

Exercise C.6. Prove (29).

Solutions for the exercises

Exercise C.1. Prove that $\mathbf{A}_{n+1} = \dot{\mathbf{A}}_n + \mathbf{A}_n\mathbf{L} + \mathbf{L}^T\mathbf{A}_n$.

$$\mathbf{A}_n = [\dot{\overline{(\mathbf{F}^T)^{-1}}}]\,\overset{(n)}{\mathbf{C}}\mathbf{F}^{-1} + \mathbf{F}^{T^{-1}}\,\overset{(n+1)}{\mathbf{C}}\;\mathbf{F}^{-1} + \mathbf{F}^{T^{-1}}\,\overset{(n)}{\mathbf{C}}\dot{\overline{\mathbf{F}^{-1}}}\ .$$

But $\mathbf{1} = \mathbf{F}^{-1}\mathbf{F}$, so $\mathbf{0} = \dot{\mathbf{F}}^{-1}\mathbf{F} + \mathbf{F}^{-1}\dot{\mathbf{F}}$. Hence, $\dot{\mathbf{F}}^{-1} = -\mathbf{F}^{-1}\dot{\mathbf{F}}\mathbf{F}^{-1}$ and

$$\begin{aligned}\dot{\mathbf{A}}_n &= -\mathbf{F}^{T^{-1}}\dot{\mathbf{F}}^T\mathbf{F}^{T^{-1}}\,\overset{(n)}{\mathbf{C}}\mathbf{F}^{-1} + \mathbf{A}_{n+1} - \mathbf{F}^{T^{-1}}\,\overset{(n)}{\mathbf{C}}\mathbf{F}^{-1}\dot{\mathbf{F}}\mathbf{F}^{-1} \\ &= -\mathbf{F}^{T^{-1}}\dot{\mathbf{F}}^T\mathbf{A}_n + \mathbf{A}_{n+1} - \mathbf{A}_n\dot{\mathbf{F}}\mathbf{F}^{-1}\ .\end{aligned}$$

When $\tau{=}t$, $\dot{\mathbf{F}}{=}\mathbf{L}$, $\mathbf{F}{=}\mathbf{1}$.

Exercise C.2. Prove that $\mathbf{F}_t(\tau) = \mathbf{F}(\tau)\mathbf{F}^{-1}(t)$. It is proved by inspection

$$(\mathrm{F}_t)_{ij} = \frac{\partial\chi_i}{\partial x_j} = \frac{\partial\chi_i}{\partial X_\ell}\frac{\partial X_\ell}{\partial x_j} = [\mathbf{F}(\tau)\mathbf{F}^{-1}(t)]_{ij}\ .$$

It follows that

$$\mathbf{C}_t(\tau) = \mathbf{F}_t^T(\tau)\mathbf{F}_t(\tau) = \mathbf{F}^{T^{-1}}(t)\mathbf{F}^T(\tau)\mathbf{F}(\tau)\mathbf{F}^{-1}(t)$$
$$= \mathbf{F}^{T^{-1}}(t)\mathbf{C}(\tau)\mathbf{F}^{-1}(t)$$

Exercise C.3. Prove that $\overset{(n)}{\mathbf{C}}_t(\tau) = \mathbf{F}_t^T(\tau)\mathbf{A}_n(\tau)\mathbf{F}_t(\tau)$. From

Exercise C.2, we have

$$\overset{(n)}{\mathbf{C}}_t(\tau) = \mathbf{F}^{T^{-1}}(t)\overset{(n)}{\mathbf{C}}(\tau)\mathbf{F}^{-1}(t)$$

and $\mathbf{F}^{-1}(t) = \mathbf{F}^{-1}(\tau)\mathbf{F}_t(\tau)$ where $\mathbf{A}_n(\tau) = \mathbf{F}^{T^{-1}}\overset{(n)}{\mathbf{C}}\mathbf{F}^{-1}$.

Exercise C.4. Prove (19). In a rigid motion, each and every material line moves without changing its length. In general,

$$|d\boldsymbol{\chi}|^2 = (\mathbf{C}_t(\tau)d\mathbf{x})\bullet d\mathbf{x}$$

where $\mathbf{C}_t = \mathbf{F}_t^T\mathbf{F}_t$ is positive definite and symmetric.

Refer $\mathbf{C}_t(\tau)$ to principal axis

$$|d\boldsymbol{\chi}|^2 = \lambda_1 dx_1^2 + \lambda_2\, dx_2^2 + \lambda_3 dx_3^2$$

where λ_1's are not negative. Moreover, $|d\mathbf{x}|^2=|d\boldsymbol{\chi}|^2$ because the motion is rigid. Choose $d\mathbf{x}=\mathbf{e}_1 dx_1$. Then $dx_1^2=|d\boldsymbol{\chi}|^2$, and it follows that $\lambda_1=1$. Similarly, $\lambda_2=\lambda_3=1$. So $\mathbf{C}_t=\mathbf{1}$. Now, we have

$$\frac{\partial\chi_m}{\partial x_i}\frac{\partial\chi_m}{\partial x_j} = \delta_{ij}\,,$$

$$\frac{\partial^2\chi_m}{\partial x_i\partial x_k}\frac{\partial\chi_m}{\partial x_j} + \frac{\partial\chi_m}{\partial x_i}\frac{\partial^2\chi_m}{\partial x_j\partial x_k} = 0\,, \tag{30}$$

$$\frac{\partial^2\chi_m}{\partial x_k\partial x_j}\frac{\partial\chi_m}{\partial x_i} + \frac{\partial\chi_m}{\partial x_k}\frac{\partial^2\chi_m}{\partial x_i\partial x_j} = 0\,, \tag{31}$$

$$\frac{\partial^2 \chi_m}{\partial x_j \partial x_i} \frac{\partial \chi_m}{\partial x_k} + \frac{\partial \chi_m}{\partial x_j} \frac{\partial^2 \chi_m}{\partial x_k \partial x_i} = 0 , \qquad (32)$$

$$(30) + (31) - (32) \text{ gives } \frac{\partial^2 \chi_m}{\partial x_k \partial x_i} \frac{\partial \chi_m}{\partial x_j} = 0 ,$$

that is,

$$2 \frac{\partial \mathbf{F}_t^T}{\partial x_i} \mathbf{F}_t = 0 .$$

Since $\mathbf{F}_t$ is not singular, $\partial \mathbf{F}_t^T / \partial x_i = 0$ and

$$\frac{\partial^2 \chi_p}{\partial x_i \partial x_k} = 0 , \quad \chi_p = A_{pq} x_q + C_p , \quad A_{pq} = \frac{\partial \chi_p}{\partial x_q} .$$

$\boldsymbol{\chi} = \mathbf{A}\mathbf{x} + \mathbf{C}$, $\mathbf{A}(\tau)$ is independent of $\mathbf{x}$. Choose an origin so that $\boldsymbol{\chi} = \mathbf{0}$. Then $\mathbf{C} = 0$. We find that

$$\mathbf{A}\mathbf{A}^T = \mathbf{1} , \quad \text{hence, } \mathbf{A} = \mathbf{Q} .$$

Exercise C.5. Prove (20).

$$\frac{d\boldsymbol{\chi}_t}{d\tau} = \boldsymbol{\Omega} \wedge \boldsymbol{\chi}_t , \quad \boldsymbol{\Omega} \text{ constant}$$

Choose coordinates

$$\boldsymbol{\Omega} = \mathbf{e}_3 \Omega .$$

Then,

$$\boldsymbol{\chi}_t = \mathbf{e}_1 \chi_1 + \mathbf{e}_2 \chi_2 + \mathbf{e}_3 \chi_3 , \quad \mathbf{e}_1 \text{ and } \mathbf{e}_2 \text{ depend on t but not } \tau$$

$$\frac{d\chi_1}{d\tau} = -\Omega \chi_2 , \quad \chi_1 = x_1 \quad \text{at} \quad \tau = t$$

$$\frac{d\chi_2}{d\tau} = \Omega \chi_1 , \quad \chi_2 = x_2 \quad \text{at} \quad \tau = t$$

$$\frac{d^2 \chi_j}{d\tau^2} + \Omega^2 \chi_j = 0 , \quad j = 1,2 .$$

After integrating and applying the condition $\boldsymbol{\chi}(t) = \mathbf{x}$, we get

$$\chi_t = \mathbf{e}_1(t)[x_1 \cos\Omega s + x_2 \sin\Omega s]$$
$$+ \mathbf{e}_2(t)[-x_1 \sin\Omega s + x_2 \cos\Omega s] + \mathbf{e}_3 = \mathbf{Q}(\Omega s)\mathbf{x} .$$

Exercise C.6. Prove (29). First compute

$$\mathbf{A}_2 = (\mathbf{u}\bullet\nabla)\mathbf{A}_1 + \mathbf{A}_1\mathbf{L} + \mathbf{L}^T\mathbf{A}_1$$

using (28), $(\mathbf{u}\bullet\nabla)\mathbf{A}_1$, $\mathbf{A}_1\mathbf{L} = \frac{1}{2}\mathbf{A}_1^2$. Then use recursion formula for $\mathbf{A}_3=\mathbf{A}_1^3$, etc., to get $\mathbf{A}_n=\mathbf{A}_1^n$. The formulas $(29)_{2,3}$ are the immediate consequence of (27) and $\mathbf{C}_t = \mathbf{F}_t^T\mathbf{F}_t$.

APPENDIX D

Stream Function–Angular Momentum, Decomposition for Axisymmetric Flow

The stress in an incompressible fluid is $\mathbf{T}=-p\mathbf{1}+\mathbf{S}$. The equations of motion are

$$\rho \frac{d\mathbf{U}}{dt} = -\nabla\Phi + \operatorname{div}\mathbf{S} \tag{1}$$

where $\Phi=p+\rho gz$. We use cylindrical coordinates (r,θ,z) with velocity components (u,v,w). We define

$$\begin{aligned} \mathbf{u} &= \mathbf{e}_r u + \mathbf{e}_z w\,, \\ \nabla_2 &= \mathbf{e}_r \frac{\partial}{\partial r} + \mathbf{e}_z \frac{\partial}{\partial z}\,. \end{aligned} \tag{2}$$

Hence

$$\begin{aligned} \mathbf{U} &= \mathbf{e}_\theta v + \mathbf{u}\,, \\ \nabla &= \frac{\mathbf{e}_\theta}{r}\frac{\partial}{\partial\theta} + \nabla_2\,. \end{aligned} \tag{3}$$

The equations of motion in cylindrical coordinates are given in Exercises for Chapter 10. Axisymmetric flows are independent of θ. Hence θ derivatives do not appear in the continuity equation and one may satisfy the continuity equation identically with a stream function $\Psi(r,z)$

$$(u,w) = \frac{1}{r}(-\partial_z\Psi,\partial_r\Psi)\,. \tag{4}$$

It is also convenient to introduce a rotation function Ω

$$v = \frac{\Omega(r,z)}{r} = r\omega(r,z) \tag{5}$$

where $\Omega=r^2\omega$ and ω is the angular velocity function. Then

$$\mathbf{U} = \Omega\boldsymbol{\lambda} + \nabla_2\Psi\wedge\boldsymbol{\lambda}\,,\quad \boldsymbol{\lambda} \overset{\text{def}}{=} \frac{\mathbf{e}_\theta}{r}\,. \tag{6}$$

We may compute successive curls of $\mathbf{U}$ using (15) and (17).

$$\boldsymbol{\zeta} = \text{curl}\,\mathbf{U} = \nabla_2\Omega\wedge\boldsymbol{\lambda} - \boldsymbol{\lambda}D^2\Psi\,, \tag{7}$$

$$\text{curl}\,\boldsymbol{\zeta} = \nabla_2 D^2\Psi\wedge\boldsymbol{\lambda} - \boldsymbol{\lambda}D^2\Omega\,, \tag{8}$$

$$\text{curl}^2\,\boldsymbol{\zeta} = -\nabla_2 D^2\Omega\wedge\boldsymbol{\lambda} + \boldsymbol{\lambda}D^4\Psi\,. \tag{9}$$

We may write the equations of motion as

$$\rho\left[\frac{\partial\mathbf{U}}{\partial t} + \boldsymbol{\zeta}\wedge\mathbf{U}\right] = -\nabla\left(\Phi + \frac{\rho}{2}|\mathbf{U}|^2\right) + \mu\nabla^2\mathbf{U} + \text{div}\,\mathbf{S}_2 \tag{10}$$

where

$$\mathbf{S} = \mu\mathbf{A}_1[\mathbf{U}] + \mathbf{S}_2\,. \tag{11}$$

The vorticity equation is

$$\rho\left[\frac{\partial\boldsymbol{\zeta}}{\partial t} + \text{curl}\,\boldsymbol{\zeta}\wedge\mathbf{U}\right] = -\mu\,\text{curl}^2\,\boldsymbol{\zeta} + \text{curl div}\,\mathbf{S}_2\,. \tag{12}$$

Using (6) and (7) we compute

$$\begin{aligned}\boldsymbol{\zeta}\wedge\mathbf{U} = &-\boldsymbol{\lambda}\left[\frac{\partial\Omega}{\partial r}\frac{\partial\Psi}{\partial z} - \frac{\partial\Omega}{\partial z}\frac{\partial\Psi}{\partial r}\right]\frac{1}{r}\\ &-\mathbf{e}_r\left[\frac{\Omega}{r^2}\frac{\partial\Omega}{\partial r} + \frac{D^2\Psi}{r^2}\frac{\partial\Psi}{\partial r}\right]\\ &-\mathbf{e}_z\left[\frac{\Omega}{r^2}\frac{\partial\Omega}{\partial z} + \frac{D^2\Psi}{r^2}\frac{\partial\Psi}{\partial z}\right].\end{aligned}$$

The equation governing Ω is the θ component of (10)

$$\rho\left[\frac{1}{r}\frac{\partial\Omega}{\partial t} + \frac{1}{r^2}\left(\frac{\partial\Omega}{\partial z}\frac{\partial\Psi}{\partial r} - \frac{\partial\,\Omega}{\partial r}\frac{\partial\Psi}{\partial z}\right)\right] = \frac{\mu}{r}D^2\Psi + \mathbf{e}_\theta\cdot\text{div}\,\boldsymbol{\tau} \tag{13}$$

where $\mathbf{e}_\theta\cdot\text{div}\,\boldsymbol{\tau}$ is given by (19). The equation governing Ψ is the θ component of (12)

$$\begin{aligned}&\rho\left\{\frac{1}{r}\frac{\partial D^2\Psi}{\partial t} + \frac{1}{r^2}\frac{\partial\Psi}{\partial r}\frac{\partial D^2\Psi}{\partial z} - \frac{\partial\Psi}{\partial z}\frac{\partial}{\partial r}\left(\frac{1}{r^2}D^2\Psi\right)\right\}\\ &\quad + \frac{\rho}{2r^3}\Omega\frac{\partial\Omega}{\partial z} = \frac{\mu}{r}D^4\Psi - \mathbf{e}_\theta\cdot\text{curl div}\,\boldsymbol{\sigma}\end{aligned} \tag{14}$$

where $\mathbf{e}_\theta\cdot\text{curl div}\,\boldsymbol{\sigma}$ is given by (20).

Exercises

Exercise D.1. Suppose G(r,z), $F_r(r,z)$, and $F_z(r,z)$ are arbitrary sufficiently smooth functions where $\mathbf{F}=\mathbf{e}_r F_r+\mathbf{e}_z F_z$. Then

$$\nabla\wedge\boldsymbol{\lambda}G = \nabla_2 G\wedge\boldsymbol{\lambda}\ , \tag{15}$$

$$\text{curl}\left(\mathbf{F}\wedge\boldsymbol{\lambda}\right) = -\boldsymbol{\lambda}\left[r\frac{\partial\frac{1}{r}F_r}{\partial r}+\frac{\partial F_z}{\partial z}\right] \tag{16}$$

where $\boldsymbol{\lambda}=\mathbf{e}_\theta/r$. If $\mathbf{F}=\nabla_2\Psi$, then

$$\text{curl}\left(\nabla_2\Psi\wedge\boldsymbol{\lambda}\right) = -\boldsymbol{\lambda}D^2\Psi \tag{17}$$

where

$$D^2 = \left(\frac{\partial^2\Psi}{\partial r^2}-\frac{1}{r}\frac{\partial\Psi}{\partial r}+\frac{\partial^2\Psi}{\partial z^2}\right).$$

Exercise D.2. Define

$$\boldsymbol{\sigma} = \mathbf{e}_r\mathbf{e}_r S_{rr} + [\mathbf{e}_z\mathbf{e}_r + \mathbf{e}_r\mathbf{e}_z]S_{rz} + \mathbf{e}_z\mathbf{e}_z S_{zz} + \mathbf{e}_\theta\mathbf{e}_\theta S_{\theta\theta}\ ,$$
$$\boldsymbol{\tau} = [\mathbf{e}_r\mathbf{e}_\theta + \mathbf{e}_\theta\mathbf{e}_r]S_{r\theta} + [\mathbf{e}_z\mathbf{e}_\theta + \mathbf{e}_\theta\mathbf{e}_z]S_{z\theta}$$

where

$$\mathbf{S}_2 = \boldsymbol{\sigma} + \boldsymbol{\tau}\ . \tag{18}$$

Show that

$$\mathbf{e}_\theta\bullet\text{div}\ \mathbf{S} = \mathbf{e}_\theta\bullet\text{div}\ \boldsymbol{\tau} = \partial_r S_{r\theta} + \partial_z S_{z\theta} + \frac{2}{r}S_{r\theta} \tag{19}$$

and

$$\mathbf{e}_\theta\bullet\text{curl div}\ \mathbf{S} = \mathbf{e}_\theta\bullet\text{curl div}\ \boldsymbol{\sigma}$$

$$=\left[\frac{\partial^2}{\partial z^2}-\frac{1}{r}\frac{\partial}{\partial r}\left(r\frac{\partial}{\partial r}\right)\right]S_{rz} + \frac{1}{r}\frac{\partial^2}{\partial r\partial z}\left[r(S_{rr}-S_{zz})\right]\ . \tag{20}$$

APPENDIX E

Domain Perturbation

This appendix is taken from the paper, "Free Surface Problems in Rheological Fluid Mechanics," by D.D. Joseph and G.S. Beavers. That paper presented a theory of domain perturbations for application to rheological problems involving free surfaces. A sequence of ever more involved model problems which exhibit all the details of the theory in a simple way was presented. These model problems are presented below. They are applicable to all sorts of problems and the allusion to rheological problems can be entirely suppressed by readers interested in other applications. Two types of problems will be considered: prescribed domain deformations and free surface problems.

We suppose that there is an analytic function $F(\phi)$ which is not known that in a neighborhood of $\phi=0$ has a Taylor series

$$F(\phi) = \frac{1}{2}\,a\phi^2 + \frac{1}{3!}\,b\phi^3 + \dots\,, \tag{1}$$

where $a=F_{\phi\phi}(0)$ and $b=F_{\phi\phi\phi}(0)$. We may think that $F(\phi)$ is representative of the nonlinear part of the stress in some fictitious material. Our idea is to find the Taylor coefficients for (1) by measuring the free surface induced by a dynamical process

$$\mathrm{F}(\mathbf{x}) = \nabla^2\phi + F(\phi) = 0 \quad \text{in} \quad V_\delta\,, \tag{2}$$

subject to the condition that

$$\mathrm{G}(\mathbf{x},\varepsilon) = \phi(\mathbf{x}) - \mathrm{g}(\mathbf{x},\varepsilon) = 0 \quad \text{on} \quad \partial V_\delta\,, \tag{3}$$

where V_δ is a bounded region of space depending on a parameter δ and $\mathrm{g}(\mathbf{x},\varepsilon)$ are given data depending on a parameter ε.

It is instructive to carry out our analysis in easy stages. First we perturb ε, leaving δ fixed and assume that

$$\phi(\mathbf{x},\varepsilon) = \sum_{\ell=0} \frac{\varepsilon^\ell}{\ell\,!}\,\phi_\ell(\mathbf{x})\,, \tag{4}$$

where, in V_δ,

$$\nabla^2\phi_0 + F(\phi_0) = 0\ ,$$
$$\nabla^2\phi_1 + F_\phi(\phi_0)\phi_1 = 0\ , \tag{5}$$
$$\nabla^2\phi_2 + F_\phi(\phi_0)\phi_2 + F_{\phi\phi}(\phi_0)\phi_1^2 = 0\ ,$$

and, on ∂V_δ ;

$$G_n = \phi_n(\mathbf{x}) - g_n(\mathbf{x}) = 0\ , \tag{6}$$

where $g_n(\mathbf{x})$ is nth derivative of $g(\mathbf{x},\varepsilon)$ at $\varepsilon=0$. If we could solve (5) and (6) for ϕ_0 we could find ϕ_1, ϕ_2, etc. as the solution of linear boundary value problems. It would be hard to solve $(5)_1$, even if we knew $F(\phi_0)$, because it is nonlinear. But we cannot solve $(5)_1$ at all because we don't know $F(\phi_0)$ except as a power series in a small neighborhood of zero.

The "rest state" is now defined as the state for which $g_0(\mathbf{x})|_{\partial V_\delta}=0$. We couple this definition with the assumption that $(5)_1$ has no solutions $\phi_0\neq 0$ when $\phi_0(\mathbf{x})|_{\partial V_\delta}=0$. Then $\phi_0(\mathbf{x})\equiv 0$ in V_δ and, replacing $(5)_{2,3}$ we get

$$\nabla^2\phi_1|_{V_\delta} = 0\ ,\quad \phi_1|_{\partial V_\delta} = g_1\ ;$$
$$\nabla^2\phi_2 + a\phi_1^2|_{V_\delta} = 0\ ,\quad \phi_2|_{\partial V_\delta} = g_2\ , \tag{7}$$

etc. These problems are linear and easy to solve even when a is unknown (in fact, $\phi_2=\phi_{21}+a\phi_{22}$, where ϕ_{21} and ϕ_{22} are independent of a). So we can use the solution which perturbs the rest state to find a and, with more work, we can get expressions which involve the other Taylor coefficients of (1) as well.

Now suppose that ε is fixed and δ varies. And suppose further that V_0 is some convenient domain whose point of convenience is, say, that V_0 has a high degree of symmetry. We are going to try to solve the dynamic problem in V_δ as a series whose coefficients can be determined from boundary value problems posed on the symmetric domain V_0.

First we map V_δ into V_0 with an invertible one-to-one mapping, which is analytic in δ and which takes points on ∂V_δ into ∂V_0:

$$\begin{gathered}
\mathbf{x} = \mathbf{x}(\mathbf{x}_0,\delta) = \sum_0 \frac{\delta^n}{n!}\mathbf{x}^{(n)}(\mathbf{x}_0) \quad \text{(analytic in } \delta) , \\
\mathbf{x}_0 = \mathbf{x}(\mathbf{x}_0,0) \quad \text{(identity)} , \\
\mathbf{x}_0 = \mathbf{x}^{-1}(\mathbf{x},\delta) \quad \text{(inverse)} , \\
\partial V_\delta \leftrightarrow \partial V_0 .
\end{gathered} \tag{8}$$

Let $H(\mathbf{x},\delta)$ be any function defined in the family of domains V_δ and introduce the notation:

$$\begin{gathered}
\tilde{H}(\delta) = H(\mathbf{x}(\mathbf{x}_0,\delta),\delta) , \\
H^{[n]}(\mathbf{x}_0) = \frac{d^n}{d\delta^n}\tilde{H}(0)
\end{gathered}$$

and

$$H^{<n>}(\mathbf{x}_0) = \left.\frac{\partial^n H(\mathbf{x},\delta)}{\partial\delta^n}\right|_{\substack{\delta=0\\ \mathbf{x}=\mathbf{x}_0}} .$$

Connection formulas (the chain rule) connect $H^{[n]}$ and $H^{<n>}$:

$$\begin{gathered}
H^{[0]}(\mathbf{x}_0) = H^{<0>}(\mathbf{x}_0) = H(\mathbf{x}_0,0) , \\
H^{[1]}(\mathbf{x}_0) = H^{<1>}(\mathbf{x}_0) + \mathbf{x}^{[1]}\cdot\nabla H^{<0>} , \\
H^{[2]}(\mathbf{x}_0) = H^{<2>}(\mathbf{x}_0) + 2\mathbf{x}^{[1]}\cdot\nabla H^{<1>} \\
+ \mathbf{x}^{[2]}\cdot\nabla H^{<0>} + (\mathbf{x}^{[1]}\cdot\nabla)^2 H^{<0>},
\end{gathered} \tag{9}$$

etc. It follows from the equations that

$$\tilde{F}(\delta) = F(\mathbf{x}(\mathbf{x}_0,\delta),\delta) = 0 ,$$

when $\mathbf{x}\in V_\delta$, $\mathbf{x}_0\in V_0$, and that

$$\tilde{F}^{[n]}(0) = F^{[n]}(\mathbf{x}_0) = 0 \quad \text{in } V_0 . \tag{10}$$

We may easily establish by mathematical induction, using the connection formulas, that

$$F^{<n>}(\mathbf{x}_0) = 0 \quad \text{in } V_0 . \tag{11}$$

For example, $F^{<0>}(\mathbf{x}_0) = 0$ in V_0 and $F^{[1]}(\mathbf{x}_0) = F^{<1>}(\mathbf{x}_0)+\mathbf{x}^{[1]}\cdot\nabla F^{<0>}(\mathbf{x}_0) = F^{<1>}(\mathbf{x}_0) = 0$. It is a bit more complicated at the boundary. Since $\tilde{G}(\delta)=0$ on ∂V_δ, $G^{[m]}(\mathbf{x}_0)=0$ on ∂V_0 and tangential derivatives of $G^{[m]}$ on V_0 must vanish but normal derivatives need not vanish. It follows that on ∂V_0

$$G^{[m]}(\mathbf{x}_0) = \left\{\left(\frac{\partial}{\partial\varepsilon} + v_n \frac{\partial}{\partial n}\right)^m G(\mathbf{x},\varepsilon)\right\}_{\substack{\varepsilon=0\\ \mathbf{x}=\mathbf{x}_0}} = 0\,, \tag{12}$$

where $\mathbf{n}$ is the outward normal to V_δ, $v_n=\mathbf{x}^{[1]}\cdot\mathbf{n}$ and $\mathbf{n}\cdot\nabla=\partial/\partial n$.

We may seek the solution of (2) and (3) in V_δ as a series defined in V_0

$$\phi(\mathbf{x}(\mathbf{x}_0,\delta),\delta) = \sum_0 \frac{\delta^n}{n!}\,\phi^{[n]}(\mathbf{x}_0)\,, \tag{13}$$

where, in V_0,

$$\begin{gathered}\nabla^2\phi^{<0>} + F(\phi^{<0>}) = 0\,,\\ \nabla^2\phi^{<1>} + F_\phi(\phi^{<0>})\phi^{<1>} = 0\,,\\ \nabla^2\phi^{<2>} + F_\phi(\phi^{<0>})\phi^{<2>} + F_{\phi\phi}(\phi^{<0>}){\phi^{<1>}}^2 = 0\,,\end{gathered} \tag{14}$$

etc., and on ∂V_0

$$G^{[n]} = \phi^{[n]}(\mathbf{x}_0) - g^{[n]}(\mathbf{x}_0)\,. \tag{15}$$

The problems (14), (15) are like (5) and (6). We can't solve them because $F(\phi)$ is given only as a Taylor series (1) with unknown coefficients and an unknown circle of convergence.

The "rest state" for the domain perturbation, like the "rest state" for the perturbation of the boundary data, may be defined by the condition $g^{[0]}(\mathbf{x}_0)|_{\partial V_0}=0$. This condition implies that $\phi^{[0]}=\phi^{<0>}=0$ on ∂V_0, hence, $\phi^{<0>}\equiv 0$ in V_0 and

$$\nabla^2\phi^{<1>}|_{V_0} = 0\,,\quad \phi^{<1>}|_{\partial V_0} = g^{[1]}(\mathbf{x}_0)\,, \tag{16}$$

$$\nabla^2\phi^{<2>} + a\phi^{<1>^2}|_{V_0} = 0 \,,$$
$$\phi^{[2]}|_{\partial V_0} = \phi^{<2>} + 2V_n \frac{\partial}{\partial n}\phi^{<1>} = g^{[2]}(\mathbf{x}_0) \,, \tag{17}$$

etc. The linear problems (16), (17) and higher order problems are solvable and not too hard to actually solve, even when a is unknown.

In our rheological problems the boundary data (ε in our first example) perturb the boundary (δ in our second example). So we may put $\varepsilon=\delta$ and construct a simple example of a domain perturbation of the rest state with a free surface. By a free surface we understand that there is a one (ε) parameter family of domains V_ε which are unknown. Supposing now that our dynamical processes (1) and (2) hold in V_ε we might expect solutions in each and every V_ε interval of the origin. But no, this will not be possible because in addition to (2) we pose an additional boundary condition, which is analogous to, but much simpler than, the condition that the jump in the normal component of stress is balanced by surface tension times mean curvature. Because we have this extra condition, we can't solve (1) and (2) in every V_ε; the extra condition can be satisfied only when V_ε is properly chosen.

As an example of the foregoing, consider the two dimensional problem specified in polar coordinates (r,θ) in Figure E.1 where the boundary data $g(\theta,\varepsilon)$ are given and correspond to a rest state $g(\theta,0)=g^{<0>}(\theta)=0$. The dynamical process $\phi(r,\theta,\varepsilon)$ and the function $f(\theta,\varepsilon)$ which gives the boundary $r=1+f(\theta,\varepsilon)$ of V_ε are unknown. We remind the reader that our aim is to show how to find $F(\phi)$; that is, the Taylor coefficient in (1) by (fictitious) experimental measurements of the (made up) boundary $r=1+f(\theta,\varepsilon)$.

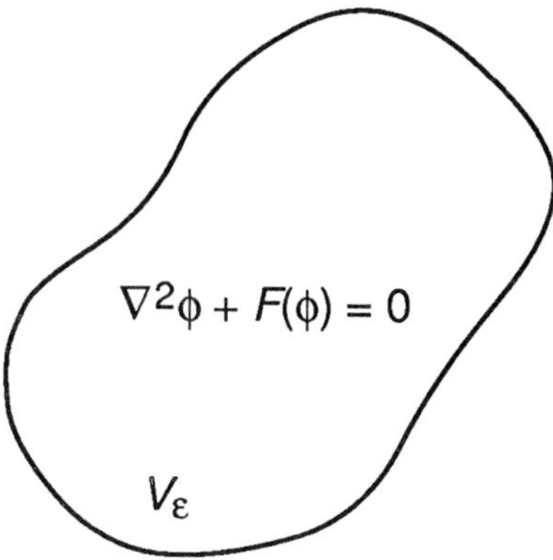

Figure E.1. Model of a free surface problem.

$$G_1(r,\theta,\varepsilon) = \phi(r,\theta,\varepsilon) - g(\theta,\varepsilon) = 0\,, \quad G_2(r,\theta,\varepsilon) = F(\phi) - f(\theta,\varepsilon)\,,$$
$$F(\phi) - f(\theta,\varepsilon) = 0 \quad \text{on} \quad r = 1+f(\theta,\varepsilon)\,. \tag{18}$$

First we solve (18) when $\varepsilon=0$ and we find that $\nabla^2\phi+F(\phi)=0$ in V_0 with $\phi(r,\theta,0)=0$ on $r=1+f(\theta,0)$ has $\phi\equiv 0$ in V_0. Then, since $F(0)=0$, $f(\theta,0)=0$ so that the reference configuration V_0 is the unit circle $r_0=1$. We seek the solution of (18) in powers of ε

$$\begin{pmatrix}\phi(r,\theta,\varepsilon)\\ f(\theta,\varepsilon)\end{pmatrix} = \sum_0 \frac{\varepsilon^n}{n!}\begin{pmatrix}\phi^{[n]}(r_0,\theta_0)\\ f^{[n]}(\theta_0)\end{pmatrix}$$

where $f^{[0]}(\theta_0)=\phi^{[0]}(r_0,\theta_0)=0$ and V_ε and V_0 are related by a shifting map

$$\theta = \theta_0\,, \quad r = r_0(1+f(\theta_0,\varepsilon))\,,$$

having all the properties required of (18). For the shifting map the deformation of V_0 is along rays and

$$\mathbf{x}^{[n]} = \mathbf{e}_r r^{[n]} = \mathbf{e}_r r_0 f^{[n]}(\theta_0)\,.$$

Note that for any function $f(\theta)$ of $\theta=\theta_0$ alone, we have $f^{<n>}(\theta)\equiv f^{[n]}(\theta)$. Using the connection formulas (9) we find that

$$\phi^{[1]}(r_0,\theta_0) = \phi^{<1>}(r_0,\theta_0)$$

and

$$\phi^{[2]}(r_0,\theta_0) = \phi^{<2>}(r_0,\theta_0) + 2r_0 f^{[1]}(\theta_0)\frac{\partial\phi^{<1>}}{\partial r_0}(r_0,\theta_0)\ .$$

On the boundary of $V_0, r_0=1$ we have, from $(18)_3$ that $F_\phi(0)\phi^{<1>}(1,\theta_0)-f^{[1]}(\phi_0)=0$. Since $F_\phi(0)=0$, we find that $f^{[1]}(\theta_0)=0$. The boundary value problems satisfied by $\phi^{<1>}(r_0,\theta_0)$ and $\phi^{<2>}(r_0,\theta_0)$ are given by Figures E.2 and E.3.

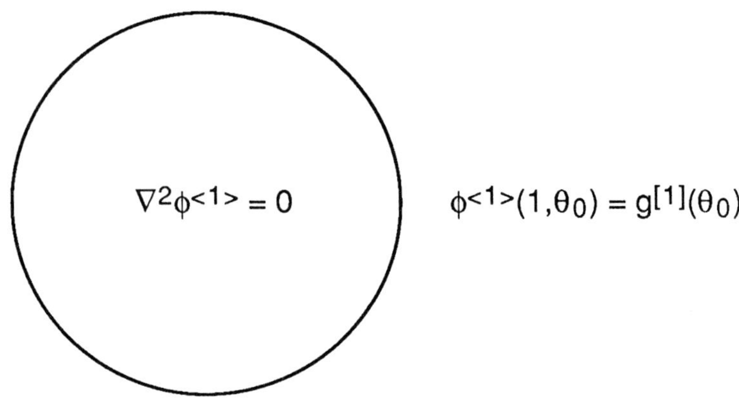

Figure E.2. The problem satisfied by $\phi^{<1>}(r_0,\theta_0)$.

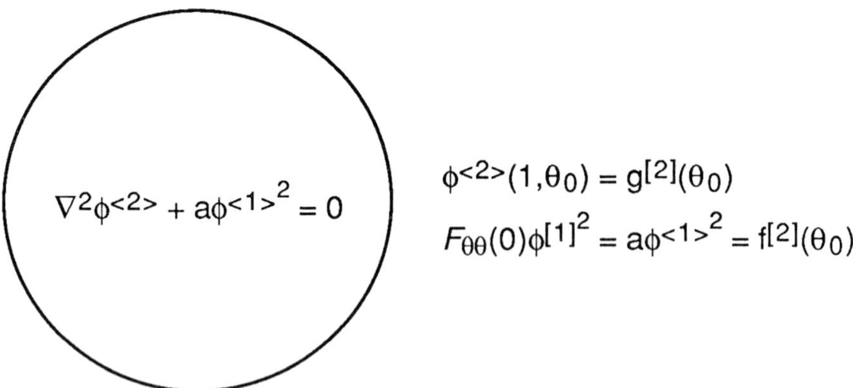

Figure E.3. The problem satisfied by $\phi^{<2>}(r_0,\theta_0)$.

The problems are easy to solve. We find from Figure E.3 that

$$f^{[2]}(\theta_0) = a\phi^{<1>^2}(1,\theta_0) ,$$

where $\phi^{<1>}(r_0,\theta_0)$ is the solution of the problem shown in Figure E.2. It follows that the first approximation to the shape of V_ε is given by

$$r = 1 + f(\theta,\varepsilon) = 1 + a\phi^{<1>^2}(1,\theta)\varepsilon^2 + O(\varepsilon^3) .$$

The next approximation depends on b as well as a. We may therefore deduce the values of derivatives of $F(\phi)$ at $\phi=0$ by monitoring the changes in the shape of V_ε when ε is near to zero.

APPENDIX F
The Wave Speed Meter

F.1 Introduction

The wave-speed meter is a Couette type device which is used to measure the speed of impulsively generated waves of vorticity into rest. These waves are usually called shear waves, in a loose terminology. No serious progress can be made in the exploration of hyperbolic phenomena in viscoelastic fluids without some devices to measure the speed of these waves. Further details about the wave speed meter can be found in papers by Joseph, Riccius, and Arney [1986] and in the Ph.D. thesis of O. Riccius [1989].

The first attempt to measure the speed of shear waves was by E.B. Lieb [1975] who measured streak lines on photographs of a tracer-laden shear flow. His measured value of 8 cm/sec for a 1% solution of CMC in 49% water and 50% glycerin is perhaps six to ten times less than values measured on the wave speed meter (see the table on carboxy-methyl-cellulose in this appendix). Lee and Fuller [1986] measured wave speeds using a phase-modulated-birefringence technique. This technique only works for the small class of liquids which are polarized. Their measurements of wave speeds in 0.75% and 1% Polyox WSR-301 in water and polybutene agree well with values obtained on the wave speed meter, but later unreported measurements using 0.5% polystyrene in tri-cresyl phosphate and in di-methyl phathalate did not agree, differing by factors of two and three respectively, with higher speeds on the wave speed meter. S.R. Burdette [1989] measures wave speeds with a laser doppler velocimeter. He measured speeds in liquids for which we have no data, but his results seem consistent with ours to the extent to which they can be compared. A fuller discussion of these comparisons is in the 1989 thesis of Riccius.

F.2 The wave-speed meter

The wave-speed meter (U.S. Patent 4,602,502) utilizes a Couette apparatus with coaxial cylinders which may rotate independently. The radius of the outer cylinder is b and that of the inner cylinder is a; b–a=d is the gap size. Liquid of height L is placed in the gap. The outer cylinder is moved impulsively. A shear wave propagates into the interior, toward the cylinder at r=a. After a certain time, the transit time, the inner cylinder is set into motion. The transit time would be the time of first reflection if the fluid was elastic or the diffusion time if the fluid was inelastic (Newtonian). We say that the fluid is elastic if the transit time is proportional to the gap size, with the same constant of proportionality for all sufficiently small gap sizes. The reciprocal of this constant is the wave speed, c.

F.2.1 The apparatus

A sketch of the wave speed apparatus is shown in Figures F.1 and F.2. A rigid structure is mounted on a vibration insulated table. It supports the moving parts of the apparatus and consists of upper and lower plates and a supporting collar. The outer cylinder is suspended on open ball bearings lying in grooves on the lower and upper plate. The inner cylinder is designed to rotate easily. It is mounted on hardened steel points which rest in jewel bearings. This provides for negligible frictional resistance to rotational motion. The inner cylinders are built such that the lowest moment of inertia possible is obtained.

The test liquid is pressed into the gap through a plunger and enters the apparatus from the bottom close to the lower jewel bearing. This loading technique suppresses the formation of air bubbles in the test liquid. The inner cylinder is buoyed up by the liquid. It is seated properly in the bottom jewel bearing by adding an appropriate weight on top of the upper bearing. The weights are evident on Figures F.1 and F.2.

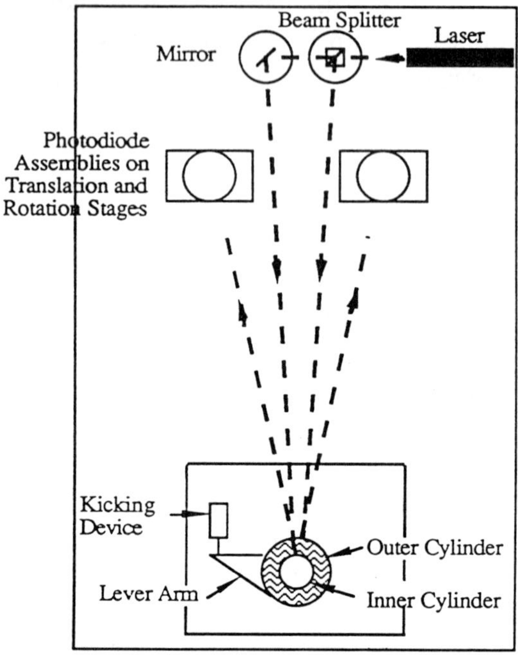

Figure F.1. Sketch of the wave-speed meter.

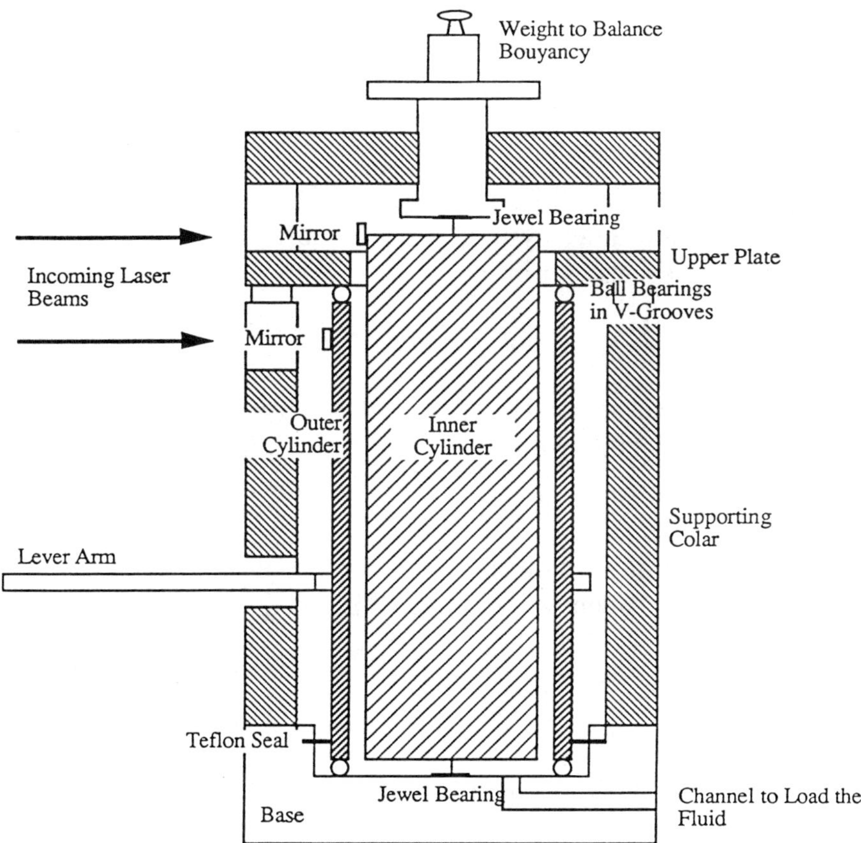

Figure F.2. Details of the cylinders' assembly.

F.2.2 Theoretical model for the wave-speed meter

We consider two long concentric cylinders of length L and radii b and a, b>a. Fluid fills the gap b–a. At some instant the outer cylinder is forced to rotate with an angular velocity given arbitrarily as $\Omega(t)$, $t \geq 0$. The inner cylinder is free to rotate on its axis, without friction. Eventually, the shear induced by the outer cylinder causes the inner cylinder to move.

The linearized problem for the device, neglecting end effects, can be expressed relative to polar cylindrical coordinates (r,z,θ)

$$\rho\frac{\partial w}{\partial t}(r, t) = \frac{\partial T_{r\theta}}{\partial r} + \frac{2}{r} T_{r\theta} , t>0 \tag{1}$$

in $a\leq r\leq b$ with $w(r, t)=0$ for $t\leq 0$. The input at the outer cylinder $r=b$ is prescribed

$$w(b, t) = b\Omega(t), t>0 . \tag{2}$$

The output at the inner cylinder is that the fluid and the cylinder velocities are the same at $r=a$

$$a\ddot{\theta} = \frac{\partial w}{\partial t}(a, t) , \tag{3}$$

and that the shear stress $T_{r\theta}(a, t)$ on the cylinder gives rise to the torque which turns the cylinder

$$I\ddot{\theta} = 2\pi a^2 L T_{r\theta}(r, t), \tag{4}$$

where L is the filling height and I the moment of inertia.

The relation between w and $T_{r\theta}$ shall be given after Boltzmann as

$$T_{r\theta}(r, t) = \int_0^\infty G(s) \left[\frac{\partial w}{\partial r} - \frac{w}{r}\right](r, t-s)ds . \tag{5}$$

We may combine (3), (4) and (5) to find for the inner cylinder

$$\frac{\partial w}{\partial t}(a, t) = 2\pi a^3 L \int_0^t G(s) \left[\frac{\partial w}{\partial r} - \frac{w}{r}\right](r, t-s)ds . \tag{6}$$

The same problem but with $w(a, t)=0$ has been solved by Narain and Joseph [1982]. We can reduce the new problem to the old one in the following way.

Since the gap $d/b \ll 1$ in the wave-speed meter, we may seek a simplified problem for small d/b as a narrow gap approximation. When a wave hits the inner cylinder, the velocity at $r=a$ is initially zero, the shear velocity is annihilated by the reflected wave at the time of first reflection, but the shear stress doubles and this provides the torque that turns the inner cylinder. This leads to a problem which satisfies $w(a, t)=0$ for the short time of first reflection.

We define v(y, t)=w(r, t), y=b–r, and seek v(y, t) as a series in δ=d/b, valid for short times in the neighborhood of the first time of reflection

$$v(y, t) = v_0(y, t) + \delta v_1(y, t) + \delta^2 v_2(y, t) + \ldots \tag{7}$$

To derive the problems satisfied by v_0, v_1, etc. we introduce the fast dimensionless times τ and σ

$$t = \frac{\delta\tau}{\Omega_0}, \; s = \frac{\delta\sigma}{\Omega_0}, \tag{8}$$

and

$$\begin{gathered} y = xd\,, \\ v(y, t) = b\Omega_0 u(x, \tau)\,, \end{gathered} \tag{9}$$

where Ω_0 is the largest value of $\Omega(s)$. We first determine that

$$\rho(b\Omega_0)^2 \frac{\partial u}{\partial \tau} = \int_0^\tau \hat{G}(\sigma) \frac{\partial^2 u}{\partial x^2}(x, \tau-\sigma)\, d\sigma - \frac{\delta}{1-\delta}$$

$$\int_0^\tau \hat{G}(\sigma) \frac{du}{dx}(x, \tau-\sigma)\, d\sigma - \left[\frac{\delta}{1-\delta}\right]^2 \int_0^\tau \hat{G}(\sigma) u(x, \tau-\sigma) d\sigma,$$

$$0 \le x \le 1$$

$$\left.\begin{aligned} u(x,\tau) &= 0, & \tau &\le 0, \\ u(0, \tau) &= \begin{cases} \hat{\Omega}(\tau), & \tau > 0, \\ 0, & \tau < 0, \end{cases} \end{aligned}\right\}$$

$$\frac{\partial u(1, \tau)}{\partial \tau} = -\frac{2\pi a^3 L}{bI\Omega_0^2}\left[\delta \int_0^\tau \hat{G}(\sigma) \frac{\partial u}{\partial x}(1, \tau-\sigma)\, d\sigma \right.$$

$$\left. - \frac{\delta^2}{1-\delta} \int_0^t \hat{G}(\sigma) u(1, \tau-\sigma)\, d\sigma \right], \tag{10}$$

where $\tilde{\Omega}(\tau)=\Omega(t)/\Omega_0$ and $\hat{G}(\sigma)=G(s)$. Short times are time td<δ/Ω_0. The main effect of this scaling is to introduce δ as a factor of some terms on the right of (10). We now expand

$$u(x, \tau) = u_0 + \delta u_1 + \delta^2 u_2 + \dots , \tag{11}$$

and identify the coefficients of powers of δ. At zeroth order

$$\rho(b\Omega_0)^2 \frac{\partial u_0}{\partial \tau} = \int_0^\tau \hat{G}(\sigma) \frac{\partial^2 u_0}{\partial x^2}(x, \tau-\sigma)d\sigma,\ 0\le x\le 1 , \tag{12}$$

with $u_0(x,\tau)=0$ for $\tau\le 0$, $u_0(1,\tau)=0$ for $\tau\ge 0$, and $u_0(0,\tau)=H(\tau)\tilde{\Omega}(\tau)$. This is a generalization of Stokes' first problem in which arbitrary initial data replace a unit step jump. At first order, we find

$$\rho(b\Omega_0)^2 \frac{\partial u_1}{\partial \tau} = \int_0^\tau \hat{G}(\sigma) \left[\frac{\partial^2 u_1}{\partial x^2} - \frac{\partial u_0}{\partial x}\right](x, \tau-\sigma)d\sigma, \tag{13}$$

and $u_1(x, \tau)=0$, $\tau\le 0$ in $0\le x\le 1$, while $u_1(0, \tau)=0$ for all $\tau\ge 0$. The motion of the inner cylinder is then governed by

$$\frac{\partial u_1}{\partial \tau}(1, \tau) = -\frac{2\pi a^3 L}{Ib\Omega_0^2} \int_0^\tau \hat{G}(\sigma) \frac{\partial u_0}{\partial x}(1, \tau-\sigma) . \tag{14}$$

Equations (13) and (14) determine $u_1(1,\tau)$. This perturbation can be established rigorously using Laplace transforms. The perturbation is not valid for long times.

In case $\tilde{\Omega}(\tau)=1$ for $\tau\ge 0$ is a step function we use the solution by Narain and Joseph

$$u_0(x, \tau) = f(x, \tau)H(\tau-\alpha x) \tag{15}$$

where

$$\frac{1}{\alpha} = \frac{1}{b\Omega_0}\sqrt{\frac{\hat{G}(0)}{\rho}}$$

is a dimensionless wave speed and

$$f(x, \alpha x^+) = \exp\left(\frac{\alpha x \hat{G}'(0)}{2\hat{G}(0)}\right) . \tag{16}$$

The solution is a valid solution of our problem only at early times near to the time $\tau=\alpha$ of first reflection, that is when

$$t = \frac{b-a}{c}\ , \quad c = \sqrt{\frac{G(0)}{\rho}}\ . \tag{17}$$

To find the motion of the inner cylinder at x=1 for short times after the first reflection, we evaluate (14), using (15).

$$\frac{\partial u_0}{\partial x}(1, \tau) = -2\alpha f(1, \tau)\delta(\tau-\alpha) + 2\frac{\partial f}{\partial x}(1,\tau)H(\tau-\alpha)\ , \tag{18}$$

where $\delta(\tau-\alpha)$ is Dirac's δ-function. Inserting this into (14) we find

$$\frac{\partial u_1}{\partial \tau}(1, \tau) = \frac{4\pi a^3 L}{Ib\Omega_0^2}\left[\alpha\hat{G}(\tau-\alpha)(1, \alpha^+) - \int_0^{\tau-\alpha}\hat{G}(\sigma)\frac{\partial f}{\partial x}(1, \tau-\sigma)d\sigma\right] \tag{19}$$

This equation governs the motion of the inner cylinder. The wave-speed meter requires that we know $u_1(1, \tau)$ for small values $\tau-\alpha\geq 0$. In this time interval we may write

$$\frac{\partial u_1}{\partial \tau}(1, \tau) = A_0 + A_1(\tau-\alpha) + O(|\tau-\alpha|^2)\ , \tag{20}$$

where

$$A_0 = \frac{4\pi a^3 L}{Ib\Omega_0^2}\alpha\hat{G}(0)\exp\left(\frac{\alpha x\hat{G}'(0)}{2\hat{G}(0)}\right),$$

$$A_1 = \frac{4\pi a^3 L}{Ib\Omega_0^2}\left[\alpha\hat{G}'(0)\exp\left(\frac{\alpha x\hat{G}'(0)}{2\hat{G}(0)}\right) - \hat{G}(0)\frac{\partial f}{\partial x}(1, \alpha^+)\right],$$

and $\frac{\partial f}{\partial x}(1, \alpha^+)$ is given in Narain and Joseph [1982].

Returning to physical variables one writes

$$\frac{\partial w}{\partial t}(a, t) = \frac{b^2}{d}\Omega_0^2\frac{\partial u}{\partial t}(1, \tau)\ .$$

Hence, using $\hat{G}(\sigma)=G(s)$ and (11)

$$\frac{\partial w(a,t)}{\partial(t-d/c)} = \frac{4\pi a^3 L b \Omega_0 \sqrt{\rho G(0)}}{I} \exp\left(\frac{dG'(0)}{2cG(0)}\right) + 0\left(\left|t-\frac{d}{c}\right|\right) + 0\left(\left|\frac{d}{b}\right|^2\right). \tag{21}$$

Since $w(a,t)=a\dot{\theta}$, we have

$$\theta(t) = B\tau^2 + 0\left(\tau^3 + \frac{d}{b}\tau^2\right) \tag{22}$$

where

$$\tau = \left|t - \frac{d}{c}\right|$$

and

$$B = \frac{2\pi a^2 L b \Omega_0 \sqrt{\rho G_0}}{I} \exp\left(\frac{dG'(0)}{2cG(0)}\right).$$

This formula shows how the inner cylinder moves after being hit by a wave.

One can compare it with response of the inner cylinder to step data in a Newtonian fluid. One writes an expression for the torque of the inner cylinder at r=a using the solution for Stokes' problem and integrate the equation of motion for the inner cylinder using these two solutions. At small times this gives

$$\theta(t) = \frac{8a^2 L b \Omega_0 \sqrt{\pi \nu}}{3I} t^{1.5} + 0\left(t^{1.5}\right) \tag{23}$$

The displacement starts at the instant that the outer cylinder is moved. There is no transit time, no delay, under the assumptions of this analysis.

The kicking device used in the measurements does not provide a step jump of velocity at the outer cylinder as assumed above. Instead there is a smooth rise which is recorded on the oscilloscope diagrams. We have solved the problem for arbitrary smooth inputs. We modelled the actual input and showed that the response for this model differs only slightly from the response for the step jump when $\frac{d}{b} \ll 1$ and $\frac{\beta}{\lambda_u} \ll 1$ where β is the time the outer cylinder needs to accelerate to a constant angular velocity and λ_u is an effective time of relaxation for an effective modulus (see chapter 19) which we modeled with a single exponential.

F.2.3 Measurements of transit times

A transit time measurement is started by applying an impulsive force to the outer cylinder. This is done by a simple kicking device or by a hammer which hits a lever arm which is rigidly attached to the outer cylinder. The kick induces an impulse, which propagates inward, and moves the inner cylinder. The onset of motion for each cylinder is determined with an optical method.

2.3.1 The optical system

The optical apparatus is the heart of the measuring system. It uses two laser beams which are directed to mirrors on the outer and inner cylinder. The reflected beams are focused onto two photodiodes which are located at a distance of 1 meter from the Couette cell. Voltages generated by the irradiated photodiodes are monitored with two voltmeters and a storage oscilloscope. When the outer cylinder moves, its laser beam is directed off the corresponding photodiode which causes the photovoltage to drop and triggers the oscilloscope. When the inner cylinder moves, its beam is removed from the second photodiode and this photovoltage drops. Both voltage versus time diagrams are recorded on the oscilloscope and later evaluated.

A polarized helium-neon laser is used to generate the two laser beams. It is aligned such that a prism-beam splitter divides the incoming beam into two rays of equal intensity. The beam splitter directs one of them onto the outer cylinder; the other is deflected by a mirror onto the inner cylinder. In front of the cylinders the divergence of the beams is limited by two pin holes of 1mm diameter. The beams are then reflected from mirrors on each cylinder and directed to the photodiodes. In front of the photodiodes each reflected beam passes a blade edge and a lens (f=2.54 cm) which focuses the beam onto its photodiode. Every set of blade, lens and photodiode is mounted on a translation and rotation stage which allows exact focusing before a measurement is started.

2.3.2 *Transit times*

To obtain transit times, the times at which each cylinder begins to move must be known. These are equal to the times at which the photovoltage begins to drop and the time difference can be computed from the oscilloscope traces. In order to interpret the plots correctly one has to relate the drop in photovoltage to the motion of the cylinders.

The voltage V is a function of the intensity I of the laser light. The intensity depends on the power P of the laser and the deviation x from the position x=0 of maximum intensity I_0 The function dependence of V(I) and I may be measured from monitoring V(t) as I(t) decays after a sudden blocking of the laser light. We fit V(I) to an exponential

$$V(1) = V_0 e^{-\beta(I_0-I)} . \tag{24}$$

When the laser power is fixed the intensity depends on the deviation x, I=f(x), I_0 =f(0) for some to-be-determined f(x). f(x)=f(–x) because the intensity cannot depend on whether the center of the laser light is to the left or right of the photodiode center. Clearly f(x) is a decreasing, even function of x and we assume that

$$f(x) = f(0) - \left(k^2x^2 + 0(x^4)\right) . \tag{25}$$

Then, up to order x^4 we have

$$V(I) = V_0 e^{-\beta k^2 x^2} . \tag{26}$$

We next observe that x=Rθ where θ is angle of rotation from the line between the cylinder and photodiode centers and R is the distance between the photodiode and cylinder center. The angle $\theta(\tau), \tau = t - \frac{d}{c}$ is given by (22). Hence

$$V = V_0 e^{-\beta k^2 R^2 B^2 \tau^4} . \tag{27}$$

where the exponent is given up to terms of smaller order $0(\tau^4)$. We can detect the start time for the inner cylinder by comparing with the oscilloscope traces of V(t). Typical oscilloscope graphs are displayed in Figure F.3. They show both voltages before the outer cylinder is moved abruptly. The transit time measurement is started when the voltage corresponding to the outer cylinder drops. This is a rapid drop over about 150 μsec due to the impact of the kicking device at the outer cylinder. The voltage which corresponds to the inner cylinder remains unchanged although it shows vibrations due to the impact. To find the onset of rotation for the inner cylinder one fits (27) to the portion when the voltage begins to drop and identifies the vertex of (27) which gives the time of onset of rotation.

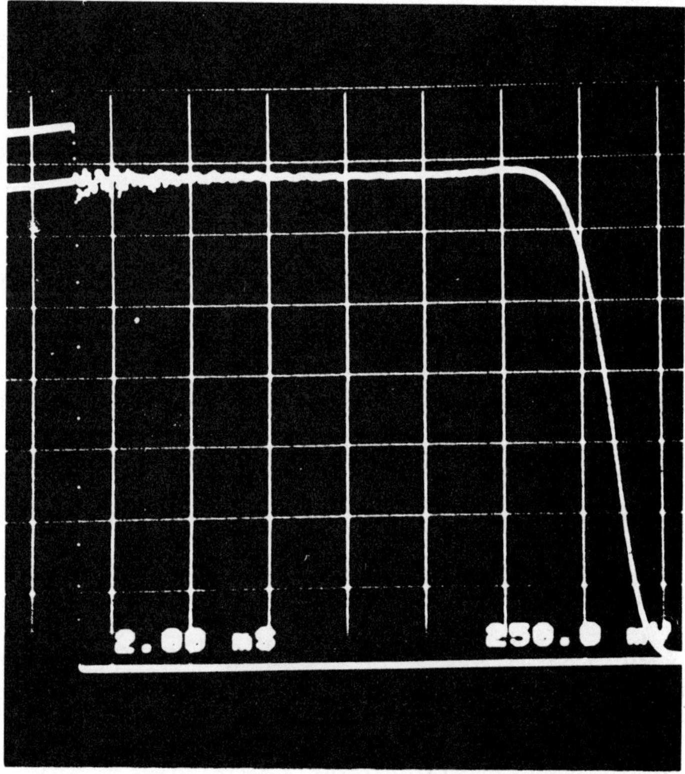

Figure F.3. 1% Polyox (coag.) in water; 2mm gap; dt=11.1 msec; c=18.0 cm/sec.

F.3 Criteria for waves

The criteria that we use to determine the existence of a shear wave are listed below.

1. We have to find one wave speed for varying small gap sizes. The gap sizes can be varied between 0.25 and 6.75 mm.

2. The measured transit times are repeatable, i.e. with a standard deviation of at most 15% over 10 readings.

3. The effective shear modulus $G_c=\pi\bar{c}^2$, where $\bar{c}$ is the average wave speed taken over small gap sizes, is greater or equal than the values of the relaxation function obtained in a stress relaxation experiment and greater or equal than the value of the storage modulus $\tilde{G}'(\omega)$, obtained in oscillatory experiments from standard rheometers.

Point (3) from above needs some motivation. The effective modulus, G_c, which can be calculated from an average wave speed corresponds to a certain value of the relaxation function G(s). It is a value at a relatively short time when fast relaxing modes such as glassy modes have already relaxed while slower "effective" modes have not. The idea of the effective modulus is explained in chapters 18 and 19. We estimate $s=0(10^{-4}$ sec) for the wave-speed meter. Common rheometers provide information only at larger times, at most for $s>10^{-2}$ sec, and measurements of the storage modulus, $\tilde{G}'(\omega)$, are limited to frequencies $\omega\leq 500$ Hz. The requirements in 3) mean that from comparison of the time scales of common rheometers with the wave-speed meter we must have $G_c\geq G(x\geq 10^{-2}\text{sec})$ and $G_c\geq\tilde{G}'(\omega\leq 500 \text{ Hz})$.

F.4 Errors

Errors in wave speed measurements can arise in reading off transit times on the oscilloscope, from inaccuracies in gap size, and during preparing and handling of the test fluids.

The variations in gap size are given through the accuracy in machining parts of the apparatus and depend on the gap size. Large errors can occur in small gaps while the error is insignificant in large gaps. Typical values are 10% for the 0.38 mm gap, 8% for the 1 mm gap, 3% for the 2 mm gap and 1% in the 6.12 mm gap. The error in reading of transit times is estimated from one standard deviation over a set of eight to ten transit time readings. We neglect error from preparing the test fluids. All three errors are discussed in the Ph.D. thesis of Riccius [1989]. The table in F.5 contains errors due to one standard deviation from the transit time measurements and the inaccuracies in gap size.

F.5 Data on shear-wave speeds

We measured transit times in different liquids:

1. aqueous glycerin in various degrees of dilution.
2. vegetable oils and miscellaneous lubricating oils.
3. silicone oils of different molecular weight at different temperatures.
4. polymer solutions in various degrees of dilution and molecular weight.

Ten different polymers were used in solvents of water, water and glycerin, decalin, toluene, tri-cresyl phosphate, di-methyl phthalate, petroleum oil and others. The main groups of polymers were poly(ethylene oxide) (WSR-301 and coagulant), poly(iso-butylene), poly(styrene), poly(acryl amide), carboxy-methyl cellulose, poly(methyl methacrylate) and several co-polymers. Most of the solvents were measured as well which are listed at the end of this group. Among the poly(iso-butylene) solutions are the calibration fluids M-1, D-1 and D-2; Fluorinert coolant; and honey and fruit jam.

The table presented in this section gives all values of shear-wave speeds, c (cm/s), measured up to December 1988. The data are obtained from oscilloscope traces except those which are marked with "C" in the rightmost column. Those transit speeds were measured on an electronic counter, prior to May 1985. After May we realized that the counter values of c contain a systematic error and could be as much as 50% lower than oscilloscope values. Oscilloscope data are considered "true" data since the oscilloscope traces allow us to calculate the time of onset of rotation of each cylinder, while the counter is triggered only when the photovoltage has dropped past a certain level and the cylinders have already moved. We refer to table 2 in Joseph, Riccius and Arney [1986] for a comparison between transit times obtained on counter and oscilloscope. Data marked with "C" are those for aqueous solutions of glycerol, soybean oil, STP, TLA 227, Amoco oil #140, SAE30 motor oil, solutions of coagulant poly(ethylene oxide) in concentrations between 1.7% and 0.075%, poly(acryl amide) Separan AP30 solutions of 1.5%, 1.25%, 0.75% to 0.05% and Honey. Oscilloscope data for soybean oil are included for comparison. We were not systematic in recording temperatures at the time of measurement prior to May 1985. These data were taken at temperatures between 22° and 24°C. Gap sizes, d, for various measurements of the transit speed are listed in mm. The fact that c was measured for many d-values in some liquids and for few d-values in others is due either to the number of gap sizes that were available at a given time, to experimental problems in loading very viscous liquids into small gaps, to other experimental problems, or because nothing new would emerge from changing the gap size further. In general, the data for different gap sizes were collected on the same day and within one hour for each gap. Data that were collected on different days are marked with "*". The table also includes an effective shear modulus G_c in Pa which is computed from the average, $\bar{c}$, of c over d as

$$G_c = \rho\bar{c}^2, \tag{28}$$

where ρ is the density. ρ is given in kg/m^3 and was obtained through volumetric measurements and weighing or from published data*. The table also lists values of $\tilde{\mu}$ in Pas, the static or zero-shear viscosity, which were taken as extrapolated values of viscosity

* Published data were used for silicone oils.

versus shear rate measurements on a Rheometrics System Four rheometer, from a Rheometrics Fluid rheometer, both in a cone-and-plate configuration, or from published data. We define an effective relaxation time λ_c (10^{-3}s) as the ratio of zero-shear viscosity and effective shear modulus. λ_c corresponds to relaxing effective modes in the fluid. It is not comparable to the relaxation times computed from conventional measurements which do not detect fast relaxing elastic responses.

The data in the table show that c is independent of d for small d in most liquids. This independence is unambiguous in most cases. We note that rapid spatial decay is expected in fluids with short memories, e.g. "Newtonian" fluids, and the response will appear to be diffusive in the larger gaps. At early times diffusion (or dispersion) of the wave could give rise to shorter transit times because the wave spreads as it propagates. Eventually diffusion manifests itself as a decrease of c with d. Such decreases are evident in aqueous glycerin solutions with even small amounts (≥20%) of water, 20 cs silicone oil, 5% poly(styrene) in decalin, decalin, and tri-cresyl phosphate. We cannot conclude that these fluids are elastic. On the other hand, all fluids will appear to be diffusive when the gaps are sufficiently large (see chap. 1.1). We require that c is independent of d for small d.

Table F.1. Aqueous glycerin in various degrees of dilution

Fluid	$\tilde{\mu}$ (Pa s)	$\bar{c}$ (cm/s)	G_c (Pa)	λ_c (10^{-3}s)	gap (10^{-3}m)	c (cm/s)	T (°C)	ρ (kg/m^3)
Glycerol	0.69	209	5471	0.12	0.25	219 ±91	24	1255
					0.5	199 ±54	25	
					1	227 ±33	25	
					2	219 ±42	25	
					3	180 ±18	25	
					6.75	157 ±12	24	
90% aqueous Glycerol	0.15	43.5	233	0.64	1	46.7 ±6.6	23	1230 C
					2	40.5 ±2.9		
					3	43.4 ±2.8	23	
80% aqueous Glycerol	0.04	18.1	39.3	1.02	0.25	18.3 ±9.5	23	1200 C
					0.5	18.2 ±4.0	-	
					1	17.8 ±2.1	24	
					2	15.4 ±0.9	24	
					3	12.2 ±1.0	24	
70% aqueous Glycerol	0.03	9.7	11.1	2.7	0.25	9.6 ±4.3	*	1180 C
					0.5	9.8 ±2.4	*	
					1	9.6 ±1.1	-	
					2	6.9 ±0.4	-	
					3	6.6 ±0.6	-	

Table F.1. Aqueous glycerin in various degrees of dilution (continued)

Fluid	$\tilde{\mu}$ (Pa s)	$\bar{c}$ (cm/s)	G_c (Pa)	λ_c (10^{-3}s)	gap (10^{-3}m)	c (cm/s)	T (°C)	ρ (kg/m³)
50% aqueous Glycerol	0.01	4.7	2.5	4	0.25	4.7 ±2.2	-	1120 C
					0.5	5.9 ±1.1	-	
					1	3.5 ±0.6	-	
					2	0	-	
					3	0	-	

Table F.2. Vegetable oils and miscellaneous lubricating oils

Fluid	$\tilde{\mu}$ (Pa s)	$\bar{c}$ (cm/s)	G_c (Pa)	λ_c (10^{-3}s)	gap (10^{-3}m)	c (cm/s)	T (°C)	ρ (kg/m³)
Olive oil (brand name "Olio Sasso")	0.06	31.9	93	0.65	0.25	30.6±10.5	25	914
					0.5	33.9±10.1	25	
					1	40.1 ±8.3	26	
					2	30.7 ±3.1	26	
					3	24.0 ±1.3	26	
Soybean oil (plus additives, brand name "Crisco")	0.046	19.9	36.5	1.26	0.25	19.2 ±9.6	-	922 C
					0.38	18.4 ±3.9	-	
					0.5	20.9 ±5.4	-	
					1	21.2 ±3.9	-	
					1.38	22.0 ±1.6	-	
					2	17.8 ±2.2	24	
					3	15.5 ±1.4	24	
Soybean oil (plus additives, brand name "Crisco") (oscilloscope)	0.046	33	100.4	0.46	0.25	34.2±15.6	24	922
					0.38	31.8 ±5.6	24	
STP (motor oil additive)	14.3	286	7050	2.03	1	277 ±64	-	858 C
					2	279 ±18	-	
					3	304 ±15	-	

Table F.2. Vegetable oils and miscellaneous lubricating oils (continued)

Fluid	$\tilde{\mu}$ (Pa s)	$\bar{c}$ (cm/s)	G_c (Pa)	λ_c (10^{-3}s)	gap (10^{-3}m)	c (cm/s)	T (°C)	ρ (kg/m^3)
TLA 227	22.3	234	4840	4.61	1	211 ±33	-	884 C
					2	246 ±10	-	
					3	245 ±11	-	
Amoco oil # 140 (gear lubricant)	1.63	368	12500	0.13	1	384 ±39	-	924 C
					2	352 ±21	-	
SAE 30 motor oil (Amoco)	0.098	76.3	516	0.19	1	81.7±21.9	-	886 C
					2	68.3 ±8.6	*	
					3	78.9 ±7.4	*	

Table F.3. Silicone oils of different molecular weight at different temperatures

Fluid	$\tilde{\mu}$ (Pa s)	$\bar{c}$ (cm/s)	G_c (Pa)	λ_c (10^{-3}s)	gap (10^{-3}m)	c (cm/s)	T (°C)	ρ (kg/m^3)
600,000 cs PDMS (Poly(di-methyl siloxane)) MN=110,000	586	1471	211000	2.78	3	1270 ±242	24	977
					6.75	1671 ±113	24	
100,000 cs PDMS MN=75,000	98	1372	184000	0.53	1	1320 ±487	23	977
					2	1423 ±183	23	
					3	1373 ±163	23	
60,000 cs PDMS MN=65,000	58	965	90700	0.64	1.5	1018 ±167	24	974
					2.5	912 ±236	26	
12,500 cs PDMS MN=41,000	12.2	598	34900	0.35	1	490 ±82	25	975
					2	636 ±91	26	
					3	668 ±56	25	

Table F.3. Silicone oils of different molecular weight at different temperatures (continued)

Fluid	$\tilde{\mu}$ (Pa s)	$\bar{c}$ (cm/s)	G_c (Pa)	λ_c (10^{-3}s)	gap (10^{-3}m)	c (cm/s)	T (°C)	ρ (kg/m³)
12,500 cs PDMS	25.0	613	37500	0.67	2	613 ±73	0	999
12,500 cs PDMS	19.8	621	38100	0.52	2	621 ±44	10	989
12,500 cs PDMS	12.2	599	34900	0.35	2	599 ±45	25	975
12,500 cs PDMS	6.6	494	23000	0.29	2	494 ±32	59	942
12,500 cs PDMS	5.6	445	18500	0.30	2	445 ±78	70	932
1,000 cs PDMS	1.09	200	3870	0.28	1	216 ±26	25	967
MN=16,500					2	170 ±17	25	
					3	214 ±32	25	
1,000 cs PDMS	1.7	257	6570	0.26	2	257 ±34	0	995
1,000 cs PDMS	1.2	241	5710	0.21	2	241 ±35	12	983
1,000 cs PDMS	0.67	172	2840	0.24	2	172 ±10	35	961
1,000 cs PDMS	0.57	176	2940	0.19	2	176 ±14	48	949

Table F.3. Silicone oils of different molecular weight at different temperatures (continued)

Fluid	$\tilde{\mu}$ (Pa s)	$\bar{c}$ (cm/s)	G_c (Pa)	λ_c (10^{-3}s)	gap (10^{-3}m)	c (cm/s)	T (°C)	ρ (kg/m^3)
20 cs PDMS	0.02	-	-	-	0.25	54.7±26.6	26	949
MN=1,500					0.5	24.0 ±7.1	25	
					1	12.7 ±2.7	25	

Table F.4. Polymer solutions of different dilution and molecular weight
F.4.A. Poly(ethylene oxide) WSR-301 and coagulant

Fluid	$\tilde{\mu}$ (Pa s)	$\bar{c}$ (cm/s)	G_c (Pa)	λ_c (10^{-3}s)	gap (10^{-3}m)	c (cm/s)	T (°C)	ρ (kg/m^3)
5.0% PO WSR-301	45	124.7	1812	24.8	1	122.6±10.7	27	1165
in water					2	119.7 ±4.9	24	
(Poly(ethylene oxide))					3	131.8 ±4.6	25	
MN=4,000,000								
4.5% PO WSR-301	31	85.7	853	36.3	1	82.7±15.3	25	1163
in water					3	87.0±13.9	25	
					6.75	87.3 ±2.3	24	
2.5% PO WSR-301	62.8	51.2	302	228	1	48.6 ±7.4	24	1150
in water					3	53.8 ±2.3	25	
					6.75	51.2 ±0.6	24	
1.3% PO WSR-301	11.9	24.7	60.8	196	0.5	24.8 ±5.4	25	999
in water					1	24.3 ±2.7	25	
					2	23.2 ±2.4	26	
					3	26.3 ±1.4	23	

F.4.A. Poly(ethylene oxide) WSR-301 and coagulant (continued)

Fluid	$\tilde{\mu}$ (Pa s)	$\bar{c}$ (cm/s)	G_c (Pa)	λ_c (10^{-3}s)	gap (10^{-3}m)	c (cm/s)	T (°C)	ρ (kg/m^3)
0.9% PO WSR-301 in water	6.23	18.3	33.4	187	0.25	16.6 ±6.4	24	995
					0.5	19.1 ±4.4	25	
					1	17.8 ±2.5	25	
					2	20.9 ±1.6	25	
					3	17.2 ±0.9	25	
0.75% PO WSR-301 in water	1.4	12.4	15.4	90.9	1	11.6 ±1.9	23	1000
					2	13.2 ±1.5	*23	
					3	12.4 ±0.8	*23	
0.5% PO WSR-301 in water	0.14	8.33	6.91	20.3	2	9.63±1.67	23	997
					3	7.02±0.75	23	
0.25% PO WSR-301 in water	0.001	4.29	1.84	0.71	0.25	4.45±1.79	23	996
					0.5	4.46±1.05	23	
					1	4.21±0.61	23	
					2	4.05±1.01	22	
50ppm PO WSR-301 in water	0.001	2.48	0.62	1.94	0.25	2.40±1.41	25	1000
					0.38	2.44±0.57	25	
					0.5	2.59±1.83	25	

F.4.A. Poly(ethylene oxide) WSR-301 and coagulant (continued)

Fluid	$\tilde{\mu}$ (Pa s)	$\bar{c}$ (cm/s)	G_c (Pa)	λ_c (10^{-3}s)	gap (10^{-3}m)	c (cm/s)	T (°C)	ρ (kg/m^3)
3% PO coagulant in water (Poly(ethylene oxide)) MN=5,000,000	181	71.7	499	363	2	71.7 ±3.2	28	970
					3	71.6 ±4.6	25	
1.7% PO coag. in water	58.4	25.7	68	859	0.25	22.3±10.2	22	1003 C
					0.5	24.4 ±6.2	22	
					1	27.0 ±2.8	*23	
					2	28.9 ±2.2	*23	
1% PO coag. in water	6.12	14.7	21.5	285	0.25	12.2 ±5.2	22	997 C
					0.5	14.3 ±3.5	22	
					1	14.5 ±1.7	23	
					2	16.1 ±5.3	23	
					3	16.0 ±0.8	23	
0.75% PO coag. in water	0.37	9.8	9.55	38.7	0.38	9.8 ±1.6	22	994 C
					0.5	9.8 ±2.1	*24	
0.5% PO coag. in water	0.1	6.9	4.69	21.3	1	6.8 ±0.7	26	999 C
					2	6.9 ±0.5	26	
					3	6.9 ±0.4	26	

F.4.A. Poly(ethylene oxide) WSR-301 and coagulant (continued)

Fluid	$\tilde{\mu}$ (Pa s)	$\bar{c}$ (cm/s)	G_c (Pa)	λ_c (10^{-3}s)	gap (10^{-3}m)	c (cm/s)	T (°C)	ρ (kg/m^3)
0.25% PO coag. in water	0.03	3.4	1.12	26.8	0.25	3.22±1.77	22	1000 C
					0.5	3.34±0.80	22	
					1	3.39±0.59	23	
					2	3.42±1.09	23	
					3	3.37±1.78	23	
0.075% PO coag. in water	0.01	1.87	0.35	28.6	0.25	1.91±1.11	22	1000 C
					0.5	1.83±0.37	22	

Table F.4. Polymer solutions of different dilution and molecular weight
F.4.B. Poly(iso-butylene)

Fluid	$\tilde{\mu}$ (Pa s)	$\bar{c}$ (cm/s)	G_c (Pa)	λ_c (10^{-3}s)	gap (10^{-3}m)	c (cm/s)	T (°C)	ρ (kg/m^3)
9.5% PIB Vistanex L100 in Decalin (Poly(iso-butylene)) MN=1,000,000	139	162	2388	58.2	1	157 ±16	25	910
					2	179 ±11	26	
					3	151 ±18	25	
6% PIB Vistanex L100 in Decalin	11	90.1	730	15.1	1	88.9±12.9	23	899
					3	93.6 ±6.5	23	
					6.75	87.6 ±3.6	23	
6% PIB Vistanex L100 in Toluene (Poly(iso-butylene)) MN=1,000,000	41	116	1160	35.3	1.38	114.2 ±7.5	24	868
					2.38	116.9 ±4.9	24	
5.5% PIB Vistanex L100 in Toluene	5.44	60.6	332	16.4	1.38	53.3 ±7.0	23	903
					2.38	67.9 ±2.6	23	
4% PIB Vistanex L100 in Toluene	0.88	53.5	244	3.6	1	48.4±10.8	24	854
					2	56.2 ±5.1	24	
					3	55.9 ±3.0	27	

F.4.B. Poly(iso-butylene) (continued)

Fluid	$\tilde{\mu}$ (Pa s)	$\bar{c}$ (cm/s)	G_c (Pa)	λ_c (10^{-3}s)	gap (10^{-3}m)	c (cm/s)	T (°C)	ρ (kg/m³)
0.25% PIB in Poly-(butene)	12.3	1250	139000	0.09	1	1580 ±726	-	893 C
					2	995 ±188	-	
					3	1170 ±125	-	
M-1, (Poly(iso-butylene) in Hexane-Poly(butene))	3.0	404.7	14067	0.21	6.12	378.7±12.2	23	859
					9.38	403.4±33.6	*23	
					12.2	430.5±10.5	*23	
2% B200 in Decalin (Boger fluid, D-1)	10	38.7	131	76.3	0.25	40.0±18.3	28	876
					0.25	41.7±20.6	28	
					0.5	43.1±13.1	28	
					1	35.3 ±5.5	*28	
					2	35.9 ±3.3	29	
					3	36.0 ±2.0	30	
10% B50 in Decalin (Boger fluid, D-2)	1.7	102.2	918	1.85	0.25	90.0±45.6	28	878
					0.5	110.1±40.0	28	
					1	94.2±18.1	28	
					2	109.6±15.3	*26	
					3	107.3±10.5	*26	

F.4.B. Poly(iso-butylene) (continued)

Fluid	$\tilde{\mu}$ (Pa s)	$\bar{c}$ (cm/s)	G_c (Pa)	λ_c (10^{-3}s)	gap (10^{-3}m)	c (cm/s)	T (°C)	ρ (kg/m³)
1% PIBM in BIS 2 (Poly(iso-butyl-methacrylate) in bis(2-ethyl-butyl-acrylate)) MN=5,500,000	0.065	20.3	38.3	1.7	0.38	21.2 ±6.7	23	930
					1	21.2 ±4.0	24	
					2	18.5 ±3.3	24	

Table F.4. Polymer solutions of different dilution and molecular weight
F.4.C. Poly(styrene)

Fluid	$\tilde{\mu}$ (Pa s)	$\bar{c}$ (cm/s)	G_c (Pa)	λ_c (10^{-3}s)	gap (10^{-3}m)	c (cm/s)	T (°C)	ρ (kg/m^3)
10% PS in Decalin	0.011	7.74	5.53	1.99	0.25	7.39±3.79	30	890
(Poly(styrene))					0.5	8.26±2.81	28	
MN=22,000					1	7.57±1.36	28	
5% PS in Decalin	0.004	6.87	4.18	0.96	0.25	11.29±7.0	27	885
MN=22,000					0.5	7.49±1.27	27	
					1	4.47±0.85	27	
					2	4.22±0.92	28	
1% PS in Decalin	0.003	-	-	-	0.25	5.64±3.26	27	873
MN=22,000					0.5	4.27±1.08	27	
					1	3.55±1.07	26	
20% PS in Decalin	2.3	153	2120	1.09	1	172 ±29	25	908
(Poly(styrene))					2	139 ±14	25	
MN=2-3,000,000					3	146 ±4.7	26	
10% PS in Decalin	0.087	41	151	0.58	0.25	39.4±18.6	26	900
MN=2-3,000,000					0.5	50.1±13.0	26	
					1	33.5 ±4.6	26	

F.4.C. Poly(styrene) (continued)

Fluid	$\tilde{\mu}$ (Pa s)	$\bar{c}$ (cm/s)	G_c (Pa)	λ_c (10^{-3}s)	gap (10^{-3}m)	c (cm/s)	T (°C)	ρ (kg/m^3)
5% PS in Decalin	0.015	-	-	-	0.25	18.07±10.2	27	885
MN=2-3,000,000					0.5	11.0 ±2.36	28	
					1	9.83±1.61	25	
0.5% PS in TCP	1.199	52.3	313	3.83	0.38	52.2±38.7	25	1148
(Poly(styrene) in					1	46.9 ±9.7	27	
tri-cresyl phosphate)					1.38	53.5 ±7.5	25	
MN=20,000,000					2.38	63.8 ±8.3	25	
					3	44.9 ±5.9	25	
					6.12	34.9 ±2.4	25	
0.5% PS in DMP	-	21.6	-	-	0.38	23.3 ±4.0	25	
(Poly(styrene) in					1	20.3 ±3.9	25	
di-methyl phthalate)					1.38	21.2 ±5.9	25	
MN=20,000,000								

Table F.4. Polymer solutions of different dilution and molecular weight
F.4.D. Poly(acryl amide)

Fluid	$\tilde{\mu}$ (Pa s)	$\bar{c}$ (cm/s)	G_c (Pa)	λ_c (10^{-3}s)	gap (10^{-3}m)	c (cm/s)	T (°C)	ρ (kg/m^3)
1.5% PA AP30 in 48.5% water and 50% glycerin (Poly(acryl amide)) MN=4,000,000	160	38.4	172	930	1	37.5 ±4.2	-	1170 C
					2	38.4 ±4.4	-	
					3	39.3 ±2.5	-	
1.25% PA AP30 in 48.75% water and 50% glycerin	112	34.3	137	818	1	33.8 ±4.5	-	1170 C
					2	33.3 ±2.8	-	
					3	35.8 ±1.9	-	
1% PA AP30 in 49% water and 50% glycerin	53.8	51.7	310	174	1	58.4 ±6.5	23	1160
					1.38	58.6 ±2.9	24	
					2.38	52.0 ±5.4	*24	
					3	46.5 ±2.7	23	
					6.12	42.8 ±3.3	24	
0.75% PA AP30 in 49.25% water and 50% glycerin	26.4	25.3	137	193	1	25.5 ±3.6	-	1160 C
					2	24.7 ±2.1	-	
					3	25.7 ±1.9	-	

F.4.D. Poly(acryl amide) (continued)

Fluid	$\tilde{\mu}$ (Pa s)	$\bar{c}$ (cm/s)	G_c (Pa)	λ_c (10^{-3}s)	gap (10^{-3}m)	c (cm/s)	T (°C)	ρ (kg/m^3)
0.5% PA AP30 in 49.5% water and 50% glycerin	10.5	22	55.9	188	1 2 3	23.5 ±3.1 22.1 ±1.9 20.4 ±0.8	- - -	1160 C
0.25% PA AP30 in 49.75% water and 50% glycerin	4.5	18.9	41.1	110	1 2 3	18.5 ±2.7 18.0 ±1.5 20.2 ±2.6	- - -	1150 C
0.1% PA AP30 in 49.9% water and 50% glycerin	0.3	8.3	7.6	39.5	1 2 3	7.7 ±1.4 7.8 ±0.8 9.3 ±1.4	- - -	1120 C
0.05% PA AP30 in 49.95% water and 50% glycerin	0.11	6.8	5.2	21.2	1 2 3	7.5 ±1.3 6.1 ±0.9 9.6 ±3.5	- - -	1130 C
0.5% PA AP30 in ethylene glycol with 2ppm Al2O3 flakes (Poly(acryl amide)) MN=4,000,000	1.4	28.1	88.4	15.8	0.38 1 2	25.8 ±3.0 27.5 ±5.4 31.1 ±4.8	25 27 26	1117

Table F.4. Polymer solutions of different dilution and molecular weight
F.4.E. Carboxy-methyl-cellulose

Fluid	$\tilde{\mu}$ (Pa s)	$\bar{c}$ (cm/s)	G_c (Pa)	λ_c (10^{-3}s)	gap (10^{-3}m)	c (cm/s)	T (°C)	ρ (kg/m³)
1.3% CMC in 48.7% water and 50% glycerin (Carboxy-methyl - cellulose) MN=160,000	129.8	80.8	744	74	1.38	81.2 ±8.8	23	1140
					6.12	80.3 ±5.5	23	
1.2% CMC in 48.8% water and 50% glycerin	84.5	55.4	347	244	1.38	50.0 ±5.0	26	1130
					3	58.8 ±9.2	24	
					6.12	55.7 ±1.6	25	
					6.75	57.1 ±1.6	25	
1.0% CMC in 49% water and 50% glycerin	44.8	50.6	289	155	1	47.3 ±6.3	-	1132
					2	51.4 ±3.0	-	
					3	53.1 ±2.6	-	
0.8% CMC in 49.2% water and 50% glycerin	5.77	48.8	269	21.2	1	49.0 ±8.5	22	1130
					2	47.2 ±3.3	23	
					3	50.1 ±3.1	22	

F.4.E. Carboxy-methyl-cellulose (continued)

Fluid	$\tilde{\mu}$ (Pa s)	$\bar{c}$ (cm/s)	G_c (Pa)	λ_c (10^{-3}s)	gap (10^{-3}m)	c (cm/s)	T (°C)	ρ (kg/m^3)
1.3% CMC in in water (Carboxy-methyl cellulose) MN=160,000	10.8	48.1	231	46.8	3	48.0 ±2.4	24	999
					6.75	48.1 ±1.9	27	
0.7% CMC in in water	0.598	30.2	91	6.6	1	32.6 ±5.5	23	1000
					2	29.8 ±2.6	24	
					3	28.1 ±3.2	23	

Table F.4. Polymer solutions of different dilution and molecular weight
F.4.F. Poly(methyl-methacrylate) and a copolymer

Fluid	$\tilde{\mu}$ (Pa s)	$\bar{c}$ (cm/s)	G_c (Pa)	λ_c (10^{-3}s)	gap (10^{-3}m)	c (cm/s)	T (°C)	ρ (kg/m^3)
9.8% PMMA Elvacite in DEM (Poly(methyl-methacrylate) in di-ethyl malonate) MN=400,000	0.71	67.7	503	1.41 3	0.25	66.8±34.6 68.5 ±8.1	26 *25	1100
2% PMMA in DEM (Poly(methyl-methacrylate) in di-ethyl malonate) MN=1,000,000	2.58	25	66.8	38.6	1 2 3	25.8 ±4.1 26.6 ±2.1 22.5 ±2.0	22 23 23	1071
1% PMMA in DEM	0.426	12.9	17.4	24.5	1.38 2.38	14.0 ±1.72 11.7 ±1.3	23 24	1050

F.4.F. Poly(methyl-methacrylate) and a copolymer (continued)

Fluid	$\tilde{\mu}$ (Pa s)	$\bar{c}$ (cm/s)	G_c (Pa)	λ_c (10^{-3}s)	gap (10^{-3}m)	c (cm/s)	T (°C)	ρ (kg/m³)
12.1% K125 in DEM (copolymer of 80% poly(ethyl-butyl acrylate) in di-ethyl malonate) MN=1,900,000	90.1	138	2161	41.7	1	140.8±16.1	25	1140
					3	140.1 ±5.6	25	
5% K125 in DEM	1.17	46.7	231	5.06	1	45.4 ±7.0	26	1060
					2	46.1 ±3.3	25	
					3	48.5 ±5.4	26	

Table F.4. Polymer solutions of different dilution and molecular weight
F.4.G. Solvents

Fluid	$\tilde{\mu}$ (Pa s)	$\bar{c}$ (cm/s)	G_c (Pa)	λ_c (10^{-3}s)	gap (10^{-3}m)	c (cm/s)	T (°C)	ρ (kg/m^3)
Decalin	0.003	5.22	2.41	1.25	0.25	6.07±2.93	27	883
					0.38	4.92 ±.92	28	
					0.5	4.67±1.29	25	
Tri-cresyl Phosphate (TCP)	0.07	-	-	-	0.38	62.5±28.9	26	1110
					1	49.4±11.0	25	
					2	36.0±14.0	25	

Table F.5. Fluorinert coolant

Fluid	$\tilde{\mu}$ (Pa s)	$\bar{c}$ (cm/s)	G_c (Pa)	λ_c (10^{-3}s)	gap (10^{-3}m)	c (cm/s)	T (°C)	ρ (kg/m^3)
Fluorinert FC 5312	2.43	12.5	30.2	80.6	0.25	12.8 ±6.6	25	1930
					0.5	12.2 ±6.1	24	

Table F.6. Honey and fruit jam

Fluid	$\tilde{\mu}$ (Pa s)	$\bar{c}$ (cm/s)	G_c (Pa)	λ_c (10^{-3}s)	gap (10^{-3}m)	c (cm/s)	T (°C)	ρ (kg/m^3)
Honey	249	1350	256000	0.97	1	1671 ±632	-	1400 C
					2	1403 ±311	-	
					3	977 ±196	-	
Cherry EP258 #1 (Pillsbury)	474	117	1490	0.32	6.75	117.2 ±5.7	25	1090
Cherry EP258 #4 (Pillsbury)	456	117	1740	0.26	3	121.2±19.0	26	1270
					6.75	113.0 ±5.9	26	
Cherry EP258 #5 (Pillsbury)	423	124	1620	0.26	3	131.4 ±8.2	25	1050
					6.75	117.1 ±4.6	24	

References†

AHRENS, M. 1989 Studies of hyperbolicity and change of type in flowing viscoelastic fluids, Ph.D. Thesis, University of Minnesota. **8** 17, 25, 29–32.

AHRENS, M., JOSEPH,. D.D., RENARDY, M., AND RENARDY, Y. 1984 Remarks on the stability of viscometric flow, *Rheol. Acta* **23**, 345–354. **4** 32; **16** 36, 47–48.

AHRENS, M., YOO, J.Y., AND JOSEPH, D.D. 1987 Hyperbolicity and change of type in the flow of viscoelastic fluids through pipes, *J. non-Newtonian Fluid Mech.* **24**, 67–83. **11** 10; **12** 2; **13** 22.

AKBAY, U., BECKER, E., KROZER, S., AND SPONAGEL, S. 1980a About a possible cause of viscoelastic turbulence, in *Rheology* (ed. Astarita, *et al.*), Plenum Press: New York, vol. 2, 79–84. **4** 31–32; **16** 41.

AKBAY, U., BECKER, E., KROZER, S., AND SPONAGEL, S. 1980b Instability of slow viscometric flow, *Mech. Res. Comm.* **7**, 199–204. **4** 31–32; **16** 41.

ALVAREZ, G., LODGE, A., AND CANTOW, H. 1985 Measurement of the first and second normal stress differences: correlation of four experiments on polyisolutylene/decalin solution "D1", *Rheol. Acta* **24**, 368–376. **17** 37, 39.

AMBARI, A., DESLOUIS, C., AND TRIBOLLET, B. 1984 Coil-stretch transition of macromolecules in laminar flow around a small cylinder. *Chem. Eng. Commun.* **29**, 63–78. **7** 18, 50, 59–60.

*ASTARITA, G. AND MARRUCCI, G. 1974 *Principles of Non-Newtonian Fluid Mechanics.* McGraw-Hill: New York. **7** 13; **9** 4.

BARLOW, A. J., HARRISON, G., AND LAMB, J. 1964 Viscoelastic relaxation of polydimethyl-siloxane liquids, *Proc. R. Soc. Lond.* **A-282**, 228–251.

BARR, G., *Viscometry*, London, 1931. **18** 28.

BATEMAN, H., DRYDEN, H.L., AND MURNAGHAN, F. Bulletin of the National Research Council, Report of the Committee on Hydrodynamics, Washington, 1932, reprinted: Dover, New York, 1956. **18** 28.

BEAVERS, G.S. AND JOSEPH, D.D. 1975. The rotating rod viscometer, *J. Fluid Mech.* **69**, 475–511. **17** 32–35.

BEAVERS, G.S. AND JOSEPH, D.D. 1977. Novel Weissenberg effects. *J. Fluid Mech.* **81**, 265–272. **17** 32, 49–50.

† General books of reference are marked with an asterisk.

BEAVERS, G.S., YOO, J., AND JOSEPH, D.D. 1980 The free surface on a liquid between cylinders rotating at different speeds, Part III, *Rheol. Acta* **19**, 19–31. **17** 32.

BECHTEL, S.E., LIN, K.J., AND FOREST, M.G. 1988 On the behavior of viscoelastic free jets with elliptical cross section, *J. non-Newtonian Fluid Mech.* **27**, 87–126. **13** 2.

BECKER, E. 1980 Simple Non-Newtonian Flows, in *Advances in Applied Mechanics* **20**, Academic Press: New York. **16** 41.

BERNSTEIN, B., KEARSLEY, E.A., AND ZAPAS, L.J. 1963 A study of stress relaxation with finite strain. *Trans. Soc. Rheol.* **7**, 391–410. **3** 16; **15** 18.

BINDING, D.M., JONES, D.M., AND WALTERS, K. 1989 The shear and extensional flow properties of M1, *J. non-Newtonian Fluid Mech.*, to appear. **17** 45.

*BIRD, R.B., ARMSTRONG, B., AND HASSAGER, O. 1977a *Dynamics of Polymeric Liquids*. Wiley: New York. **16** 23, **18** 7.

*BIRD, R.B., ARMSTRONG, B., AND HASSAGER, O. 1977b *Dynamics of Polymeric Liquids, vol. 2: Kinetic Theory*. Wiley: New York. **16** 23, **18** 8–9, 34.

BIRKHOFF, G. 1954 Classification of partial differential equations, *J. Soc. Indus. Appl. Math.* **2**(1), 57–67. **4** 19.

BÖHME, G. 1974 Eine Theorie für sekundäre Strömungserscheinungen in Nicht-Newtonschen Fluiden. *Deutsche Luft- and Raumfahrt Forschungsbericht* 74–24. **16** 41; **17** 11, 32.

*BÖHME, G. 1987 *Non-Newtonian Fluid Mechanics*. North Holland: Amsterdam. **17** 18; **19** 14.

BOLTZMANN, L. 1874 Zur Theorie der elastischen Nachwirkung sitzungber. Kaiserl-Akad. Wiss. (Wien), *Math. Naturwiss classe* **70** (II), 275–306. **1** 22, **18** 17, 30.

BOOR, C.D. 1971 CADRE: an algorithm for numerical quadrature, in *Mathematical Software* (ed. J. Rice). New York: Academic Press, 417–449.

BRENSCHEDE, E. AND KLEIN, J. 1970 Druckverluste und instabiles Fließen elastischer Flüssigskeiten im Hockdruck-Kappilarviskosimeter, *Rheol. Acta* **9**, 130–136. **13** 13, 15.

BROADBENT, J., KAY, A., LODGE, A., AND VALE, D. 1968 Possible systematic error in the measurement of normal differences in polymer solutions in steady shear flow. *Nature* **217**, 55–56. **17** 31.

BROYER, E., AND MACOSKO, C.W. 1975 Comparison of cone and plate, bicone and parallel plates geometries for melt rheological measurements. *S.P.E. Tech. Papers* **21**, 343–345. **11** 9, 16.

BURDETTE, S.R. 1989 Development of the velocity field in transient shear flows of viscoelastic liquids. *J. non-Newtonian Fluid Mech.* **32**, 269–294.

BURGERS, J.M. 1939 Mechanical Considerations, Model Systems, Phenomenological Theories of Relaxation and of Viscosity. 1st Report on Viscosity and Plasticity, Committee for the Study of Viscosity of the Academy of Sciences at Amsterdam, II (5). **18** 18.

BUTCHER, J.C. 1876 On the formation of the equation of motion. *Proc. London Math. Soc.* **10**, 103–135. **18** 22.

CALDERER, M.C., COOK, L.P., AND SCHLEINIGER, G. 1989 An analysis of the Bird-Deaguiar Model for Polymer Melts, *J. non-Newtonian Fluid Mech.* **31** (2), 209–225. **6** 30.

CARLSLAW, H.S. AND JAEGER, J.C. 1963 *Operational Methods in Applied Mathematics*, 2nd ed. Dover. Pg. 207. **8** 5, **19** 14.

CASWELL, B. 1976 The stress singularity in a linear viscoelasti fluid at the leading edge of a flat plate. Chalmers University of Technology: *Proceedings of VIIth International Congress on Rheology*, 432–33. **8** 5.

CHOI, H.C., SONG, J.H., AND YOO, J.Y. 1988 Numerical simulation of the planar contraction flow of a Giesekus fluid, *J. non-Newtonian Fluid Mech.* **29**, 347–380. **6** 33.

*CHRISTENSEN, R.M. 1982 *Theory of Viscoelasticity: An Introduction.* Second Edition. Academic Press: New York.

CHU, B.T. 1962 Stress waves in isotropic linear viscoelastic materials, *J. Méc.* **1**(4), 439–462. **19** 14.

COCHRANE, T., WALTERS, K., AND WEBSTER, M. F. 1981 On Newtonian and non-Newtonian flow in complex geometrics, *Phil. Trans. Roy. Soc.* **301**, 163–181. **17** 30.

COLEMAN, B.D. AND GURTIN, M.E. 1968 On the stability against shear waves of steady flows of nonlinear viscoelastic fluids, *J. Fluid Mech.* **33**, 165–181. **20** 3.

COLEMAN, B.D., GURTIN, M.E., AND HERRERA, R.I. 1965 Waves in materials with memory, I, *Arch Rat. Mech. Anal.* **19**, 1–19. **7** 51; **19** 14.

COLEMAN, B.D. AND MARKOVITZ, H. 1964 Normal stress effects in second-order fluids, *J. Appl. Phys.* **35**, 1–9. **16** 23.

*COLEMAN, B.D., MARKOVITZ, H., AND NOLL., W. 1966 Viscometric flows of non-Newtonian fluids; Theory and experiment, *Springer Tracts in Natural Philosophy* **5**. **16** 32.

COLEMAN, B.D. AND NOLL, W. 1960 An approximation theorem for functionals, with applications in continuum mechanics, *Arch. Rat. Mech. Anal.* **6**, 355–370.

COLEMAN, B.D. AND NOLL, W. 1961 Foundations of linear viscoelasticity, *Rev. Modern Phys.* **33**, 239–249. **15** 8, 16; **16** 5–6, 17; **20** 9.

COLIN, M.J. 1893 Sur la Rigidité des Liquides, *Comptes rendus* **116**, 1251–1253. **18** 26–27.

CRAIK, A.D. 1968 A note on the static stability of an elastico-viscous fluid, *J. Fluid Mech.* **33**, 33–38. **16** 28–29.

*CROCHET, M.J., DAVIES, A.R., AND WALTERS, K. 1984 *Numerical Simulation of non-Newtonian Flow.* Elsevier: Amsterdam.

CROCHET, M.J. AND DELVAUX, V. 1990 Numerical simulation of inertial viscoelastic flow, with change of type, in *Nonlinear Evolution Equations that Change Type,* B. Keyfitz and M. Shearer, eds. IMA Volumes in Mathematics and its Applications. New York: Springer-Verlag.

DELVAUX, V. AND CROCHET, M.J. 1990 Numerical simulation of delayed die swell. *Rheologica Acta* **29**, to appear.

DELVAUX, V. AND CROCHET, M.J. 1990 Inertial viscoelastic flow past circular cylinder. *J. non-Newtonian Fluid Mech.,* to appear.

DE METZ, G. 1888 Uber die temporäre Doppelbrechung de Lichtes in totierenden Flüssigkeiten. *Ann. Phys. Chem.* **35**, 497. **18** 25.

DUNWOODY, J. AND JOSEPH, D.D. 1984 Systematic linearization for stability of shear flows of viscoelastic fluids, *Arch. Rat. Mech. Anal.* **86**, 65–84. **16** 36, 41.

DUPRET, F., MARCHAL, J.M., AND CROCHET, M. 1985 On the consequence of discretization errors in the numerical calculation of viscoelastic flow, *J. non-Newtonian Fluid Mech.* **18**, 173–186. **6** 33.

EDWARDS, B.J. AND BERIS, A. 1990 Remarks concerning compressible viscoelastic models, *J. non-Newtonian Fluid Mech.* (in press)

ERICKSEN, J.L. 1975 Equilibrium of bars, *J. Elasticity* **5**, 191–201. **4** 25.

EVANS, D., HANLEY, H., AND HESS, S. Jan. 1984 Non-Newtonian phenomena in simple fluids, *Physics Today* **37**, 26–33. **16** 3, 34.

FERGUSON, J., HUDSON, N.E., AND WARREN, B.C.H. 1987 Structural changes during elongation of polymer solutions, *J. non-Newtonian Fluid Mech.* **23**, 49–72. **10** 2–3, 5, 18.

*FERRY, J. D. 1980 *Viscoelastic Properties of Polymers.* Wiley: New York. **18** 35; **19** 8.

FOSDICK, R. AND KAO, B.G. 1980 Steady flow of a simple fluid around a rotating sphere, *Rheol. Acta* **19**, 675–697. **17** 18.

FRAENKEL, L.E. 1988 Some results for a linear, partly hyperbolic model of viscoelastic flow past a plate, in *Material Instablities in Continuum Mechanics and Related Mathematical Problems* (ed. J.M. Ball). Clarendon Press: Oxford. **8** 2, 7–8, 21, 23, 25, 27.

GIESEKUS, H. 1963 Die simultane Translations-und Rotations bewegung einer Kugel in einer elastovisken Flussigkeit, *Rheol. Acta* **3**, 59–71. **17** 6, 54.

GIESEKUS, H. 1967 Die sekundarströmung in einer Kegel-Platte-Anordnung: Abhängigkeit von der Rotationsgeschwindigskeit bei verchiedenen Polymersystemen, *Rheol. Acta* **6**, 339–353. **17** 18.

GIESEKUS, H. 1968 Verschiedene Phänonmene in Strömungen viskoelastischer Flüssigkeiten durch Düsen, *Rheol. Acta*. **8**, 411–421. **10** 2; **13** 13, 15; **17** 18.

GIESEKUS, H. 1982 A simple constitutive equation for polymer fluids based on the concept of deformation-dependent tensorial mobility, *J. non-Newtonian Fluid Mech.* **11**, 69–109. **1** 21; **2** 9; **6** 12.

GLEISSLE, W. 1982 Stresses in polymer melts at the beginning of flow instabilities (melt fracture) in cylindrical capillaries, *Rheol. Acta* **21**, 484–487. **16** 48.

GODDARD, J.D. AND MILLER, C. 1967 Technical Report 06673-8-T. University of Michigan. **16** 40.

GREEN, A.E., AND RIVLIN, R. 1957 The mechanics of non-linear materials with memory—part 1, *Arch. Rat. Mech. Anal.* **1**, 1–21. **16** 5, 6.

GUILLOPÉ, C. AND SAUT, J.C. 1987 Résultats d'existence pose des fluides élastiques à loi de comportment de type differentiel. *C.R. Acad. Sci. Paris* **305**, 489–492. **6** 36.

GUILLOPÉ, C. AND SAUT, J.C. 1990 Global existence and one-dimensional nonlinear stability of shearing motions of viscoelastic fluids of Oldroyd type. *Mathematical Modeling and Numerical Analysis*, to appear. **6** 36.

HARDER, K.J. 1989 Influence of wall strain rate, polymer concentration and channel height upon drag reduction and turbulent structure. Master's thesis in Mechanical Engineering, Purdue University. **7** 206

*HARRISON, G. 1976 *The Dynamic Properties of Supercooled Liquids*. Academic Press: London and New York. **18** 12, 32; **19** 4.

HERMES, R.A. 1965 Flow of viscoelastic fluids past a flat plate, Ph.D. thesis, University of Minnesota. **8** 33–34, 36.

HERMES, R.A. AND FREDERICKSON, A.G. 1967 Flow of viscoelastic fluids past a flat plate, *Amer. Inst. Chem. Eng. J.* **13**, 253–259. **8** 33–35.

HESS, W.R. 1920 Die innere Reibung gelatinierender Lösungen. *Koll. Zeitschrift* **27**, 154. **18** 28.

HILL, C.T. 1972 Nearly viscometric flow of viscoelastic fluids in the disk and cylinder system. II: Experimental, *Trans. Soc. Rheol.* **16**, 213–245. **17** 21.

HOFFMAN, A.H. AND GOTTENBERG, W.G. 1973 Determination of the material functions for a simple fluid from a study of the climbing effect, *Trans. Soc. Rheol.* **17**, 465–485. **17** 32.

HOOPER, A. AND BOYD, W.G. 1983 Shear flow instability at the interface between two viscous fluids, *J. Fluid Mech.* **128**, 507–528. **4** 20.

HU, H., RICCIUS, O., CHEN, K., ARNEY, M., AND JOSEPH, D.D. 1990 Climbing constants, second order corrections of Trouton's viscosity, wave speeds and delayed die swell for M1. *J. non-Newtonian Fluid Mech.*, to appear. **17** 40–41.

*HUILGOL, R.R. 1975 *Continuum Mechanics of Viscoelastic Liquids.* Halstead Press: Delhi, and Wiley: New York. **2** 10; **16** 40.

HULSEN, M.A. 1986 Analysis of the equations for viscoelastic flow: type boundary conditions and discontinuities. Tech. report WTHD 185, Mech. Eng. Delft Univ. of Technology. **5** 14; **6** 33–35; **20** 4, 35.

HUNTER, J.K. AND SLEMROD, M. 1983 Viscoelastic fluid flow exhibiting hysteric phase changes, *Phys. Fluids* **26**, 2345–2351. **4** 29.

HUPPLER, J.D., ASHARE, E., AND HOLMES, L.A. 1967 Rheological properties of three solutions. Part I: Non-Newtonian viscosity, normal stresses, and complex viscosity. *Trans. Soc. Rheol.* **11**, 159–179. **18** 9.

JAMES, D.F. 1967 Laminar flow of dilute polymer solutions around circular cylinders, Ph.D. Thesis. Cal. Inst. Tech., Pasadena. **7** 16, 58.

JAMES, D.F. AND ACOSTA, A.J. 1970 The laminar flow of dilute ploymer solutions around circular cylinders, *J. Fluid Mech.* **42**, 269–288. **7** 16–18, 49–50, 60–61.

JAMES, D.F. AND GUPTA, O.P. 1971 Drag on circular cylinders in dilute polymer solutions, *Chem. Eng. Progr.* **67** № 111, 62–73. **7** 13, 61.

JEFFREYS, H. 1929 *The Earth.* Cambridge University Press. **2** 3.

JOHNSON, M. AND SEGALMAN, D. 1977 A model for viscoelastic fluid behavior which allows non affine deformation, *J. non-Newtonian Fluid Mech.* **2**, 255–270. **20** 30.

JOSEPH, D.D. 1974 Slow motion and viscometric motion; stability and bifurcation of the rest state of a simple fluid, *Arch. Rat. Mech. Anal.* **56**, 99–157. **13** 9; **17** 20.

*JOSEPH, D.D. 1976 *Stability of Fluid Motions II*, Springer-Verlag: New York. **3** 19; **16** 15, 29.

JOSEPH, D.D. 1977 Rotating simple fluids, *Arch. Rat. Mech. Anal.* **66**, 311–344. **16** 11, 13, 16.

JOSEPH, D.D. 1980a An integral invariant for jets of liquid into air, *Arch. Rat. Mech. Anal.* **74**, 389–393. **13** 3.

JOSEPH, D.D. 1980b Short communication: Some inequalities in the theory of simple fluids, *Rheol. Acta* **19**, 389–390. **16** 24.

JOSEPH, D.D. 1980c Instability of the rest state of fluids of arbitrary grade larger than one, *Arch. Rat. Mech. Anal.* **75**, 251–256. **16** 24.

JOSEPH, D.D. 1985 Hyperbolic phenomena in the flow of viscoelastic fluids, in *Viscoelasticity and Rheology* (eds. A.S. Lodge, J. Nohel, and M. Renardy). Academic Press. **3** 16; **8** 7; **12** 42.

JOSEPH, D.D. 1989 Remarks on inertial radii, persistent normal stresses, secondary motions, and non-elastic extensional viscosities, *J. non-Newtonian fluid mechanics* **32**(1), 107–114. **17** 3.

JOSEPH, D.D. AND BEAVERS, G.S. 1976 The free surface on a simple fluid between cylinders undergoing torsional oscillations, *Arch. Rat. Mech. Anal.* **62**, 323–352. **17** 33.

JOSEPH, D.D. AND BEAVERS, G.S. 1977a Free surface problems in rheological fluid mechanics, *Rheol. Acta* **16**, 169–189. **16** 5, 23; **17** 47, 51; **App. E** 1*ff*.

JOSEPH, D.D. AND BEAVERS, G.S. 1977b Free surfaces induced by the motion of viscoelastic fluids. In *The Mechanics of Viscoelastic Fluids,* Appl. Mech. Div. ASME (ed. R. Rivlin) **22**, 1977. **16** 5, 23; **17** 34.

JOSEPH, D.D., BEAVERS, G.S., CERS, A., DEWALD, C., HOGER, A., AND THAN, P. 1984 Climbing constants for various liquids, *J. Rheology* **28**, 325–345. **17** 33–36.

JOSEPH, D.D., BEAVERS, G.S., AND FOSDICK, R. 1973 The free surface on a liquid between cylinders rotating at different speeds. Part II, *Arch. Rat. Mech. Anal.* **49**, 381–401. **17** 32, 34.

JOSEPH, D.D. AND CHEN, K. 1988 Anomalous elongational flows and change of type, *J. non-Newtonian Fluid Mech.* **28**, 47–67. **10** 2.

JOSEPH, D.D. AND FOSDICK, R. 1973 The free surface on a liquid between cylinders rotating at different speeds. Part I, *Arch. Rat. Mech. Anal.* **49**, 321–380. **17** 32–33.

JOSEPH, D.D., MATTA, J., AND CHEN, K.. 1987 Delayed die swell, *J. non-Newtonian Fluid Mech.* **24**, 31–65. **10** 2; **12** 23, 37, 40–42; **13** 2*f*, 13.

JOSEPH, D.D., NARAIN, A., AND RICCIUS, O. 1986 Shear-wave speeds and elastic moduli for different liquids. Part 1: Theory, *J. Fluid Mech.* **171**, 289–308. **19** 2.

JOSEPH, D.D., NGUYEN, K., AND MATTA, J. 1983 Jets into liquid under gravity, *J. Fluid Mech.* **128**, 443–468. **13** 3.

JOSEPH, D.D. AND PREZIOSI, L. 1989 Heat waves. *Rev. Mod. Phys.* **61**, 41–73.

JOSEPH, D.D. AND PREZIOSI, L. 1989 Heat waves—addendum. *Rev. Mod. Phys.* **62**, to appear.

JOSEPH, D.D., AND RENARDY, M. 1990 Stokes' first problem for viscoelastic fluids with finite memory, *Rheol. Acta* **28**, 453–456.

JOSEPH, D.D., RENARDY, M., AND SAUT, J.C. 1985 Hyperbolicity and change of type in the flow of viscoelastic fluids, *Arch. Rat. Mech. Anal.* **87**(3), 213–251. **3** 16; **6** 33; **10** 6; **12** 6, 38, 42; **15** 15, 18.

JOSEPH, D.D., RICCIUS, O., AND ARNEY, M. 1986 Shear wave speeds and elastic moduli for different liquids; Part II: Experiments, *J. Fluid Mech.* **171**, 309–338. **7** 58, 60; **10** 3, 16; **12** 38, 40; **13** 13; **18** 32; **19** 2; **App. F** 2, 18.

JOSEPH, D.D. AND SAUT, J.C. 1986 Change of type and loss of evolution in the flow of viscoelastic fluids, *J. non-Newtonian Fluid Mech.* **20**, 117–141. **4** 19; **10** 6; **20** 3, 28.

JOSEPH, D.D. AND SAUT, J.C. 1989 Short wave instabilities and ill-posed problems, *Theoretical and computational fluid dynamics*, to appear. **4** 22.

JOSEPH, D.D. AND STURGES, L. 1980 A normal stress amplifier for the second normal stress difference, *J. non-Newtonian Fluid Mech.* **6**, 325–331. **17** 54.

KAMATH, V.M. AND MACKLEY, M.R. 1989 Redetermination of polymer relaxation moduli and memory functions using integral transforms, *J. non-Newtonian Fluid Mech.* **32**, 119–144.

KAPLUN, S. 1954 The role of coordinate systems in boundary-layer theory. *Z. Angew. Math. Phys.* **5**, 111–135. **8** 7.

KAYE, A. 1962 Non-Newtonian flow in incompressible fluids. College of Aeronautics, Cranfield, England, Tech Note. **15** 18.

KAYE, A. 1973 The shape of a liquid surface between rotating concentric cylinders, *Rheol. Acta* **12**, 206–211. **17** 32.

KAZAKIA, J.Y. AND RIVLIN, R.S. 1981 Run-up and spin-up in a viscoelastic fluid I, *Rheol. Acta* **20**, 111–127. **19** 14.

KEENTOK, M., GEORGESCU, A.G., SHERWOOD, A.A., AND TANNER, R. 1980. The measurement of the second normal stress difference for some polymer solutions, *J. non-Newtonian Fluid Mech.* **6**, 303–324. **17** 36–37, 39, 47.

KOHLRAUSCH, F. 1863 Über die elastische Nachwirkung bei der Torsion. *Ann. Phys. Chem.* **119**, 337. **18** 30.

KOHLRAUSCH, F. 1866 Beiträge zur Kenntniss der elastischen Nachwirkung. *Ann. Phys. Chem.* **128** (1), 207 and 399. **18** 30.

KOLPIN, B., BEAVERS, G.S., AND JOSEPH, D.D. 1980 Free surface on a simple fluid between cylinders undergoing torsional oscillations. IV: Oscillating rods, *J. Rheology* **24**, 719–739. **17** 33.

KONIUTA, A., ADLER, P.M. AND PIAU, J.M. 1980 Flow of dilute polymer solutions around circular cylinders. *J. non-Newtonian Fluid Mech.* **7**, 101–106. **7** 1, 55–57, 61.

KRAMER, J.M. AND JOHNSON, M. 1972 Nearly Viscometric Flow in the Disk and Cylinder System. I: Theoretical, *Trans. Soc. Rheol.* **16**, 197–212. **16** 41; **17** 21.

KREISS, H.O. 1978 *Numerical methods for Solving Time-Dependent Problems for Partial Differential Equations.* Les presses de l'Université de Montréal, Montréal. **4** 23.

KUNDT, A. 1881 Über die Doppelbrechung des Lichtes in bewegten reibenden Flüssigkeiten. *Ann. Phys. Chem. (Wiedemann)* **8**, 110. **18** 25.

KUO, U. AND TANNER, R. 1974 On the use of open-channel flows to measure second normal stress difference, *Rheol. Acta* **13**, 443–456. **17** 47.

LAKLALECH, H. 1985 Thèse troisième cycle. Sur une classe d'equations integrodifférentielles liées à la mécanique des fluides non newtoniens, Mathématique, Université de Paris Sud, Orsay. **16** 15.

LARMOR, J. 1907 *Memoir and Scientific Correspondence of the Late Sir George Gabriel Stokes.* Cambridge University Press. **18** 20.

*LARSON, R.G. 1988a *Constitutive Equations for Polymer Melts and Solutions.* Butterworths: Boston. **18** 8.

LARSON, R.G. 1988b Viscoelastic inertial flow driven by an axisymmetric accelerated surface. *J. Fluid Mech.* **196**, 449–465. **11** 2*ff*, 8, 16.

LEE, J.S. AND FULLER, G. G. 1987 The spatial development of transient Couette flow and shear wave propagation in polymeric liquids by flow birefringence. *J. non-Newtonian Fluid Mech.* **26**, 57–76. F

LEWIS, J.A. AND CARRIER, G.F. 1949 Some remarks on the flat plate boundary layer. *Quart. Appl. Math.* **7**, 228–234. **8** 2, 7.

LIEB, E.B. 1975 Viscoelastic flow behavior in accelerating shear fields. PhD Thesis, University of Rochester. F

LODGE, A.S. 1956 A network theory of flow birefringence and stress in concentrated polymer solutions. *Trans. Faraday Soc.* **52**, 120–180. **15** 18.

*LODGE, A.S. 1964 *Elastic Liquids.* Academic Press: London and New York.

*LODGE, A.S. 1974 *Body Tensor Fields in Continuum Mechanics (With Applications to Polymer Rheology).* Academic Press: New York. **1** 11.

*LODGE, A.S., RENARDY, M., AND NOHEL, J.A. 1985 *Viscoelasticity and Rheology.* Academic Press: New York.

LODGE, A.S., SCHIEBER, J.D., AND BIRD, R.B. 1987 The Weissenberg effect at finite rod-rotation speeds, *J. Chem. Phys.* **88**, 4001–4007. **17** 40.

MARCHAL, J.M. AND CROCHET, M. 1987 A new fixed finite element for calculating viscoelastic flow, *J. non-Newtonian Fluid Mech.* **26**, 77–114. **6** 33.

MARKOVITZ, H. 1968 The emergence of rheology. *Physics Today* **21**(4), 23–30. **18** 28.

MARKOVITZ, H. 1977 Boltzmann and the beginnings of linear viscoelasticity. *Trans. Soc. Rheol.* **21**(3), 381–397. **18** 30.

MASON, W.P. 1947 Measurement of viscosity and shear elasticity of liquids by means of a torsionally vibrating crystal. *Trans. ASME* **69**, 359–367. **18** 12.

MASON, W.P., BAKER, W.O., MCSKIMIN, H.J., AND HESS, J.H. 1949 Measurement of shear elasticity and viscosity of liquids at ultrasonic frequencies, *Phys. Rev.* **75** № 6, 936–946. **18** 12, 14, 31.

MAXWELL, J.C. 1873 On double refraction in a viscous fluid in motion, *Proc. R. Soc. Lond.* **22**, 46–47. **18** 17, 19.

MAXWELL, J.C. 1891 *Theory of Heat.* IVTA Edition, 264–265. **4** 4.

MCDONOUGH, R.N. AND HUGGINS, W. 1968 Best least-squares representation of signals by exponentials. *IEEE Trans on Aut. Control.* Vol. AC-**13**(4), 408–412. **18** 34.

MERRINGTON, A.C. 1943 Flow of viscoelastic materials in capillaries, *Nature*, London **152**, 663. **13** 13.

METZNER, A.B. 1967 Behavior of suspended matter in rapidly accelerating viscoelastic fluids: the Uebler effect. *AIChE J.* **13**, 316–318. **9** 19.

METZNER, A.B. AND ASTARITA, G. 1967 External flows of viscoelastic materials: Fluid property restrictions on the use of velocity-sensitive problems. *AIChE J.* **13**, 550–555. **7** 54, 58.

METZNER, A.B., UEBLER, E.A., AND FONG C.F.C.M. 1969 Converging flows of viscoelastic materials, *AIChE J.* **15**, 750–758. **9** 1*ff*, 13, 15, 21.

METZNER, A.B., WHITE, J.L., AND DENN, M.M. 1966 Behavior of viscoelastic materials in short time processes, *Chem. Eng. Prog.* **62**, 81–92. **13** 13–14.

MICHAUD, F. 1922 Rigidity of Gels. *Comptes Rendus* **175**, 1196 and *Ann. Phys.* **19**, 63. **18** 28.

MIDDLEMAN, S. 1968 *The Flow of High Polymers.* Wiley: New York. **13** 13.

MORRISON, J.A. 1956 Wave propagation in rods of Voigt materia and viscoelastic material with three-parameter models. *Q. Appl. Math.* **14**, 153–169. **19** 14, 22.

MORSE, P.M. AND FESBACH, H. 1953 *Methods of Theoetical Physics*, McGraw-Hill, New York.

NARAIN, A. 1986 On K-BKZ and other viscoelastic models as continuum generalizations of the classical spring-dashpot models, *Rheol. Acta* **25**, 1–14. **15** 18.

NARAIN, A. AND JOSEPH, D.D. 1982 Linearized dynamics for step jumps of velocity and displacement of shearing flows of a simple fluid, *Rheol. Acta* **21**, 228–250; Corrigendum, 1983, *Rheol. Acta* **22**, 906. **19** 14, 17, 21, 24; **App. F** 7, 10.

NARAIN, A. AND JOSEPH, D.D. 1983a Linearized dynamics of shearing deformation perturbing rest in viscoelastic materials. *Equadiff* **'82**. Lecture Notes in Mathematics, vol. 1017, Springer. **8** 5; **18** 7; **19** 14.

NARAIN, A. AND JOSEPH, D.D. 1983b Remarks about the interpretation of impulse experiments in shear flows of viscoelastic liquids, *Rheol. Acta* **22**, 528–538. **18** 7, 15; **19** 14.

NIGGEMAN, M. 1981 A model equation for non-Newtonian fluids, *Math. Meth. Appl. Sci.* **3**, 200–217. **16** 32.

NOLL, W. 1958 A mathematical theory of the mechanical behavior of continuous media. *Arch. Rat. Mech. Anal.* **2**, 197–226. **15** 2.

OLDROYD, J.G. 1950 On the formulation of rheological equations of state, *Proc. R. Soc., London* A**200**, 523–541. **2** 6.

OLMSTEAD, W.E. 1975 A homogeneous solution for viscous flow around a half-plane, *Quart. Appl. Math.* **33**, 165–169. **8** 7.

*PAPANASTASIOU, A.C., SCRIVEN, L.E., AND MACOSKO, C.H. 1983 An integral constitutive equation for mixed flows: viscoelastic characterization, *J. Rheol.* **27**. 387–410.

PASCAL, H. 1989 Propagation of shear stress disturbances in a non-Newtonian fluid. *Physica D* **39**, 262–266. **19** 27.

*PEARSON, J.R.A. 1985 *Mechanics of Polymer Processing.* Elsevier: London.

*PETRIE, C.J.S. 1979 *Elongational Flows.* Pitman: London.

*PETRIE, C.J.S., AND DENN, M.M. 1976 Instabilities in polymer processing, *AIChE J.* **22**, 209–236.

PHAN-THIEN, N. 1983a Coaxial-disk flow and flow about a rotating disk of a Maxwellian fluid, *J. Fluid Mech.* **128**, 427–442. **11** 2*ff*, 6, 8.

PHAN-THIEN, N. 1983b Coaxial-disk flow of an Oldroyd-B fluid: exact solution and stability, *J. non-Newtonian Fluid Mech.* **13**, 325–340. **11** 2 *ff*, 6, 8.

PHAN-THIEN, N., AND TANNER, R. 1977 A new constitutive equation dreived from network theory, *J. non-Newtonian Fluid Mech.* **2**, 353–365. **1** 21.

PHELAN, F.R., MALONE, M.F. AND WINTER, H.H. 1989 A purely hyperbolic model for unsteady viscoelastic flow. *J. non-Newtonian Fluid Mech.* **32**(2), 197–224. **6** 35.

PIPKIN, A.C. 1964 Small finite deformations of viscoelastic solids, *Rev. Mod. Physics.* **36**, 1034–1041. **16** 5.

*PIPKIN, A.C. 1972 Lectures on viscoelasticity theory, *Springer Applied Math. Sci.*, **7**.

PIPKIN, A. AND OWEN, R. 1967 Nearly viscometric flows, *Phys. Fluids* **10**, 836–843. **16** 35.

POISSON, S.D. 1829 Sur les Equations Generale de l'Equilibre et du Mouvement des Corps Solides Elastiques et des Fluides. *Journal de l'Ecole Polytechnique Tome* **13** (cah. 20). **18** 17.

PREZIOSI, L. 1989 On an invariance property of the solution to Stokes' first problem for viscoelastic fluids, *J. non-Newtonian Fluid Mech.* **33**(2), 219–224. **19** 26.

PREZIOSI, L. AND JOSEPH, D.D. 1987 Stokes' first problem for viscoelastic fluids, *J. non-Newtonian Fluid Mech.* **25**, 239–259. **19** 2, 14, 26.

PRONY, R. 1795 Essai expérimental et analytique, etc., Paris: *J. de l'Ecole Polytechnique* **1**, cahier 2, 24–76. **18** 34.

REGIRER, S.I. AND RUTKEVICH, I.M. 1968 Certain singularities of the hydrodynamic equations of non-Newtonian media. *J. Appl. Math. Mech. (PMM)* **32**, 962–966. **4** 30.

REINER, M. 1949 *Twelve Lectures on Theoretical Rheology.* Amsterdam. **18** 29.

RENARDY, M. 1982 Some remarks on the propagation and non-propagation of discontinuities in linearly viscoelastic liquids. *Rheol. Acta* **21**, 251–254. **18** 15; **19** 14.

RENARDY, M. 1988a Inflow-boundary conditions for steady flows of viscoelastic fluids with differential constitutive law, *Rocky Mt. J. Math.* **18**, 445–453. **6** 33–34, 36.

RENARDY, M. 1988b Recent advance in the mathematical theory of steady flows of viscoelastic fluids, *J. non-Newtonian Fluid Mech.* **29**, 11–24. **6** 33, 35–36.

RENARDY, M. 1989a Mathematical analysis of viscoelastic flow. *Ann. Rev. Fluid Mech.* **21**, 21–36. **6** 36; **20** 39.

RENARDY, M. 1989b A well-posed boundary value problem for supercritical flow of viscoelastic fluids of Maxwell type. In *Nonlinear Evolution Equations That Change Type* (eds. B. Keyfitz and M. Shearer), Springer, to appear. **6** 36.

RENARDY, M. 1989c On Rankine-Hugoniot conditions for Maxwell liquids, *J. non-Newtonian Fluid Mech.*, **32**(1), 69–78. **20** 36.

*RENARDY, M., HRUSA, W., AND NOHEL, J. 1987 *Mathematical Problems in Viscoelasticity*. Longman Scientific and Technical: Burnt Mill, Harlow, Essex, England. **Intro.** iii; **5** 11; **19** 22; **20** 3, 28.

RICCIUS, O. 1989 Shear wave speeds and elastic moduli for different liquids, Ph.D. Thesis, University of Minnesota. **App. F**, 2, 17.

RICCIUS, O., JOSEPH, D.D., AND ARNEY, M. 1987 Shear-wave speeds and elastic moduli for different liquids. Part III, Experiments—Update, *Rheol. Acta.* **26**, 96–99. **8** 33; **10** 16.

RICHTMEYER, R.D. AND MORTON, K. 1967 *Difference Methods for Initial Value Problems*. Interscience, New York. **4** 22.

RIVLIN, R.S. 1955 Further remarks on the stress-deformation relations for isotropic materials, *Arch Rat. Mech. Anal.* **4**, 681–672. **App. A** 7.

RIVLIN, R.S. 1982 Run-up and spin-up in a viscoelastic fluid, II, *Rheol. Acta* **21**, 107–111. **19** 14.

RIVLIN, R.S. 1983 Run-up and spin-up in a viscoelastic fluid, 3, *Rheol. Acta* **22**, 213–222. **19** 14.

RIVLIN, R.S. AND SAWYERS, K.N. 1971 Nonlinear continuum mechanics of viscoelastic fluids. Annual Review of Fluid Mechanics **3**, 117–146. **15** 18.

ROBINSON, I. 1976 An algorithm for automatic integration using the adaptive Gaussian technique. *The Australian Computer Journal* **8**, 106–115.

ROSCOE, R. 1965 The steady elongation of elasto-viscous liquids. *Brit. J. Appl. Phys.* **16**, 1567–1571. **17** 8.

RUTKEVICH, I.M. 1969 Some general properties of the equations of viscoelastic incompressible fluid dynamics, *J. Appl. Math. Mech. PMM* **33**, 30–39. **20** 4, 35.

RUTKEVICH, I.M. 1970 The propagation of small perturbations in a viscoelastic fluid, *J. Appl. Math. Mech.(PMM)* **34**, 35–50. **4** 13.

SAUT, J.C. AND JOSEPH, D.D. 1983 Fading memory, *Arch. Rat. Mech. Anal.* **81**, 53–95. **15** 13, 16; **19** 14.

*SCHOWALTER, W.R. 1978 *Mechanics of Non-Newtonian Fluids*. Pergamon Press: New York.

SCHWEDOFF, T. 1889 Recherches expérimentales sur la cohésion des liquides, *J. Phys.* **8**, 341–359. **18** 26.

SLEMROD, M. 1976 A hereditary partial differential equation with applications in the theory of simple fluids, *Arch. Rat. Mech. Anal.* **62**, 303–321. **16** 15, 29.

SLEMROD, M. 1978 Instability of steady shearing flows in a non-linear viscoelastic fluid, *Arch. Rat. Mech. Anal.* **68**, 3, 211–225. **20** 3, 21–22, 28.

SLEMROD, M. 1983 Admissibility criteria for propagation of phase boundaries in a Van der Waals fluid, *Arch. Rat. Mech. Anal.* **81**, 301–315. **4** 25.

SLEMROD, M. 1985 Breakdown of smooth shearing flow in viscoelastic fluids for two constitutive equations: The vortex sheet vs. the vortex shock. Appendix to the paper Joseph, D.D. 1985, Hyperbolic phenomena in the flow of viscoelastic fluids. *op.cit.*. **5** 11; **20** 3, 21–22.

SONG, J.H. AND YOO, J.Y. 1987 Numerical simulation of viscoelastic flow through a sudden contraction using B type dependent difference method. *J. non-Newtonian Fluid Mech.* **24**, 221–243. **6** 32.

SPERA, F.J., BORGIA, A., AND STRIMPLE, J. 1988 Rheology of melts and magmatic suspensions. *J. Geophys. Res.* **93**, 10,273–10,294. **17** 33.

STEWARTSON, K. 1953 On the flow between two rotating coaxial disks. *Proc. Camb. Phil. Soc.* **49**, 333–341. **11** 11.

STRANG, G. 1966 Necessary and insufficient conditions for well-posed Cauchy problems. *J. Diff. Eq.* **2**, 107–114. **4** 23.

STURGES, L. 1977 Secondary motion induced by the rotation of a wavy rod, *Rheol. Acta* **16**, 476–483. **12** 10; **17** 13, 16–17.

STURGES, L. 1981 A theoretical study of extrudate swell, *J. non-Newtonian Fluid Mech.* **9**, 357–378. **17** 24, 28.

STURGES, L. AND JOSEPH, D.D. 1975 Slow motion and viscometric motion. Part V: The free surface on a simple fluid flowing down a tilted trough, *Arch. Rat. Mech. Anal.* **59**, 359–387. **17** 48.

*STURGES, L. AND JOSEPH, D.D. 1980 Short communication: A normal stress amplifier for the second normal stress difference, *J. non-Newtonian Fluid Mech.* **6**, 325–331.

TAMMANN, G. AND FAUST, O. 1909 Rigidity and its connection with inner friction. *Z. physik Chem.* **71**, 51–58. **18** 26, 28–29.

TAMMANN, G. AND LAUER, L. 1908 Rigidity in liquids. *Z. physik Chem.* **63**, 141–150. **18** 26, 28–29.

TANNER, R. 1962 Note on the Rayleigh problem for a viscoelastic fluid, *Z. angew. Math. Phys.* **13**, 573–579. **19** 14.

TANNER, R. 1966 Plane creeping flows of incompressible second-order fluids, *Phys. Fluids* **9**, 1246–1247. **17** 56.

TANNER, R. 1970 Some methods for estimating the normal stress function in viscometric flows, *Trans. Soc. Rheol.* **14**, 483–507. **17** 47.

*TANNER, R.I. 1985 *Engineering Rheology.* Clarendon Press: Oxford.

TANNER, R. AND PIPKIN, A. 1969 Intrinsic errors in pressure-hole measurements, *Trans. Soc. Rheol.* **13**, 471–484. **17** 31–32.

TIEU, H.A. AND JOSEPH, D.D. 1983 Extrudate swell for a round jet with large surface tension, *J. non-Newtonian Fluid Mech.* **13**, 203–222. **17** 24, 27–28.

TIEU, H.A., JOSEPH, D.D., AND BEAVERS, G.S. 1984 Interfacial shapes between two superimposed rotating simple fluids, *J. Fluid Mech.* **145**, 11–70. **17** 52.

TORDELLA, J.P. 1969 Unstable flow of molten polymers in F. Eirich, ed., *Rheology: Theory and Applications*, vol. 5, Academic Press: New York, London. **4** 29.

TROGDON, S. AND JOSEPH, D.D. 1980 The stick-slip problem for a round jet, I. Large surface tension, *Rheol. Acta* **19**, 404–420. **17** 24, 28.

TROUTON, F.T. AND ANDREWS, E.S. 1904 On the viscosity of pitchlike substances, *The London, Edinburgh, and Dublin Philosophical Magazine* **7**, 347. **18** 26, 28.

*TRUESDELL, C. AND NOLL, W. 1965 The non-linear field theories of mechanics, *Flugge's Handbuch der Physik*, **III**, 3, Springer. **App. A** 2, 4, 7.

ULTMAN, J.S. AND DENN, M.M. 1970 Anomalous heat transfer and a wave phenomenon in dilute polymer solutions, *Trans. Soc. Rheol.* **14**, 307–317. **7** 1, 51, 58–61.

VAN DYKE, M. 1975 *Perturbation Methods in Fluid Mechanics.* Parabolic Press: Stanford. **8** 7.

VERDIER, C. AND JOSEPH, D.D. 1989a Similarity solutions that give rise to hyperbolicity and change of type in steady flow of viscoelastic fluid, *J. non-Newtonian Fluid Mech.* **31**, 301–323. **11** 2, 8.

VERDIER, C. AND JOSEPH, D.D. 1989b Change of type and loss of evolution of the White-Metzner model, *J. non-Newtonian Fluid Mech.* **31**,325–343. **4** 15.

*WALTERS, K. 1975 *Rheometry*. Wiley: New York.

WALTERS, K. 1983 The Second-Normal-Stress Difference Project, *IUPAC 29th Symposium on Macromolecules 83*, Bucharest, Romania: Plenary and Invited Lectures, part 2, pp. 227–237. **17** 39.

WALTERS, K. AND WATERS, N. 1968 On the use of a rheogoniometer. Part 1: Steady shear. In *Polymer Systems—Deformation and Flow* (eds. R. Wetton and R. Whorlow), pp. 211–235. London: Macmillan. **17** 19.

WEBER, W. 1835 Über die Elastizität der Seidenfäden. *Ann. Phys. Chem.* **34**, 247. **18** 30.

WHITHAM, G.B. 1974 *Nonlinear waves.* Wiley-Interscience, New York. **20** 2, 8, 36.

WINEMAN, A. AND PIPKIN, A. 1966 Slow viscoelastic flow in tilted troughs, *Acta Mechanica* **2**, 104–115. **17** 47.

WINTER, H. 1987 Evolution of rheology during chemical gelation. *Prog. Colloid and Polymer Sci.* **75**, 104–110.

YIN, W.L. AND PIPKIN, A.C. 1970 Kinematics of viscometric flow, *Arch. Rat. Mech. Anal.* **37**, 111–135. **16** 32, 35.

YOO, J.Y., AHRENS, M., AND JOSEPH, D.D. 1985 Hyperbolicity and change of type in sink flow, *J. Fluid Mech.* **153**, 203–214. **9** 1.

YOO, J.Y. AND JOSEPH, D.D. 1985 Hyperbolicity and change of type in the flow of viscoelastic fluids through channels, *J. non-Newtonian Fluid Mech.* **19**, 15–41. **11** 10; **12** 2.

YOO, J.Y., JOSEPH, D.D., AND BEAVERS, G.S. 1979 Higher-order theory of the Weissenberg effect, *J. Fluid Mech.* **92**, 529–590. **17** 32.

Author Index

Subject Index

Permissions

Springer-Verlag and Dr. Daniel D. Joseph would like to gratefully acknowledge the following for granting permissions:

Foreword Figures 1–3 from Joseph, D.D., O. Riccius, and Arney M. © 1986. Shear wave speeds and elastic moduli for different liquids: part II: experiments. *J. Fluid Mech.* 171, 309–338. Reprinted with the permission of Cambridge University Press.

Figures 7.2a–c, 7.3a–c, 7.6a,b from Anacosta, J., © 1970. *J. of Fluid Mech.* 42, 269–288. Reprinted with the permission of Cambridge University Press.

Figure 7.5 from Gupta, O.P, D.F. James © 1971. Drag on circular cylinders in dilute polymer solutions. *Chemc. Eng. Progress Symposium Series* 67, 62–73. Reproduced by permission of the American Institute of Chemical Engineers.

Figures 7.7–8 from Koniuta, A., P.M. Adler, and J.M. Piau © 1980. Flow of dilute polymer solutions around circular cylinders. *J. Non-Newtonian Fluid Mech.* 7, 101–106, Reprinted by permission of Elsevier Science Publishers.

Figures 7.11–15 from Delvaux, V., M.J. Crochet © 1990. Numerical simulation of delayed die swell. *Rheol. Acta* 29. Reprinted by permission of Dr. Dietrich Steinkopff Verlag.

Figure 8.10 from Hermes R.A., A.G. Fredrickson © 1967. Flow of viscoelastic fluids past a flat plate. *Amer. Inst. Chem. Eng. J.* 13, 253–259. Reproduced by permission of the American Institute of Chemical Engineers.

Figures 9.1, 9.3–5 from Yoo, J.Y., M. Ahrens, D.D. Joseph © 1985. Hyperbolicty and change of type in sink flow. J. Fluid Mech. 153, 203–214. Reprinted with the permission of Cambridge University Press.

Figure 9.2 from Metzner, A.B., E.A. Uebler, C.F.C.M. Fong © 1969. Converging flows of viscoelastic materials. *Amer. Inst. Chem. Eng. J.* 15. 750–758. Reproduced by permission of the American Institute of Chemical Engineers.

Figure 10.3 from Joseph, D.D., K. Chen © 1988. Anomalous elongational flows and change of type. *J. of Non-Newtonian Fluid Mech.* 28, 47–67. Reprinted by permission of Elsevier Science Publishers.

Figures 11.1–21c from Verdier, C., D.D. Joseph © 1989. Change of type and loss of evolution of the White-Mertzner model. *J. of Non-Newtonian Fluid Mech.* 31, 325–343. Reprinted by permission of Elsevier Science Publishing.

Figures 12.1–9 from Yoo, J.Y., D.D. Joseph © 1985. Hyperbolicty and change of type in the flow of viscoelastic fluids thru channels. *J. of Non-Newtonian Fluid Mech.* 19, 15–41. Reprinted by permission of Elsevier Science Publishers.

Figures 12.10–15 from Ahrens, M., J.Y. Yoo, D.D. Joseph © 1987. Hyperbolicty and change of type in the flow of viscoelastic fluids through pipes. *J. of Non-Newtonian Fluid Mech.* 24, 67–83. Reprinted by permission of Elsevier Science Publishers.

Figures 13.1–6, 13.9–20 from Joseph, D.D., J. Matta, K. Chen © 1987. Delayed die swell. *J. of Non-Newtonian Fluid Mech.* 18, 31–65. Reprinted by permission of Elsevier Science Publishers.

Figure 13.7 from Brenschede, E., J. Klein © 1970. Druckverluste und instabiles Flieben elastischer Flüssigskeiten im Hockdruck-Kappilarviskosimeter. *Rheol. Acta* 9, 130–136. Reprinted by permission of Dr. Dietrich Steinkopff Verlag.

Figure 13.8 from Giesekus, H. © 1968. Verschiedene Phänonmene in Strmungen viskoelastischer Flüssigkeiten durch Düsen. *Rheol. Acta* 8, 411–421. Reprinted by permission of Dr. Dietrich Steinkopff Verlag.

Figures 17.1–3 from Sturges, L. © 1977. Secondary motion induced by the rotation of a wavy rod. *Rheol. Acta* 16, 476–483. Reprinted by permission of Dr. Dietrich Steinkopff Verlag.

Figure 17.4 from Walters, K. N. Water © 1968. On the use of a rheogoniometer; part I: steady shear. *Polymer Systems— Deformation and Flow.* Reprinted by permission of Macmillan Publishers.

Figures 17.8a–f from Tieu, H.A., D.D. Joseph © 1983. Extrudate swell for a rojnd jet with large surface tension. *J. Non-Newtonian Fluid Mech.* 13, 203–222. Reprinted by permission of Elsevier Science Publishers.

Figure 17.9 from Cochrane, T., K. Walters, M.F. Webster © 1981. On Newtonian and non-Newtonian flow in complex geometrics. *Phil. Trans. Roy. Soc.* 301, 163–181. Reprinted with the permission of Cambridge University Press.

Figure 17.13a–f from Ju, H., O. Riccius, K. Chen, M. Arney, D.D. Joseph © 1990. Climbing constants, second order corrections of Trouton's visc., wave speeds. . . . *Non-Newt. Fluid Mech.* to appear. Reprinted by permission of Elsevier Science Publishers.

Figures 18.1–6 and 19.1–2 from Joseph, D.D., O. Riccius, M. Arney © 1986. Shear wave speeds and elastic moduli for different liquids; part II: experiments. *J. Fluid Mech.* 171, 309–338. Reprinted with the permission of Cambridge University Press.

Figures 19.3 from Preziosi, L., D.D. Joseph © 1987. Stokes' first problem for viscoelastic fluids. *J. Non-Newtonian Fluid Mech.* 25, 239–259. Reprinted by permission of Elsevier Science Publishers.

Figures 20.1–2 from Whitham, G.A. © 1974. *Linear and Non-linear Waves.* Reprinted by permission of John Wiley & Sons, Inc.

Applied Mathematical Sciences

55. *Yosida:* Operational Calculus: A Theory of Hyperfunctions.
56. *Chang/Howes:* Nonlinear Singular Perturbation Phenomena: Theory and Applications.
57. *Reinhardt:* Analysis of Approximation Methods for Differential and Integral Equations.
58. *Dwoyer/Hussaini/Voigt (eds.):* Theoretical Approaches to Turbulence.
59. *Sanders/Verhulst:* Averaging Methods in Nonlinear Dynamical Systems.
60. *Ghil/Childress:* Topics in Geophysical Dynamics: Atmospheric Dynamics, Dynamo Theory and Climate Dynamics.
61. *Sattinger/Weaver:* Lie Groups and Algebras with Applications to Physics, Geometry, and Mechanics.
62. *LaSalle:* The Stability and Control of Discrete Processes.
63. *Grasman:* Asymptotic Methods of Relaxation Oscillations and Applications.
64. *Hsu:* Cell-to-Cell Mapping: A Method of Global Analysis for Nonlinear Systems.
65. *Rand/Armbruster:* Perturbation Methods, Bifurcation Theory and Computer Algebra.
66. *Hlaváček/Haslinger/Nečas/Lovíšek:* Solution of Variational Inequalities in Mechanics.
67. *Cercignani:* The Boltzmann Equation and Its Applications.
68. *Temam:* Infinite Dimensional Dynamical System in Mechanics and Physics.
69. *Golubitsky/Stewart/Schaeffer:* Singularities and Groups in Bifurcation Theory, Vol. II.
70. *Constantin/Foias/Nicolaenko/Temam:* Integral Manifolds and Inertial Manifolds for Dissipative Partial Differential Equations.
71. *Catlin:* Estimation, Control, and the Discrete Kalman Filter.
72. *Lochak/Meunier:* Multiphase Averaging for Classical Systems.
73. *Wiggins:* Global Bifurcations and Chaos.
74. *Mawhin/Willem:* Critical Point Theory and Hamiltonian Systems.
75. *Abraham/Marsden/Ratiu:* Manifolds, Tensor Analysis, and Applications, 2nd ed.
76. *Lagerstrom:* Matched Asymptotic Expansions: Ideas and Techniques.
77. *Aldous:* Probability Approximations via the Poisson Clumping Heuristic.
78. *Dacorogna:* Direct Methods in the Calculus of Variations.
79. *Hernández-Lerma:* Adaptive Markov Control Processes.
80. *Lawden:* Elliptic Functions and Applications.
81. *Bluman/Kumei:* Symmetries and Differential Equations.
82. *Kress:* Linear Integral Equations.
83. *Bebernes/Eberly:* Mathematical Problems from Combustion Theory.
84. *Joseph:* Fluid Dynamics of Viscoelastic Liquids